MULTICHIP MODULE TECHNOLOGY HANDBOOK

Electronic Packaging and Interconnection Series

Charles M. Harper, Series Advisor

ALVINO ▪ *Plastics for Electronics*

CLASSON ▪ *Surface Mount Technology for Concurrent Engineering and Manufacturing*

G. DI GIACOMO ▪ *Reliability of Electronic Packages and Semiconductor Devices*

GINSBERG AND SCHNOOR ▪ *Multichip Module and Related Technologies*

HARMON ▪ *Wire Bonding in Microelectronics,* 2/e

HARPER ▪ *Electronic Packaging and Interconnection Handbook,* 2/e

HARPER AND SAMPSON ▪ *Electronic Materials and Processes Handbook,* 2/e

HARPER AND SEYERT ▪ *Hybrid Microelectronics Handbook,* 2/e

HWANG ▪ *Modern Solder Technology for Competitive Electronics Manufacturing*

KONSOUSK AND HOLLARD ▪ *Electronic Packaging of High-Speed Circuitry*

LAU ▪ *Ball Grid Array Technology*

LAU ▪ *Flip Chip Technologies*

LAU ▪ *Solder Joint Reliability of BGA, CSP, Flip Chip, and Fine Pitch SMT Assemblies*

LICARI ▪ *Multichip Module Design, Fabrication, and Testing*

Related Books of Interest

BOSWELL ▪ *Subcontracting Electronics*

BOSWELL AND WICKAM ▪ *Surface Mount Guidelines for Process Control, Quality, and Reliability*

BYERS ▪ *Printed Circuit Board Design with Microcomputers*

CHEN ▪ *Computer Engineering Handbook*

CHRISTIANSEN ▪ *Electronics Engineers' Handbook,* 4/e

COOMBS ▪ *Electronic Instrument Handbook,* 2/e

COOMBS ▪ *Printed Circuits Handbook,* 4/e

GINSBERG ▪ *Printed Circuits Design*

JURAN AND GRYNA ▪ *Juran's Quality Control Handbook*

JURGEN ▪ *Automotive Electronics Handbook*

MANKO ▪ *Solders and Soldering,* 3/e

RAO ▪ *Multilevel Interconnect Technology*

SZE ▪ *VLSI Technology*

VAN ZANT ▪ *Microchip Fabrication,* 3/e

MULTICHIP MODULE TECHNOLOGY HANDBOOK

Philip E. Garrou
Dow Chemical Company

Iwona Turlik
Motorola

McGRAW-HILL

New York San Francisco Washington, D.C. Auckland Bogotá
Caracas Lisbon London Madrid Mexico City Milan
Montreal New Delhi San Juan Singapore
Sydney Tokyo Toronto

McGraw-Hill

A Division of The McGraw-Hill Companies

1 2 3 4 5 6 7 8 9 0 DOC/DOC 9 0 2 1 0 9 8 7

ISBN 0-07-022894-9

The sponsoring editor for this book was Stephen S. Chapman and the production supervisor was Pamela Pelton. It was set in Times by J. K. Eckert & Company, Inc.

Printed and bound by R. R. Donnelley & Sons Company.

To our children

Marshall Philip Blair, Christopher Whitney,
Daniel Thomas, and Karin

In remembrance

During the writing of this book we lost a colleague, a friend and a mentor. George Messner, former Senior Scientist with Photocircuits, PCK, Kollmorgen, and AMP AKZO, and IDC, passed away January 20, 1996. During his distinguished career, George served as the President of IEPS, member of the Steering Committee of the IEEE Computer Packaging Committee, and member of ISHM and IPC. With his extensive background in laminate technology, he was, in recent years, concerned with the fabrication of MCMs—primarily, MCM-Ls. This interest in MCMs led George to organize and co-edit the Thin Film Multichip Module treatise published by ISHM in 1992. When we were both newcomers to microelectronic packaging, George was always there, available to offer advice or share experiences. We, and the industry as a whole, miss him very much.

CONTENTS

PREFACE

This handbook is intended to serve as a detailed introduction and guide to MCM technology. This book was conceived with the aim of providing a comprehensive source of information in this burgeoning area.

Chapter 1 examines the technical drivers for MCM technology; Chapters 2–4 review the various materials and technology options for MCM-C, MCM-D and MCM-L, giving commercially relevant examples. Chapter 5 introduces the recent technology of large area processing (LAP) where printed wiring board (PWB) and flat panel display technology (FPD) are used to produce low-cost thin film MCM modules. Chapter 6 details the advances in 3D MCM technology for highly volume-constrained systems. Chapters 7 and 8 examine MCM package design and chip to module assembly, while Chapter 9 examines module to board assembly. Chapter 10 examines MCM substrate design, and Chapters 11 and 12 examine MCM electrical performance. Thermal issues are examined in Chapter 13. The need for known good die (KGD) is detailed in Chapter 14, and module test and inspect are examined in Chapter 15. Together, these chapters present a complete look at the state of the art in multichip packaging technology.

While every attempt has been made to keep the material presented as current as possible, it should be remembered that MCM technology is still in a stage of rapid development and change.

ACKNOWLEDGMENTS

Completing a book of this scope and complexity required the talents and skills of many people. Special thanks to the chapter authors who, as experts in their fields, made the chapters possible. In writing and editing this handbook, the continuing support and interest of many of our colleagues inside as well as outside of Dow Chemical and Motorola have been appreciated. We especially acknowledge the support of Don Dix (Dow Chemical) and Mauro Walker (Motorola). Iwona Turlik thanks Stephanie Germack, Brenda Lane, and Julie Ehardt for their indispensable help with the manuscripts.

CONTRIBUTORS

Robert Darveaux is currently employed at Amcor. Previously, he was a senior staff engineer in the Land Mobile Products Sector of Motorola in Plantation, Florida. Dr. Darveaux received a BS in nuclear engineering from Iowa State University and a PhD in materials science and engineering from North Carolina State University. His research and development activities have been in the areas of thermal performance of multichip modules, mechanical characterization of solders, finite element modeling, power device packaging, failure analysis, and reliability prediction. His modeling and fatigue analysis work has been integral to Motorola's BGA implementation process. He is author of more than 20 technical articles and holds three patents.

Robert C. Frye received a PhD in electrical engineering from M.I.T. in 1980. Since then, he has been with Bell Laboratories. In recent years, his work has focused on advanced electronic interconnection technology and multichip modules. He is currently a distinguished member of technical staff in the Electronic Packaging Research Department. He is a member of the IEEE, IMAPS, and the Materials Research Society.

Philip E. Garrou is currently Dow Chemical's chief scientist for microelectronics. During his 23 years at Dow, he has been involved with many microelectronic technology areas, including AlN electroceramics and polymers for microelectronics packaging applications. Dr. Garrou is program director for the MMS/Dow large area processing line, the DARPA Consortium for Large Area Processing (CILAP), and the DARPA Seamless High Off Chip Connectivity (SHOCC) consortium.

Dr. Garrou has authored more than 50 articles on microelectronics and co-edited the 1992 book *Thin Film Multichip Modules.* He is associate editor of the *International Journal of Microcircuits and Microelectronic Packaging.*

Dr. Garrou is a senior member of IEEE. He was elected to the board of governors of the CPMT society of IEEE (1994–1996, 1997–1999) and served as chairman of the IEEE CPMT technical committee on materials (1992–1994).

Dr. Garrou was elected president of IMAPS (ISHM) in 1997, where he previously served as technical VP (1995--96); Chairman of the Advanced Packaging subcommittee (1992–1994) and chairman of the Material Division (1992). Dr. Garrou is co-founder of the ISHM/IEEE International Symposium on Advanced Packaging Materials and was technical chair of the 4th International MCM Conference (Denver, 1995).

Dr. Garrou earned a BS and PhD in chemistry from North Carolina State University and Indiana University, respectively.

Barry K. Gilbert received a BS in electrical engineering from Purdue University and a PhD in physiology and biophysics from the University of Minnesota. He is a staff scientist and director of the Special Purpose Processor Development Group at the Mayo Foundation, Rochester, Minnesota. His research interests include the development of algorithms for the real-time transmission and analysis of wide-bandwidth image and signal data; the design of specialized signal-processing computers, using advanced chip technologies such as gallium arsenide and indium phosphide to execute these tasks; the development of CAD tools to allow the timely design of high-complexity digital signal processors; and the development of advanced electronic packaging technologies such as multichip modules which will be capable of supporting digital integrated circuit-based processors operating at gigahertz system clock rates.

John K. Hagge is a senior principal engineer with Rockwell International's Advanced Technology Center in Avionics and Communications, assigned to help develop next-generation interconnection and miniaturized packaging technologies. He has held various management and development positions with Rockwell since 1966 and has authored many technical papers and book chapters. He has also served as a consultant to the Air Force, reviewer for various journals and NSF, lecturer in packaging courses, and senior science master at Kagumo Higher School in Kiganjo, Kenya. He is a member of ASME, IMAPS, Pi Tau Sigma, Tau Beta Pi, Phi Kappa Phi, and Sigma Xi.

R. Wayne Johnson is a professor of electrical engineering at Auburn University. At Auburn, he has established teaching and research laboratories for advanced packaging and electronics assembly. Research efforts are focused on materials, processing, and modeling for multichip packages, electronics assembly and high-temperature electronics. He has worked in MCM design, MCM-L, -C and -D substrate technology as well as SMT, wire bond, and flip chip assembly techniques. He has published and presented many papers at workshops and conferences and in technical journals. He has also co-edited one IEEE book on MCM technology and written two book chapters in the areas of silicon MCM technology and MCM assembly. Dr. Johnson is a senior member of the IEEE and a member of the CPMT Society Board of Governors. He was the 1991 president of the International Society for Hybrid Microelectronics (ISHM). He received the 1993 John A. Wagnon, Jr., Technical Achievement Award from ISHM and was named a fellow of the Society in 1994.

Dr. Johnson received BE and MSc degrees in 1979 and 1982 from Vanderbilt University, Nashville, Tennessee, and a PhD in 1987 from Auburn University, Auburn, Alabama, all in electrical engineering. He has worked in the microelectronics industry for DuPont, Eaton, and Amperex.

Wulf Knausenberger has a BS in engineering science and a PhD in solid state science from Penn State University. For 25 years, he worked as a member of technical staff and as a technical manager with AT&T Bell Labs in Whippany, New Jersey. Among his major achievements was initiating and leading the development of the AT&T standardized electronic packaging hardware system that is presently being manufactured worldwide (by Lucent Technologies) in large volumes. Since retiring from AT&T in 1995, he has been active as a consultant and lecturer specializing in interconnection technology and electronic packaging issues.

Dr. Knausenberger has extensive experience in the modeling and optimization of interconnection systems and strategic planning for interconnection technology. He regularly organizes workshops and sessions for technical society meetings. His September 1984 technical paper (with co-author Len Schaper), "The Myth of Cheap Wiring," was published in the *IEEE CHMT Transactions* and has become widely recognized as a major contribution to the fundamental understanding of interconnection technology.

He is a fellow of the IEEE and IMAPS. He has served as the president of IEPS, a member of the advisory council of IMAPS, and a member of the steering committee of the IEEE Computer Society's Computer and Systems Packaging Committee, a member of the board of governors of the IEEE CPMT Society, and an associate editor of the IEEE CPMT Transactions on Packaging. He received the 1993 Achievement Award for his outstanding contributions to the field of electronic packaging from the International Electronics Packaging Society.

John U. Knickerbocker is a distinguished engineering manager at IBM Microelectronics in East Fishkill, New York. He manages the Advanced Packaging Product Development organization responsible for packaging and interconnection products for IBM internal and OEM customer requirements. He received BS and MS degrees from Alfred University and a PhD from the University of Illinois, all in ceramic engineering. Dr. Knickerbocker has worked in packaging microelectronics for over 14 years at IBM. He holds 27 patents and patent applications, is responsible for 15 technical publications and numerous technical presentations, and is a fellow of the American Ceramic Society.

Mr. Knickerbocker serves as FTAB with Sematech Assembly and Packaging and is a member of IMAPS. With an expertise in packaging and assembly, his skills include materials, process and structural development of multilayer ceramic packages, along with associated C-4 interconnection, surface mount technology and assembly. He has reached IBM's 7th Invention Plateau, and IBM awards including the Corporate Award, three Division awards, Outstanding Technical Achievement award, Patent Portfolio award and Top Contributor Award.

Ananda H. Kumar is presently a member of the technical staff at Applied Materials. He has worked in the areas of ceramic and thin film packaging for nearly 20 years at IBM Microelectronics and David Sarnoff Research Labs. Dr. Kumar has written many papers and has earned 30 patents related to packaging technology. Dr. Kumar was the recipient of the top IBM Corporate Technology award in 1991 for his contributions to the development of glass-ceramic multilayer substrates. He has a PhD in chemical engineering from the University of Illinois.

Victor Nelson received a BSEE from the University of Kentucky and MS and PhD degrees from Ohio State University in 1977 and 1978. Since 1978, he has been a member of the electrical engineering faculty at Auburn University. His primary research interests include computer architecture, fault-tolerant computing, multiprocessor system design, microprocessor applications, and computer-aided design and testing of digital systems, including test methods for multichip modules. The latter includes work on low-cost test methods for automotive MCMs, in conjunction with the Consortium for Vehicle Electronics. At Auburn, he developed the Fault Tolerance/Distributed Computing Lab, the Whirlpool Home Appliance Network, the VLSI CAD and Testing Lab, and the MCM Test Advisor Program (MTAP). In addition to authoring a number of technical papers, he is co-author of one textbook on fault-tolerant computing, and another on digital logic system analysis and design.

Theodore G. Tessier has a BS from Laurentian University in Canada and an MSc in applied polymer science from the University of Ottawa. He is principal staff engineer and manager of new process and product development within the Advanced Interconnect Systems Laboratory of Motorola's semiconductor products sector. He currently has responsibility for advanced semiconductor packaging programs including flip chip, chip scale packaging and advanced display assembly technologies. He has previously worked at Motorola's Corporate Manufacturing Research Center, the Microelectronics Center of North Carolina, and Northern Telecom.

Mr. Tessier has published more than 50 professional publications and book chapters and holds six patents. He is a senior member of IEEE and IMAPS, chairman of IEEE CPMT TC-5 Materials Subcommittee, and a member of the technical programs committees of ECTC, IMAPS and the MCM-Denver Conference.

Jim Trent is a product manager for automotive power train applications in Motorola's Transportations System Group, located in Austin, Texas. He has been with Motorola for 12 years, with 3 years in product engineering, 3 years in semiconductor packaging, and 6 years at the Strategic Electronics Division performing circuit design, system design, and project management.

He transferred into the Motorola's Semiconductor Products Sector (SPS) in June of 1991. Since joining SPS, he has had various assignments ranging from design, development, test, research/development, and management. Jim's main emphasis throughout his career has been in the design and test of system and/or multi-chip modules. He has authored multiple MCM articles ranging from MCM Test and Design Philosophies to MCM-L Manufacturing Issues. He graduated in 1985 with a BS in electrical and computer engineering from Kansas State University.

Rao R. Tummala is petit chair professor and director of PRC, Georgia Tech, Atlanta, Georgia. Dr. Tummala received a BS in physics, mathematics, and chemistry from Loyola College, India; a BE in metallurgical engineering from the Indian Institute of Science, Bangalore, India; an MS in metallurgical engineering from Queen's University; and a PhD in materials science and engineering from the University of Illinois. He joined the faculty at Georgia Tech in 1993 as a petit chair professor in electronics packaging and as Georgia State research scholar. He is also the director of the Low-Cost Electronic Packaging Research Center, funded by NSF (as one of its Engineering Research Centers), the state of Georgia, and U.S. electronics industry. Before joining Georgia Tech, he was an IBM fellow at IBM Corporation, where he invented a number of major technologies for IBM's products for displaying, printing, magnetic storage, and multichip packaging. He is both a fellow of IEEE and the American Ceramic Society, a member of the National Academy of Engineering, 1996 general chair of IEEE-ECTC, and 1996 president of ISHM. He was recently named by *Industry Week* as one of the 50 stars in the U.S., for improving US competitiveness.

He is co-editor of widely used *Microelectronics Packaging Handbook.* He has published 90 technical papers, holds 21 U.S. patents, and is responsible for 44 other inventions. He has received a number of awards, including the David Sarnoff award, the Sustained Technical Achievement award from IEEE, John Wagnon's award from ISHM, the Materials Engineering Achievements award from ASM-I, the Distinguished Alumni award from University of Illinois, and the Arthur Friedberg Memorial award from American Ceramic Society.

Dr. Tummala's current research interests include packaging materials (metals, ceramics, and polymers) and processes, mechanical properties of materials, thin and thick MCMs, thermal and electrical designs, and integrated passive components.

Iwona Turlik received her MS in electrical engineering and PhD in technical science from the Technical University of Wroclaw, Poland. She has worked as the manager of exploratory processing with Bell Northern Research, where she was involved in process and device development for several Si and GaAs technologies. She was the director of advanced packaging technology programs, MCNC, Research Triangle Park, NC, and a professor with the Electrical Engineering Department, University of North Carolina a Charlotte, where she conducted research in the material science and process technology of high-performance systems. Currently, she is vice president and director of the Corporate Manufacturing Research Center at Motorola in Schaumburg, Illinois.

Dr. Turlik is an IEEE fellow, serves on the board of directors of IEPS, and is a member of the Surface Mount Council.

Christian M. Val, PhD (physics), was employed at Thomson-CSF from 1968 to 1995 in the Active Components and Military Computer Departments and the Thomson Central Lab (LCR/Orsay). Since that time, he has served as chairman and CEO of 3D PLUS Company. He has been a member of the Scientific and Technical Council of the Thomson Group since its creation in 1989 and is the technical director of ISHM France. He was named a life member of the International Microelectronics and Packaging Society (IMAPS) in 1997.

Dr. Val is the recipient of the IEPS Achievement Award for Outstanding Contributions to the Field of Electronics Packaging (Marlborough, Massachusetts, 1990) and the John A. Wagron, Jr., Technical Achievement Award, given by the ISHM (Boston, Massachusetts, 1994). He has published some 70 technical papers in international publications, and 20 have been presented at ISHM conferences in the U.S.A., Japan, and Europe, and IEEE symposia in the U.S.A. He holds 60 corporate and 3 personal patents.

Russell J. Wagner is manager of advanced operations engineering for Rockwell Collins Inc. His current responsibilities involve the development of next-generation electronics manufacturing technology and processes. Mr. Wagner received a BS in electrical engineering technology from Milwaukee School of Engineering in 1980. In his 16 years with Rockwell, he has undertaken various engineering assignments involving the research, development, application, and management of advanced electronic packaging design and test technology. He is the author or co-author of more than a dozen technical publications on advanced packaging design and test, including several book chapters. He has been an invited instructor, technical panelist, chairman, and speaker on the subjects of KGD, MCM, and test technology at many industry conferences and workshops. Mr. Wagner holds awards from Rockwell and the industry in general for distinguished technical achievement. He serves on the SEMATECH Low Cost Project Technical Advisory Board; Rockwell Corporate Technical Panel on Electronic Packaging, Interconnect and Assembly; and the IEEE Test Technology Technical Committee. He is a member of the IEEE and IMAPS.

CHAPTER 1

TECHNICAL DRIVERS

Dr. Iwona Turlik
Motorola

1.1 INTRODUCTION

The semiconductor industry has been providing integrated circuits (ICs) in standard packages for more than 25 years. However, it has become increasingly clear that the performance advantages inherent in VLSI/ULSI semiconductor devices will not be realized at the system level with conventional packaging.[1] Consequently, many new packaging technologies have emerged, ranging from ball grid arrays (BGAs), which combine a dense area array of contacts with a surface mount assembly process, to multichip modules (MCMs).[2] The challenge is to simultaneously meet three system metrics: size, performance, and cost. The main drivers behind MCMs stem from different segments of the market. At one end, markets such as electronic data processing and networking are driving increasing lead counts and improved electrical performance as they move toward greater computational and transmission/switching capabilities. On the other end, package size reductions and cost are being driven by markets moving toward smaller portable products such as digital cameras, personal computers, and cellular phones.

Multichip modules have features that enable smaller, lighter systems and higher speed performance to be obtained by eliminating individual packages and their parasitics. Multichip module technology is one of the most efficient packaging technologies. The bare devices are attached to the substrate by using one of the interconnection method such as wire bonding, tape automated bonding (TAB), or flip chip. In this approach, the size of the package is dictated by the size of IC devices and a number of input and output connections per device. Therefore, the package is of minimum size and has superior electrical performance. To date, depending on the type of substrate used or the method of fabrication, MCMs have been divided into four basic groups:

- MCM-C are multichip modules that are built using either cofired ceramic substrates or thick film interconnection technology on ceramic substrate.
- MCM-D are multichip modules that have interconnection structure built by deposition of thin film metals and dielectric materials over silicon, diamond, ceramic, or metal substrate.
- MCM-L are multichip modules that have substrates built by using the most advanced PWB technology.
- 3D MCMs are multichip modules that employ vertical stacking of IC devices.

These definitions for MCMs are now accepted and frequently used worldwide in the electronic industry, because they cover all commonly used manufacturing methods of MCMs. However, the industry has begun to realize that there is a large amount of process and assembly related commonality between these basic fabrication technologies and all of the hybridized analogs such as MCM-D/L or MCM-C/D and so on have emerged. This synergy is just now beginning to be exploited with interconnect processes and tools that are compatible "across the board."[3] It is still to be determined what fabrication processes and interconnect structures will be the industry's preferred method, but the substrate chosen will reflect the application and the technologies that will maximize performance, size, and cost effectiveness.

1.2 SYSTEM PACKAGING CHALLENGES

Electronic industry depends on and marches in step with integrated circuit (IC) developments. As higher levels of integration are achieved, new demands are being placed on circuit packaging and system engineers (Fig. 1.1) Currently available VLSI and future ULSI circuits require ever increasing input-output (I/O) count and constant improvements in thermal and electrical performance from the circuit package. These demands must be met if the full potential of these complex high-performance circuits is to be realized. The circuit package that appears to be most suited to fill these high-performance needs is the MCM.

As shown in Fig. 1.2, MCM construction requires multidisciplinary skills. Proficient application of these skills provides a solution to the driving forces of packaging application. Given the diversity of available technologies and the interrelations between the package metrics and system performance, it is important that the package architect be able to quickly quantify or rank the technology options with respect to system requirements.[6]

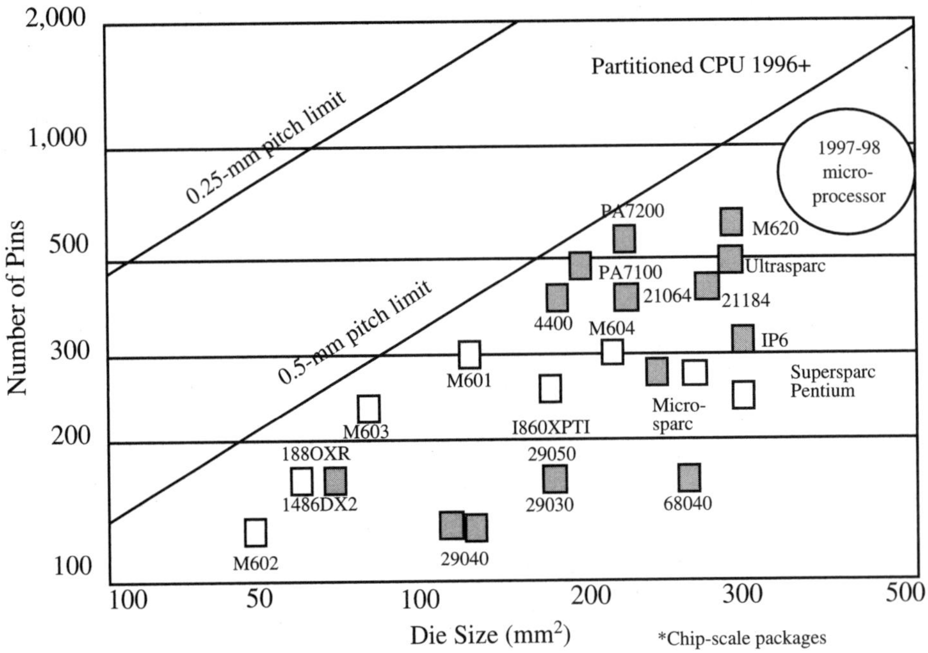

FIGURE 1.1 Processor requirements met by 0.5-mm pitch limit (next generation will require 0.25-mm-pitch packaging).

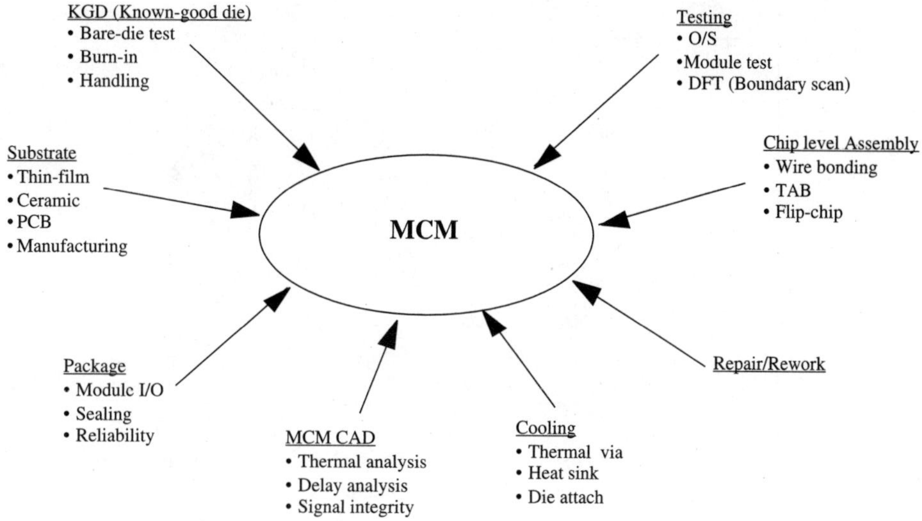

FIGURE 1.2 Various aspects of MCM packaging.

System limits depend strongly on the specific application or end use of the chip. The key issues to be determined are the impact of a proposed device and packaging technology on overall system performance. Figure 1.3 shows the basic relationships among the parameters contributing to system performance. As the figure shows, the package effect on system cycle time is a measure of package time-of-fight characteristics. The circuit density that can be supported affects the overall cycle time in terms of both circuit and package delay.[7]

The package materials and mechanical structure dictate the power distribution and thermal capacity, which affect the circuit density at the package level. These considerations influence the choice for circuit power level and package density. In addition, the package level geometrics that can be used for line and via/size spacing influence the number of chip I/O; all of this can further constrain the realizable circuit density for multichip designs. Table 1.1 shows typical design rules of various interconnect substrates.[8]

1.3 PACKAGING EFFICIENCY

The main benefit provided by MCMs is their ability to form extremely dense conductor matrix for the interconnection of bare chips. Therefore, that capability should become a part of their definition. To define this high-density capability of multichip modules, a suitable metric had to be established. Knausenberger and Schaper[9] proposed substrate efficiency, which was defined as:

$$F_x = \frac{A_{si}}{A_s} \tag{1.1}$$

where F = substrate efficiency
A_{si} = active silicon area (cell)
A_s = total substrate area

Because data about the active cell area in a chip are usually not readily available, a better measure is the use of actual silicon chip area of the devices mounted on the substrate. Such a metric was used by Reche[10] where he defines packaging efficiency as:

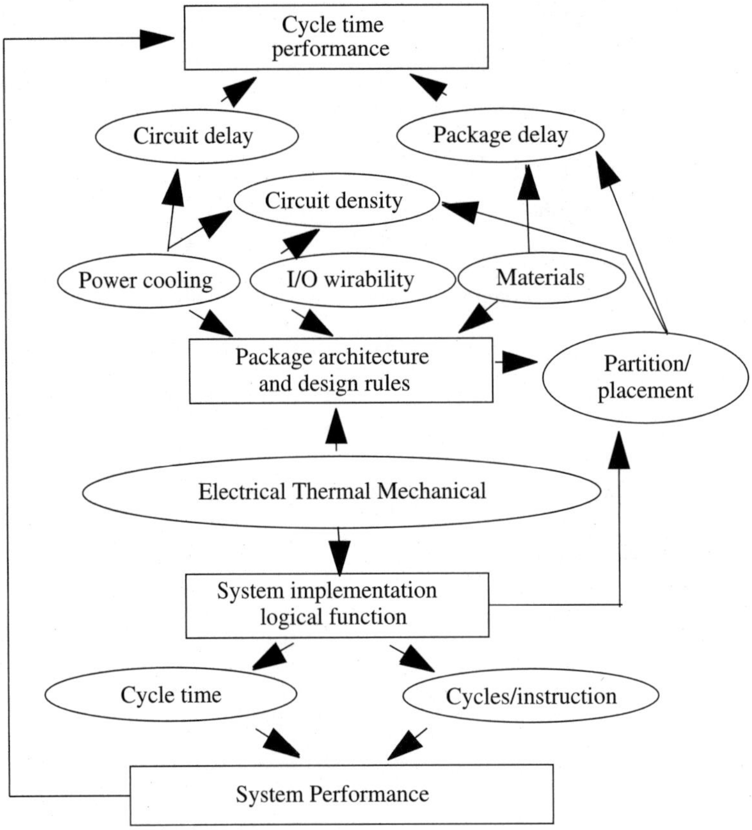

FIGURE 1.3 System performance parameters.

Table 1.1 Carrier Attributes Comparison

	MCM-L	MCM-C	MCM-D
Substrate	Printed wiring boards	Cofired ceramic low-E ceramics	Low-E polymers, silicon on base
Material	Epoxy glass	Al2O3	Polyimide, BCB
Dielectric constant	2.8–4.5	4.5–10	2.6–4
Thermal dissipation	Poor–medium	High	Moderate
Line width, μm	75–750	90–125	12–50
Line pitch, μm	Low	Moderate	High
I/O density	Low	Moderate	High
Substrate cost	Low	Moderate	High

$$F_p = \frac{A_{sc}}{A_p} \tag{1.2}$$

where F_p = packaging efficiency
A_{sc} = silicon chip area
A_p = package area

The result of this packaging density analysis of various substrates is given in Fig. 1.4. This plot indicates how the conductor widths and the fan-out requirements affect the packaging density of all commonly used substrate categories.

Chao and coworkers[11] compare the density available from typical production PWB and multilayer ceramic technologies to the density expected from an average thin film MCM construction. For a significant reduction in the number of layers, notable improvements are made in overall connection and via density which would result in a significant size reduction for a given system. This is shown in Table 1.2.

Rockwell's evaluation of the relative efficiency for several packaging technologies compared to MCM-D (here called *silicon-on-silicon*) is shown in Table 1.3.[12]

In the now classic paper by Neugebauer, et al.,[13] comparison is made between four high-density packaging approaches to implement a hypothetical array processor consisting of 160 VLSI chips comprising 64 microprocessors, 64 RAMs, and 16 glue chips for a total signal I/O count of 11,920. The interconnection of 160 single chips mounted on a double-sided PWB is compared by way of 16, 4, 1 MCM. Performance advantages are seen in Table 1.4. The 16 MCM approach shows 15× improvement, the 4 MCM approach shows a 30× improvement. The 50× improvement from the 160-chip MCM is discounted because it is projected to be offset by lower module yields.

1.4 MINIATURIZATION

Hagge and coworkers at Rockwell have presented numerous examples of military system miniaturization using MCM fabrication techniques.[14–16] When the 82-chip PWB/SMT board was replaced by thin film MCM-D, a 4:1 size reduction was achieved.

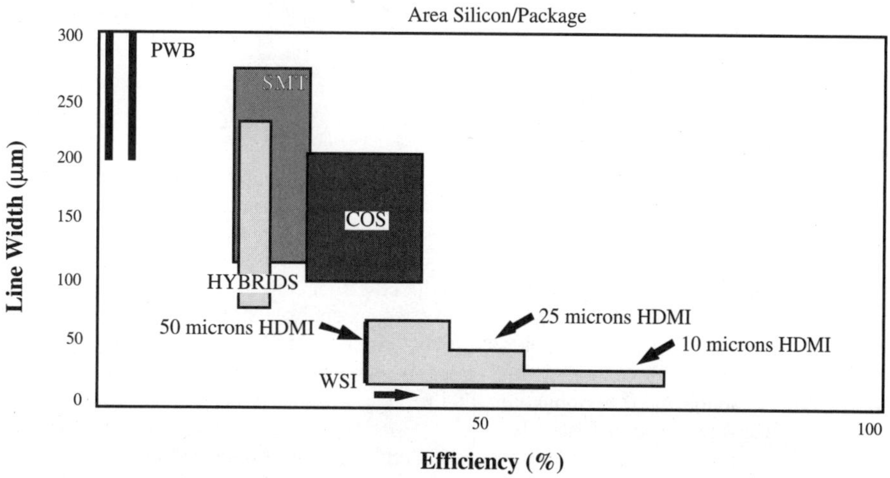

FIGURE 1.4 Packaging efficiency: area silicon/package.

Table 1.2 Comparison of interconnection densities

Technology	Connection density (in/in^2/layer)	Signal layers (total layers)	Total length (in/in^2)	Via density per in^2
PC boards (3 tracks, 0.100" pitch)	30	12 (12)	360	100
Multilayer ceramic substrate (1 track, 0.020" pitch)	50	20 (42)	1,000	2,500
Cu/PI thin film (2 tracks, 140 μm pitch)	350	4(8)	1,400	33,000

Table 1.3 Comparison of several packaging technologies

Packaging technology	Line/space size (μm)	Packaging efficiency	Typical ICs/in^2
Integrated circuit	0.3–3	100	
Silicon-on-silicon	10–50	30–60	8–20
Chip-on-board	100–200	15–30	5–9
Hybrid thick film	125–250	6–14	2–4
Surface mount PWB	125–250	6–14	2–4
VHSIC packages	100–200	5	3–1
DIP PWB	200–300	1–3	1–2

The size reduction possible using MCM technology can also be observed in the reported Hughes example of reduction of 28 PWBs assembled with DIP single chip packages to five PWBs using SMT technology to five MCM with 3.4× reduction in the substrate area (Fig. 1.5).[17]

1.5 RELIABILITY

Improvements in electronic system reliability is, of course, a benefit to all system users. In the conventional single chip package, surface mount design, each chip-to-chip signal must travel from chip to package, to lead, to solder joint, to circuit board, to solder joint, to lead, to package, and to chip.[18] The leads, solder joints, and circuit boards are all exposed to potential environmental degradation. In the MCM design, the signal travels directly from chip-to-substrate-to-chip, all within the protection of a package environment. Many of the passive components of the system also move from the exposed environment of the circuit board into the MCM. This is depicted in Fig. 1.6.

In another example,[12] two circuit boards were replaced by one MCM, resulting in a 3.5× size reduction and the elimination of 94 percent of solder joints and 84 percent of the hermetic I/O. The data are shown in Table 1.5.

The impact of fewer solder joints on reliability is best understood by examining Figure 1.7, which compares the relative reliability of different contact joints. It is clear, from a system reliability standpoint, that interconnection made on silicon is preferable.

Table 1.4 Comparison of interconnections for three high-density packaging approaches for 160-chip 7-Mgate array processor system

	Packaging parameter	100 SCMs on PWBs	16 MCMs on PWBs	4 MCMs on PWBs	Single MCM
Density	Total package footprint area, cm^2	1,202	370	280	207
	Functional density, 1,000 gates/cm^2	6	19	25	34
I/Os	No. on MCM	—	8,384 (72%)	9,900 (84%)	11,000 (97%)
	No. on PWB	11,000 (97%)	3,216 (25%)	1,600 (13%)	—
	No. on backplane	400 (3%)	400 (3%)	400 (3%)	400 (3%)
Interconnections	Total length, m	1,750	883	684	684
	Avg. signal path length between I/Os, cm	15.2	7.68	5.95	5.6
	MCM area, in^2	—	2.25	9.0	36
	PWB area, in^2	210	64	49	—
	MCM layers	—	2	2	2
	PWB layers	6	26	4	—
Package performance	Avg. propagation delay between I/Os, ns	1.08	0.48	0.34	0.22
	CV^2 power dissipated for avg. signal path, mW	25	13	10	9
	Power × delay for avg. signal path	27	6	4	3
Figure of merit	Density × performance, gates/cm^2-pJ	220	3,220	6,250	11,300
	(normalized)	(1)	(15)	(30)	(50)

1.6 FUTURE CHALLENGES

As described above, electronic systems designed with MCM can count on:

- Improved electrical performance
- Increased IC packaging density
- Miniaturization
- Reduced power consumption
- Increased reliability
- Lower system cost

The multichip market has evolved from demonstrating technical feasibility to use in a variety of applications. The technology exists to build MCMs for mainframe computers, missiles, switching systems, camcorders, and pagers (Fig. 1.8).[20] For many users, there are still barriers to MCM use.

Table 1.5 Comparison of surface mount and silicon-on-silicon packaging

Design	Area (in^2)	Number of solder joints	Number of board-to-board I/O	Number of hermetic I/O	Number of connector contacts
Two SMT PWBs	13.40	496	15	194	56
MCMs	3.84	30	0	30	0

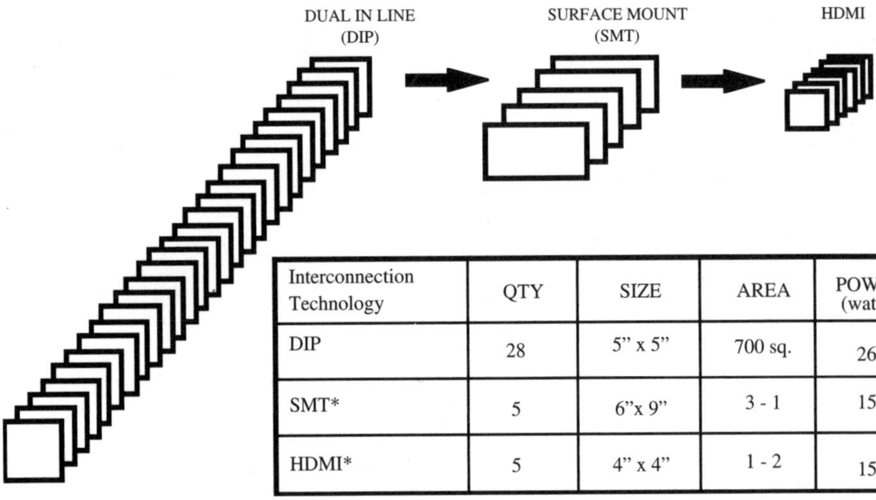

Interconnection Technology	QTY	SIZE	AREA	POWER (watts)
DIP	28	5" x 5"	700 sq.	268
SMT*	5	6"x 9"	3 - 1	158
HDMI*	5	4" x 4"	1 - 2	158

FIGURE 1.5 Performance improvements from 0.6 MIPs to 4.5 MIPs over DIP version.

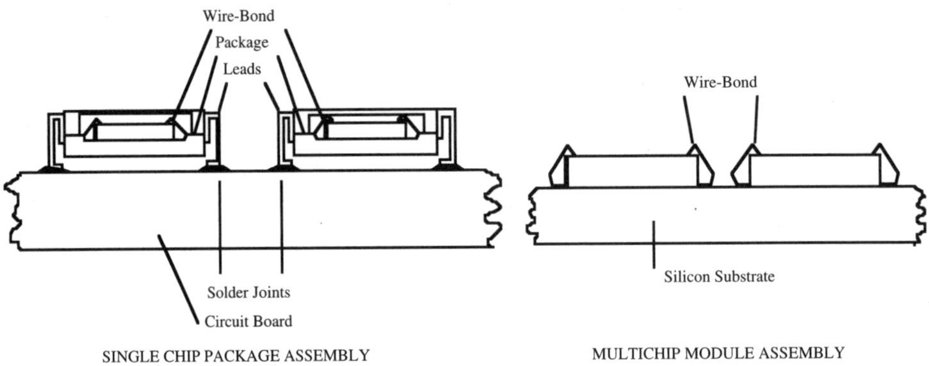

FIGURE 1.6 Reduction in electrical path "segments" between two ICs with multichip module technology: (left) nine segments with conventional through hole or surface mount technology using single-chip packages and (right) three segments in multichip interconnection.

Expanded adoption depends on the introduction of low-cost MCMs. This includes introduction of low-cost substrate materials or low-cost fabrication methods, the availability of known-good dice (KGD), and improved test methods. MCM design methodology must be further improved to cope

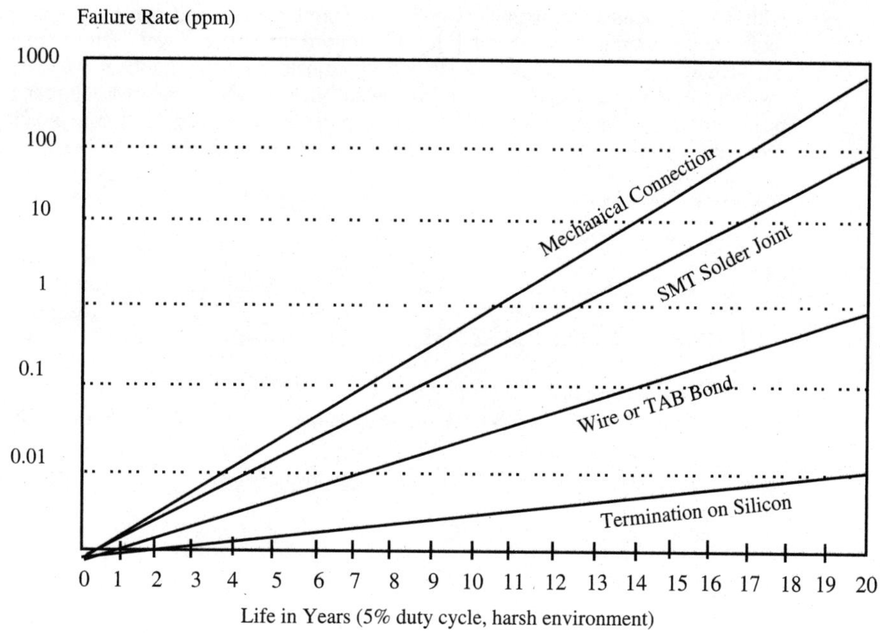

FIGURE 1.7 Relative reliability of different contact joints.

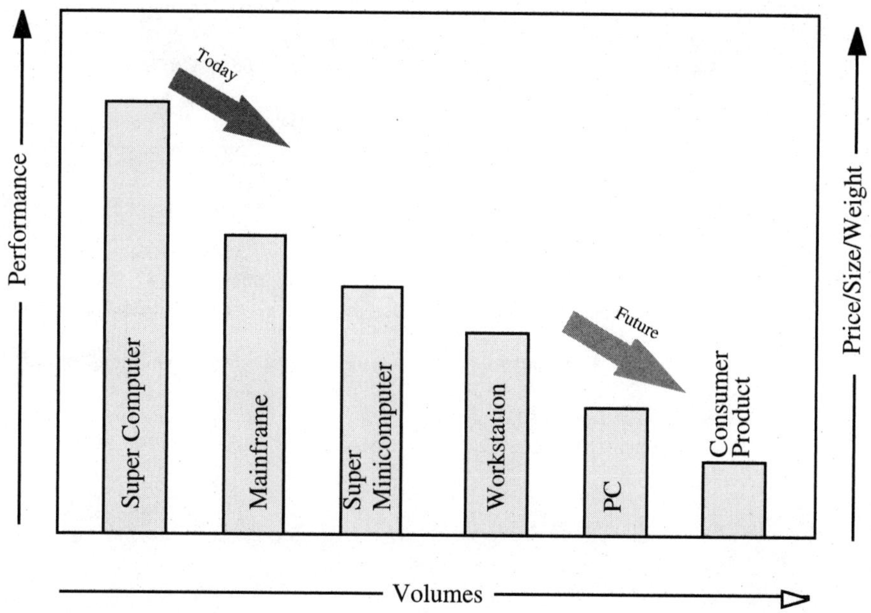

FIGURE 1.8 MCM Applications

with design change quickly and easily by integrating thermal and electrical analysis and verification tools more smoothly, from the chip level design to the system level design. Moving to MCMs too soon could result in unworkable products and projects, but delaying the move to MCMs could mean losing the competitive race. Considering the advances in packaging over the past 15 years, from the beginning of standardization of chip carriers to the level of assemblies now produced worldwide, it seems realistic to expect continued progress with newer and better MCM technology solutions.

REFERENCES

1. C. Neugebauer, R.O. Carlson, R.A. Fillion, "High Performance Interconnections between VLSI Chips," *Solid State Technology,* 31 (6)93 (June 1988).
2. IBIS Associates, Inc., *Prospectous,* 1990.
3. T. Tessier, "The Future of MCMs," *Advancing Microelectronics,* 6 (March/April 1995).
4. V.J. Biancomano, "Interconnects & Packaging," 43, *Electronic Engineering Times* (June 24, 1996).
5. T. Sudo, "Present and Future Directions for Multichip Module Technologies," *IEE Trans, Electron.,* vol. E78-C, No. e, 684 (June 1995).
6. D.P. Seraphim, R. Lasky, and C.-Y. Li, *Principles of Electronic Packaging,* McGraw-Hill, New York, 1989.
7. D.P. LaPotin, T.R. Mazzawy and M.L. White, "Early Package Analysis: Consideration and Case Study," *Computer,* 30 (April 1993).
8. J. Lisowskiand D. Clocher, "Multi-Chip Module Development Process," *APCON '95 Papers,* Part 2, 1995.
9. W.H. Knausenberger and L.W. Schaper, "The Myth of Cheap Wire," *Proc. NEPCON West,* Anaheim, (Feb. 1984).
10. J.Reche, "High Density Interconnect for Advanced Packaging," *Proc. NEPCON West,* Anaheim, (March 1989).
11. C.C. Chao, et al., "Multilayer Thin-Film Substrate for Multichip Packaging," *Proc. 38th EIA/IEEE Electron. Compon. Conf.,* Los Angeles, 276 (May 1988).
12. J.K. Hagge, "Ultra-Reliable Packaging for Silicon-on-Silicon WSI," *Proc. 38th EIA/IEEE Electron. Compon. Conf.,* Los Angeles, 282 (May 1988).
13. C.A. Neugebauer, R.O. Carlson, R.A. Fillio, T.R. Haller, *Proc. NEPCON West,* Anaheim (1989).
14. J.K. Hagge, "Ultra-Reliable HWSI with Aluminum Nitride Packaging," NEPCON West, Anaheim, (March 1989).
15. J.K. Hagge, "Applications for Multichip Modules in Avionics," *ISHM Works,* Multichip Modules, Nantucket (June 1990).
16. A.A.Evans, J.K. Hagge, "Advanced Packaging Concepts-Microelectronics Multiple Chip Modules Utilizing Silicon Substrates," *Proc. 1st SAMP Int. Electron. Conf.,* 37 (1987).
17. R.P. Himmel and J.J. Licari, "Fabrication of Large-Area, Thin-Film Multilayer Substrates," *Proc. ISHM Int. Symp. Microelectron.,* Baltimore, 454 (Oct. 1989).
18. D.J. Pedder, "Flip Chip Solder Bonding for Microelectronic Application," *Hybrid Circuits,* (15)4 (Jan. 1988).
19. BPA Report, System 2000.
20. J. Vardaman, "Multichip Packages," *Advanced Packaging,* 36 (Jan./Feb. 1995).

CHAPTER 2

MCM-C MATERIALS, PROCESSES, AND APPLICATIONS

Dr. Rao Tummala
Georgia Institute of Technology
Dr. Philip Garrou
Dow Chemical Company
Dr. John Knickerbocker
IBM Corporation
Dr. Andy Kumar
Applied Materials Corporation

2.1 INTRODUCTION

Commercial ceramic technology for multichip modules can be divided into four major categories:

- Thick film hybrid processes
- High-temperature cofired alumina processes (HTCC)
- Low-temperature cofired ceramic/glass based processes (LTCC)
- High Tc aluminum nitride cofired substrates (AIN)

Thick film hybrids were first introduced into high-end computers in 1980 by UNIVAC. Although generally not considered capable of generating the density required for MCM technology, recent technologies such as photoimageable thick film processing appear to be significant advancements.

HTCC alumina technology for multichip modules has been commercialized for computer, telecommunications, and aerospace applications. It was first applied by IBM for its mainframe applications in early 1980s based on a 33-layer alumina-molybdenum substrate (90 × 90 mm in size) capable of interconnecting, powering, cooling, and protecting 100 bipolar chips.[1] The technology subsequently evolved to a size of 127 mm square, with 63 cofired layers, and supported up to 121 chips with thermal dissipation by air of 600 W for the module.[2–4] At about this time, the industry began to assess three competing technology options for increased mainframe and supercomputer performance. These options include

- Thin film wiring based on copper-polymers (see Chap. 3)
- Better ceramic-metal multilayer systems
- Redistribution of thin film layers on top of either alumina or the new glass-ceramic systems (discussed in detail in Chap. 3)

All three approaches were pursued by the industry, NEC having pursued thin film on alumina substrate and IBM and Fujitsu having pursued low-temperature glass-ceramic-copper cofired multilayer technology. IBM, in addition, pursued the use of one layer of thin film on both alumina and glass-ceramic systems. Although alumina and glass-ceramic systems offered excellent electrical properties including relatively low dielectric constant (down to 5.0) and copper conductivity, as well as excellent thermal expansion match, they have one major drawback: the thermal conductivity of HTCC packages are limited to 20 W/m° C and the thermal conductivity of LTCC is limited to approximately 2 W/m° C less. The recent development of AlN is aimed at applications that require maximum thermal capability. This chapter reviews the multichip ceramic developments in thick film hybrids, high-temperature cofire alumina, HTCC, low-temperature cofire alumina, LTCC, and aluminum nitride ceramics (AlN), as they are used in the fabrication of ceramic MCMs. We will examine materials, processes, and properties and examine some of the specific applications that have been commercialized in MCM-C.

2.2 THICK FILM HYBRIDS

Thick film technology has recently been covered in depth[5] and as such will only be touched on in this chapter.

2.2.1 Thick Film Technology

Thick film hybrid technology produces interconnect by the successive deposition of conductors, dielectric, and/or resistor patterns onto a base substrate. The thick film material, in the form of a paste, is screen printed onto the underlying layer (as shown in Fig. 2.1) then dried and fired.

Lipka has studied the effects of wire surface quality, wire diameter, mesh thickness and operator reproducibility on fine line screen printing and concluded that the limiting factor on fine conductor printing is a complex issue.[6]

The metallurgy chosen for a particular hybrid construction depends on a number of factors including cost sensitivity, conductivity requirements, solderability, wire bondability, and so on. Typical thick film materials are listed in Table 2.1. A comparative summary of typical ceramic interconnect properties is given in Table 2.2.

In 1980, Sperry UNIVAC (now Unysis) introduced the system 80, based on thick film ceramic MCM technology. The substrates usually consisted of three to five conductor layers with integral resistors.[8]

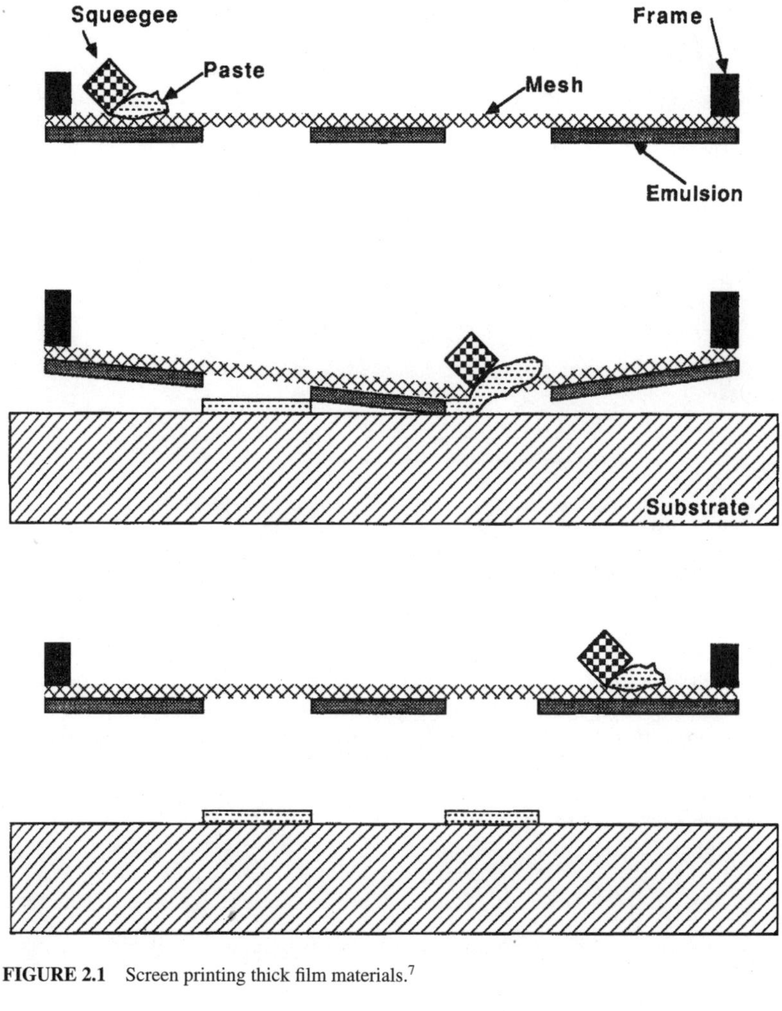

FIGURE 2.1 Screen printing thick film materials.[7]

Table 2.1 Typical Thick Film Materials

Base substrates	Al_2O_3, AlN, BeO
Conductors	Au, Ag, Pd/Ag, Pt/Pd/Au, Cu
Dielectrics	glass ceramics, recrystallizing glasses
Resistors	RuO_2 doped glasses

Table 2.2 Summary of Ceramic MCM Technology Properties[7]

	Thick Film	HTCC	LTCC
Line width (μm)	125	100	100
Via diameter (μm)	250	125	175
Avg. no. conductor layers	1–6	1–75	1–75
Conductor resistance (mΩ/sq)	2–100	8–12	3–20
ε' dielectric	5–9	9–10	5–8
CTE	4–7.5	6	3–8
Tc dielectric (W/m°C)	2	15–20	1–2
Relative cost*	medium	high	high
Relative cost†	medium	medium	medium
Tooling costs	low	high	high
Capital outlay	low	high	medium

* low volume
† high volume

2.2.2 Photoimageable Thick Film Processing

A photoimageable thick film process, Fodel®, has been developed which is reportedly capable of producing 1 mil lines and spaces and 3-mil vias. As shown in Fig. 2.2, a photoactive paste is placed on a substrate and exposed through a mask to define lines or vias. The materials are developed using an aqueous process and fired using typical thick film methods. Both Cu and Au metallizations and up to ten layers have been demonstrated with successful performance under MIL-883.[9]

During the UV exposure, the photoinitiators in the pastes absorb light and initiate radical polymerization of the acrylic monomers in the paste. The organic materials are removed during the binder burnout and firing processes.

2.2.3 Diffusion Patterning

To overcome the limitations of traditional screen printing a higher resolution *diffusion* patterning technique has been developed.[10] The process involves printing and drying an unpatterned dielectric

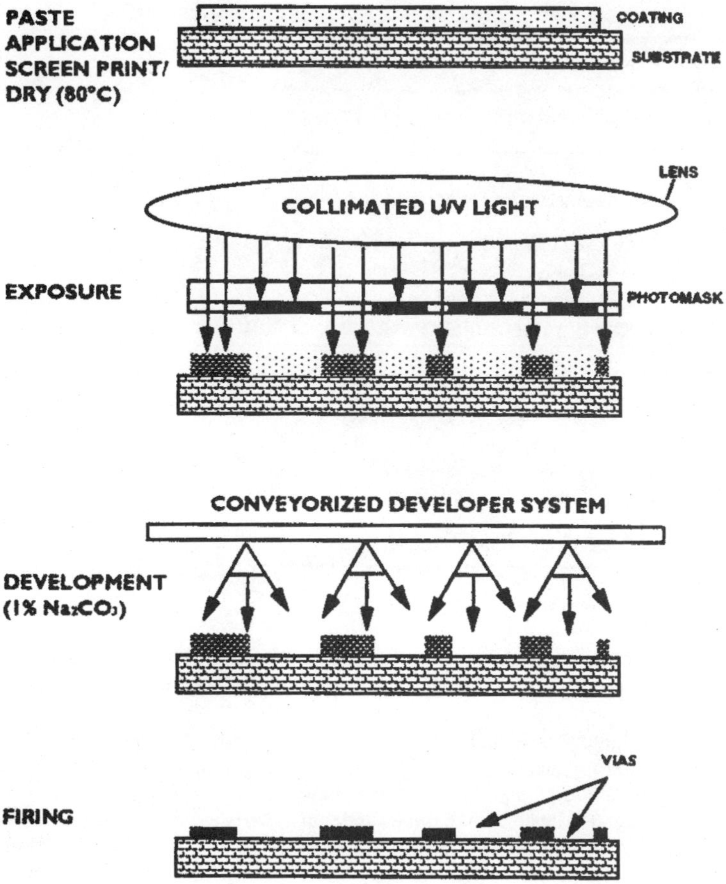

FIGURE 2.2 Fodel® photoimageable thick film processing.[9]

layer, then overprinting a via pattern. The via pattern paste contains an agent that causes the dielectric film beneath to become water soluble. The agent is transferred to the dielectric film as the paste dries. The vias are then generated by rinsing the dielectric layer with water. Reportedly, 150 μm vias are routinely achieved. The basic process is shown in Fig. 2.3.

2.3 HIGH-TEMPERATURE COFIRED ALUMINA (HTCC)

2.3.1 Introduction

HTCC alumina packages are defined as having >85 percent alumina content in the dielectric material, with standard commercial products being approximately 92 percent alumina. Although pure alumina can be used and provides higher mechanical strength, a second phase of glass is most often used at 4 to 12 percent to aid densification at lower sintering temperatures. Alumina (with glass) packages provide a high mechanical strength (~420 MPa), thermal expansion of $\sim 7 \times 10^{-6}$/°C,

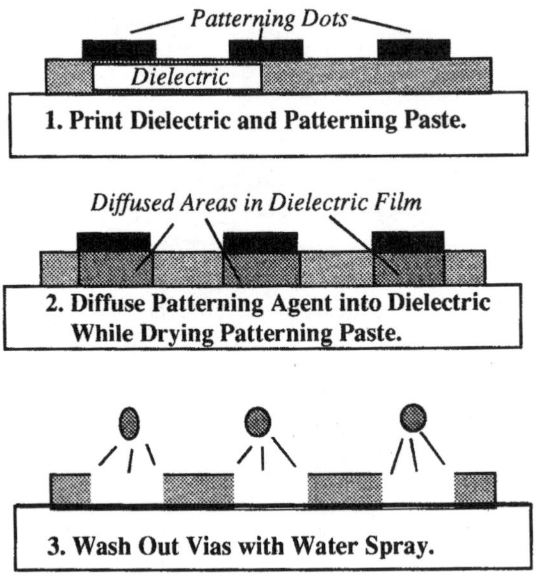

FIGURE 2.3 Diffusion patterning process flow.[10]

dielectric constant ~9.5 and thermal conductivity of ~20 w/m°K. Multilayering of the ceramic provides high density wiring for area array first (chip to package) and second level (package to board) interconnection.

Design of alumina multichip modules should be concurrent with system and chip designs to optimize system performance and minimize cost without compromising reliability. Extensive detail of package electrical design and thermal design have been described elsewhere.[4, 11–13] The significance of package integrated design concurrent with chip, board, and system design has been shown to optimize cost and performance.[14] For 0.5 micron silicon components, package options such as area array interconnection and mixed voltage offer system benefits. Area flip chip technology maximizes die-area utilization and boosts chip I/O density. Surface mount area array packaging such as ball grid array (BGA) or area array pin grid array (PGA) packages provide higher I/O densities which can reduce board size. Multivoltage package capability can offer two or more voltages such as 3.3 and 2.5 V and therefore support reduced chip power consumption. Reduced chip power consumption benefits include longer battery life for portable systems and reduced power dissipation (cooling). Multichip modules with tightly packed chips offer significantly reduced chip-to-chip net length, enabling high-speed buses to run at processor speeds with lower noise, smaller capacitive loading, and shorter time of flight. MCMs can have higher reliability at the system level and lower cost by using smaller, less complex boards with fewer components attached.

2.3.2 Materials for HTCC Alumina Packages

Typical alumina ceramic compositions are given in Ref. 4. Metal conductors of tungsten or molybdenum are used for compatibility in cofiring to temperatures of 1600° C. Materials compatibility during cofiring dictates raw materials of alumina with glass used for densification, and any conductor metal powders (W or Mo) must be designed to closely match onset, rate, and volume shrinkage; promote adhesion; and minimize thermal expansion mismatch between conductor and dielectric. In addition to cofire requirements, green tape and substrate fabrication processes require organic materials such as binders and plasticizers for consistent layer processing through each operation prior to sintering (see below in processing). Metal powders, additives, and organic materials are optimized to ensure conductor paste consistency for via fill and patterning on the ceramic tape or greensheet. Tape and paste organics must be removed by pyrolysis and/or carbon oxidation during cofiring. More details on process compatibility issues are described elsewhere.[4] After sintering, finishing metals of nickel and gold are typically plated onto metal features to provide an interconnection adhesion/diffusion barrier and to enhance wettability and bondability characteristics, respectively. Design of the substrate structure and finishing metallurgies must be consistent with the interconnection technology to be employed.

2.3.3 Processing HTCC Ceramics

An overall process flow for multilayer ceramic substrate fabrication is shown in Figure 2.4. Although the processes described below are specific for IBM, they are generally similar for all ceramic processors. Key process technologies are discussed in the following sections: alumina raw material and green sheet technology, metal powder and paste, thick film metallization inspection, stacking and lamination, sintering, plating, and electrical test.

Alumina Raw Material and Greensheet Technology. The raw materials used in fabrication of alumina substrates include aluminum oxide, glass, binder, plasticizer, and solvent. Materials specifications are used to control alumina particle size, surface area, impurity, and agglomeration. Glass frit is controlled through composition, glass transition and softening point, particle size, and surface area. Molecular weight, group chemistry, and viscosity controls are used for the binder and plasticizer. The raw materials, along with solvent, are homogenized in a ball milling operation. The dispersed ceramic slurry is passed through a filter to remove contaminates and agglomerates. Next, the slurry is cast into thin sheet form where the solvent is removed (Fig. 2.5). The resulting product, called a *greensheet,* is used as the basic building block in the fabrication of substrates. The choices of material and processes are controlled to provide structural uniformity in greensheet microstructure and binder/plasticizer distributions. Greensheets are also designed to maintain dimensional control and to provide good mechanical properties, including, toughness, modulus, yield strength, compressibility, and laminated bond strength. Consistency in greensheet microstructure forms the foundation for consistency in thick film metallization and ceramic sintering.

Metal Powder and Paste. Metal powders cofired with aluminum oxide in volume production are molybdenum- and/or tungsten-based metals due to the ~1600° C temperature used during sintering. In addition to a conductive metal powder, glass powders can be used to aid in densification, shrinkage matching to the dielectric, chemical and mechanical bonding to the ceramic, and for thermal expansion adjustment. A thick film paste often uses metal powder, glass powder, organic resins and

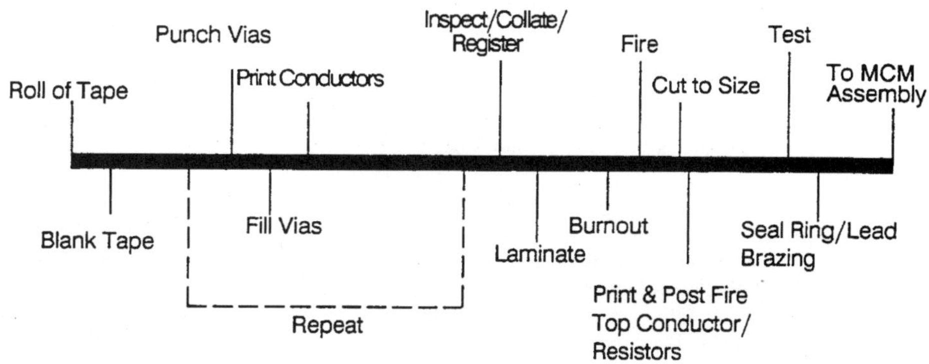

FIGURE 2.4 MLC process flow.

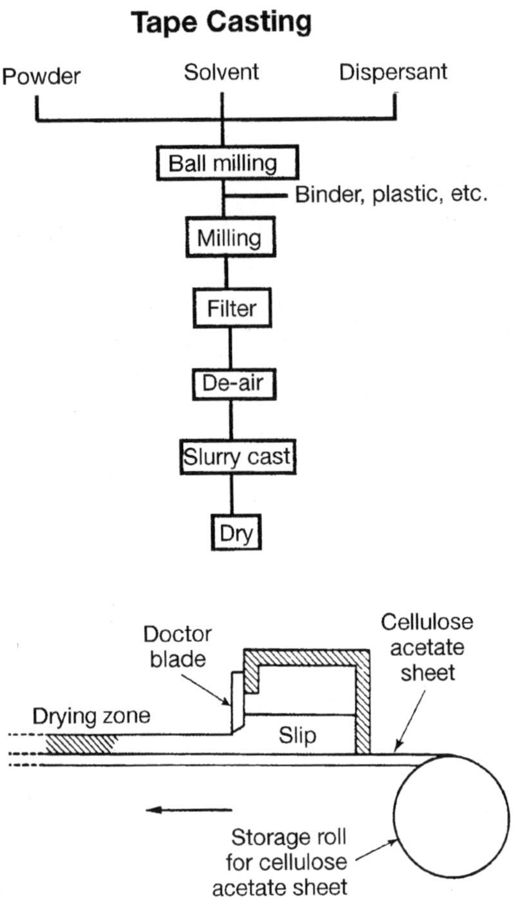

FIGURE 2.5 Doctor blade and slip casting.

solvents. Compositions are varied to control screening properties, metal shrinkage, and conductivity. Paste fabrication begins with batch mixing, dispersion, and deagglomeration, which are completed on a three-roll mill. Solids content, composition, particle size distribution, and rheological properties are controlled for paste reproducibility.

Thick Film Metallization. The greensheet is cast, dried, stripped from the carrier film, and blanked into defect-free sheets, typically ~200 mm sq. The greensheet is then processed through punching, screening, and inspection operations.

Mechanical punching is accomplished by using a punch and die bushing to pattern a greensheet with its unique pattern of holes, for Z metallization and corner locating or reference holes for subsequent alignment and stacking. For larger vias (200 μm) with less dense patterns, a fixed-head punch is most often employed in high-volume production. For smaller vias (75 to 150 μm) and higher densities, a programmable automated punch is most often used for production. In either case, the punch is pushed through the greensheet, which is supported by a die bushing of larger size than the punch diameter. For small vias (25 to 50 μm), laser formation of the via or photo patterning of the greensheet is employed. A schematic of a mechanical punch is illustrated in Fig. 2.6. Positional accuracy, verification of hole formation, and absence of debris are control parameters.

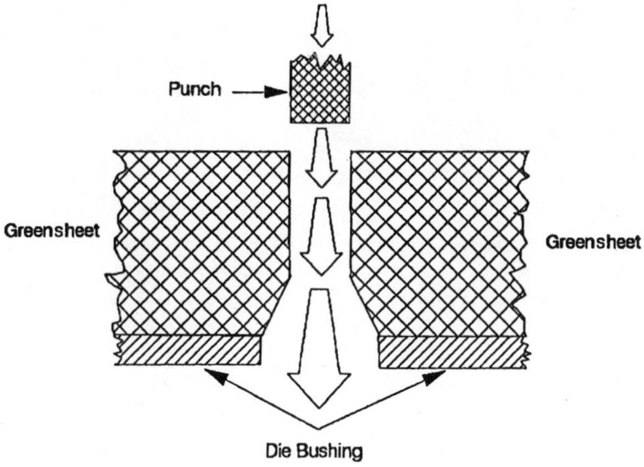

FIGURE 2.6 Schematic of greensheet punching.

Screening of thick film conductive metal pastes can use either a one or two step operation. The two step process first uses a mask patterned with holes for via fill followed by a drying operation (to remove paste solvents) and second screening and drying step to pattern the surface of the greensheet with signal and voltage wiring. In these operations a metal mask is often used for via fill and an emulsion coated metal screen (mesh mask) or etched metal mask is used for patterning. The one-step process used at IBM[1,3] employs an extrusion of metal paste through a etched metal mask to fill vias and form thick film wiring patterns simultaneously. The patterned greensheet is then dried. In both processes, it is desirable to balance the interaction between paste and greensheet to optimize paste adhesion and control geometrical line formation without causing significant dimensional instability.

Inspection. Following greensheet screening, an inspection of patterns can be used to ensure that no opens or shorts exist and to verify complete via fill. The operation uses optical contrasting between paste and greensheet. Inspection is used for patterning with very high wiring density where the cost of operation is warranted by yield benefits at final electrical test.

Stacking, Lamination, and Sizing. In stacking and lamination, each individual layer type is precision aligned using location holes and pins for alignment or individually aligned and cut greensheets using the edge for alignment. Precise registration of each layer must be made to ensure good electrical contact between layers and avoid shorting. The operation of stacking, followed by an elevated temperature and pressure at lamination, will consolidate the individual sheets into a single unit. A binder such as polyvinyl butyryl provides good bonding using temperatures just above the glass transition temperature. Other organic systems have been developed that require the application of an adhesive between each layer to avoid delamination in subsequent operations. In any organic system, a balance must be struck between having adequate material flow or greensheet compressibility to ensure complete encapsulation of metal features (signal lines, voltage planes) without having too much lateral flow, causing excessive distortion. For smaller multichip modules processed multi-up through lamination, individual substrates can be diced using a hot guillotine knife or green sizing saw. In addition, edge and corner chamfers can be introduced at this point prior to sintering by use of a routing, machining, or cutting operation.

Sintering. At sintering, the multilayer ceramic composite of organics, metal powder, ceramic, and glass powders is cofired into a dense insulating ceramic with embedded 3D networks of electrically conductive wiring. In the stages of the cofire process with heating to 600° C, the organic materials are removed by means of decomposition or pyrolysis. Any carbon residue is removed with a wet hydrogen or wet forming gas atmosphere at temperatures up to 1200° C. The thermodynamics of the Mo-O-C system or W-O-C system must be controlled through the pH_2/pH_2O ratio to control reaction of carbon oxidation (shown in Eq. 2.1) without oxidizing either molybdenum or tungsten.

$$C + H_2O \rightarrow CO_2 + H_2 \tag{2.1}$$

With continued heating to temperatures in the range of 1200 to 1600° C, densification, often aided by liquid phase sintering, occurs where both powder metal and ceramic powder/glass powder coalesce into a dense substrate. Control of temperature and atmosphere is used to provide consistency in shrinkage rates, which must be maintained for final substrate and metal feature dimensional control. As reported elsewhere,[1,15,16] total linear shrinkages of 14 to 17 percent are common. In addition to total volumetric shrinkage, matching onset and rate of metal and ceramic shrinkages are important for consistent dimensional control. Control of lot-to-lot variations in powder characteristics, greensheet properties, paste properties, laminate properties, and sintering must be continuously maintained to control the final substrate shrinkage and distortion. Improved methods for atmosphere and temperature control have also enhanced the ability to maintain dimensional controls where shrinkage tolerance of ±0.3 to ±0.1 percent are required.

Plating and Electrical Test. The final substrate processing is plating and electrical test. Metal surface features are plated (typically with nickel and gold) to provide a wettable surface for connection. Chip-to-package interconnections use either chip solder connections (flip chip) or wire bond pads. Package-to-board interconnections using input/output (I/O) pads are prepared for I/O pins or surface mount attach such as ball grid arrays or column grid arrays.

The plating process uses either an electrolytic nickel bath and electrolytic gold or an electroless nickel bath and electroless gold bath, or combinations as required by the product. The electrolytic processes use an electrical contact to each feature to be plated. The plating bath provides nickel or gold source during plating. For electroless plating, a bath is chosen that selectively plates all surface metal features without depositing nickel or gold on the ceramic. After depositing electroless nickel, the substrate is heat treated to improve adhesion, and thin gold is plated to prevent oxide formation and improve wettability. For processing of wire bond pads and seal bands, a thick gold layer is selectively plated to ensure robust connections. The electroless processes are often required for multichip modules because many surface features that interconnect one chip to another do not also connect to an I/O pin, thereby eliminating the opportunity for a frequently used electrolytic connection methodology of connecting to pins. Alternatively internal connections can be made for electrolytic plating in some cases but is limited in application due to wireability.

Substrates are electrically tested by contacting top and bottom surface pads to ensure the integrity of the signal and voltage wiring connections. Chip-to-chip as well as from chip-to-board wiring circuits are evaluated. Parametric testing can ensure that net resistances and capacitances are within desired specification limits. Subsequent to electrical substrate test, I/O pins are brazed on the surface using solder, copper/silver, silver, or gold/tin metallurgies. The pins are selectively plated with nickel and gold to improve wettability and prevent corrosion.

Module Assembly. The preferred chip assembly process with ceramic multichip modules consists of taking both known good die (Chap. 14) and known good packages (Chap. 7), chip joining (Chap. 8), testing (Chap. 15), and chip rework (when needed), followed by encapsulation.

For high-density interconnections, flip chip or C-4 (controlled collapse chip connection) technology is used to mount silicon die to the substrate surface. After placing all chips on the surface (often with flux to improve wettability to surface connections) the substrate and flip chips are heated to simultaneously make all solder connections. Chip in-place tests can check for interconnection integrity. If required, chips can be removed and replaced by either furnace reflow or a localized heat source. Capping or cover assembly mounting is used to enhance the multichip module system life through either hermetic encapsulation or module nonhermetic enhancements.

2.3.4 HTCC Alumina MCM Conclusions

The use of alumina multilayer ceramics for multichip modules applications have been proven in both industry and the military.[1–4,13,17–20] IBM realized the benefits of improved performance, high reliability, and low cost in a compact size when it introduced an alumina MCM in the 1970s (4300 series mainframe).[2–17]

HTCC Ceramic MCM packages are also finding application in telecommunications.[16] MCM-C can provide a cost/performance solution for a wide range of applications based on the following merits:

1. Electrical performance characteristics including 50 Ω impedance, low conductor resistance, ability to integrate passive components such as capacitors and inductors, ability to achieve high wiring density (ease of increasing the number of wiring layers at low cost), ability to support high-speed simultaneous switching drivers, and ease of supporting multiple reference voltages[14,16,21]
2. Inherent thermal performance characteristics superior to MCM-L and MCM-D[16]
3. Time demonstrated reliability[2,4,22]

2.4 LOW-TEMPERATURE COFIRED CERAMIC (LTCC) SUBSTRATES

2.4.1 Introduction

The use of glass and glass-ceramics in electronic packaging dates back to the invention of semiconductors. Glasses are used for sealing T-O type packages and CERDIPs, as crossover and interlevel dielectrics in hybrid substrates, and as sintering additives in alumina substrates. The first reported development of crystallizing glasses to form self-supporting glass-ceramic substrates with cofired conductors took place in IBM in the late 1980s, soon after this company's pioneering launch of the cofired alumina substrates into commercial use. Japanese computer manufacturers Fujitsu and NEC soon followed suit. The very success of the cofired alumina substrate spurred the development of the multilayer glass-ceramic substrates to obtain certain performance and reliability advantages over similar alumina substrates such as those listed in Table 2.3.[24] These advantages are derived from the higher electrical conductivity lines of copper, silver, or gold; lower dielectric constant of the glass-ceramic; and closer CTE match of the substrate to silicon. These advantages were important in developing the fastest mainframe and super computers.

2.4.2 Glass-Ceramic Compositions

Two approaches have been used to obtain glass-ceramic compositions suitable for fabricating self-supporting substrates. In the first, fine powder of a suitable glass composition is used that has the ability to sinters well in the glassy state and simultaneously crystallize to become a glass-ceramic. More commonly, mixtures of ceramic powders are used, such as alumna and a suitable glass in nearly equal proportions, to obtain densely sintered substrates. In both cases, the low sintering temperature (850 to 950° C) allows for cofiring thick film inks of copper, silver, or gold with the ceramic. Because many glass and ceramic powders can be used to obtain densely sintered glass-ceramics, the actual choice is often made on the basis of other desirable attributes in the resulting glass ceramic such as low dielectric constant for lowering the signal propagation delay and coefficient of thermal expansion (CTE) closely matched to the CTE of silicon to improve the reliability of solder interconnections. Unlike the case of a crystallizing glass, the mixed glass and ceramic approach allows much wider choice of materials. Thus, many companies have developed their own particular LTCC compositions. Table 2.4 presents a partial list of companies and their LTCC compositions.

Table 2.3 Comparison of Alumina and Glass-Ceramic Multilayer Substrates

Attribute	Multilayer Alumina Substrates (HTCC)	Multilayer Glass/Ceramic Substrates (LTCC)
Metallization	(Mo thick film)	(Cu thick film)
Dielectric constant (k)	9.4	5.0
Signal propagation delay (ps/cm)	131	95
Capacitance (crosstalk)	2.61	1.16
Line conductivity (0.0025 × 0.01 cm)	0.46	0.14
DC losses		60% lower
Number of layers	33	6 3
Impedance (Ω)	60	75
Ceramic CTE ($\times 10^{-6}$/°C)	6.0	3.0
Reliability of flip chip	No fails in >20 billion solder interconnections	Extend similar reliability connection-hours record to larger chips
Manufacturing process	Greentape process, 17% lateral firing shrinkage	Greentape process, 0% lateral firing shrinkage

The industry's first glass-ceramic copper multilayer substrate was developed by IBM for use in its fastest mainframe computers. These 127.5 × 127. 5 mm, 63-layer substrates, shown in Fig. 2.7, have now been surpassed both in size and number of circuit layers by glass-ceramic substrates made by Fujitsu and NEC.

IBM used a glass in the MgO-Al_2O_3-SiO_2 system to fabricate the glass-ceramic substrate having cordierite as the principal crystalline phase. Finely ground glass powder constituted the only inorganic material in the green tape. The green tape blanks were provided with screen printed thick film copper conductor patterns and copper paste feed via holes. During the firing process, the glass powder sinters and crystallizes to give the body sufficient rigidity to retain its shape. Figure 2.8 shows the differential thermal analysis curve of this glass to illustrate the sequence of sintering and crystallization events taking place.[26] Upon completion of the firing cycle, the glass fully sinters and crystallizes to cordierite with small amounts of binary magnesium silicates such as enstatite. Several others have also successfully developed LTCC substrates using crystallizable MgO-Al_2O_3-SiO_2 glasses and other systems.[27,28]

Fujitsu and NEC use mixtures ceramic and glass to fabricate large, multilayer LTCC substrates. Fujitsu uses a mixture of alumina with 40–50 percent borosilicate glass, choosing the composition carefully to match the shrinkage characteristics of thick film copper and gold.[29] To promote good sintering, the glass and ceramic were ground to particle size of 5 microns or less. The dielectric con-

Table 2.4 Examples of Glass-Ceramics Used in Substrate Applications[2]

	Material	Firing Temp. (°C)	TCE ($\times 10^{-6}$/°C)	Dielectric Constant (1 MHz)	Bending Strength (Mpa)	Conductor Material	Company
Cordierite-type crystallizable glasses	FC-01	900–1050	2.2–3.5	4.9–5.6	170	AuCu	NTK
	Zn-cordierite					Ag/Pd	Fujitsu
	Cordierite	850–950	2.4–5.5	5.3–5.7	180–230	Cu	IBM
	β-Spodumene		2.83	5.65		Cu	IBM
Glass-filled composites	Boro-silicate alumina	1000	4.0	5.6	240	Cu	Fujitsu
	Pb-borosilicate + alumina	900	4.2	7.5	300	Au	NEC
	CaO-A_2O_3-SiO_2-B_2O_3 + alumina	880	5.5	7.7	200	Au, Ag/Pd	Narumi China
	MgO-Al_2O_3-SiO_2-B_2O_3 + silica	850–950	3–7.9	4.25–5	150	Ag, Ag/Pd	Hitachi
	Glass + Al_2O_3 + $CaZrO_3$	850	7.9	8.8	210	Au, Ag	DuPont
	SiO_2-B_2O_3+ alumina+ $2MgOSiO_2$	850–950	6.0	6.5	200	Ag/Pd	Asahi Glass
Crystalline phase ceramics	$BaSn(BO_3)_3$	960–980	5.5	8.5	170	Ag/Pd	Toshiba
	CaO-Al_2O_3-SiO_2-B_2O_3	950	4.8	6.7	250	Cu, Ni	Taiyo Yuden
	BaO-CaO-Al_2O_3-SiO_2-B_2O_3	950–1000	8.0	6.1	200	Ag, Ag/Pd	Murata

stant of the resulting glass-ceramic was 5.7, which is significantly less than that of alumina. NEC successfully used a mixture of 55 wt.% alumina and 45 wt.% lead boro-silicate glass to fabricate its glass-ceramic of dielectric constant 7.8.[30] These substrates had thick film lines of either silver-palladium or gold. Fujitsu's 245 × 245 mm, 61-layer substrates with 30 signal layers, is still the largest multilayer substrate ever used. The thick film copper lines in this substrate added up to nearly a kilometer in length. NEC's 225 × 225 mm, 78-layer substrate with 13 signal layers is also quite impressive. Both Fujitsu and NEC developed their substrates for use in supercomputer applications where high signal propagation speed is of paramount importance. Very low dielectric constant (k = 4.0–4.5)

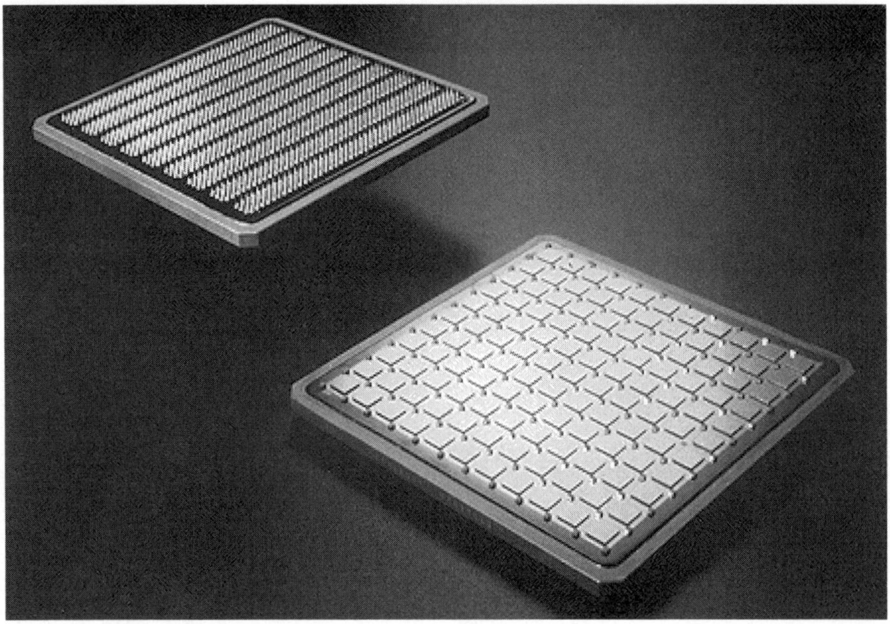

FIGURE 2.7 IBM glass-ceramic/Cu multilayer substrates with brazed I/O pins and flip chip soldered chips.

values have been obtained in LTCC substrates either by deliberate introduction of porosity[31] or by rendering certain high silica compositions sinterable at about 1000 C.[32]

When mixtures of glass and ceramic are fired together, the glass softens, flows, and wets the surface of the ceramic particles, causing the consolidation of the body by capillary forces. In the absence of the ceramic phase, the glass would ball up due to surface tension. In most cases, the glass and the ceramic surface react to form new crystalline phases, affecting the properties of the resulting glass-ceramic substrate—usually in the desired direction. For example, when mixtures of alumina and lead boro silicate are fired together, the reaction product is anorthite, a calcium-alumino-silicate compound having low dielectric constant, low CTE, and high strength. When a mixture of cordierite and borosilicate glass was used to obtain a low CTE glass-ceramic, the reaction between the two produced a result quite opposite what was expected—a CTE of nearly $17 \times 10^{-6}/°C$ instead of the expected value of $2\text{–}3 \times 10^{-6}/°C$ due to the formation of cristobalite. Addition of alumina to the composition was found to inhibit this reaction.[30,32]

2.4.3 Glass Ceramic Substrate Fabrication

Shrinkage Control. The fabrication of multilayer glass-ceramic substrates follows substantially the same processing sequence as used for alumina substrates. However, there were at least three unique and difficult challenges to overcome. The first challenge, especially in the production of the large, multilayer glass-ceramic substrates, was shrinkage control. A tolerable degree of shrinkage variation cannot exceed few tenths of a percent over large substrates without seriously affecting the positioning of the bonding pads.

Two methods for achieving such high degree of shrinkage control have been successfully employed. The first is the traditional method of shrinkage control by green density control. Due to the presence of large volume fraction glass in the composition and very fine particle sizes of the ceramic and glass powders used, the sintered body easily achieves full densification. Hence, the only

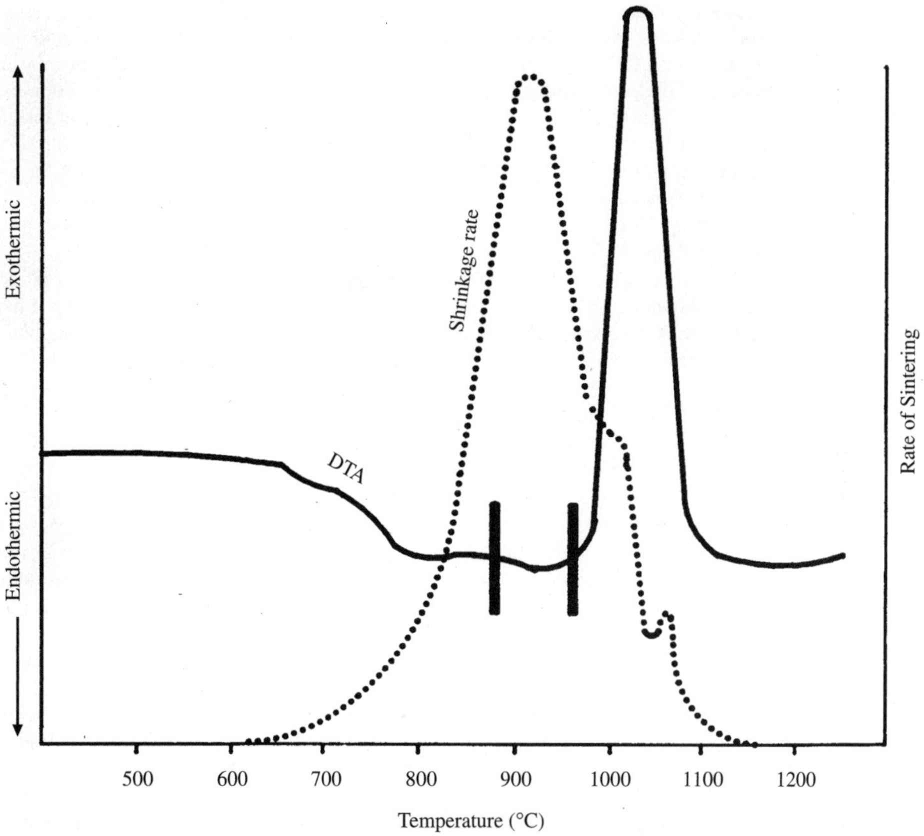

FIGURE 2.8 Differential thermal analysis (DTA) and shrinkage rate curves for a sinterable $MgO\text{-}Al_2O_2\text{-}SiO_2$ glass powder.

factor controlling the shrinkage is green density. This, in turn, is controlled by the volume fraction solids in the slurry, the degree of dispersion of the particle in the green tape, and the lamination uniformity. In practice, the particle sizes and slurry composition are generally first optimized for lowest shrinkage level, good slurry dispersion, and casting characteristics, as well as green tape properties such as flexibility, ability give clean punched via holes without cracking, and stability. Then, shrinkage control is achieved solely through control of lamination pressure.

The other method for shrinkage control (successfully used by IBM in fabricating their large glass-ceramic substrates) is to totally constrain the lateral shrinkage. The glass-ceramic system in question had the ability to sinter to full density despite this constraint by accommodating the needed shrinkage in thickness alone. By this method, they achieved unprecedented low shrinkage variation of 0.1 percent. Another novel method for suppressing the lateral shrinkage, at least for green laminates of 15 layers or less is to fire the glass-ceramic substrates on suitably prepared, thin metal plates of matched expansion.[33] Early in the sintering process, the glass ceramic layers adhere to the Cu-Mo-Cu metal plate which then acts to totally suppress the lateral shrinkage. The integral metal plate of the substrate can be used as heat spreader, ground plane, or simply as a reinforcing member. Also, single- or double-sided substrates can be produced in large format on these metal plates. Control of shrinkage and distortion in multilayer substrates also requires careful formulation of thick film inks for matched shrinkage.

Binder Burnout. The organic binders used for forming the green tape normally burn off in air at around 500° C without any carbon residue. Thus, when thick film silver or gold inks are used in conjunction with the glass-ceramic, the substrates can be fired in air without any problem of carbon removal. NEC used gold thick film pastes to fabricate their large glass ceramic substrates. Others have successfully produced multilayer substrates with silver inks. With silver inks, a yellow discoloration is generally observed around the conductors due to the formation of a volatile silver oxide at the beginning of the firing cycle. There is still a considerable reluctance on the part of some to use silver inks because of the early history of catastrophic failures due to silver migration in hybrid substrates that was eventually solved by alloying with palladium. Time and again, tests have shown that no silver migration occurs in densely sintered substrates.

The application of LTCC to consumer applications requires the use of cost-effective metallurgy. The gold conductors used in typical LTCC constructions are highly reliable but present a cost disadvantage. Silver inner layers would have a significant cost advantage over gold; however, practitioners have worried about Ag migration under THB (biased 85° C/85 percent humidity) conditions.

Recent studies have shown that silver inner layers are reliable under a variety of accelerated aging conditions.[34,35,37] However, all surface metallization layers except gold migrated. This is resolved by using a barrier via fill material[36] that connects the silver inner layer conductors to the gold surface metallurgy. This system passed extensive reliability testing.

Due to higher the cost of gold pastes, and because of the concerns of silver migration, the use of copper thick film inks for glass-ceramic substrates looks very attractive. Copper pastes also have better electrical conductivity than either silver or gold. However, the use of copper thick film inks needs an inert or slightly reducing sintering ambient in which the charred binder residues are not removed.

Three distinct approaches to dealing with the binder removal problem have been used successfully. The simplest is that used by Fujitsu, where the binder used was selected on the basis of its ability to completely decompose (unzip) to monomers before any charring takes place in neutral ambient. Others have found that while these so-called *unzippable polymers* are effective in removing more than 99.99 percent of carbon, small amounts of carbon residues remain behind due to the difficulty of removing the adsorbed binder molecules on the glass powder surface. Even very small amounts of residual carbon (~100 ppm) can completely destroy the insulation resistance of the ceramic [see Ref. 30].

The second approach to binder removal is to fire the substrates in an ambient that is oxidizing to carbon while remaining reducing to copper.[38] Such a regime exists in the required temperature range. The needed very small oxygen potentials are achieved through mixing small amounts of hydrogen in steam. In a typical firing cycle, the green laminates are heated in nitrogen to the chosen temperature for binder burnout. They are held at this temperature long enough to completely remove the carbon residues of the binder by gasification. The bisque laminate is then heated further to the final sintering temperature in nitrogen or hydrogen so as to drive out any absorbed moisture. Without this drying, the water dissolved in the glass will come out when it crystallizes, causing the substrate to bloat.

A novel approach to overcoming the binder burnout problems is to use copper oxide thick film inks in place of copper thick film inks for the internal layers of the glass ceramic substrates.[39] The green laminate is first fired in air to 500° C to completely remove the binder. It is then cooled and refired in a mixture of nitrogen and hydrogen at 300° C to reduce the copper oxide in the thick film lines to porous copper. Subsequent heating to 900° C in nitrogen completely consolidates the copper lines and sinters the glass/ceramic.

Ceramic-Metal Interface. The large difference in the coefficients of thermal expansion of the glass-ceramic and the thick film lines of copper, silver, or gold poses a significant problem if it becomes necessary to obtain a hermetic glass-ceramic substrate. At via locations, the larger contraction of the metal filling the vias compared to the glass-ceramic will cause it to separate from the sidewall. To overcome this problem, it is common to add significant quantities of glass or ceramic fillers to the thick film inks—sometimes exceeding the percolation limit for conduction. Fortunately, the segregation of the metal particles to the grain boundaries between the glass and ceramic fillers allows for sufficient electrical conduction even at these very low levels of metal loading in the via inks.

Because high conductivity, rather than adhesion, is important in the buried conductor lines, the thick film inks used for them are either only lightly loaded with fillers or not at all.

Surface Metallization. The adhesion of bonding pads and other thick film features such as seal bands on the surface of the glass-ceramic substrate surface is problematic. Although these can be cofired, it is often found necessary to fire these surface patterns in a separate step following sintering to obtain better feature location accuracy or to form surface metallization with a different thick film ink. When copper thick film ink is used for the internal lines in a substrate, gold or silver-palladium inks are used for surface metallization. If silver thick film lines and features are fired on the substrate, they need to be covered over with electroless nickel to prevent silver migration. If gold dots are to be printed on top of copper or silver filled vies, it becomes necessary to interpose another metal such as palladium or rhodium between them to prevent the formation of the so-called *Kirkendall voids* at their interface due to vastly differing diffusion rates of one into the other.[40]

2.4.4 Future of Glass-Ceramic Substrate Technology

In the nearly ten years since their first commercial use, glass-ceramic substrates appear to have found little use beyond the few multichip substrate applications in high-performance computers. Among the reasons may be that (a) most were developed and manufactured for use in captive situations in large computer companies, and (b) they have some important limitations such as low thermal conductivity, low strength, poor thick film pad adhesion, or high cost compared to alumina substrates. Hence, these could not find use in single-chip packages where alumina is dominant. Even in the special use packages such as microwave packages, mixed signal modules, and so forth, the design rules are invariably geared towards the use of alumina ceramic packages.

The greatest potential for glass-ceramic substrates may lie in the ability to incorporate resistor and capacitors in the internal layers of the glass-ceramic.[41] This ability is derived from the fact that, when thick film inks of silver or gold are used, the firing temperatures (800–950° C) and the air ambient are similar to the firing conditions needed for many thick film inks used for resistors and capacitors. Buried resistors, capacitors, and inductors can be formed in the internal layers of glass-ceramic substrates with high yield, good tolerances, and high stability. Such substrates, when available commercially, are expected to find application in telecommunication and other portable electronic applications.

2.5 ALUMINUM NITRIDE

2.5.1 Introduction

In the 1990s, aluminum nitride (AlN) products are being used in a variety of commercial and military applications. Increasing frequency, increasing circuits per device, and smaller component sizes are continuing to drive increased power requirements per unit area, even with lower and multiple voltages. Thermal management with solutions such as AlN can provide superior cooling to ensure reliable system operation. A global AlN infrastructure continues to expand with improved quality and consistency in powder and components. Applications of AlN in electronics range from commercial and military production levels to expanding prototype assessments in data processing, communications, and automotive requirements, each with production-level applications.

Aluminum nitride (AlN) packages typically offer a thermal conductivity of 150–200 W/m°K, a level which can be compared with many metals or other high thermal conductive materials such as berillia (BeO) or silicon carbide (SiC). AlN has a thermal coefficient of expansion of 4.4 ppm, which is better matched to silicon than alumina or plastics. Other characteristics are comparable with alumina ceramics as shown in Table 2.5 and Figs. 2.9 and 2.10.

One should take advantage of AlN for thermal management through thermal design at the system level. At this level, temperature control can be critical to system performance optimization and affect reliability. For example, increased CMOS operating temperatures can lead to decreasing performance (0.25 percent circuit delay per °C rise).[42] Device failure is also accelerated with increased thermal cycling in an Arrhenius relationship. For higher operating temperatures, device failure accelerates through failure mechanisms such as corrosion and electromigration.[43] Therefore, to optimize performance and ensure reliability, thermal dissipation becomes important.

Table 2.5 AIN Properties Comparison

	AIN	HTCC	LTCC	BeO	Mullite	Si	Cu
Thermal conductivity, W/m°K	175	25	2	260	5	150	394
Density, g/cm	3.3	3.9	2.6	2.8	2.3		8.9
CTE, RT–300° C × 10^{-6}/°C	4.4	7.1	3	7.2	4.2	3	18
Dielectric constant, 1 MHz	8.9	9.5	5.0	6.7	6.4		
Dissipation factor, 1 MHz	0.0004	0.0004	0.0002	0.0004	0.0004		
Bending strength, MPa	320	420	210	220	380		

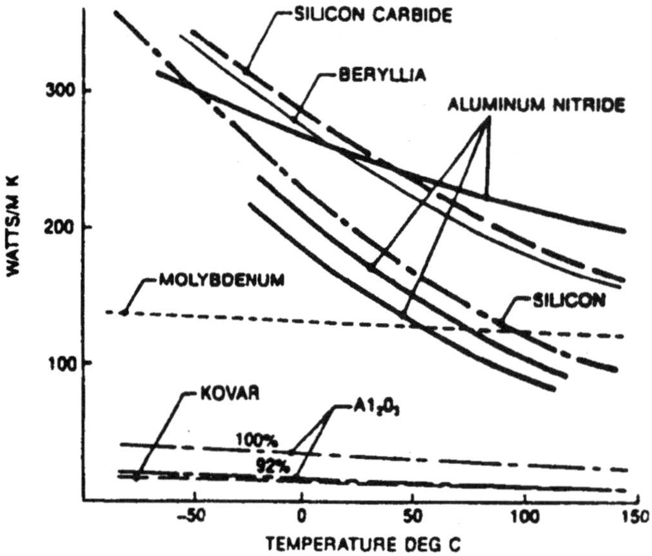

FIGURE 2.9 Thermal conductivity of three grades of AlN vs. other packaging materials.[63]

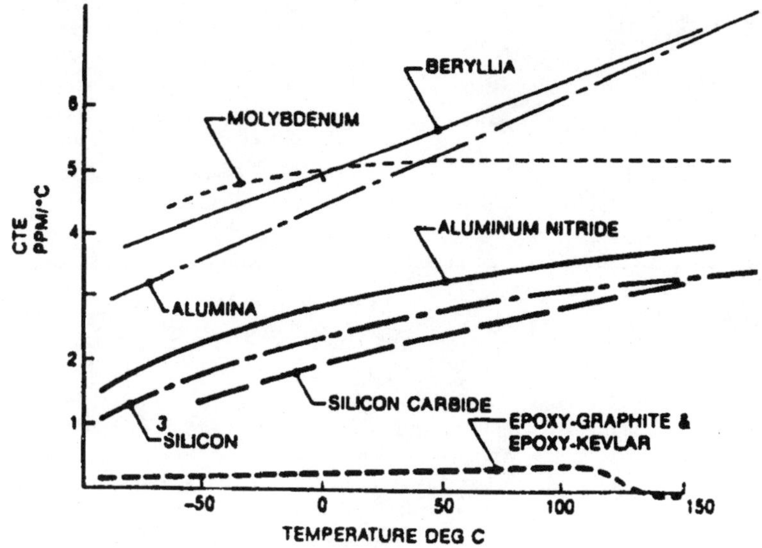

FIGURE 2.10 TCE of AlN vs. other packaging materials.[63]

This is clearly shown in Fig. 2.11[62] where a Si chip (1.2 × 0.55 × 0.16 mm) that dissipates 2 W is attached to an HTCC alumina or AlN substrate (50 × 50 × 0.635 mm). The thermal conductivity of the AlN substrate was 190 W/m°C. The resulting thermograms shown in Fig. 2.11 clearly demonstrates that the area under the Si chip attached to the AlN substrate is cooler (47 vs. 64° C) and that the edge temperature of the AlN is higher since the heat is being dissipated.

For thermal dissipation, the thermal resistance is proportional to thickness and inversely proportional to thermal conductivity. Additionally, the resistance can be dramatically influenced at all inter-

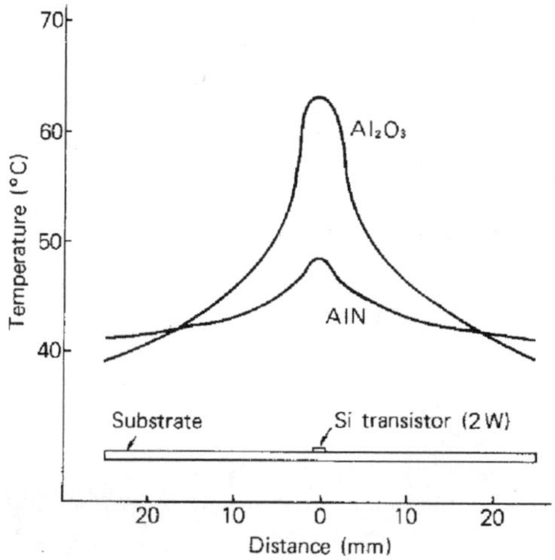

FIGURE 2.11 Heat dissipation characteristics for HTCC alumina and AlN substrates.[62]

face junctions. For an increasing number of applications, AlN can provide the most cost-effective system solution for thermal dissipation.

2.5.2 Materials for Multilayered AlN Packages

Aluminum nitride is a synthetic compound manufactured by two processes: the carbothermal reduction of alumina (Eq. 2.2), and/or the direct nitridation of aluminum metal (Eq. 2.3).

$$Al_2O_3 + 3C + N_2 \rightarrow 2AlN + 3CO \tag{2.2}$$

$$2Al + N_2 \rightarrow 2AlN \tag{2.3}$$

Although single-crystal AlN has been shown to have a theoretical thermal conductivity of 320 W/m°C, polycrystalline AlN typically exhibits much lower thermal conductivities due to oxygen and metallic impurities in the Wurtzite-like lattice structure.[45,46] In practice, sintering aids such as yttrium oxide (Y_2O_3), calcium oxide (CaO), and/or glass frit additions are made to promote densification, getter oxygen from the surface and lattice of AlN, and reduce oxygen diffusion into the AlN grains. Cofiring AlN with densification aides and a conductor metal of tungsten or molybdenum can achieve thermal conductivities of 150 to over 200 W/m°K.

As discussed in the alumina materials sector, cofiring requires materials compatibility to achieve shrinkage matching for onset, rate, and volume shrinkage of both metal and ceramic. In addition, adhesion and thermal expansion between conductor and dielectric must be controlled. Similarly, organic compatibility with fabrication of greensheets and conductor paste must provide both process compatibility prior to sintering as well as provide ease of removal during cofiring. Finishing metallurgies of nickel and gold are typically plated onto the metal features.

2.5.3 AlN Materials and Greensheet Technology

Aluminum nitride powders produced from either the carbothermal or the direct nitride process are utilized in the fabrication of multilayer ceramic substrates most often with yttria or calcium oxide and possibly a glass. AlN powders from the carbothermal process can provide a narrower particle size distribution, a smaller particle size, and lower oxygen content, which attributes are beneficial for controlled sintering. Optimization of the AlN formulation must be made for compatibility with sintering parameters, cofired metallurgy, process parameters, and the desired product characteristics. For example, three cofiring approaches have been commonly practiced,[47–52] which include

- Hot pressing or pressure-assisted sintering
- Pressureless high-temperature sintering (>1800° C)
- Pressureless low-temperature sintering (<1650° C)

For hot pressing and high-temperature sintering, alkali earth or rare earth oxides can adequately aid sintering. For low-temperature sinter, either organometallic precursors or, more commonly, sintering additives are used to initiate liquid phase or viscous sintering.

For cofiring, compatibility with processing (i.e., greensheets through sintering) and metallization with tungsten or molybdenum are required. This is similar to alumina and low-temperature cofire ceramic (LTCC) systems discussed previously. Factors such as organics used in casting for greensheets and those used in metal paste fabrication must be able to meet process requirements and subsequently be removed in cofiring without AlN package electrical and mechanical characteristics.[53]

Metal Powder and Paste. Metal powders cofired with aluminum nitride are most often tungsten or molybdenum due to the high cofire temperatures of 1450 to 1900° C. The metal paste should be optimized for consistent via fill and patterning on greensheets without causing distortions due to paste solvent greensheet interactions. In addition, the metal system should be optimized to match AlN shrinkage (onset, rate, volume) and provide good adhesion characteristics.[54,55] Paste additives include glass frit, ceramic additives, organic resins, and solvents. Paste processing is similar to alumina paste processing as described above.

Inspection, Stacking, Lamination and Sizing. Aluminum nitride multilayered packages can use the same tooling and processing as alumina-molybdenum or tungsten and LTCC-copper or silver cofired modules.

Sintering. Cofiring aluminum nitride substrates requires that several critical processing parameters be met to achieve success. As prerequisites, the potential compatibility of AlN powder composition and W metallurgical paste should be able to achieve full density with near matched shrinkages and good interfacial adhesion. Tungsten and AlN shrinkage is a near match.[55]

During sintering, too aggressive a pyrolysis can cause delamination. Control of oxygen partial pressure must be at a level to avoid oxidation of the aluminum nitride powder and the metallurgical conductor. The cofiring conditions should provide means of organic removal adequate to avoid a decrease in the ceramic electrical resistance and must provide a conductor with high electrical conductivity. The cofire process and equipment must also be able to achieve the desired thermal performance required in the package, typically 120 to 200 W/m°K.

Achievement of high thermal conductivity in AlN is dependent on a number of factors. As discussed above, the AlN powder and its inherent impurity levels is important. Oxygen and metallic impurities can impact thermal conductivity if incorporated in the lattice structure. Sintering aides such as Y_2O_3 and CaO getter oxygen form the AlN surface and/or lattice structure. They form another phase (Y-Al-O and Ca-Al-O), which tends to concentrate at grain boundaries and triple points and can migrate to the surface. This can reduce oxygen diffusion into the AlN, which aids the attainment of high thermal conductivities.[48] The Y_2O_3-CaO-Al_2O_3 phase diagram shows that a number of phases are possible. Y_2O_3 has higher eutectic temperatures and has a higher affinity for oxygen. CaO can react with Al_2O_3 or AlON to form eutectics below 1500° C in low-temperature sintering. Combinations of Y_2O_3 and alkaline earths (such as CaO), or modifications with fluorides, has led to improved low-temperature sintering compositions.[50] A nitrogen atmosphere with low hydrogen additions is often used for sintering. Sintering in a carbon containing atmosphere can also lead to high thermal conductivity AlN. Carbon present during sintering from the atmosphere or small residual levels from organic removal can help to reduce AlON to AlN. This is derived from the carbothermal reduction of oxide on the surface of AlN particles during sintering.

Plating and Electrical Test. Substrate finishing is completed by deposition of metal features with metals such as nickel and gold and subsequent verification of product electrical specifications using electrical test. Similar to alumina, nickel and gold can be electrolytically plated on the surface features of the AlN package to meet the needs of the interconnection specifications.

AlN is resistant to water, acid solutions, and ethanol but is attacked by alkaline solution.[56,58] Therefore, caution must be used in selection of electrolytic or electroless plating compositions that utilize alkaline reagents. Resistance degrades with increasing pH, however, AlN samples have been shown to be stabilized by pretreatment processes.[56,57,59] Electrical testing and module assembly are similar to alumina testing described above.

AlN Package Performance. Toshiba researchers have compared the thermal resistance of Al_2O_3 and AlN packages when a 4 W gate array is packaged in them. It is clear in Fig. 2.12 that the alumina PGA with heat sink can be replaced by the AlN PGA without a fin.[77]

Figure 2.13 shows a modeled comparison of the thermal resistance for a face-up TAB back bonded 1 sq. cm chip attached to multilayer HTCC alumina or AlN. The resistance of the chip and a 2 mil filled epoxy die attach are included (these contribute 0.3°C-sq cm/W). It is clear that the greater the number of layers, the greater the impact of the higher Tc AlN.

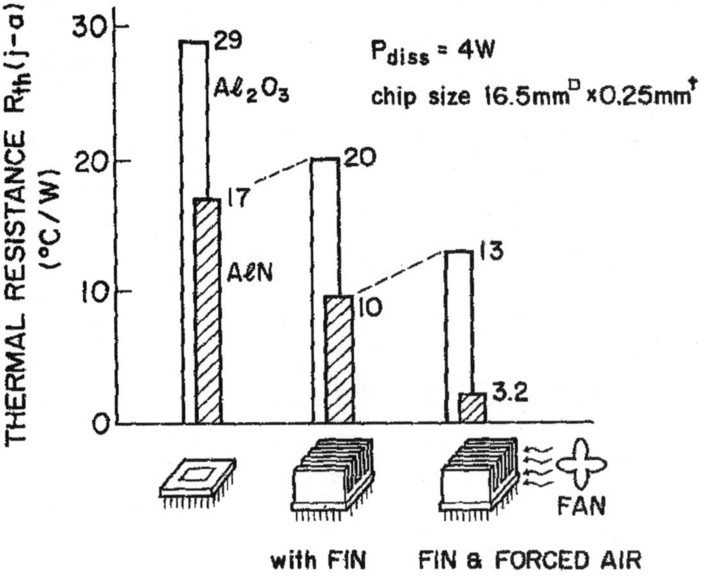

FIGURE 2.12 Thermal resistance changes in Al_2O_3 and AlN packages.[77]

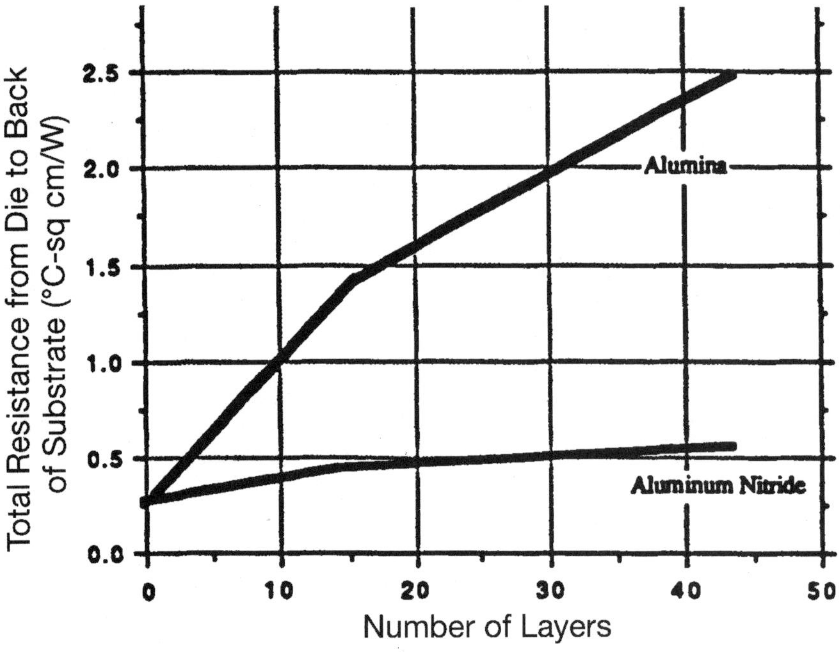

FIGURE 2.13 Comparison of the thermal resistance of AlN and Al_2O_3 modeled MCM packages.[78]

Aluminum nitride packages are finding increased applications in communications, military, and data processing applications, especially where thermal dissipation requirements are high or where thermal expansion match to silicon is desired. Fujitsu has used an AlN single-chip package in its VP 2000 series supercomputer for a number of years.[60] Multichip modules have also been reported for multiple programs focused on double-sided packages[47] and, for monolithic microwave integrated circuits (MMICs),[61] double-sided packages offer high-density, three-dimensional interconnection structures for short interconnections. The MMIC packages provide miniaturization, lighter weight, and lower cost by approximately 20 to 40 percent compared to an Al_2O_3 package with heat sink.[61]

Aluminum nitride will continue to expand into markets previously dominated by BeO for both cost and environmental benefits. As the AlN infrastructure continues to grow, AlN powder and production costs will be reduced, and AlN multilayer package technology will expand.

2.6 TYPICAL MCM MERCHANT DESIGN RULES

2.6.1 Kyocera

Kyocera offers MCM-C packages based on the set of materials detailed in Table 2.6.

2.6.2 NTK

NTK offers MCM-C packages based on the set of materials shown in Table 2.7.

2.6.3 IBM

IBM HTCC and LTCC technologies are shown in Table 2.8. The HTCC technology is available in sizes up to 127 mm sq. and 63 layers of Mo wiring. The LTCC technology uses a lower dielectric constant glass ceramic material, copper thick film wiring, and copper thin film pad metallurgy. It is also available in sizes up to 127 mm and 63 layers of thick film copper wiring.

Table 2.6 Kyocera Commercial Material Sets

	Alumina	AlN	Mullite	LTCC	LEC
Dielectric composition	Approx. 92% Al_2O_3	Approx. 90% AlN	mullite	alumina, quartz, lead borosilicate glass	alumina, quartz, borosilicate glass, cordierite
Thermal conductivity (W/m°C)	19	150	5	2.2	2
Dielectric constant	10	8.5	6.4	7.9	5.0
Conductor	W/Mo	W	W/Mo	Au/Ag	Cu
Sheet resistivity, m-Ω/sq	8	10	8	5	2
TCE (10^{-6}/°C)	7	4.6	4.2	7.0	4.0

Table 2.7 NTK Commercial Material Sets

	HTCC HA-921 Alumina (>92%)	LTCC (FC-01)	Mullite	AlN
Primary crystalline phase	corondum	cordierite	mullite	
Dielectric constant	9.4	5.1	6.6	8.9
CTE (10^{-6}/°C)	6.8	3	4	4.3
Thermal conductivity (W/m°C)	17	3	5	170
Firing temperature (°C)	>1500	>1700	1500	950
Breakdown voltage (kV/mm)	15	15	15	15

Table 2.8 Comparison of IBM Mainframe Technologies

	HTCC	LTCC
Material properties		
Conductor	molybdenum	copper
Dielectric	92% Al_2O_3	cordierite
Dielectric constant	9.5	5.0
Thermal conductivity (W/m°C)	20	5
TCE (10^{-6}/°C)	6.5	3
Design rules		
Min. via pitch (mm)	0.225	0.225
Min. via diameter (mm)	0.10	0.09
Line width (mm)	0.1	0.075
Electrical characteristics		
Impedance (Ω)	50	60
Line capacitance (pF/cm)	2.4	1.3
Line resistance (Ω/cm)	0.32	0.22
Line propagation (pF/cm)	118	80
Via capacitance (pF/cm)	3.5	2.0
Via resistance (Ω/cm)	0.07	0.05

2.7 MCM-C APPLICATION EXAMPLES

2.7.1 IBM

As described in earlier sections, the IBM 3090 and ES 9000 mainframes were developed using their HTCC and LTCC technologies (see Table 2.8).

One application is a RISC 6000 central processing unit upgrade from single-chip module (SCM) on a printed circuit board (PCB) packaging to an enhanced function with multichip module (MCM) to package the central processing unit.[13] Shown in Fig. 2.14, the R-6000 module is manufactured for IBM internal PC and workstation applications. It is an HTCC-based module that is 7.85 mm thick, and consists of 44 layers of alumina. The conductor is 0.100 mm wide Mo. There are 10 signal redistribution layers and 30 signal layers with a wiring density of 40 cm/cm sq. per layer. Nine chips are attached to the module by C4 joining, the balls on 250 μm pitch. I/O is achieved through an interstitial pin grid array on the bottom side. The module is designed to dissipate 55 W and incorporates an under-chip encapsulant for top surface passivation. Superior system characteristics based on benefits of MCM technology were proven. Multilayer ceramic was selected based on lower cost, lower wiring resistance, lower connector inductance, and availability.

2.7.2 Kyocera

Kyocera ceramic MCM technology has been detailed extensively.[64,65] Kyocera has been a supplier of the two-chip "P6" module for Intel. It is an HTCC, 387 I/O ceramic module with W metallization. It is shown in Fig. 2.15.

Figure 2.16 shows a four-chip MCM built for LSI logic. This 383-pin LTCC EWS (engineering workstation) graphics module allows a 20 to 30 percent operational speed increase over its single-chip analogs.

FIGURE 2.14 IBM HTCC R-6000 MCM.[13]

FIGURE 2.15 Kyocera P6 MCM-C for Intel.

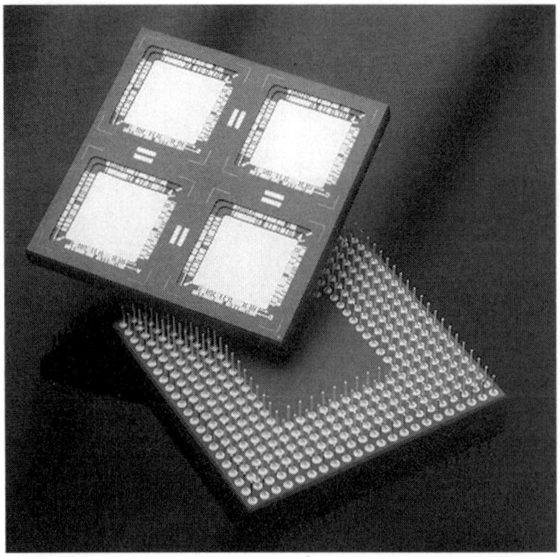

FIGURE 2.16 LSI Logic MCM-C EWS graphics module.

2.7.3 Unysis

To achieve their goal of a producing an entire 48-bit mainframe computer on a single personal computer-size PWB, Unysis used HTCC MCM-C technology.[66,67] The 5.0 × 5.84 cm cavity-up ceramic PGA SCAMP (single-chip A-series mainframe processor) package is shown in Fig. 2.17. It contains one ASIC CMOS processor and ten 32k × 8 CMOS SRAMs. The ceramic package contains 10 ceramic layers and 4033 buried vias, and it has 255 pins on a 2.54 mm grid. By locating the SRAMs in the same package with the processor chip, the access time to the control store was reduced. The

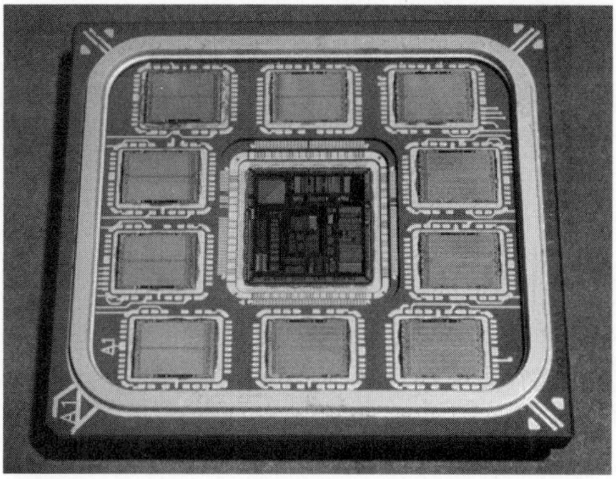

FIGURE 2.17 Unysis SCAMP HTCC MCM[67]

ability to locate the power and ground pins together with the planes in the ceramic eliminated the need for discrete capacitors on the package.

2.7.4 Fujitsu

Fujitsu introduced its copper conductor multilayer glass/ceramic substrates in the VP 2000 supercomputer and the M-1800 mainframe.[68] The glass ceramic is a mixture of alumina and borosilicate glass that sinters at <1000° C.

Figure 2.18 shows the 52 layer LTCC substrate for the M-1800 mainframe computer. Table 2.9 lists the properties of this ceramic circuit board.

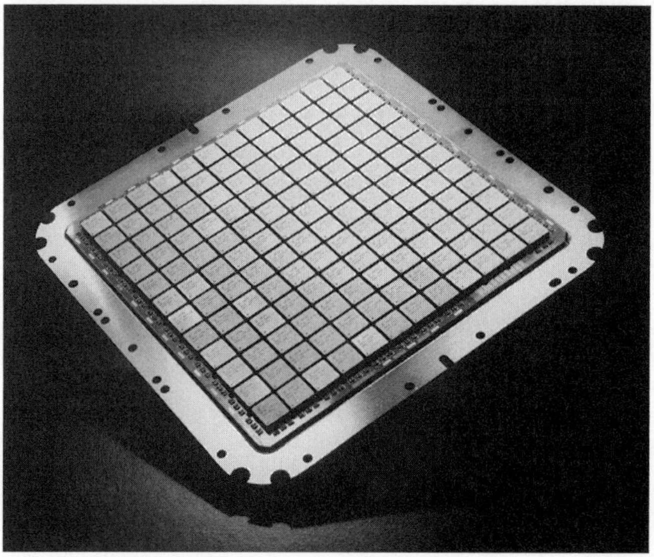

FIGURE 2.18 Fujitsu 52-layer LTCC substrate for the M-1800 mainframe.[68]

Table 2.9 Properties of the Fujitsu M-1800 LTCC MCM

Dielectric material	alumina/borosilicate glass
Dielectric constant	5.7
TCE ($\times 10^{-6}$)	4
Flex strength (MPa)	200
Thermal conductivity (W/m°C)	2.5
Conductor	Cu
Sheet resistivity (mΩ/sq)	1.2
Propagation delay (ns/m)	8.0
Line width (μm)	95
Line pitch (μm)	450
I/O pins	
signal	4320
power	4320
Size (mm)	245 × 245

2.7.5 NEC

NEC developed ceramic MCM technology for their SX-2 and SX-3 mainframes[69] and their ACOS system 3900.[70] They chose 55 percent alumina, 45 percent lead borosilicate glass compositions which are sinterable at 900° C as the dielectric material. The substrate size is 225 × 225 mm with 11,540 pins. The SX-3 module described in Chapter 3 (see Fig. 3.37) has 78 glass-ceramic dielectric layers of which 13 are conductive layers. The sintered Au has an electrical resistivity of 3 mΩ-cm. This information is summarized in Table 2.10. For the final packages, this ceramic base is capped with Cu/PI interconnect as described in Chapter 3.

Table 2.10 NEC Ceramic for MCM-C Mainframe Applications[70]

Substrate size (mm)	225 × 225
Dielectric material	alumina/Pb borosilicate glass
Number of layers	78
Via hole diameter (μm)	200
Conductor	gold
Resistivity	3.0 μΩ-cm
Line width (μm)	200
conductor layers	13

2.7.6 Honeywell

The Honeywell Advanced Spaceborne Computer Module (ASCM) program is described more fully in Chapter 3. The MCM shown in Fig. 2.19 contains three CPUs and two FPPs. The MCM-C memory module shown in Fig. 2.20 contains nine memory ICs and a memory line driver IC. Table 2.11 compares the design geometries for the MCM-D and MCM-C builds. They concluded that the MCM-D technology was capable of much higher density bonding pads and much higher interconnect density per layer but that the MCM-C technology has a more established vendor base, and the MCMs could be manufactured more easily and reliably at lower cost.[71]

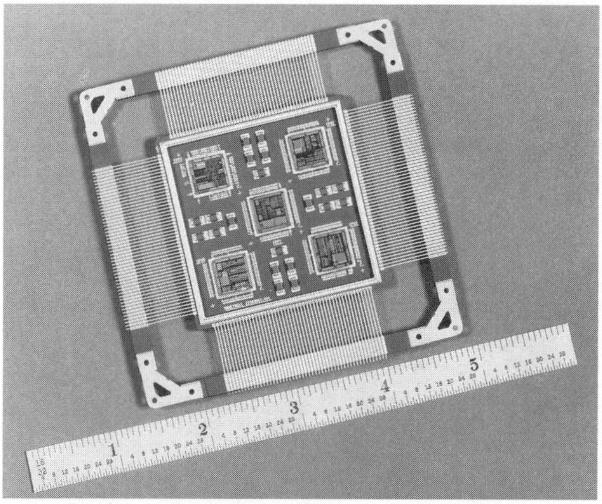

FIGURE 2.19 Honeywell generic VHSIC Spaceborne Computer MCM.[71]

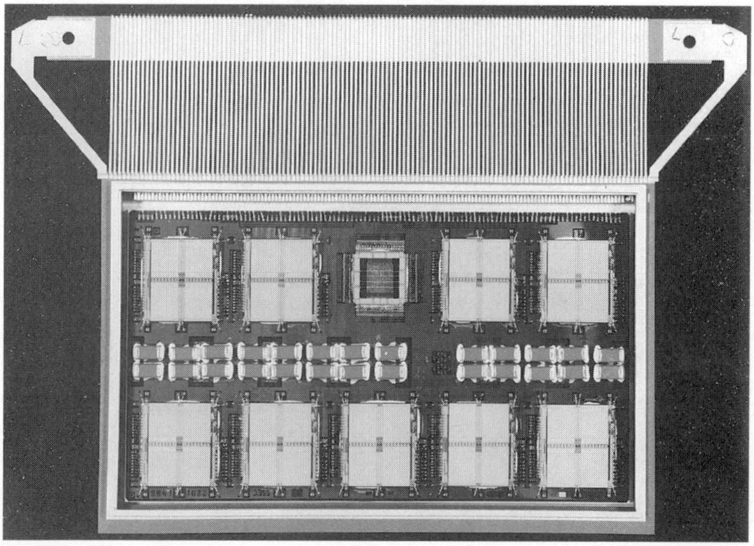

FIGURE 2.20 Honeywell MCM-C memory module.[71]

Table 2.11 Design Geometries for MCM-D and MCM-C Modules Used in the Honeywell ASCM Program[71]

	Generic VHSIC Module		Memory Module	
	MCM-D	MCM-C	MCM-D	MCM-C
Signal layers	2	6	1	4
Power/ground layers	1	6	1	6
Bond pad layers	1	1	1	1
Line width (mil)	1	4.7	1.5	4.7
Routing pitch (mil)	4	12	6	12
Bond pad pitch (mil)	6 (one row)	12 (two rows)	6 (one row)	12 (two rows)

2.7.7 Hughes

Hughes has applied LTCC since 1991 in many packaging areas including high-speed digital and microwave circuits.[72,73] The LTCC process provides 15 layers of metallization with no base substrate, using instead the ceramic as both dielectric and structural support for the package. Buried and top conductors were gold, although silver buried conductors are available. Table 2.12 shows the characteristics of the Hughes LTCC process for the HAC-32 general purpose processing element.

2.7.8 Lockheed

The AlN serial processor module shown in Fig. 2.21 was produced by the joint effort of Lockheed Missiles Systems, Ferro-EMD, and Carborundum.[75] The circuit was built on a 3 × 4-inch substrate. The circuit has eight metal layers, five mil lines, and ten mil vias. It required 46 separate firings.

Table 2.12 Characteristics of Hughes LTCC Substrates for HAC-32 Processor[72]

	LTCC
Conductor	
Material	gold
Thickness (μm)	12
Line width (μm)	150
Line pitch (μm)	300
Catch pad (μm)	200
Dielectric	
Thickness (μm)	85
Via diameter (μm)	150
Dielectric constant	7.9
Electrical	
Line resistivity (mΩ/sq.)	5
Line capacitance (pF/cm)	<3.0

FIGURE 2.21 Lockheed AlN processor module.[75]

2.7.9 Dassault

Dassault has reported on a photoimageable thick film process to commercially produce MCMs such as the one shown in Fig. 2.22.[76] This 10 × 4.3-cm, six conductor layer substrate with 125 μm lines and spaces and 125 μm vias was fabricated with Fodel dielectric and conventional screen printed Au thick film conductor. The circuit contains 10 ASICs and operates at 40 MHz. The module is sealed in a hermetic aluminum package and is used in an aeronautical application.

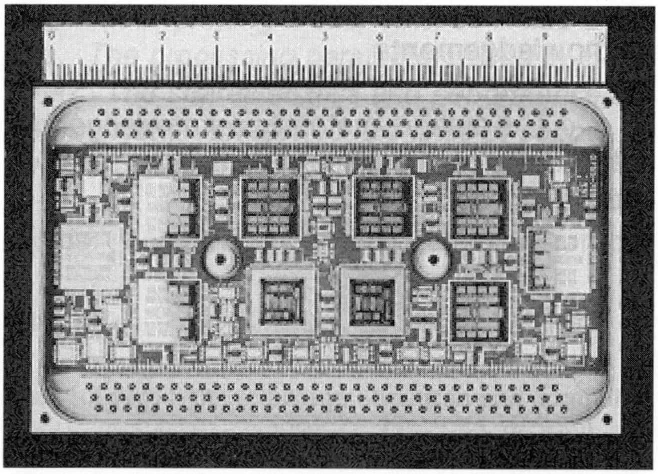

FIGURE 2.22 Dassault phase data multiplexer.[76]

2.8 CERAMIC MCM FUTURE DIRECTIONS

Advancements in materials, processes, and productivity will continue to benefit applications for which ceramic multichip modules provide the most cost-effective solution. Materials enhancements, such as integration of inductors, capacitors, and filters into multichip modules, will lead to increased reliability, lower costs, and smaller sizes for packages addressing the expanding communications market.

Process and productivity enhancements for both substrates and interconnections, along with finer pitch wiring in ceramic substrates, will lead to package solutions at much higher silicon to package I/O levels and package-to-board I/O. Multiple voltages will allow for increased portability due to reduced power consumption. Flip chip and area array interconnections will support the ever increasing I/O requirements for silicon.

Thick film, HTCC, LTCC, and AlN will be used in both SCM and MCM applications based on the best cost performance system-level solutions (MCM-C). For the highest I/O level requirements, single-layer or few-layer thin film will be placed on MCM-C ceramic packages to provide the best solution for some applications.

REFERENCES

1. Blodgett, A.J., "Microelectronic Packaging," *Sci. Amer.,* vol. 249, 1983, p. 86.
2. Tummala, R., "Multi-chip Packaging in IBM—Past Present & Future," *Proceed. Int. MCM Conf.,* Denver, 1993, p. 1.
3. Young, W.S., "Multi-layer Ceramic Technology," in *Ceramic Materials for Electronics,* R.C. Buchanan, Ed., p. 403, Marcel Dekker, New York, 1986.
4. *Microelectronics Packaging Handbook,* R. R. Tummala and E. Rymaszewski, Eds., Van Nostrand Reinhold, New York, 1989.
5. Hankey, D., Shaikh, A., Vasudevan, S., and Khadilkar, C., "Thick Film Materials and Processes," in *Hybrid Microelectronics Handbook,* J. Sergent and C. Harper, Eds., McGraw Hill, 1995, chapter 3.
6. Lipka, K., "Limits of Fine Conductor Line Widths: Today's Screen Printing Technology," *Proceed. ISHM,* 1993, p. 475.
7. Johnson, R.W., "Introduction to Hybrid Microelectronics" chapter 1 in *Hybrid Microelectronics Handbook,* J. Sergent and C. Harper, Eds., McGraw Hill, 1995.
8. Amey, D., "A Thick Film Hybrid Logic Module for Commercial Computer Applications," *Proceed. Int. Microelectronics Conf.,* 1981.
9. Hagen, N., Henderson, J., Nebe, W., and Osborne, J., "High Resolution Thick Film Material System," *Int. J. Hybrid Microelectronics,* vol. 12, 1989, p. 175.
10. Felten, J, Collins, J., and Wang, C., "High Resolution Via Patterning by Aqueous Diffusion Patterning," *Proceed. ISHM,* 1991, p. 545.
11. Chambers, B., et al., "Evaluation of A Multi-chip Module (MCM) Thermal Design Tool: A Three Chip MCM Thermal Mock-up As A Case Study," *Int. J. Microcircuits and Elect. Pkg.,* vol. 16, 1993, p. 249.
12. Chong, F.C., et al., "The Evolution of MCM Test From High Performance Bipolar Mainframe Multi-chip Modules To Low Cost Workstation Multi-chip Modules," *Proceed. Int. Conf. MCMs,* Denver, 1993, p. 404.
13. Sechler, R., "RISC System/6000 Central Processing Unit (CPU) MCM," *Proceed. Int. MCM Conf.,* Denver, 1993, p. 22.
14. Nowak, M., *EE Times,* June 26, 1995 p. 50.
15. Blodgett, A.J., and Barbour, D.R., "Thermal Conduction Module: A High Performance Multi-layer Ceramic Package," *IBM J. Res. Develop.,* vol. 26, 1992 p. 30.
16. Fukui, M., et al., "Multi-layer Ceramic Substrates (MLCS) The Alternative Printed Circuit Board Technology," *Proceed. ISHM,* 1993, p. 172.
17. Blodgett, A.J., "A Multi-layer Ceramic Multi-chip Module," *IEEE Trans. CHMT,*1980, p. 634

18. Knickerbocker, J., et al., "IBM System/390 Air-cooled Alumina Thermal Conduction Module," *IBM J. Res. Develop.,* vol. 35, 1991 p. 330.

19. Chopra, M., and Sechler, R., "Comparison of MCM's for A High Performance Workstation," *Proceed. Int. MCM Conf.,* Denver, 1993, p. 457.

20. Hargis, B., "Multi-chip Module Technology Comparison," *Proceed. IEPS,* 1993, p. 959.

21. Mecca, J.P., et al., "Characterization of Simultaneous Switching Noise in Multi-layer Ceramic Multi-chip Modules," *Proceed. IEPS,* 1993, p. 872.

22. Blake, B., and Ahmed, S., "Developing Manufacturing Products," *Proceed. NEPCON West,* 1991. p. 265.

23. Perfecto, E., et al., "Multi-level Thin Film Packaging Technology at IBM," *Proceed. Int MCM Conf.,* Denver, 1993, p. 474.

24. Kumar, A.H., and Tummala, R.R., "State of the art, Glass-ceramic/copper Multilayer Substrate for High Performance Computers," *Int. J. Hybrid Microelectronics,* vol. 14, 1991, p. 137.

25. Garrou, P., and Turlik, I., "Materials of Construction: Substrate, Dielectric, Metallization" in *Thin Film Multichip Modules,* G. Messner, I. Turlik, J. Balde, P. Garrou, Eds., ISHM Press, 1992, p. 92.

26. Knickerbocker, J., Kumar, A.H., and Tummala, R.R., "Sinterable, Cordierite Glass-ceramics," *Bull. Am. Ceram. Soc.,* vol. 72, 1993, p. 91.

27. Holleran, L.M., MacDowell, J.F., and Martin, F.W., "Glass-ceramics for Electronic Packaging," European patent application, no. 0 289 222 A1(1988).

28. Kondo, K., Okuyama, M., and Shibata, Y., "Low Firing Temperature Ceramic Material for Multilayer Ceramic Substrates," in *Multilayer Ceramic Devices, Advances in Ceramics,* vol. 19, J.B. Blum and R. Canion, Eds., Am. Ceramic Society Publications, Westerville, Ohio, 1986.

29. Niwa, K., Takeda, Y., "Multilayer Glass-Ceramic Composite Circuit Board for Fujitsu VP 2000 Series," *Fujitsu Sci.Tech. J.,* vol. 27, no, 2, pp. 179–186, 1991.

30. Shimada, Y., Kobayashi, Y., Kata, K., Takamizawa, H., and Kurano, M., "Large multilayer glass-ceramic substrate for supercomputer," in *Proceed. 40th Electronic Components Conference,* Las Vegas, 1990, p. 76.

31. Shimada, Y., et al., "High Speed Computer Packaging with Multilayer Glass-Ceramic Systems," *Proceed. ISHM,* 1991, p. 176.

32. Gupta, T.K., and J.H. Jean, "Principles of Development of a Silica Dielectric for Microelectronic Packaging," *J. Mat. Sci.,* vol. 11., 1996, p. 243.

33. Kumar, A H., Prabhu, A.N., and Thaler, B., "Low-Temperature-Cofired-Ceramic on Metal Substrate Technology," *Proceed. Int. MCM Conf.,* Denver, 1995, p. XX.

34. Gallo, S., Mones, A., Hormadaly, J., Cicognani, J., Brown, R., Massiot, M, and Perchais, E., "LTCC Circuits for MCM-C Applications: A Cost Effective Packaging Solution," *Proceed. ISHM,* 1993, p. 178.

35. Martin, T., and Schroeder, D., "Reliability Analysis of LTCC MCM's Utilizing Silver Conductors," *Proceed ISHM,* 1994, p. 295.

36. Drozdyk, L., Getty, R., Hormadaly, J., Needes, C., and Mones, A., "Transition Via Fill Material for Gold-Silver Interconnections," *Proceed ISHM,* 1994, p. 301

37. Williams, T., and Shaikh, A., "Silver Via Metallization for Low Temperature Cofired Ceramic Tape," *Proceed. ISHM,* 1994, p. 324.

38. Herron, L.W., Master, R., and Tummala, RR., "Cofiring Process for Glass-Ceramic/Copper Multilayer Substrates," *Proc. 41st. Electronics Components Conference,* Atlanta, 1991, p. 5.

39. Baba, Y., et al.," Co-fireable Copper Multilayer Ceramic Substrates," *Proceed. ISHM,* Seattle, 1988, p. 405.

40. Page, J., and Webster, D., "Transitioning from Gold to Copper in a Multilayer System," *Proc. 30th. Electronics Components Conference,* San Francisco, 1980, p. 294.

41. Brown, R.L., and Shapiro, A.W., "Integrated Passive Components and MCM's: The Future of Microelectronics," *Proceed. Int. MCM Conf.,* Denver, 1993, p. 487.

42. Gillett, J., "New Packaging Challenges Posed by High Power CMOS," *Proceed Int. MCM Conf.,* 1993, p. 97.

43. Hannemann, R., "Electronic System Thermal Design for Reliability," *IEEE Trans. Reliability,* R-26, 1977.

44. Slack, G.A., et al., "The Intrinsic Thermal Conductivity of AlN," *J. Phys. Chem. Solids,* vol. 48, 1987, page 641.

45. Kuramoto, N., et al., "Sintering Process of Translucent AlN and Effect of Impurities on Thermal Conductivity of AlN Ceramics," *Togto-Ktokai-Shi,* vol. 93, 1985, p. 41.

46. Knudsen, A.K., et al., "The Influence of Impurities on The Optical Thermal and Electrical Properties of Sintered AlN," *Proceed. Int. Microelectronics Conf.,* Yokohama, 1992, p. 270.

47. Minehan, W., et al., "Current Technology For Hot Pressed, Co-fired AlN Electronic Packages," *Advancing Microelectronics,* vol. 21, 1994, p. 31.

48. N. Kuramoto, N., et al., "Sintering and Properties of High Purity AlN Ceramics," *Advances In Ceramics,* vol. 26, 1989, p. 107.

49. Kuramoto, N., "Thin Film and Co-fired Metallization on Shapal Aluminium Nitride," *Advancing Micro-Electronics,* vol. 21, 1994, p. 15.

50. Chiano, Y. H., and Copus, Y. A. "Low Temperature Sintering and Co-sintering of Aluminium Nitride Substrates," *Proceed. ISHM,* 1993, p. 407.

51. Iwase, N., et al., "AlN Substrates and Packages," *Advancing Microelectronics,* vol. 21, 1994, p. 24.

52. Horn, D., et al., "Development of Metallized Aluminum Nitride For Low Temperature Sintering of Electric Packages," *Advancing Microelectronics,* vol. 21, 1994, p. 29.

53. Yan, H., et al., "Polyvinyl Butyryl Pyrolysis, "Interactions With Plasticizer and AlN Ceramic Powder," *Proceedings Material Research Society,* 1991.

54. Chiao, Y.H., et al., "Interfacial Bonding In Brazed and Co-fired Aluminum Nitride," *Proceed. ISHM,* 1991, p. XX.

55. Chiao, Y. H., and Yang, K., "Co-fired Metallization of Aluminum Nitride Substrates," *Proceed. ISHM,* 1992, p. 88.

56. Garrou, P.E., "Aluminum Nitride For Microelectronic Packaging," *Adv. Microelectronics,* vol. 21,1994, p. 6.

57. Smith, S., and Hazen, B., "Thin Film Protective Coatings for Processing AlN Substrates," *Proceed. ISHM,* 1991, p. 48.

58. Kurokawa, Y., et al., *Proceed. ECC,* 1986, p. 412.

59. Kurihara, Y., et al., IEEE Trans CHMT, vol. 12, 1989, p. 330.

60. Sugimoto, M., "AlN Applications," *Adv. Microelectronics,* vol. 21, 1994, p.27.

61. Imura, R., "Development of Aluminum Nitride Multi-layer Packages," *Adv. Microelectronics,* vol. 21, 1994, p. 39.

62. Kurokawa, Y., Hamaguchi, H., Shimada, Y., Utsumi, K., and Takamizawa, H., "Highly Thermal Conductive AlN Substrate," *NEC Journal of R&D,* no. 85, 1987, p. 15.

63. Hagge, J., "Ultra Reliable HWSI With AlN Packaging," *Proceed NEPCON West,* 1989.

64. Fukui, M., Hori, M., Makihara, C., and Terasawa, M., "Multilayer Ceramic Substrates: The Alternative PWB Technology," *Proceed ISHM,* 1993, p. 172.

65. Sigliano, R., and Gaughan, K., "Ceramic Material Options for MCMs," *Proceed. Int. MCM Conf.,* Denver, 1992, p. 291.

66. Nelson, J., and Cheves, C., "MCM for a Cost Driven Mainframe," *Proceed IEPS,* 1989, p. 230.

67. Rhodes, R., "Multichip Packages at Unysis: Co-fired Ceramic to thin Film," *Proceed NEPCON West,* 1991, p. 477.

68. Yokouchi, K., Kamehara, N., and Niwa, K., "Packaging Technology for High Speed Computers—Multilayer Glass/Ceramic Circuit Board," *Proceed. ISHM,* 1991, p. 183.

69. Shimada, Y., Koybayashi, Y., Kata, K., Kurano, M., and Takamizawa, H., "Large Scale Multilayer Glass-Ceramic Substrate for Supercomputers" *Proceed. ECTC '90,* p. 76.

70. Yamada, M., Nishiyama, M., Tokaichi, T., and Okano, M., "Packaging Technology for the NEC ACOS System 3900," *Proceed. ECTC,* 1992, p. 745.

71. Jacobsen, W., Jensen, R., Loy, J., and Speerschneider, C., "Application of MCM-D and MCM-C for Spaceborne Data Processors," *Proceed. Int. MCM Conf.,* Denver, 1993, p. 525.

72. Brown, R., and Bone, R., "A Comparison of MCM Packaging Technologies for a 32-bit High Performance Processor," *Proceed. ISHM,* 1993, p. 562.

73. Brown, R., and Ehlert, M., "A Flexline Manufacturing Approach for Low Temperature Co-fired Ceramic MCM," *Proceed. ISHM,* 1992, p. 7.

74. DiLazzaro, J., and Nauman, D., "Low Temperature Co-fired Multilayer Ceramics for High Performance Packages," *Proceed ISHM,* 1991, p.408.
75. Shaikh, A., "Thick Film Pastes for AlN Substrates," *Advancing Microelectronics,* Special AlN edition, P. Garrou, Ed., 1994, p. 18
76. Cazenave, J.P., and Suess, T., "Fodel® Photoimageable Materials—A Thick Film Solution for High Density MCMs," *Proceed. ISHM,* 1993, p. 483.
77. Miyashiro, F., et al., "High Thermal Conductivity AlN Ceramic Packages and Substrates," *IEEE Trans. CHMT,* vol. 13, 1990, 313.
78. Dolbear, T., "Thermal Management of Multichip Modules," *Proceed. NEPCON West,* 1990, p. 1186.

CHAPTER 3

MCM-D THIN FILM MATERIALS, PROCESSES, AND APPLICATIONS

Dr. Philip Garrou
Dow Chemical Company

3.1 INTRODUCTION

This chapter covers MCM-D materials, processing, and applications. Because of the evolution that has occurred over the last decade in MCM technology, exact definitions are difficult to come by, as was discussed in Chap. 2. Figure 3.1 shows the overlap that has occurred between classical MCM-C, MCM-D, and MCM-L. Because MCM-D is basically defined by the use of thin film materials, a case can be made for the inclusion of MCM-D/L and MCM-D/C in this chapter. The former is covered in Chap. 4, MCM-L, and Chap. 5, on low-cost large-area processing (LAP). The latter is covered in this chapter inasmuch as the ceramic is serving mainly as the base support for the thin film interconnects, and in Chap. 2, MCM-C. Therefore, the reader is also directed to those chapters for further information on laminate/thin film and ceramic/thin film material properties, processing, and applications.

3.2 CONSTRUCTION MATERIALS

This section focuses on the materials used to fabricate the high-density MCM-D interconnect. The materials of construction can be classified into three distinct categories: the thin film dielectric, the substrate, and the conductor metallization. Although the pad level metallurgy is also part of the final MCM interconnect substrate, its composition is determined by the choice of assembly technique and thus will be covered more thoroughly in Chap. 8.

3.2.1 Thin Film Dielectrics

Dielectrics for thin film packaging can be classified into two categories: polymeric and inorganic. Although ICs have traditionally used SiO_2 as the dielectric layer, MCMs and other thin film packages have evolved to a much greater extent with polymeric materials. The impetus for this has been

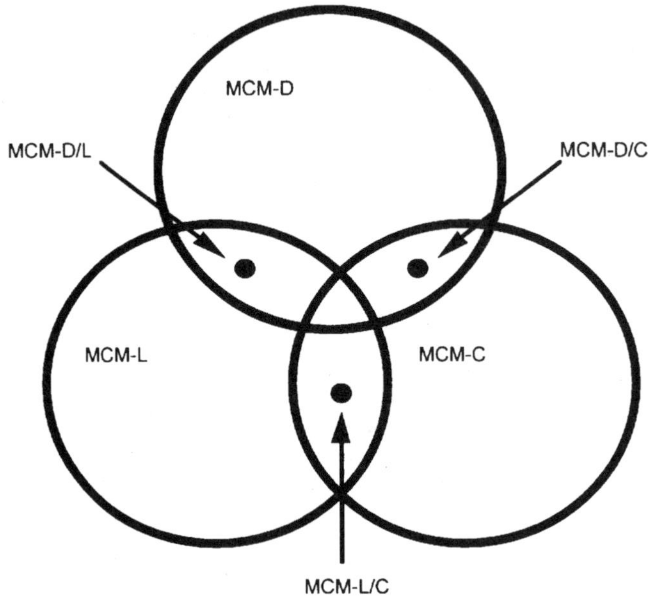

FIGURE 3.1 MCM overlapping technologies.

the capability of polymers to form thicker layers (with much lower stress), the speed and cost of deposition (versus typical CVD SiO_2 techniques), and the better planarization (SiO_2 is conformal) and lower dielectric constant offered by polymers. On the other hand, SiO_2 unquestionably offers better thermal conductivity, thermal stability, and chemical inertness.

Polymer Dielectrics. The use of polymers as insulating layers in microelectronic packaging is relatively recent.[1] NEC was the first to use a polymer dielectric in a commercial product—the SX-2 supercomputer—in the late 1980s.[2]

The suitability of a specific polymeric material is highly dependent on the process chosen to fabricate the structure, the intended application of the module, and the environment in which the module will be expected to operate. Commercially available materials are usually categorized by the via formation technique with which they are compatible, i.e., dry etching, wet etching or photosensitive.

In general, polyimides (PIs), benzocyclobutenes (BCBs), and cyanate ester formulations (CEs) are the only polymeric materials used commercially to fabricate thin film substrates. As the use of thin films has spread to redistribution layers on laminates, alternate epoxy-based materials have been examined for lower performance requirements. Such materials are described and compared in Chap. 5.

Polyimides (PIs). Polyimides contain the imide functional unit shown below:

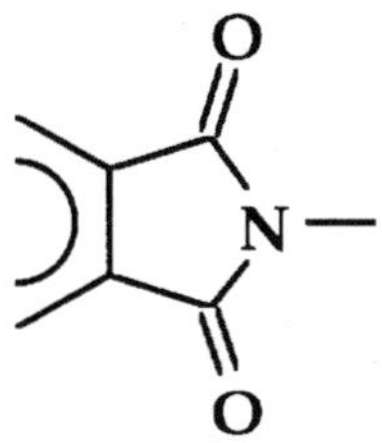

Structure 3.1

Commercially viable materials are derived from the reaction of an aromatic dianhydride with an aromatic diamine[3,4] as shown in Eq. 3.1. Most products are supplied as soluble polyamic acid (PAA) intermediates, which evolve water upon curing to yield the final PI structure. Some are supplied as preimidized materials. The traditional solvent is NMP. To obtain good film quality, high molecular weight (Mw) PAAs are needed that result in low-solubility, high-viscosity systems. PAA solutions are stored at 0° C. The "curing" reaction typically requires 350–400° C or higher.

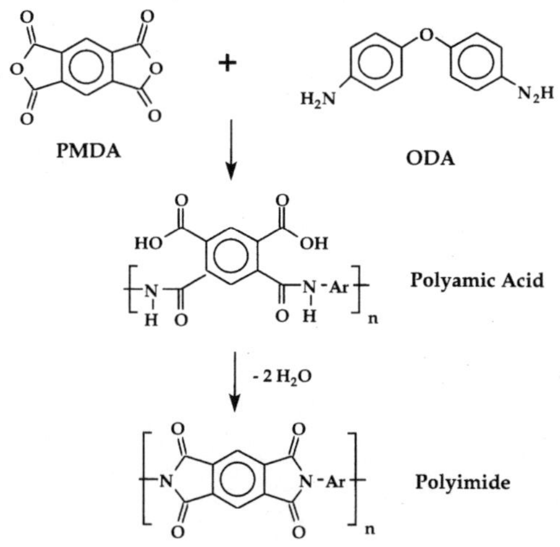

(3.1)

More detail on the chemistry of Polyimides is available in Refs. 5–8.

Low-CTE Polyimides. Detailed studies have been carried out to understand the relationship between PI structure and coefficient of thermal expansion (CTE).[10,11] Most PIs have an aligned structure and thus unequal X-Y (in the plane) versus Z (out of plane) CTE values as well as anisotropic electrical and mechanical properties.[5,9]

PIs having rigid backbones reveal very high modulus and low biaxial (in plane) CTE values. These CTE values are highly anisotropic, thus Z-axis CTEs reportedly are >> 100 ppm.[12] "Low CTE" PI products are exemplified by Pyralin™ type V products, PIQ-L100 (IV) and the OCG 500 series.

Polyamic Esters. Recent IBM studies have identified polyamic ester polyimides (Structure 3.2) that do not have some of the undesirable characteristics of polyamic acids such as their high curing temperatures, reactivity with copper metallizations, and poor planarization capabilities.[13] Such materials eliminate (produce) alcohols, not water, during their curing reactions. Production of these materials requires many more processing steps than for standard PAAs. Such materials include the OCG 600 series.[14,198]

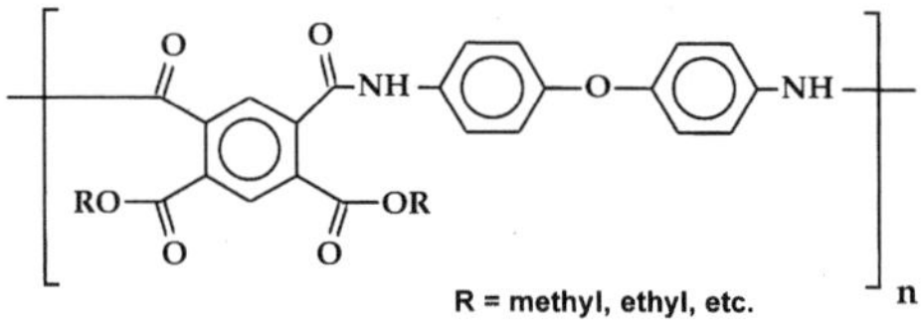

Structure 3.2

Commercially Available Polyimides. PIs must contain the imide structural unit (I). The PIQs and PPQs discussed below are referred to as PIs by the user community but, strictly speaking, are not polyimides.

- Pyralin™

  The Pyralin series of PI products from DuPont are based on several amine/dianhydride chemistries as shown in Table 3.1. Type I chemistry is based on PMDA/ODA, type III on BTDA/ODA/m-PDA, and type V on BPDA/p-PDA.[15]

- PIQ™

  PIQ polyimides, commercialized by Hitachi Chemical, have structures exemplified by PIQ-L100 (Structure 3.3). These structures show increased thermal stability due to the polyisoindoloquinazolinedione structure.[16]

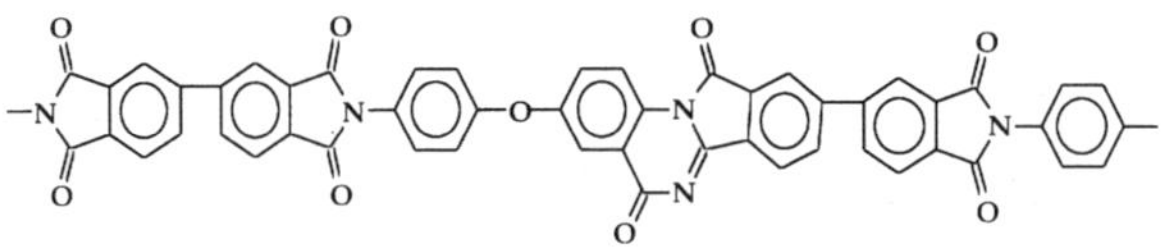

Structure 3.3

- Syntorg™

  Polyphenylquinoxalines (PPQs) (Structure 3.4), commercialized by Cemota under the trade name Syntorg, are fully cyclized and stable at room temperature.[17] The solvent system is m-cresol/xylene.

Table 3.1 Pyralin™ PI Chemistries

	Chemistry Type		
	I	III	V (low CTE)
Standard	PI 2545	PI 2525	PI 2611
Photosensitive		PI 2722	PI 2730
Wet etch			WE 1111

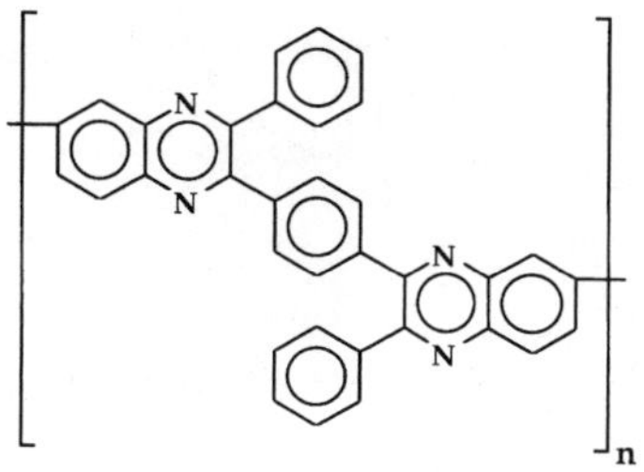

Structure 3.4

- Probamide™

  Polyimides, commercialized by Olin/Ciba Geigy (OCG) under the trade name Probamide™, encompass several structural varieties. The 400 series is based on BTDA and ortho alkyl substituted diamines and is inherently photosensitive due, it is thought, to the presence of the benzophenone moiety. These materials are not as oxidatively stable as other PI materials.[18] The 500 series is based on BPDA/PPD chemistry, and the 600 series is based on polyamic acid esters.[14]
- Ultradel™

  Polyimides commercialized by Amoco carry the trademark Ultradel™. The Ultradel™ 4000 series is a wet processable PI.[19] Enhanced process control reportedly comes from separate solutions for the etching of the polymer and the dissolution of the photoresist mask. Ultradel 4212 (Structure 3.5), is shown below.

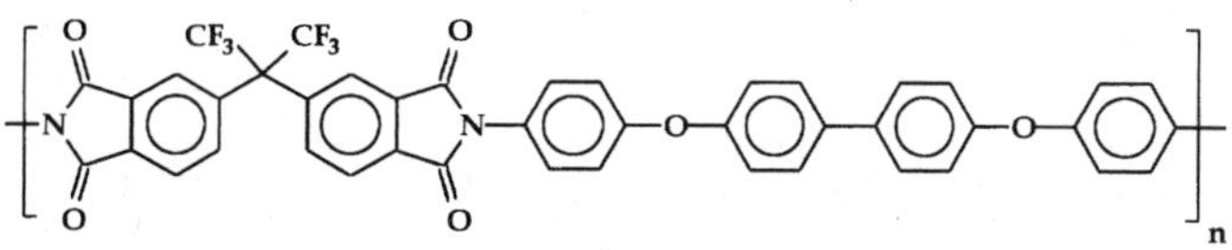

Structure 3.5

The Ultradel™ 5000 series are auto-photosensitive PIs similar to Probamde™ 400 (see above) but partially fluorinated to give a lower dielectric constant.[20]

Benzocyclobutenes (BCBs)

- Cyclotene™

  Benzocyclobutenes (BCBs),[21,22] commercialized by Dow Chemical under the trade name Cyclotene®, are a family of thermoset resins containing the ring structure shown in Structure 3.6. The 3000 series is dry etchable, and the 4000 series is photoimageable. The chemical structure of the commercial siloxy-containing DVS-BCB (Structure 3.7) is shown below. The resins are supplied partially polymerized ("B-staged") to obtain appropriate viscoelastic handling properties. Dry etchable BCB solutions are stable at room temperature. The polymerization (curing) of BCB proceeds through a two-step process—a thermally driven ring opening followed by a 4 + 2 Diels Alder reaction producing a polymer matrix based on a structure similar to Structure 3.8.

Structure 3.6

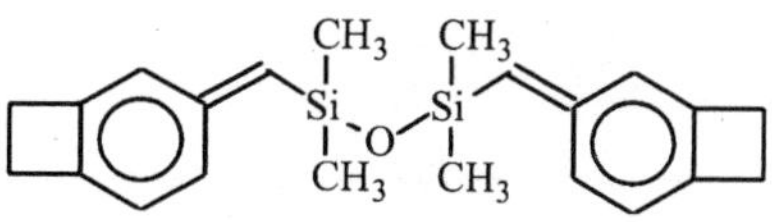

Structure 3.7

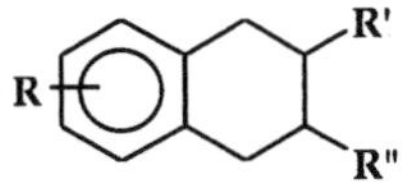

Structure 3.8

Benzocyclobutenes have totally different chemical structures when compared to PIs and therefore have significantly different physical, electrical and processing characteristics. Curing is typically carried out at 210 to 250° C, without the evolution of water or other by-products.

Photosensitive Dielectrics. Photosensitive dielectrics can be categorized as *negative* or *positive* working. For positive working materials, the irradiated area undergoes photolytic cleavage and is removed upon development. No positive working materials are currently commercially available, although it is an area of intense research.

Negative working materials contain photoreactive groups that lead to photo-cross-linking with adjacent polymer chains when exposed to UV light. This leads to a differential solubility between the exposed and unexposed portions of the film and, thus, development like a negative photoresist. The three common frequencies of exposure are 365 nm (I line), 405 nm (h line), and 436 nm (g line). Negative working materials swell due to absorption of organic solvent during development, resulting in limited resolution. Such materials are available from a number of sources. All currently available photodefinable materials require refrigeration.

In 1976, Rubner and co-workers at Siemens reported on a PMDA/ODA based negative working photosensitive polyimide.[23] This technology was widely licensed and formed the foundation of many of today's photosensitive products. The first generation systems suffered from poor shelf life, poor photospeed, and cracking problems during development.[24]

Although photosensitive materials are inherently more expensive to manufacture, they are generally chosen to lower overall processing costs. Several processing steps can be omitted when using photoimageable polymers, as shown in Fig. 3.2.

Photoimageable Polyimides. Commercial materials are limited to negative working systems. Negative working PSPIs can be further categorized by the source of photoactivity. PIs that are not inherently photosensitive are modified by "covalent attachment" or "ionic" attachment of photosensitive groups to the parent PAA molecules. These are referred to as *ester-type* and *salt-type* PSPIs, respectively. The ester materials are of the same overall chemical structure as the polyamic acid esters described by structure 3.2, except the R group contains a photo-cross-linkable species. The salt-based materials are derived from the reaction of polyamic acids with bases (usually amine compounds containing the photo-cross-linking species).

Mechanical Properties. Kojima and co-workers[26] have compared the properties of "ester-based" and "salt-based" PSPIs for the same diamine/diacid combinations as shown in Table 3.2. While the ester-based materials show better resolution in thicker layers, the salt-based materials reveal vastly superior mechanical properties. Of course, such results are highly dependent on the specific chemical backbones that were chosen for comparison.The presence of oxygen in the curing environment has been shown to drastically reduce the percent elongation in several families of PIs.[27]

Table 3.2 Comparison of Properties for Cured PSPI Films[26]

TSPI type	Elongation (%)	Tensile strength (Mpa)	Peel strength (g/cm)*
Ester	<1	<10	0
Salt	10	116	260

* adhesion to Si wafer after 20 hr of PCT

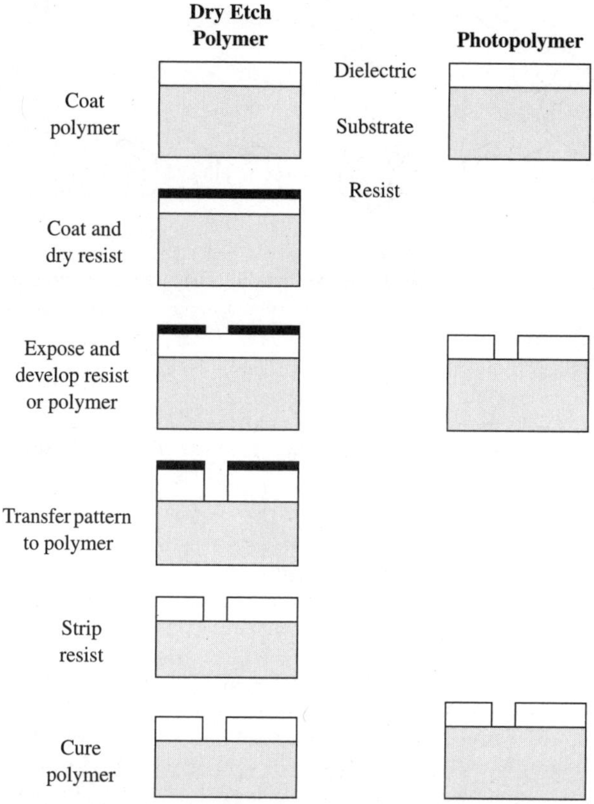

FIGURE 3.2 Processing steps for photosensitive dielectrics.

Inherently Photosensitive Materials. PIs derived from BTDA and certain ortho-alkyl substituted aromatic amines are inherently photosensitive.[18] Inherently photosensitive PIs have longer shelf life and shrink far less than the ester- or salt-based materials. The OCG 400 series Probamides and the Amoco 5000 series are of the inherently photosensitive type. The Amoco material is partially fluorinated to lower the dielectric constant.

Positive Acting PSPIs. Although laboratory studies have shown that positive acting photosensitive PIs (PSPIs) offer better resolution and can be developed with aqueous base, such materials are not commercially available, and experimental materials have been limited to very thin films.[28,29]

The use of PSPIs has been extensively detailed by NTT,[30] NEC,[31] Boeing,[32] and AT&T.[33]

Photoimageable Benzocyclobutene (PS-BCBs). Photoimageable Cyclotene™ formulations commercialized as the 4000 series are capable of generating approximately 3–12 μm thick cured films.[34–36,202] Processing details are given in Sec. 3.3. The use of PS-BCB has been detailed by NEC,[37,38] Siemens,[39] and Motorola.[130,199]

Photoimageable Cyanate Ester. A proprietary cyanate ester resin, PHP 92, is used by AT&T for their PolyHIC technology.[40] The formulation consists of a blend of triazine resin (see Structure 3.9) (40–65 percent), acrylated acrylonitrile-butadiene rubber, acrylated epoxy, and other minor components such as hardeners, photo-cross-linkers, adhesion promoters, photoinitiators, pigments, surfactants, and thermal stabilizers. These photoactivated materials cure at <300° C. These formulated products are manufactured for internal use only. This dielectric formulation is used in the AT&T PolyHIC process described in Sec. 3.4.1.

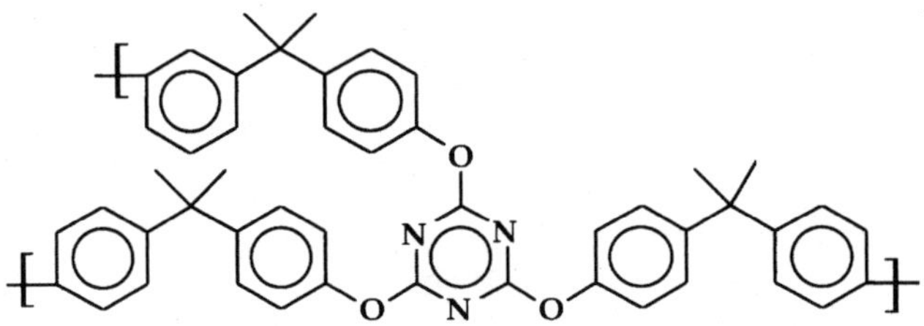

Structure 3.9

General properties for the aforementioned commercial dielectrics are compared in Table 3.3 and are discussed below.

Polymer Properties

Electrical Properties

Dielectric Constant. Dielectric constant (ε_r) is the most critical electrical parameter for a microelectronic polymer. The lower the dielectric constant, the faster the signal propagation velocity, as described by Eq. 3.2.

Table 3.3 Properties of Commercially Available Thin Film Polymer Dielectrics

Vendor	Product designation	Photo	ε'	Tg (°C)	CTE (ppm)	Tensile modulus (GPa)	Tensile strength (MPa)	Elongation (%)
Amoco	Ultradel 4212		2.9	295	50	2.8	101	30
	5106	+	2.8	400	24	1.3	122	70
Ciba-Geigy	Probamide 400	+	3.0	350	39		140	56
	500		3.2	400	6–7	11.6	444	28
	600		3.2	366	23	2.7	194	104
Cemota	Syntorg IP-2000		2.9	350	55		119	4
Dow	Cyclotene 3022		2.65	>350	52	2.9	85	9
	4026	+	2.65	>350	52	2.9	85	9
Dupont	Pyralin 2540		3.5	400	26	1.3	160	60
	2611		2.9 (x,y)	350	5	6.6	600	60
	WE-1111			385	19	4.4	300	55
	2730	+	2.9 (x,y)	350	15		170	
Hitachi	PIQ-13	–	3.4	290	50		116	10
	L100	–	3.2	360	3		320	22
	PL-2135	+	3.3	270	40	3.6	124	10
Toray	Photoneece UR-3800	+	3.3	280	45	3.4	145	30
	UR-5100	+			25		200	20

$$Vp = c\sqrt{\frac{1}{\varepsilon_r}} \tag{3.2}$$

where Vp = velocity of propagation
ε_r = dielectric constant of the insulator
c = speed of light

A lower dielectric constant allows signal traces to be designed wider and the dielectric thickness decreased. It also allows one to maintain the same characteristic impedance while lowering the line resistance and crosstalk. Good dielectric performance is obtained from a material whose dielectric constant is invariant with frequency and temperature.

Dielectric constants for typical MCM polymers are shown in Table 3.3. Polyimides have dielectric constants in the range 3.2 to 3.4. Low CTE PIs reveal highly anisotropic dielectric constants [estimated 3.0(z), 3.8–4.0(x,y)].[9] DVS-BCB has an isotropic dielectric constant of 2.65. CEs have an inherent dielectric constant of 2.8, but this is raised considerably (3.4–3.8) with the incorporation of 35 to 60 percent additives which all have considerably higher ε_r.[40]

Dissipation Factor. Dissipation factor (ε'') and loss tangent ($\varepsilon_r/\varepsilon''$) are also important electrical parameters, especially at high frequencies. Low values are indicative of minimal conversions of electrical energy to heat and little overall power loss.

Dielectric Strength. Dielectric strength/breakdown voltage define the voltage that causes current to flow through an insulator. The materials listed in Table 3.3 have values in the range of 10^6 V/cm.

Polymer Mechanical Properties. The mechanical properties of a polymer are determined in part from stress strain curves such as the one shown in Fig. 3.3. Stress is the force per unit area that is being exerted on the material. Strain is the change in dimension that occurs in response to the stress. Point 1 is the yield point and is a measure of the strength of the material and its resistance to deformation. Point 2 is the break point. Between point 1 and point 2 is a region where the material undergoes plastic deformation.

The ratio of stress to strain is the *elastic modulus*. The greater the elastic modulus, the greater the force needed to deform the polymer. Modulus is thus a measure of the stiffness or hardness of the polymer. The ratio of tensile stress to tensile strain (in the linear region at small strain values) is called Young's modulus (Eq. 3.3).

$$E = \frac{\sigma}{\varepsilon} \tag{3.3}$$

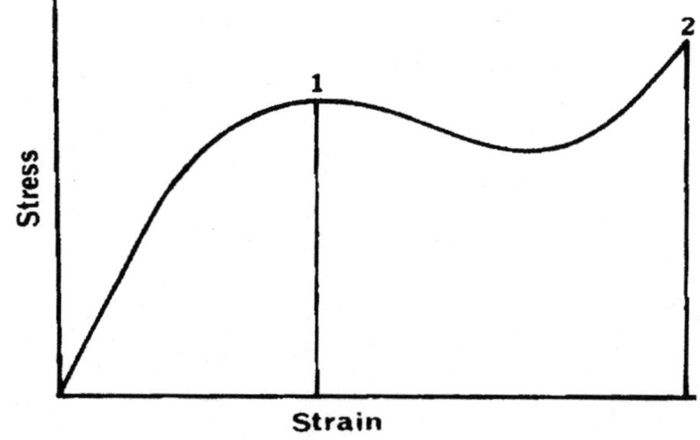

FIGURE 3.3 Typical polymer stress strain curve.

where E = Young's modulus
σ = tensile stress
ε = tensile strain

Hard brittle polymers have high moduli of elasticity, no well defined yield point, and low strain at break. Hard strong polymers have high moduli of elasticity and yield points and moderate stress at break. Hard tough polymers have high moduli of elasticity, moderate yield points, and high stress and strain at break.

Glass Transition Temperature, Tg. Another important mechanical property is the glass transition temperature, Tg. This is the temperature at which the polymer will deform in response to an external load. A thermoset polymer above its Tg takes on the properties of a cross-linked rubber, it will deform, but it cannot flow. A thermoplastic material above its Tg will flow. If processing of a thin film structure is to occur at a temperature above the Tg of the polymer dielectric, the mechanical integrity of the structure may be compromised. Thermoplastic materials are generally not used to fabricate thin film structures because such materials flow when taken above their Tg. BCBs and CEs are thermosets. PIs are thermoplastics that cross-link during the final curing reaction.

Water Absorption. Water absorption impacts both the electrical properties and the processibility of MCM dielectrics. Absorption of water, with a dielectric constant of 78, raises the ε_r of an insulating polymer and thus affects circuit performance. Absorbed moisture can also cause severe damage to structures during fabrication. Uncontrolled outgassing of moisture from underlying layers of dielectric during subsequent high-temperature processing can result in blistering and delamination of the thin film structures being fabricated.[42–43] Bakeout cycles are routinely employed during the fabrication of MCMs with polymers that exhibit significant moisture absorption. Gridded ground planes and, in some cases, gridded signal traces are used to allow moisture to escape.[44] It has been reported that PI moisture absorption increases after aging at 85°C/85% RH.[45]

Moisture uptake is specific to the polymer in question and the conditions of the experiment. Data in the open literature has been reported for exposure of varying thicknesses of polymer film for varying times at varying RH. Literature values for a variety of materials are given (with experimental conditions where available) in Table 3.4.

Jensen has noted that the ε_r of PMDA/ODA-based PI increases linearly from 3.1 at 0 percent RH to 4.1 at 100 percent RH.[46] He concluded that the electrical design tolerances must accommodate such variations in electrical characteristics, or the MCM must be hermetically sealed to control the humidity level.

Table 3.4 Water Uptake by Polymer Dielectrics under Varying Conditions

Polymer	Condition	Water Uptake (%)
Cyclotene 3022	24 hr water boil	0.2
	80% RH	0.1
Cyclotene 4026	24 hr water boil	0.3
IP 200	unknown	0.8
PI 2555	1 hr @ 50% RH	2–3
PI 2611	2 hr water boil	0.4
PIQ 13	unknown	2.3
PIQ I100	2 hr water boil	0.5
Prob 400	100% RH	5.0
Prob 500	50% RH	0.75
Prob 600	50% RH	1.3
Ultradel 5106	unknown	3.5
UR 3800	24 hr immersion @ RT	1.1

Kobayashi, et al., have detailed the moisture bake-out procedures necessary to reliably deposit aluminum onto PIQ dielectric.[47]

Kambe and co-workers have noted that the ε_r of PI increased after exposure to a pressure cooker test (PCT) from 3.1 to >3.5 after 25 hr. Moisture was subsequently released after heating 17 hr in air at 200° C.[48]

Heacock has studied the hydrolysis of PMDA/ODA, the reverse of the imidization reaction, at elevated temperature and humidity.[49] The reaction proceeds according to Eq. 3.4 At 80° C and 60 percent RH, $K = 2.66 \times 10^{-5}$. PI hydrolysis has also been discussed in detail by Volksen.[13]

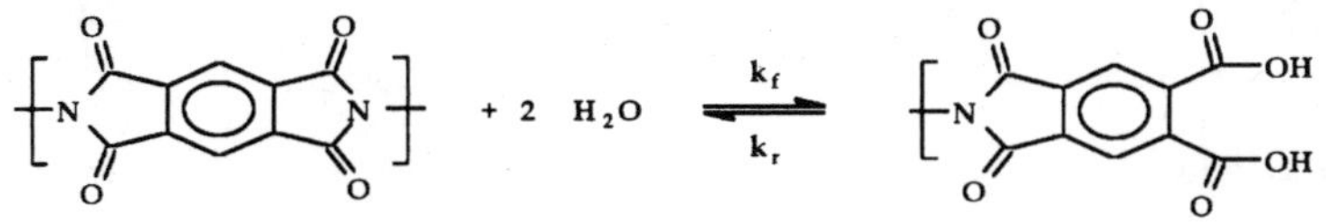

(3.4)

Figure 3.4 shows moisture absorption and desorption rates for DVS-BCB when humidity is increased to 81 percent and then decreased to 6 percent at ambient temperature.[50] Such low moisture absorption and rapid equilibration allows the elimination of traditional bakeout steps.

Planarization. Planarization by polymeric materials has recently been reviewed by Schiltz.[51] Relief existing on a surface prior to photolithography will affect the resolution and quality of subsequent conductor traces. One of the main functions of the insulator layer is to planarize the underlying topography and thus provide a flat focal plane for the next layer. Lack of topographic planarity can lead to subsequent nonuniform metal thickness (poor step coverage). These thinned and weakened areas are susceptible to cracking.[52] The degree of planarization (DOP) for a polymer dielectric is defined for lines and vias in Fig. 3.5.[53,54]

Thin film coatings of BCB exhibit excellent planarization properties, thus eliminating photolithographic resolution issues caused by the generation of relief when poorly planarizing materials are used to fabricate multiple levels of interconnect. The degree of planarization is dependent on the film thickness versus the feature thickness; the number of coatings; and the feature width, height, and spacing.[55] Typical planarization versus feature width is shown in Fig. 3.6 for single coatings of varying thicknesses.

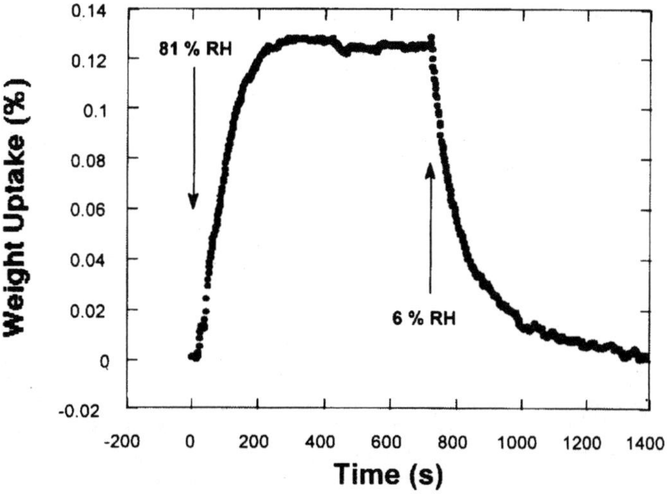

FIGURE 3.4 Water absorption/desorption for Cyclotene 3022.[50]

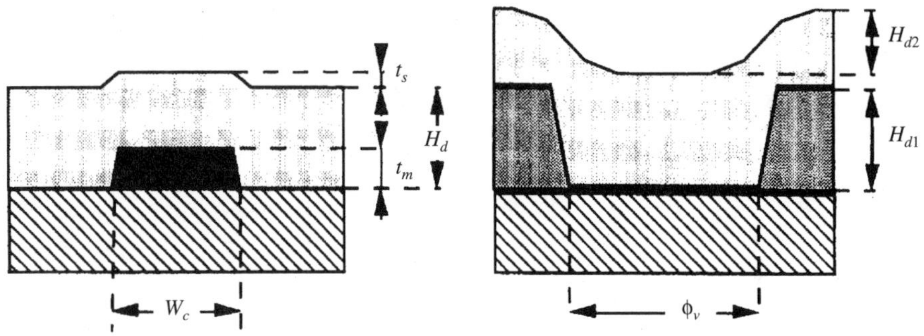

Planarization over conductor tracks:

$$\text{D.O.P.} = 1 - \frac{t_s}{t_m}$$

Planarization over vias:

$$\text{D.O.P.} = 1 - \frac{H_{d2}}{H_{d1}}$$

FIGURE 3.5 Planarization of an isolated feature.

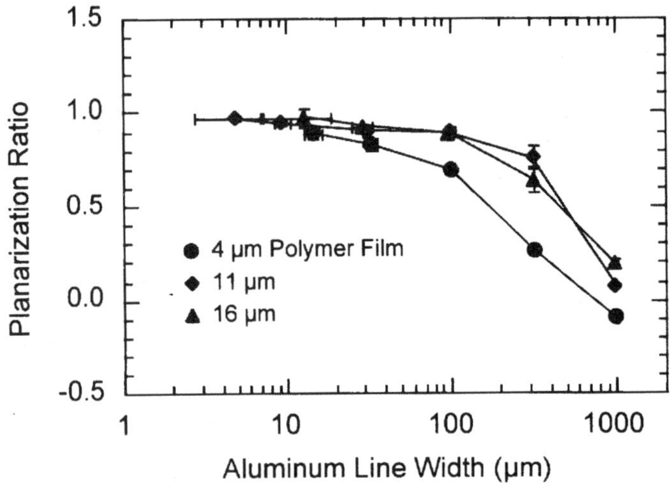

FIGURE 3.6 DVS-BCB planarization over 4 μm thick aluminum lines.[55]

In conformal via structures, BCB dielectrics not only planarize underlying conductor layers but also planarize the conformal vias so that the next orthogonal layer of conductor can be routed directly over the vias. Under identical conditions, BCB showed 65–70 percent planarization of via features versus 10–15 percent for typical polyimide dielectrics.[54]

Senturia and co-workers[56] have shown that the DOP obtained by fully cured PI films is the composite of two processes: drying and curing. During drying, loss of planarization occurs from shrinkage when the polymer can no longer flow. For polyimides, imidization and its associated water loss, which occur during curing, cause further shrinkage, which adds to planarization loss. Shrinkage is reported to be as high as 50+ percent depending on the polyimide. Mw also plays a role in planarization: the lower the Mw, the better the planarization.

Figure 3.7 reveals the degree of planarization versus the number of coating applications needed if the initial coating gives DOP = P_1 as calculated by Chao.[57] DOP for PIs can be reportedly be improved by optimizing spin conditions.[57]

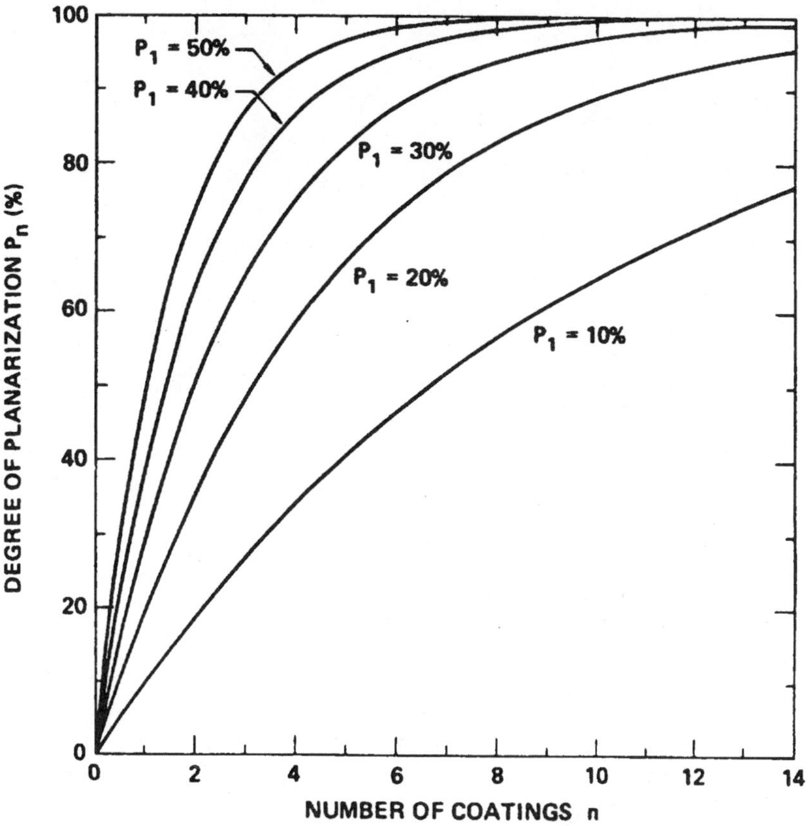

FIGURE 3.7 Degree of planarization as a function of the number of coatings.[57]

In general, thin lines and narrow spaces are easier to planarize.[53] This is depicted in Fig. 3.8, where loss of planarization with increasing conductor width is clearly evident for Pyralin™ 2611.[58]

Planarization is shown for 5 μm of dielectric over 3 μm conductors (Ti/Cu and Ti/Cu/Ti) patterned on SiO_2. There appears to be a surface dependence, i.e., higher planarization over Ti than Cu.

A conformal via/photo PI process has been described by Perfecto et al.[59] The photo PI is patterned by full field projection lithography. The degree of planarization is 35 percent for 7 μm thick PI over 5 μm Cu traces on 25 μm pitch.[60]

Etchback Planarization Process. Oki has developed an etchback planarization process similar to that used to planarize conformal spin on glass or CVD oxide dielectric layers.[61] In the described process, after sputtering patterning and plating up Cu posts, features are covered with PI dielectric and cured. The cured PI is coated with a sacrificial positive resist layer which is hard baked. Dry etching of the structure is then carried out under conditions where the etch rate of PI and the etch rate of the resist are approximately equal. Etching continues until the tops of the via posts emerge. These steps are depicted in Fig. 3.9. Planarization for a variety of dielectric materials is given in Table 3.5.

Chemical Resistance. Polymer dielectrics must be resistant to the processing and cleaning chemicals used in the fabrication of the structures. The usual manifestation of poor chemical resistance is swelling, cracking, or crazing of the polymer layer. Thermosets are generally much more resistant than thermoplastic materials. Most of the materials listed in Table 3.3 are reported to be resistant to normal processing chemicals.[43,62]

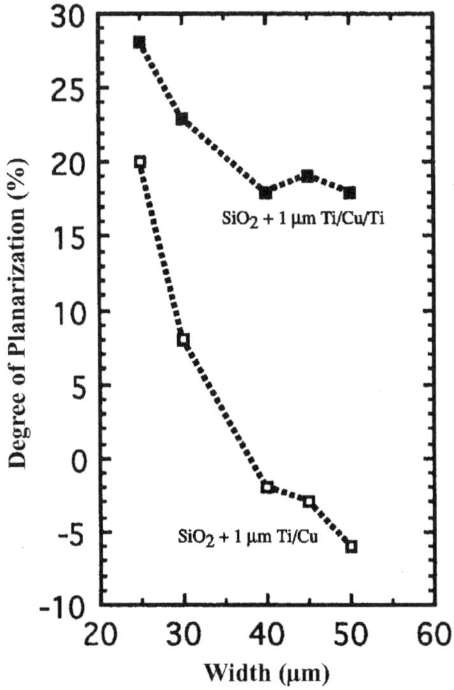

FIGURE 3.8 Planarization vs. conductor line width for 5 microns of PI 2611.[58]

Table 3.5 Planarization of Typical Thin Film Polymer Dielectrics

Polymer	Planarization (%)
Cyclotene 3022	90–95
Cyclotene 4026	70–90
IP 200	25
PI 2525	30
PI 2611	18–28
PI 2722	28
PIQ 13	25
PIQ L100	25–28

PIs are not stable to strongly basic solutions, which causes problems with typical electroless plating solutions.[63] BCB films are not stable in the presence of strongly oxidizing acid solutions.

Preimidized PIs are soluble before curing and after solvent removal. As such, solvent resistance is poor unless they are taken to an elevated temperature to induce cross-linking of the linear molecules. Poor solvent resistance is manifested by crazing (microcracking) of the polymer surface. For example, Hodge et al.[64] have reported the following data for OCG Probamide 293 (PMDA-ODA-DAPI) (see Table 3.6).

Thermal Stability. A material is stable at a given temperature when its properties at that temperature do not change. Thermal stability for an MCM polymer dielectric needs to be in excess of the temper-

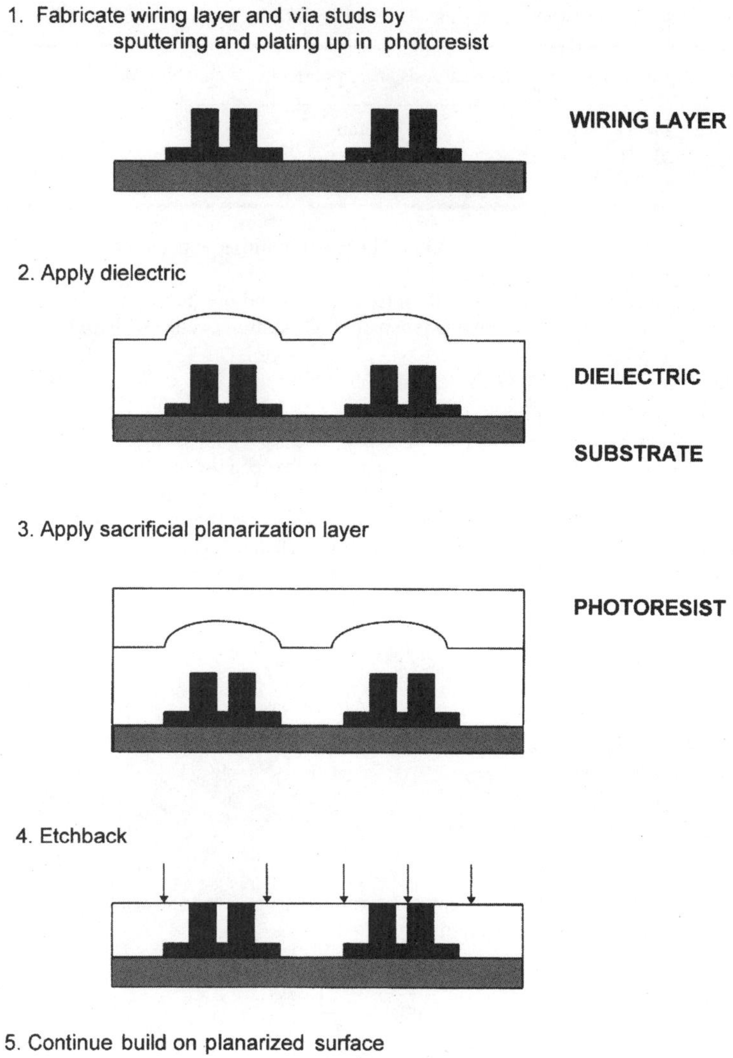

FIGURE 3.9 Oki etchback planarization process.[61]

atures experienced during subsequent processing and/or repair steps. In general, the dielectric will be subjected to temperatures <350° C during chip attach or other assembly processes.[42] Thermal stability is best determined from isothermal thermogravametric analysis, TGA, on thicknesses similar to those used to fabricate the structures. Isothermal data is required for meaningful results because thermal degradation is limited by solid state diffusion and therefore is thickness dependent.[21]

Most fully aromatic PIs are stable in air and nitrogen at 400° C. The Hitachi PIQ materials were specially designed to produce the most thermally stable materials available. DVS-BCB stability is limited to 350° C under nitrogen.

Polyimides that are not fully aromatic, such as the preimidized materials and BCBs, are inherently more susceptible to oxidation at elevated temperatures due to the presence of benzylic and/or

Table 3.6 Solvent resistance of Probamide 293 under different "curing conditions."[64]

Cure ("x-link") conditions	Solvent resistance, γ-butyrolactone	Solvent resistance, NMP
Hot plate, 2 min @ 300° C	1 min	4 min
Hot plate, 30 s @ 350° C	5+ min	5+ min
Furnace cure, 350° C	2 min	5 min

aliphatic functionality in the cured films. The use of antioxidants increases the stability of DVS-BCB in air at high temperatures.[65]

The thermal stability (determined under nitrogen and the thermo-oxidative stability (determined in air) of BCBs have been studied extensively.[66] Oxidation occurs with an activation energy of 25 ± 1.5 Kcal per mole.

The PPQ/Cu interface reveals degradation in air above 200° C. This has been interpreted as copper catalyzed degradation.[68]

Stress. Stress is a reliability issue because excessive stress buildup can exceed the adhesive strength of an interface resulting in delamination or exceed the tensile strength of the polymer, resulting in cohesive fracture of the dielectric layers. Intrinsic stress is built into a film due to the shrinkage and contraction that can occur after a film has adhered to the underlying surface. Thermal stress is generated, after cure, as the polymer cools down (from its Tg) as result of the mismatch in CTE between the polymer film and the substrate. This is described by Eq. 3.5 and shown in Fig. 3.11.

$$\Delta\sigma_f = \frac{E_f(\alpha_s - \alpha_f)\Delta T}{1 - \nu_f} \tag{3.5}$$

where σ_f = film stress
E_f = Young's modulus
α_s = CTE substrate
α_f = CTE film
ΔT = temperature difference
ν_f = film Poisson ratio

The biaxial stress (σ_f) is generally determined from substrate warpage measurements. Stress is related to substrate warpage, according to Eq. 3.6.

$$\Delta\sigma_f = \frac{E_s t_s^2}{6(1 - \nu_s)Kt_f} \tag{3.6}$$

where E_s = Young's modulus of substrate
t_s = substrate thickness
t_f = film thickness
K = substrate curvature ($1/R_u - 1/R_c$), where R = radius of curvature for coated and uncoated samples, respectively
ν_s = Poisson's ratio for the substrate

For silicon, $E/1 - \nu_c = 1.8 \times 10^{11}$ Pa.

Soane[69] has compared the stressed developed in PMDA-ODA (Pyralin™ 2545) and DVS-BCB (Cyclotene™ 3022). When both materials were subjected to two hours of curing at 260° C (full cure for BCB, partial cure for PI), the PI films developed substantial intrinsic stress on heating due to film contraction after solvent evaporation, whereas BCB stress was mainly developed during cooling of the film from 260° C to RT. This is depicted in Fig. 3.10. Stress develops in PI during heating *and* cooling, but stress in BCB develops only during the cooling cycle. The author concludes that PMDA/

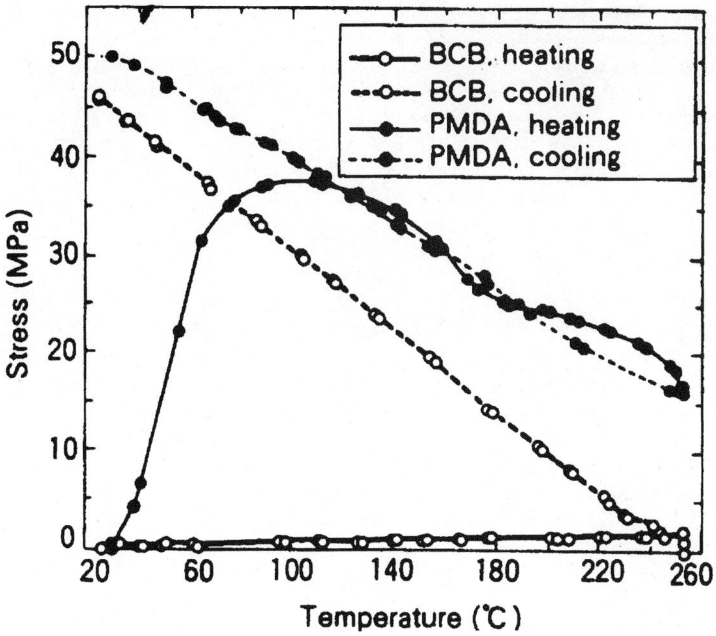

FIGURE 3.10 Stress-temperature profiles for DVS-BCB and PMDA/ODA PI.[69]

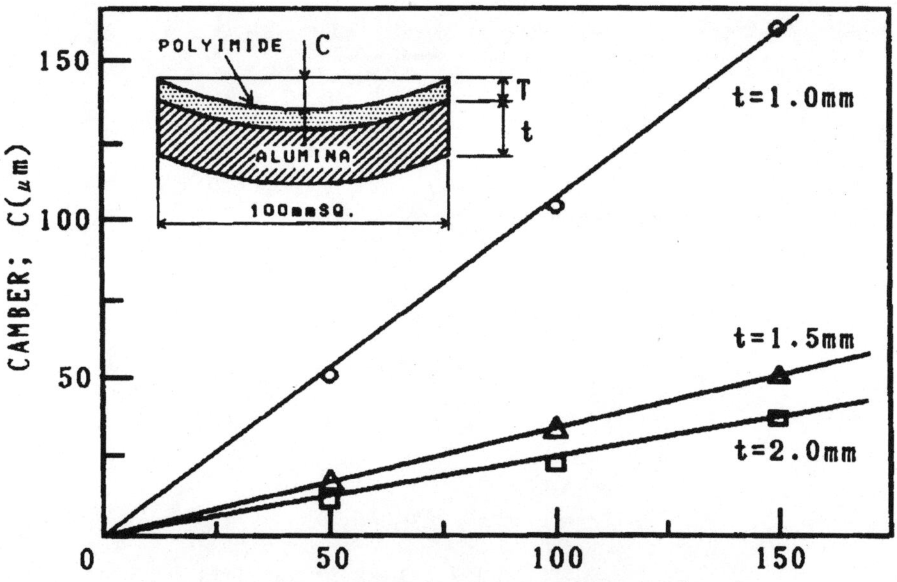

FIGURE 3.11 Camber in alumina substrate caused by PI stress.[48]

ODA has higher cumulative stress at 25° C, in spite of its lower calculated thermal stress, due to the intrinsic stress induced during solvent evaporation.

Film stresses are biaxial in the plane of the substrate, except near the film edges, where the biaxial stresses are compounded by shear stresses at the film substrate interface. Biaxial stresses are independent of film thickness. Shear stress is proportional not only to the CTE of the two interface materials and the temperature excursion that they have seen, but also to the modulus of the materials and inversely proportional to the thickness of the layers. Lower moduli and thicker layers translate into lower shear stress.

The most pronounced effects of stress are observed at stress concentrators such as the corners or edges of patterned features.[21] It is for this reason that round vias are favored over square vias, and sloped via bottoms are favored over sharp undercutting.

Stresses that are generated by polymer dielectrics can lead to fabrication problems from substrate warpage, especially when using silicon substrates.[70] If wafer bow becomes too severe, vacuum fixtures cannot hold the wafers down flat enough to allow sharp focusing over the entire wafer diameter. Problems have also been encountered in automated cassette-to-cassette handling equipment due to wafer sticking if the warpage is too severe. Warpage problems increase with substrate dimensions and the total number of dielectric layers. Ceramic substrates, with higher moduli, are more resistant to such bowing. Tensile stresses are also imparted by the metallization layers.

Polymer stresses for a series of PIs have been correlated with relative humidity.[71] As shown in Table 3.7, humidity has a tendency to swell the polyimide layers and thus somewhat relax the stress. This effect of RH may account for some of the differences in published stress data.

The camber created by PI coatings (CTE 45, cure temperature 400° C) on differing thicknesses of alumina ceramic is shown in Fig. 3.11.[48] The camber is <50 μm when the total PI thickness is <100 μm and the alumina thickness is more than 1.5 mm. Published stress values for a variety of dielectric materials are given in Table 3.8.

Adhesion. Polymer adhesion is needed at four interfaces: polymer to substrate, polymer to metal, metal to polymer and polymer to polymer. Adhesion is intimately related to other issues such as

Table 3.7 Stress as a Function of Relative Humidity

PI	0% RH	50% RH	100% RH
PI 2525	41	31	18
PIQ L110	8	5	0
PIQ 2611	4	0	–5

Table 3.8 Thin Film Polymer Stress Values

Polymer	Stress (Mpa)
Cyclotene 3022	24–38
Cyclotene 4026	28–30
PI 2525	40–54
PI 2611	4–11
PIQ 13	28–30
PIQ L100	4–5
Prob 400	55
Prob 500	<5
Prob 600	20
UR 3800	45

water absorption and stress. Temperature and humidity are known to degrade adhesion. Adhesive failure usually manifests itself as a delamination between interfaces.

Various measurement techniques have been used to assess adhesive strength. A meaningful test should duplicate the stress that the MCM will be subjected to in fabrication and service. Tape pull,[72] stud pull,[73] peel tests,[74] and microindentation[75] tests do not adequately simulate the stresses imparted on an MCM structure but do offer a somewhat quantitative means of comparing the adhesion of differing dielectric materials to different interfaces. In general, a good indication of adequate adhesion can be obtained by crosshatching a 1 mm grid into the cured film with a razor blade. Adhesive tape is applied to the interface and is pulled off. If none of the film has been pulled off, the test structure is subjected to boiling water, 85° C/85% RH and/or PCT conditions (121° C/2 atm) and retested with the tape process.

Adhesion is very sensitive to surface preparation. Cleaning of the surface prior to polymer deposition removes reactive layers such as oxides and/or particulate contamination. Surface treatment by plasma or ion beam techniques and/or by "adhesion promoters" are used to induce better adhesion

Adhesion Promoters. Adhesion promoters are formulated to chemically react with both the substrate surface and the deposited film. PI adhesion promoters are usually amino silanes such as APS[76] while BCB adhesion promoters include APS or MOPS[77] depending on the carrier substrate composition. The adhesion promoters are either included in the dielectric formulation—"self-priming" (some PIs)—or applied as a separate solution prior to the dielectric (BCB and some PIs).

It has been noted that the thickness of the adhesion promoter layer for PIs should be carefully monitored because each amine reaction into a PI linkage causes cleavage of a PI chain, and excessive chain scission is actually detrimental to adhesion and the mechanical properties of the resultant film.

The adhesion of Cu deposited onto PI is reportedly increased 25 fold by oxygen RIE of the PI surface.[78,79] Studies have detailed the adhesion of Cr/Cu/Cr to a PMDA/ODA film that has been exposed to either Ar backsputter or O_2 RIE. Peel strength increases from 80 g/cm to approximately 500 g/cm. It is shown that the treatments change the surface free energy rather than roughening the surface as was previously thought.[80]

Inorganics. Far less has been published detailing the use of inorganics as thin film dielectrics for MCMs. SiO_2 is the only inorganic to receive any real attention in the commercial arena. Thick SiO_2 layers are difficult to deposit crack free due to the normally high stress that CVD techniques impart. Patented technology has been described for the controlled compressive stress PECVD deposition of >10 μm coatings.[81,82] It is claimed that SiO_2-based modules are more rugged, offer a harder surface for wire bonding, have a higher tolerance for high temperature IC rework, and are more reliable under thermal cycling and thermal shock. SiO_2-based processes have been described by n-Chip (Sec. 3.4.5) and Toshiba (Sec. 3.4.13).

3.2.2 Carrier Substrates

The thin film interconnect carrier (substrate) must present a flat and highly polished surface to build upon, since the subsequent thin film fabrication requires the use of lithographic techniques. The substrate should be inert to the process chemicals, gas atmospheres, and temperatures used during the fabrication of the interconnect. Mechanical properties such as modulus are important, because the substrate must be strong enough to withstand handling, thermal cycling, and thermal shock.

The substrates must meet certain CTE constraints, because it interfaces with very large silicon chips on one side and the package on the other. Thermal conductivity is also important, because there needs to be a mechanism to conduct the heat away from the closely spaced, heat generating semiconductor chips. Table 3.9 presents some relevant electrical, physical, and mechanical properties of substrate materials now in use or being considered.

Laminate carriers are detailed in Chaps. 4 and 5. Basically, three classes of materials are being used today: silicon wafers, metals, and ceramics. Each carrier material represents a unique response to the required surface roughness, flatness, mechanical and thermal stability and/or compatibility with thick and/or thin film processing. The recent drive for *high density large area processing* has generated interest in glass as a carrier material. The properties of this material are detailed in Chap. 5.

Table 3.9 Properties of MCM Carrier Materials

Substrate	CTE (ppm/°C)	TC (W/mK)	ε′	Flex mod (kpsi)
Si	2.7–3.5	150	11.7	2
Al	22–24	238		
Cu	17	400		
Cu/Mo/Cu	5.1	166 (x,y) 151 (z)		
Al_2O_3 (96%)	6.0–7.7	20	9	30–50
Al_2O_3 (99.6%)	6.5	27	10	30–40
Mullite	4–5	4–7	5.5–6.5	18–40
BeO	8	250	6.7	20
AlN	4.1	175–200+	8.5	40–50

Silicon. Silicon as a substrate material has many advantages.[83] In terms of CTE matching, the substrate is an exact match to the silicon chips that will be mounted on it. The thermal conductivity of silicon is 10 times higher than conventional alumina. Silicon wafers are extremely flat, smooth, and dimensionally stable. Silicon wafers allow the fabrication of integral decoupling capacitors.[84] A unique advantage of silicon is that active circuitry (transistors) can be incorporated into the substrate.[85,86] Pearson, of Lockheed, has proposed that interconnect can be built on top of active circuitry such as microprocessors, and support chips can be attached to the interconnect over the substrate by solder bumps[87] as shown in Fig. 3.12.

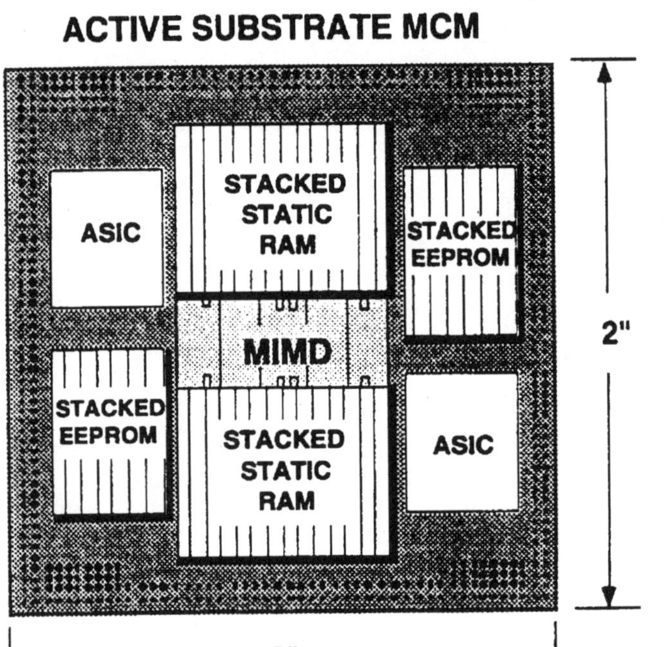

FIGURE 3.12 Interconnect created over active microprocessor wafers.[87]

Low-cost, highly polished wafers are compatible with automated IC processing equipment. When silicon wafers are used as the substrate, the power, ground, and signal layers are built on top of the substrate. The resulting circuitized substrate must be placed into a package for connection to the outside world. To realize the full potential of a silicon carrier, the packaging of the module has to be optimized with respect to electrical, thermal and mechanical performance (see Chap. 7).

Problems associated with the use of silicon wafers include poor flexural strength, a relatively high dielectric constant, and excessive warpage caused by the stresses imposed by the dielectric and metallization layers. In general, the lower Young's modulus of a silicon carrier results in three times the warpage versus the same thickness ceramic substrate when a dielectric/metal thin film structure is deposited on it.[88]

Metallic Substrates. Metallic substrates have been used to optimize the thermal and mechanical requirements while minimizing substrate raw material and processing cost. Metal substrates can be manufactured easily in large panels size or cut to the shape of a silicon wafer to be accepted on IC processing equipment. Metallic carriers have excellent thermal properties but can react with the common wet etchants used to define conductor patterns.

Metallic sandwiches such as Cu/Mo/Cu can be tailored to control CTE and thermal properties. 5%Cu/Mo/5%Cu is reported to have a TC of 135 W/m°K, a CTE of 5.1 ppm, an as-manufactured surface finish of 0.813 μm (32 μin) and camber of 0.0005 in/in.[89]

DEC and MMS have commercialized technology based on aluminum carriers.[90,91]

Ceramic Carrier Technology. Many ceramic materials have been examined or are under consideration as carriers. Ceramics can be used in two distinct ways: simply as a mechanical support upon which the power, ground, and signal layers are fabricated by thin film techniques or as a cofired ceramic structure that contains power and ground distribution planes and some of the signal layers, and upon which some or all of the thin film signal layers and/or chip signal redistribution layers are fabricated.

Ceramic processing can be categorized in several ways, including thick versus thin film and high- versus low-temperature cofire. Thick film processes such as doctor blading of dielectric and conductor layers, and cofiring processes such as lamination of individual green sheets, utilize materials with coarse granularity that are not easily adapted to micron-size dimensions.[92] Thin film processes require much smoother surface finishes to match up with the requirements of photolithography.

High-temperature cofired (HTCC) parts require firing at high temperatures due to their high ceramic content. This, in turn, necessitates the use of refractory metallizations (such as W) that are inherently more resistive. Low-temperature cofire has a higher glass content and thus can be fired at lower temperatures, with higher conductivity metallizations.

To apply thin film to metallized ceramic substrates, the carrier must present a flat, smooth surface for subsequent thin film fabrication. The ceramic carriers have rough topography due to distortions that occur during sintering, inherent ceramic porosity, and the protruding top surface metallurgy. This may exceed the photolithographic depth of focus. The relation between line width and projection aligner focus depth (with different numerical apertures) is shown in Fig. 3.13.[93] To define patterns on rough surfaces, a larger focus depth is required. This in turn inhibits fine line definition. As a rule of thumb, the substrate should have a roughness of less than 1000 Å and should have a flatness of 10 μm or less to be suitable for thin film lithography.[94]

The lapping and polishing needed for the desired post-fired surface finish in most part determines the cost of the cofired carrier. In some cases, the rougher finish of the cofired ceramic versus silicon carrier can be circumvented by an initial polymer planarization coat prior to the deposition of thin film metallization.[70] The material properties and processing of ceramics are covered in greater detail in Chap. 2.

3.2.3 Conductor Metallizations

Table 3.10 presents the resistivity (ρ), coefficient of thermal expansion (CTE) and thermal conductivity (TC) of the metals typically used in the fabrication of thin film MCM structures. The specific

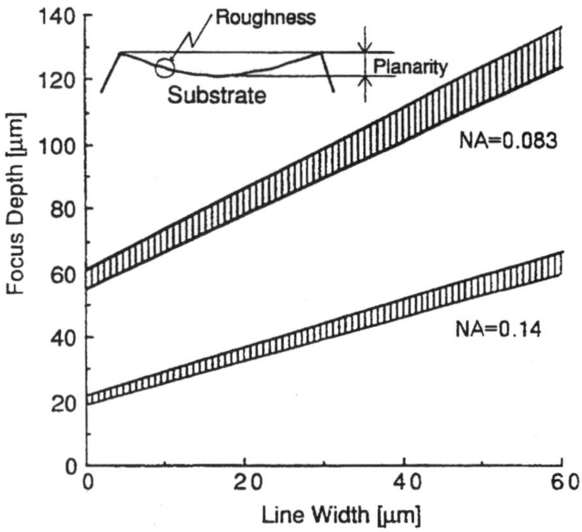

FIGURE 3.13 Relationship between line width and focus depth.[93]

Table 3.10 Physical Properties of Metals Used in the Fabrication of Thin Film Structures

Metal	ρ (μΩcm)	CTE (10^{-6}/°C)	TC (W/m°C)
Cu	1.67	18	approx. 400
Al	2.6–4.3	23	240
Au	2.3	14	300
Ni	6.9	13	90
W	5.5	45	200
Cr	13–20	63	66
Ti	55	90	22
Pt	10.5	90	70
Pd	11	110	70

conductor material chosen will depend on the design, electrical requirements, and process chosen to fabricate the MCM. Metal deposition and definition techniques are detailed in Sec. 3.3.3. The most critical requirements for conductor metallizations are conductivity and reliability.

Aluminum. Aluminum is in use by a number of commercial organizations (see Table 3.14 in Sec. 3.4) for thin film interconnect. It is a low-cost material that has adequate conductivity and can be deposited and patterned by typical IC techniques. It is also relatively resistant to further oxidation. It can be sputtered or evaporated but cannot be electroplated.

Copper. Copper offers significant conductivity improvements over aluminum and is more electromigration resistant. Copper can be deposited by sputtering, evaporation, electroplating, or electroless plating. Copper rapidly oxidizes, forming a variety of oxides which have poor adhesion to polymer dielectrics and copper itself. A barrier/adhesion layer of Cr or Ti is required when interfacing PI dielectrics, whereas this is reportedly not necessary with BCB (see Sec. 3.5).

Gold. Au is used in thin film structures to minimize via contact resistance problems caused by oxidation of Cu and Al. Au can also be deposited by sputtering, evaporation, electroplating, or electroless plating. Although its resistivity is excellent, its cost is high. Adhesion to Au is usually poor, and it thus requires an adhesion layer (typically 50–200 nm of Ti or Ti/W).

Pad-level metallization must be conductive, show good adhesion to the dielectric, and be compatible with the chosen assembly process. Pad level metal is usually finished in gold because of its corrosion resistance. It normally requires adhesion/barrier layers such as Ni, Cr, or Ti/W. Metallization thickness of 1–3 μm of Au is needed for wire bonding or TAB, whereas only a flash is required for a soldering surface. In general, the pad level requirements are dictated by the chip attach method of choice. Further details can be obtained in Chap. 8.

3.3 THIN FILM PROCESSING

3.3.1 Dielectric Application and Curing

Analogous to photoresist application in the IC industry, the application of organic dielectrics have evolved as a spin coating technology. One needs to achieve the desired thickness in the fewest coating applications possible; exhibit thickness uniformity across the substrate; and produce a smooth, planar, pinhole-free film. The need for planarization has been described separately (Sec. 3.2.1). Cured film thickness is a function of both spin speed and spin time. Nonuniformity across the substrate can result if the spin speed is either too slow or too fast.[41]

IBM has reported that PI coating uniformity across a ceramic substrate is typically 8 percent.[95] Coating uniformity for BCB has been reported as <2 percent for the dry etchable Cyclotene 3022. Variations in dielectric thickness can affect subsequent etching steps where one may overetch certain areas while just clearing others. A 10 percent overetch during processing is typical for reported PI process sequences. It is important to remove organic contaminants from any surface prior to film deposition, because spin on polymers tend to pull away from these areas leaving *pinholes*.[96]

Spin coating of polymer dielectrics usually involves the application of an adhesion promoter to the substrate surface followed by the dielectric it self. Some polymers are self-priming (i.e., contain adhesion promoter in the dielectric solution).

Spray Coating. There is very little published literature on spray coating of dielectric layers. The thickness, uniformity, and texture of the sprayed surface are influenced by the viscoelastic properties of the polymer, the concentration of the polymer in solvent, and the critical surface tension at the interface.[97,98] AT&T PolyHIC substrates are manufactured using a spray coating process (see Sec. 3.4.1).

Other Coating Techniques. As the emphasis in thin film processing shifts to lowering costs, practitioners are more closely examining alternative coating technologies. This is discussed in detail in Chap. 5.5.

Curing

PI. During polyamic acid-based PI curing, two processes take place: loss of NMP solvent and release of water. Improper curing can have a pronounced effect on the properties of the final PI film. For example, electromigration has been reported to occur between the leads of PI overcoated chips caused by outgassing of NMP and water from the film which had been insufficiently cured at 300°C.[99] IR monitoring of films undergoing cure has demonstrated that the rate of reaction is dependent on both temperature and thickness. Thicker PI films, or thicker regions of nonuniform films, tended to cure to a greater extent than thinner films under identical thermal processing.[100] A detailed IR analysis of the thermal curing of BTDA/ODA/MPDA and BPDA/PDA has been reported.[101] It has also been reported that PI curing can be monitored by microdielectrometry.[102] It has been shown for PIs that tensile stress increases linearly with cure temperature.[103] Thus, undercured films will show unrealistically low stress values.

BCB. Extensive kinetic studies have revealed a curing activation energy, Ea, of 36 ± 0.9 Kcal/mol.[104] No volatiles are evolved upon cure, so shrinkage is low (<5 percent). Curing is best monitored by FT-IR spectroscopy.[67,105] Curing must be carried out in an inert (nitrogen) atmosphere (<100 ppm O_2 recommended). Curing can take place in a typical box oven, on a nitrogen purged hotplate or in a controlled atmosphere belt furnace. Time/temperature conversion curves (see Fig. 3.14) that describe the curing process have been developed from the kinetics of the reaction. At vitrification (solidification), the reaction slows down an order of magnitude but continues in the solid state. Processing to partial (approx. 70–80 percent) cure is recommended to achieve the best layer-to-layer adhesion.[38,106] Full cure can be obtained on a time scale of minutes to hours by controlling the curing temperature. In a typical convection oven, "soft" curing is achieved in 30 min at 210° C, and a full "hard" cure is obtained in approximately 1 hour at 250° C. Typical cure schedules with ramp up and cooldown take approximately four hours to complete.

Rapid Thermal Curing [RTC]. Traditionally, polymer dielectric curing has been carried out in a box oven configuration in a batch mode. The heat-up, hold, and cool-down cycles reported by most dielectric manufacturers can occupy 4–12 hours of processing time and create logistical work flow and work-in-progress problems. In addition, an oven loaded with hundreds of wafers puts a lot of product at risk at once, especially when the builds are near completion. To circumvent such issues, rapid thermal curing (RTC) was developed for BCB.[107]

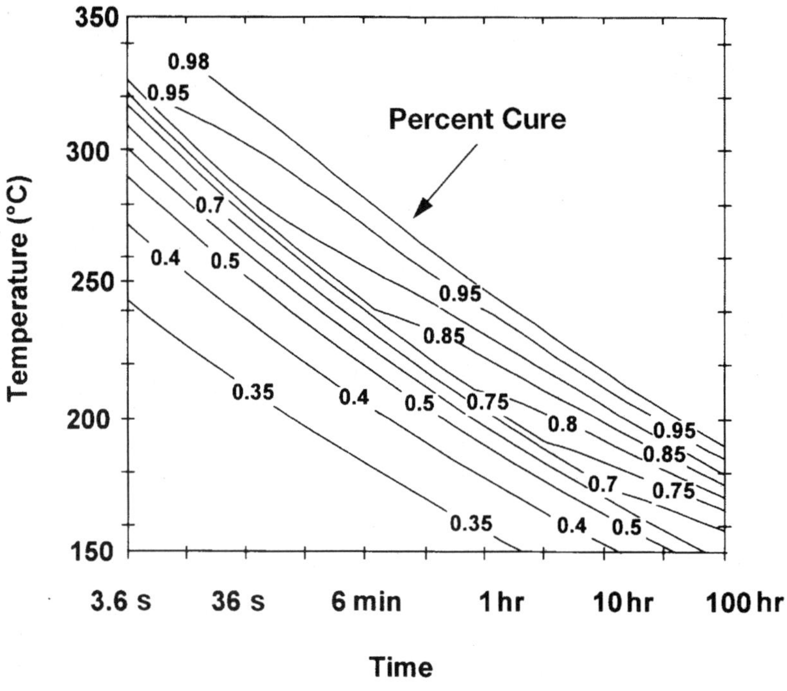

FIGURE 3.14 BCB time-temperature curing curves.[106]

Rapid thermal curing in a belt furnace (IR or convection) decreases typical cure times from 4–5 hours to <15 min, as shown in Fig. 3.15, and shows no deleterious effect on stress, adhesion, or planarization.[107]

Hot plate curing has also been described.[67] Kohl and co-workers extended this concept in their investigation of the curing of various PI and BCB materials using RTC hot plate processing.[64] They concluded that RTC is a viable process that resulted in polymer films with equivalent mechanical and electrical properties when compared to standard furnace curing for BCB and preimidized PIs. Polyamic acid-based PIs were shown to be incompatible with RTC techniques due to inferior properties developed when compared to standard furnace curing. Higher dielectric constant and much higher stress were observed.

Rapid curing of the polyamic acid ethyl ester of PMDA-ODA has been examined in a continuous-mode furnace.[198] Cure time cycles were reduced to 3.5 hr with no impact on mechanical properties or film stress.

3.3.2 Via Formation

Via formation in the dielectric layer is necessary to interconnect the signal, ground, power, and pad layers. The main techniques of via formation have included

- Dry etching
- Wet etching
- Use of photosensitive dielectrics
- Laser ablation

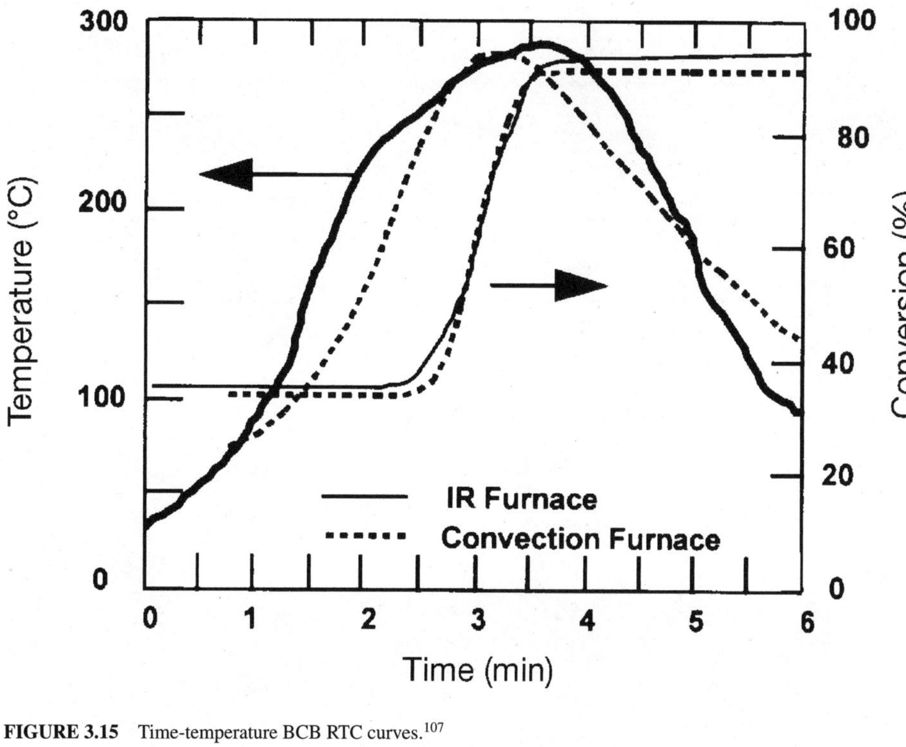

FIGURE 3.15 Time-temperature BCB RTC curves.[107]

Laser ablation can be further defined as scanning laser ablation (SLA) or projection laser ablation. These techniques are discussed below.

Irrespective of the creation technique, good vias are synonymous with low-resistance vias. The key to fabrication of low-resistance vias, by any technique, is minimizing the amount of residue to ensure a clean metal-to-metal contact area. Via resistance is determined by measuring the voltage drop across the via, and the two contact interfaces above and below it, and dividing this voltage by the interjected current level. Normally, a linear relationship is seen between current and voltage typical of an ohmic contact. If contamination of the via has occurred, a nonlinear C/V curve will result.

For PIs, Myszka has defined general via resolution limits for photosensitive, RIE, and SLA versus dielectric thickness[108] as shown in Table 3.11.

Stacked Stud Vias. As discussed further in Sec. 3.3.3, metal features can be defined in photoresist and plated up. The features are coated with PI, and the dielectric is cured. Excess PI is then removed by chemical/mechanical polishing of the structure. Scratches in the PI generated during the polishing process can lead to line opens at subsequent operations. Slurry aggregates, filtration, tool cleaning and maintenance, and pad material are the key factors in controlling such scratch defects.[109] The uniformity of the PI removal during the planarization process determines the uniformity of the dielectric thickness. In subsequent layers, the vias may be built directly on top of the previous layer due to the flatness uniformity that has been achieved.

Conformal Vias. If subtractive processing is used, vias cannot be vertically stacked (unless they are filled by plating them up) but rather must be staggered or staircased in a so-called *conformal* via process. In such processing, blanket metal deposition conformally follows the sidewall angle created by the dielectric via etching process and therefore is filled in by the next layer of dielectric, as is depicted in Fig. 3.16. The next layer of vias must therefore be offset or "staircased." When using highly planarizing materials such as Cyclotene™ 3022, the vias may be fabricated in folded staircase technology, i.e., layer 3 may be on top of layer 1.

Table 3.11 Typical Resolution Limits of Via Generation Approaches for PI Dielectrics[108]

	Dielectric Thickness		
Via approach	10 μm	15 μm	20 μm
Photo	25	50	50
RIE	5	10	10
Laser	5	10	10

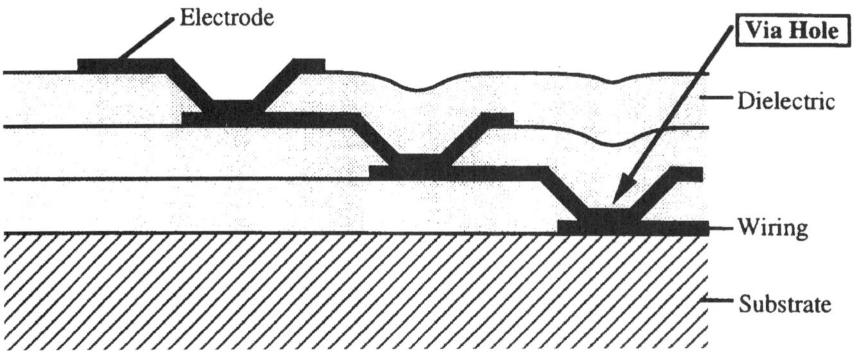

FIGURE 3.16 Conformal via structure.

Dry Etching. The technique of plasma processing has been thoroughly reviewed elsewhere.[110] Both plasma and reactive ion etching take advantage of the presence of plasma activated ionic species to remove material. The difference between the two etching processes is that plasma etching uses chemical etching to remove material, whereas both chemical reactions and ion bombardment are employed in reactive ion etching.

The final shape profile and size of the etched via is determined by the masking technique. Sloped sidewall vias are obtained with plasma etching, whereas tailored slopes or vertical sidewalls are more easily attainable with reactive ion etching. It has been reported that RIE processes show etch nonuniformity of approximately 15 percent.[95]

Dry Etch "Soft" Masking. To use a soft mask such as a photoresist to pattern a polymer dielectric, there must be a measurable difference in the resist to dielectric etch rates. BCB shows poor resist selectivity to most photoresists in oxygen/fluorine plasmas. A commercial resist, AZP 4620, has been identified that results in a 1.5 BCB/resist selectivity at an etch rate of 0.6 μm/min when etched in a 1:4 O_2 to CF_4 plasma.[111] A thick positive photoresist is usually used to mask cured PI. The thickness of the photoresist must be substantially thicker than the PI because the etch rate of photoresists is approximately the same as that of PIs.

Dry Etch "Hard" Masking. If the thickness of the dielectric precludes the use of a photoresist mask, a variety of metal or inorganic (SiO_2) masks can be used.[112,113] Hard mask deposition at or above the Tg of the dielectric has resulted in cracking of the mask and the underlying dielectric layers.[112] Rutter has compared the use of photoresist, SiO_2, and aluminum etch masks and has related the sidewall angle to the etch mask used.[114] Typical mask thicknesses used for RIE processing are in the range of 3000 to 5000 Å.[108] For scanning laser ablation processes, 2–4 μm thick metal layers are required to withstand the fluences typically used for scanning laser ablation (see below). Depositing, patterning, and etching a hard mask is a costly and time-consuming process. Hard mask processing of BCB has been detailed.[21,115]

Etch Gas Compositions. Although PIs can be etched in an oxygen plasma, many practitioners have preferred to etch in CF_4/O_2 or SF_6/O_2 to obtain better control of the sidewall slope for subsequent metallization[42] and approximately five times higher etch rates.[116] It has been reported that RIE of dielectrics in pure O_2 plasma results in large columnar Cu grains during subsequent metal deposition.[117]

Due to its silicon content, DVS-BCB cannot be etched in pure O_2 but rather requires fluorinated etchant gasses. The process has been studied in detail.[21] Etch rates of >1 μm per minute have been reported. Formation of surface SiO_2 has been detected at some O_2/CF_4 gas compositions,[111] better results being obtained with SF_6/O_2.[97]

The effects of low-power and high-power O_2 RIE and O_2/CF_4 RIE on PI via formation have been detailed by Shih and co-workers.[118] Low-power RIE has a narrow process window. High-power RIE gives very poor results due to heavy redeposition. O_2/CF_4 has the largest process windows and is able to consistently open up clean trenches and via holes on PI and metal surfaces.

X-ray photoelectron spectroscopy (XPS) of oxygen plasma etched PI surfaces reveal a dramatic increase in surface oxidation, while surfaces etched in SF_6/O_2 or CF_4/O_2 reveal incorporation of fluorine.

Wet Etching. Wet etching of PIs is carried out by developing areas of the partially imidized films in aqueous alkaline solution. Tight process control of the thickness of the coatings is necessary. The isotropic nature of the etch process usually results in shallow wall profiles. Final via definition occurs subsequent to curing, thus further blowout of the top dimension occurs. For instance, WE-111 is reported to give a 45° post-cure wall profile.[119]

Strong base etchants for wet via processing of partially cured PIQ PIs have been described.[120] It has been reported that wet etching of 1–2 μm PI results in wall slopes of 55°, whereas unacceptably shallower slopes are obtained for thicker layers.

Inoue et al.[121] have detailed a wet etch process for 8 μm PI. Although well controlled, the process becomes complex and includes the need for Cr barrier metallization and argon backsputter descumming.

Others have reported that wet etching of PI is very sensitive to temperature variations during pre-imidization. Using 12 μm cured coatings, the via forming process is limited to > 50 μm.[122]

Rucker and co-workers report the fabrication of an i486™ based MCM using Ultradel™ 4212 wet etchable PI.[123] The softbake step was shown to be critical in defining the etch rate and the adhesion of the film to underlying layers. Key process control parameters included softbake time and temperature. A reproducible etch rate of 15 μm/min could be achieved, which resulted in an undercut of 9 ± 1 μm for 14 μm uncured film (10 μm cured). After curing, the via wall profile shows rounded top and bottom and an approximately 70° sidewall angle, which results in excellent step coverage during subsequent metal deposition. Oxygen descum was necessary to achieve low-resistance vias.

Ori et al. have described in detail the via overhangs that develop during the base etching of PI.[124] To eliminate this overhang, which was found to impact overall via reliability, the PI is applied thicker than required, and subsequently the vias with overhang are blanket wet etched using dilute base to cause smoothing.

Photosensitive Via Formation. The creation of vias in a photosensitive polymer simply entails exposure and development of the polymer before polymer curing. The areas that are not exposed are washed away during the development for a negative working photo polymer.

Storage. Negative working PSPIs and PSBCBs require refrigeration (freezer) and have a finite shelf life. Viscosity tends to increase with time at elevated temperatures (RT) due to premature cross-linking, causing thickness variations upon application, and exposure variations, because the concentration of cross-linker is changing with time.

Application and Prebake. PSPIs can be applied by various techniques, although spin coating has been the traditional technique. Some PSPIs require multiple spin coats to obtain the desired thickness. There is considerable "art" involved with obtaining good photolithography on dielectric layers produced from multiple coats, each of which require prebake and exposure.

PSBCB (Cyclotene™ 4000 series) can be applied by a number of coating techniques (see Chap. 5.5). High-efficiency coating techniques—including spray coating, extrusion coating, and meniscus coating—have been successfully evaluated on glass, aluminum, and laminate substrates.[130] Shrinkage on cure is typically <5 percent, thus the ultimate thickness needed for each coating is achievable in one spin coat operation by choice of the appropriate resin concentration. This can significantly affect high-density substrate fabrication costs.[36]

Prebake is used to optimize the solvent content of the PSPI and PSBCB films. To much solvent can result in delamination, too little solvent can result in cracking.[200]

Exposure. The most important factor in obtaining the desired feature profile is the exposure process. Exposure can be performed with (a) projection steppers or scanners or (b) printers in the hard contact, soft contact, or proximity mode. Contact exposure will produce steeper wall profiles than projection exposure because of the reduction in focus effects.[28] The use of contact exposure in manufacturing results in yield difficulties due to mask cleaning considerations.

Line width and via definition are usually dependent on precise control of the UV-exposure and the thickness of the dielectric. When variations in the thickness of the dielectric are large, due to poor planarization (see Sec. 3.2.1), the photolithographic process becomes unacceptable.

In general, for PSPIs, the mask dimension defines the bottom via dimension, while the top dimension will change (blow out) upon development and cure. For photo BCBs, the mask dimension defines the top via dimension and the bottom via dimension is determined by the sidewall angle.

The feature profile for most PSPIs can be modified by varying the exposure wavelength and or dose. Increasing the dose will result in steeper wall profiles. It has been reported that increased dose does not affect the mechanical properties of the PSPI upon cure,[125] however, an increase in dose has been reported to increase the amount of residue (scum) in the vias.[33]

BCB exposure has been reported for both projection stepper and contact printer in the proximity mode.[202] At low exposure doses, liftoff of the film during processing can result. At excessive exposure doses, the vias begin to swell shut. Typical doses for 5 μm and 10 μm films are 0.3 and 0.8 Joules, respectively.

Develop and Post-Develop Processes. Traditional developing techniques include

- Dip-tank immersion, which can be enhanced by stirring or ultrasonics
- Puddle developing, usually on an automated track coater
- Spray developing

One should examine the quantity of the development and rinse solvents required as well as the time required per unit for the development process. While several minute development times would be acceptable in a dip tank process that is batch loaded, long development times would be unacceptable on a track coater that processes individually.

PSPIs. To decrease via blowout upon shrinkage, several post-exposure stabilization methods for PSPIs have been suggested, such as blanket exposure, which enhances the cross-linking prior to thermal cure.[126]

PSBCBs. Similar to the dry etchable materials, the photodefinable BCBs show low (approximately 5 percent) shrinkage. Thickness is lost, however, during the develop step when low Mw fragments are "extracted" from the exposed film. Film retention is typically 75 to 85 percent, depending on exposure and develop conditions. Such loss of deposited thickness during the development does not distort the via shape.[202] Figure 3.17 shows an SEM cross section of a 35 μm via in a 12 μm cured film.

Curing and Descum

PSPIs. Significant film shrinkage occurs when negative working PSPIs are cured (up to 60 percent, depending on the specific chemistry of the material and the thickness of the layer).[25] During curing, the cross-linkers and any remaining unreacted photogroups are volatilized along with the water that is generated during the curing process. This film shrinkage has been reported to produce "cusping" or "crowning" at via edges[32,33,126,128] as shown in Fig. 3.18.

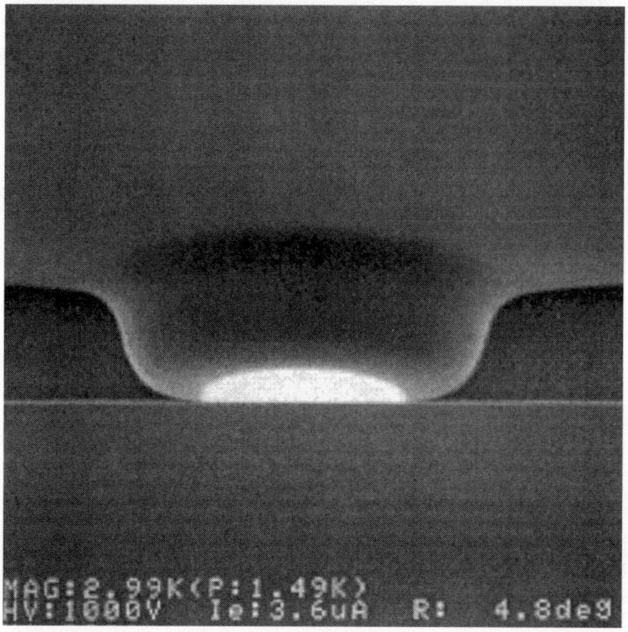

FIGURE 3.17 Cross section of a cured BCB via.

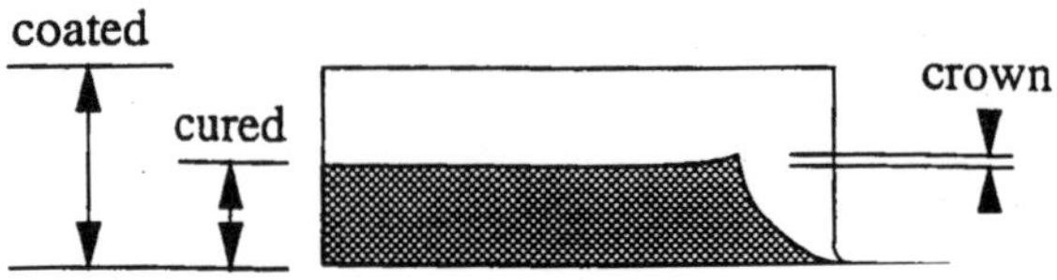

FIGURE 3.18 Crowning effect in some PSPIs.[128]

This crowning effect has been reported to result in as large as a 2 micron ridge. Subsequent layers of photoresist tend to be thinned in these areas, and photoresist breakdown in these thinned areas has been reported to be a significant cause of electrical opens.[129]

Shrinkage affects not only the film thickness but also the developed feature wall profile. Figure 3.19 shows the affect of shrinkage on isolated, semi-dense, and dense features. Note the top via dimension is significantly smaller for the semi-isolated features.[126]

Although the goal of the development process is to remove all unexposed material, current manufacturing processes require a so-called "descum" to remove the thin polymer residue that is left behind and ensure good electrical contact between metal layers. This process is usually carried out with an oxygen plasma source such as an asher or barrel etcher. The process sequence shown in Fig. 3.20 has been suggested by Kotzias[33] for the fabrication of Al/PSPI layers.

PS-BCB. Pre- and (optional) postbake conditions are typically 60–80° C for 15–30 min in an oven or 80–100° C for 1–2 min on a hot plate for a 3 to 12 μm final cured thicknesses.[200] Exposure by projection or proximity (125 to 1000 mJ/cm^2, i-line) techniques, followed by development and cure, produces smooth, sloped sidewall vias as shown in Fig. 3.17. Of course, the specific exposure and development conditions used are film thickness dependent. Via sidewall angles can be varied from 50 to 65°, depending on exposure and cure conditions. The 5 and 10 micron formulations can produce 2.5 aspect ratio vias in the final cured films. Standard oven curing and RTC have both been applied to photo BCB. Via descum in oxygen/CF_4 or forming gas is necessary to remove the last traces of organic contamination and achieve minimum via resistances.[34] The full process sequence for photo BCB is shown in Fig. 3.21 and detailed in Refs. 200–202.

Laser Processing. Laser ablation has been shown to be a viable technique for via formation. The mechanism of ablation is the thermally induced breaking of polymer chains and the subsequent volatilization of the chain fragments. The thermal effect is extremely localized due to the collimated nature of the laser beam and the short pulse duration.

Materials that strongly absorb at wavelengths of radiation emitted by excimer lasers 193 nm (ArF), 248 nm (KrF), 308 nm (XeCl) show the highest ablation efficiency. The use of 308 nm light significantly simplifies optical design and optical material choice and reduces the cost of tooling, compared to 248 and 193 nm light sources.[131]

The primary variables in the ablation process are the fluence, the number of pulses, and the focus of the image on the surface. A further consideration in the ablation process is the control and removal of debris generated by the process. Processing time depends on pattern geometry and area. Equipment down time in production is reportedly 10 percent.[132]

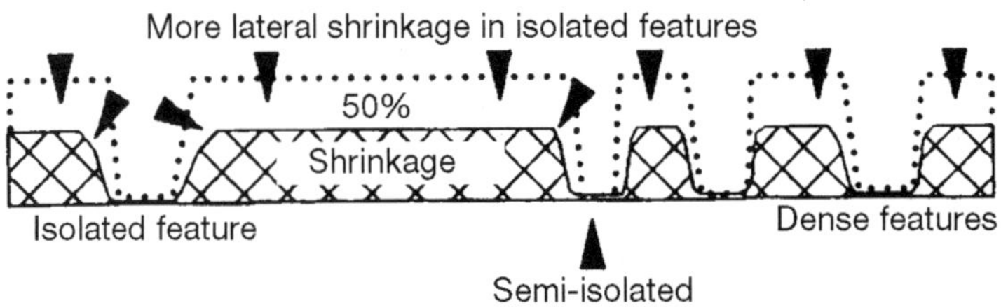

FIGURE 3.19 Effect of pattern density of wall profile.[126]

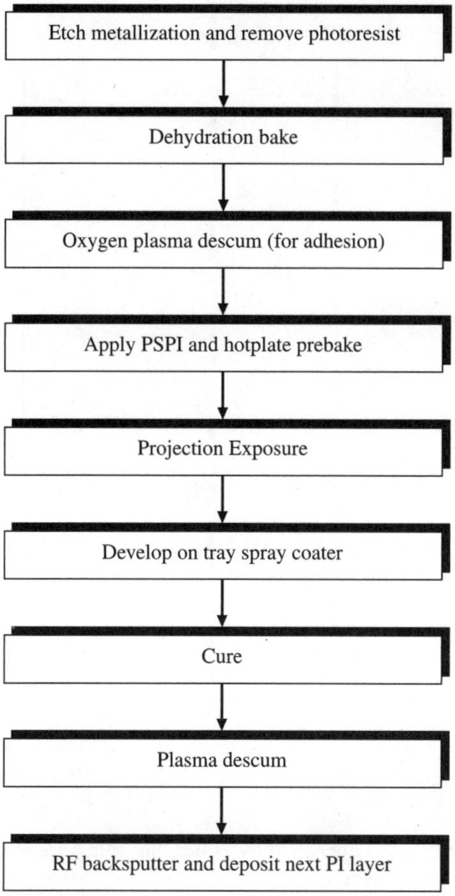

FIGURE 3.20 Process sequence for photo PI MCM fabrication.[33]

Scanning Laser Ablation. Scanning laser ablation involves the rastering of a laser across a hard mask patterned dielectric layer to form vias. Mask undercutting, an issue with RIE via generation processes, does not occur for laser processing. The wall angles are also much more near vertical.

Tessier has detailed the excimer scanning laser ablation (SLA) of a variety of dielectrics, including low-stress PI, fluorinated PI, PPQ, and BCB.[133] At low fluences (<200 mJ/cm^2), debris from the ablation process deposits on top of the polymer being ablated and obstructs further incident laser light. Ultradel 4212 shows consistently higher etch rates over the range of fluences studied. BCB is found to ablate more readily at 248 nm (KrF). Etch rates versus fluence are shown in Fig. 3.22.

Projection Laser Ablation. Projection laser ablation involves the use of a discrete dielectric on quartz mask to project the via ablation pattern onto the substrate surface, analogous to semiconductor photolithographic steppers.

The glass ceramic/PI thin film packages used in the IBM ES 9000 (see Sec. 5.12) is the first commercial application of projection laser ablation.[132,142] A 150 W XeCl (308 nm) gas laser is used. Etch rates of approximately 0.1 μm of PI can be obtained per 308 nm laser pulse at energy densities of 200mJ/cm^2. Debris is minimized by atmosphere control and air flow over the part. RF-induced plasmas have been shown to remove the debris on a 127 mm sq. substrate in under 5 min.[139] In the IBM ES 9000 mainframe, 100,000 75 μm vias are ablated per layer in 18–20 μm PI. This is accomplished in a step-and-repeat mode with one chip site being ablated at a time. IBM reports >99 percent yields on layers with >100,000 vias when laser ablating through a mask using an excimer laser at 308 nm.[60]

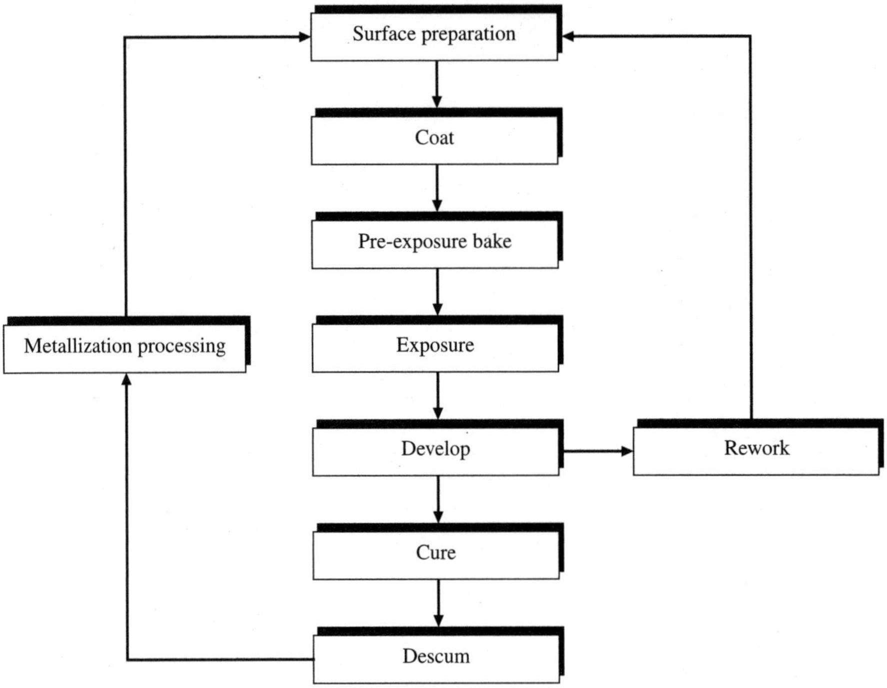

FIGURE 3.21 Processing sequence for Photo BCB.

3.3.3 Metal Deposition and Definition Techniques

Metallization for thin film-based MCMs is deposited by either thin film vacuum processes such as evaporation or sputtering or by wet processes such as electrolytic or electroless plating. Film thickness uniformity is key to repeatable manufacturing of thin film structures. In general, vacuum deposition processes result in more uniform films than do electroplating processes.

Darrow has reported comparisons between Cu, Al, and Au deposited by evaporation, sputtering, and electroplating.[134] In general, electroplating is used for copper and gold, and sputtering or evaporation are used for aluminum.

Metal Deposition Techniques

Evaporation. Metal deposition by evaporation involves the thermal heating of the metal in a vacuum and subsequent condensation of the metal onto a substrate. Conventional thermal evaporation is a low-energy process. Impact onto the substrate is mild, and therefore adhesion to smooth surfaces is usually poor. Since the deposition is a direct line-of-sight process, the metal is deposited with a near 90° angle of incidence. This is advantageous for lift-off processing, as will be discussed below. For common evaporation sources such as E-beam guns, deposition occurs in a upward vertical direction, requiring substrates to be suspended in a tooling fixture. Evaporators usually involve manual loading and are not easily adapted to automated wafer handling.

Sputtering. Metal deposition by sputtering involves bombardment of a metal target source by energetic particles, typically Ar^+ in a plasma. Metal atoms are dislodged from the target and deposited onto a substrate. Substrate heating can be a problem because of short target to substrate distances, but adhesion to a variety of substrates is inherently better. DC magnetrons are common sputtering sources that reduce the problem of substrate heating by magnetic confinement of electrons in the plasma.

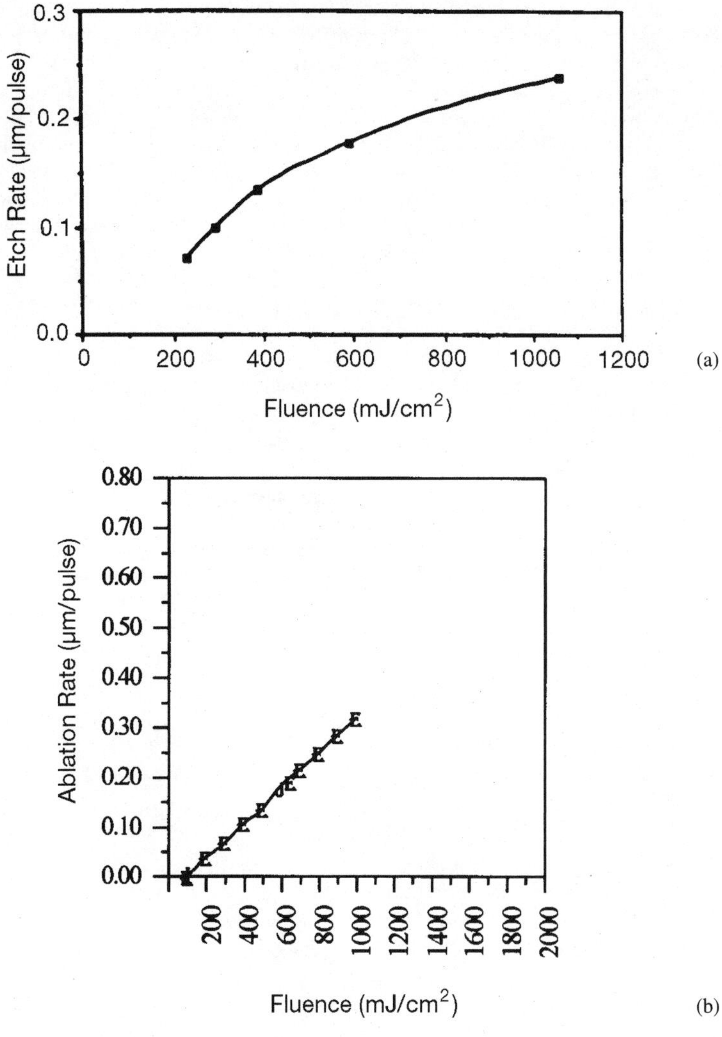

FIGURE 3.22 Etch rates vs. fluence for (a) low-stress PI and (b) Cyclotene 3022.

Sputtering suffers from low deposition rates and the poor utilization of targets due to uneven flux distribution (important for gold deposition). Sputtering is adaptable to automation. It has been reported that stress in sputtered metals can be controlled by control of the process variables.[134,135,181] Sputtering tends to be a conformal process that results in very good step coverage of features.

Contamination must be removed from the substrate surface to ensure metal adhesion at the interface. Ion milling or RF sputtering is used to remove several hundred Å of the surface of the substrate prior to coating. This process removes organic contaminants and, in some cases, is thought to cause surface roughening which aids in adhesion.

Electroplating. Electroplating involves the selective deposition of ions from a plating solution. An electrical potential is applied to a previously deposited conductive seed layer to initiate the electrochemical reaction. The main advantage of electroplating is high deposition rate, typically >1 μm/min.

The substrate acts as the anode, and metal ions are reduced and deposited on exposed regions of the seed layer (through a patterned photoresist). This is a disadvantage in that it increases the processing steps by first requiring a seed layer to plate upon and then having to remove it from all areas where there is to be no metal pattern.

Electroplating processes are dynamic in nature and therefore require tight process control. Chemical depletion of the bath must be continuously monitored to maintain reproducible plating conditions. Impurities in the plating baths have been reported to result in poor adhesion, pitting, stress, cracking, and other undesirable attributes. The plating rate on an electrode surface depends on the flow conditions of the plating bath, because this changes the concentration of metallic ions near the electrode surface.[108]

Nonuniform deposition rates can present a processing problem. Variations in current density resulting from different dimensions on the pattern being plated will result in greater deposition on finer features, which results in thickness variations across the pattern.

Since Cr, Cu, Au, and Ni can be electroplated, surface pad metallurgy such as Cu-Ni-Au can be created by plate-up technology. Thin Au (1500 Å) can be used for soldered components, but thicker Au (>1 μm) is needed for bonded interconnects.

Electroless Plating. Unlike electroplating, which requires a continuous metallic base held at a given current or voltage, electroless plating requires only a catalytic seed layer without electrical contacts. The metal is built up by selective deposition onto a catalytic seed layer by the reduction of metallic ions in aqueous solution.

Electroless plating solutions for Cu, Au, and Ni are available. Ni gives a high deposition rate but results in conductors or reduced conductivity. Au usually exhibits a slow deposition rate, approximately <2 μm/hr.

The electroless Cu deposition reaction is based on the reduction of Cu ions in an alkaline solution of formaldehyde and hydroxyl ions. The overall reaction can be described by Eqs. 3.7 and 3.8.

$$\text{Anodic reaction: } 2\text{HCHO} + 4\text{OH}^- \rightarrow 2\text{HCOO}^- + 2\text{H}_2\text{O} + \text{H}_2 \tag{3.7}$$

$$\text{Cathodic reaction: } \text{Cu}^{++} + 2\text{e}^- \rightarrow \text{Cu} \tag{3.8}$$

The overall reaction requires a solution pH in the range 11.5–13 at a plating temperature between 40 and 70° C. Because PIs are not chemically stable in such highly alkaline solutions, bulk stability and interface stability problems have been reported.[63] The deposition process is favorable at surfaces such as Au and Ni but is poor at inert surfaces like polymer dielectrics and SiO_2. These latter surfaces, therefore, need to be activated (usually with a Pd salt) to initiate selective deposition on them.

Metal Stress. Total metal film stress, like total dielectric film stress, is composed of both thermal and intrinsic stress. Thermal stress results from the coefficient of thermal expansion (CTE) mismatch between the deposited film and the substrate. The level of intrinsic stress, and whether it is tensile or compressive, is influenced by system geometry and operating conditions as much as by process factors.[134] Stress increases as the coating gets thicker. It has been noted that refractory metal adhesion/barrier layers can be the major contributor to composite film stress.

Windischmann[135] has pointed out that ion bombardment during sputter deposition can alter microstructure, resulting in coating density changes and, under specific conditions, reversal from tensile to compressive stress. Metal stress can be addressed through process parameters such as gas type and pressure, temperature, substrate composition, and deposition rate.

Subtractive Definition. In subtractive definition, a blanket layer of metal is deposited by one of the above techniques. The metal layer is then masked, and exposed areas are removed by wet chemical etching. Since wet etching is isotropic, the technique works best for patterns where the width of the feature is several times the thickness. When common metal etchants, high in acid concentration, are used for subtractive etching, the fine lines created for the interconnect are severely undercut and might be completely removed. For this reason, less aggressive etchants such as ammonium persulfate are used, which result in approximately 4 percent reduction in line width for 25 μm lines.[108] For high aspect ratio features, anisotropic dry etching such as RIE would be necessary. Because of cost and the lack of suitable processing for Cu metallurgies, there are no examples of RIE etched metallization in thin film packages in commercial production.

Due to the isotropic nature of subtractive etching, the resist mask is sized smaller than the desired metal feature dimensions to compensate for the undercutting during etching. Subtractive etching of aluminum has less undercut than copper, resulting in better line profile.

Differences in center-to-edge dimensions across a substrate are usually compensated for by overetching. This becomes more important as the substrate size increases. The primary defect type observed during subtractive etching of copper and aluminum is shorts, due to resist flaws and contamination.

Additive Definition

Additive Electroplating. Additive electroplating is feasible for both copper and gold wiring. Additive electroplating allows high aspect ratio wiring, because feature tolerance is determined by the photoresist lithography tolerance. Thickness uniformity is determined by uniformity of the electroplating process. Many factors can affect the plating uniformity such as current density, bath agitation, anode/cathode spacing, rack design, and pattern density. Using tight process control, Motorola has reported <10 percent thickness variation over 16 in^2 substrates.[108]

In the conventional conformal via electroplating process, a *blanket layer* of metal is deposited as the plating base. The metal base is patterned using a photoresist, and the desired features are plated up in the exposed regions of the photoresist. Electrical isolation of the features is achieved by subsequent wet etching of the plating seed layer. This is depicted in Fig. 3.23.

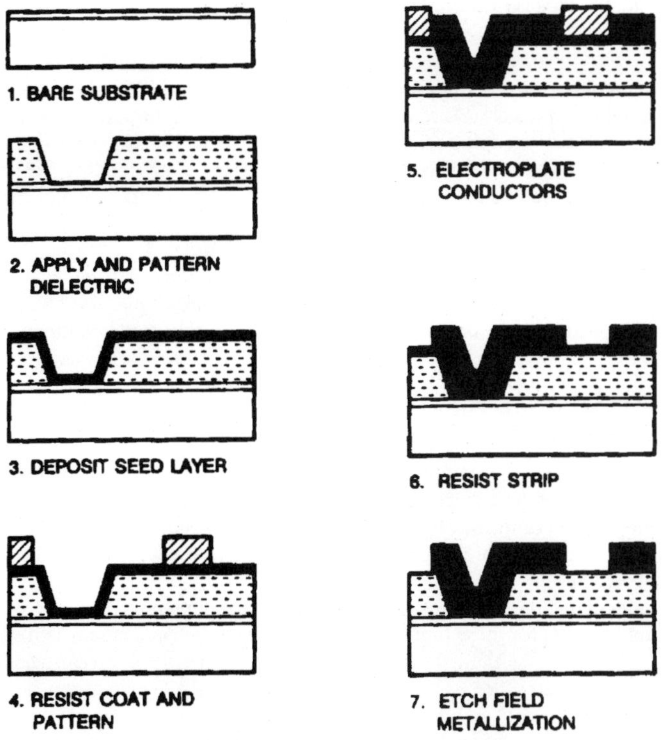

FIGURE 3.23 Process flow for conventional additive electroplating process.

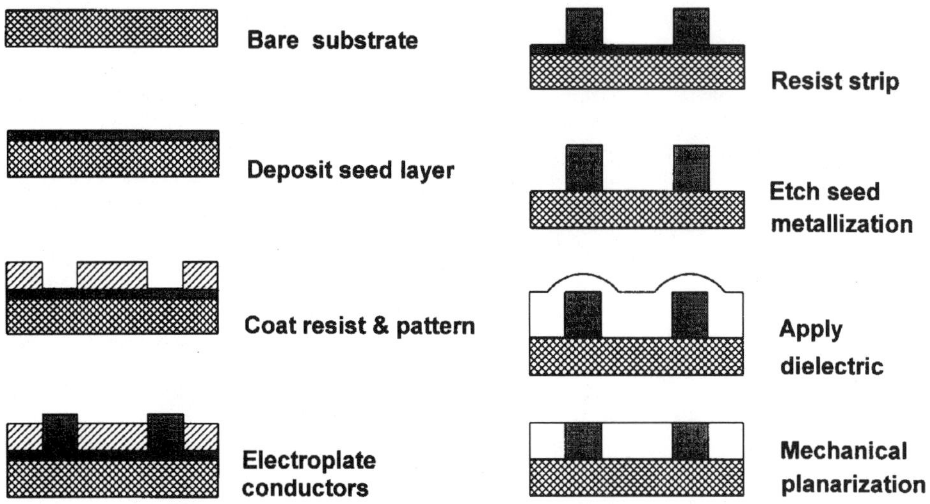

FIGURE 3.24 Process flow for stacked additive electroplating.

In the conventional stacked via process, a seed is similarly deposited on the substrate. Resist is coated and patterned, and the conductor is plated up in the exposed regions of the photoresist. The resist is stripped, the seed is metallization etched away, and the dielectric is applied and mechanically planarized to expose the stud tops as depicted in Fig. 3.24.

Lift-Off Processing. Lift-off processing, developed for high-definition intrachip interconnect, has been practiced commercially for thin film packaging by IBM. This technique involves evaporation of metal through a defined mask.[17] The mask is fabricated such that the metal deposited on top of the mask, and the metal deposited through the mask openings onto the base substrate, are discontinuous, as shown in Fig. 3.25. The mask is then dissolved, which "lifts off" the metal deposited on the mask surface. The process gives better feature definition but is more complex and expensive than other metal definition processes.

The abnormal angle of deposition and the overhang that develops during evaporation results in trapezoidal shaped features.[95] Once the mask is lifted off, a gap is left between the dielectric layer and the metal. If the subsequent layer of dielectric does not "flow" into this gap (i.e., gap fill), chemicals can be entrapped in this area and/or shorts can occur during metallization.

3.4 THIN FILM MCM PROCESSES

Since the initial references to thin film interconnect first started appearing in the early 1980s, a plethora of process sequences have been strung together to fabricate test structures and "demonstration vehicles." Many of the companies involved with the early manufacturing of MCM-D substrates are no longer active, for a variety of reasons. (Many of these were described by Turlik and Adema[136] and will not be covered here.)

This section will detail the processes developed by current merchant suppliers. Section 3.6 details commercial applications for these thin film structures in various market segments.

3.4.1 AT&T

PolyHIC. AT&T Polymer Hybrid Integrated Circuit (PolyHIC) technology has been in production for its switching and transmission equipment in their Merrimack Valley facility since 1987.[137] Its

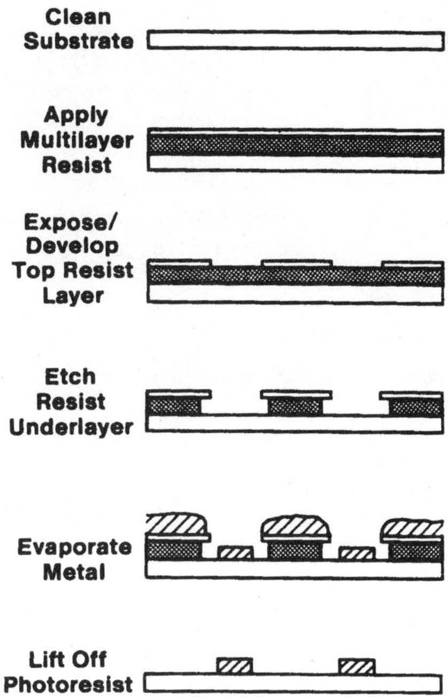

FIGURE 3.25 Lift-off processing.

standard technology uses copper conductors and TaN resistors on a thin film alumina substrate. The cyanate ester based dielectric is described in Sec. 3.2. The low curing temperature allows the use of precision TaN terminating resistors which provide near lossless transmission line capability at Gb/s frequencies.

Process flow is shown in Fig. 3.26.[91,138] Laser-drilled through holes are generated in the alumina substrate for designs requiring double sided structures.

The first level of metal is created by sputtering a Ti adhesion layer and plating up copper metal. TiN resistors are fabricated and laser trimmed. The dielectric is applied by spray deposition, exposed, developed, and cured. Conformal, staggered 100 μm square vias are normally developed in the nominal 50 μm dielectric thickness. A second level can be created by repeating this process sequence. Line widths and spaces range from 50 to 100 μm. Top-level metallurgy is Ti/Cu/Pd/Ni/Au. The Pd passivates the Ti/Cu interface. The Ni acts as a barrier layer between the Cu and the Au. Au is used to permit soldering and wire bonding of devices.

Si-on-Si. The initial AT&T silicon-on-silicon process initially known as Advanced VLSI Process (AVP).[84] It was based on Ni passivated, electrodeposited Cu metallization and incorporated a complex multilevel metal scheme. More recently, an Al/PI process was developed that reportedly requires much simpler processing.[139]

The Al is deposited by DC magnetron sputtering. RF backsputtering is used to enhance the adhesion between PI and the deposited Al. The aluminum is photolithographically patterned and defined by wet chemical etching. Typical conductor dimensions are 10–40 μm wide and 1.5–2.5 μm thick.

The process was developed using Hitachi PIX-L112 PI (20 μm between power and signal and 5 μm between the other layers). The via levels are defined by RIE etching through an Al mask. Ta-Si thin film resistors are deposited by sputter deposition. SiO_2 and Si_3N_4 are used to form thin film integrated capacitors. In applications that require thick (>5 μm) dielectric an electroless Ni plating process is used to provide a planar surface for subsequent processing. A cross section of the process is shown in Fig. 3.27. Later commercial implementation of this process as the MicroInterconnect Tech-

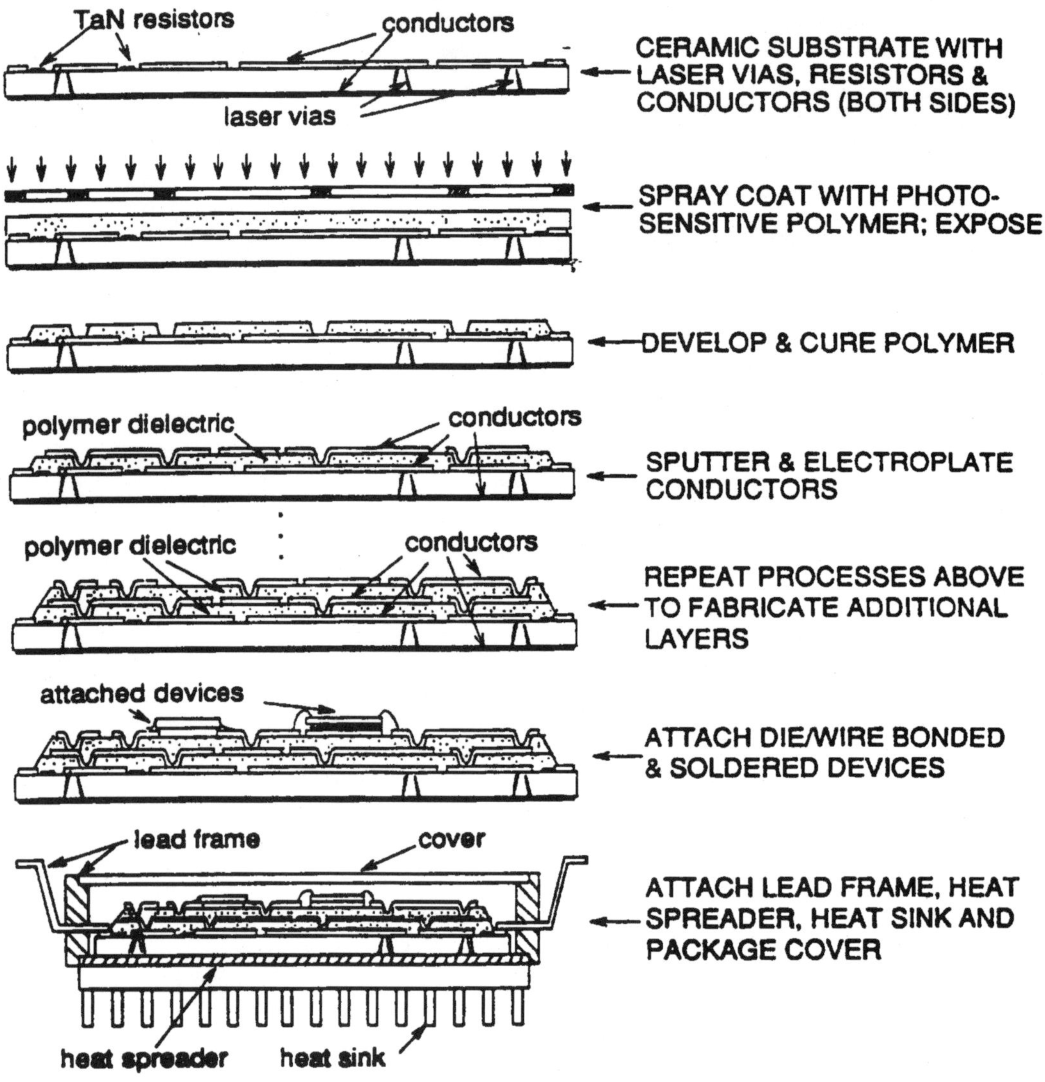

FIGURE 3.26 Schematic of PolyHIC MCM fabrication process.

nology process is detailed in Sec. 3.6. This implementation uses PSPI and the process sequence shown in Fig. 3.20.[85,86]

3.4.2 IBM

Since the early 1980s, IBM has developed and implemented single chip and multichip multilayer thin film technology for both ends of the cost-performance continuum.[60] Cu/PI and Al/PI thin film structures have been implemented on silicon, alumina and glass ceramic carriers. Several of these technologies are detailed below. Commercial applications are discussed in Sec. 3.6.

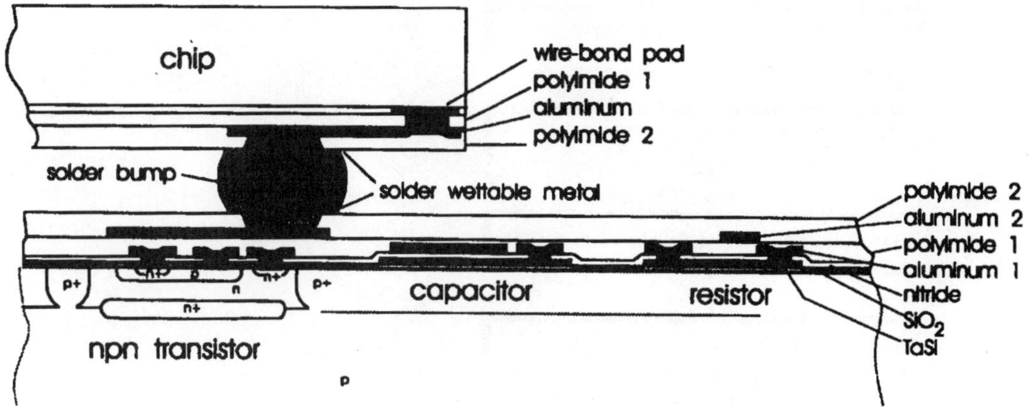

FIGURE 3.27 Cross section of AT&T Si-on-Si process.[85]

Thin Film for Redistribution on Cofired Ceramic. Thin film technology was developed for signal redistribution on systems such as the ES 9000.[140,141] The thin film structure for the ES9000 processor uses a planarized stud process for high wiring density and Cu/PI thin film layers, projection laser ablation for via formation, chemical mechanical polishing for planarization, and laser CVD for thin film repair.[59]

The first thin film layer is composed of pads to capture the ceramic vias, which are nonplanar due to sintering of the cofired base structure. This metallization layer [Cr (200 Å), Cu (2 µm), and Cr (200 A)] is defined by sputtering, photoresist definition, and subtractive etching. PI is coated and cured at 400° C. Vias are ablated in the PI using a projection laser ablation process.[142] The via studs are formed by sputter deposition of a Cr adhesion layer followed by a thick layer of Cu, which is chemically mechanically polished to form the studs. This polishing is required due to the poor planarization achieved by typical PIs (see Sec. 3.2.1). Such planarization allows precise control of the deposited PI thickness which, in turn, results in better electrical performance.]

A redundant metal layer is formed next by sputtering Cr/Cu/Cr, creating a lift-off stencil using RIE techniques and evaporating down the surface pad layer of Cr/Cu/Ni/Au. The redundant metal level is then removed from unwanted areas by photoresist application and subtractive etching. A PI layer is applied and cured, and pads are opened (for soldering) by laser ablation.

Process flow is detailed in Fig. 3.28. A cross section of the ES 9000 is shown in Fig. 3.29.

Thin Film Al/PI on Silicon. Al/PI thin film technologies were developed to package CMOS chips in MCMs.[44,143,144] The thin film structure is defined by a conformal via process. The PI is patterned by RIE through photoresist. The wiring is patterned by subtractive etching of aluminum. The process and materials were chosen to be compatible with a CMOS manufacturing line.[44] Al (4–9 percent Cu) conductors are 3 µm thick and typically 13 µm wide on 25 µm pitch. Minimum wiring dimensions are 8 µm. The dielectric is 3–6 µm PI, vias are dry etched. Via size is 8 µm for wiring and 50 µm for power and ground. The top surface metallization is evaporated Cr/Cu/Au for standard IBM C4 joining technology.

3.4.3 Hughes–HDMI

The Hughes HDMI-1 process is based on dry etch, low-CTE PI and aluminum interconnect. Substrates have been fabricated on 150 mm silicon and/or ceramic (alumina or aluminum nitride) wafers.[88] Several structures are available with differing degrees of complexity as shown in Fig. 3.30.

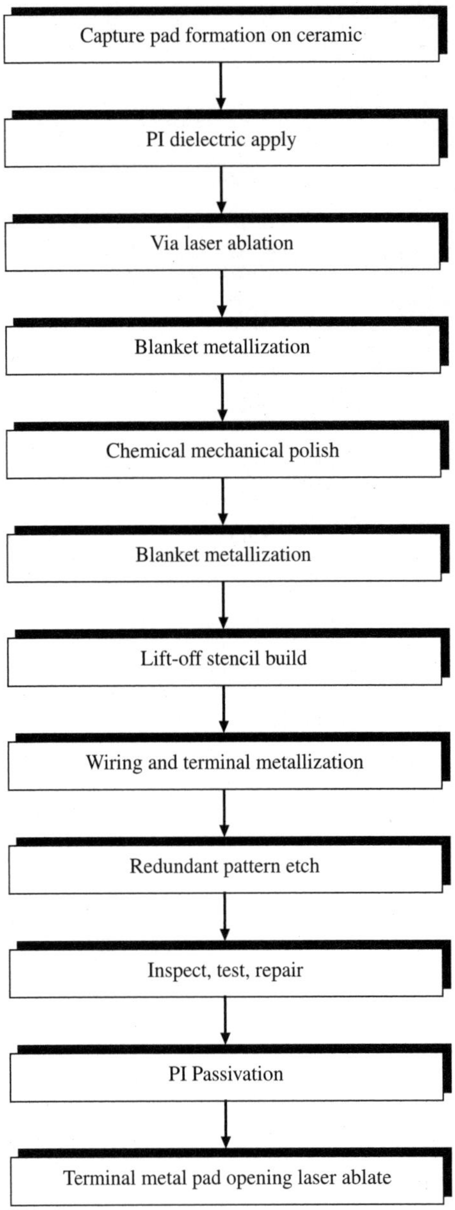

FIGURE 3.28 Process flow used in the fabrication of thin films for ES 9000.[59]

HDMI-1 consists of power, ground, signal 1, signal 2, and chip attach metal layers that are sputtered in 5 μm Al and wet etched. The top Al layer is sputtered with a thin Ti/W layer and a thin Au layer, and then 5 μm of Au are plated over the Al. Vias are plasma etched after standard lithography is used to define the pattern. Vias are typically staggered to avoid the buildup of nonplanarity, because the vias are not filled.

Aluminum nitride packages, which are much more CTE compatible and deliver a much higher thermal conductivity (>200 W/cm°K versus approximately 20 for alumina or 2 for low-fire glassy

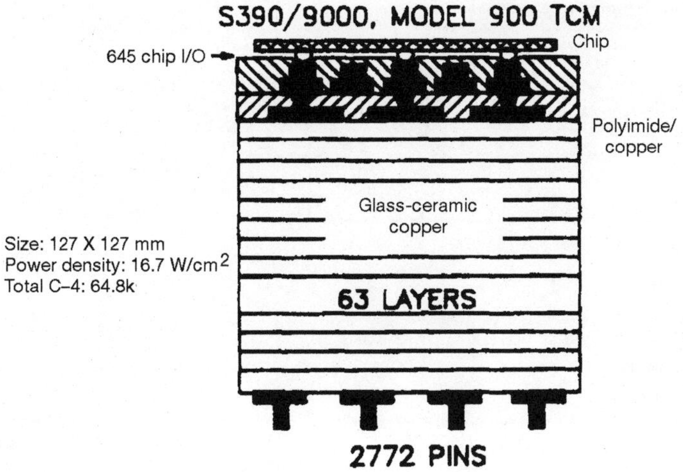

FIGURE 3.29 Cross section of ES9000 MCM-CD substrate.

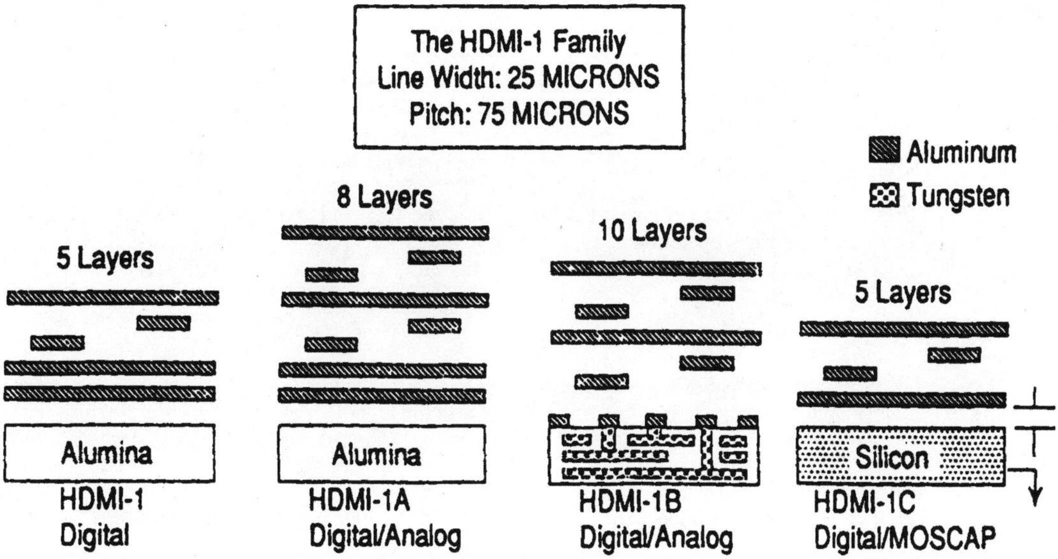

FIGURE 3.30 Cross section of Hughes product family.[145]

alumina tape) have been developed in as large as 3.8 × 3.8-in sizes under a NOSC contract by Hughes, Coors, and Grace.[146,147] A cross section is shown in Fig. 3.31.

3.4.4 MMS

The MMS D5000 process retains many of the processing steps of the initial DEC process for the VAX 9000 (see Sec. 5.6.1) while substantially reducing the manufacturing costs inherent to such a large-CPU system. 150 mm Al wafers are used as substrates. The wafers are coated on the back side

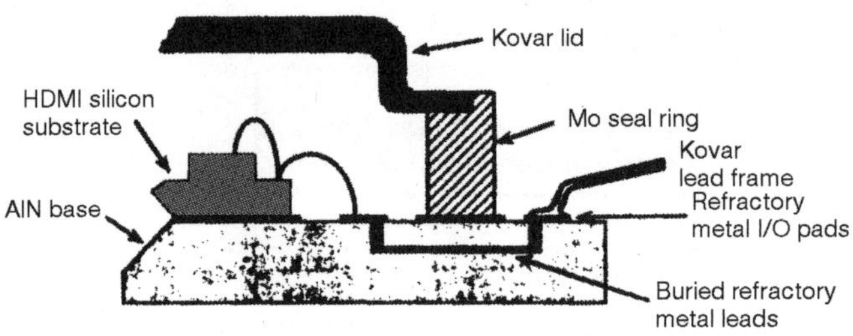

FIGURE 3.31 NOSC AlN package.

and edges to protect the wafers during processing. Both BCB and PI dielectrics are offered. Cr is used as an adhesion/barrier layer, and copper is electroplated up. It is no longer necessary to remove the interconnect from the substrate, because the aluminum substrate itself is now used as one of the power distribution layers as well as one electrode of a decoupling capacitor. Because the electrical requirements (versus the VAX 9000) are also relaxed, the stripline structure of the multichip unit can be replaced by a simpler microstrip configuration. MMS design rules of 15 μm lines and 50–70 μm line pitches are typical.[90,91]

3.4.5 n-Chip SiCB™

n-Chip fabricates thin film interconnect on a silicon substrate. The n-Chip technology uses proprietary plasma-enhanced CVD to deposit undoped SiO_2 dielectric ($\varepsilon_r = 3.75$). SiO_2 is deposited by a high-rate PECVD process. The film is preloaded under compressive stress of <100 Mpa at room temperature. Control of stress during the deposition allows layers as thick as 10.5 μm to be deposited.[149,150] Characteristics of the three main technologies are shown in Table 3.12.[171] A cross section of the nC2000 is shown in Fig. 3.32. Integral decoupling capacitors are common to all structures. Resistor layers and electroplated copper are used in the 2000 and 3000 series for higher performance and higher-frequency operation. A typical 15 μm wide line exhibits a capacitance of 1.2 pF/cm (Cu-based structures). Interconnect density of >1000 in/ in sq. is reported to be easily achievable.

3.4.6 General Electric–HDI

The GE high-density thin film interconnect process (HDI) is a chips-first approach.[151] A schematic cross section of the process is shown in Fig. 3.33. In the nominal process, bare chips are inserted top down in cavities that have been created in a ceramic substrate, and the interconnect structure is formed over the chips and substrate. The interconnect is formed by lamination of a thin PI sheet over the substrate to form the first dielectric layer. Via holes are made through the film by laser ablation down to the I/O pads. Connections to the chip pads and the interconnect traces are formed by a 4 μm

Table 3.12 SICB™ Technologies

	nC1000	nC2000	nC3000
Target speed (MHz)	0–100	100–400	400–1000+
IC technology	CMOS	BiCMOS, ECL, GaAs	BiCMOS, ECL, GaAs
Signal/power layers	2/2	2 + pad/2	2 + pad/3
Min. pitch/width (mm)	25/10	36/16	36/16
Signal layer metal	Al	Cu	Cu
TF resistor option	no	10 W/cm	10 W/cm

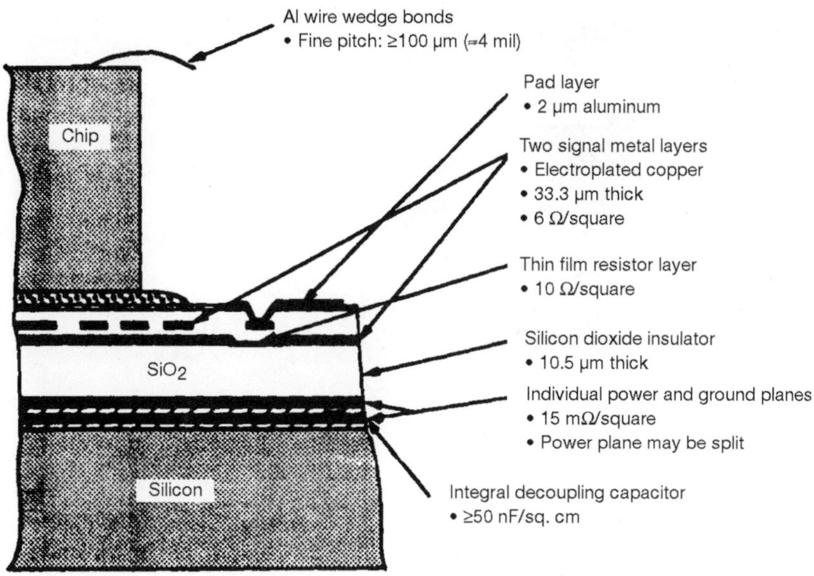

FIGURE 3.32 Schematic of nC2000 cross section.

Ti/Cu seed sputtering and plate-up process. The metal interconnect pitch is typically 4 mil. Multiple layers are formed by repeating the PI film lamination, via formation, and metallization processes. Typically, two orthogonal signal layers and power and ground layers are processed. Substrate variations include TI's MCM-L/O laminate version[148] (Fig. 3.34) and GE's metal matrix and direct bond copper (DBC) versions.[153]

A high-Tg thermoplastic is used for die attach. Ultem™ 1000, a 217° C Tg thermoplastic, is then used as overlay adhesive for a Kapton film on which the interconnect is created. The Ultem is reported[154] to produce void-free lamination at approximately 300° C. Via formation and metal patterning is accomplished by argon ion laser (351 nm). Subsequent layers of dielectric are limited by the melt flow temperature of the Ultem overlay adhesive. A siloxane polyimide/cycloaliphatic epoxy resin formulation was specifically formulated for this application. Actual dielectric layers therefore consist of 1 mil Kapton/0.5 mil adhesive layers. Top-level pads are opened by ablation of the overlay dielectric layer.

3.4.7 NTK

Kambe and co-workers have described a process based on a cofired alumina substrate. TaN thin film termination resistors are formed on the top layer of the cofired ceramic, then five Cr/Cu/Cr metal layers and five PSPI layers are deposited. Vias are stacked plated-up posts. Ground rules of 25 μm line width and 50 μm vias on 150 μm pitch are typical. Such structures have passed extensive reliability testing.[48]

3.4.8 Kyocera

Kyocera has divulged very little about their thin film MCM processing.[155] The substrate tends to be cofired multilayer mullite. A special blend of PI has been developed to give reliable adhesion after high-temperature storage and pressure cooker testing, although it is not known whether this material is being used commercially.

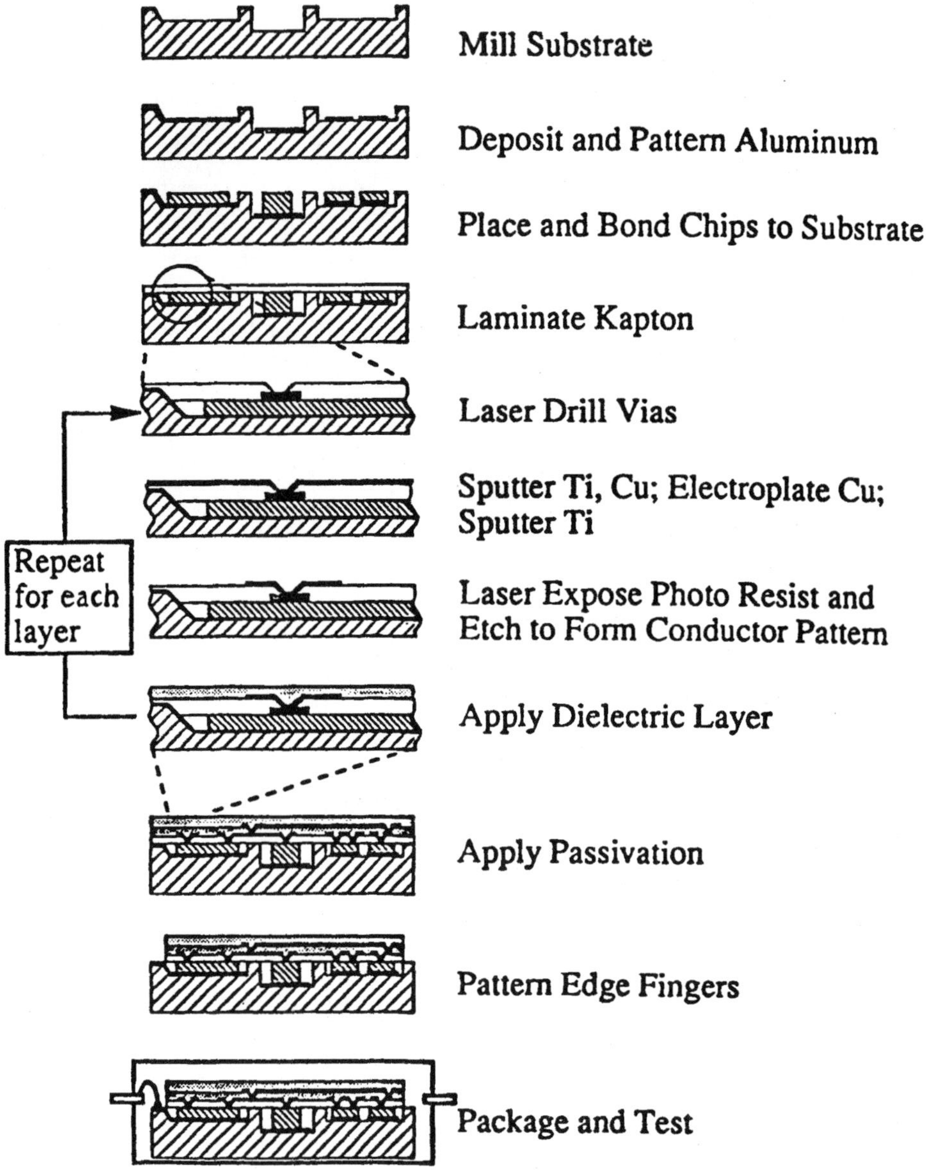

FIGURE 3.33 Cross section of GE chips first MCM technology.[154]

3.4.9 OKI

Oki has described an MCM-D/C process using LTCC cofired glass-ceramic substrate and Cr/Cu/Cr/PI thin film interconnect on a 40 mm sq. format.[156] The 5 μm copper conductors are plated up over a sputtered Cu seed layer. There are four conductive layers: two for signal, one for ground, and one containing chip pads. The LSI pads are perforated to avoid blistering of the Cu pads during PI curing.

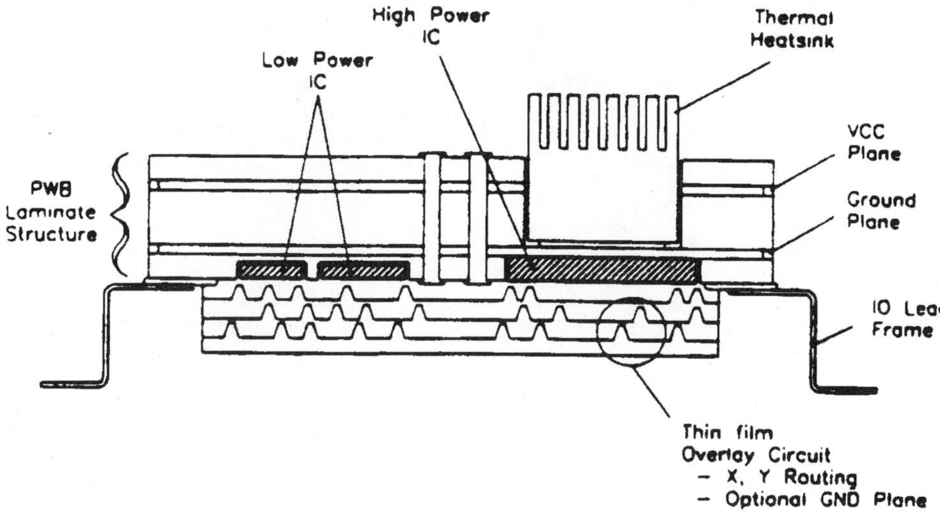

FIGURE 3.34 Cross section of MCM L/O.[152]

Multiple coat and cure steps for the PI were needed to achieve a 14 μm cured thickness. In 14 μm of BPDA/PDA PI they were able to produce 40 μm dry etched vias (70 μm top diameter, 40 μm bottom diameter) with reproducibility. The developed MCM technology runs at 1 Gb/s with satisfactory signal integrity.

3.4.10 IMC

In the commercial IMC thin film process,[157] interconnects are manufactured with four layers with sputtered Al and BCB dielectric. Thin film resistors and capacitors are fabricated in a TiN layer. The process is MIL qualified. The process is suitable for digital clock frequencies up to 500 MHz. Die are typically wire bonded to the substrate.

3.4.11 Thomson

The Thomson MCM line manufactures copper/PI structures on 100 to 150 mm sq. ceramic substrates.[158] Five design rule sets are available, as listed in Table 3.13. The typical five-layer structure consists of power, ground, x, y, and pad layers. They use a photosensitive PI dielectric with thickness ranging from 5 to 15 μm. Both Al and Au wire bonding are used, and a hot gas chip removal process has been developed.[159]

Table 3.13 Thomson MCM-D/C Ground Rules

	A	A′	B	C	D
Conductor width (μm)	75	60	38	25	20
Conductor pitch (μm)	200	150	100	75	50
Via diameter (μm)	60	50	30	20	10

3.4.12 Fujitsu

Fujitsu has developed MCM technology[160] for applications such as their commercial business server (see Sec. 3.6). Multilayer thin films (five layers max.) are built of Cu (5 μm thickness max.; 12 μm width min.)/PI (10 μm max.) with 20 μm min. vias on an AlN substrate. The maximum size of the AlN substrate is 135 mm sq. The chips are area array connected (100 μm 95/5 solder bumps on 210 μm pitch). Thermal vias are arranged to conduct heat from the chip through the thin film through the bumps to the substrate.

3.4.13 Toshiba

Toshiba has described technologies based on MCM-D/C and MCM-D thin film technologies.[161,162] The physical and electrical parameters of their MCM-D technology using SiO_2 dielectric has been described.[162] A cross section is shown in Fig. 3.35. The SiO_2 dielectric is formed by TEOS CVD deposition. The thickness of the silicon dioxide dielectric significantly influences the line capacitance per unit length. It is difficult to deposit thick (approximately 10 μm) dielectric layers, thus thinner layers positively affect the manufacturing costs. Vias are etched in the dielectric by wet etching. The aluminum interconnect is also patterned by wet etching.

Toshiba has commercialized MCM-D/C technology is based on copper/PI thin film on alumina or aluminum nitride cofired ceramic.[161] It consists of five conductor layers, including a chip mount layer, two signal layers, and two ground layers. Signal lines are sandwiched between ground to optimize the characteristic impedance. The pad layer is plated with 1.5 μm of gold for wire bonding. The total PI thickness is 90 μm. The cofired alumina includes two conductor layers for power supply and ground. Chips are mounted with silver-filled epoxy. Thermal vias formed in the thin film structure effectively reduced the thermal resistance of the structure. A cross section is shown in Fig. 3.36. Several of the processes described above are compared in Table 3.14.

3.5 RELIABILITY

3.5.1 Copper/Polymer Interfaces

The Copper/ PI Interface. Copper easily corrodes at Cu/PI interfaces and the oxide produced has poor adhesion.[163] The mechanical and electrical properties of PI can degrade when cured in the pres-

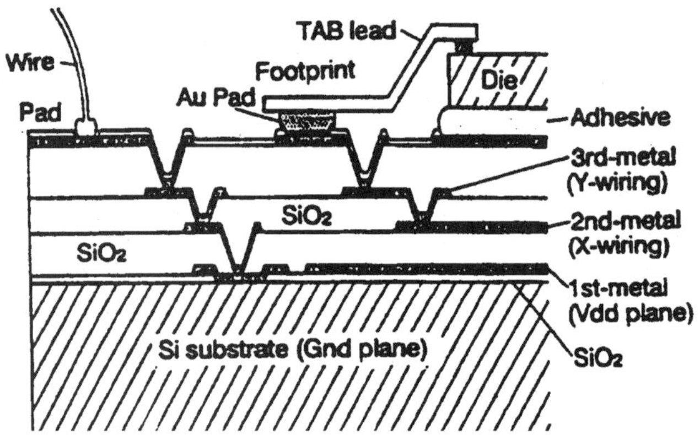

FIGURE 3.35 Cross section of Toshiba MCM-D technology (SiO_2 dielectric).[162]

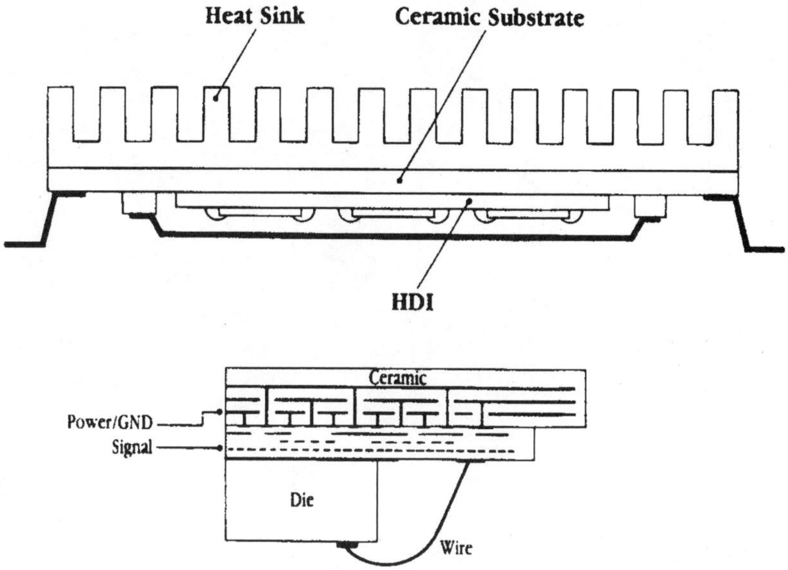

FIGURE 3.36 Cross section of Toshiba MCM-D/C structure (TAM10Z series).

ence of copper.[164,165] Copper has been shown to diffuse into PI by oxidation at the Cu/PI interface, migration of copper ions into the PI matrix and, finally, reduction of the ionic copper to small nodules of copper metal by the matrix. The incorporation of copper reduces the breakdown voltage and increases the dielectric constant of the PI. Thin diffusion barrier layers of passivating metals stop this diffusion of metal into the PI (see Ref. 95 and references therein).

As copper is converted to its oxides in the PI driven corrosion process, the resistance of the copper interconnects increases because copper oxides have higher resistivity. Niwa et al.[127] have shown that Cu (II) diffuses approximately 8 μm from a copper/PSPI interface after a 1 hr prebake at 80° C. This copper-rich "scum" is difficult to remove with a typical oxygen descum process. Amino silane adhesion promoter decreased the Cu concentration and increased adhesion to the copper after 800 hr of PCT.

Cech et al.[166] have shown that the corrosion rate for protected lines is eight times higher than for unprotected lines. Resistance increases after 85%RH/85° C testing were four to five times lower for fully encased lines than for 60 percent encased (top and bottom only) lines.

Rao[167] has used TEM to examine of PI-Cr and PI-Cu interfaces on a completed MCM substrate that has been subjected to 120° C, 85 percent RH, and 40V bias for 856 hr. While the Cr-PI interfaces appear clean, Cu nodules, the largest of which is 25 nm, can be observed out to 100 nm from the PI-Cu interface.

Skin effects for high-frequency transmission are expected to be adversely affected by these highly resistive barrier layers. The degradation is expected to be most severe for thin conductor lines with relatively thick passivation/barrier layers.[96]

Adema et al.[168] have shown that sputtered Si_3N_4 and PECVD SiO_xN_y behave as diffusion barriers between Cu/PI interfaces, which result in improved performance at high frequencies over unprotected transmission lines. Copper was also shown to diffuse through sputtered SiO_2. BCB was shown to be impervious to copper migration as expected.

Barrier metallization such as Cr or Ti is required for structures containing Cu conductor and polyimide dielectrics to achieve good adhesion between the Cu and the PI and to prevent chemical attack on the Cu by the PI, or more specifically the polyamic acid precursor (see Sec. 3.6.1). The barrier layer also serves to protect the underlying copper during subsequent O_2 RIE.

Table 3.1 Comparison of Merchant Thin Film MCM Technologies

	AT&T PolyHIC	IBM MLTF	Hughes HDMI-1	MMS	n-Chip nC1000	TI/GE	NTK	Kyocera	IMC	Thomson
Conductor	Cu	Cu	Al	Cu	Al	Cu	Cu	Cu	Al	Cu
Adhesion/barrier		Cr		Cr		Ti	Cr	Cr		Cr
Thickness (μm)	8	5	5	2–8	2	4	10	3, 5	3	5
Width (μm)	50–100	13–25	25	20	15	40	25	30	20	50–200
Pitch	125–225	25–100	75	75		100	75	76.2	50	50–200
Pad level	Ti/Pd/Cu/Ni/Au	Cr/Cu/Ni/Au	TiW/Ni/Au	Cu/Ni/Au	Al	TiW/Au	Au, Sn/Pb	Ni/Au	Al, Ni/Au	Cu/Au
Substrate	alumina	ceramic or silicon	silicon, alumina, or AlN	aluminum	silicon	ceramic or laminate	ceramic	ceramic	silicon	ceramic
Dielectric	CE	PI	PI2611	PI, BCB	SiO_2	Kapton	PI	PI	BCB	PI
Photosensitive	+	–	–	–	NA	–	–	±	±	+
Thickness (μm)	50	5.5–12.5	5–10	3.5–12	7	35	25	20	7	5–15
Via dia (μm)	100	10–20	35	30–50		25	50	30	30	15–60
Via type	staggered conformal		conformal	conformal	conformal	conformal staggered	stacked	conformal	conformal	conformal
Chip attach		Al, Au wire, TAB	WB	WB, FC	WB	direct contact	WB, FC, TAB	WB, FC, TAB	Al & Au WB	Al, Au WB

Cr is known to be a strong oxygen getter, and thus it is necessary to reduce the Cr thickness to reduce interface contact resistance. A layer of at least 100 Å needed to get continuous film coverage.

During any etching step, the Cu/PI nodules or deposits impede clearing of the vias and thus the formation of low-resistance contacts. Cr/Cu/Cr is subtractively etched in separate Cr and Cu etching solutions. In this process, the sidewalls of the Cu are left bare (i.e., not covered with Cr), and thus the copper is still subject to interaction with the PI and/or oxidation. When subsequent layers are deposited, a short ion beam etching step can be employed to remove the few hundred Å Cr layer.[95]

The Copper/BCB Interface. Copper diffusion is not observed when BCB is deposited directly on copper metallization[106,168,169] even after 1500 hours at 85% RH/85° C, in contrast to results with polyamic acid-based polyimides.[168]

3.5.2 Environmental Reliability Testing

MMS has described the reliability testing performed on an assembled R3000 workstation module.[170] MMS has also detailed the reliability of their Cu/BCB Twinstar™ product.[202] Samples were electrically tested at discrete intervals during the stress conditions outlined in table 3.14. Testing was done on whole wafers, diced substrates, and fully assembled parts.

Okajima has evaluated the environmental reliability of IBM Yasu Si-on-Si technology.[171] Packages passed the following stresses: 300 hr of 85% RH, 85° C at 7.5 V, 2000 cycles of –40 to +125° C. Kambe[48] has detailed reliability studies on the NTK process (see Sec. 5.4.7).

Reliability of via continuity was investigated by thermal cycle testing (MIL 883C, condition C). No opens were found, and there was no resistance change after 1000 thermal cycles. Metal adhesion was studied by pad pulls after thermal cycling. No drop in adhesion strength was observed after 1000 thermal cycles. Insulation resistance between a pair of traces is studied before and after pressure cooker testing (PCT), (120° C/85% RH). Parts made from n-Chip SiCB™ technology have been subjected to standard Mil 883C testing.[82] IMC Al/BCB technology has been qualified under MIL 883B.[157]

NEC has shown that RISC 3000 prototype modules made of Cu/BCB have passed 500 thermal cycles (–45 to 125° C), high-temperature exposure (125° C for 3200 hr), and high-temperature/high-humidity environments (85° C/85% RH for 3200 hr).[37]

TI has examined the reliability of the GE HDI technology.[172] Thermal cycling was performed air-to-air and liquid-to-liquid from –65 to 150° C and –55 to + 125° C. In general, via resistance increased with thermal cycles. After 250 thermal cycles, via resistance increased 71 percent after cycling with a ΔT of 215° C compared to 33 percent increase at ΔT of 175° C. Via resistance increases in a manner consistent with low cycle via fatigue.

Table 3.14 MMS Twinstar™ Reliability Testing

Stress condition	Electrical parametric testing, whole wafers	Device functional performance, assembled parts	Mechanical performance, substrates
High-temp storage (150° C, air)	500 hr		100 hr
Temp and humidity, 85° C/85% RH	168 hr		
Temp cycle, –55 to 125° C	500 cycles		1000 cycles
Thermal shock, –55 to 125° C	500 cycles		
HTOL, 125° C, 3.3 V bias		1000 hr	
THB, 85° C/85% RH, 3.3 V bias		1000 hr	

3.6 APPLICATION AREAS

Applications for thin film packaging have traditionally been performance driven. The demands of high-performance systems, such as ECL-based mainframe computers and military and space applications, have thus led the way as thin film market applications. Applications have been dominated by high-clock-rate and I/O-intensive requirements. As we pass the mid 1990s, it has become increasingly clear that thin film technology will have to deliver system cost savings as well as performance benefits to expand market acceptance. As long as thin film MCM technology stays focused on low-volume, high-performance applications, costs will remain high.

It is expected that thin film MCMs will see acceptance in "interconnect rich" architectures such as tightly coupled multiprocessors, crossbar switches, multiple bus architectures, and cache memory systems. Interconnect rich applications should include military applications such as electronic radar and image processing. Digital telecom switches and digital TV processors will also require elaborate coding and decoding schemes to transport massive quantities of data at GHz frequencies.

3.6.1 Mainframe Computers

Thin film MCM technology has been implemented by mainframe manufacturers in systems such as the IBM 390/ES9000,[140] the NEC SX3,[173] the Hitachi M880,[121] and the DEC VAX-9000.[90] It should be noted that, in Japan, they have never used bare die in commercial mainframe applications. The lack of bare die infrastructure led them to use very small "chip carriers" that could be burned in and tested before assembly. At NEC, they were called flipped TAB carriers (FTCs); at Hitachi, they were called micro chip carriers (MCCs). These carriers clearly are the precursors of today's chip-scale packages (CSPs).[203]

IBM 390/ES9000. In 1989, IBM announced the introduction of the ES9000 processor modules for their 390 mainframes.[140,174] Chips are solder bonded to the substrate. The backsides of the chips are interfaced with compliant pistons to remove the heat. Each substrate has 63 metal layers, is 127.5 mm sq., and supports 121 chips.

Signal and power distribution are achieved through copper signal paths and vias in the cofired structures. The copper conductor exhibits a threefold improvement in conductivity over the Mo metallurgy used in the earlier systems.

Pins are brazed to the underside of the substrate. Six of these modules are mounted on the 22-layer glass/epoxy PWB with zero insertion force (ZIF) connectors. The thin film located on the top side of the cofired ceramic substrate redistributes connections between the chip and substrate grids.

NEC SX-3. The NEC SX-3 Supercomputer is an evolutionary extension of the SX-2. Highly advanced bipolar technology is used for the basic VLSI chips. The key packaging elements of the NEC SX-3 supercomputer include TAB'ed chips, flipped TAB chip mounting (FTC, flipped TAB carrier), a 22.5 cm sq. thin film-ceramic substrate upon which the 100 FTCs are mounted, and the 12,000 chip terminations are connected.[2,31,173]

The face-down TAB packaging was chosen because of the ease of LSI cooling. Figure 3.37 is a cross-sectional view of the FTC. The I/O are arranged on staggered grid, which allows the use of high-pin-count VLSI chips. For cooling, the chip is die bonded directly to the FTC cap, and it is possible to obtain a very low thermal resistance by using copper-tungsten or, more recently, aluminum nitride cap material.

The substrate has to accommodate a large number of wires and at the same time connect the wires with the least signal propagation delay. The cross section of this PI/alumina multilayer substrate (MLS) is shown in Fig. 3.38. Alumina is used as the base substrate with embedded power supply and ground wiring to lower the impedance of the power distribution circuits. I/O pins are brazed to the bottom surface in a staggered grid. The PI/wiring layers are formed on top of this cofired ceramic structure. The Au signal layers are 25 μm wide on 75 μm pitch. The Photoneece™ photosensitive PI is 20 μm thick. Square vias are 60 μm wide. Vias are created by projection exposure. Development of the vias takes 200 s in a ultrasonic dip tank.[175] This technology is practiced internally to supply requirements for the SX-3 and the ACOS computer systems.

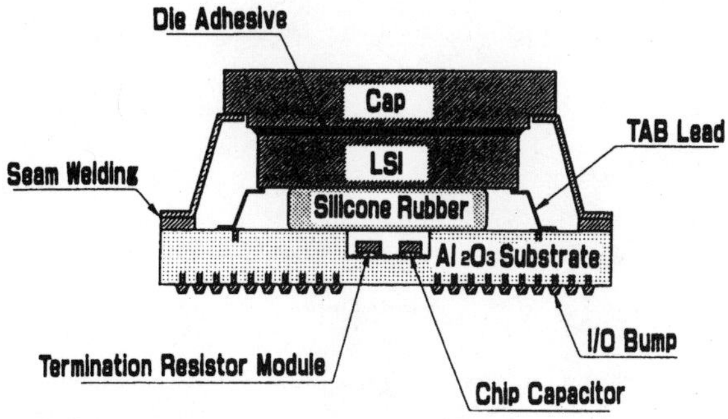

FIGURE 3.37 Cross section of the NEC FTC

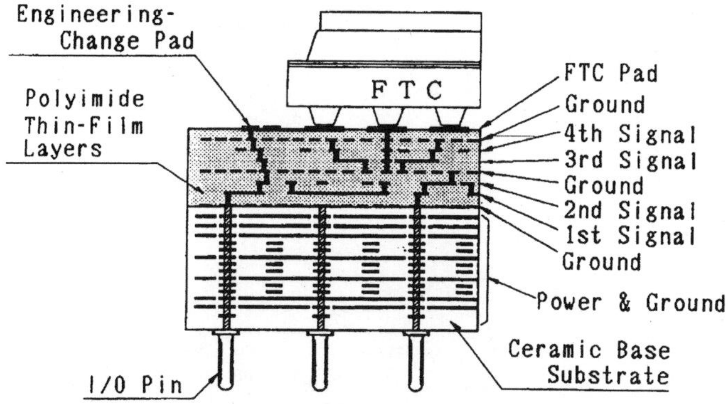

FIGURE 3.38 Cross section of the SX-3 Pl-alumina substrate.

Since each chip has a maximum of 20,000 gates, the package has a maximum of 2 million gates. One complete processor unit is mounted on the MCP.

ACOS. The NEC general-purpose computer, ACOS System 3900, uses much of the same thin film packaging technology as the SX-3.[176] Table 3.16 compares the technology parameters associated with the SX-3 and the IBM ES-9000.

DEC VAX-9000. The VAX-9000 CPU was partitioned into 20 different high-speed MCM-D modules each using different module designs.[90,91] An exploded view of the multichip unit (MCU) is given in Fig. 3.39.

Table 3.16 Comparison of SX-3 and ES 9000

	NEC	IBM
Model	SX-3/SX-X	ES 9000
Material	Alumina Au/PI thin film	Glass/ceramic Cu/PI thin film
Dimensions (mm)	225 × 225	127.5 × 127.5
Conductor layers (thick film/thin film)	15/8	63/2

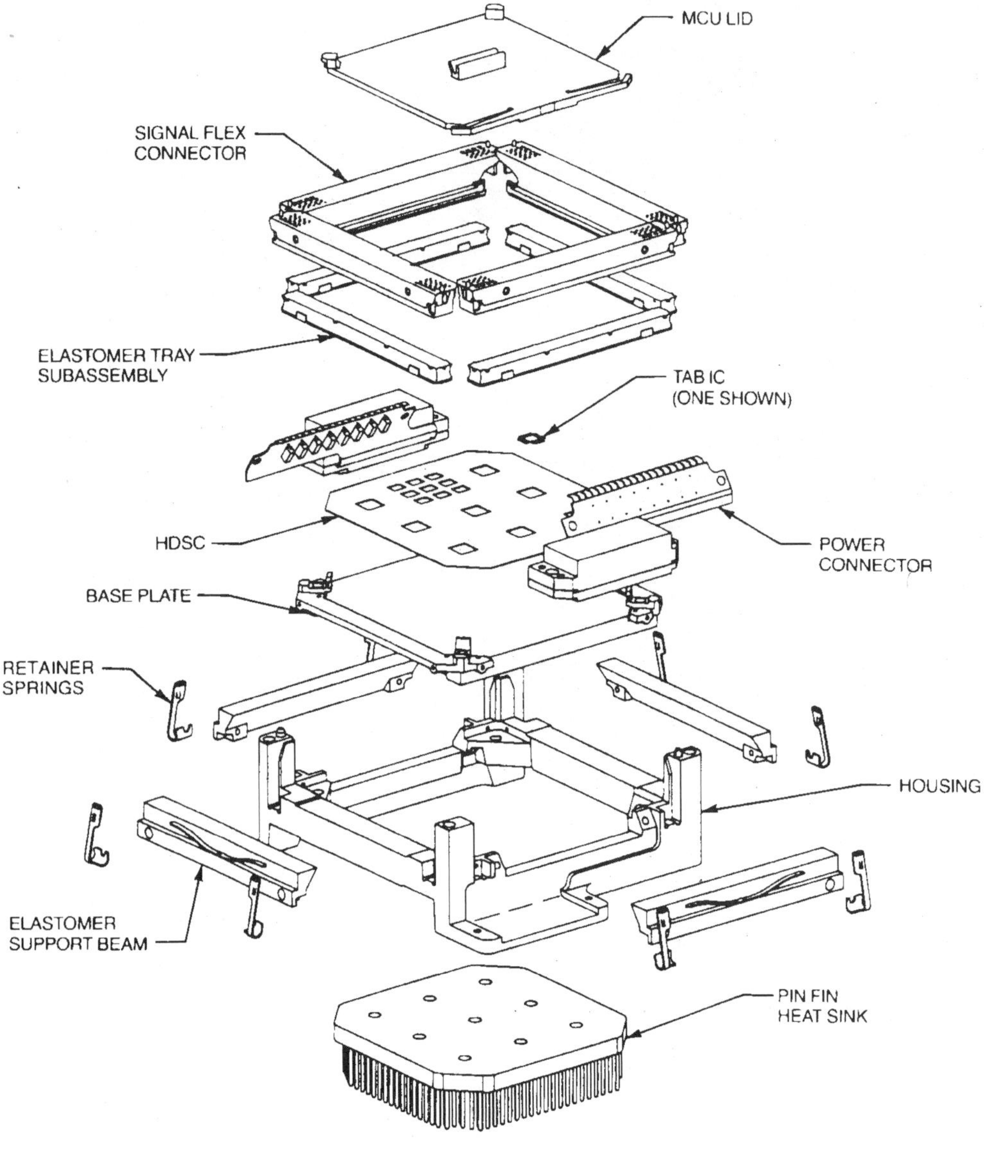

FIGURE 3.39 An exploded view of the Vax 9000 MCU.

The high-density signal carrier (HDSC) or MCM interconnect consists of nine Cu conducting layers separated by PI dielectric (four for power, two for x and y signals, tow for ground, and one for surface connection). 150 mm aluminum wafers are used as carriers for the fabrication of the HDSC. They are coated on the back sides and edges with PI to protect the Al during processing. Cr is used as adhesion/barrier layer, copper signal lines are electroplated up. Oxygen/SF_6 is used to etch the vias. The interconnect is fabricated in two parts: the signal core and the power core, as shown in Fig. 3.40. Before the two structures are mated, PI is dry etched off the backside of the aluminum plates, and the

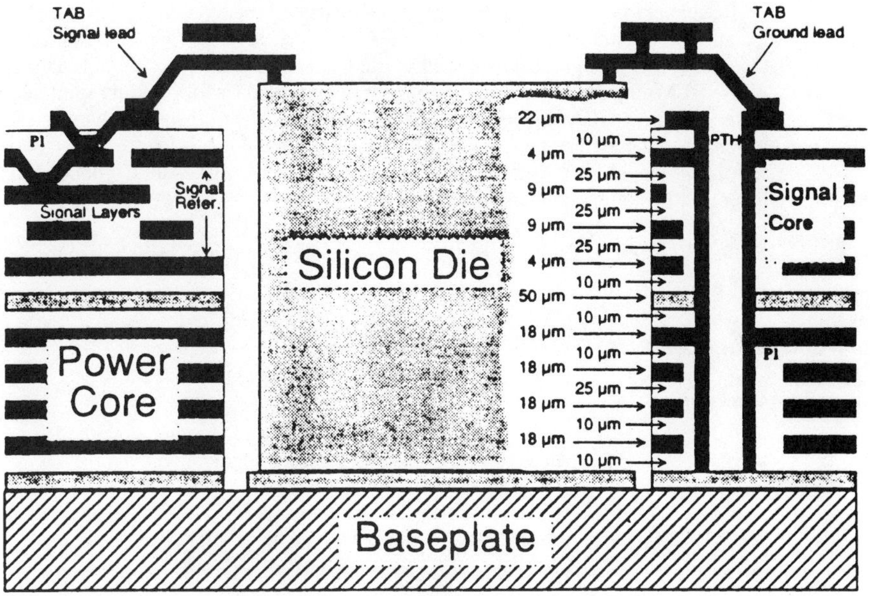

FIGURE 3.40 Cross section of the high-density signal carrier.

aluminum is etched off the signal core. They are then drilled, plated, and mated to interconnect the power core to the signal core.

Die are TAB attached in cavities etched into the HDSC, to the metallic baseplate with diamond filled die attach resin which provides extremely low thermal resistance path. The modules are manufactured to handle 270 W of power.

HITACHI M-880. Figure 3.41 shows a cross section of the Hitachi M-880. Each 0.8-μm ECL process logic and logic/memory LSI chip is packaged in a *micro chip carrier* (MCC). The MCC consists of five thin film Al/PI layers and seven mullite glass/tungsten thick film layers.

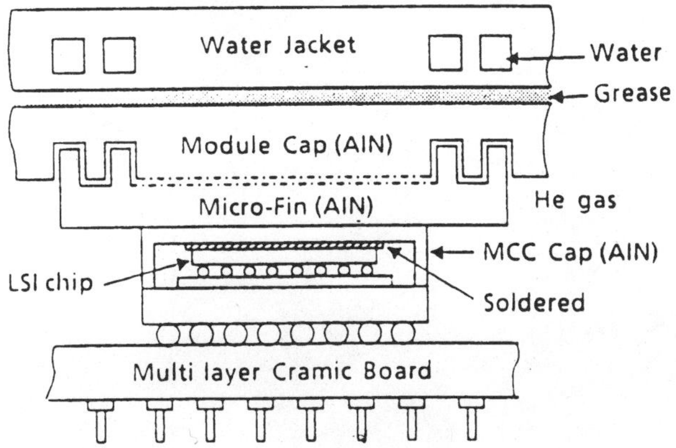

FIGURE 3.41 Cross-sectional view of MCC for Hitachi M-880.

Roughness in the cofired ceramic is compensated for by a layer of wet etchable PI that smooths the surface for subsequent thin film layers. The next layer is a Cr-Si-O sputtered resistor layer. For the wiring layers, a photosensitive PI is used to form 30 μm vias in 9 μm thick PI. Ball level metallurgy (BLM) for the BGA package consists of sputtered Cr followed by sputtered Ni/Cu alloy on top of the thin film. They are patterned by wet etching. The Cr acts as a barrier layer and an adhesion promoter to the PI. The Ni/Cu acts as a Ni plating seed and a solder diffusion barrier. Ni is plated up, followed by gold, which ensures solder wettability. The MCCs are C4 bonded to the surface of a 44 layer mullite cofire ceramic mother board.[93,121]

Fujitsu VP-2000. The Fujitsu VP 2000 (Fig. 3.42) and the M 1800 utilize the technology discussed in Sec. 3.4.12.

3.6.2 Workstations, Servers, Desktops, and Portables

Workstations

Ross HYPERSPARC. The modules, shown in Fig. 3.43, are manufactured for Sun microsystems' workstations. These MCMs are also sold by Ross as upgrades to existing Sun SPARCstation™ and

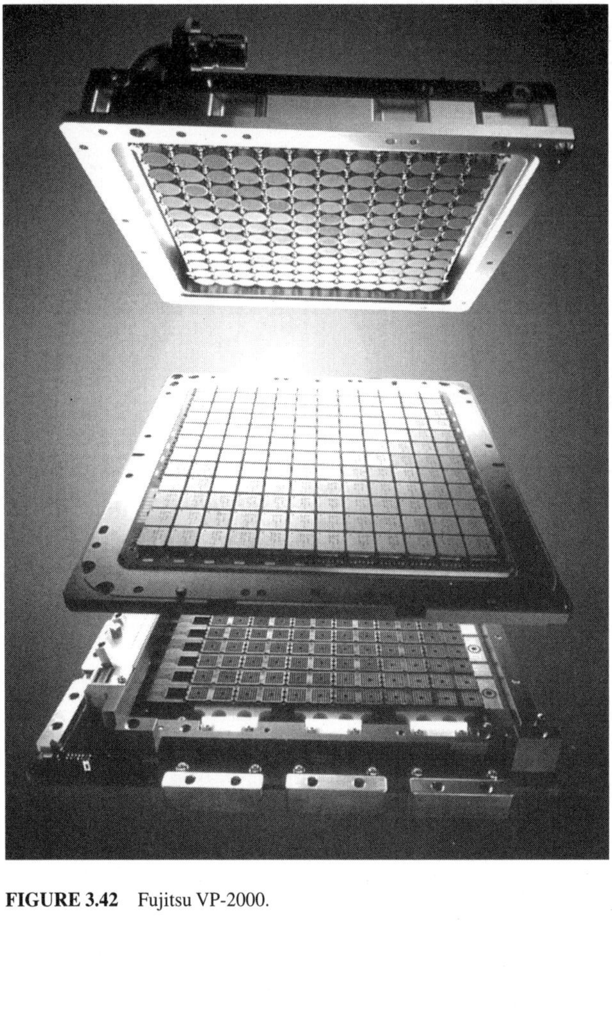

FIGURE 3.42 Fujitsu VP-2000.

FIGURE 3.43 HyperSPARC module fabricated by SiCB™ technology.[178]

SPARCserver™ computer systems. The MCMs operate at higher clock frequencies than would be possible if the die were packaged in traditional single-chip packages. In addition the I/O buffer design of an IC going into an MCM need not be as robust, because the electrical loads being driven by the IC are much lower. This, in turn, results in lower power consumption and less heat to dissipate in the system.

n-CHIP. n-Chip has produced a 40-MHz SPARC CPU module that has been commercialized in the UNIX based workstation made by Tadpole.[178] The small footprint of the module was important in this laptop application, as was the 20–30 percent reduction in power consumption afforded by the low-capacitance interconnect. The 90-MHz hyperSPARC module is manufactured using n-Chip's nC1000 technology (see Sec. 3.4.5). The four to six chips are more manufacturable than a monolithic version yet are partitioned in such a way as not to reduce overall system performance.

MMS. An electrically equivalent product is manufactured by MMS using the technology described in Sec. 3.4.4.

MMS Twinstar. The MMS Twinstar MCMs, which contain two Pentium™ microprocessors, are designed for use in workstations and servers. The Twinstar offers a 70 percent increase in processor application performance over standard single Pentium processors. The Twinstar is delivered in a 347-pin PGA with aluminum baseplate and heat fins. At 100 MHz, the module dissipates 12 W typical (18 W maximum) power. The module in Fig. 3.44 was fabricated in Cu/BCB thin film.[202]

NEC RISC-3000 Prototype. Fig. 3.45 shows a prototype of a high-density RISC 3000 module fabricated with Cu/BCB technology. Good reliability was reportedly obtained without barrier/adhesion layer metals such as Ti, Cr, or Ti/W. The direct Cu/BCB interface resulted in a 30 percent reduction in signal transmission speed versus the Cr/Pd/Au/PI technology used in previous products.[37]

FIGURE 3.44 MMS Twinstar MCM.

3.6.2.2 Servers

Fujitsu K6000. Figure 3.46 shows the MCM used in the commercial Fujitsu K6000 business server.[160] Chips for the server are 0.5 μm CMOS LSIs. For further cooling, a heatsink is glued to the back of the AlN substrate. Technology for this server is discussed in Sec. 3.5.12.

SUN SPARC-MMS CACHE SRAM. MMS has commercialized several large secondary cache implementations such as the four-chip MCM shown in Fig. 3.47. The module was originally routed using laminate technology in a 3-mil line and space, four-layer design. However, the 1.5 W SRAMs produce a 6 W package, which required thermal chip sites to be cut in the laminate, which increased the size of the module to that of the single-chip equivalents. The current cost premium for the thin film module was reported to be approximately 10 percent.[197]

Desktops, Portables, and Notebooks

MMS Chipframe™ 80486 Upgrade Products. MMS has developed Chipframe™, a PC booster 80486 upgrade family of modules for 80286 and 80386 desktop and notebook systems. The module,

FIGURE 3.45 Prototype of R 3000 module.[37]

FIGURE 3.46 Fujitsu K6000 business server.

shown in Fig. 3.48, consists of a TI 486SLC CPU, 16 kB cache controller, and 25 MHz clock generator with the same dimensional area and pinout as the original CPU.[179]

MMS Chipframe™ Northstar Processor Modules. Each MMS Northstar 64 or 128 contains an Intel Pentium™ CPU chip, a cache controller, and 18 dual-port SRAM chips (512 kB of secondary cache). The modules are designed for servers and high-end desktop systems using NT™ and

FIGURE 3.47 MMS SRAM cache product.[197]

FIGURE 3.48 MMS Chipframe™ CPU upgrade.

UNIX™. They are packaged in a 349-pin PGA.[180] An example of a Northstar module is shown in Fig. 3.49.

3.6.3 Military and Space

Each succeeding generation of military and/or space product tends to have higher performance and smaller form factor than its predecessor. Performance has traditionally come from increasing device integration. However, growing device size and increasing I/O count have caused the ever larger chip package to increase in volume and decrease in performance. Military applications for MCMs often require nonstandard form factors tailored to fit the configuration of the military system. Size and weight requirements can restrict the use of standard SMT single-chip packages mounted on PWBs and backplanes. The most common approach to resolving this dichotomy is to minimize the ratio of device package area to active silicon area by MCM approaches. It is for this reason that MIL and space applications have driven the leading edge of MCM technology over the last decade.

Although some of the MIL and space vendors have substrate fabrication and MCM assembly in house, most have relied on outside vendors for one or both of these activities. The following descriptions are not comprehensive but, rather, attempt to give a flavor of some of these activities.

Hughes. Using their HDMI technology, Hughes has developed many military MCM applications. As an example, Fig. 3.50 shows the HAC-32™ processing element of Hughes data processors.[182] It is a 20-MIPS general-purpose processing element based on a 32-bit JIAWG processor, an Intel I960™ MX RISC microprocessor, and 194 MB SRAMS in a 2 × 4-in footprint. Physically, the HAC-32 is approximately one-tenth the size and weight of the comparable elements if individually packaged. The HAC32 supports a range of military applications which include the F-22 Advanced Tacti-

FIGURE 3.49 MMS Northstar processor module.[180]

FIGURE 3.50 Hughes HAC processor.[182]

cal Fighter and Navy electronic warfare systems such as the F/A-18, F-14, A-6, and the Harrier aircraft.

Raytheon. Raytheon has developed a thin film on cofired ceramic architecture for a variety of radar and IR imaging applications.[183] AT&T PolyHIC technology was subcontracted to create the thin film portion of the substrate. An example is shown in Fig. 3.51.

Rockwell

Dual 1750A Processor. The dual 1750A, Fig. 3.52, consists of two 1750A processors with memory management sharing 512 kB of RAM, 16 kB of EEPROM, direct memory access, and parallel and RS232 serial interfaces. The entire processor, which uses direct die attach, has 140 components including 33 ICs. It is packaged on a single 2.5 × 2.5-in silicon substrate. The processors operate at 40 MHz with a combined throughput of 5 million instructions per second (MIPS). The silicon substrate is mounted into an integral housing/connector assembly. Module dissipation is 8 W with a maximum junction temperature rise of 4° C.

Avionics Memory Module. The memory module shown in Fig. 3.53 is packaged in an advanced 2.2 × 2.2-in cofired hermetic AlN package. The memory module contains 64 passive devices and 32 ICs, including 16 RAM devices on two silicon substrates. The 3 in^2 of silicon replaces 24 in^2 of circuit board assembly real estate in a previous design. The AlN package has 136 I/O with separate electrical, thermal, and mechanical interfaces to the circuit board.

Matra. Matra has described its military digital signal processor (DSP) based on a TI 320 C30 processor, SRAM, and memory buffer. The 32-MHz module shown in Fig. 3.54 is built on an 80× 53 mm ceramic substrate and uses Cu/PSPI interconnect. It is packaged in a 240-I/O format. The processor generates 6 W total power. It is used for image processing, signal processing, spectrum analysis, radar, sonar, and robotics applications.[185]

FIGURE 3.51 Raytheon module.[183]

FIGURE 3.52 Rockwell dual 1750A processor with on-board memory.[184]

FIGURE 3.53 Rockwell avionics memory module.[184]

TI. TI has developed several thin film MCM technologies internally. Their work in the military arena can be exemplified by the Aladdin Program,[186] which seeks to package mainframe processor performance in a soup can sized volume. The resulting 2-GFLOPS processor occupies a cylindrical volume of 4.5 inches in diameter and 3.1 inches in height. The technology used to fabricate these parts was silicon-on-silicon, Al/PI technology. The six layers of metal and five layers of BPDA/PDA PI were processed on 125-mm wafers. Ti/W was used as a adhesion layer for the 8-μm PI layers. The 25-μm vias were dry etched on 50-μm pitch.

Logic chips are TAB'ed and tested, excised, and mounted in an inverted orientation using gold bumps. Memory chips are tested and assembled into a 3D stack and mounted on end using extended tabs as pins that are soldered to the substrate.

After the substrate is populated with capacitors, ICs, and 3D memory, it is mounted into its housing by use of gold compression bonds from the substrate pads to the housing. The Aladdin consists of five basic processor modules connected by compressible connectors. This module is shown in Fig. 5.55.

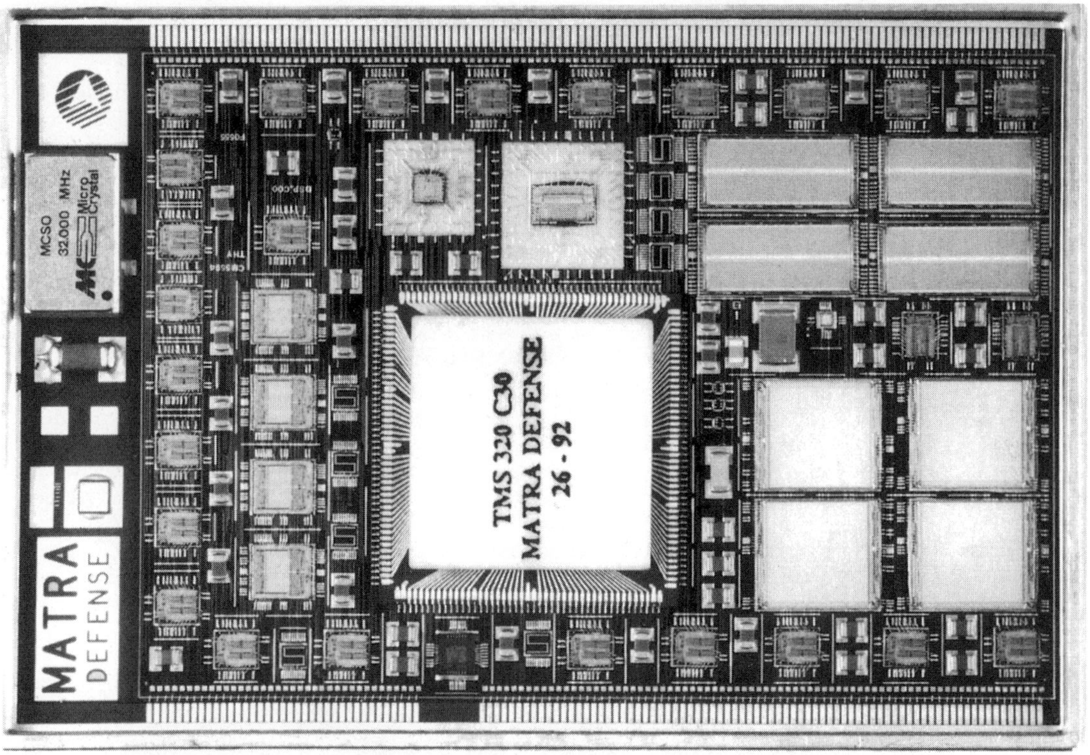

FIGURE 3.54 Matra signal processing module.[185]

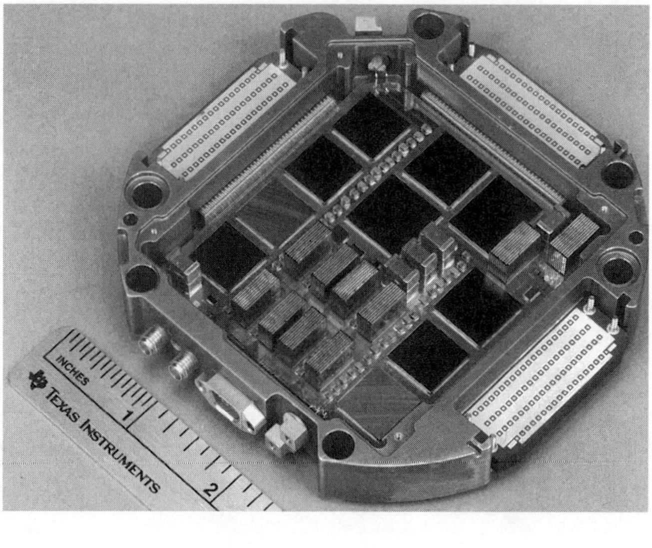

FIGURE 3.55 Aladdin microprocessor.

Saab/Combitech. The Saab/Combitech 486 Computer Module System fabricated by IMC is shown in Fig. 3.56. It contains an Intel486™ microprocessor, 1 MB flash PROM, 0.5 MB SRAM, and ASIC devices for memory management and I/O handling. It has ben qualified by MIL H 38534.[187]

Honeywell. Honeywell's program is exemplified by the Advanced Spaceborne Computer Module (ASCM) program, which is a general-purpose, space-qualified unit based on 1750A processor design. An overall view of the ASCM module is shown in Fig. 3.57. The chassis contains four elongated SEM-E PWBs. Both single-chip and multichip modules are surface mounted to both sides of the boards. The overall assembly consists of on processor subassembly, two memory subassemblies, and an I/O subassembly.

There are three types of MCMs in the ASCM system. The GVSC (Generic VLSI Spaceborne Computer) module shown in Fig. 3.58 contains three CPUs and two floating point processors (FPPs). The memory MCM, shown in Fig. 3.59, contains nine 8k × 8 SRAMs and a memory line driver. The third MCM contains cache memory (six 8K × 8 SRAM die and a cache controller chip). The system has been fabricated in cofire MCM-C and in MCM-D technology.[188]

Deutsche Aerospace. Microwave transmit/receive (T/R) radar modules have been fabricated using thin film technology on ceramic substrates in combination with a thin film logic control MCM.[189] Microstrip lines with integrated NiCr thin film resistors and dielectric spaced ground layer defined on the opposite side of an alumina substrate define the RF layer on top of the substrate. Metallized through holes are used to connect the front pattern (RF) and the substrate spaced ground plane.

The logic control unit, which is situated between the RF striplines, is constructed from three layers of metallization and PI or BCB insulation. GaAs MMICs and SPDT devices (switches) are

FIGURE 3.56 Saab Combitech 486 computer module system.

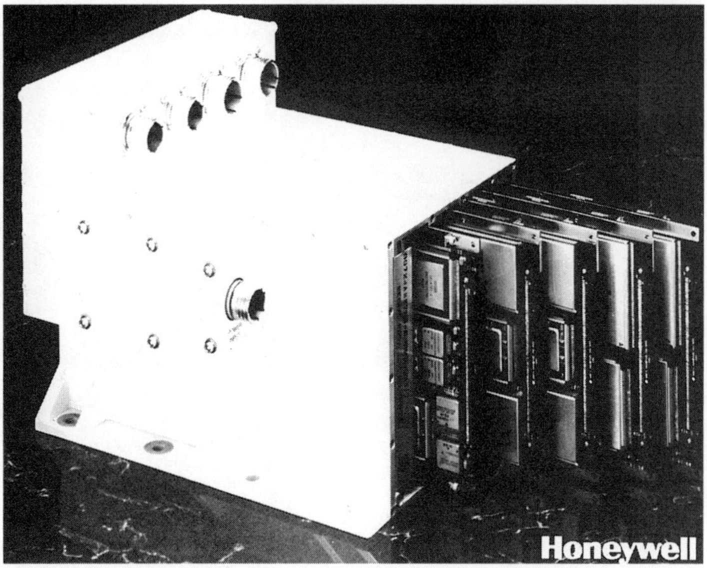

FIGURE 3.57 The Advanced Spaceborne Computer Module.

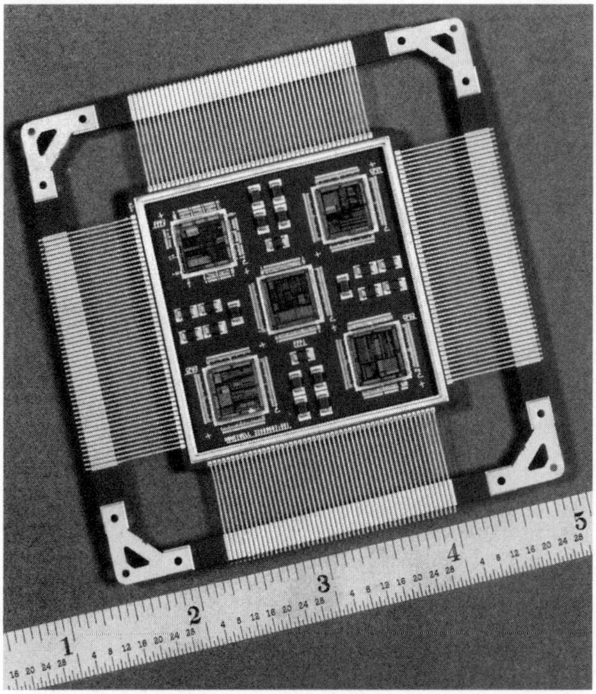

FIGURE 3.58 The Generic VHSIC Spaceborne Computer MCM.

FIGURE 3.59 ASCM memory board with memory MCMs.

mounted on the logic control substrate as shown in Fig. 3.60. The face of an active phased array radar antenna contains as many as 2000 such elements.

GE. The GE HDI technology described in Sec. 3.4.6 has been used to prototype and manufacture many military products. Figure 3.61 depicts a 44-channel A/D engine control module for military jet aircraft. It contains 13 IC and 16 precision resistor network chips embedded below the Cu/PI interconnect, as well as 82 surface mounted signal decoupling capacitors.

3.6.4 Telecommunication

Demand is growing for a broadband integrated services digital network (B-ISDN) that can handle high-speed digital communications services such as data, video, and high-definition TV. Such large-capacity, high-speed switching and transmission systems will require MCM packaging technologies to implement signal switching.[190,191] Saving space, due to the elimination of multiple single-chip packages and passive components, can also be of great value in mobile telecommunications.[192]

NTT ATM Switch. NTT has described thin film technology for ATM (asynchronous transfer mode) switching.[193] Although demand for the broadband integrated services digital network (B-ISDN) has been slower that initially projected, such high-speed digital signal services are on the horizon. The asynchronous transfer mode (ATM) is expected to be a key technology for this system. The main uses of ATM switching systems will be high-speed communication including voice, video, and data. Future ATM switching systems will be built of many ATM switching elements with high-speed interconnects between them. Thin film MCM technology is expected to allow the fabrication of entire ATM switching elements within a single package.[193]

Figure 3.62 shows a cross section of an ATM switch using thin film technology. For a 4× 4 ATM shared memory switch, the MCM contains 4 BiCMOS LSIs, 16 bipolar LSIs for conversion between TTL and ECL, 2 8:1 multiplexer bipolar SSIs, and 2 1:8 demultiplexer LSIs. The MCM has six lay-

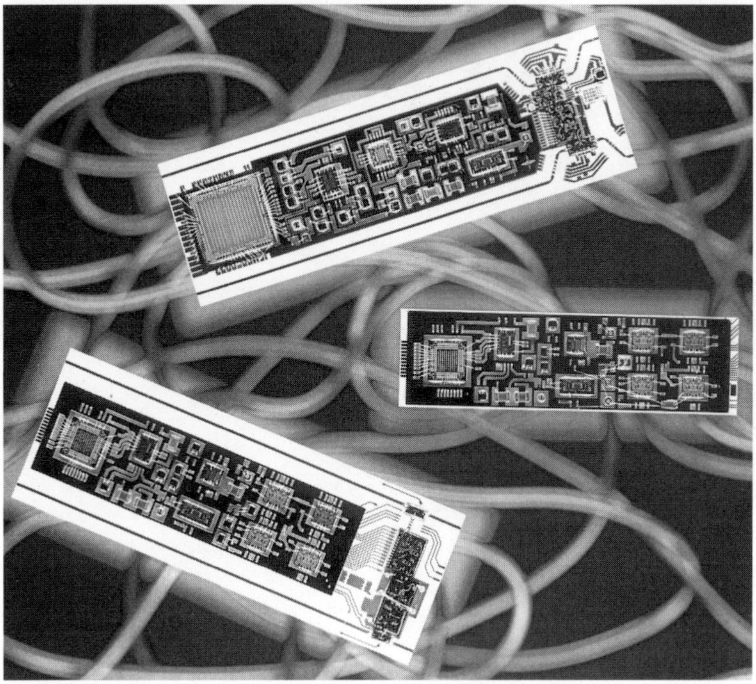

FIGURE 3.60 Deutche Aerospace T/R MCM radar module.[189]

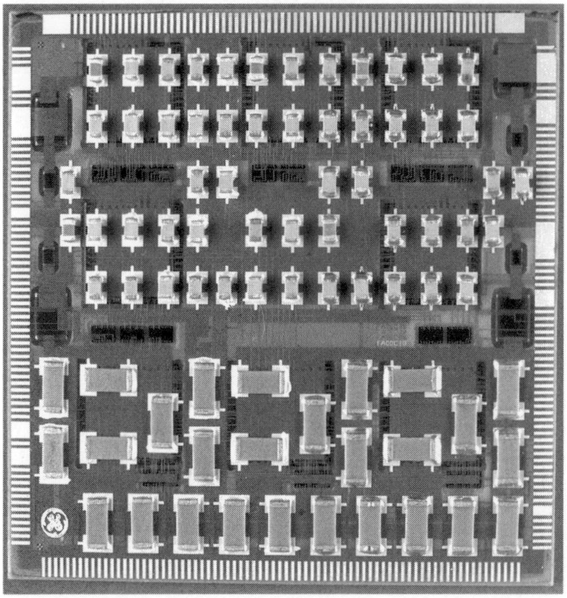

FIGURE 3.61 GE engine control module.[151]

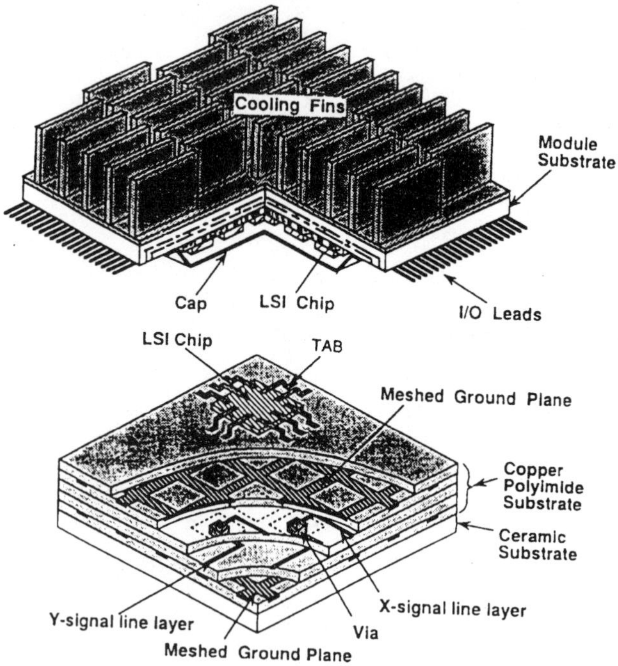

FIGURE 3.62 Cross section of NTT switch.[194]

ers of Cu/PI thin film, two of which are meshed over supply layers. To increase the yield of the MCM ATM switches, TAB technology is used to allow pretesting of the LSI devices.

AT&T Timing Generator. Figure 3.63 depicts a timing generator that operates above 300 MHz. The circuit, fabricated using AT&T PolyHIC technology, contains 7 chips, more than 100 thin film resistors, and 30 capacitors. The circuit operates with controlled impedance lines, tightly controlled delays, and terminating resistors.[194]

3.6.5 Consumer

Inherently, MCM-D thin film technology does not contain any elements that should be considered exotic or inherently expensive. The high cost of the initial mainframe and military and space applications was, and is, caused by low volume and low throughput per unit operation and low yield due to large x-y dimensions. It has not been uncommon to see the 250 × 250-mm MCM-CD (for mainframes) and the 100 × 100-mm MCM-D on silicon or ceramic (for military applications). Such structures have suffered from substrate yield problems due to killer defects in the same way that ICs have had to deal with such issues as they increase in x-y dimension. This has caused designs to incorporate costly rerouting pathways and has demanded costly rework to yield such structures.

The approach in considering consumer applications has been to consider the cost savings that can be achieved at the system level by using thin film technology to replace existing components.[195,196] The lower parasitic loading results in less power dissipation and thus, at the system level, reduced power supply costs, less demand for cooling, and denser packaging of boards in cabinets. The smaller footprint of the thin film MCM can reduce the size and thus cost of mother boards, and most importantly of all, it allows one to incorporate large numbers of passive components directly into the substrate.[195]

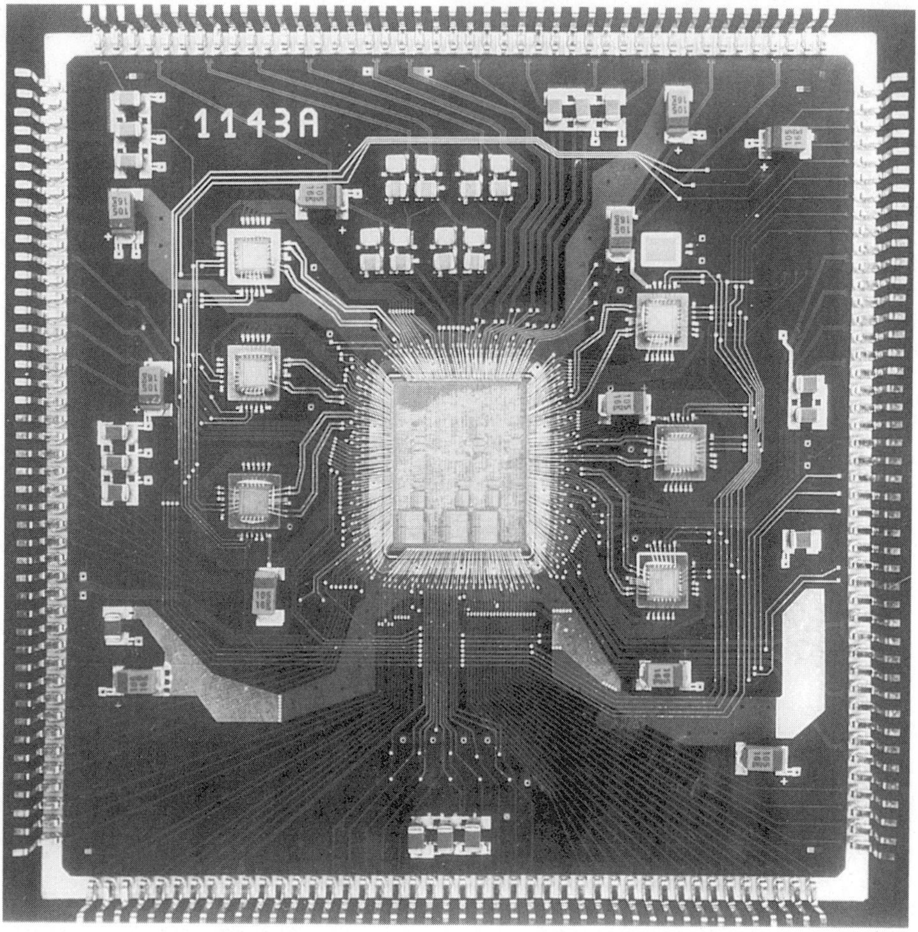

FIGURE 3.63 PolyHIC timing generator.

AT&T Partner Telephone. AT&T has described the premanufacturing of silicon based MCM for its Partner™ telephone, a relatively large volume consumer business.[85,86] Key to the low cost are the fabrication of the needed passive components directly on the carrier and the adoption of existing SMT assembly infrastructure to the chip mounting on the carrier wafers. The leadless, bare die are furnished with solder wettable flip chip pads in place of conventional wire bond pads. The I/O bond pad array is rerouted from the conventional perimeter positions to area array flip chip under the chip. This rerouting increases the pitch from as little as 5 mil (0.13 mm) to 16 mil (0.4 mm), which then allows for fine pitch SMT assembly. Rerouting involves the addition of an additional Al conductor layer, an additional dielectric layer with I/O vias, and solder wettable pads. The "active" MCM interconnect is a three-level, 10 μm interconnect with containing 119 resistors, 8 capacitors, 7 diodes, and 152 transistors.[86] Figure 3.28 shows a cross section of the final structure.

Because each MCM interconnect substrate is only 9.5 × 6.65 mm in size, this permits patterning of 148 per 125 mm wafer. Figure 3.64 shows the populated MCMs on the silicon carrier. It has been noted that the cost savings from mounting and stocking the passive components is greater than the cost of the components themselves.

FIGURE 3.64 Populated six-chip MCM wafer.

REFERENCES

1. Evans, A., Hagge, J., "Advanced Packaging Concepts—Microelectronics Multiple Chip Modules Utilizing Silicon Substrates," *Proc. 1st Int. SAMPE Elect. Conf.,* 1987, p. 37.
2. Murano, H., Watari, T., "Packaging Technology for the NEC-SX3 Supercomputer," *IEEE Trans CHMT,* vol. 15, 1992, p. 411.
3. Bower, G. W., Frost, L. W., *J. Polym. Sci. A,* vol. 1, 1963, p. 3135.
4. Frost, L. W., Keese, I., *J. Appl. Polym. Sci.,* vol. 8, 1964, p. 1039.
5. Ferger, C., Ferger, C., "Selection Criteria for MCM Dielectrics," in *MCM Technologies and Alternatives,* D. Doane, P. Franzone, Eds., Van Nostrand Reinhold, 1993, p. 311.
6. Ferger, C., et al., *Polyimides, Materials, Chemistry and Characterization,* Elsevier, 1989.
7. Polyimides, K.L. Mittal, Ed., *Plenum Press,* 1984, p. 353.
8. *Recent Advances in PI Science and Technology,* W. Weber, M. Gupta, Eds., SPIE, 1987.
9. Deutsch, A., et al., "Measurement of Dielectric Anisotropy of BPDA-PDA PI in Multilayer Thin Film Packages," *IEEE Trans. CHMT,* vol. 17, 1994, p. 486.
10. Numata, S., Oohara, S., Fujisaki, K., Imaizumi, J., Kinjo, N., "Thermal Expansion Behavior of Various Aromatic Polyimides," *J. Appl. Polym. Science,* vol. 31, 1986, p. 101.
11. Numata, S., Fujisaki, K., Makino, D., Kinjo, N., "Chemical Structures and Properties of Low Expansion Polyimides," *Proc. 2nd Int. Symp. Polyimides,* 1985, p. 164.

12. Boese, D., Herminghaus, S., Yoon, D.Y., Swalen, J.D., Rabolt, J., Smith, B. A., *App. Phys. Letters,* vol. 59, 1991, p. 1043.

13. Volksen, W., Yoon, D. Y., Hendrick, J., "Polyamic Alkyl Esters: Versatile Polyimide Precursors for Improved Dielectric Coatings," *Proc. ECTC, 1991,* p. 572.

14. Breen, M., Weber, W., "Applicability of Polyamic Aester Precursors to High Frequency, High Density Interconnect Fabrication," *Proc. IEPS,* 1994, p. 177.

15. Merriman, B. T., Craig, J. D., Nadar, A.E., Goff, D. L., Pottiger, M.T., Lautenberger, W.J., "New Low CTE Polyimides for Inorganic Substrates," *Proc. ECTC,* 1989, p. 155.

16. Saiki, A., Mukai, K., Harada, S., Miyadera, Y., "Development of Polyimide Isoindoloquinazolinedione in Multilevel Interconnections for Large-Scale Integration (LSI)," in "Polymeric Materials for Electronic Applications," *ACS Symp. Series #184,* E. Feit, C. W. Wilkens, Eds., 1982, p. 123.

17. Schiltz, A., Bouzid, M.J., Jourdain, D., Torres, J., Palleau, J., Obwerlin, J.C., Laviale, D., Nechtschien, J., "Lift-off Techniques for Multichip Modules," *Proc. NEPCON West,* 1990, p. 975.

18. Pfeifer, J., Rohde, O., "Direct Photoimaging of Fully Imidized Solvent-Soluble Polyimides," *Proc. 2nd Int. Conf. on Polyimides,* 1985, p. 130.

19. Neuhaus, H., "A High Resolution, Anisotropic Wet Patterning Process Technology for MCM Production," *Proc. Int. Conf. MCMs,* 1992, p. 256.

20. Neuhaus, H., "Photosensitive PI for MCM-D/L," *Proc. IEPS,* 1994, p. 457.

21. Burdeaux, D., Townsend, P., Carr, J., Garrou, P.E., "Benzocyclobutene (BCB) Dielectrics for the Fabrication of High Density Thin Film Multichip Modules," J. Electronic Materials, vol. 19, 1990, p.1357.

22. Heistand, H.R., DeVellis, R., Manial, T.A., Garrou, P.E., Stokich, T.M., Townsend, P.H., Adema, G.M., Berry, M.J. and Turlik, I., "Advances in MCM Fabrication with Benzocyclobutene Dielectric," *Proc. ISHM,* 1991, p. 96.

23. Rubner, R., *Siemens Forsch-u. Entwicki-Ber.,* vol. 5, 1976, p. 232.

24. Pottinger, M., *Solid State Technology,* 1989, p. 51.

25. Ahne, H., Kruger, H., Pammer, E., Rubner, R., in "Polyimides," K.L. Mittel, Ed., *Plenum Press,* New York, 1984, p. 933.

26. Kojima, J., Sekine, H., Suzuki, H., Satou, H., Makino, D., Kataoka, F., Tanaka, J., Shouji, F., "Photosensitive Polyimide for IC Devices," *Proc. ECTC,* 1989, p. 920.

27. Maw, T., Hopla, R.E., *Proc. MRS,* vol. 203, 1991, p. 71.

28. Perfecto, E.D., Osborn, C., Berger, D., "Factors That Influence Photosensitive PI Lithographic Performance," *Proc. Int Conf MCMs,* Denver, 1993, p. 40.

29. Kubota, S., Tanaka, Y., Moriwaki, T., Eto, S., "Positive Working Photosensitive Polyimide," *J. Electrochem. Soc.,* vol. 138, 1991, p. 1080.

30. Sasaki, S., Kon, T., Ohsaki, T., Yasuda, T., "A New Multichip Module Using a Copper/PI Multilayer Substrate," *IEEE CHMT,* vol. 12, 1989, p. 658.

31. Watari, T., Murano, H., "Packaging Technology for the NEC SX Supercomputer," *IEEE Trans. CHMT,* vol. 8, 1985, p. 462.

32. Chakravorty, KK., Chien, C.P., Cech, J.M., Tanniellian, M.H., Young, P., *IEEE Trans CHMT,* vol. 13, 1990, p. 200.

33. Kotzias, B.D., Murray, D.O., "Review of PSPI Process for Manufacturing Multichip Modules," *Int. J. Microcircuits and Elect. Pkging.,* vol. 16, 1993, p. 339.

34. Rutter, E., Moyer, E., Harris, R., Frye, D., St. Joer, V., Oaks, F., "A Photodefinable BCB Resin for Microelectronic Applications," *Proc. Int. MCM Conf.,* Denver, 1992, p. 394.

35. Moyer, E.S., Rutter, E.W., Bernius, M.T., Townsend, P.H., Harris, R.F., "Photodefinable BCB Formulations for Thin Film Microelectronics Applications Part II," *Proc. IEPS,* 1992, p. 37.

36. Garrou, P.E. et al., "Cost Modeling and Fabrication of MCMs using Photosensitive Cyclotene™ (BCB), *Proc. Int. Microelect. System Tech. Symp.,* Berlin, 1994, p. 569.

37. Shimoto, T., Matsui, K., Utsumi, K., "Cu/PhotoBCB Thin Film Multilayer Technology for High Performance Multichip Modules," *Proc. Int. Conf. MCMs,* Denver, 1994, p. 115.

38. Shimoto, T., Matsui, K., Kimura, M., Utsumi, K., "High Density Multilayer Structure using BCB Dielectric," *Proc. 7th Int. Microelectronics Conf.,* 1992, p. 325.

39. Radlik, W., et al., "MCM-D Technology for a Communication Application," *Proc. Int Conf. MCMs,* Denver, 1994, p. 402.
40. Sweetman, E., "Characteristics and Performance of PHP-92: AT&T's Triazine Based Dielectric for Polyhic MCMs," *Proc. Int. Conf. MCMs,* Denver, 1992, p. 401.
41. Clatterbaugh, G. V., Charles, H.K., "The Application of Photosensitive Polyimide Dielectrics in Thin Film Multilayer Hybrid Circuit Structures," *Proc. ISHM, 1988,* p. 320.
42. Chao, C., Scholz, K.D., Liebovitz, J., Cobarruviaz, M., Chang, C.C., "Multilayer Thin Film Substrates for Multichip Packages," *IEEE Trans. CHMT,* vol. 12, 1989, p. 180.
43. Tessier, T.G., Adema, G.M., Turlik, I., "Polymer Dielectric Options for Thin Film Packaging Applications," *Proc. ECC,* 1989, p. 127.
44. Kimura, A., Tsujimura, T., Saitho, K., Kohno, Y., "Si-on Si Technology: Process and Characteristics of Si Carrier Substrate," *Proc. Int. Microelect. Conf.,* Yokohama, 1992, p. 337
45. Denton, D., Pranjoto, H., "Gravimetric Measurement of Moisture Uptake in Polyimide Films Used in Integrated Circuit Packaging," *MRS Symp. Proc.,* vol. 154, 1989, p. 97.
46. Jensen, R.J., Douglas, R.B., Smeby, J.M., Moravec, T.J., "Characteristics of Polyimide Material for Use in Hermetic Packaging," *Proc. VHSIC Pkging Conf.,* Houston, 1987, p. 193.
47. Kobayashi, S., et al., "Aluminum Sputter Deposition onto PI Coated Substrates," *Proc. ECTC,* Las Vegas, 1990, p. 383.
48. Kambe, R., Kuroda, M., Imai, R., Kimura, Y., "Copper PI Multilayer Substrate for High Speed Signal Transmission," *Proc. ECTC.,* 1991, p. 14.
49. Heacock, J.F., "A Kinetic Study of the Hydrolysis of Polyimide Film," in *Polyimides,* Webber, W.D., Gupta, M.R., Eds., Society of Plastics Engineers, 1987, p. 174.
50. Projanto, H., Denton, D., "Moisture Uptake in BCB Films for Electronic Packaging Applications," *MRS Symp. Proc.,* Boston, vol. 203, 1990, p. 295.
51. Schiltz, A., 'A Review of Planar Techniques for Multichip Modules," *IEEE Trans. CHMT,* vol. 15, 1992, p. 236.
52. McDonald, J.F., Lin, H.T., Grueb, H.J., Philhower, R.A., Dabra L.S., "Techniques for the Fabrication of Wafer Scale Interconnections in Multichip Packages," *IEEE Trans. CHMT,* vol. 12 1989, p. 195.
53. Rothman, L.B., "Properties of Thin Polyimide Films," *J. Electrochem. Soc.,* Solid State Science and Tech.,1980, p. 2216.
54. Tessier, T., "A Comparison of Common MCM-D Dielectric Material Performance," *Proc. 6th SAMPE Electronics Conf.,* 1992, p. 347.
55. Stokich, T.M., Fulks, C.C., Bernius, M.T., Burdeaux, D.C., Garrou, P.E., Heistand, R.H., "Planarization with Cyclotene 3022 (BCB) Polymer Coatings," *Material Research Soc. Symp. Proc.,* vol. 308, 1994, p. 517.
56. Day, D.R., Ridley, D., Mario, J., Senturia, S.D., "Polyimide Planarization in Integrated Circuits," in *Polyimides,* K.L. Mittal, Ed., Plenum Press, 1984, p. 767.
57. Chao, C.C., Wang, W.V., in *Polyimides,* K.L. Mittal Ed., Plenum Press, 1984, p. 783.
58. Beyne, E., Van Hoff, R., Pieters, F., Lenaerts, S., Achen, A., "The Use of BCB and Photo BCB Dielectric in MCM-D for High Speed Digital and Microwave Applications," *Proc. Int Conf. MCMs,* Denver, 1995 p. 513.
59. Perfecto, E., Prasad, K., Wilkens, P., White, G., Tummala, R., Prasad, C., Redmond, T., "Multi-level Thin Film Packaging Technology at IBM," *Proc. Int. MCM Conf.,* Denver, 1993, p. 474.
60. Prasad, K., Perfecto, E., "Multilevel Thin Film Packaging: Applications and Processes for High Performance Systems," *IEEE Trans. CHMT,* Part B, vol. 17, 1994, p. 38.
61. Kasuya, Y., Takahashi, Y., Uno, Y., Iguchi, Y., Kanamori, T., "Planarized Cu/PI Thin Film Substrates," *Proc. ISHM,* 1993, p. 193.
62. Fong, S.O., Keister, F.Z., Peters, J.W., "Polybenzocyclobutene resins in Multilayer Thin Films for Electronic Packaging," *Proc. SAMPE,* 1990, p. 602.
63. Kersten, P., Schammler, G., Reichl, H., "An Additive Approach to Multichip Modules, Using Electroless Metallization Processes," *Int. J. Microcircuits and Electronic Packaging,* vol. 17, 1994, p. 73.
64. Hodge, T.C., Landmann, B., Bidstrup, S.A., Kohl, P.A., "Rapid Thermal Curing of Polymer Interlayer Dielectrics," *Int. J. Microcircuits and Electronic Packaging,"* vol. 17, (1994), p. 10.

65. Stokich, T., Burdeaux, D.C., Mohler, C., Townsend, P.H., Warrington, S., Tou, J., Han, B.J., Pryde, C., Bair, H., Johnson, G., "Thermal and Oxidative Stability of Polymer Thin Films Made from DVS-BCB," *MRS Symposium Proc.,* MRS, Anaheim, 1991.

66. Stokich, T. et al., "Thermal and Oxidative Stability of Polymer Thin Films Made From DVS-BCB," MRS, Anaheim, 1991.; Stokich, T. et al., "Advances in the Thermooxidative Stabilization of DVS-BCB Polymer Coatings," *MRS Symposium Proc.,* vol. 265, 1992, p. 275.

67. Dibbs, M.G., Townsend, P.H, Mohler, C.E., Stokich, T.M., Huber, B.S., Heistand, R.H., Garrou, P.E., Adema, G., Berry, M., Turlik, I., "Cure Management of BCB Dielectric for Electronic Applications," *Int. SAMPE Elect. Conf.,* vol. 6, 1992, p. 1.

68. Evan, R., Torres, J., Palleau, J., Oberlin, J.C., Lavial, D.,Templia, F., Pantel, R., "Physiochemical Stability of Copper/Polyphenylquinoxaline Interfaces," *Proc. of IEMT,* 1990, p. 240.

69. Birnath, R.W., Soane, D.S., *Characterization of Stresses in Polymer Films for Microelectronics Applications,* chapter 29, American Chemical Soc., 1989, p. 356.

70. Himmel, R., Licari, J., "Fabrication of Large-Area, Thin-Film Multilayer Substrates," *Proc. ISHM,* 1989, p. 454.

71. Pan, J.T., Poon, S., "Film Stress in High Density Thin Film Interconnect," *MRS Symp. Proc.,* vol. 154, 1989, p. 27.

72. *ASTM D3359-78,* American Society for Testing and Materials.

73. Stokich, T., Dow Chemical, private communication.

74. DeLeo, J.A., Gupta, J.R., "Thermal Stability of Polyimide Adhesion with Various Coupling Agents," *Proc. 2nd Int. Conf. on Polyimides,* W.D. Webber, M.R. Gupta, Eds., 1985, p. 399.

75. Conway, H.D., Thomson, J.P., *J. Adhesion. Science & Tech,* vol. 3, 1988, p. 227.

76. Greenblatt, J., Araps, C.J., Anderson, H.R., "Aminosilane Polyimide Interactions and their Implications in Adhesion," in *Polyimides,* K.L. Mittal, Ed., Plenum Press, 1984, p. 573.

77. Townsend, P., et al., "Adhesion of Cyclotene Coatings on Silicon Substrates," *MRS Symp. Proc.,* Boston, vol. 323, 1993, p. 365.

78. Sotou, H., "PIQ PI for MCM Applications," *Proc. NEPCON West,* 1989, p. 921.

79. Paik, K.W., Ruoff, A.L., *J. Adhesion Science Tech, A,* vol. 6, 1988, p. 1004.

80. Flitsch, R., Shi, D-Y., "A Study of Modified Polyimide Surfaces as Related to Adhesion," *J. Vac. Sci. Tech.,* vol. A8 1990, p. 2376.

81. Horton, T., "MCM Driving Forces, Applications and Future Directions," *Proc. NEPCON West,* 1991, p. 487.

82. Marella, P.F., Tuckerman, D.B., "Prequalification of n-Chip SiCB Technology," *Proc. Int. Conf. MCM,* Denver, 1992, p. 376.

83. Hagge, J., "Ultra-reliable Packaging For Silicon-on-Silicon WSI," *IEEE Trans CHMT,* 12, 1989, p. 170.

84. Bartlett, C., Segelken, J., Teneketges, N., "Multichip Packaging Design for VLSI Based Systems," *Proc. ECC,* 1987, p. 518.

85. Dudderar, T.D., Degani, Y., Spadafora, J.G., Tai, K.L., Frye R.C., "AT&T μSMT Assembly: A New Technology For Large Volume Fabrication of Cost Effective Flip-Chip MCMs," *Proc. Int. Conf. MCMs,* Denver, 1994, p. 266.

86. Cokely, D., Strittmatter, C., "Redefining the Economics of MCM Applications," *Proc. Int. Conf. MCMs,* Denver, 1994, p. 306.

87. Pearson, R.E., "Active Silicon Substrate MCM Technology for Sensor Signal Processing and Control," *Proc. Int Conf. MCMs,* Denver, 1994, p. 618.

88. Shambrook, K, Trask, P., *Proc. ECTC,* 1989, p. 656.

89. S. Jain, Climax Specialty Metals Corp., personal communication.

90. Deshpande, U., Shamouilian, S., Howell, G., "High Density Interconnect Technology For VAX 9000 System," *Proc. IEPS,* 1990, p. 46.

91. Abbasi, S. et al. "The Technology and Manufacture of the Vax-9000 Multichip Unit," in *Multichip Modules: The Basics,* Doane and Franzione, Eds., Van Nostrand, 1993, p. 769.

92. Tummala, R., in *Microelectronic Packaging Handbook,* R. Tummala, E. Rymaszewski, Eds., Van Nostrand Reinhold, 1989.

93. Takeda, K., Naritsuka, Y., Hirada, M., "High Density Module with Small Packaged LSI's for Mainframe Computers," *Int. Conf. MCMs,* Denver, 1993, p. 260.
94. Leung, G.B., Sands, S.A., "A Thin Film on MLC Application," *Proc. ECTC,* 1991, p. 10.
95. Paraszczak, J., et al., "Fabrication and Performance Studies of Multilayer Polymer/Metal Interconnect Structures for Packaging Applications," *Proc. ECTC,* 1991, p. 362.
96. Wilson, A.M., "Use of Polyimides in VLSI Fabrication," in *Polyimides,* K.L. Mittal, Ed., Plenum Press, 1984, p. 715.
97. Berry, M.J., Tessier, T.G., Turlik, I., Adema, G. M.,Burdeaux, D.C., Carr, J.N., Garrou, P.E., "Benzocyclobutene as a Dielectric for MCM Fabrication," *Proc. ECTC,* 1990, p. 746.
98. Rickeri, P., Stephanie, J., Slota, P., "Evaluation of Photosensitive Polyimides for Packaging Applications," *Proc. ECC,* 1987, p. 220.
99. Kohara, M., Mashiko, Y., Nakazaki, K., Nunoshita, M., "Mechanism of Electromigration in Ceramic Packages Induced by Chip Coating PI's," *Proc. ECTC,* Las Vegas, 1990, p. 894.
100. Ginsburg, R., Susko, J.R., in *Polyimides,* K.L. Mittel, Ed., Plenum Press, New York, 1984, p. 237.
101. Pryde, C.A., "IR Studies on Polyimides I: Effects of Physical and Chemical Changes During Cure," *J. Polym. Science,* Part A, vol. 27, 1989, p. 711.
102. Day, D.R., Senturia, S.D., "In-Situ Monitoring of Polyamic Acid Imidization with Microdielectrometry," in *Polyimides,* K.L., Mittal Ed., Plenum Press, 1984, p. 249.
103. Elsner, G., *J. Applied Polymer Science,* vol. 34, 1987, p. 815.
104. Stokich, T.M., Lee, W. M., Peters, R.A., "Real Time FT-IR Studies of the Reaction Kinetics for the Polymerization of Divinyl Siloxane Bis-Benzocyclobutene Monomers," *MRS Symp. Proc.,* vol. 227, 1991, p. 103.
105. Mohler, C.E., Strandjord, A.J., Castillo, D.J., Stachowiak, M.R., Heistand, R.H., Garrou, P.E., Tessier, T.G., "Micro ATR as a Probe of BCB Layers for MCM-LD Applications," *Proc. MRS Symp.,* 1993, vol. 323, p. 295.
106. Heistand, R., DeVellis, R., Manial, T., Kennedy, A., Stokich, T., Garrou, P., Takahashi, T., Adema, G., Berry, M., Turlik, I., "Advances in MCM Fabrication With Benzocyclobutene Dielectric," *Proc. Int. Microelectronics Conf.,* Yokohama, 1992, p. 320.; *Journal of Microcircuits and Microelectronic Packaging,* vol. 15, 1992, p. 183.
107. Garrou, P., Heistand, R., Dibbs, M., Manial, T., Stokich, T., Mohler, C., Townsend, P., Adema, G., Berry, M., Turlik, I., "Rapid Thermal Curing of BCB Dielectric," *IEEE Trans. CHMT,* vol. 16, p. 46.
108. Myszka, E., Demet, G., Fuerhaupter, H., Tesier, T., Hoffman, W., Mukerji, P., Stafford, J., Lindsey, S., "The Development of A Multilayer Thin Film Cu/PI process for MCM-D Substrates," *Int. Conf. MCMs,* Denver, 1992, p. 1.
109. Perfecto, E., Saji, K., Desai, K., Saltarelli, J., "Thin Film Learning in Manufacturing," *Proc. IEPS,* 1994, p. 481.
110. Mlynko, W.E., Cain, S.R., Egitto, F.D., Emmi, F., in *Principles of Electronic Packaging,* Seraphim, Lasky, and Li, Eds., McGraw Hill, New York, 1989, p. 425.
111. Rogers, B., Berry, M.J., Turlik, I., Garrou, P.E., Castillo, D., "Soft Mask for Via Patterning in Benzocyclobutene," *Proc. ISHM,* 1993, p. 187; *Journal of Microcircuits and Microelectronic Packaging,* vol. 17, 1994, p. 210.
112. Tessier, T.G., Turlik, G.M., Sivan, D., Yung, E.K., Berry, M.J., "Process Considerations in Fabricating Thin Film MCMs," *Proc. IEPS,* 1989, p. 294.
113. Tessier, T.G., Stafford, J.W., Hoffman, W.F., "Via Processing Options for MCM-D Fabrication: RIE vs. Laser Ablation," *Proc. ECTC,* 1991, p. 827.
114. Rutter, J.J., "Via Formation in Thick PI Layers for Si Hybrid MCMs," *Proc. Electronic Materials Process. Cong.,* 1989, p. 351.
115. Johnson, R. W., Phillips, T.L., Weidner, W. K., Hahn, S.F., Burdeaux, D.C., Townsend, P., "Benzocyclobutene Interlayer Dielectrics for Thin Film Multichip Modules," *IEEE Trans. CHMT,* vol. 13, 1990, p. 347.
116. Tessier, T.G., Hoffman, W.F., Stafford, J.W., "Via Processing Options for MCM-D Fabrication: Excimer Laser Ablation vs. Reactive Ion Etching," *Proc. ECTC,* 1991, p. 827.

117. McDonald, J.F., Lin, H.T., Grueb, H.J., Philhower, R.A., Dabra L.S., "Techniques for the Fabrication of Wafer Scale Interconnections in Multichip Packages," *IEEE Trans. CHMT,* vol. 12 1989, p. 195.

118. Shih, D.Y., et al., "Factors Affecting the Interconnection Resistance and Yield in the Fabrication of Multilayer Polyimide/Metal Thin Film Structures," *Proc. ECTC,* 1992, p. 1002.

119. Summers, J., Murray, D., Mocella, M., Springman, M., Howe, J., Lautenberger, W., "Wet Etching Polyimides for MCM Applications," *Proc. Int. Conf. MCMs,* Denver, 1992, p. 247.

120. Saiki, A., Mukai, K., Nishida, T., Suzuki, H., Mikino, D., in *Polyimides,* K.L., Mittel Ed., Plenum Press, NY, 1984, p. 827.

121. Inoue, T., Matsuyama, H., Matsuzaki, E., Narizuka, Y., Ishino, M., Tanaka, M., Takenake, T., "Microcarrier for LSI Chip Used in the HITAC M-880 Processor Group," *IEEE Trans CHMT,* vol. 15., 1992, p. 7.

122. Schammler, G.K.H., Glaw, V., Chmiel, G., "Comparison of the Metallization of Chemically and Laser-Etched Structures in BPDA-PDA Polyimide," *IEEE Trans. CHMT,* vol. 16, 1993, p. 720.

123. Rucker, T., Murali, V., Shukula, R., "A Wet Etch PI Process for MCMs," *Proc. Int. Conf. MCMs,* Denver, 1992, p. 251.

124. Ori, T., Shibuya, H., Ohnuki, Y., Utsunomiya, J., Iida, S., "Development of PI Substrate for Gb/sec Glass MCM-D/C," *Proc. ISHM,* 1994, p. 518.

125. Ree, M., Chen, J.R, *Polymer Preprints,* 31(2), 1990, p. 594.

126. Perfecto, E.D., Berger, D.G. Osborn, C.T., White, G., "Engineering PSPI's for MCM-D Applications," *Int. J. Microcircuits & Elect. Pkging.,* vol. 16, 1993, p. 319.

127. Niwa, K., Eguchi, M., Asano, M., "Patterning Techniques of PSPI on Copper," *Proc. Int. Microelect. Conf.,* Yokohama, 1992, p. 51.

128. Hubbard, R.L., Lehman-Lamer, G., "Very High Speed Multilayer Interconnect Using Photosensitive PI," *Proc. ISHM,* 1988, p. 374.

129. Clatterbaugh, G.V., Charles, H.K., "The Application of Photosensitive PI Dielectrics in Thin Film Multilayer Hybrid Circuit Structures," *Proc. ISHM,* 1988, p. 320.

130. Strandjord, A., Garrou, P., Heistand, R.H., Tessier, T.G., "MCM-LD: Large Area Processing Using Photo BCB," *IEEE Trans CPMT,* Part B, Advanced Packaging, vol. 18, 1995, p. 269.

131. Patel, R., Redmond, T., Tessler, C., Tudryn, D., Pulaski, D., "Laser Via Ablation Technology for MCM Thin Film Packaging—Past, Present and Future at IBM Microelectronics," *Proc. ISHM,* 1994, p. 31.

132. Wolbold, G.E., Tessler, C.L., Tudryn, D.J., "Polymer Ablation with a High Power Excimer Laser Tool in Manufacturing," *Proc. IEMT,* Kanazawa, Japan, 1993, p. 22.

133. Tessier, T.G., Chandler, G., "Compatibility of Common MCM-D Dielectrics with Scanning Laser Ablation Via Generation Processes," *Proc. ECTC.,* 1992, p. 763.

134. Darrow, D., Vilmer-Bagen, S., "Comparative Analysis of Thin Film Metallization Methodologies for High Density Multilayer Hybrids," *Proc. Int Conf. MCMs,* Denver, 1992, p. 56.

135. Windischmann, H., "Intrinsic Stress in Sputtered Thin Films," *J. Vac. Sci. Technol.,* A9, 1991, p. 4.

136. Turlik, I.,and Adema G.M., "Main Construction Approaches for MCM-D," in *Thin Film Multichip Modules,* G. Messner, I. Turlik, J. Balde and P. Garrou, Eds., ISHM Press, 1992.

137. Shiflett, C., Bucholtz, D., Faudskar, C., Small, R., Markham, J., "High Density Multilayer Hybrid Circuits Made with Polymer Insulating Layers (Polyhics)," *Proc. ISHM,* 1986, p. 481.

138. Shah, A., Sweetman, E., Hoppes, C.,"A Review of AT&T's PolyHIC MCM Technology," *Proc. NEPCON West,* 1991, p.850.

139. Lau, M.Y., Tai, K.L., Frye, R.C., Saito, M., Bacon, D.D., "A Versatile, IC Process Compatible MCM-D for High Performance and Low Cost Applications," *Proc. Int. Conf. MCMs,* Denver, 1993, p. 107.

140. Tummala, R., Potts, H., Ahmed, S., "Packaging Technology For IBMs Latest Mainframe Computers (S/390/ES9000)," *Proc. ECTC,* 1991, p. 682.

141. Redmond, T., et al., "Polyimide Copper Thin Film Redistribution on Glass Ceramic/Copper Multilevel Structures," *Proc. ECTC,* 1991, p. 689.

142. Redmond, T.F., Lankard, J.R., Balz, J.G., Proto, G.R., Wassick, T.A., "The Application of Laser Process Technology to Thin Film Packaging," *IEEE Trans CHMT,* vol. 16., 1993, p. 6.

143. Bregman, M.F., Kimura, A., Matsui, T., Nishida, H., Nishiyama, K., Ohkuma, H., Tanaka, A., Kovac, C., McQueeney, D., "A Thin Film MCM for Workstation Applications," *Proc. ECTC,* 1992, p. 968.

144. Kimura, A., Tsujimura, T., Kohno, Y., "Fabrication and Characteristics of Silicon Carrier Substrates for Silicon on Silicon Packaging," *Proc. Int. Conf. MCMs,* Denver, 1992, p. 23.

145. Bone, R.L., et al. "High Performance Aerospace MCM Technology Development at Hughes," in *MCM Technology and Alternatives,* D. Doane and P. Franzone, Eds., Van Nostrand Reinhold, 1993, p. 695

146. Kovacs, A.L., Ehlert, M.R., McDonald, R.B., "Mechanical Aspects of AlN packages for VLSI Packaging," Proc. IEPS, 1990, p. 199.

147. Luh, E., Enloe, J., Kovacs, A., Lucernoni, R., "Metallization of AlN Packages," *Proc. IEPS, 1990,* p. 208.

148. Walter, D., "Flexibility and Performance of TI's Multilayer TF/laminate Interconnect Technology," *Proc. Int. MCM Conf.,* Denver, 1995, p. 553.

149. McWilliams, B., "Comparison of Multichip Interconnect Technologies," *Proc. IEPS,* 1991, p. 63.

150. Pierson, D., Drobac, S., Parry, D., "An 80 Mhz MIPS R6000 CPU Using Multichip Module Technology," *Proc. NEPCON West,* 1993, p. 954.

151. Fillion, R.A., Wojnarowski, R.J., Bicknell, W.H., Daum, W., Forman, G., "Non Digital Extensions of Embedded Chip MCM Technology," *Proc. Int. Conf. MCMs,* Denver, 1994, p. 464.

152. Walter, D., Roszel, L., "MCM-L/O—Thin Film Interconnect on a Printed Wiring Board (Laminate) Substrate," *Proc. 7th Int. SAMPE Electronics Conf.,* 1994, p. 99.

153. Fillion, R., et al., "Development of a Plastic Encapsulated Power Multichip Technology for High Volume, Low Cost Commercial Electronics," *Proc. ISHM,* 1994, p. 84.

154. Gorczyca, T.B., Cole, h.s., Gorowitz, B., "Advanced Packaging Applications for the GE-HDI Multilamination Process," *Proc. IEPS,* 1993, p. 633.

155. Tanaka, J., Tanahashi, S., Terasawa, M., Sigliano, R., "A Multichip Module-PI on a Mullite Ceramic," *Proc. IEPS,* 1990, p. 421.

156. Ohnuli, Y, Ori, T., Yoshihara, K., Nonaka, Y., Senda, M., Shibuya, H., Takahashi, Y., "Development of Cu/PI Thin Film Multilayer on LTCC Substrate," *Proc. IMC,* Omiya, 1994, p. 215.

157. Strandberg, J., Theide, H., Karlsson, A., Weiland, A., "High Reliability 4 Metal Layer MCM-D Structure with BCB as Dielectric," *Proc. Int. MCM Conf.,* Denver, 1995, p. 223.

158. Droguet, J.P, "Apilot Line for MCM Substrates," *Proc. 8th ISHM Europe,* Rotterdam, 1991, p. 280.

159. Droguet, J.P, "An MCM-D Foundry From Copper/PI TFML Bare Substrates to MCMs," *Proc. Int Conf. MCMs,* Denver, 1994, p. 384.

160. Kikuchi, S., Yamamoto, H., Seyama, K, Hirano, M., Moriizumi, K., "High Performance MCM-D Technology," *Proc. IEEE MCM Conf.,* Santa Cruz, 1994, p. 9.

161. Nii, M., Omori, J., Nakazawa, T., Sasaki, T., Kobayashi, J., Kimijima, S., Fukuoka, Y., "Evaluation of Cu/PI Thin Film Substrate for MCMs," *Proc. IMC,* Yokohama, 1992, p. 259.

162. Okada, T., et al., "RISC Based MCMs for Workstations—Performance Evaluation Among Types of MCMs," *Proc. Int. Conf MCMs,* Denver, 1993, p. 230.

163. Chambers, S.A., Chakravorty, K.K., "Oxidation at the PI/Cc Interface," *J. Vacuum Science Tech.,* vol. A6, 1988, p. 3008.

164. Burrell, M.C., Fontana, J., Chera, J.J., "Study of the Enhanced Oxidative Degradation of Polymer Films at Polymer/Copper Oxide Interfaces," *J. Vacuum Science Tech.,* vol. A6, 1988, p. 2893.

165. Shih, D.Y., Paraszczak, J., Klymko, N., Flitch, R., Nunes, S., Lewis, J., Yang, C., Cataldo, J., McGouey, R., Graham, W., Sarrino, R., Galligan, E., "A Study of the Chemical and Physical Interaction between Copper and Polyimide," *J. Vacum Science Tech.,* vol. A7, 1989, 1402.

166. Cech, J.M., Burnett, F., Chien, C.P., "Reliability of Passivated Copper Multichip Module Structures Embedded in PI," *IEEE Trans CHMT,* vol. 16, 1993, p. 752.

167. Rao, S., "TEM Examination of MCM-D Substrate with Copper Conductors," *Proc. Int. Conf. MCMs,* Denver, 1992, p. 286.

168. Adema, G.M., Hwang, L.T., Rinne, G.A., Turlik, I., "Passivation Schemes for Cu/PI Thin Film Interconnections Used in MCMs," *Proc. ECTC,* 1992, p. 776.

169. Manial, T., Heistand, R.H., Garrou, P., "BCB for Advanced Packaging Applications," *Proc. Int. SMT Conf.,* San Jose, 1992, p. 39.

170. Teoh, H., Shen, N.T., Yang, J., "Reliability Performance of an R3000 Based MCM For Desktop Workstations," *Proc. IEPS,* 1993, p. 545.

171. Okajima, K., "Reliability Evaluation of Si-on-Si multichip modules," *Proc. Int. Microelect. Conf.,* Yokohama, 1992, p. 533.
172. Samith, E., "Reliability Assessment and Growth of a MCM High Density Interconnect Overlay Structure," *Proc. ISHM,* 1994, p. 372.
173. Akihiro, D., Toshihiko, W., Hideki, N., "Packaging Technology for the NEC SX-3/SX-X Supercomputer," *Proc. ECTC,* 1990,p. 525.
174. Tummala, R., Ahmed S., "Overview of Packaging Technology for the IBM Enterprise System/9000 Based on the Glass-Ceramic Copper/Thin Film Thermal Conduction Module," *IEEE Trans CHMT,* vol. 15, 1992, p. 426.
175. Inoue, T., Seino, K., Tamura, K., Hasegawa, S., Kimbara, K., "Photosensitive PI Application for Multichip Packages," *J. Photopolymer Science & Tech.,* vol. 5., (1992), p. 385.
176. Yamada, M., Nishiyama, M., Tokaichi, T., Okano, M., "Packaging Technology for the NEC ACOS System 3900," *Proc. ECTC,* 1992, p. 745.
177. Niwa, K., Fujitsu, private communication.
178. Tuckerman, D.B., Benson, D., Moore, H., Horner, J., Gibbons, J., "A High Performance Second Generation Sparc MCM," *Proc. Int. Conf. MCMs,* Denver, 1994, p. 314.
179. Griswald, B., Deshpande, U., "Building High Volume MCM-D for the PC Industry," *Proc. IEPS,* 1993, p. 628.
180. Private communication, Howard Green, MMS, Cupertino CA.
181. Denman, J., "nChips Silicon Circuit Board Technology," *Proc. NEPCON West,* 1994, p. 2039.
182. Brown, R., Boone, R., "A Comparison of MCM Packaging Technologies for a 32 Bit High Performance Processor," *Proc. ISHM,* 1993, p. 562.
183. Arndt, L., "A Large Area, 3-D Stackable, MCM-D on C Signal Processor," *Proc. Int. Conf. MCMs,* Denver, 1994, p. 603.
184. Hagge, J., "State-of-the-Art Multichip Modules for Avionics," *IEEE Trans. CHMT,* vol. 15, 1992, p. 29.
185. Michaud, M., Rames, J.C., "Development of a DSP Function Using MCM Technology," Proc. Int. Conf. MCMs, Denver,1994, p.378.
186. Terrill, R.E., "Aladdin: Packaging Lessons Learned," *Int. MCM Conf.,* Denver, 1995, p. 7.
187. Carlson, Jan, "486 Computer Module System," *Int. MCM Conf.,* Denver, 1995, p. 18.
188. Jacobsen, W.F., Jensen, R.J., Loy, J.M., Speerschneider, C.J., "Application of MCM-C and MCM-D for Spaceborne Data Processors," *Proc. Int. Conf. MCMs,* Denver, 1993, p. 525.
189. Feurer, E., Oppermann, M., Holl, B., "M4CMs for Microwave Applications," *Proc. ISHM,* 1994, p. 49.
190. Ohsaki, T., "Electronic Packaging in the 1990's—A Perspective From Asia," *IEEE Trans CHMT,* vol. 14, 1991, p. 254.
191. Kishimoto, T., Yusada, K., Oka, H., Sasaki, S., Kaneko, Y., Kawauti, M., Harada, A., Sugiura, N., Nakano, K., "Small Planar Packaging System Combined with Card on Board Packaging for High Speed, High Density Switching Systems," *Proc. ISHM,* 1994, p. 253.
192. Noren, J., Brofman, P., "Interconnection Considerations for a Hybrid MCM," *Proc. Int. Conf. MCMs,* Denver, 1994, p. 299.
193. Doi, Y., Yamada, H., Sasaki, S., Kishimoto, T., Tomimuro, H., "An ATM Switch Hardware Technology Using Multichip Packaging," *IEEE Trans CHMT,* vol. 16, 1993, p. 60.
194. Goldman, S., "AT&T's MCM-D Process, Design and Manufacturing," *Int. Conf. MCMs,* Denver, 1993, p. 435.
195. Frye, R.C., Tai, K.L., Lau, M.Y., Lin, A.W., "Silicon-on-Silicon MCMs with Integrated Passive Components," *Proc. IEEE MCM Conf.,* Santa Cruz, 1992, p. 155.
196. Frye, R.C., Shah, A.V., "Targeting Low Cost, High Volume MCM Applications," *Proc. Int MCM Conf.,* Denver, 1993, p. 12.
197. Mok, S., "Volume Implementation of a MCM-D Based Cache SRAM Product for Workstation and PC Application," *Proc. Int. Conf. MCMs,* 1994, p. 320.
198. McHerron, D., et al., "Applications of PMDA-ODA PI in MCM-D," *Proc. Int. MCM Conf.,* Denver, 1995, p. 410.
199. Tessier, T., Myszka, E., "Approaches to Cost Reducing MCM-D Substrates," *Proc. ECTC,* 1993, p. 570.

200. Strandjord, A., Ida, Y., Rogers, B., Garrou, P., Cummings, S., Kisting, S., "Photo BCB, MCM-D Processing and Area Array Assembly," *Int. Journal Microcircuits and Electronic Packaging*, vol. 19, 1996, p. 260.

201. Czornyj, G., Asano, M., Beliview, R., Garrou, P., Hiramoto, H., Akida, A., Kruez, J., Rhode, O., chap. 11, "Polymers in Packaging," in *Microelectronics Packaging Handbook,* R. Tummala, E. Rymaszewski, and A. Klopfenstein, eds., Chapman and Hall, New York, 1997.

202. Skinner, M., Castillo, D., Cummings, S, Garrou, P., Chazen, D., Liu, K., Reinschmide, R., Westbrook, S., Ho, C., Rogers, B., "Twinstar™ Dual-Pentium Processor Module," *Int. Journal Microcircuits and Electronic Packaging*, vol. 19, 1996, p. 358.

CHAPTER 4

MCM-L MATERIALS, PROCESSES, AND APPLICATIONS

Theodore G. Tessier
Motorola Inc.
Dr. Phillip E. Garrou
Dow Chemical Company
George Messner
(deceased)

4.1 INTRODUCTION

The third broad category of MCMs has been dubbed the MCM-L, where *L* represents the laminate substrate upon which the interconnect is fabricated. As explained in Chap. 1, the major difference between an MCM-L and a traditional PWB being that the chips are directly attached to the laminate board and are not prepackaged, and the laminate is of much higher density. Direct chip attach to PWB has been around for decades, especially in the Far East, as *chip-on-board* (COB). Such structures were typically low-cost, low-density subsystems used in consumer products such as hand-held electronic games and inexpensive watches. MCM-L as envisioned in the mid to late 1980s is a low-cost, high-density interconnect solution for direct chip attach, as will be discussed further in Sec. 4.1.1.

This chapter does not seek to be a primer on PWB fabrication, information about which is available elsewhere,[1] but rather discusses the wide range of high-density MCM-L technologies that have been developed throughout the industry over the past decade. We will, as well, discuss some of the applications where this form of high-density interconnection technology has been successfully used. There is some intentional overlap with Chap. 5, where sequential dielectric on laminate technologies and materials are described as part of the large area processing discussion.

4.1.1 Definition of MCM-Ls

There are varying opinions as to what substrate types should be included within the MCM-L category. Most agree that the interconnection of bare chips on a high-density laminate substrate using traditional PWB processes and materials should be classified as an MCM-L. However, even such simple classifications are gradually collapsing as the enabling aspects of MCM-D and MCM-C technologies are being added to the technologies available to fabricate laminate based structures. The best example of this has been the emergence of the *large area processing* based MCM-D/L subclassification (see Sec. 4.2.3 and Chap. 5). This variation involves the application of one or more sequentially built thin film layers, generally fabricated using an inherently photodefinable dielectric material deposited as a liquid onto a laminate-based substrate core. For the most part, these photovia PWB technologies have significant similarities in process flow to the MCM-D technologies described in detail in Chaps. 3 and 5.

Another approach is to include any type of PWB substrate that interconnects unpackaged, bare chips, regardless of the number of chips or the substrate density, i.e., to include all chip on board (COB) applications. Most practitioners feel that it is more reasonable to establish some arbitrary demarcation line between "normal" COBs and "true" MCM-Ls. As the use of bare die assemblies becomes increasingly prevalent in the industry (including widespread usage of these technologies in high-volume consumer products such as portable communication and computing products including PCMCIA cards), any view of a clean MCM-L definition breaks down. Perhaps the most useful characteristic to use in assessing the relative potential of a given MCM-L technology is interconnection density.

Early on, a packaging efficiency was defined as a metric to evaluate the density capability of various packaging technologies.[2]

$$\text{Packaging efficiency (\%)} = \frac{\text{Silicon chip area}}{\text{Package area}}$$

In other words, this measures how much of the surface of the board can be used for chip mounting pads versus how much has to be avoided because of the interconnect traces and holes/pads. The result of this packaging density analysis on various packaging technologies is given in Fig. 4.1.

In Fig 4.2, one sees the impact of feature reduction on board size. A fourfold reduction in board features will have a 16× effect on board size reduction. As we shall see later, the key to achieving this overall size reduction is the reduction in hole and pad size—a much more difficult task than reduction in size the size of the traces. Figure 4.3 shows that reduction in the drilled hole size from 30 to 13 mils allows double the interconnect density to be routed between the pads for wide (12 mil) lines, but the pads become the limiting density issue at smaller linewidths.

4.1.2 Advantages and Disadvantages of MCM-Ls

The economic advantage that MCM-Ls have over other MCM technologies is based on the ability to fabricate these circuits on large panels with a multiplicity of identical patterns. This capability not only reduces the manufacturing cost of individual units but provides an important advantage when rapid ramp-up to high-volume production is required. Quick response to volume orders is critical if any MCM packaging solution is to become a viable option in the explosive consumer electronics

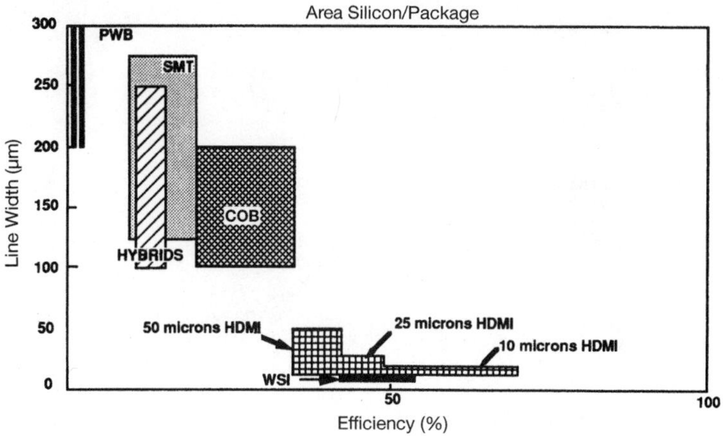

FIGURE 4.1 Packaging efficiency plot for various packaging technologies.

PWB parameter	Easy to achieve	Desired value
Component pitch (mil)	50	12.5
Line width (mil)	8	2
Pad diameter (mil)	40	10
Hole diameter (mil)	20	5
Hole density (per sq. in)	35	560
Trace density (in/in sq.)	20	80

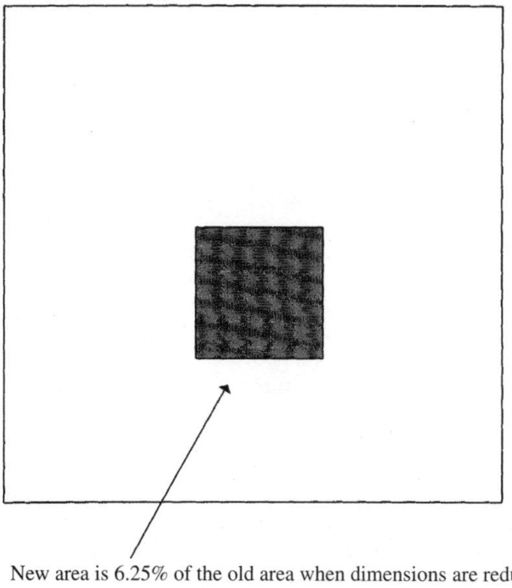

FIGURE 4.2 Impact of PWB geometries on board size.

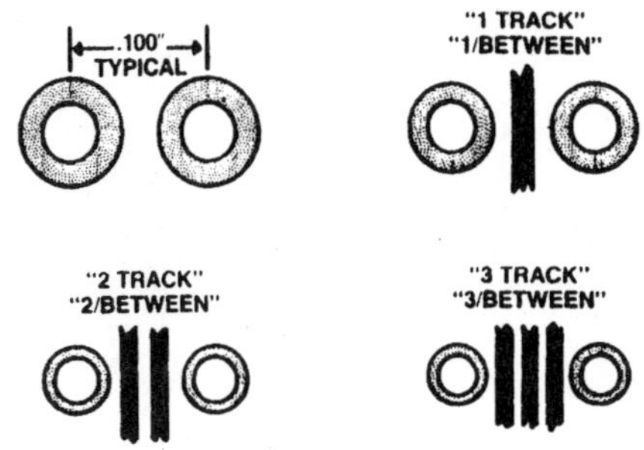

"Tracks between" technologies for holes on 0.100 centers.

Line/space width, inches	Hole dia.: 0.030 inch PTH Hole c-c: 0.100 in	0.013 inch PTH (SM) 0.100 inch	0.013 inch PTH (SM) 0.050 inch
0.012	10	20	
0.010	10	30	
0.008	20	40	
0.007	20	40	20
0.006	30	50	40
0.005	40	60	40
0.004	50	70	60

FIGURE 4.3 Board density inhibited by holes and pads.

segment of the industry, which includes workstations and portable communication products. The sheer magnitude of the existing PWB infrastructure provides a number of key advantages to laminate-based high-density interconnection technologies that are not generally available to the other MCM formats. MCM-Ls can be batch produced in large quantities and at high yields using a "tried and true" technology base. In the case of modest-density MCM-Ls, sourcing from several worldwide sources is possible. In the case of the higher-density MCM-Ls, there is a considerable narrowing of supplier options but still much more available than with either of the other two flavors of multichip modules. Table 4.1 highlights the level of activity underway in the industry to develop higher density interconnection technologies.[3]

There is a general trend toward extending the use of proven materials and processes to ever higher densities wherever possible and a general reluctance to implement newer, unproven materials. The vast PWB infrastructure assures relative ease of prototyping especially for lower complexity MCMs while having the built in capacity required to satisfy high-volume demand when required. By far the greatest bottle neck to achieving significantly higher-density MCM-Ls is the via/though hole and associated capture pads required for z-axis connection between layers. For the most part, technological extensions of existing print and etch conductor generation technologies are more than adequate for satisfying the bulk of the high-volume MCM-L substrate needs.

MCM-L solutions are usually more limited in interconnect density relative to advanced MCM-C and MCM-D technologies. Interconnection densities for these MCM options are compared in Table 5.1, in Chap. 5. PWB laminates are poor heat conductors, making heat extraction from small, powerful chips difficult. Heavy copper planes and solid thermal vias can be used to minimize this issue. In some cases, embedded metal planes have served as effective heat spreaders. Also, copper slugs and cut-outs have been used in MCM-Ls for direct heat transfer from the chips to the heat sinks. Although such an approach can substantially degrade the interconnection density of these substrates, sufficient channel capacity still remains to satisfy most MCM-L requirements. These options are shown in Fig. 4.4.

Table 4.1 A Synopsis of the Various High-Density PWB Technologies under Development within the Electronics Industry as of 1995.[3]

Company	Via technology	Metallization process*
Cannon	photovia	PPE
CMK/Toa Gose	photovia	PPE
Continental Circuits	photovia	PPE
HADCO	photovia	PPE
HP-Dycostrate	plasma	PPE
Hitachi	plated via post	PPE
Ibiden	photovia	FBE
IBM Yasu (SLC)	photovia	PPE
IBM Endicott	photovia	FBE
IBM Singelfingen (Ger)	plasma	PPE
Lares Cozzi	laser ablate	PPE
Matsushita	laser ablate	E
Mommers	laser ablate	PPE
NEC	photovia	PPE
Paralex/Allied Signal	wet or plasma	PPE
Shendahl	laser ablate/punch	PPE
3M	laser ablate or wet etch	PPE
Zycon	laser or photovia	PPE

* PPE = panel plate and etch, FBE = full-build electroless, E = etch

As flip chip on board technologies and flip chip on package technologies become more prevalent, an alternative thermal solution presents itself where the bulk of the heat can be extracted from the backside of the die.[4] This, in turn, is expected to further diminish the thermal dissipation requirements of the interconnection medium.

Another issue that must be accommodated in using laminate-based MCM substrates is the mismatch between the inherent high coefficient of thermal expansion (CTE) of the substrates (about 15 to 17 ppm) and the low CTE of silicon chips (3.2 ppm). In the early days of MCM development, a widely held notion was that a substrate with a CTE matched to silicon would ultimately be required to achieve acceptable levels of reliability for flip chip assemblies. This largely resulted in a general bias toward MCM technologies based on relatively expensive, low-CTE substrate technologies including silicon, aluminum nitride, and glass ceramic. More recently, it was discovered that the use of a suitable underfill material placed between the die and the substrate could compensate for CTE mismatches, providing adequate reliability for most high-volume MCM applications.[5] This discovery has been the single most important factor in the evolution of MCM technologies to date, as it served to eliminate one of the major reservations to MCM-L usage.

4.2 BASIC MCM-L CONSTRUCTIONS

As indicated previously, this chapter is not intended to be a primer on PWB fabrication, but some exposure to traditional PWB processing is necessary.

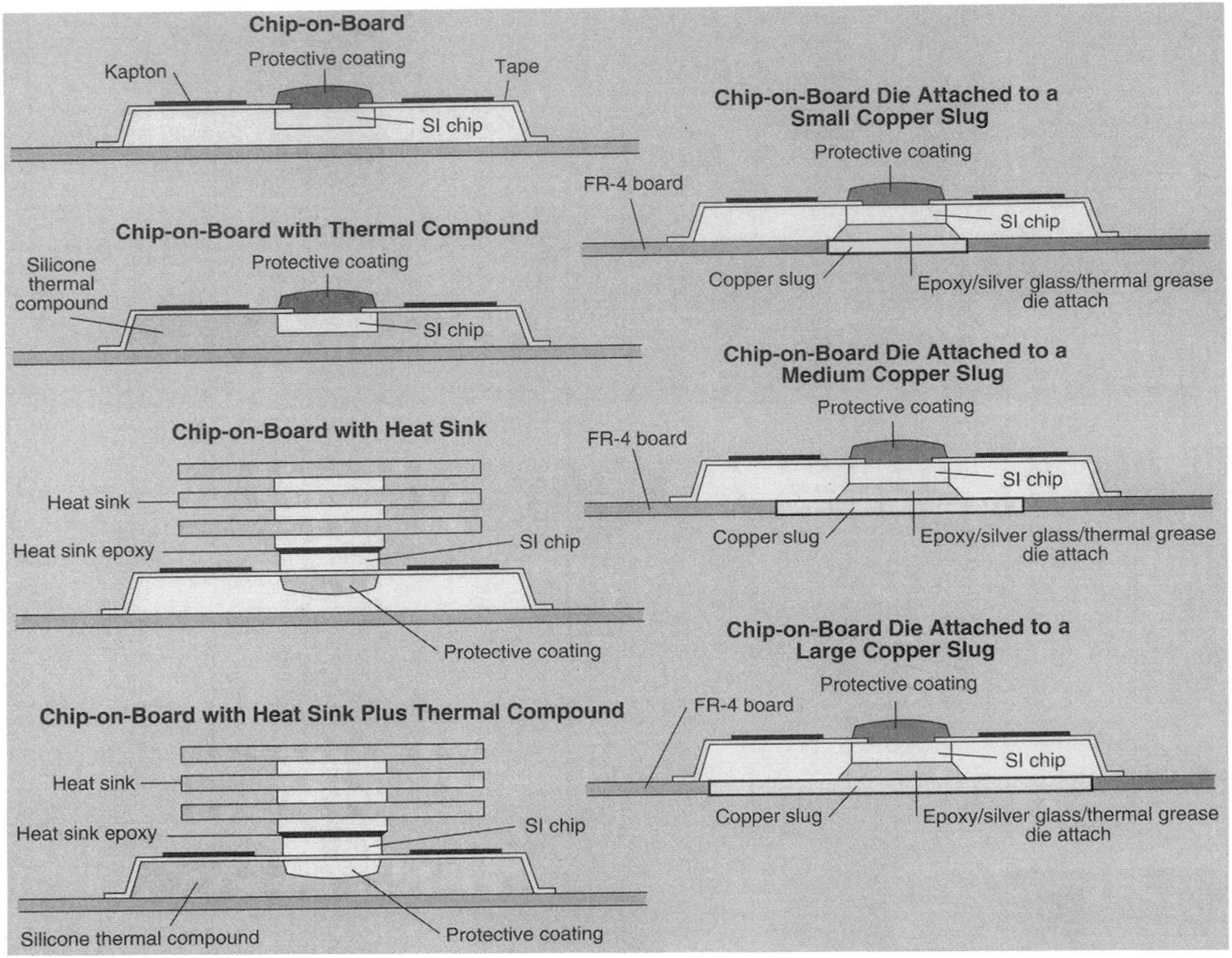

FIGURE 4.4 Thermal mounting methods for chip on board applications.

A traditional PWB process flow is outlined in Fig. 4.5. A schematic of the lamination and drilling processes is shown in Fig. 4.6.

Until recently, MCM-L development has involved evolutionary technological advances to shrink the dimensions of interconnect lines (traces) and vias (drilled holes). More recently, a heightened need for higher-density PWB technologies to satisfy the substrate needs of single-chip semiconductor packages—most notably fine pitch ball grid arrays (BGAs)—has developed. This demand, in turn, has fueled the development of alternative blind via and through hole generation technologies as replacements to conventional mechanical drilling as well as enhanced photolithographic and metallization definition processes. The examples of MCM-L technologies selected for discussion here indi-

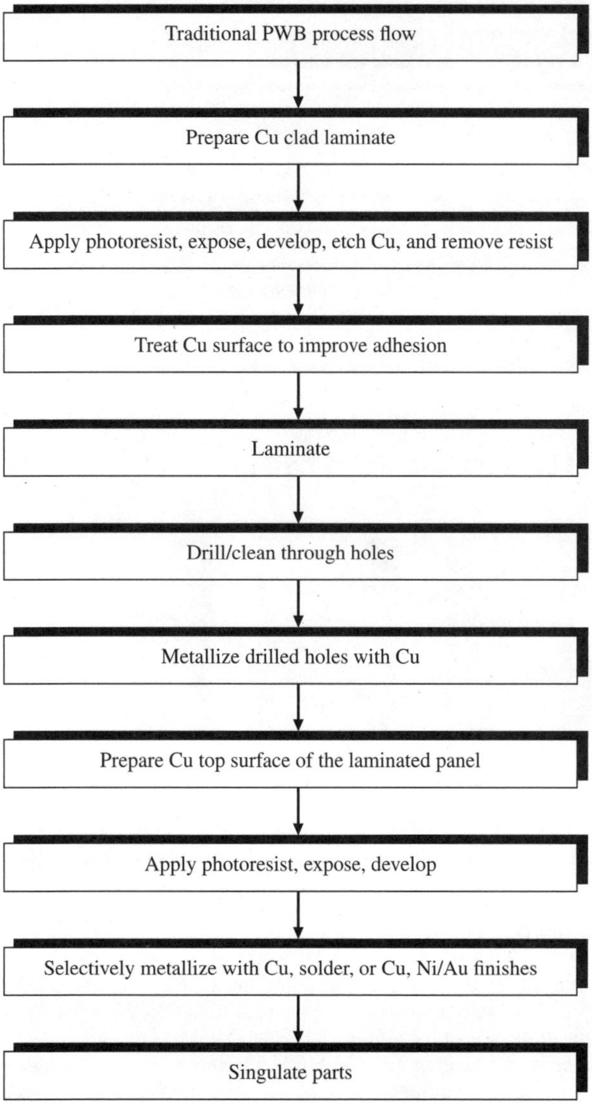

FIGURE 4.5 Process sequence for photo PI MCM fabrication.[33]

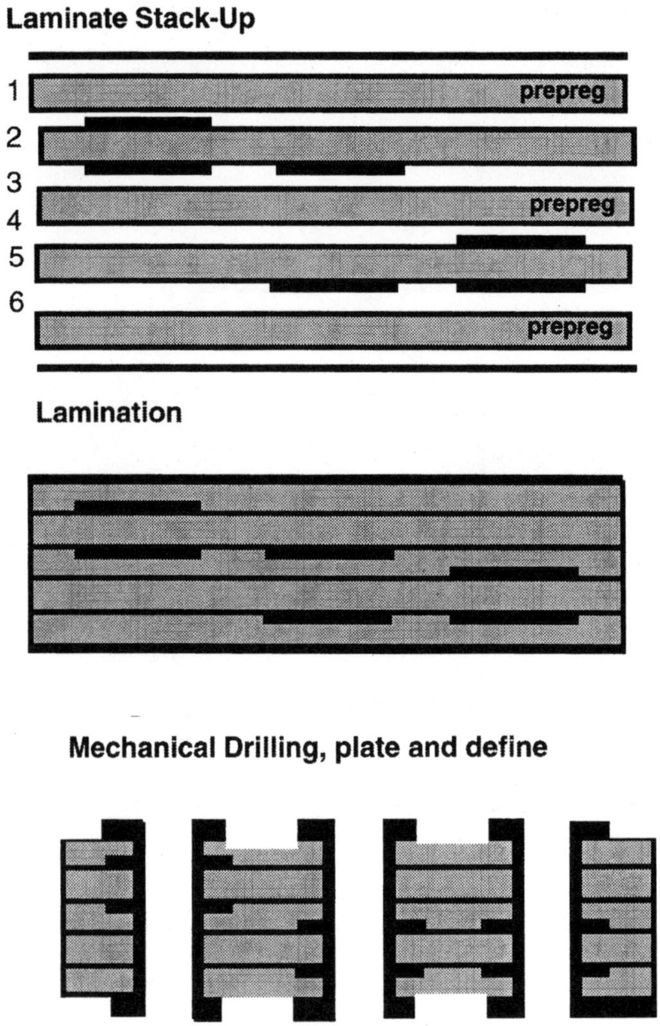

FIGURE 4.6 Lamination and drilling process.

cate that application of printed wiring board technology, whether based on extensions of conventional processes or enhanced with new processing alternatives, can produce dense multichip modules.

4.2.1 MCM-Ls Using Conventional PWB Constructions

From a cost perspective, wherever possible, it is desirable to use conventional PWB technologies for MCM-L fabrication. This is becoming more and more difficult as the need for multichip modules with higher interconnection density continues. Typical MCM-L design rules are summarized in Table 4.2 and compared to the leading edge of MCM-L requirements. Although there is a dramatic increase in the number of I/O to be interconnected on some die for high-performance applications, an equally daunting challenge is encountered in accommodating the fanning out of I/O from small die with much fewer I/O.

Table 4.2 Comparison of Leading Edge Requirements to High-Volume Laminate Capability

	High volume	Leading edge requirements
Average number of layers	4–12	6–8
Line width and spacing (μm)	75–125	50–75
Via dimensions (μm)	200–300	100–150
Via capture pads (μm)	300–450	150–250
Surface pad pitches (μm)	200–450	100–150
Board finish	Ni/Au	Ni/Au, solder
Board thickness (6 layers, μm)	500	300

The need for higher-density substrate technologies is growing rapidly. Currently, the size of through holes and blind vias and their corresponding capture pads are the single most challenging obstacle to achieving dramatic gains in interconnection density. This has focused the PWB industry on the development of alternative blind and buried via board constructions. As MCM technologies are being considered for high-volume consumer product applications, a focus on containing the cost of high-density MCM-Ls is critical.

The use of mechanically drilled blind and buried via board constructions, though costly, has emerged as a means of significantly extending the interconnect density of conventional PWB fabrication methods. Mechanically drilled through holes and blind vias remain the most costly process steps in the fabrication of PWBs today, with 340 μm (13.5 mil) hole diameters being most common. This is especially so for blind and buried via constructions, because either costly multiple drilling and lamination steps or slower throughput controlled depth drilling processes must be used. Schematic representations of a five-layer FR-4 board fabricated using these two processes are compared in Fig. 4.7.

Of the two techniques used, sequential lamination continues to be the most common approach for fabricating these blind and buried via constructions, because it is most compatible with the existing PWB infrastructure. In this process flow, a number of conventional drilling steps are performed on partial layer stackups to generate the required z-axis connections between specific layers. In the particular example shown in Fig. 4.7*a*, two double-sided copper clad laminate sheets are drilled, plated, and patterned in parallel to form layer pairs 1–2 and 4–5. A prepreg layer and additional copper foil layer are laminated to one side of layer pair 4–5. Through holes are then drilled through this partial stack, plated, and patterned to form the necessary connections between layers 5 and 3. A prepreg layer is then used to laminate layer pair 1–2 to substack 3–5. An additional drilling, copper plating,

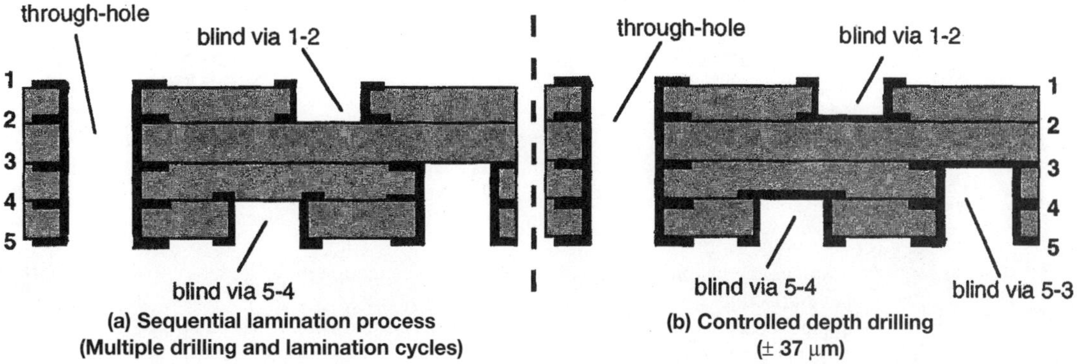

FIGURE 4.7 Blind and buried PWB constructions based on extensions of conventional PWB processes.[6]

and patterning step is then required to generate through holes between layers 1 and 5. At the end of the day, this sequential build strategy results in a total of four separate drilling cycles and two lamination cycles. This significant number of additional processing steps account for the higher average cost per square inch for comparable five-layer boards of approximately \$1.00, compared to approximately \$0.40 to \$0.50 for a conventional board construction (without blind vias).

The alternative approach that has been used by other PWB suppliers (Fig. 4.7*b*) in fabricating these blind and buried via boards has involved the use of controlled depth drilling. This construction strategy involves printing and etching of inner layers and a single lamination step. In a single drilling step, controlled depth holes are drilled down so as to contact buried metal pads within the inner layers of the multilayer stack from one or the other side of the board. The depth control of the drilling process provided by mechanical means is limited to ±35 μm. This practical limit to z-axis control limits the usefulness of this type of approach as laminate thicknesses decrease below 50 μm. The major advantage of this approach is that only a single lamination and plating process is required, thereby significantly shortening the process flow through the circuit board shop. On the other hand, implicit in control depth drilling is a single panel stack height, thereby imposing significant additional demand for drilling capacity on more sophisticated and costly single-spindle drilling machines. For a recent typical leading edge four-layer, 2.54 × 2.54 cm ball grid array, multichip module, a total drilling time of 450 minutes per 18 × 24 in panel for control depth drilling from one or the other side was estimated. *This involves drilling in excess of 85,000 holes per panel!*

Regardless of which of the two traditional PWB approaches is selected for fabricating these blind and buried via boards, drilling accounts for more than 50 percent of MCM-L fabrication costs today. There are also some noteworthy and alarming trends in the number of the blind vias required for leading-edge MCM-L and high-density substrates for chip-scale package attach in portable communication products. As a general rule, the number of blind vias in boards in these applications is doubling every three years. This trend is shown graphically in Fig. 4.8. Due to the onset of chip-scale packaging and flip chip packaging technologies, this trend if expected to continue, if not worsen, beyond the year 2000, resulting in a near quadrupling in the cost of a 10 in^2, five-layer substrate with 340 μm (13.5 mil) holes/vias. In addition, as shown in Fig. 4.9, the need for smaller blind and buried via technologies will continue driving drilling costs almost exponentially from approximately \$0.99 per 1000 holes for 13.5 mil holes up to approximately \$18.00 per 1000 holes for 4 mil mechanically drilled holes—almost an 18× increase in drilling costs.

For higher-density MCM-L applications, there is a trend toward the use of 25 μm to 50 μm wide conductors, smaller grids, and pad pitches with 75 to 100 μm diameter landless vias. Examples of using this approach to collapse layers of interconnect into a high-density layer pair has been proposed.[38] The desire to accommodate greater interconnection density by designing ever finer conductors is rather misguided. It is much more productive to increase the channel capacity, and thus the layer density, by minimizing the pad sizes around vias or even totally eliminating them.

Based on these observations, it is clear that a paradigm shift in the way that higher-density substrates are manufactured will be required to maintain the cost structure of conventional PWBs. Consequently, the development of alternative mass via generation technologies has become critical. There is currently a tremendous amount of activity underway in the industry to develop alternative small via generation processes. An overview of the various options available has been described in some detail.[1] A generic test vehicle proposed by workers at Motorola Inc.[6,7] has been widely adopted by the PWB industry as a de facto standard for the assessment of high-density PWBs. This board, called MRTV-2, and its predecessor, MRTV-1, are shown in Fig. 4.10. All of the various technologies under development can be categorized into one of three main via-generation classifications, namely, photodefined vias involving (a) inherently photosensitive dielectric materials,[8] (b) plasma,[9] and (c) laser processing.

Photovia-based technologies have received the most attention in the literature to date, since IBM-Yasu first described its *surface laminar circuitry* (SLC) technology.[10]

For the most part, the bulk of the U.S. PWB industry has taken a "wait and see" attitude toward small via generation. A number of issues have limited the spread of photovia technologies from OEMs to a broader segment of the PWB industry. Undoubtedly, the biggest hurdles involved in implementing a photovia-based technology are the capital equipment costs of setting up a facility,

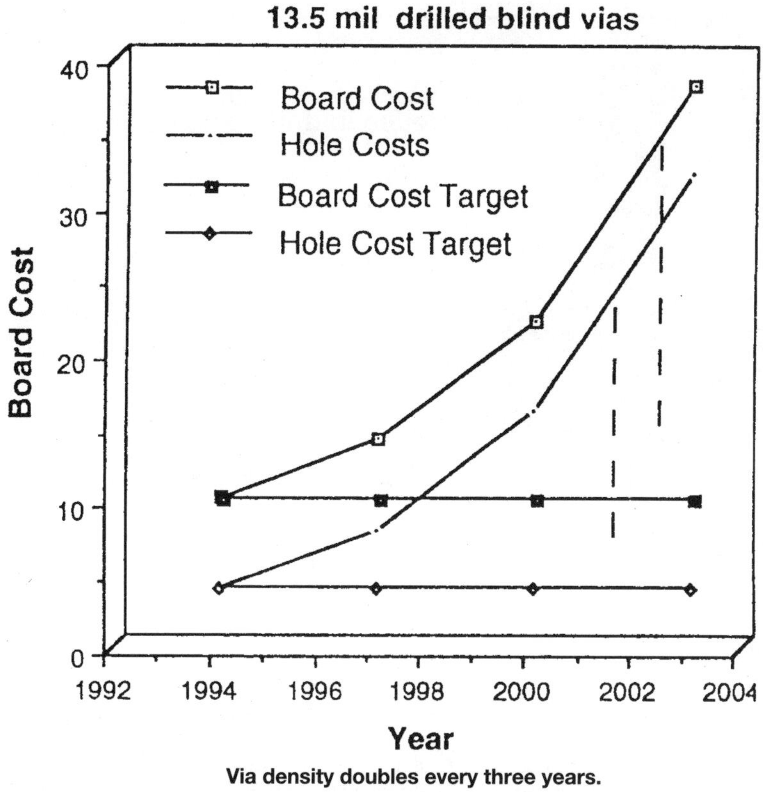

FIGURE 4.8 Trends in blind via densities and their impact on board cost.[8]

estimated to be in the range of \$5–10 million, depending on throughput needs and the technical complexity and yield risks associated with these sequential board fabrication processes. The financial tolls associated with overcoming these hurdles could lead to the demise of most of the manufacturers in the chronically low-margin PWB industry if not executed flawlessly. Laser- and plasma-based via technologies, on the other hand, are more readily implementable small-via approaches and are highly compatible with the existing PWB infrastructure at a cost comparable to the cost of a single multihead drilling machine (\$350–500 thousand).[11] Consequently, laser processing allows a more gradual and lower-risk evolutionary migration from current PWB fabrication technologies toward higher-density interconnects.

As the need for finer and finer lines and spaces continues, several approaches to fabricating higher-yielding, reliable interconnections have been developed. For the most part, the conventional PWB industry has been based on the use of copper clad laminates with copper foil thicknesses of 17 and 35 μm (0.5 to 1.0 oz). Inner layers are patterned using straightforward print and etch techniques. Copper foil is laminated to outer layers of a board stackup and laminated to the inner layers using prepreg laminate. The through holes and/or blind vias are seeded using either electroless copper or a carbon or graphitic based activating solution. This provides electrical continuity allowing plating up of copper in the z-axis connections.

As a means of achieving higher interconnect densities, a number of variations on the theme have been developed, most of which are based on semiadditive plating. In semiadditive plating, a seed layer of copper is deposited either in the form of a thin copper foil or by electroless processing. A

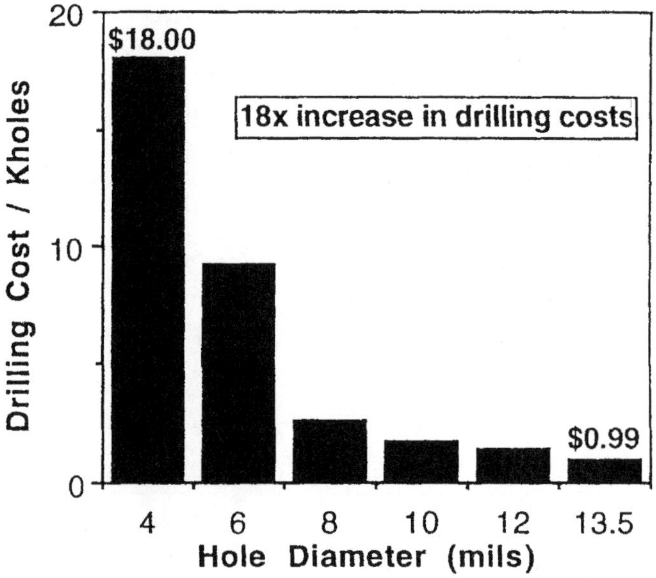

FIGURE 4.9 Exponential increase in the cost of mechanical drilling small through holes and vias.[7]

patterned photoresist layer on the copper is used to define where the conductors are to be plated up onto the surface of the panel. The photoresist is stripped in a subsequent process and the bus layer etched to allow isolation of the individual conductors in the pattern being defined. In addition to the processes mentioned above, fully additive plating processes are gradually gaining some attention, particularly for the new high-density interconnection technologies. In this case, electroless plating is used, which enables the selective deposition of copper without the need for subsequent etchback steps. Although fully additive technologies hold the potential of providing the highest interconnection density achievable, significant hurdles remain including the relatively slow deposition rate associated with electroless copper plating, the relatively poor stability of these baths relative to electrolytic plating baths, and associated operating costs. In the case of photovia-based PWB technologies, there is an added degree of complexity related to the roughening of the photo-hardened dielectric surface to allow adequate adhesion of electroless copper. This dielectric roughening process is critical to ensuring the adhesion and hence the reliability of resultant sequentially built thin film structures fabricated in this manner, and it usually involves a combination of mechanical and chemical roughening. Permanganate etchants have been commonly used for this chemical roughening processes.[12] The resultant complexity of photovia technologies has limited the interest with which the bulk of the PWB industry have embraced this approach to higher-density interconnects. For the most part, to date, only broadly diversified OEMs have risked the investment in photovia-based PWB technologies.

4.2.2 Few-Chip Package Constructions

A popular approach has been to build very small simple PWBs that connect just a few (typically two) bare die with some discretes. After component assembly, these boards are placed over the leadframe paddle or package base, resulting in incorporation of simple MCMs into DIP, PQFP, or BGA package formats.[42] Such substrates are usually manufactured using hundreds of boards on a panel or mass produced in tape form. A schematic cross section of such a four-die MCM-L packaged in a

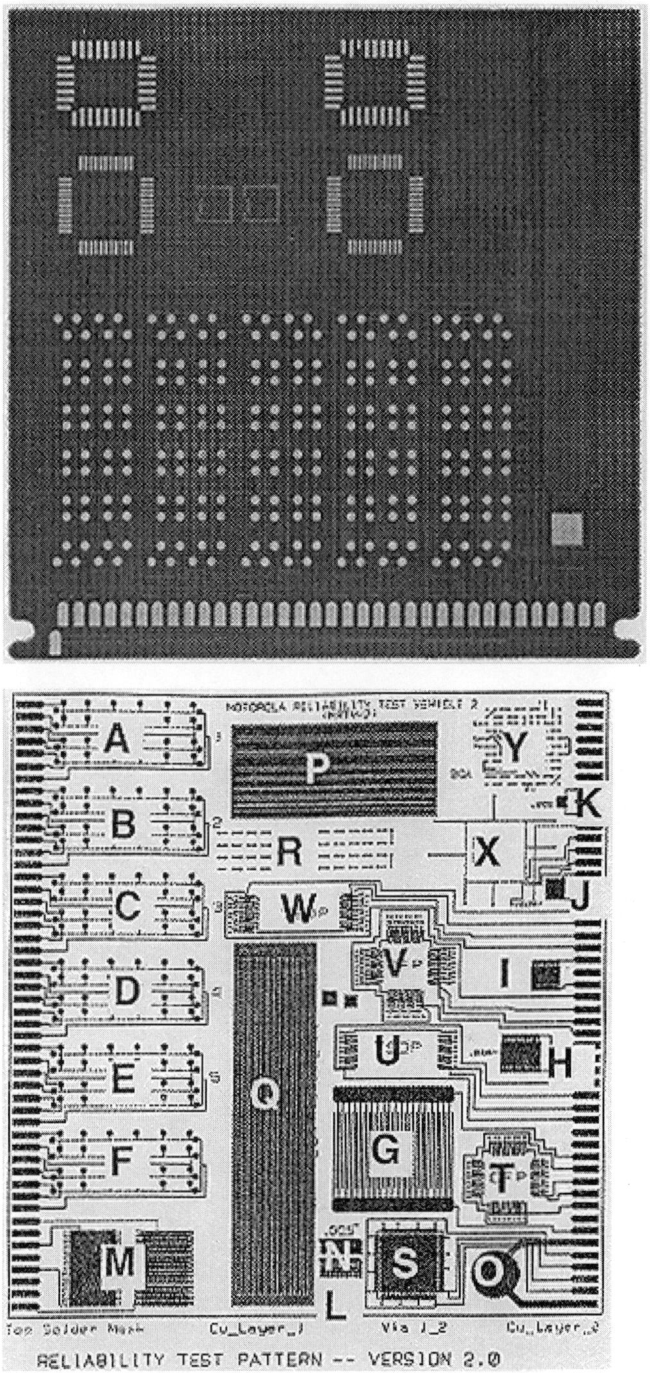

FIGURE 4.10 Motorola reliability test vehicles used for high-density PWB technology assessment.[6]

PQFP is shown in Fig. 4.11. A representation of this approach, illustrated in Ibiden's Packthol technology, is shown in Fig. 4.12. Although technologically unsophisticated relative to some of the high-performance MCM applications described recently, there are millions of these few-chip modules being manufactured in the world each year, hidden within conventional single-chip package form factors.

4.2.3 MCM-D/L: Thin Film Additive

Conventional PWB board fabrication methods cannot support the fanout requirements of the very dense area array I/O patterns encountered in fully populated ball grid array packages, chip-scale packages, and flip chip on board assemblies. The need to provide an escape route from such densely spaced terminals had led toward the development of special conductor redistribution layers placed between the chip and the conventional PCB substrate with inherent small and mass via generation capabilities. As highlighted earlier in this chapter, drilling of very small holes is expensive. This has led to interest in the use of inherently photodefinable dielectric materials capable of providing via generation *en masse*. Because of the implicit elegance of these photovia-based technologies, they have been gathering most of the early interest in the industry; however, their complexity has gated their broad implementation in the industry.

IBM Yasu—Surface Laminar Circuit. The addition of thin-film redistribution layers to an FR-4 laminate base to interconnect high I/O area array chips was first publicly disclosed by IBM-Japan,[13–16] which call it the surface laminar circuit (SLC) technology. By defining vias in photosensitive epoxy instead of drilling holes for signal line interconnection, up to 300 cm/cm^2 can be manu-

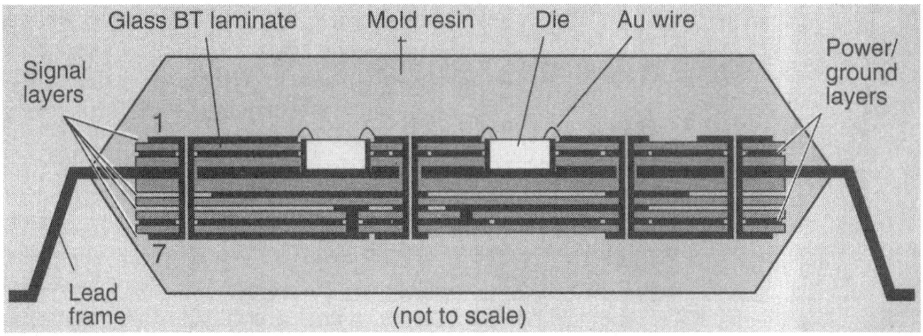

FIGURE 4.11 Few-chip package concept in a traditional PQFP form factor.

FIGURE 4.12 Mass-produced Ibiden "Packthol" substrates for few-chip applications.

factured when four layers of thin film are used.[15] Low cost is achieved by using existing materials and tooling from the printed circuit board industry. The SLC process sequence is detailed in Fig. 4.13, and a cross section of this structure is shown in Fig. 4.14.

Ordinary glass-reinforced FR-4 laminate is used as the substrate. The high-density interconnect is sequentially built up by deposition of photosensitive epoxy soldermask and plated copper. The photo epoxy is applied to the 500 × 600 mm panels by a curtain coating process. In adopting this application technique, refinements had to be made to minimize the occurrence of pinholes and other catastrophic defects for damaging the finished circuit. The panel runs under the application curtain at a speed of 3–4 m/s. Signal via holes are formed by photo exposure and development of the photo epoxy dielectric. A micrograph of typical via structures is shown in Fig. 4.15. The initial via profile obtained after development of the solder mask dielectric is in act concave. An aggressive mechanical grinding process removes approximately 50 percent of the dielectric thickness, thereby removing the upper retrograde portion of the via profile, which results in a positive slope. Final cure of the epoxy is achieved thermally.

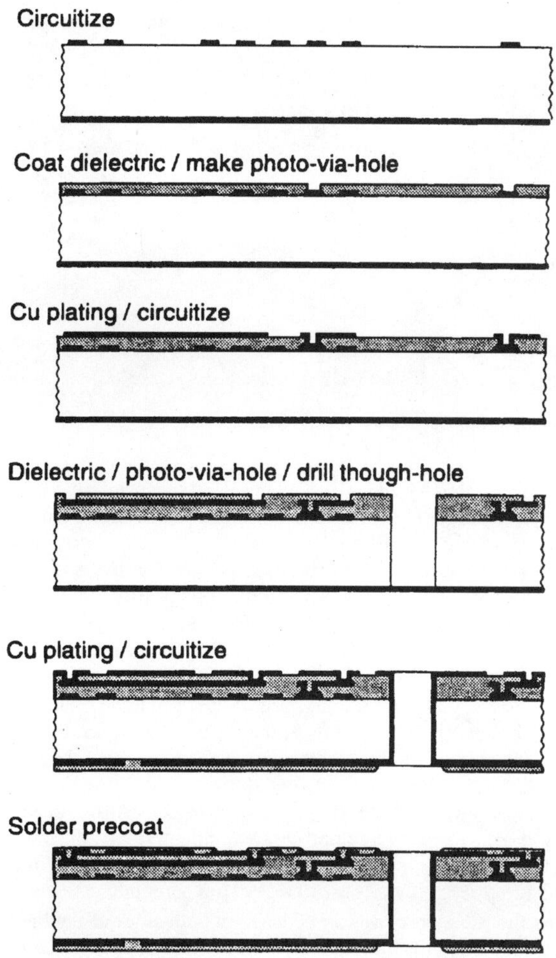

FIGURE 4.13 SLC manufacturing process steps.

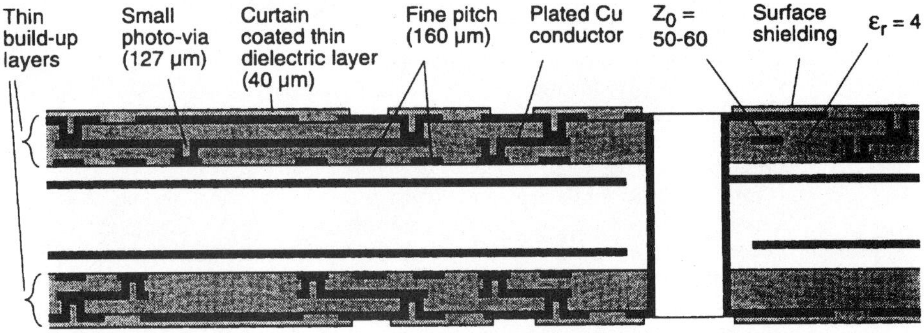

FIGURE 4.14 Schematic cross section of the IBM Yasu SLC technology.

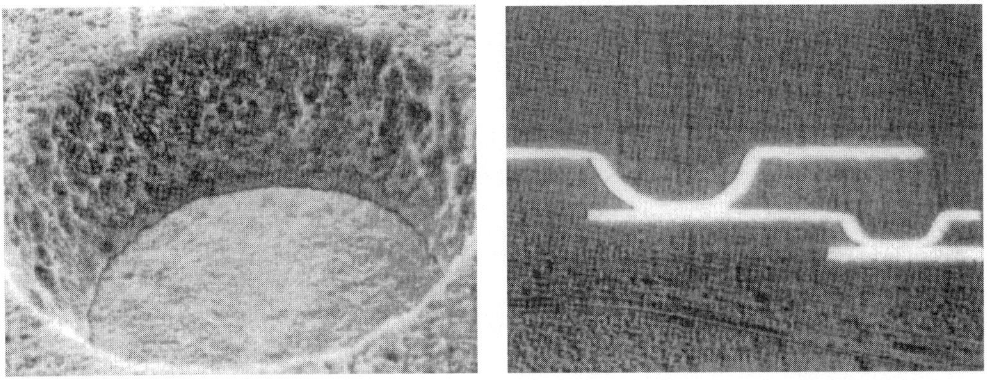

FIGURE 4.15 SEM (left) and cross section (right) of SLC via.

The conductor layer is generated by electroless plating and electroplating to obtain the necessary thickness of copper. The circuits are then defined by subtractive etching. Before the electroless plating, the surface of the epoxy must be roughened by a permanganate etch to achieve the desired copper adhesion. The chemical roughening process used is quite aggressive and provides an additional planarization function. In a typical application, several microns of fully cured dielectric are removed using an in-line process. Under- or overetching is reported to give poor adhesion.[14] Properties of the commercial epoxy soldermask (Probimer™ 52) used as dielectric are detailed in Sec. 5.4.2.

Sequential buildup of SLC thin-film layers have been described in the literature with up to six high-density layers having been applied,[16] although it is thought that the bulk of commercial product today is one to two layers. In the original versions of the SLC design rules, the high-density thin film layers were processed on the base substrates and through holes to connect the outer high-density layers to the core layers in the base substrate fabricated subsequently (see Fig. 4.13). More recently, alternative SLC constructions have been developed that allow for drilling of vias prior to thin film processing. By plugging these vias with an appropriate conductive or nonconductive material, circuitry on the overlying high-density circuits is possible. Feature dimensions in production are typically 75 to 100 μm lines; 125 μm photovias on 250 to 300 μm capture pads.

It should be noted that the SLC structure can be built on both sides of the base substrate and that, in fact, a balanced construction of thin film layers on either side of the base substrate is required to reduce substrate distortion and warpage, especially for thin base substrates such as those used in low profile packaging applications, e.g., PCMCIA cards and portable product applications.[17] Figure 4.16

FIGURE 4.16 Workstation adapter card manufactured in SLC technology.

shows a 292 × 88 mm workstation adapter card manufactured at Yasu. It is a 2-D high-speed graphics card having seen 376 (maximum) I/O bare area array chips attached. It consists of two layers of buildup on a five-layer FR-4 board. Figure 4.17 shows a 49 × 77 mm, 5250 local area network adapter PCMCIA card containing two bare chips (maximum 273 I/O). It consists of two buildup layers on a four-layer FR-4 board. The IBM SLC process has been licensed by AMP–AKZO.[18]

The key technology for implementation of such structures is the under chip epoxy "buffer" layer that absorbs the stress created by the CTE differential between epoxy board and the area array direct chip attach (DCA) solder bumps. This attach technology is discussed further in Chap. 8. As of early 1996, SLC technology was in volume manufacturing (>500 panels per day).

IBM-Endicott (FRL, SLC-E, SLC-A). A photovia technology analogous to the SLC process was developed at IBM-Endicott, called the *film redistribution layer* (FRL).[19] In this case, however, an internally developed dry film dielectric material called *additive solder mask* (ASM) has been used. In the case of the FRL process, the original intent of this technology was to fabricate the bulk of the board layers using conventional PWB technologies and only use the FRL thin film technology, as the name implies, for direct chip attach (DCA) redistribution in a single surface layer. This philosophy was somewhat different from SLCs in that this technology was intended to be used for the fabrication of multilayer structures from the start, thereby targeting significant layer reduction as a fundamental goal.

A summary of the FRL process is shown schematically in Fig. 4.18. Prior to building up the high-density layer, the plated-through holes in the base substrate are plugged with a copper-filled conductive paste. The dry film dielectric used in the FRL process is very similar to process as Probimer 52 with N-Butyrolactone used as the developer.

Current FRL design rules allow 50 μm line widths, 75 μm spaces, photovia diameter is typically 100 μm with a land diameter of 150 μm. The minimum center-to-center distance for the photovias is 238 μm, and the dielectric thickness is approximately 70 μm. In the case of the dry film dielectric, a slight retrograde via profile is obtained that does not appear to adversely impact the quality of the subsequent metal deposition processes. Typical via profiles of an FRL thin film structure are shown in Fig. 4.19.

The plated-through holes can be filed with copper paste or dielectric. To achieve 3 to 4 lb/in peel adhesion of plated copper to the photo dielectric, it was necessary to roughen the surface by "dry processing." Disadvantages to dry film dielectric are reported to be relatively poor uniformity over uneven topography, design limitation due to the available commercial film thicknesses, and higher cost.

Preliminary reliability results have been published and indicate that for 20 mm chips, flip chip attached to these substrates (–65 to 150° C, 100 cycles; –55 to 125° C, 300 cycles; 50 hr of HAST at 130° C, 80 percent RH) test conditions can be passed.[20]

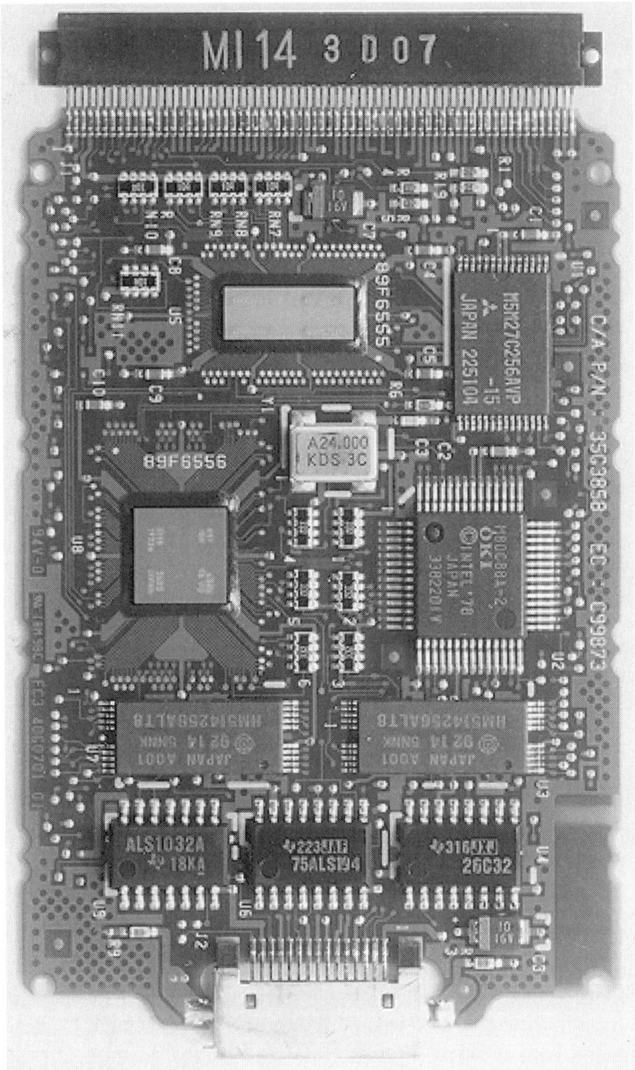

FIGURE 4.17 Workstation adapter card manufactured in SLC technology.

Ibiden Photo-AAP10. The Ibiden Photo-AAP10 process is the least sophisticated of two photovia technologies that Ibiden practices. It is based on inherently photodefinable dielectric materials and full-build electroless plating. It uses thin film polymer technology similar to SLC to produce high-density redistribution layers on top of a classic PWB structure. The full-build electroless copper plating process (Fig. 4.20) allows for fully additive processing and resultant higher density capabilities.[21] Additionally, a considerable amount of effort has been expended in developing a higher deposition rate electroless copper plating bath. Photo-AAP10 was initiated approximately five years ago to extend the capabilities of conventional PWB constructions to higher densities.

The thin film D/L process involves a two-resin photodielectric system. Resin 1 in this two-dielectric system is a two-component formulation of an epoxy resin and a thermoplastic rubber. The resin 1

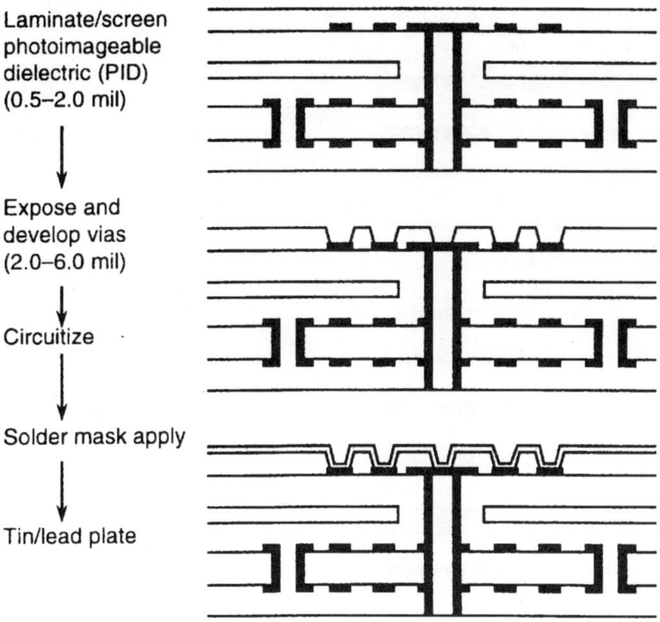

FIGURE 4.18 Process sequence for IBM FRL.[5]

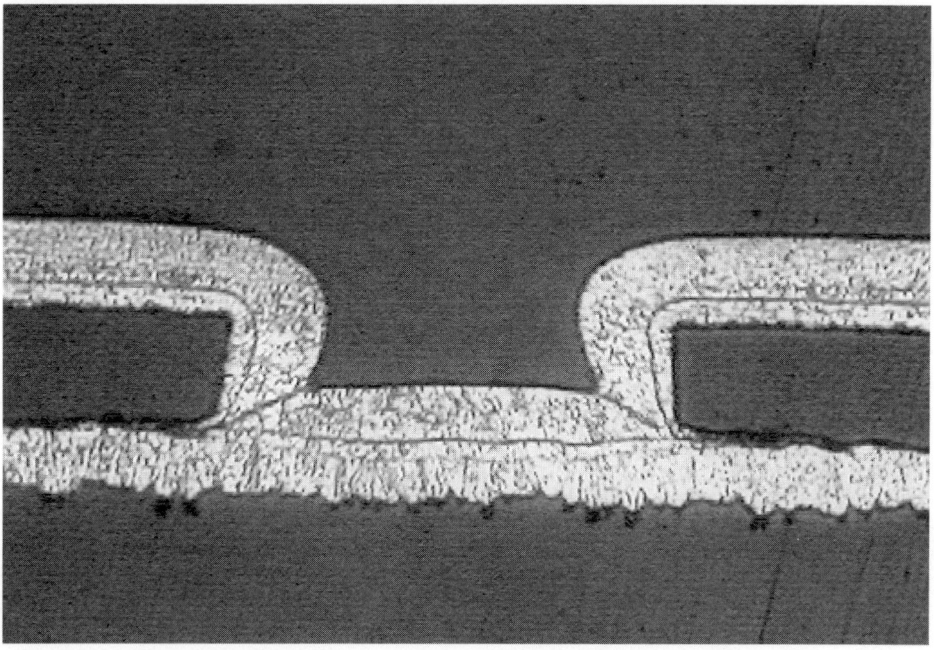

FIGURE 4.19 FRL via profile indicating slightly retrograde nature of dry film generated photovias.

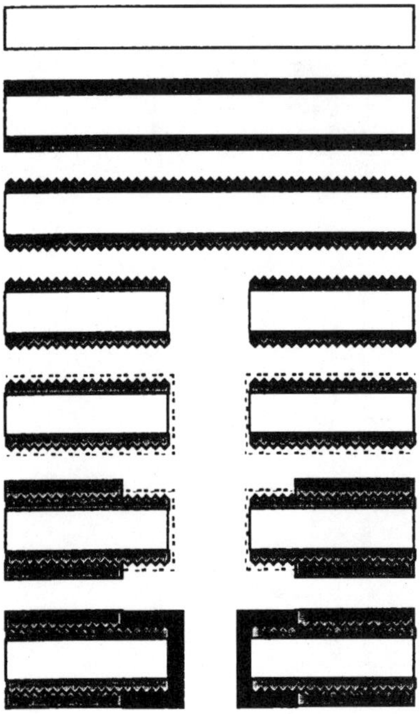

FIGURE 4.20 Ibiden AAP10 additive PWB processing.

layer is applied to the base substrate, baked, and photolithographically patterned to define vias down to metal pads on the layer below. Typical nearly vertical via profiles are reported. To enhance the adhesion of the electroless copper to the dielectric layer, a chemical roughening process based on chromic acid or permanganate wet etching is used. The second layer of photosensitive dielectric (resin 2) is applied and imaged to form the trenches into which is plated the copper conductors using a full-build electroless copper plating process. The exposed surfaces of roughened resin 1 are selectively activated to define the sites of subsequent electroless copper deposition. Additional layers of interconnect can be applied sequentially. Figure 4.21 shows typical dimensions reported for this technology.

Dielectric thickness is typically 20 to 40 μm, and 50 micron lines and 100 μm vias on 150 to 200 μm capture pads are reported as typical. Electroless gold plating is used to form the wire bondable top level metallurgy.

Ibiden IBSS Process. Another variation on their photoimageable buildup method evolved out of their semiconductor packaging fabrication facility for high-density BGA and multichip module applications. This process, generically referred to as the *build up* technology and called IBSS, is currently in the pilot plant production phase of development. In this case, a two-component, proprietary, photoimageable dielectric based on interpenetrating polymer networks (IPN) is applied by curtain coating over the base substrate. This two-layer dielectric scheme allows the selective roughening of the first dielectric layer, allowing selective seeding and subsequent deposition of electroless copper. The upper dielectric layer remains unaffected by this selective roughening process, thereby serving as a permanent resist that defines the dimensions of the conductor layers. After photo formation of vias, the surface is processed to etch out special filler particles and thus provide additional adhesion. After catalyzing the surface, the plating resist pattern is applied and vias and pattern metallized by fully additive copper deposition. Figure 4.22 provides an illustration of the IPN resin components. A cross section of the IBSS technology is shown in Fig. 4.23. Design rules for IBSS are shown in Fig. 4.24.

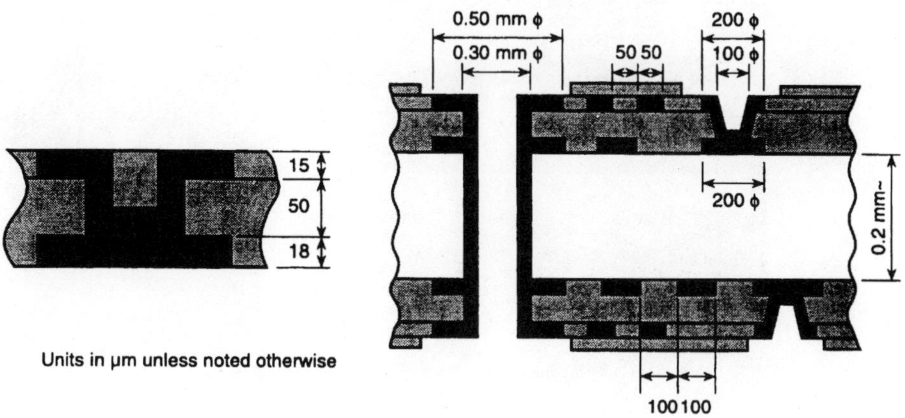

FIGURE 4.21 Typical ground rules for AAP10.

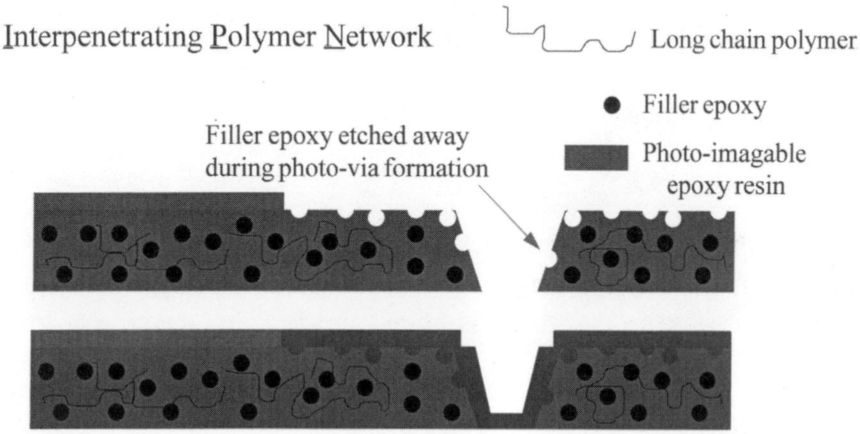

FIGURE 4.22 IPN resin components.

NEC. NEC has described a D/L additive process using Cu/photo BCB thin film and conventional glass-epoxy PWB.[44] Figure 4.25 shows a cross section of the structure. The PWB through holes were epoxy filled and polished back to allow the thin film fabrication. Figure 4.26 shows a prototype of a high-density RISC module. Table 4.3 details ground rules and reliability results.

Other Buildup Processes. A number of materials suppliers, including Dow Chemical (BCB), Amoco Chemical (preimidized Ultradel™ 5010), and Shipley (FP-91111), have announced similar process sequences that can be used to form sequential buildup layers. These are discussed in Chap. 5.

4.2.4 MCM-L Structures Using Flexible Materials

The increased use of flexible, predominantly polyimide, thin film materials for manufacture of dense patterns of MCM-Ls is largely due to the fact that these substrate materials allow the simultaneous production of a large numbers of microvias in large panel or, in some cases, even in roll-to-roll pro-

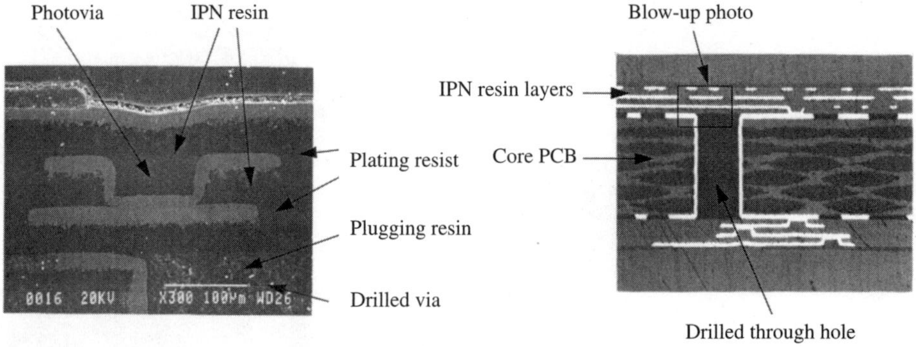

FIGURE 4.23 Cross sections of Ibiden's buildup photovia technology.

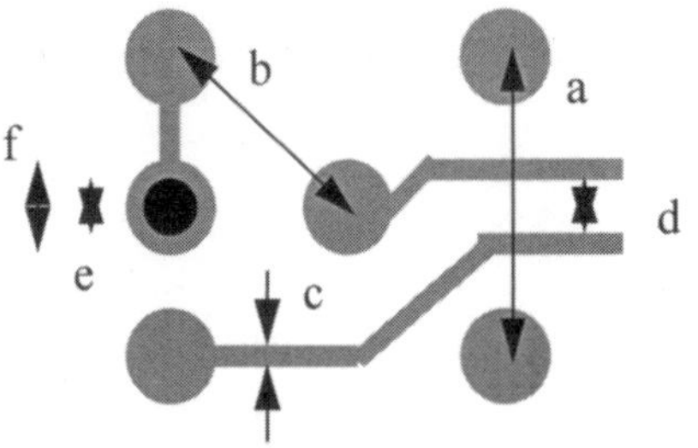

a. Pad pitch with two lines
b. Pad pitch with no lines
c. line width
d. line space
e. via diameter
f. land diameter

IBSS	a	b	c	d	e	f	IPN thickness
1H 96	425	300	50	50	100	150	35
2H 96	350	250	40	40	85	125	35
1H 97	325	225	35	35	80	125	30
2H 97	300	215	30	30	75	125	35

Core	Line	Space	Land dia.	TH dia.
1996	100	100	500	300
1997	75	75	500	300

FIGURE 4.24 Design rules for IBSS technology.

cessing formats by means of either laser ablation through an appropriate etch mask or plasma etching, providing considerable economic advantages over alternative glass reinforced laminates. This trend is expected to continue with the development of alternative flexible or semirigid laminate materials, including thin FR-4, aramid laminate, and liquid crystalline polymers.

Recently introduced thin, two-layer (i.e., adhesiveless) polyimide materials enhanced the quality of formed holes and patterns and had contributed to increased interest in application of such materials to MCM-L manufacture. Because these polyimide films are thin, in the range of 25 to 50 μm, it is possible to form in them an acceptably shaped through vias, as small as 50 or even 25 μm at very high throughputs. The major advancements that continue to be made in the use of lasers in PWB via fabrication are highlighted toward the end of this section. In the remainder of this section are highlighted some of the most promising polyimide film based MCM-L approaches that have appeared in the literature, as well as some key older technologies that have formed the industry in this area of advanced thin film processing.

General Electric's MCM-F Process. Over the past decade or more, engineers at GE developed what has become known as *high-density interconnect* (HDI) multichip module technology. In this

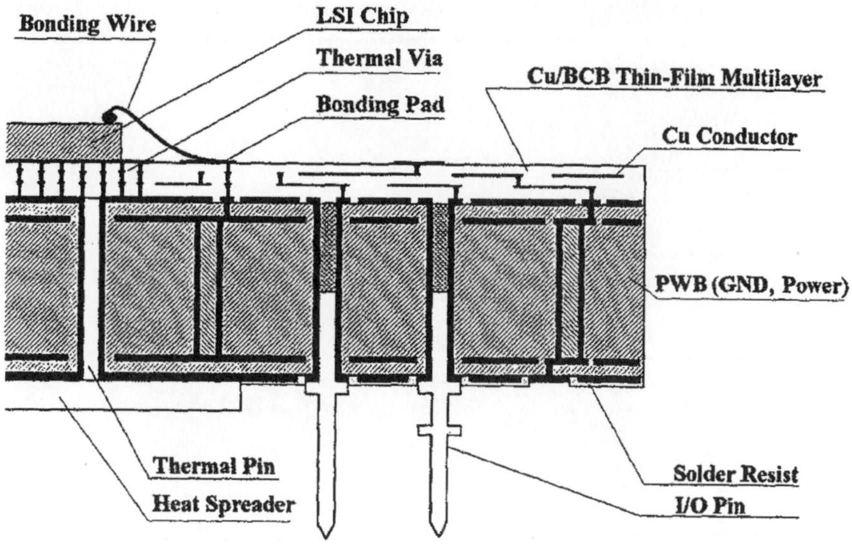

FIGURE 4.25 Cross section of NEC MCM D/L structure.[44]

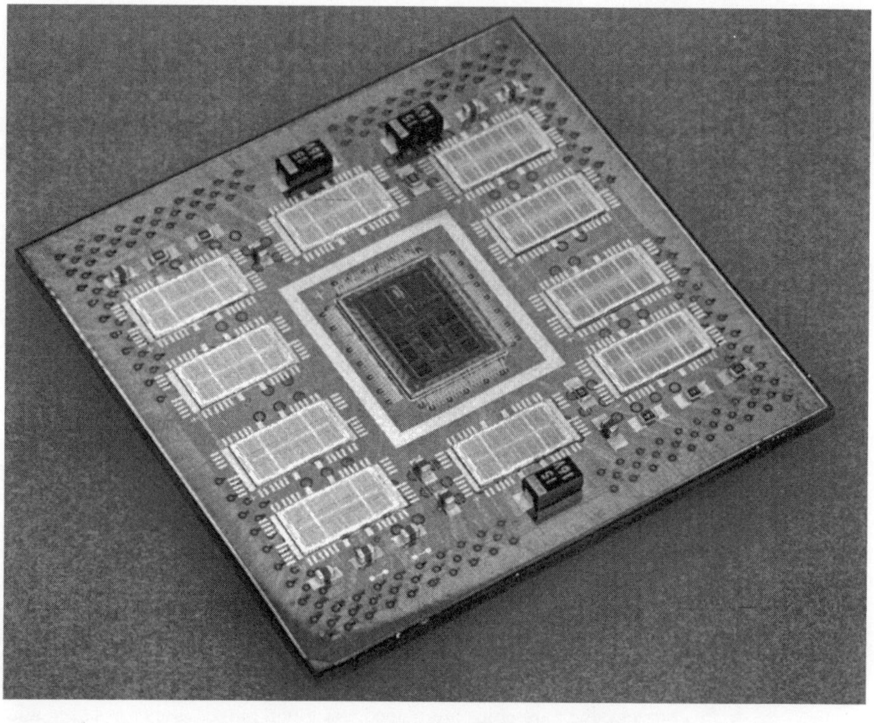

FIGURE 4.26 Photograph of NEC high-density RISC module.[44]

Table 4.3 Design Rules and Reliability Results for NEC D/L Technology.

Signal line material	Cu		
Conductivity (mΩcm)	1.8		
Width (μm)	40		
Pitch (μm)	108		
Impedance (Ω)	50		
Delay time (ps/cm)	55		
Dielectric			
Via hole dia. (μm)	40		
Dielectric constant	2.7		
PWB			
Via hole dia. (mm)	0.1		
Through hole dia.	0.5		
Reliability	Shorts/opens, 113 nets	Resistance change (%)	Number of cycles
Thermal (–65 to 125° C)	none	<1	1000
High-temp aging (125° C)	none	<1	3000
85° C/85% RH	none	<1	3000

"chips first" MCM technology, integrated circuits are bonded face side up to a ceramic base substrate with milled out cavities and very precisely controlled dimensions. A polyimide film was used in this technology to serve as the first layer of interconnect to the underlying die. Although the HDI technology has been successfully used in several military applications and demonstrated to be very reliable, the limited repairability and inherent high cost of this assembly technology prevents it from receiving broad acceptance in the MCM segment of the industry. More recently, efforts have been made to extend this technology toward a simpler and less costly process by adopting the idea of a separately prefabricated two-sided flexible circuit overlay.[22,23]

Two-sided interconnection patterns of the overlay are formed by laser ablation and patterning techniques using thin film deposition and etching techniques that are readily available within the flex circuit segment of the PWB industry. The overlay film can be fabricated in roll-to-roll format and subsequently panelized and coated on one side by a thermoset adhesive. The chips and other components are then placed face down and allowed to cure. From then on, there are two paths to complete such an MCM. For MCMs that can be configured for nonplanar surfaces, the assembly can be flipped over with additional thin film processing possible if needed. This film-based MCM-L fabrication scheme has been dubbed *multichip-on-flex* (MCM-F) technology. For more conventional MCMs where resistance to mechanical damage is required, a mold frame is placed around the MCM perimeter and the whole structure encapsulated, cured, and then flipped over and moved toward additional interconnect formation as required. This further extension of the HDI technology is called *encapsulated multichip-on-flex* (MCM-E/F) and is shown in Fig. 4.27.

For both of these MCM types, the final interconnect structure is formed by laser ablation of carrier and adhesive materials to expose the chip pads, then the interconnects are built up by sputter deposition of the full thickness of TI and Cu and then pattern etching, thus forming a three-layer thin film interconnect structure. Despite these significant efforts to reduce the cost of this technology, it is believed by these authors that it will not be able to compete with some of the emerging alternative high-density interconnection technology options based on "chips last" approaches.

MCM-L/O of Texas Instruments. Texas Instruments developed an analogous film overlay type of approach to the GE technology described in the previous section. From the start, TI opted to use a separately prepared flexible film involving a single-layer pair in its MCM-L/O method.

In this process, the bare chips are placed and glued into cavities formed in a laminate-based substrate containing power and ground planes in such a way that their surfaces are coplanar with the sur-

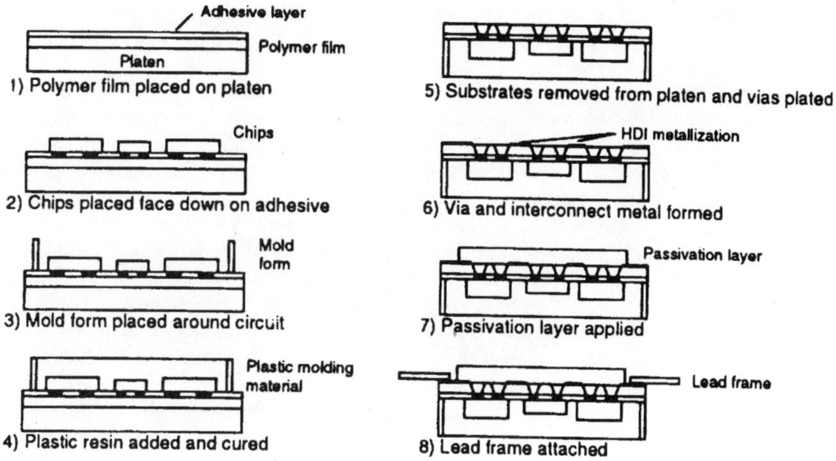

FIGURE 4.27 GE plastic encapsulated chip on flex process flow.

face of the board. Then, a polyimide overlay containing signal interconnects on both sides is glued over the chips and the board. The connection between embedded chips and the overlay is made by laser etching vias to the chip terminals with subsequent metallization of the vias and patterning of the metal layer. A schematic representation of the MCM-L/O technology is shown in Fig. 4.28.

Dycostrate. The Dycostrate technology was developed more than five years ago by Dyconex Ltd., a Swiss company[9] and has been licensed for use by H-P in Boblingen, and IBM in Singelfingen. This higher-density substrate technology was based on the use of thin flexible, single or double metal layer polyimide films. A plasma etch-based mass via generation process has been used in this process.

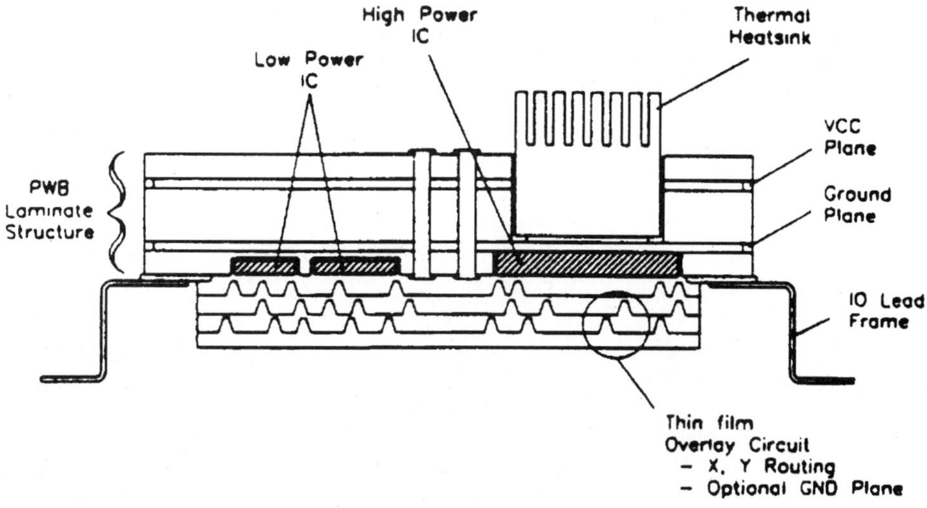

FIGURE 4.28 Cross section of MCM L/O.

The standard Dycostrate process involves the plasma etching of small (75–100 μm) blind vias in 25 or 50 μm thick copper-clad polyimide film through windows etched in the copper foil. This mask is patterned into one of the two copper cladding layers using conventional print and etch technology. After the via generation process, electroless copper seeding of the through holes with subsequent electroplated copper panel plating is used to build up the desired copper thickness and generate the z-axis connections. The resultant layer pair is then patterned using conventional print and etch techniques. This process has been demonstrated to be capable of producing 100 μm lines and spaces and 150 μm vias on 250 μm pads. Once fabricated, the interconnect can be applied to a stiffener PWB, which can contain etched power and ground distribution layers on its surfaces.

The connection between the flex circuit patterns and the power and ground can be made through special, larger, 200–400 μm holes dispersed throughout the substrate. Solder paste can be screened into these holes and reflowed producing a reliable connection to the underlying layers of metal in the base substrate.[9] Alternatively, a four-layer thin film circuit can be made by laminating single-sided flex laminates on both sides of the competed two-sided circuit. The processing of these outer layers is identical to the process sequence described above.

Multichip modules and other forms of high-density interconnection substrates have been fabricated using the Dycostrate technology. The Dycostrate technology as practiced by HP is capable of supporting both wire bond chip to substrate interconnection as well as flip chip on board. The potential use of the Dycostrate technology in the fabrication of a wide range of higher density PWB applications, including cellular phone boards with multiple flip chip attachment sites, has been investigated and described elsewhere.[25]

Plasma-based via generation is a particularly attractive technology, because it is compatible with most PWB facilities capabilities. In fact, most PWB facilities use plasma desmearing processes after conventional mechanical drilling. The general availability of suitable plasma processing tools and relatively low capital equipment costs make the Dycostrate technology particularly attractive. The biggest limitation of plasma etched vias relative to laser processing is the inherent isotropic nature of the plasma etching process. This inherent lack of directionality of the plasma limits the resolution achievable with this mass via generation technology to approximately a 2:1 aspect ratio. Although reactive ion etching is an alternative form of plasma etching that provides substantially improved directionality to the etching process,[26] typically by means of strong electromagnets which contain the plasma, it is not a viable option for high-volume reel-to-reel processing. Substantial undercutting of the copper mask can occur, resulting in overhanging of the copper foil around the via opening. A copper etchback process has been developed to reduce the extend of this undercut. Recently, work has also been underway to cost reduce the Dycostrate process by replacing the polyimide film originally specified with an epoxy based resin coated copper material which is etchable under the plasma etching conditions described above. Preliminary results of these alternative thin film materials have been described elsewhere.[25,27] The Dycostrate process flow is detailed in Fig. 4.29.

4.2.5 Historically Important Laminate-Based Technologies

For the sake of completeness, two MCM-L manufacturing methods will be described that have substantially impacted more recent technology development. Although these technologies are no longer in practice, they have had considerable influence on the development of some of the more recent MCM-L and MCM-D technologies that followed.

PACTEL's Microtec Technology. In the early 1970s, Sanford Lebow of PACTEL manufactured dense multilayer polyimide substrates that were used to interconnect a variety of packaged and bare die. This "Microtec" process used the basic elements of PWB manufacture, namely sequential lamination of polyimide prepregs, conventional photoprinting, and electroless and electroplating of copper for the manufacture of dense MCM substrates.[28]

A schematic representation of this process flow is shown in Fig. 4.30. It consists of a flat and polished panel over which a photomask was applied and the signal and pad patterns of the first metal layer defined. Copper was plated up through the openings in the temporary photoresist layer to a required height. This patterning and plating process was repeated to form via posts that were used to

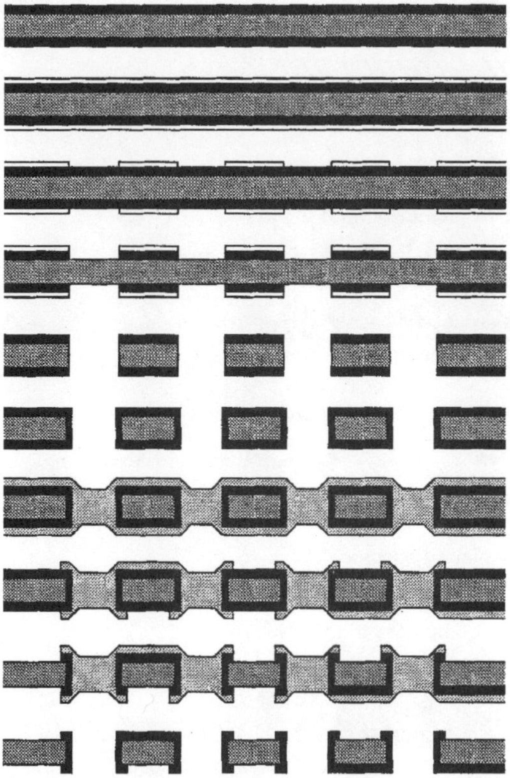

FIGURE 4.29 Schematic of Dycostrate process flow.

provide z-axis connection between adjacent metal layers. A polyimide prepreg was laminated over the plated pattern and via posts, resulting in planarization of the metal conductor and via post topography. The polyimide material covering the vias was removed and planarized in a mechanical lapping process, at the time involving the use of a sanding block leaving the tops of the via posts exposed and the cured polyimide dielectric suitably roughened for the application of the next conductor layer. By repeating this process sequence, multilayer structures with up to five metal layers were produced. The resultant MCM-L panel was then peeled off its temporary substrate and cut into individual circuits for assembly.

Features as small as 50 μm lines and 75 μm spaces and 75 μm sq. vias were demonstrated using this technology. In addition to these very impressive density capabilities, novel (for the time) structures were also demonstrated using this fabrication technology including coaxial via structures. This substrate technology was truly ahead of its time and, as a result, served as the starting point for a number of higher-density MCM technologies that followed, including the MCC[29] and IBM stacked via MCM-D technologies[30] and alternative MCM-L approaches.

Siemens Multiwiring Substrate. The most complex application of PWB technologies to the manufacture of MCM-Ls to date was carried out by Siemens in Germany in a mainframe computer applications during the 1980s. These MCMs were 115 mm sq., each containing 16 VLSI chips on 13–16 conductive layers. The conductors were 80 μm thick, and the vias were 0.9 mm on a 0.5 mm grid. In the fabrication of these substrates, Siemens employed more or less conventional PWB fabrication technologies with the addition of excimer laser based blind via generation and the use of a nonrein-

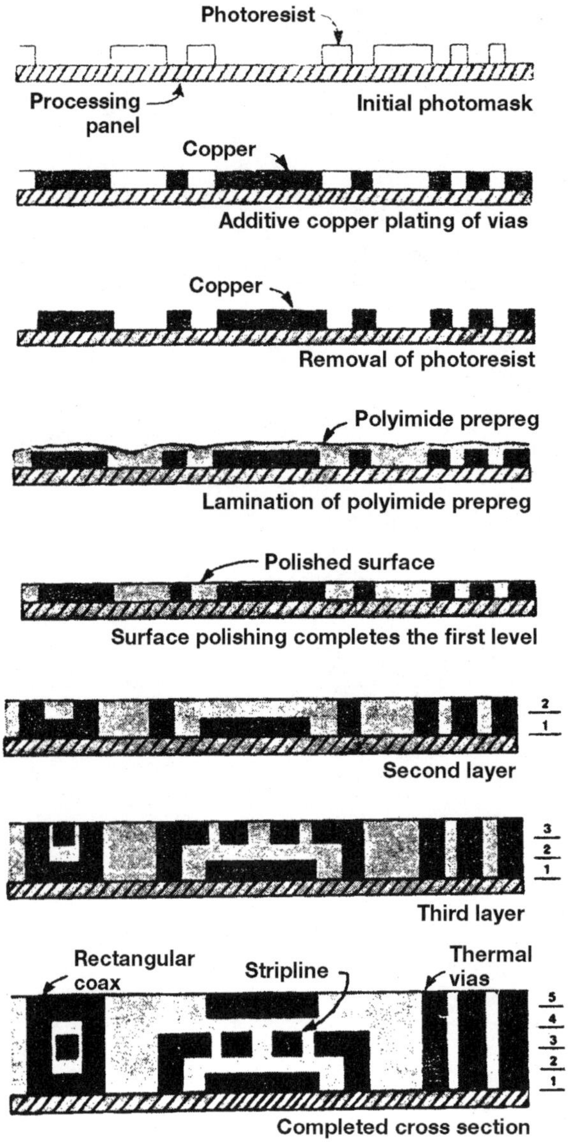

FIGURE 4.30 Process flow of PACTEL's Microtec technology.

forced polyimide film with acrylic adhesive layers.[31] Due to the high degree of complexity of these boards (it is reported that it took four days per layer for build, test, and repair), electrical testing was carried out after the formation of each conductive layer—before proceeding with the processing of the next metal layer. Figure 4.31 shows an individual *multiwiring module.* Although this substrate fabrication technology was largely too complex and costly for most other electronic applications, it served to nurture the development of the laser industry, as did the early work in lasers for the fabrication of MCM-D/Cs at IBM.[31] A number of the major strides made recently in the use of laser pro-

FIGURE 4.31 Siemens multiwiring module.

cessing for small blind via generation, which are highlighted in the remainder of this section, owe their origins to these technologies.

4.2.6 Emergence of Laser-Based Technology Options

Lasers emitting light from the UV to the infrared have been shown to be capable of generating high-density arrays of holes well under 100 μm in diameter. The spectrum of lasers that have been used for processing polymer-based materials have included excimer lasers, Nd-YAG solid state lasers, and CO_2 lasers. Above a dielectric material's specific ablation threshold, the energy that laser beams impinge onto organic films causes a vaporization or etching of the film with every pulse of light. As mentioned in the previous section, excimer laser processing has been used in the fabrication of high-density PWBs at Siemens[26] and in the fabrication of high-density multichip modules at IBM[31] and Motorola.[33] In the past, the major impediment to more widespread implementation of these laser processing technologies has been the capital and operating costs associated with these systems, their poor beam utilization, and resultant throughput and overall poor system reliability.[32] Additionally, because laser machining technologies evolved from applications involving small substrate sizes (i.e., MCM-D), systems available on the market were for the most part incompatible with the large format requirements (e.g., 450 mm × 600 mm) of the PWB industry. These issues had stunted the spread of laser-based via ablation technologies into the PWB industry. Major advancements have been made in the past few years both in laser processing equipment and processes that have dramatically improved its feasibility for PWB fabrication.

A number of different masking technologies have been developed for laser processing. The major masking approaches can be classified into one of three general classes (see Fig. 4.32): projection imaging, contact masking, and conformal masking. The particular masking scheme selected is based on a number of factors, including minimum feature size, throughput required, laser power, wavelength of the laser, and so on. Projection ablation technologies have been further subclassified into either point-to-point or multiple via projection technologies. Point-to-point projection imaging has been the most common masking technology used to date, as it provides the most flexibility. Because CNC control is possible with point-to-point projection imaging, this approach allows production

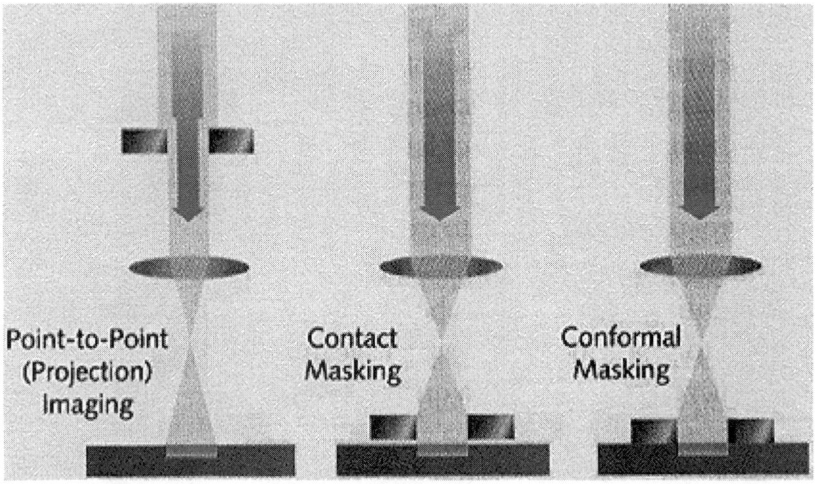

FIGURE 4.32 Masking alternatives for laser ablation.

direct from CAD/CAM data. Although this technique has been shown to be throughput limited for most laser technologies, the relatively low output power and high repetition rate (5,000 pulses/s) of Nd-YAG lasers makes point-to-point imaging the only viable option.[34] IBM–East Fishkill has developed a very sophisticated dielectric on quartz projection ablation masking technology which has been used for several years for high-density MCM-D fabrication with XeCl (308 nm) excimer lasers.[32] Although this technology is particularly effective, it is a very wasteful process with respect to beam utilization. Litel has also developed a holographic masking technology compatible with short-wavelength excimer laser sources which serves to concentrate light to specific areas on the substrate surface, in parallel, resulting in ultra-high volume via ablation.[35] This technology has been actively investigated by ARPA's MCM-L and MCM-D consortia.

Contact masks involve the patterning of a metal layer applied directly onto the surface of the laminate or dielectric material that is to be ablated. The openings in this metal layer are used to define the location where the via is to be generated. Most CO_2 laser based processes that have been used in the flexible circuit, and PWB industries are based on contact masks patterned in copper-clad foils. Although a conformal mâsk provides the highest resolution via capability of all of the available masking technologies, it is also the most process intensive option. Contact masking approaches involve the use of a metal mask or stencil which is fabricated in thin 50–125 µm thick stainless steel or brass. Although this approach is limited to via sizes of approximately 75 µm in dimension, it allows the use of a reusable robust mask, thereby significantly reducing the via generation cost. The laser technology used and in particular the selection of masking technologies adopted is based on the process flow and material combinations used. An comparison of the status of the various laser via processing technologies available is summarized in Table 4.4.

The TEA-CO_2 and Nd-YAG lasers are compatible with both glass reinforced and non-glass-reinforced laminates. The key to the dramatic improvement in via ablation performance of the TEA-CO_2 laser over previous generations has been the high peak power (around 5 kW) and short pulse duration (40 µs), which results in instantaneous vaporization of both glass fabric and the polymeric matrix resin, in turn resulting in dramatically reduced charring and clean holes with minimal glass fibers and reflowed glass residue. This made these laser processing technologies particularly attractive for implementation into the existing rigid PWB segment of the industry. The Nd-YAG laser is also able to machine through copper if required, thereby allowing the easiest integration into an existing PWB environment. Control depth drilling with the Nd-YAG laser tools is possible by controlling the number of pulses and the peak pulse power. On the other hand, the particular wavelength of the TEA-CO_2 laser is nearly completely reflected from copper surfaces thereby providing very precise z-axis

Table 4.4 Laser Via Ablation Technology Comparison

Laser type	Excimer laser	Traditional CO_2	TEA-CO_2	ND-YAG
Ablation tool technology	x-y scan galvanometer	x-y scan	x-y scan	x-y scan
Masking technologies	Projection mask, holographic	Conformal or shadow mask	Conformal or shadow mask	Point-to-point projection
Material/laser interaction	–	–	+	+
Glass reinforced laminates	+	–	+	+
Nonreinforced laminates	–	–	–	+
Copper metallization				
Wavelength (nm)	308	10,600	10,600	266 and 355
Pulse duration (ms)		1.0	0.001–0.1	0.00005
Typical laminate thick. (μm)	50	100	100	100
Minimum via dia. (μm)	25	50–100	50	25
Throughput (vias/min)*	–	–	1,200–3,000	600–720
Throughput (vias/min)†	187,500		1,200–3,000	600–720

* 100 μm thick FR-4 equivalent
† 100 μm thick aramid/polyimide film equivalent (no copper foil)

depth control when ablating blind vias down to copper pads on inner layers. Based on the rate of via generation described above, laser processing has reached a stage in its development at which it competes with mechanical drilling. The choice of one of these laser ablation options over another is largely based on the existing interconnect fabrication flow from which the particular PWB supplier is migrating. In the remainder of this high-density PWB section, examples of the successful implementation of these three classes of laser via generation are highlighted.

Sheldahl Via Grid Technology. Over the past few years, with the assistance of the ARPA-funded MCM-L consortia, Sheldahl has been establishing a capability for roll-to-roll manufacturing of high-density, flexible circuits for use in MCM-L and semiconductor packaging applications.[36,45] This highly automated substrate fabrication process is expected to be in full production as this book goes to press. A process flow is shown in Fig. 4.33.

Sheldahl has been implementing an ultra-high throughput excimer laser based via ablation technology based on Litel's proprietary masking technology.[35] Sheldahl has also been evaluating and implementing other laser technologies for other parts of their process, including circuit excision and generating openings in their cover layers. The Litel mass via generation process has been shown to be particularly compatible with Sheldahl's proposed process flow and planned product offerings where holes are patterned in unclad polyimide film, prior to vapor deposition of a Cu seed layer and a subsequent electroplated copper process.[36]

Their so-called *Via Grid* strategy is to provide metallized polyimide film with filled 25 μm diameter vias on a fixed grid and preapplied photoresist ready for photopatterning and conductor customization. The development of such a fixed via grid strategy is crucial in overcoming the high costs inherent in Litel's masking technology for moderate volume and high product mix applications. via grid could enable the trailing edge of the PWB industry to migrate toward higher-density interconnection capabilities with minimal or no capital investment or risk.

In addition to Sheldahl's fixed via grid material, a custom via grid option will also be available for high-volume applications where the NRE costs of tooling up a custom phase mask can be justified. One such example was successfully demonstrated recently for a high-power application. In this particular implementation, dense arrays of 25 μm diameter vias on a 50 μm pitch were used to extract heat away from a die assembly. The near planar surface topography obtained with these microvias was substantially better than an alternative single massive via, thereby enabling more reliable component attach, while providing comparable thermal performance.

To provide additional interconnection capability, the roll-to-roll process manufactured two-sided layers could be stacked one above another to form a multilayer structure. Sheldahl proposed that the vertical interconnections between these stacked layers could be made using an anisotropic conductive film (ACF) material called *Z-Link.* This ACF consists of solder particles embedded in a polymer matrix and are interleaved with patterned flex material layers. The preregistered, stacked assemblies are then laminated, and the dispersed conductive particles are fused to the pads. Meanwhile, the thermosetting matrix polymerizes, thus forming layer-to-layer permanent connections. This multilayering approach is demonstrated schematically in Fig. 4.34. Considerable obstacles to the acceptance of this type of PWB construction remain, including the minimum pad size required to achieve an acceptable contact resistance and the material cost.

The migration to higher-density PWB technology is expected to be slow, largely being paced by the availability of higher-density substrate technology at affordable cost. It is also clear that, for quite

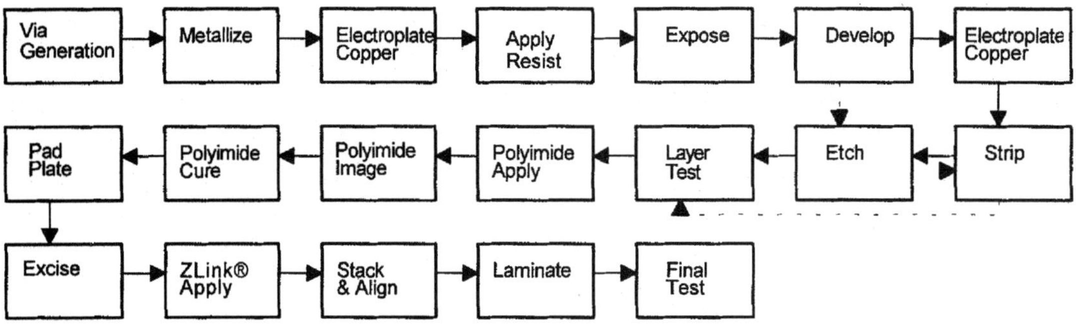

FIGURE 4.33 Sheldahl process flow.

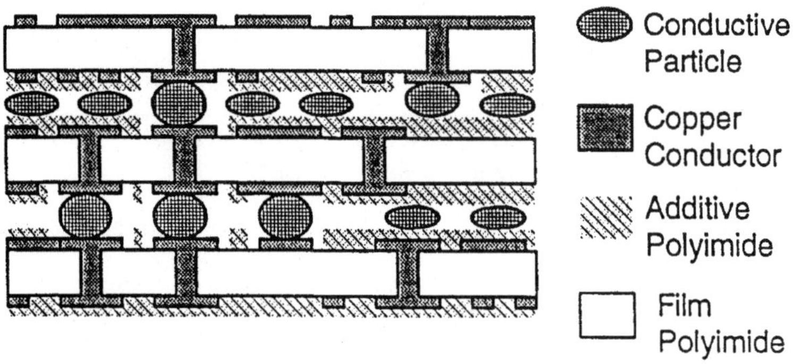

FIGURE 4.34 Sheldahl's flexible circuit and Z-Link-based multilayer construction.

some time, there will be an imbalance in the degree of interconnection required over the surface of an MCM or high-density PWB, with some areas of these substrates requiring higher density, for example, around the flip chip site of a high I/O die, with substantially lower density required where conventionally packaged surface mount packages may continue to be used. Another opportunity for the use of high-density flexible circuits with dense redistribution and interconnecting patterns is to apply them selectively to lower-density PWB, thereby providing "patches" of high-density interconnection.[37] Provided a suitably automated process for patch alignment and fixing were to be developed, this approach would allow very selective use of higher-density interconnect, thereby containing the board cost over an approach where the more costly HDI technology is applied over the whole board surface. This patch would be connected to the main board by using Z-Link to make connection from I/O pads on the perimeter of the patch to complementary pads on the rigid board, as shown in Fig. 4.35.

A method of substantially reducing the number of layers in MCMs has been developed by workers at the University of Arkansas, called the *interconnection mesh power system* (IMPS). The IMPS process is particularly applicable to the fine pitch capabilities of the Sheldahl technology.[38] This approach involves the placement of power and ground distribution on an orthogonal grid on both sides of a flexible circuit based layer pair with vias used where needed. In the spaces between these power and ground structures, signal traces can be placed forming a dense conductor matrix (see Fig. 4.36). This allows the collapse of a conventional four- or six-layer board into a high-density two-layer structure.

NCI Electronics HDI-Buildup Technology. NCI is a large and growing Japanese board manufacturer that has been migrating over the past couple of years from a low layer count, low-complexity PWB focus toward higher value-added interconnection technologies. Recently, it has been focusing on PWB applications requiring higher densities of blind and buried vias, including portable communication, high-density ball grid array, and multichip module applications.[34] NCI has been the most active PWB supplier in evaluating all the various via generation options available, including laser ablation, mechanical punching, and photovia technologies. Based on extensive studies, they have come to the conclusion that a high-throughput TEA-CO_2 laser-based via technology is marginally more expensive than a photovia approach, but the reliability of the resultant laser via based interconnect is much greater. In addition, laser-based via technology is more readily implementable into their existing factories. As a result, NCI has chosen to pursue laser-based via processing as their primary vehicle to higher-density PWBs. A schematic of a typical buildup board structure for a high-density BGA application is shown in Fig. 4.37. This trend toward laser processing appears to be taking hold in Japan, with substantial capital investment in laser based via generation technologies having been made over the past couple of years. In this construction, thin unreinforced epoxy adhesive (50 μm thick) preapplied to copper foil is laminated to one or both sides of an FR-4 core. Additional thin film

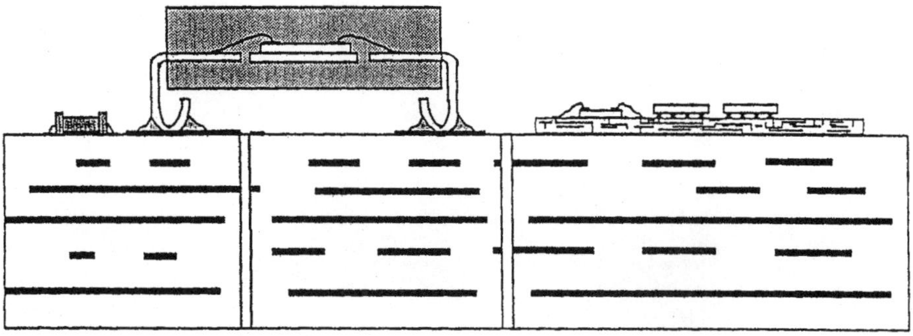

FIGURE 4.35 Schematic of localized high-density patch applications for flex circuits.

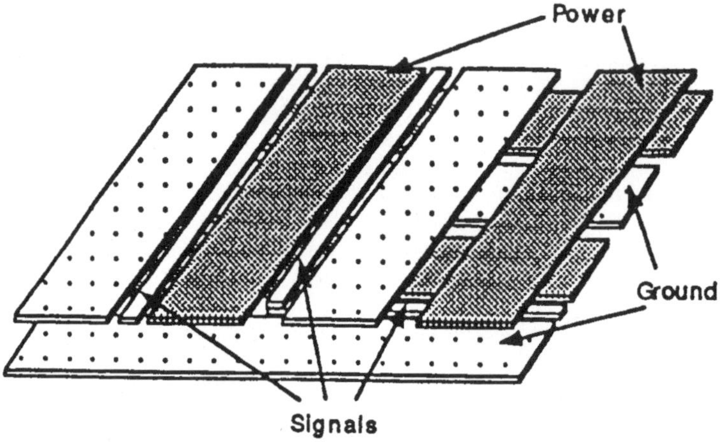

FIGURE 4.36 IMPS structure.

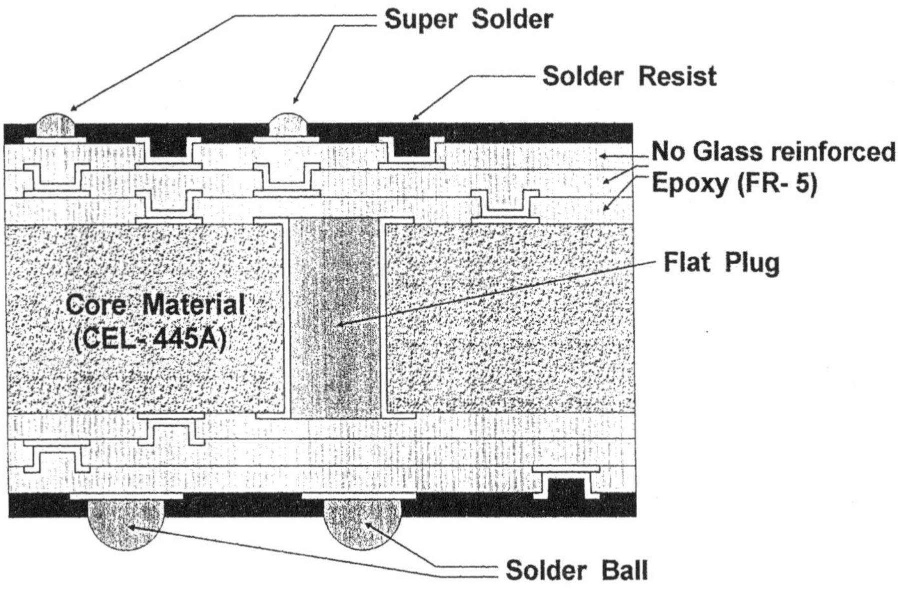

FIGURE 4.37 Cross section of a typical NCI buildup board.

layers can be deposited sequentially as needed. This approach has been used to dramatically reduce the cost of higher layer count boards by allowing significant layer reduction.

Matsushita's ALIVH Technology. Matsushita has also internally developed a high-density PWB technology based on TEA-CO_2 laser ablation using a galvanometer-based ablation tool built by Mitsubishi. This technology is referred to as *any layer interconnection via hole* (ALIVH). The galvanometer-based workstation allows the panel to remain stationary while the laser beam is being directed toward defined x-y coordinates on the panel where vias are to be located. To the best of our

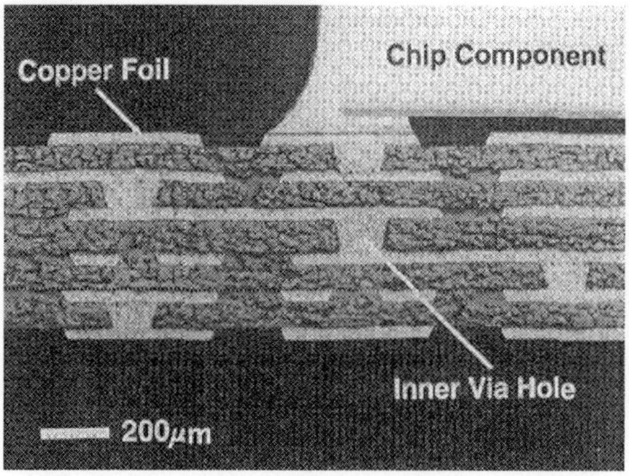

FIGURE 4.38 PWB fabricated using the ALIVH technology.

knowledge, a metal shadow mask is used to define the via dimensions. Based on the success of the Matsushita technology and throughputs of up to 6,000 vias per minute, more than a dozen of these systems were sold into Japan in 1995 alone.

A cross section of an ALIVH board is shown in Fig. 4.38. In this novel PWB fabrication method, laser vias are generated in low flow, highly dimensionally stable aramid laminate prepreg. The laser-generated holes are then filled with a copper-filled conductive paste and copper foil laminated to either side. This PWB process flow is the most environmentally benign of all of the high-density interconnect options currently in development, with the potential for copper etching being the only wet process required. In addition, the elimination of copper plating allows for a much greater control of conductor heights, thereby allowing finer pitch component attachment technologies, including Matsushita's *stud bump* flip chip attach technology.[34]

4.3 OTHER MCM-L TECHNOLOGIES

4.3.1 Hitachi Transfer Laminate Circuit Technology

The Hitachi transfer laminate circuit (TLC) technology[46] is shown schematically in Fig. 4.39. A thin barrier layer is continually formed on one side of a copper foil by electroplating (*b*). A pattern is defined using photoresist and copper is plated up to the desired thickness (*c, d*). The photoresist is stripped off, and the copper lines are subjected to an oxidation treatment to improve the adhesion upon lamination (*e*). Transfer of the conductive lines to the substrate is carried out by sandwiching prepreg between the substrate and the copper foil with the plated conductor lines and heating them under pressure (*f*). After the lamination process, the copper foil and barrier layer are selectively etched off, and the surface is polished.

4.3.2 Toshiba Buried Bump Interconnect Technology

Toshiba has developed a technology for interconnecting PWB layers with piercing bumps rather than plated-through holes, thus avoiding drilling and plating.[47] Conductive bumps are formed on copper foil by screen printing and drying silver paste to match the needed via positions. The shape of the bumps is conical, and their height is greater than the thickness of the subsequent prepreg. When combined with prepreg under heat and pressure, the bumps pierce the laminate. Another copper foil is subsequently laminated onto the prepreg in a lamination press. This is shown in Fig. 4.40 where B^{2}it processing is compared to conventional PWB processing.

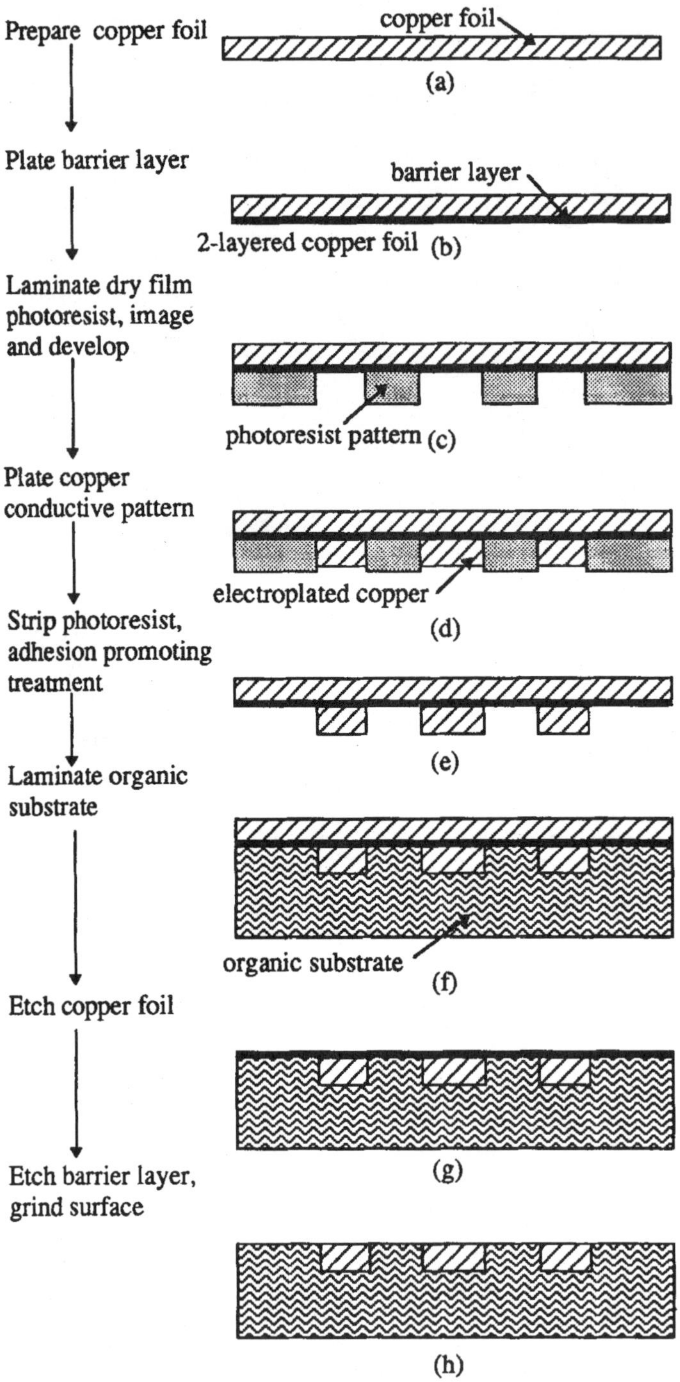

FIGURE 4.39 Hitachi TLC process flow.

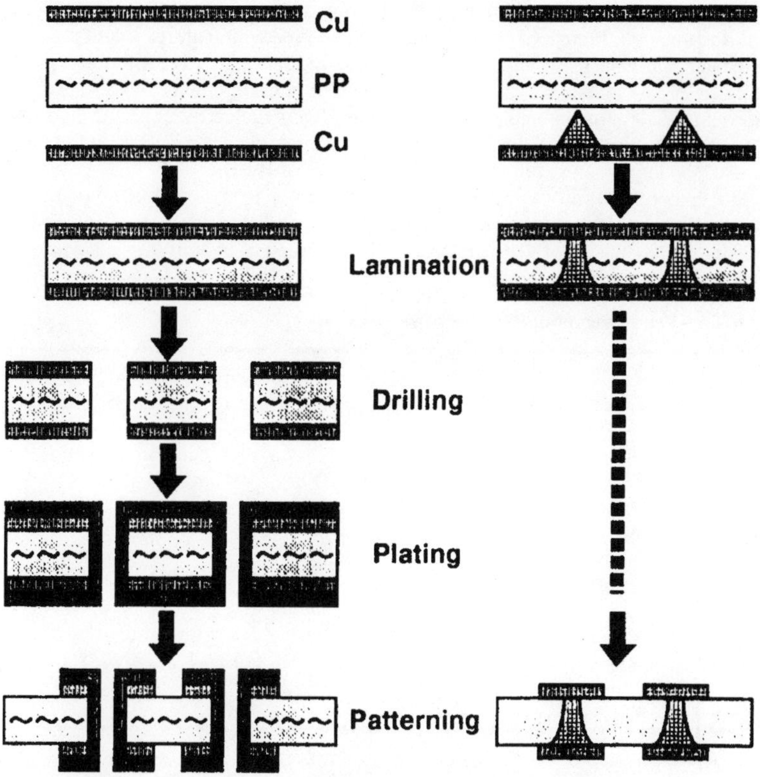

FIGURE 4.40 Comparison of conventional PWB and Toshiba B^{2}it processing.[47]

The technology has passed the following reliability tests: (a) high-temp storage 85° C for 1000 hr, (b) low-temp storage –30° C for 1000 hr, 65° C/95 percent humidity for 1000 hr, temp cycling –40 to +85° C, 1000×. Current design rules are shown in Table 4.5.

Table 4.5 Toshiba B^{2}it Ground Rules

Line and space	100/100
Bump diameter (μm)	300
Land diameter (μm)	500
Bump pitch (μm)	600
Layers	1 to 6

4.4 MCM-L SUBSTRATE MATERIALS

4.4.1 Organic Base Laminates

The majority of MCM-L substrates have been and continue to be built using conventional FR-4 epoxy/fiberglass based laminates. It is the mainstay of all the PWB manufacturing and accounts for over 85 percent of all base laminates used. Thus, its properties and performance are well know and well documented. Increased operational speed and reduced features have spawned a vigorous search for newer laminate materials that satisfy these requirements. For the most part, this improvement effort in base laminates was directed toward increasing their glass transition temperature, Tg, improving their dimensional stability, reducing their dielectric constant, and improving surface to allow the definition of smaller features. Table 4.6 provides information on the main characteristics of some of the laminate variations in the manufacture of PWBs.

Dimensional stability is currently a limiting factor in the production of MCM-L boards. Figure 4.41 graphically depicts the effect of material movement over a 10 × 10-in board. The dimensional stability limits the size of the pads needed to capture vias, even if the vias a laser or plasma etched.

Table 4.6 Key Properties of PWB Substrate Materials

Reinforcement/resin	Tg (°C)	x-y CTE (ppm)	z CTE (ppm)	ε′ @ 1 MHz	Dimensional stability (%)	Water abs. (%)
E-glass/epoxy	125	14–18	80	4.7	0.04	0.15
E-glass/PI	250	12–16	60	4.5	0.05	0.35
Woven Kevlar™/epoxy	125	6–8	105	–3.9	0.06	0.85
S-glass/cyanate ester	230	8–10	40	3.6	0.03	0.08
Quartz/PI	250	6–8	34	4.0	0.04	0.35
Thermount™/high-Tg epoxy	180	7–9	110	3.9	0.03	0.44
Thermount™/PI	230	7–9	80	3.6	0.03	0.81

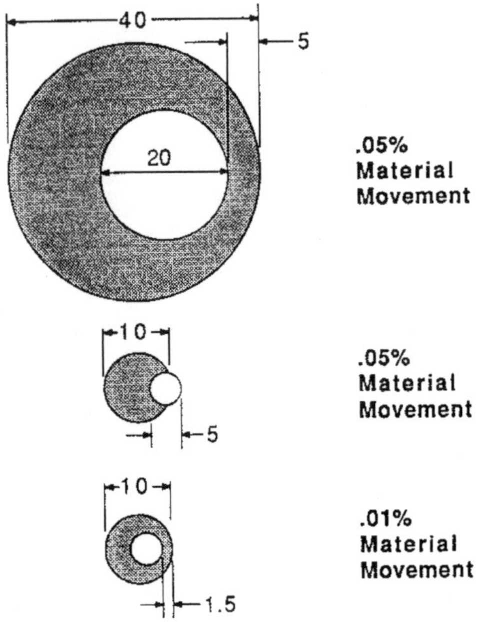

FIGURE 4.41 Effect of dimensional stability over a 10-in PWB.

Aramid "paper" reinforced laminates such as TL-01, commercialized by Tejin, have TCE of 6 ppm and a high Tg of 194° C.[48] Such Aramid-based laminates exhibit outstanding dimensional stability and a smooth surface, as shown in Fig. 4.42. It is formed by impregnating a nonwoven aramid fabric with epoxy resins. Similar materials, known as Thermount™, have been developed by DuPont.[39]

"Linear laminates" also reduce the dimensional expansion of the laminates in x and y directions as compared to conventional FR-4 materials. While glass filaments are still used for reinforcement in this case, they are not woven but oriented parallel to each other, alternating to filament layers in orthogonal fashion (see Fig. 4.43). The material can be impregnated with a variety of epoxy or polyimide resins materials. Compositech linear laminates also exhibit smooth surfaces and high dimensional stability.[43]

It should be noted that, because of their characteristics, polyimide-based laminates have been preferred for military applications, despite their significantly higher cost and hydroscopic nature. Laminates using BT-epoxy resins or cyanate esters, while more expensive than FR-4, have higher Tg, lower dielectric constant, and are also good materials for applications involving wire bonding based chip interconnection. There are also a number of multifunctional epoxy-based formulations designed to raise the Tg of the epoxy laminates. The bifunctional formulations have Tgs in the 140° C area, while multifunctional formulations have values above 170° C.

The need for thinner, more cost-effective laminates has produced a new class of laminate material called *resin-coated copper.* This unreinforced, B-stageable class of laminate/film is supplied on copper foil and is used in several of the processes mentioned in Section 4.2, including the Dycostrate technology being developed at HP and NCI.

4.4.2 Cast Films

As described in detail in Sec. 4.2.3, there are a number of MCM-L constructions that augment the conventionally built PWB structures with layers of deposited unreinforced dielectric films to serve as

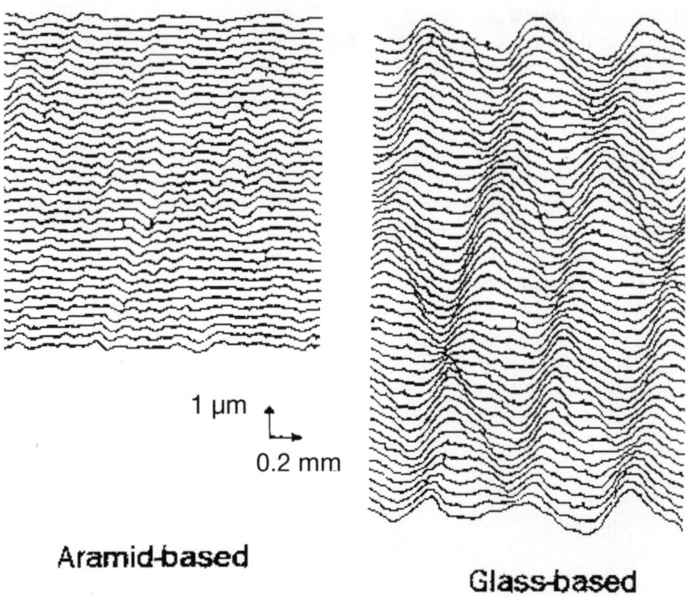

FIGURE 4.42 Surface roughness comparison between aramid and glass weave reinforced laminates.

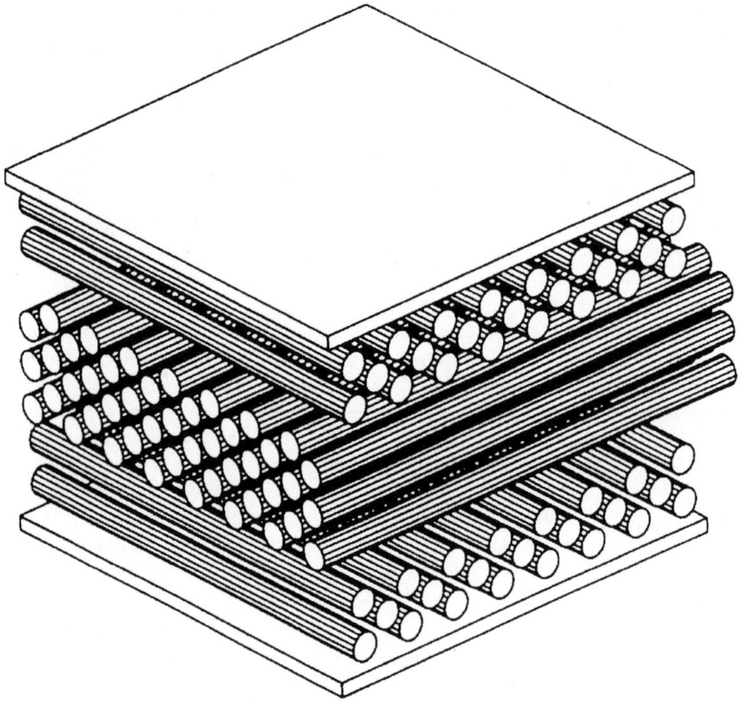

FIGURE 4.43 Linear laminates used in high-density MCM-Ls.

pattern redistribution from tight terminal arrays. Commercial materials are discussed in Chap. 5. The epoxy liquid coating films not discussed in Chap. 5 are mainly proprietary formulations. Generally, it is desired that such coatings be homogeneous, exhibit good and reproducible coating and leveling properties, be fast curing, have relatively long shelf life, and adhere to themselves and underlying conductor layers.

In some cases, they must be also photosensitive, to ease via formation and make this operation more economical. For such materials, it is important to minimize shrinkage during cure after the vias have been developed out to maintain vertical sidewall via profiles. Low shrinkage coatings are also required to reduce stresses to the underlying substrate and thin film layers. Otherwise, multiple layers of deposited dielectrics, as mentioned in Chap. 5, result in severe deformation of the base substrate, especially for unbalanced constructions.

Most flexible circuits used in MCM-L applications are based on adhesiveless polyimide films. Previous generations of polyimide film used adhesive layers, typically based on acrylate or epoxy chemistries for bonding to copper foil. Two primary approaches have been developed for the production of adhesiveless polyimide films. In one approach, a polyimide film is cast onto a copper foil and then cured. Alternatively, copper metallization can be applied directly to one or both sides of a polyimide film using a combination of vapor deposition and electroplating.

4.4.3 Metals and Surface Finishes

Most interconnections in MCM-L substrates are built with copper metallization. The particular metallization surface finish required is determined by the requirements of the specific chip attachment technology used.

For tape automated bonding (TAB) chip attachment technologies and for flip chip attach, a thickness of 50 to 100 μm of solder on the bonding pad provides good and reliable connections. Efforts have been underway to cost reduce solder based flip chip attach by eliminating the need for solder on the board side of the chip connection, and providing all of the solder needed for connection on the die side. For wire bonding applications, a clean, soft gold surface, usually plated over a layer of nickel, is used for thermosonic bonding.

4.4.4 Chip Attachment Area and Escape Method

The most area efficient chip assembly on MCMs is flip chip attachment. With such chip mounting methods, close to 90 percent of available substrate area utilization is possible on MCM-Ds and MCM-Cs, but not on conventionally built MCM-Ls, which cannot handle grid arrays below 400 μm pitches. However, in the flip chip terminations, solder connection pitches below 250 μm are becoming commonplace. This board-level routing challenge can be equally daunting for relatively low-I/O, small die as it is for large, high-I/O die as shown in Fig. 4.44. In this case, the use of an HDI technology capable of accommodating the fanout needs of a given application is compared to the fanout limitations of conventional PWB technology. Here, the capabilities of high-density interconnection technologies, including 100 μm diameter vias on a 250 μm pitch, allow efficient capture and redistribution of flip chip footprints. When combined with the use of "underfill" technologies between the chips and the substrate, to reduce the joint stress caused by CTE mismatches, it allows the reliable interconnection of bare die to the lowest-cost substrate option.

The main reason for the use of direct chip attach of bare chips to a substrate is to reduce the area needed for their interconnection. A comparison of the chip spacing possible with flip chip attachment methods is summarized in Table 4.7.

4.5 UNDERFILL

There are essentially no questions of reliability of flip chip attachment to a ceramic MCM-C substrate. Nor is there much cause for concern with flip chip attachment to an MCM-D substrate, given that the CTE of chip and substrate should match, and thus no stresses should develop that may affect integrity of joints.

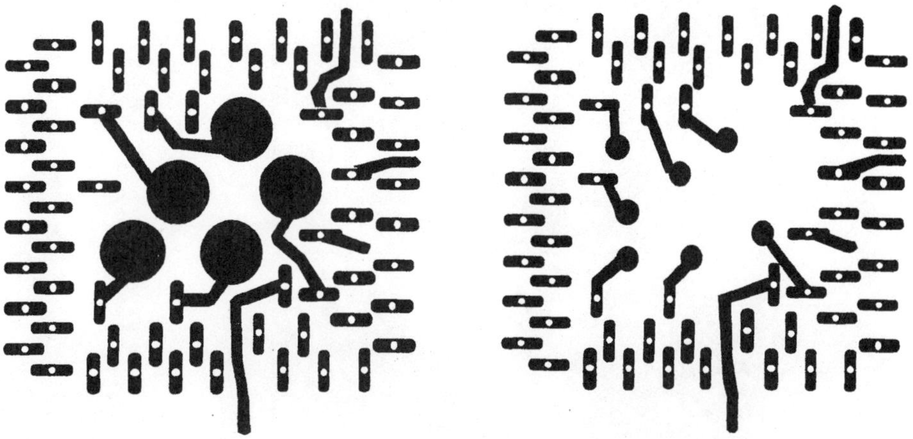

FIGURE 4.44 Fanout comparison of low-I/O die used in conventional and high-density PWB technology.

Table 4.7 A Comparison of Bare Chip Attach Space Requirements

Attachment method	Chip spacing (mil)
Flip chip	15–20
Wire bonding	70–80
Flip TAB	100–120
Regular TAB	150–400

It is a significantly different situation when integrated circuits are placed down onto an MCM-L. Due to the CTE mismatch of the ICs (3.2 ppm) with the substrate (14 to 16 ppm), an underfill material is required to compensate for any relative material movement caused by CTE mismatch. When the space between the chip and the substrate is filled with a suitable underfill material, the stresses caused by the CTE mismatch are distributed over the entire area of the chip and not concentrated on the individual solder joints. Consequently, more than a 10× improvement in flip chip reliability results. Several underfill materials have recently been evaluated for *flowability* and cure time.[49]

4.5.1 Chip-Size Packages

Recently, there has been a substantial amount of interest in the industry toward the development of chip scale package (CSP) technologies where bare die are mounted into a more robust, small package, allowing for conventional testing and burn-in with a minimal increase in size over the die footprint (no more than 1.2 × the chip area). Tessera's μ-BGA package,[40] where peripheral chip pads are connected to a flexible membrane with a redistributed area array footprint on the bottom (Fig. 4.45) and Motorola's *slightly larger than integrated circuit carrier* (SLICC) package[41] (Fig. 4.46) are two examples of such CSPs.

These CSP developments are significant to MCM-Ls for two primary reasons. First, this emerging semiconductor packaging requirement will provide additional momentum for the development of

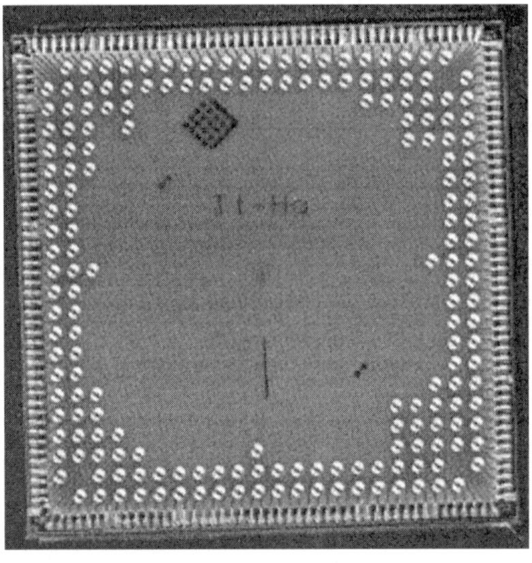

FIGURE 4.45 The Tessera Compliant Chip (TCC) package.

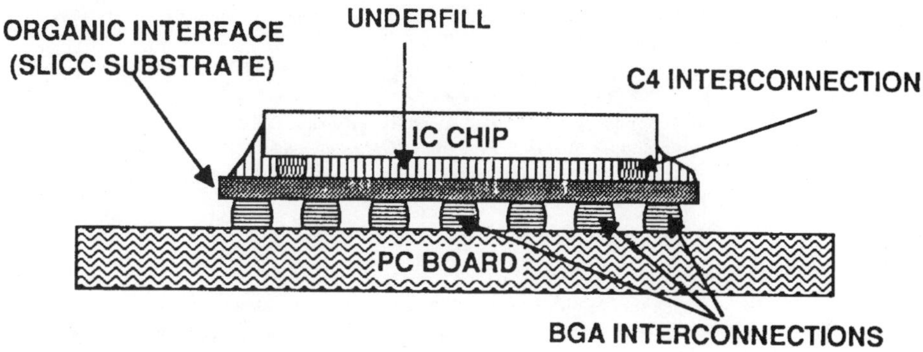

FIGURE 4.46 The Motorola SLICC package.

advanced PWB technologies that will be directly applicable to MCM-L needs. In addition, CSPs will accelerate the implementation of MCM-L technologies into segments of the industry that do not possess the assembly capabilities required for a direct transition from 0.5 mm surface mount PQFP assembly into bare die and MCM-L-based assembly.

4.6 CONCLUSIONS AND STATUS

Laminate technologies continue to evolve to meet the density requirements of today's microelectronics industry. We are clearly seeing a shift from drilled holes to the laser and photovia technologies that were described in this chapter. The analyses contained in this chapter represent a snapshot of MCM-L technologies in time, based on the information available to the authors. We are sure that the technologies described herein will most likely change as the suppliers further develop their products.

REFERENCES

1. D. P. Seraphim, D. E. Barr, W. T. Chen, G. P. Schmitt and R. R. Tummala, "Printed Circuit Board Packaging," in *Microelectronics Packaging Handbook,* Van Nostrand Reinhold, New York, pp. 853–922 (1989).
2. J. H. Reche, "High Density Multichip Interconnect for Advanced Packaging," *Proc. NEPCON West,* 1989, p. 1308.
3. T. G. Tessier, J.G. Aday and H. Crews, "Selecting Flip Chip on Board Compatible High Density PWB Technologies," *Proc. Int. MCM Conference,* Denver, 1995, p. 42.
4. N. G. Koopman, T. C. Reiley and P. A. Totta, "Chip to Package Interconnections," in *Microelectronics Packaging Handbook,* Van Nostrand Reinhold, New York, pp. 361–391 (1989).
5. D. Suranayana et al., "Flip Chip Solder Bump Fatigue Life Enhanced by Polymer Encapsulation," *Proc. 40th ECTC,* 1990, p. 338.
6. J. G. Aday, T. G. Tessier, H. Crews and J. Rasul, "A Comparative Analysis of High Density PWB Technologies," *Proc. Int. MCM Conference,* Denver, 1996, p. 239.
7. T. G. Tessier, J. Aday and H. Crews, "Selecting Flip Chip on Board Compatible High Density PWB Technologies," *Int. J. Microcircuits and Electronic Packaging,* vol. 18, Number 4, 1995, p. 319.
8. Y. Tsukada, "Low Cost Thin Film Multilayer Substrate Fill Gap with LSI," in *Nikkei Microdevices,* pp. 61–67, 1991.
9. W. Schmidt, "A Revolutionary Answer to Today's and Future Interconnect Challenges," *Proc. of Printed Circuit World Convention VI,* San Francisco, CA, 1994, p. T12–1.

10. Y. Tsukada, S. Tsuchida and Y. Mashimoto, "Surface Laminar Circuitry Packaging, *Proc. 42nd ECTC,* 1992, p. 22.
11. T. G. Tessier and J. G. Aday, "Casting Light on Recent Advancements in Laser Based MCM-L Processing," *Proc. Int. MCM Conference,* Denver,1996, p. 6.
12. M. Nakamura, M. Kato, M. Asai and H. Takenaka, "High Reliability, High Density Build Up Printed Circuit Board for MCM-L," *Proc. Int. MCM Conference,* Denver,1996, p. 36.
13. Y. Tsukada, S. Tsuchida, Y. Mashimoto, "Surface Laminar Circuitry and Flip Chip Attach Packaging," *Proc. Int. Microelectronics Conf.,* Yokohama, 1992, p. 252.,
14. Y. Tsukada, S. Tsuchida, Y. Mashimoto, "Surface Laminar Circuit Packaging," *Proc. 42nd ECTC,* 1992, p. 22.
15. Y. Tsukada, Y. Yamanaka, "Design and Electrical Performance of Circuit Laminar Circuit/Flip Chip Attach Packaging Technology," *Proc. IEPS,* 1993, p. 1111.
16. Y. Tsukada, "Design and Performance of SLC-MCM," *Proc. NEPCON West,* 1994, p. 521.
17. P. Kam and H. Lowe, "Assembly Challenges of PCMCIA Cards," presented at ISHM Advanced MCM Technology Workshop, Ogunquit, Maine, 1995.
18. C. Sullivan, R. Funer, R. Rust, M. Witty, "Low Cost MCM-L for Vehicle Applications." *Proc. Int. MCM Conf.,* Denver, 1996, p. 142.
19. C. Boyko, F. Bucek, V. Markovich and D. Mayo, "Film Redistribution Layers Technology," *Proc. 43rd ECTC,* 1993, p. 302.
20. D. O. Powell, T. Heline, R. Magnuson, V. Markovich, A. Bhatt, "Early Experiences with FRL Technology," *Proc. IEPS,* p. 649, 1994.
21. E. Enomoto, M. Asai, Y. Sakaguchi, C. Ohashi, "High Density Printed Wiring Boards Using Advanced Fully Additive Processing." *Proc. IPC,* 1989, p. 1.
22. R. Fillion, R. Wojnarowski, B. Gorowitz, W. Daum and H. Cole, "Conformal Multichip on Flex (MCM-F) Technology," *Proc. Int. MCM Conference,* Denver, 1993., p. 52.
23. R. Fillion, G. Forman, R. Kolc and R. Wojnarowski, "The Latest Development on the Low Cost Multichip-on-Flex (MCM-F), *Proc. 2nd Int. FLEXCON,* 1995, p. 97.
24. Walter, D., Roszel, L., "MCM-L/O—Thin Film Interconnect on a Printed Wiring Board (Laminate) Substrate." *Proc. 7th Int. SAMPE Electronics Conf.,* 1994, p. 99.
25. M. Moser and T. G. Tessier, "Higher Density PCBs for Enhanced SMT and Bare Chip Assembly Applications," *Proc. Int. MCM Conference,* Denver, 1995, p. 543.
26. F. Bachman, "Excimer Laser Drill for Multilayer Printed Circuit Boards: From Advanced Development to Factory Floor," *MRS Bulletin,* December 1989, pp. 49–53.
27. H. Holden, "Comparing Costs for Various PWB Build Up Technologies," *Proc. Int. MCM Conference,* Denver, 1996, p. 15.
28. S. Lebow, "High Density / High Speed Multichip Packaging," *Proc. 6th IEPS,* 1986, p. 417.
29. J. T. Pan, S. Poon and B. Nelson, "A Planar Approach to High Density Copper Polyimide Interconnect Fabrication," *Proc. 8th IEPS,* 1988, p. 174.
30. E. Perfecto, K. Prasad, P. Wilkens, G. White, R. Tummala, C. Prasad, T. Redmond, "Multilevel Thin Film Packaging Technology at IBM, *2nd Int. MCM Conference,* Denver 1993, p. 203.
31. T. F. Redmond, J. R. Lankard, J. G. Balz, G. R. Proto and T. A. Wassick, "The Application of Laser Process Technology to Thin Film Packaging," *Proc. 42nd ECTC,* 1992, p. 763.
32. C. Pico, et. al., "Micromachining of Polyimides Using Ultraviolet Solid State Lasers." *Proc. 2nd Int. Flexcon,* 1995, p. 152.
33. J. M. Morrison, T. G. Tessier and B. Gu, "A Large Format Modified TEA-CO2 Laser Based Process for Cost Effective Small Via Generation," *Proc. Int. MCM Conference,* Denver, 1994, p. 369.
34. Y. Bessho, Y. Tomura, T. Shiraishi, M. Ono, T. Ishida and K. Omoya, "Advanced Flip Chip Bonding Technique to Organic Substrates," *Proc. ISHM,*1995, p. 35.
35. A. H. Smith, "Phase Mask Machining for High Throughput Via Formation," *Proc. Int. MCM Conference,* Denver, 1993, p. 208.
36. L. Lemke, "Recent Developments in the Production and Application of Advanced Laminate Based MCMs and Chip Carriers," *Proc. Int. MCM Conference,* Denver, 1995, p. 235.
37. J. Tourne, "Laser Via Technologies for High Density MCM-L Fabrication," *Proc. Int. MCM Conference,* Denver, 1995. p. 71.

38. L.W. Shaper, S. Ang, Y. L. Low and D. R. Oldham, "Electrical Characterization of the Interconnected Mesh Power System (IMPS) MCM Topology," *IEEE Trans. CPMT.,* Part B, vol. 18, 1995, p. 99.
39. Diekman, J., Mirhej, M., "Nonwoven Aramid Papers: A New PWB Reinforcement Technology," *Proc. IEPS,* 1990, p. 123.
40. J. Fjelstad, T. DiStefano and K. Karavakis, "Multilayer Flexible Circuits with Area Array Interconnections for High Performance Electronics," *Proc. 2nd Int. FLEXCON,* 1995, pp. 110.
41. P. Lall, G. Gold, B. Miles, K. Banerji, P. Thompson, C. Koehler and I. Adhihetty, "Reliability Characterization of the SLICC Package," *Proc. 46th ECTC,* 1996, pp. 1202.
42. M. J. Begay and R. Cantwell, "MCM-L Cost Model and Application Case Study," *Proc. Int. MCM Conference,* Denver, 1994, p. 332.
43. Compositech Ltd., 120 Ricefield Lane, Hauppauge, NY 11788–2071.
44. Matsui, K., Shimoto, T., Utsumi, K., "MCM D/l Technology Based on Cu/Benzocyclobutene Thin Film Multilayer Structures." *Proc. ISHM,* 1995, p. 398.
45. Gengel, G., "A Process for the Manufacture of Cost Competitive MCM Substrates." *Proc. Int. MCM Conf.,* Denver, 1994, p. 182.
46. Tsubomatsu, Y., Yoshidomi, Y., Ohata, H., Yamazaki, T., Fukutomi, N., "Transfer Laminate Circuit Technology for Fine Pitch Wiring Technology." *Proc. 44th ECTC,* 1994, p. 786
47. Sato, Y., Shibayama, K., Hamano, H., Fukuoka, Y., "A High Density Substrate with Buried Bump Interconnection Technology." *Proc. Int. Microelect. Conf.,* Omiya, 1996, p. 345.
48. Hirakawa, *IEEE Trans. CHMT,* vol. 13, 1990, p. 570.
49. Wun, L.K., Margaritis, G., "Evaluation of fast Flow, Fast Cure Underfills for Flip Chip on Organic Substrates." *Proc. ECTC,* 1996, p. 540.

CHAPTER 5
HIGH-DENSITY, LARGE-AREA PROCESSING (LAP)

Dr. Philip E. Garrou
Dow Chemical Company
Theodore G. Tessier
Motorola Inc.

5.1 INTRODUCTION AND DEFINITIONS

5.1.1 Convergence of MCM Process Technologies

Over the past decade, the need for higher-density interconnection solutions has been accelerating. This trend has been largely driven by performance enhancement and, more recently, miniaturization. The trend was first initiated by the migration from the low semiconductor packaging efficiencies associated with through hole mounting of components onto printed circuit boards to a gradual implementation of surface mount components. Over the past 10 years, surface mount quad flat packages and thin small outline packages have evolved down to peripheral I/O pitches of 0.4 mm and less. More recently, peripheral I/O package types with ever increasing I/O requirements have been displaced by ball grid array and other area array packaging technologies. These area array technologies afford considerable loosening of I/O pitches over the peripheral package types that they have been replacing, while simultaneously serving the need for higher-density PWB interconnect.

The interest in bare die assembly and multichip modules arose largely due to the need to take advantage of the significant performance advantages that bare die assemblies could afford. These deposited dielectric (MCM-D) multichip modules were largely driven in the mid to late 1980s by the mainframe and military sectors of the electronics industry. Because maximizing performance was the overriding consideration for these types of applications, little consideration was given to the manufacturability of these technologies or their associated fabrication costs. Consequently, a focus on MCM interconnection technologies based on the use of obsolescent semiconductor processing equipment, typically small-format (100–150 mm) substrates such as silicon wafers and/or ceramic substrates prevailed for several years. Using such processes and materials, very high density MCM substrate technologies, were developed (minimum feature sizes in the 8 to 15 μm range and blind vias in the 10 μm range), but no significant gains toward MCM-D cost reduction goals were achieved.

Perhaps the most significant turning point in the evolution of MCM technology over the past 10 years was the discovery that reliable flip chip assembly to substrates with dissimilar coefficients of thermal expansion was possible provided an under-chip encapsulant was used.[1] This dismissed the previously held notion that a substrate CTE closely matched to silicon was essential for reliable flip chip attachment and broadened the scope of development activity to include work with material sets than had not been considered previously.

At the same time, the PWB industry has been striving to keep up with the ever increasing density requirements that the move to surface mount technology has brought. Until recently this has been achieved by means of evolutionary improvements in conventional PWB fabrication. Consequently, gradual improvements in interconnect densities have been realized by gradual migrations from 200 μm lines and spaces down to capabilities of 100 μm and below. Similarly, mechanically drilled holes with dimensions of 350 μm have become quite common in the industry. More recently, the use of blind and buried drilled vias with diameters down to 100 to 150 μm has had to be implemented to satisfy the localized high density requirements of *ball grid array* (BGA), μ-BGA, and *direct chip attach* (DCA) and multichip module substrates. The nearly exponential cost increase of small-hole drilling as the hole diameter drops below 350 μm has served to motivate the PWB industry to explore alternative means of fabricating blind vias, either by the use of inherently photodefinable dielectrics or by the implementation of plasma or laser based mass via generation technology.[2]

During this same time period, major investments in research and capital equipment purchases were being made by the *flat panel display* (FPD) industry, based largely in Japan. This activity was focused on the fabrication of displays for the then emerging laptop computer segment of the industry. The need to process 400 to 500 mm large glass panels (with photolithographic requirements in the 5 to 10 μm range) resulted in major developments in the area of large-panel processing technology that are directly applicable to the fabrication of MCM-D substrates in larger panel formats.

As we entered the 1990s, the simultaneous softening of both the mainframe computer and military electronics segments of the industry made it clear that manufacturability and fabrication cost had been overlooked in the obsession for MCM performance. Additionally, it became obvious that, for the foreseeable future, the bulk of the high-volume bare chip on board substrate requirements would not require the density capabilities that classical MCM-D technologies were capable of. Consequently, over the past few years, a greater willingness to make design manufacturability trade-offs has emerged, with substrate cost becoming the overriding factor for most applications. Both the MCM-D and PWB factions of the industry have been scrambling to position their technologies to meet the emerging needs of portable computing and wireless telecommunication products. This chapter will describe and compare the large-format deposited dielectric based process technologies that have emerged from the PWB industry, as well as the evolving MCM-D technology base recently supplemented by the larger panel processing activities of the LCD industry.

5.1.2 State of the Art Circuit Density vs. Processing Area

The circuit density capabilities of typical MCM-D and photovia PWB technologies in 1995-97 are compared to those of MCM-C and conventional PWB constructions in Table 5.1.

Table 5.1 Comparison of Wiring Densities Achievable with MCM-C, PWB, Photovia PWB, and Large-Format MCM-D High-Density Technologies.

	MCM-C	PWB (high-end)	Epoxy photovia PWB	Large-format MCM-D with photovias
Line width (μm)	100	75–125	50–100	10–25
Via diameter (μm)	100–125	200–300	100–125	25–50
Via pad diameter (μm)	150–200	400–500	150–300	50–100
Dielectric constant	5–9.5	4.5	3.5 (thin film)	2.6–3.5

Printed Wiring Board Evolution. As mentioned earlier in this chapter, the PWB industry has been evolving to higher interconnect densities in response to specific substrate drivers, most notably multichip module and flip chip on board requirements. As of 1996, the leading edge of high-volume PWB manufacturing was characterized by interconnect geometries of 100–125 μm lines and spaces. The greatest obstacle to achieving significantly increased interconnect densities has historically been the ongoing use of mechanically drilled holes. The majority of PWBs today are fabricated using a minimum drilled hole dimension of 350 μm. Additionally, capture pads in excess of 500 μm have been used to allow for the significant movement of materials associated with the PWB fabrication process. These dimensions are very prohibitive in attaining higher-density goals. Although some limited success has been demonstrated in extending the use of mechanically drilled holes both through hole and blind and buried to 50 μm or less,[3] this approach is not economically feasible for most applications in light of the nearly exponential increase in costs associated with drilling small holes (see Chapter 4).

This apparent limitation to the extendability of mechanical drilling has fueled activity in alternative mass via generation processes. Most of the attention to date has focused on sequential buildup of layers of photodefinable dielectrics and electroless/electroplated copper plating. In addition to these deposited dielectric-based PWB technologies, alternative thin laminate-based substrates utilizing either laser or plasma etched via generation technologies are also being developed. These thin laminate-based process technologies are discussed in detail in Chapter 4. The announcement of IBM-Yasu's *surface laminar circuitry* technology in 1991 has resulted in tremendous activity in deposited dielectric-based PWB technologies. A series of recent papers summarizing the current status of these photovia-based PWB technologies has appeared in the literature.[2,4–6] A summary of current photovia PWB technology in Japan is presented in Table 5.2.

Copper thicknesses in the 5–17 μm range are becoming common, with the general trend toward thinner metals in light of the time-dependent semiadditive plating processes that are commonly used. This is a significant change from the historical approach in the PWB industry to relatively thicker 0.5–1 oz (17–35 μm) thick copper foils.

The photovia PWB design rules forecast for the year 2000 are expected to continue to trail the evolving higher-density capabilities of large-format MCM-D processing, due primarily to the much higher dimensional stability of the base substrate used for the MCM-D structures. Although some significant improvements in laminate dimensional stability have occurred over the past few years in the area of nonwoven laminates, clearly these advances are still not adequate to compete with the more dimensionally stable glass and metal substrates that are being proposed for LAP MCM-D processing.[8,12,14,15] In addition, the PWB industry will likely remain relatively conservative in setting goals for higher-density fabrication capabilities, preferring to incrementally improve the performance of PWB fabrication tools wherever possible rather than making large investments in LAP FPD processing tools. As the MCM-D technologies adopt the LAP processing tools required to reach a panel size that is comparable to that of photovia PWBs, investment in some of these more sophisticated processing tools will be required to remain competitive (Fig. 5.1).

Table 5.2 Photovia PWB Technology in Japan

Technology	Company	Material	Via process	Via dia. (μm)	Via land (μm)	Line/space (μm)
DV-multi	NEC	epoxy film	photo	100	230	50/50
IBSS	Ibiden	epoxy liquid	photo	100	150	50/50
ALIVH	Matsushita	aramid-epoxy	laser	150	300	60/90
	Fujitsu	epoxy	photo	100	150	50/50
VB-2	Victor	epoxy	laser	100		95/75
B^2IT	Toshiba	BT laminate	paste/bump	200	300	90/90
Hitavia	Hitachi	adhesive film	drill	200	500	100/100
MSF	Shinko	PI	laser	120		20/20
SLC	IBM Yasu	epoxy liquid	photo	100	175	50/50

5.2 ECONOMIC MOTIVATIONS FOR LAP THIN FILM PROCESSING

The economic advantages of large-panel MCM-D production were first described in 1992. It was deduced that the equipment advances being implemented in the flat panel display industry could also have a major impact on reducing the price of thin film MCM-D manufacturing if such large-panel techniques could be adopted by the emerging thin film MCM industry.[7] Z-Systems and IMC Microsystems published the first work on a 350 × 350 mm FPD format (Al/BCB thin film interconnect on glass) and proposed that larger panel production was possible.[8] The very favorable economics of large-panel processing are depicted in Fig. 5.2, where 10 times more modules are fabricated on a 350 × 350 mm panel than is possible on a 150 mm diameter wafer. Even considering the greater cost of producing the larger panels due to more expensive capitalization and anticipated initial lower yields, it was predicted that the overall cost of substrate fabrication would be reduced. Several groups have addressed the economics of such processing with a more quantitative analysis.

Workers at Motorola have examined the feasibility of using large-panel processing equipment borrowed from the LCD industry for the processing of panel sizes up to 500 mm.[9] The resultant cost structure for fabrication of 50 × 50 mm modules on 100, 200, 400, and 500 mm substrates are depicted in Fig. 5.3. These results project that, for a 100 mm line operating at 50 percent cumulative yield, MCM substrate fabrication cost would range from \$20 to \$12/in^2 as annual volume production increases from 50,000 to 2 million modules. The resultant fabrication cost range for the 400 and 500 mm panels were projected to be very similar, at approximately \$3.50/in^2 for volumes of 2 million modules per year.

A collaborative effort between groups at Dow and Motorola,[10,11] using Envision™, a thin film processing cost model, have evaluated sensitivity analyses on various LAP manufacturing scenarios using the LCD equipment lists detailed above. More than a threefold cost differential exists between some of the manufacturing scenarios considered, and, a tenfold cost differential is projected for specific 150 mm IC equipment based processing versus 350 mm LAP processing.

Workers at IBM have also examined the cost impact of large-area processing.[12] Figure 5.4 compares a 125 mm small format facility to a 450 mm LAP facility running at 10 times the volume with tool throughput improvements and materials use reductions inherent to the LAP process. These results further support the advantage of larger format substrates in driving down MCM-D processing costs.

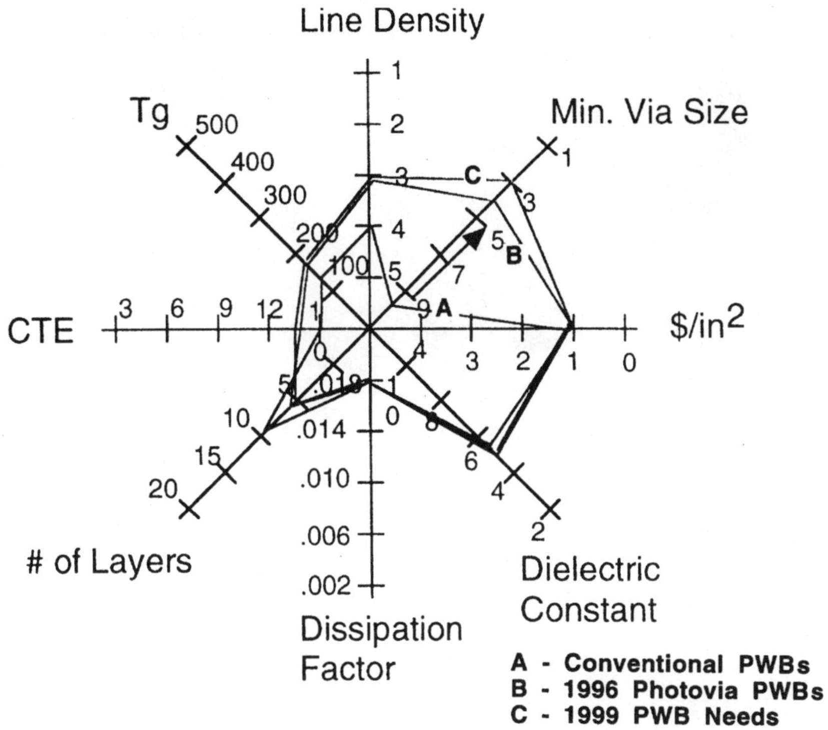

FIGURE 5.1 Trends in PWB technology needs.

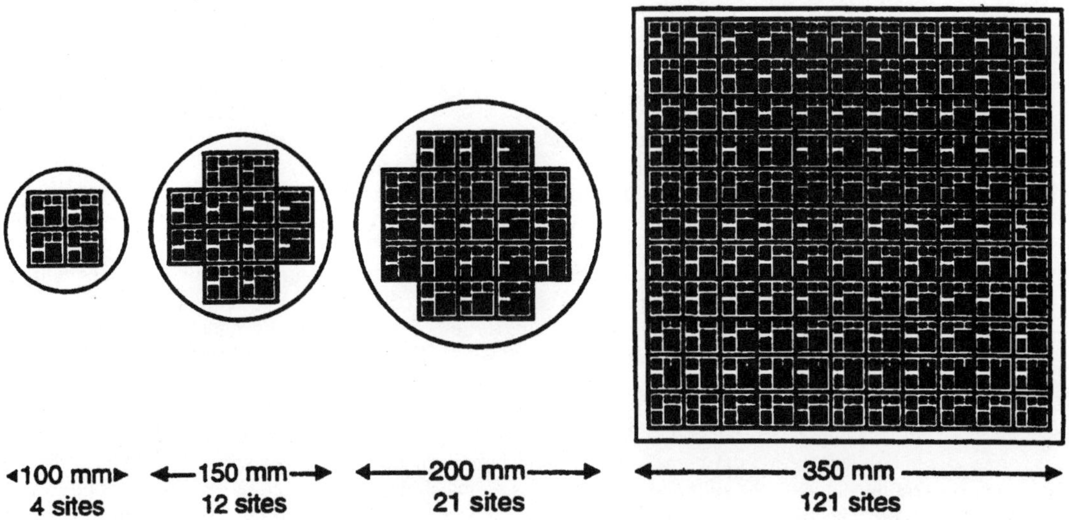

FIGURE 5.2 Trends in PWB technology needs.

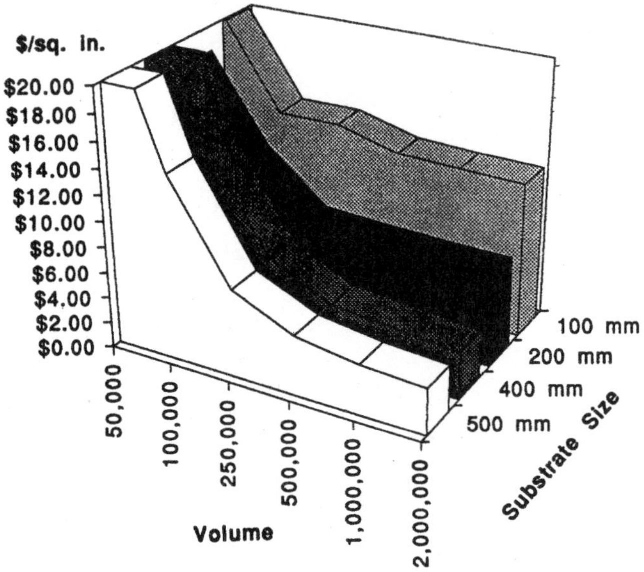

FIGURE 5.3 Motorola LAP cost projections for substrate size from 100 to 500 mm on a side square substrate.[9]

5.3 LEVERAGING THE FLAT PANEL DISPLAY INFRASTRUCTURE

Flat panel displays (FPDs) have been commercially available for more than two decades but, until approximately four or five years ago, had been restricted to relatively small sizes. Over the past few years, the much larger display requirements of laptop computers have become the predominant driving force for the manufacture of large passive and active matrix displays. The flat panel display industry is undergoing rapid growth, i.e., from 1990 levels of less than $1 billion to an anticipated $20–25 billion per year business by the year 2000.[13] This phenomenal growth has fueled a drive toward the development of large-format processing equipment for this market with the bulk of the activity concentrated in Japan.

A de facto standard in glass substrate size has developed around the size requirements of displays for laptop computers. These displays are typically 250 mm (10 in) to 300 mm (12 in) on the diagonal. For this reason, as shown in Fig. 5.5, glass substrate sizes of 350 mm × 400 mm for a two-up configuration with a move to 450 mm × 550 mm panels in a four-up configuration over the next few years is anticipated. This has led to the development of large-format processing tools (examples shown in Fig. 5.6) that are directly applicable to the fabrication of high-density thin film circuitry for large-area MCM-D processing or higher density photovia PWB processing.

For the most part, the lithographic capabilities of FPD processing tools exceed the near-term requirements of the emerging MCM industry with only a few high-performance, low-volume exceptions. The photolithographic requirements of active matrix display fabrication processes are compared with those of integrated circuit, conventional MCM-D technologies and advanced photovia PWB processes shown in Table 5.3. Based on this comparison, the LCD lithography tools are more than adequate for most interconnect applications. In fact, in many aspects, the LCD processing equipment could be simplified to more closely match the less demanding requirements of thin film interconnect fabrication. Consequently, reductions in the capital equipment costs can be expected. DARPA has funded several LAP MCM-D based research activities to enable the development of such an MCM-D processing infrastructure,[14,15] under the premise that any equipment technology

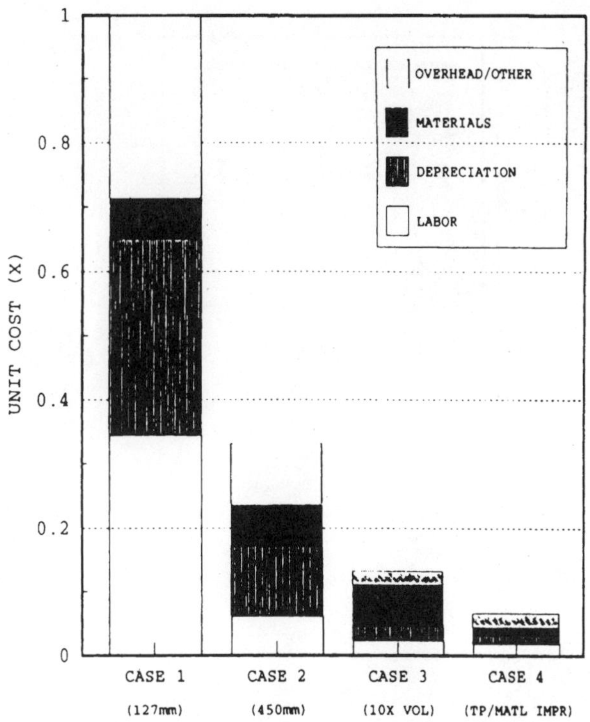

	Case 1	Case 2	Case 3	Case 4
Panel size	127	450	450	450
Substrates/panel	4	64	64	64
Panel starts/yr (k)	15	1	10	10
Panel yield (%)	90	90	90	90
Tool throughput	–	–	–	2–3×
Material usage (%)	–	–	–	75

FIGURE 5.4 IBM LAP cost projections.

developed would be applicable to a wide range of technologies from high-density PWBs for cost-sensitive consumer applications to ultra-high-performance MCM-Ds for military applications.

5.4 LAP CONSTRUCTION MATERIALS: OPTIONS AND LIMITATIONS

5.4.1 Substrates

For the most part, substrates for LAP have been limited to the material classifications of laminates, metals, and glass. The main criteria is that the substrate be inexpensive, available in LAP sizes (minimum 350 × 350 mm), and offer a usable surface finish and suitability for subsequent processing steps.

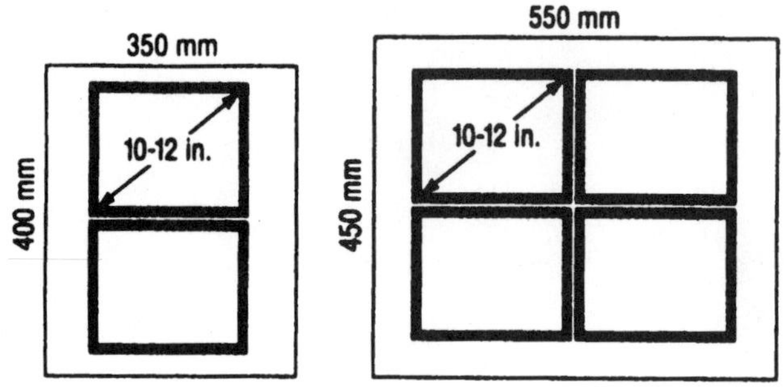

FIGURE 5.5 Driving force for increase in flat panel display manufacturing format.

Table 5.3 Comparison of Requirements for Semiconductor, MCM-D, and FPD Constructions.

	IC	MCM-D	ALMCD
Substrate	silicon	ceramic, Si	glass
Substrate size (mm)	150–200	150	400
Min. resolution (μm)	0.25	≈10	2–4
Overlay	0.25	1.0–5.0	0.5–1.
Masking levels	15–20	8–10	5–9
Line width uniformity (%)	±10	±10	±10
Throughput (parts/hr)	30–60	25–50	5–9

Irregularities in a surface may be thought of in the three categories depicted in Fig. 5.7, namely roughness, waviness, and flatness. Surface roughness (h) is reported as an average value. It is reported either as RMS (root mean square) or CLA (centerline average).[16] By taking an average, one must keep in mind that the average value could mask a single deep peak or valley that could cause discontinuities in deposited thin films. Lack of adequate flatness (h3) will interfere with thermal contact for heating and cooling and will distort patterns during exposure.

Glass Substrates. Glass has been proposed because of the flat panel display infrastructure. Display-grade glass is typified by Corning 7095F. The physical properties of this glass are shown in Table 5.4.

Table 5.4 Physical Properties of Corning 7095F Glass

Property	Value
Thickness	1.1 ± 0.1 mm
	0.7 ± 0.07 mm
Length/width	±0.2/400 mm
Squareness	≤0.10%
Flatness	3.5 μm/25 mm
Warp	≤0.65 mm

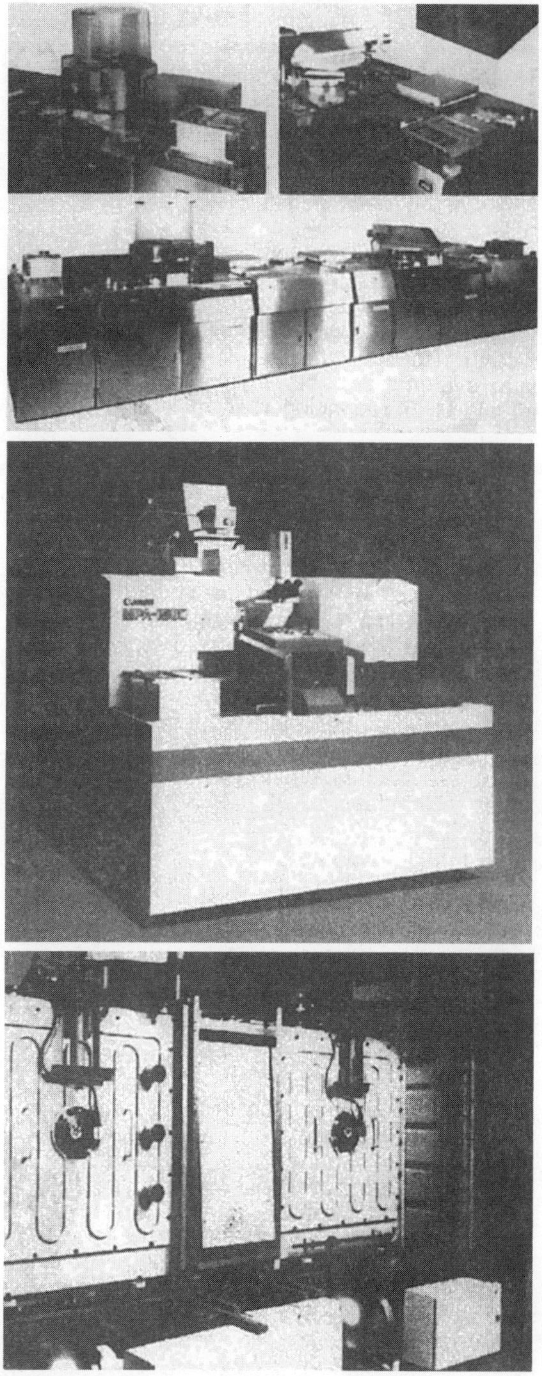

(a)

(b)

(c)

FIGURE 5.6 Photos of LAP processing equipment including (*a*) a spin coating track, (*b*) a photolithography tool, and (*c*) an in-line sputtering system.

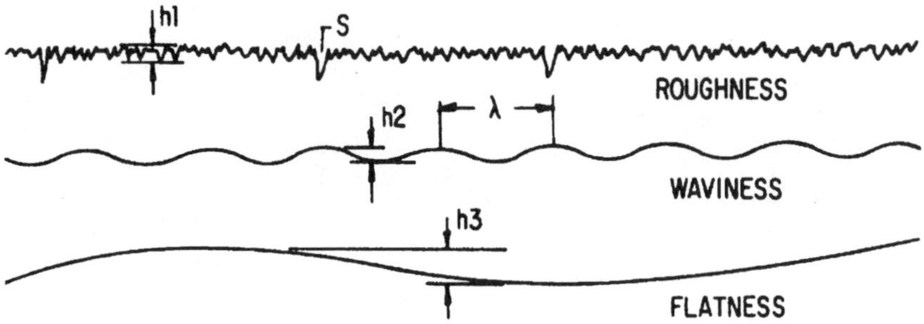

FIGURE 5.7 Characteristics of surface finish.[16]

Laminate Substrates. The copper clad glass reinforced laminates used most commonly in the PWB industry have much rougher surface finishes than is the case with conventional MCM-D substrates. This roughness is the result of two principal factors: the use of woven glass fabric in the laminate, and the imprint of the copper tooth in the laminate during lamination. Typical surface roughnesses of glass reinforced laminate after copper etch is shown in Fig. 5.8.[17] More recently, laminates have been developed that are based on nonwoven oriented glass fibers and/or chopped random fibers that exhibit considerably less surface topography than that observed with woven glass reinforced laminates. Additionally, finer-tooth copper is also being used because of the continuing need to move toward thinner foils with tighter control of thickness. In addition, the thicker copper foils (17 to 35 μm thick) used on laminates result in significant topographical variation. Of all of the dielectrics evaluated either for MCM-D or photovia PWB applications in the industry to date, BCB has been shown to provide the best planarization results.[17] For the most part, the photosensitive epoxy dielectrics, though not capable of providing the same degree of planarization as is possible with BCB, provide planarization, albeit after mechanical polishing, that is adequate for the 100 μm line and space geometries that are being targeted for the near term.

Laminate materials are commonly available in formats of 450 mm × 600 mm and larger. The degree of warpage of a PWB base substrate is dependent on several factors, including the overall thickness of the base substrate, the construction, whether balanced or unbalanced, the distribution of the copper at various layers in the board construction, and so on. Warpage values of approximately 5.0 to 7.5 μm/mm are common. To compensate for this high degree of warpage, contact lithography is most commonly used in the PWB industry, with heavy glass panels used to sandwich the substrate and artwork together, thereby providing intimate contact between the mask and substrate. Additionally, since automated in-line processing equipment is used extensively throughout the PWB fabrication process. With roller driven feeds to assist in transfer of panels through the in-line processing cells, these levels of warpage can generally be readily accommodated. It should be noted, however, that deposited dielectrics will generally increase substrate warpage when using one-sided constructions.

Metallic Substrates. MMS/Dow and the CILAP consortium have chosen aluminum as one of the substrates of choice.[15] The properties of aluminum metal are outlined in Table 3.9. The high thermal conductivity is an advantage over laminate and glass, although several layers of thin insulator do separate the metallic base from the attached chips. If thermal contact to the metal base is required for optimum power dissipation, it is generally at the expense of routing in the area of the substrate immediately around the high-power die. Surface finish requirements are the same as those described for glass. The higher modulus of metals translates into less concern over warping during processing due to the stress imparted by the thin film layers.

Ceramic Substrates. Currently, ceramic substrates generally are not available in the range of flatness and surface roughness required for MCM-D high-density processing in panel sizes larger than 250 × 250 mm. This is the size adopted by IBM, NEC, and others during the fabrication of main-

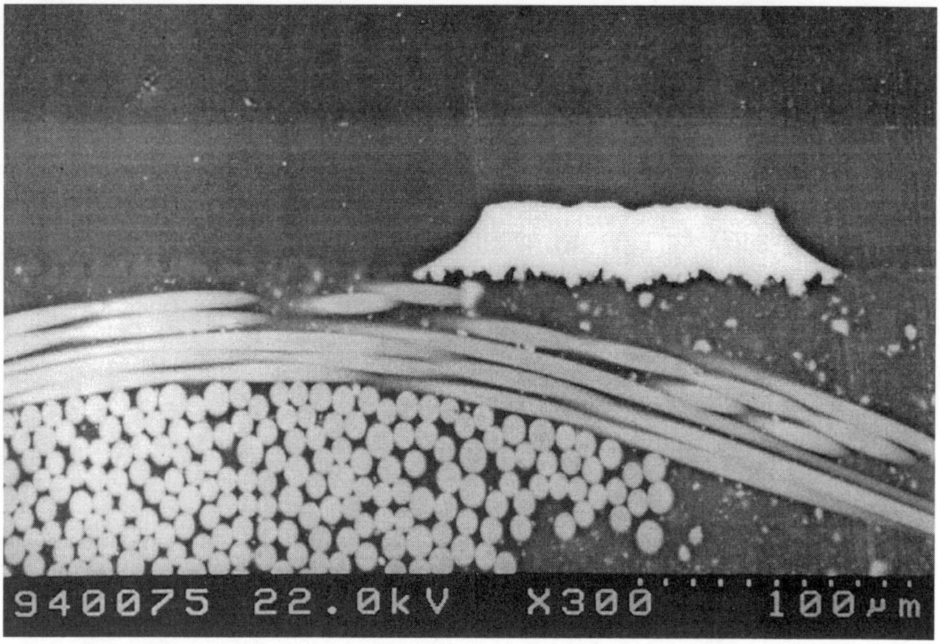

FIGURE 5.8 Cross section of underlying laminate and patterned copper features.[17]

frame MCMs of the 1980s such as the IBM *thermal conduction module* (TCM). IBM has proposed a decal approach (Sec. 5.6.2) in which the thin film interconnect can be transferred to ceramic after fabrication on a large-area substrate such as glass.

5.4.2 Dielectrics

A complete discussion of polymer dielectrics being used on MCM-D structures is given in Chap. 3. The same materials can be used as thin film dielectrics for the fabrication of lower cost MCM-D substrates on large-format glass or metallic substrates.[19] When a laminate interconnection medium is used as the base substrate, deposited dielectrics are selected that are curable at temperatures at or below the Tg of the laminate system used. Consequently, the photovia PWB processes that have been described in the literature to date have used inherently photodefinable dielectric materials. Technologies based on solder mask chemistries with relatively low cure temperatures, most notably epoxy based systems such as Ciba-Geigy's Probimer 52,[20,50–53] and Shipley's FP-9111,[21,22] have been reported. These materials can be cured readily in an in-line IR furnace. Similarly, BCB[23] can also be rapidly cured at temperatures as low as 210° C, making it compatible with most laminates including multifunctional glass reinforced epoxy laminates (Tg > 180–200° C).[17,24,25] These materials are compared in Table 5.5.

Shipley has recently developed FP-91111, a photosensitive epoxy dielectric, for use in thin film MCM D/L and photovia PWB applications. This epoxy-based system images and cures using an onium salt and a sensitizer.[22] The formulations that are currently available have been optimized for curtain coating. Detailed process flow is shown in Fig. 5.9. The described process reportedly has the ability to resolve 25 μm lines and 75 μm vias in 25 μm thick films. The development process is fully aqueous. FP-91111 is reportedly fully cured at 175° C to produce a 200° C Tg.

The Shipley recommended metallization approach involves a full-build electroless Cu process. A temporary plating resist is applied and patterned. A solvent swell and permanganate etch step is then

Table 5.5 Comparison of Properties for LAP Photodielectrics

Property	Probimer 52	Cyclotene 4026 (photo-BCB)	Shipley FP9111
Dielectric constant	3.5	2.65	3.2
Dissipation factor	—	0.0008	—
Maximum Tg (°C)	117	350	≈180
CTE (ppm/°C)	69	52	70
Modulus (GPa)	2.25	2.9	300 kpsi
Tensile strength (MPa)	—	90	≈40
Elongation (%)	—	8	6–8
Moisture absorption (%)	>1.0	<0.2	0.5

used to enhance the metal to dielectric adhesion prior to a Pd/Sn catalysis step and electroless copper plating. Copper adhesion based on peel tests is reported to be in excess of 5 lb/in (0.9 kg/cm). Similar processing approaches have been reported recently.[6]

Dry etchable and photosensitive BCBs (Cyclotene 3000 and 4000 series respectively) have been described in Chapter 3. A surface texturing of the Cu foil by either μ-etching or black oxide treatment results in significantly enhanced Cu/BCB adhesion over that obtained with as received Cu foil due to more effective anchoring (mechanical bonding) of the deposited BCB to the roughened copper surface.[17] The "black oxide" process is commonly used in PWB fabrication to enhance dielectric to metal adhesion. It involves precleaning the copper clad boards in a sulfuric acid/hydrogen peroxide solution, followed by immersion in an alkaline sodium chlorite solution. Alternatively, a copper microroughening process can be carried out by immersion in a potassium persulfate/sulfuric acid solution. Approximately 2 μm of Cu thickness is generally lost from the surface copper layer from an initial 17 μm (0.5 oz) copper foil during this roughening procedure.

5.4.3 Metallization

Large-panel sputtering equipment has allowed MCM-D LAP practitioners to continue using vapor deposition of aluminum and copper. A large-panel sputtering system is shown in Fig. 5.6c to highlight the massive dimensions of this equipment. Despite the significant gains in throughput and panel processing size that have been allowed by the implementation of large-format sputter deposition tools into MCM-D substrate fabrication applications, the cost of full-build sputter deposition remains somewhat prohibitive. There has been a trend in MCM-D approaches toward the implementation of semiadditive metallization processes where a thin seed copper metallization is deposited. In this case, the bulk of the metal thickness is built up by means of electroplating through a patterned photoresist layer with subsequent resist strip and etchback of the sputtered seed layer to isolate conductor traces. The process sequence described in Chap. 3 is analogous to that of the photovia semiadditive process shown schematically in Fig. 5.10.

In the case of the photovia PWB approaches, the most common metallization technologies are based on semiadditive copper plating where an electroless copper plating process is used to deposit a seed layer. The bulk of the copper thickness (up to about 10 μm if necessary) is deposited by panel electroplating with subsequent printing and etching of the copper layer or by semiadditive processing

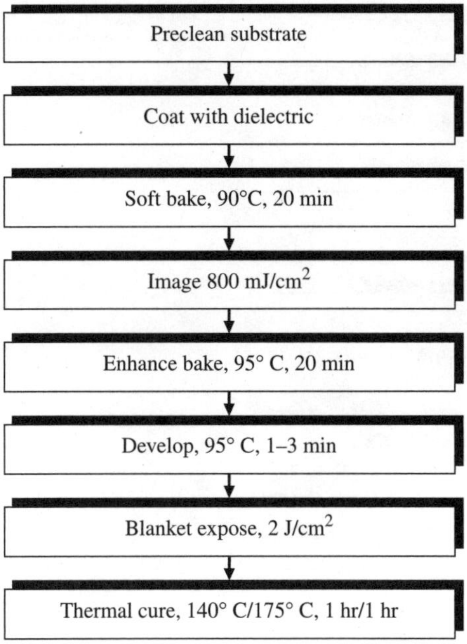

FIGURE 5.9 Process flow for Shipley FP-9111.[22]

where a patterned electroplating process is used. Due to vast electroplate and electroless copper experience that the PWB industry has acquired over the past 30 to 40 years, and its relative unfamiliarity with vapor deposition based metallization processes, it is expected that photovia PWB metallization processes will remain largely the same. Fig. 5.11 shows a cross section of such a photovia.

Some activity is underway in the industry to develop a full-build electroless copper plating process. This process, though likely to provide the greatest potential for the highest density, is typically of considerably lower throughput that is possible with electroplating. Ibiden has described an MCM-L substrate technology based on full-build electroless plating.[6]

To date, electroless/electroplating of copper to photosensitive epoxy-based dielectric materials has proven to be somewhat difficult to maintain in high-volume production. Dielectric roughening processes have been essential to maintain adequate adhesion between metallization layers and underlying dielectric layers and have included plasma as well as chemical and combinations of chemical and mechanical processing.

5.5 LAP UNIT OPERATIONS

The general trend in both photovia PWB and large-panel MCM-D processing is toward increased levels of automation. This increased automation results in dramatically increased throughput, allows minimization of manpower requirements, and results in significantly improved yields. Global defects resulting from handling damage, tool malfunction, and processing errors can be minimized with automation. Localized, random defects, on the other hand, are caused by contamination by the tools, environment, operators, and materials. These general items must also be carefully regulated to ensure high yields. The following section describes a number of the process and equipment options that are currently being investigated for use in LAP fabrication.

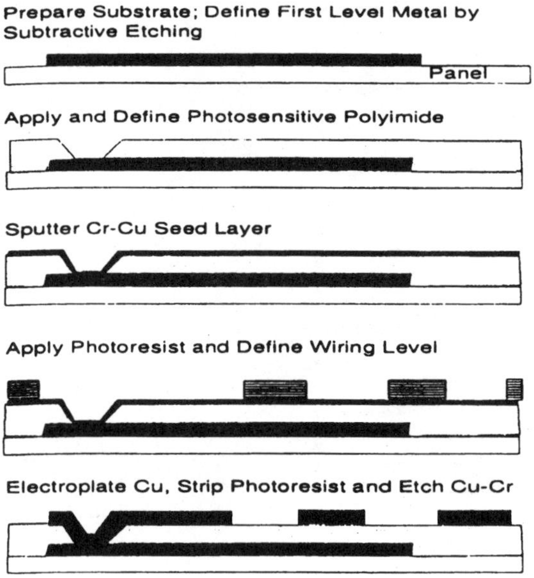

FIGURE 5.10 IBM semiadditive thin film process sequence.

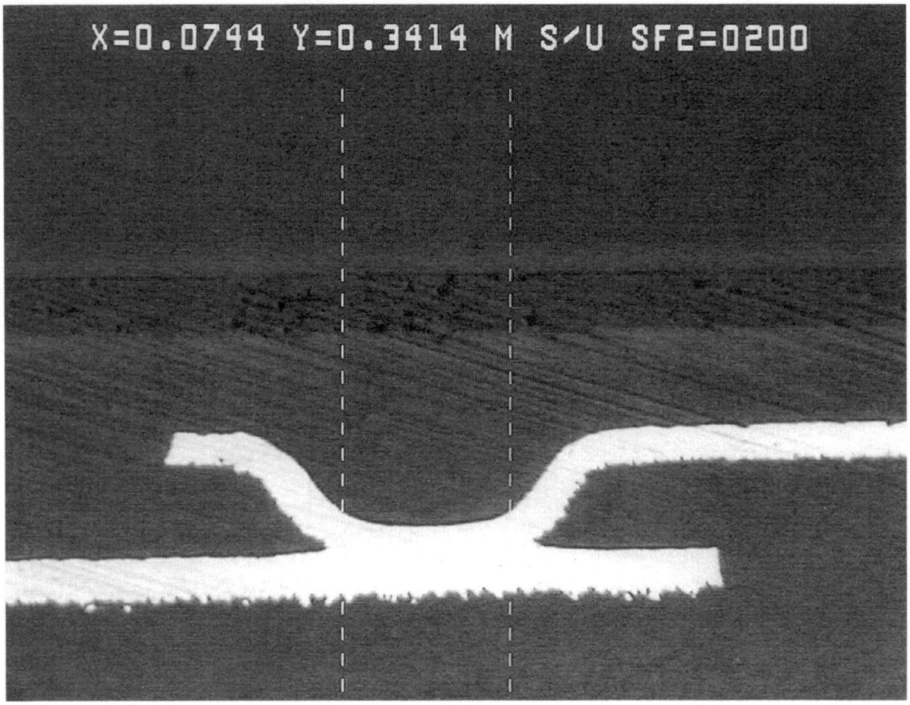

FIGURE 5.11 Cross section of metallized PWB photovia.

5.5.1 Photoresist and Dielectric—Deposition and Curing

The spin coating deposition techniques derived from IC manufacturing technology results in highly uniform coatings but are highly inefficient from a material utilization perspective, with up to 90 percent material wastage. Inherent in the concept of large-area processing is that minimal waste photoresist and polymer dielectric deposition techniques are needed for large-panel, fine line manufacturing at low cost.

The large-format spin coaters (Fig. 5.12) being manufactured by Karl Suss,[26] Steag,[27] Chuo Ryken,[28] and others for the FPD industry are similar in that the chuck, substrate, and a fully enclosed chamber spin in unison. This chuck-in-chamber system avoids air turbulence and minimizes solvent evaporation. Material utilization is improved over that observed with typical open-bowl spin coaters but cannot compete with the other high utilization coating techniques listed below.

In curtain coating, a continuous curtain of the resin dissolved in an appropriate carrier solvent is forced through a slit while the panels to be coated are passed under the curtain on a conveyor belt at high speeds. The distance from the coating head to the substrate is much larger than in other coating techniques. The curtain is established either by gravity or pressure applied to the fluid. The material flow is continuous with undeposited coating material being collected and recirculated. A typical curtain coating process is summarized schematically in Fig. 5.13. By controlling the viscosity, belt speed, and pumping speed, a specified reproducible coating thickness can be maintained over the panel (<10 percent variation in thickness). Curtain coating provides material utilization efficiencies in excess of 90 percent. A disadvantage with curtain coating is the loss of solvent as the film falls through the air. This solvent loss requires the monitoring of resin viscosity and periodic addition of casting solvent to maintain a particular viscosity range. Such systems are available from several sources, including Ciba Geigy[20] and Koating Machinery Inc.[29]

A spray coating process (Fig. 5.14) involving the deposition of a resin in an appropriate casting solvent onto a substrate also has been used. Air or nitrogen atomizes the solution, forming an elliptical spray pattern that is reciprocated over the surface of the substrate with some overlap to ensure uniformity and coverage. A metering pump is used to control volumetric flow rate into the nozzle. Other process variables that can affect the coating thickness and quality include air pressure, air-to-coating ratio, and the viscosity of the coating material. To ensure adequate edge coverage, a significant amount of overspray is required. A number of large-panel compatible spray coating systems are available from companies including Specialty Coating Systems[30] and Zicon.[31]

Extrusion coating has been used to coat substrates at high efficiency levels.[32] Figure 5.15 is a schematic diagram of an extrusion coating system. The dielectric resin is metered through a slotted bar while the substrate is translated beneath the flowing material. Process variables that affect the

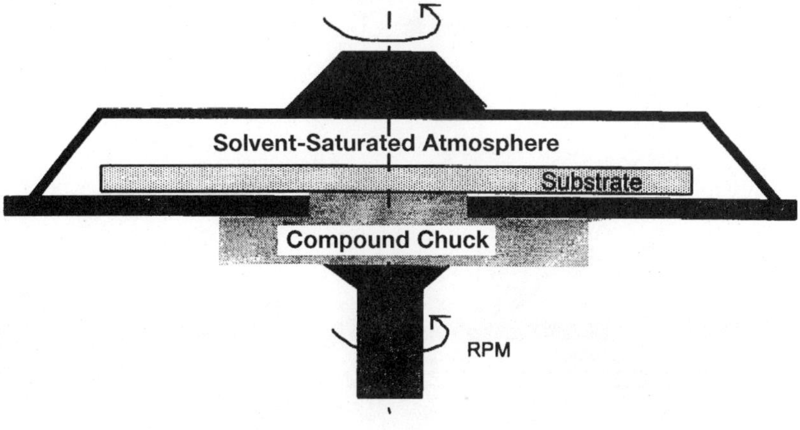

FIGURE 5.12 Schematic of a large-format spin coating process.

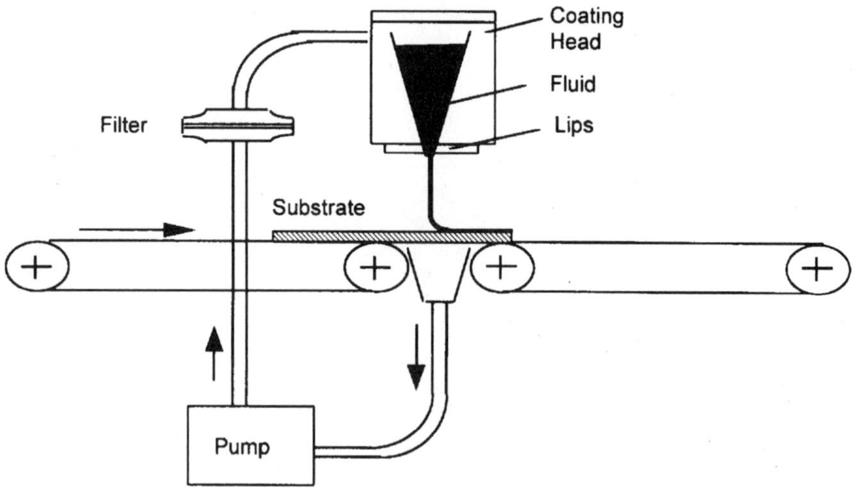

FIGURE 5.13 Schematic of a curtain coating process.

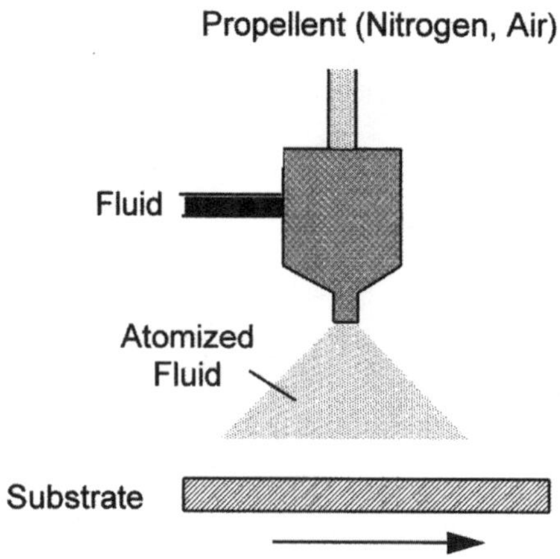

FIGURE 5.14 Schematic of a spray coating process.

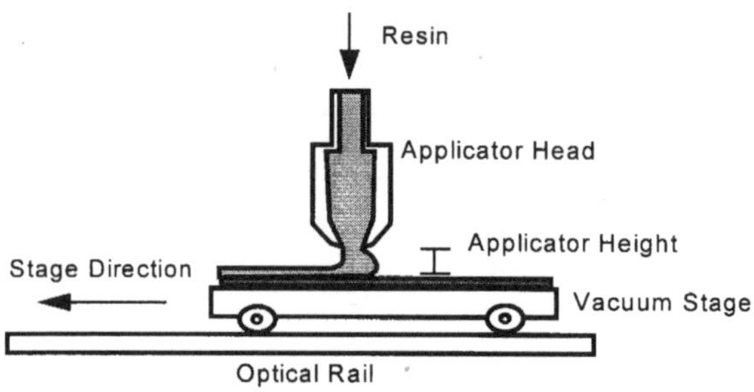

FIGURE 5.15 Schematic of an extrusion coating.

coating thickness and quality include volumetric flow rate, substrate translation speed, gap between the nozzle and the board, and resin rheology. Material utilization is in excess of 90 percent.

Meniscus coating is somewhat analogous to a continuous blade coating process. Fluid is pumped out through a porous tube (with 10 μm pores). A downward laminar flow of solution is established on the periphery of the tube. Excess fluid that runs out over the tube is collected in a reservoir, where the resin is recirculated. The substrate is inverted and translated, at a given speed, over the flowing fluid, at a given gap. This system is shown schematically in Fig. 5.16. Material utilization efficiency is estimated to be >90 percent.[24] Process parameters such as substrate gap, substrate translation velocity, percent solids in resin, and evaporation rate of the carrier solvent control the coating uniformity and film thickness. Meniscus coating is available from Specialty Coatings Systems.[30]

Many vendors of roller coating systems have arisen to satisfy the needs of the printed wiring board industry. The Gyrex system[33] is depicted in Fig. 5.17. In this roller coating process, fluid is transferred from a doctor bar to a rubber coating roll containing numerous fine grooves. The amount of fluid transferred to the coating roll and ultimately deposited on the substrate in controlled by the doctor bar pressure against the coating roll.

Curtain coating has been implemented for dielectric coating deposition in manufacturing by IBM Yasu[4] spray coating, meniscus coating, and extrusion coating have been compared in terms of coatability and coating uniformity.[24]

Uniformity of the films created using these coating techniques is an important issue. Differences in dielectric thickness can affect electrical performance, photolithography, and solvent development of the vias. Uniformity across the boards was measured to be in the range of 1 to 4 percent (1 sigma) for all of the various coating techniques as shown in Fig. 5.18. These results are typical of that achievable with most common cast dielectric materials in use. Extrusion coating[60] and meniscus coating[61] have been studied in more detail (see Sec. 5.6.2).

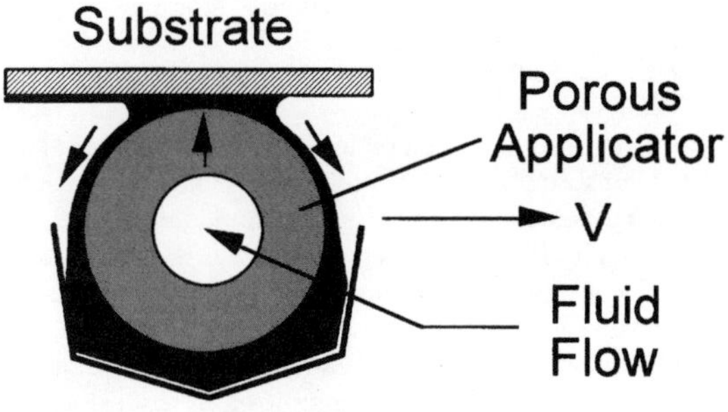

FIGURE 5.16 Schematic of a meniscus coating process.

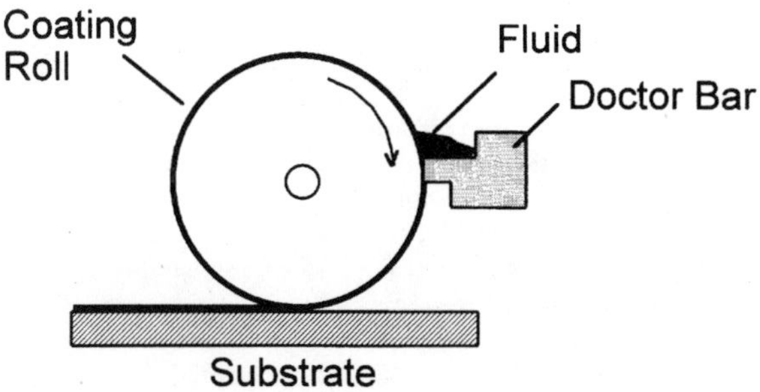

FIGURE 5.17 Schematic of a roller coating process.

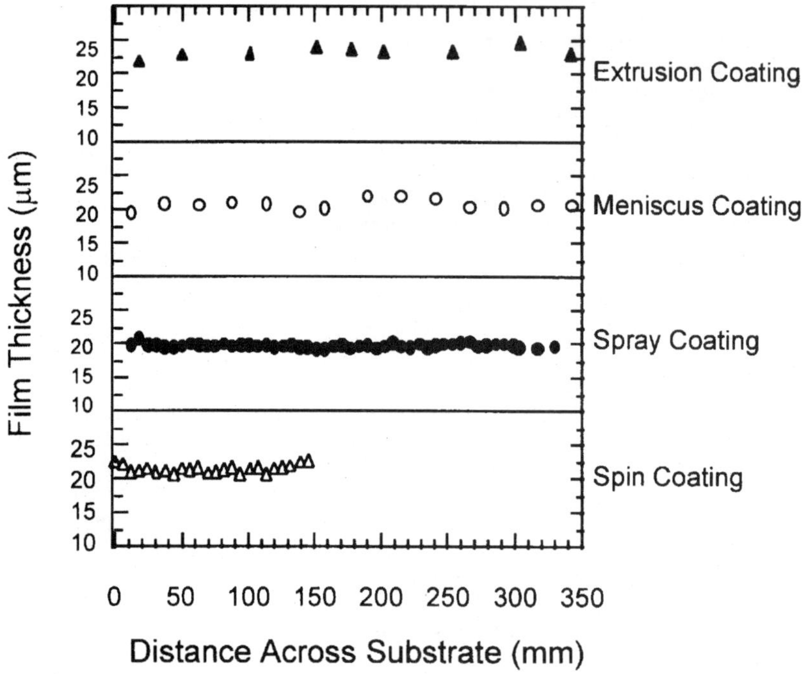

FIGURE 5.18 Uniformity of photo-BCB applied to 350 mm glass.[15]

Curing. There are two important curing parameters: temperature and time. The importance of curing temperature is related to the substrate of choice. When using laminate, the area under the curing time/temperature curve (see Chap. 3) cannot exceed what the laminate can withstand. The major factor for laminate is its Tg. For example, FR-4 laminate with a Tg of 125° C cannot withstand typical PI curing temperatures of 350° C. Cure time is important, because it has a large impact on the overall process time and therefore has a significant impact upon work in progress.

Rapid thermal curing of BCB in a belt furnace reveals typical cure times of <15 min.[34] The BCB layer needs to be soft cured at 210° C to maximize layer-to-layer adhesion. A final hard cure at 250° C is applied after all the thin film layers have been built. Photosensitive epoxy materials commonly used in the fabrication of photovia PWBs can also be cured in a belt furnace with a peak cure temperature below 180° C.

5.5.2 Metal Deposition and Definition

Metal deposition technologies used fall into one of three major classes, namely subtractive, fully additive, and semiadditive processes. In a subtractive process, the complete metal thickness is deposited over the whole surface of the panel. This is commonly carried out by vapor deposition processes, either sputtering or evaporation. Due to the size of the substrates, sputtering has become the vapor deposition method of choice for the processing of large-format substrates. Highly automated, in-line sputtering systems have been developed for these types of applications (see Fig. 5.6*c*). Alternatively, some subtractive processes have involved panel electroplating of copper. In either case, a subsequent photolithographic process is commonly used to pattern the deposited metal layer.

So far, fully additive plating processes have only been used in photovia PWB applications and involve selective electroless copper deposition onto a substrate using either a temporary masking layer to define the area of the panel where copper is to be deposited or a two-layer permanent dielectric process.[6] Although fully additive plating processes are inherently capable of providing the highest densities of all of the metal deposition processes available,[2] they are prone to a number of specific issues, including low plating rates compared to electroplated copper, significant bath stability problems, and higher processing costs. Semiadditive plating processes typically involve the deposition of a thin seed layer over the surface of the substrate either using vapor deposition or electroless copper deposition with the subsequent plating up of the bulk of the copper thickness by electroplating through a temporary patterned resist layer. Semiadditive processing is by far the most widely used of the three metal deposition and definition processes described here.

Horizontal and vertical in-line plating equipment both for electroless and electroplated copper processing is actively being developed by the PWB industry to enable in line photovia PWB fabrication and is directly applicable to the needs of large-area MCM-D fabrication as well.

5.5.3 Photolithography

Scanning and stepping projection aligners have been developed for the flat panel display industry that can currently accommodate substrates up to 550 × 600 mm. Suppliers of such systems include Nikon,[35] Cannon,[36] and MRS.[37] An overview of available FPD photolithographic equipment options has recently appeared.[38] For FPD applications, resolution of 2–4 μm is typically required. With expected MCM resolution requirements as little as 10 μm and substrate sizes of 1.5–25 mm^2 versus ≈625 mm^2 for FPD, it is expected that yields as good or better than those obtained for active matrix thin film displays should be obtainable. Recently thin film display yields have exceeded the 70 percent benchmark.

The current photolithographic limit on 450 × 600 mm (18 × 24 in) PWB panels using contact photomasking is in the 25 to 50 μm range. This limitation is due to both the local roughness of the laminates and the dimensional stability limitations of low-cost mylar masks. It is felt that noncontact projection exposure or high-speed direct write laser techniques will be needed as photovia PWBs and large-panel MCM-D substrates move below 50 μm lines and spaces.[39]

Traditionally, large-area photolithography relies either on step-and-repeat procedures, contact/proximity printing, or focused-beam direct write systems. A contact exposure printer holds the substrate in hard contact with the mask. Illumination with a high-intensity lamp transfers the mask image to the substrate. In a proximity mode, a gap is maintained between the mask and the board. Limitations include defect generation (more from contact than proximity), which causes yield losses and reduces mask life. The proximity mode can result in feature size variation issues due to variations in mask-substrate gap. In a full field projection tool, the image field of the lens is sufficient to accommodate the entire substrate. Such systems are limited by the fundamental trade-off between the desired resolution and the largest board that can be imaged. For IC substrates, full field projection exposure is a common way to achieve both high resolution and high yield at acceptable costs. However, for substrates of significantly larger dimensions, it becomes increasingly expensive to manufacture projection optics for the entire exposure field and retain tight tolerances.

In a step-and-repeat projection tool, the substrate area to be patterned is broken up into several segments which are imaged one at a time by stepping from one segment to the next. Compared to a single-field tool, a step-and-repeat tool significantly lowers the exposure throughput that can be achieved and has been known to produce stitching errors between adjacent exposure segments, and these lead to loss of yield. Direct write systems using a focused laser beam in a raster mode to expose the pixels on a substrate is also a option to satisfy future density needs. Substrate exposure times can be quite long depending on the desired resolution and the complexity of the pattern data.

Several exposure tools have recently been described that have been specifically manufactured for large-area FPD and LAP MCM manufacturing. Although their resolution is not as good as that obtained from IC grade steppers, their resolution should be adequate and their costs are projected to be 3–5 times less.

Tamarack has reported on a lithography system that can handle any substrate size up to 500 × 600 mm.[40–42] The basic idea is to continuously scan the mask and substrate relative to imaging optics and exposure unit so that the mask pattern is successively transferred to the substrate. Figure 5.19 shows a schematic drawing of such an assembly. The collimated UV light beam hits the mask first, then enters the imaging optics and finally forms the 1:1 image of the exposed portion of the mask on the substrate.

Scanning this exposure field over the entire mask area transfers the complete image to the substrate. Large-area masks have recently become available for flat panel display manufacture.[40,41] The system may be combined with an excimer laser and used as a laser ablation tool.

Anvik Corp.[43] has also reported on a projection system that operates in either an exposure or ablation mode. Figure 5.20 depicts the Anvik system. Since the mask and the substrate are mounted rigidly on the single-layer stage, issues related to relative mask and substrate movements during imaging are eliminated. The stage moves the substrate and the mask in unison in both the x and y directions. The XeCl (308 nm) excimer laser light source transfers the pattern images contained within the hexagonal region illuminated on the mask to the substrate. It is claimed that the overlap

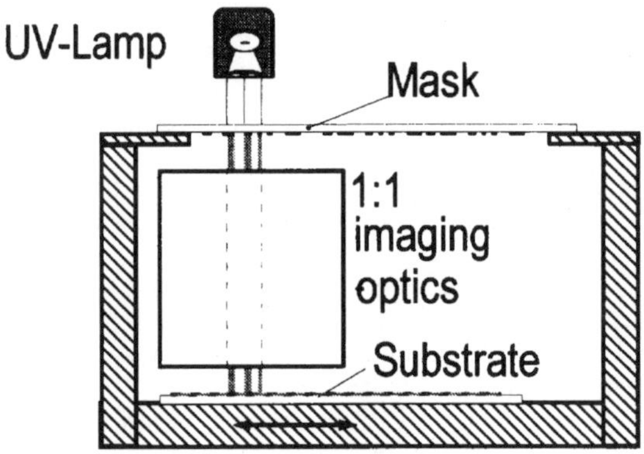

FIGURE 5.19 Schematic of scanning projection exposure.[40]

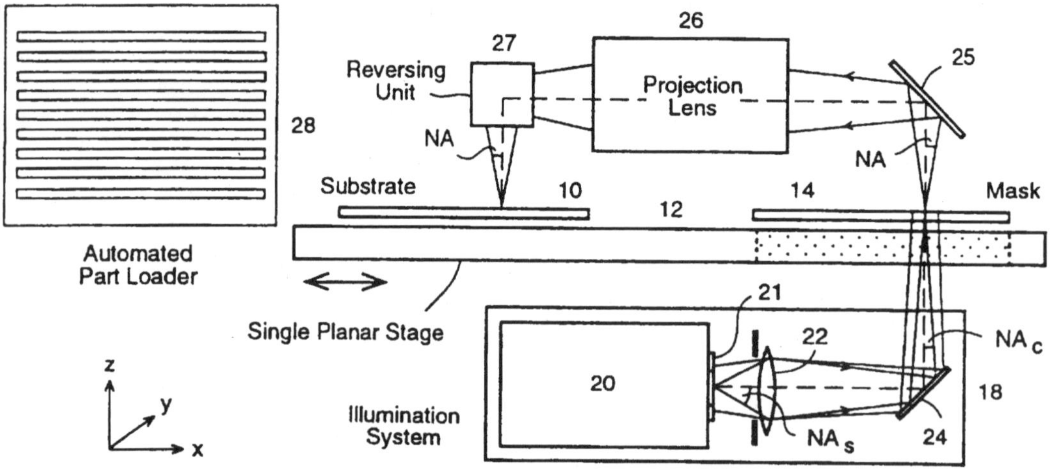

FIGURE 5.20 Schematic illustration of the Anvik LAP patterning system showing a board (10) and a mask (14) held rigidly on a single planar stage (12), an illumination system (18) having an excimer laser light source (20), a projection lens assembly (26), and an automated substrate holder (28).[43]

between adjacent scans is such that the transition from one scan to the next is totally seamless and free from exposure nonuniformity, because the triangular overlap region in the Anvik system comprises only one-sixth of the full beam cross section compared to the 50 percent overlap required by square or diamond shaped beams. The projection imaging lens provides a resolution of under 3 μm.

5.5.4 Via Generation by Laser Ablation

Laser processing can allow the generation of high-density small vias in polymer dielectric layers more efficiently than conventional mechanical drilling for PWB fabrication. Excimer laser processing has been used in the fabrication of high density PWBs at Siemens[44] and in the fabrication of high-density multichip module substrates at IBM[45] and Motorola.[46] The major impediment to more widespread implementation of these laser processing technologies has been the capital and operating costs associated with these systems, poor beam utilization with resultant low throughput, and generally poor system reliability. The many advantages of using lasers for the fabrication of small blind and buried vias, however, have fueled an interest within the electronics industry to resolve these processing and equipment issues. Up until the past few years, all of the laser processing tools that were available on the market were limited to panel sizes of approximately 150 × 150 mm sq. More recently, laser tool suppliers have begun to realize the huge potential market for large-area processing tools within the high-density printed wiring board and large-format MCM-D and have developed tools compatible with these larger formats.

Excimer, Nd-YAG[47] and modified CO_2 lasers[48] are all capable of ablating most common dielectric materials used. Typically, one of three fundamental masking technologies is used to define the ablation sites where vias are to be generated. Point-to-point drilling involves an approach whereby a panel is moved to specific x-y coordinates, and then a hole of specific diameter is drilled by firing a predetermined number of pulses. Typically, point-to-point tools use different apertures that can be switched in fractions of a second to drill different hole diameters. Point-to-point provides the ultimate flexibility in that these systems are completely data driven, eliminating the need for discrete masks. Point-to-point ablation has been found to be a particularly effective approach for high repetition rate (>2k pulses/s) Nd-YAG lasers as they allow ablation rates through 25 μm dielectric of up to 40 to 50 holes per second.

When hole densities rise over approximately 70 holes per inch square, point-to-point drilling becomes less cost effective. In the case where higher via densities are required, the average fabrication cost per via can be significantly reduced by adopting a parallel via generation technology. Conformal or contact metal masks are the simplest mask type and have been demonstrated to provide very reproducible via ablation results for vias down to approximately 25 to 50 μm. Conformal metal masks commonly used include vapor deposited aluminum or copper or electroplated copper layers, typically at least 4 μm thick. Recently, modified CO_2 lasers have been found to be particularly useful for the drilling of blind vias with end-pointing on underlying copper layers and large-panel processing tools with panel handling capabilities of up to 450 × 600 mm.[48]

IBM pioneered the use of projection excimer laser ablation tools for MCM-D fabrication.[45] In this case, a patterned dielectric on quartz mask was used to define ablation sites. Details of this ablation technology have appeared in the literature. Unfortunately, the systems that had been developed for these applications were relatively small format and dependent on costly mask making processes.

Phase mask machining is another masking option that was invented in 1990 by Litel Instruments and has since been in development. In this case, a laser beam is laterally scanned over a phase mask that alters the phase and amplitude of the light so that the desired via pattern is obtained in the workpiece plane. This masking approach provides the highest beam utilization and the greatest throughput of all of the current laser options because large arrays of vias can be drilled simultaneously. The schematic shown in Fig. 5.21 shows the mask construction and the beamlets that are generated resulting in mass via generation. Via generation rates of several thousand vias per second have been reported.[49] The major limitation of phase masking technologies are the elevated mask making costs, currently estimated at significantly greater than $10,000 per mask for a 6 × 6-in mask plate.

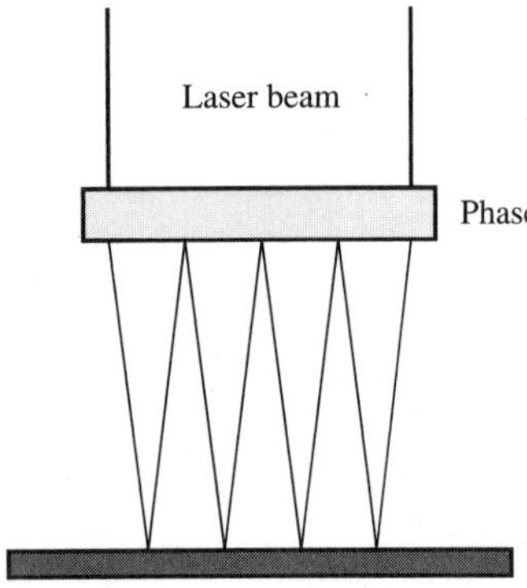

FIGURE 5.21 Phase mask via generation.

5.5.5 Inspection and Electrical Test

Due to the sequential nature of photovia LAP technologies, inspection and electrical test, both while in process and after completion, in full panel form is crucial during process development. Over the past few years, automatic optical inspection systems have been developed with the huge data processing capabilities required to map optical defects of high-resolution patterns over large areas. This has enabled a more statistical analysis of defect-contributing factors in processes and has been shown to be particularly useful in establishing new processes. The goal in LAP processing, however, for the most part must be to reduce the amount of inspection that is currently done, which is close to 100 percent inspection of every circuit layer to just a final electrical test at the end of the process.

Electrical test systems capable of handling the large-panel formats as well as to test a high mix of different products are beginning to become available.

5.6 HIGH-DENSITY LAP PROCESSES

High-density LAP processes were in production and being scaled up as of 1996. They can be divided into several categories, depending on whether the high-density wiring is added to the substrate in a sequential fashion.

5.6.1 Sequential (on Laminate)

The addition of thin film redistribution layers to a laminate base to interconnect high I/O area array chips was first publicly disclosed by IBM-Japan,[50–53] called the *surface laminar circuit* (SLC) technology. The IBM SLC process has been licensed by AMP-AKZO[58] and others. IBM Endicott has developed an analogous photovia technology using dry film dielectric called *film redistribution layer* (FRL).[25,54] Disadvantages to dry film dielectric are reported to be relatively poor uniformity over uneven topography and design limitations (due to the available commercial film thicknesses) and higher cost. These technologies are discussed in detail in Chap. 4.

5.6.2 Sequential (Nonlaminate)

IBM Fishkill. IBM Microelectronics' large-format MCM-D processing program has been described in detail.[55,56] A class 100 clean room is proposed for dielectric and photoresist application and exposure operations. The main elements are via definition by photo PI (g line) or laser ablation and wiring definition by pattern electroplating. One strategy in use by IBM involves the fabrication of thin film on temporary glass substrates with subsequent removal of the "decals" for placement onto various base substrates as shown in Fig. 5.22.

The vias are patterned by a 1X projection printer. The photo PI dielectrics are batch immersion developed. Following develop, the dielectric is cured at 350° C under nitrogen and descummed with an oxygen plasma (as little as 400 Å of residue is reported to cause electrical opens).

Although subtractive etching of a blanket Cr/Cu/Cr sputtered or evaporated metal offers the lowest cost, large linewidth variations make this process unacceptable for aggressive grounders. Patterned electroplating provides the most uniform wiring lines and equivalent thickness variations to sputtering. Vacuum deposition of a metal seed layer is followed by electrolytic plating through a resist stencil. After flash seed removal, electroless metal capping is used to prevent Cu/PI interac-

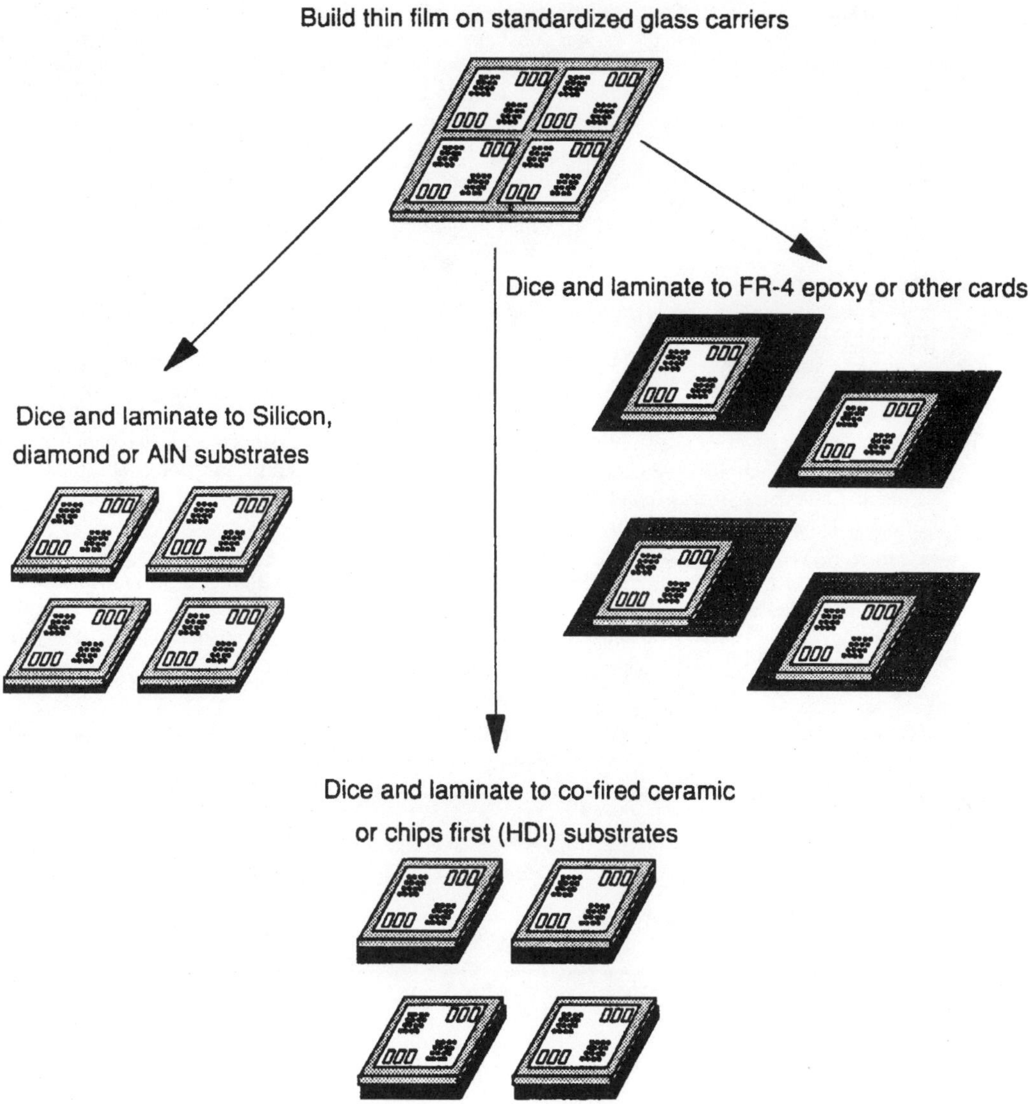

FIGURE 5.22 IBM thin film decal approach.[57]

tions (see Chap. 3). The flash etching of the seed layer is reportedly extremely critical, because shorts due to ineffective seed removal can predominate. This technology has been demonstrated on 300 × 300 mm panels.

Hughes. Hughes has announced the construction of a 600 × 600 mm format large-area processing line for the fabrication of thin film on ceramic interconnect. Published process details include the use of low-stress polyimide as the dielectric, dielectric application by extrusion coating, and phase mask machining (see Sect. 5.5.4) for via formation. It is claimed that this process will allow the production of 150 × 225 mm thin film MCMs for the military.[14]

MMS/Dow. A team of MMS and Dow engineers have been working since the fall of 1994 to build a LAP manufacturing line using various substrates and a Cu/BCB interconnect process [62]. In addition, they are both members of the DARPA-supported initiative, the "Consortium for Intelligent Large Area Processing," under which MMS and Dow [with partners FAS (extrusion coating), RTC (automated belt furnace), Johnson Mathey Electronics (laminate panels), Motorola (test vehicles), and Sandia (test)] developed LAP optimized benzocyclobutene formulations and automated, sensor controlled, coating and curing operations.[15]

The MMS/Dow team developed the MMS Cu/BCB process into a 400 mm format and began shipping LAP prototypes in late 1996, with full qualification and volume shipments scheduled for late 1997.

Panel Substrate Issues. Initial developments centered on aluminum, glass, and PWB.

As discussed in Sec. 5.4.1, substrate finish is of paramount importance for a manufacturable LAP process. Substrate bow is also a great concern. Since bow is directly related to panel dimensions, an equivalent stress will generate more bow on a larger panel. This has been studied in detail by the CILAP consortium.[63] For small deflections, Eqs. 3.5 and 3.6 an be used on 100 to 150 mm wafers to accurately calculate the bows normally encountered. For large deflections, nonlinear analysis and gravity must be taken into account. Table 5.6 shows the calculated versus experimental data for a 400 × 400 mm, 1.27 mm thick, Al panel (BCB stress of 30 MPa). Maximum deflection is observed at the corners of the panel. It was found that application of a slight vacuum (0.01 psi) would decrease the calculated deflection by approximately 50 percent.

Coating. The inherently lower efficiency of spinners as discussed in Sec. 5.5.1 compelled the LAP team to evaluate other high-efficiency coating technologies for BCB dielectric and photoresist, including extrusion[60] and meniscus[61] coating.

Extrusion Coating. Extrusion coating (see Fig. 5.15) was studied under the CILAP DARPA agreement using a FAS extrusion coater especially designed for BCB large area processing.[60] Table 5.7

Table 5.6 Comparison of Simulation and Experimental Data for Maximum Deflection (mm) in a 400 × 400 mm, 1.217 mm Thick Al Panel

BCB thickness (μm)	Linear analysis	Nonlinear analysis	Nonlinear analysis with gravity	Exp. data	Application of 0.01 psi vacuum
40	1.62	1.48	0.89	0.62	0.4
15	0.54	0.53	0.13	0.15	

Table 5.7 Intrapanel Uniformity

Panel number	Avg. panel thickness (μm)	Standard deviation (%)
30	13.78	2.25
31	13.83	1.29
32	13.91	1.93
33	13.93	2.26
34	13.94	2.34
35	13.88	1.64
36	13.91	1.94
37	13.84	1.49
Avg.	13.85	1.89

shows intrapanel BCB thickness uniformity data generated from 100 equidistant measurements on a 400 × 400 mm glass panel. Inter panel uniformity showed a standard deviation of 0.4 percent.

Meniscus Coating. Meniscus coating (Fig 5.16) was also studied in detail by the Dow/MMS team.[16] The CSC Cavex™ meniscus coating revealed inter- and intrapanel uniformity for BCB and AZ photoresists as detailed in Tables 5.8 and 5.9, respectively.

Curing. One of the inherent advantages of the Dow/MMS BCB process has been "rapid thermal curing," which offers an extremely rapid reduction in cure time when compared to other dielectric processes. For LAP processing, each substrate offered a different cure situation. For instance, polished aluminum is reflective to IR, leading to BCB temperatures 20–30° C higher than the base aluminum (due to BCB emissivity of 0.6–0.8). Glass, on the other hand, had no such reflectivity issues, and the substrate temperature exactly matched the BCB layers above it. PWB substrates required additional effort to identify an appropriate rapid thermal curing process. BCB's cure temperature of as low as 210° C allows thin film structures to be cured on PWB substrates. However, it was necessary to balance ramp rate, soak time, and peak temperature to prevent barrel cracking. The developed furnace has a throughput of one to two panels per minute, seven heating and two cooling zones, variable slew and dwell times, dielectric cure monitoring, and control by the use of noncontact IR sensors. Figure 5.23 shows a comparison of 150 mm wafer and 400 mm panel processing.

5.7 CONCLUSIONS AND FUTURE TRENDS

The trend toward large-area processing of deposited dielectric based interconnects, whether based on MCM-D or photovia PWB origins, is continuing at an accelerated pace as the need for higher density interconnect structures required to feed the emerging wireless communication and computing markets grows. As MCM-D-based technologies strive to meet the cost challenge set by MCM-L and high-density PWBs, there is a need to avoid the attraction of adopting solutions that provide interconnect density capabilities beyond the requirements of the bulk of the high-volume interconnect segments of the electronics industry to achieve cost competitiveness. On the photovia PWB side of

Table 5.8 Inter- and Intrapanel Uniformity for BCB Coatings

Panel number	Process rate (min/panel)	Avg. thickness (μm)	Variation within Panel (RSD, %)
1	1.5	6.09	1.64
2	1.5	6.15	0.76
3	1.5	0.55	1.13
4	1.5	0.65	1.13
5	1.5	0.27	1.08

Table 5.9 Inter- and Intrapanel Uniformity for AZ™ Photoresists

Formulation	Process rate (min/panel)	Thickness (μm)	Variation within panel (RSD, %)	Panel-to-panel variation (RSD, %), five panels	Temp (°C)
AZ P4110	2.75	10.26	1.3	n/a	22
	1.0	10.4	<2	0.4	38
AZ P4210	1.33	20.7	<2.1	2.8	22
	1.0	19.3	<2.1	1.0	38

FIGURE 5.23 Comparison of 150 mm and 400 mm LAP processing.

the house, especially in North America, PWB suppliers need to awaken to the fact that their inactivity in implementing these high-density capabilities into their factories has led to a significant erosion of the higher end of the PWB market to offshore competitors.

REFERENCES

1. D. Suryanarayana, R. Hsiao, T.P. Gall, and J.M. McCreary, "Flip Chip Solder Bump Fatigue Life Enhanced by Polymer Encapsulation," *Proc. 40th IEEE ECTC,* p. 338, 1990.
2. T.G. Tessier, J. Aday, and H. Crews, "Selecting Flip Chip on Board Compatible High Density PWB Technologies," *1995 Int. MCM Conf.,* p. 42, 1995.
3. D.N. Light, J.S. Kresge, and C.R. Davis, "Integrated Flex: Rigid Flex Capability in a High Performance MCM," *1994 Int. MCM Conf.,* p. 430, 1994.
4. Y. Tsukada, "Low Cost Thin Film Multilayer Substrate Fill Gap with LSI, *Nikkei Microdevices,* pp., 61, 1991.

5. C. Boyko, F. Bucek, V. Markovich, and D. Mayo, "Film Redistribution Layer Technology," *Proc. 43rd ECTC,* p. 302, 1993
6. M, Nakamura, M. Kato, M. Asai, and H. Takenaka, "High Reliability, High Density, Build Up Printed Circuit Board for MCM-L," *1995 Int. MCM Conf.,* p. 36, 1995.
7. T.G. Tessier and P.E. Garrou, "Overview of MCM Technologies: MCM-D," *Proc. 1992 Int. Symp. Hybrid Microelectronics,* p. 235, 1992.
8. C. Hodges, J. Hull, J. Strandberg, J. Hagel, H. Thiede, and A. Karlsson, "Large Panel Format MCM Substrates," *Proc. 1993 Int. MCM Conf.,* Denver, Co., p. 125, 1993.
9. T.G. Tessier and E. Myszka, "Approaches to Cost Reducing MCM-D Substrate Fabrication," *Proc. 43rd ECTC,* p. 570, 1993.
10. D. Frye, M. Skinner, M. Heistand, P. Garrou, and T.G. Tessier, "Cost Implications of Larger Area MCM Processing," *Proc. 1994 Int. MCM Conf.,* Denver, p. 69, 1994.
11. P.E. Garrou, et. al., "Cost Modeling and Fabrication of MCM's Using Photosensitive Cyclotene," *Proceed. Int. Microelect. System Tech. Symp.,* Berlin, p. 569, 1994.
12. E. Perfecto, G. White, and G. Saji, "Manufacturing Considerations for MCM-D Large Format Fabrication," *Proc. Int. Symp. Hybrid Microelectronics,* p. 208, 1994.
13. W.C. O'Mara, "Active Matrix Liquid Crystal Displays, Markets," *Solid State Technology,* p. 63, (January 1992).
14. P.A. Trask and V.A Pillai, "Large Format MCM-D Processing," *Proc. 1994 Int. MCM Conf.,* Denver, p. 94, 1994.
15. L.J. Laursen and P. Garrou, "Consortium for Intelligent Large Area Processing (CILAP)," *Proc. 1995 Int. MCM Conf.,* Denver, p. 253, 1995.
16. R. Brown, "Thin Film Substrates," in *Handbook of Thin Film Technology,* Maisel and Glang, Eds., 1970, McGraw-Hill.
17. A.J. Strandjord, R. H. Heistand, J.N. Bremmer, P.E. Garrou, and T.G. Tessier, "A Photosensitive BCB on Laminate Technology (MCM-LD)," *Proc. 44rd ECTC,* p. 374, 1994.
18. P.A. Trask, "Large Format MCM-D Processing," ISHM, Ogonquit Advanced MCM Technology Workshop, June 1995. Referenced by permission of author.
19. W. Lautenberger, B.C. Auman, J.R. Butera, D.L. Goff, C.N. Lazaridis, H.K. Seiler, and J.W. Labadie, "Polyimides for Large Area Multichip Modules," *Proc. 1995 Int. MCM Conf.,* Denver, p. 501, 1995.
20. Ciba Geigy Tarrytown, NY.
21. Shipley Company, Marlborough, MA.
22. P. Knudsen and R. Brainard, "A Photoimangeable Dielectric for Sequential PWB Fabrication," *Proc. Surface Mount Int.,* p. 351,1993.
23. Dow Chemical, Midland, Michigan.
24. A. Strandjord, P.E. Garrou, R. Heistand, and T.G. Tessier, "MCM-D/L: Large Area Processing Using PS-BCB," *IEEE Trans. CPMT,* vol. 18(2), 1995.
25. K. Matsui, T. Shimoto, Shimada, Y., K. Utsumi, "MCM D/L Technology Based on Cu/Photo-BCB Thin Film Multilayer," *Proc. ISHM,* 1995, p. 396.
26. Karl Suss, Waterbury Center, Vermont.
27. Steag Industrie Inc., 1150 Route 52, Fishkill, New York 12524.
28. Chuo Ryken Co. Ltd., 8-5-1 Higashi-Suna, Koto-ku, Tokyo 136, Japan.
29. Coates ASI, Phoenix, AZ.
30. Specialty Coating Systems, Indianapolis, IN.
31. Zicon Corp., Mount Vernon, NY.
32. T. Snodgrass and G. Blackwell, "Advanced Dispensing and Coating Technologies for PI Films," *Proc. 1992 Int. MCM Conf.,* Denver, p. 428–434, 1994.
33. Gyrex-Talus Industries, Goleta, CA.
34. P.E. Garrou, R.H. Heistand, M. Dibbs, T.A. Manial, C. Mohler, T. Stokich, P.H. Townsend, G.M. Adema, M.J. Berry, and I. Turlik, "Rapid Thermal Curing of BCB Dielectric," *IEEE Trans. CHMT,* vol. 15, 1993, p. 46–53.
35. Nikon Precision Inc., Belmont, CA.
36. Canon USA Inc., Irving, Texas.

37. MRS Technology Inc, Chelmsford, MA.

38. P. Burggraaf, "FPD Lithography: System Technology Pushed to the Limit," *Semiconductor International,* December 1992, p. 42–44.

39. D.W. Palmer and W. Worobey, "From Display and PWB to MCMs: Large Panel Manufacturing," *Proc. 1994 Int. MCM Conf.,* Denver, p. 81, 1994.

40. H. Mueller, "Large Area Fine Line Patterning by Scanning Projection Lithography," *IEEE Trans. CPMT-B,* vol. 18, p. 33, 1995.

41. G.R. Guel and D.E. Rector, "Very Large (25 in. diagonal) Precision Masks for FPD Manufacturing," in 1994 SID International Symposium, *Digest of Technical Papers, Society for Information Displays,* p. 764, 1994.

42. R.E. Sheets and H. Muller, "Scanning Projection Exposure Tools for Large Area Thin Film Displays," 1994 Display Manufacturing Conference, *Digest of Technical Papers, Society for Information Displays,* p. 97, 1994.

43. K. Jain, J. Hoffman, T. Dunn, W. Folster, and D. Panchal, "Large Area, High Throughput, High Resolution Patterning and Via Drilling System," *Proc. 1995 Int. MCM Conf.,* Denver, p. 327, 1995.

44. F. Bachmann, "Excimer Laser Drill for Multilayer Printed Circuit Boards: From Advanced Development to Factory Floor," *MRS Bulletin,* p. 49, Dec. 1989.

45. T.F. Redmond, J.R. Lankard, J.G. Balz, G.R. Proto, and T.A. Wassick, "The Application of Laser Process Technology to Thin Film Packaging," *Proc. 42nd ECTC,* p. 763, 1992.

46. T.G. Tessier, W.F. Hoffman, and J.W. Stafford, "Via Processing Options for MCM-D Fabrication: Excimer Laser Ablation vs. Reactive Ion Etching," *Proc. 41st ECTC,* p. 827, 1991.

47. C. Pico, Y. Sun, M. Owen, B. Larson, B. Kreitzberg, and E. Swensen, "Micromachining of Polyimides Using Ultraviolet Solid State Lasers," *2nd Int. Conf. on Flexible Circuits,* p. 152, 1995.

48. J.M. Morrison, T.G. Tessier, and B. Gu, "A Large Format Modified TEA CO2 Laser Based Process for Cost Effective Small Via Generation," *Proc. 1994 Int. MCM Conf.,* Denver, p. 369, 1994.

49. A.H. Smith, "Phase Mask Machining for High Throughput Via Formation," *Proc. Int. MCM Conf., Denver,* p. 208, 1993.

50. Y. Tsukada, S. Tsuchida, Y. Mashimoto, "Surface Laminar Circuitry and Flip Chip Attach Packaging," *Proc. Int. Microelectronics Conf.,* Yokohama, p. 252, 1992.

51. Y. Tsukada, S. Tsuchida, Y. Mashimoto, "Surface Laminar Circuit Packaging," *Proc. 42nd ECTC,* p. 22, 1992.

52. Y. Tsukada, Y. Yamanaka, "Design and Electrical Performance of Circuit Laminar Circuit/Flip Chip Attach Packaging Technology," *Proc. 1993 IEPS,* p. 1111, 1993.

53. Y. Tsukada, "Design and Performance of SLC-MCM," *Proc. NEPCON West,* p. 521, 1994.

54. D.O. Powell, T. Heline, R. Magnuson, V. Markovich, A. Bhatt, "Early Experiences with FRL Technology," *Proc. 1994 IEPS,* p. 649, 1994.

55. E.D. Perfecto, G. White, K. Saji, W. O'Donnell, T. Redmond, "Manufacturing Considerations for MCM-D Large Format Fabrication," *Proc. ISHM,* 1994, p. 208.

56. G. White, E. Perfecto, T. DeMercurio, T. Redmond, M. Norcott, "Large Area Fabrication—A Practical Approach to Low Cost MCM-D," *Proc. 1994 Int. MCM Conf.,* Denver, p. 37, 1994.

57. C. Narayan, S. Purushothaman, F. Doany, A. Deutsch, "Thin Film Transfer Process for Low Cost MCM Fabrication," *IEEE Trans CHMT, B,* vol. 18, p. 42, 1994.

58. C. Sullivan, R. Funer, R. Rust, M. Witty, "Low Cost MCM-L for Vehicle Applications," *Proceed Int. MCM Conf.,* Denver, 1996, p. 142.

59. E. Enomoto, M. Asai, Y. Sakaguchi, C. Ohashi, "High Density Printed Wiring Boards Using Advanced Fully Additive Processing," *Proc. IPC 32nd Annual Meeting,* 1989, p. 1.

60. Wells, G.T., Gibson, G., Newquist, R., Martin, B., Chieh, E., Lykins, J., Ho, C., "Large Area Processing: Performance Studies on Extrusion Coating," *Proc. IMAPS,* 1997, in press.

61. Garrou, P., Rehg, T., Heistand, R., Martin, B., Chieh, E., Lykins, J., Ho, C., "Large Area Processing: Meniscus Coating of Thin Film Polymer Dielectric and Photoresist," *Proc. IMAPS,* 1997, in press.

62. Chieh, E., Lykins, J., Ho, C., Moresco, L., Koblis, M., Haranghy, M., Skinner, M., Chen, T., Garrou, P., Cummings, S., Heistand, R.H., "Development of Large Area Processing for Thin Film Substrates," *Proc. Int. MCM Conf.,* Denver, 1997, p. 234.

63. Yang, K., Im, J., Heistand, R.H., "Significance of Coating Stress on Substrate Bow in Large Area Processing of MCMs," *Proc. IEEE IEMT,* Austin, 1997.

CHAPTER 6

3-D PACKAGING

Dr. Christian Val
3-D Plus S.A.

6.1 INTRODUCTION

The driving forces behind 3-D packaging are the same as those behind 2-D interconnection with the arrival of the multichip module (MCM) on the market, which then allowed

- Electronic systems density increase
- Performance increase
- Cost of packaging reduction

Since the development of integrated circuits, it has been known that densification leads to performance improvements and cost reduction. Table 6.1 presents the improvement factors that can be expected in 2-D techniques by the beginning of the year 2000. The vision involves direct flip chip attachment to a highly integrated substrate with passives and optoelectronics to achieve an order of magnitude improvement in cost, performance, size, and reliability for the future requirements of the electronic market.[1] Finally, 3-D technology permits by-passing the barrier of 100 percent of the efficiency of the packaging—in other words, to have one silicon area superior to that of the substrate area by which the chips are interconnected. The key question is, "Why bypass this ratio of 100 percent?." The main response is "savings." It can be easily verified that a drastic reduction in size, weight, or volume will result in a significant reduction in production cost. In this area, we can observe a very strong trend over the past 30 years.[2]

Table 6.1 2-D Packaging Evolution

	Current packaging	PRC packaging	Improvement factor
Size (nominal)	10	1	10×
Packaging efficiency	8%	80%	10×
Performance	50–100 MHz	200 MHz–1 GHz	5–10×
Cost of MCM	10	1	10×
Reliability (ppm)	10	1	10×

Source: Rao Tummala, Packaging Research Center, Georgia Tech, 1995

The evolution of densification is very strong at the level of two large interconnection families:

- Surface mounting techniques using packaged dice
- Multichip module techniques using bare dice

6.1.1 Surface Mounting Techniques for Packaged Dice

The trend to higher density has led to the following:

- 1-D interconnection, with two or four rows of leads, spaced from 2.54–0.5/0.3 mm up to the very small peripheral array (VSPA)
- 2-D interconnection with the coming of *pin grid array* pitch at 2.54 and 1.27 mm, then of the *pad grid array* and *ball grid array*

With those techniques, the densification at first led to a decrease of the thickness of the packages to meet the needs of portable equipment. The curve of Fig. 6.1 shows this trend where, during the past 30 years, the thickness of packages have been divided by 10. For the next 10 years, one can estimate that it will be possible to divide by 100.

6.1.2 Multichip Module Techniques for Unpackaged Dice

The trend for unpackaged dice seems to be almost the same as that for packaged ones. One observes in Fig. 6.2 that a factor of 8 to 10 has been gained in packaging efficiency in 30 years between a through hole mounted printed circuit board and the densest multichip module. Note that a ratio of 60 to 70 percent was obtained with the thinnest line width (10 to 25 μm), and a ratio of several hundred percent achieved with linewidths of 50 to 300 μm with the 3-D technique. Another view of densification is given by the diagram in Fig. 6.3, which presents the number of connected points per dm^2, with respect to surface mounting techniques, MCM techniques, and 3-D techniques. 3-D interconnection allows for ten times more points of connection per unity of surface (up to 100, 000 points/dm^2). Irvine Sensors has announced a stacking of 1000 chips in the next ten years, with almost 10^7 connected points/dm^2.[2,3] Two parameters under consideration—packaging efficiency and number of connected points per cm^2—indicate an important gap between the 2-D and the 3-D interconnection, and we know that other criteria can affect these techniques. In either case, the thermal, mechanical, and electrical behavior of the 3-D stacking has not yet limited their applications. The evolution of the technique of 2-D interconnection has been succinctly described by drawing a parallel with that of 3-D techniques. In doing so, we understand

- The stacking approach of identical chips, e.g. memory, in comparison with the single packaged die
- The stacking approach of multichip modules in comparison with 2-D/ MCM packaging

A preliminary attempt to classify the families of 3-D packaging is presented in Table 6.2.

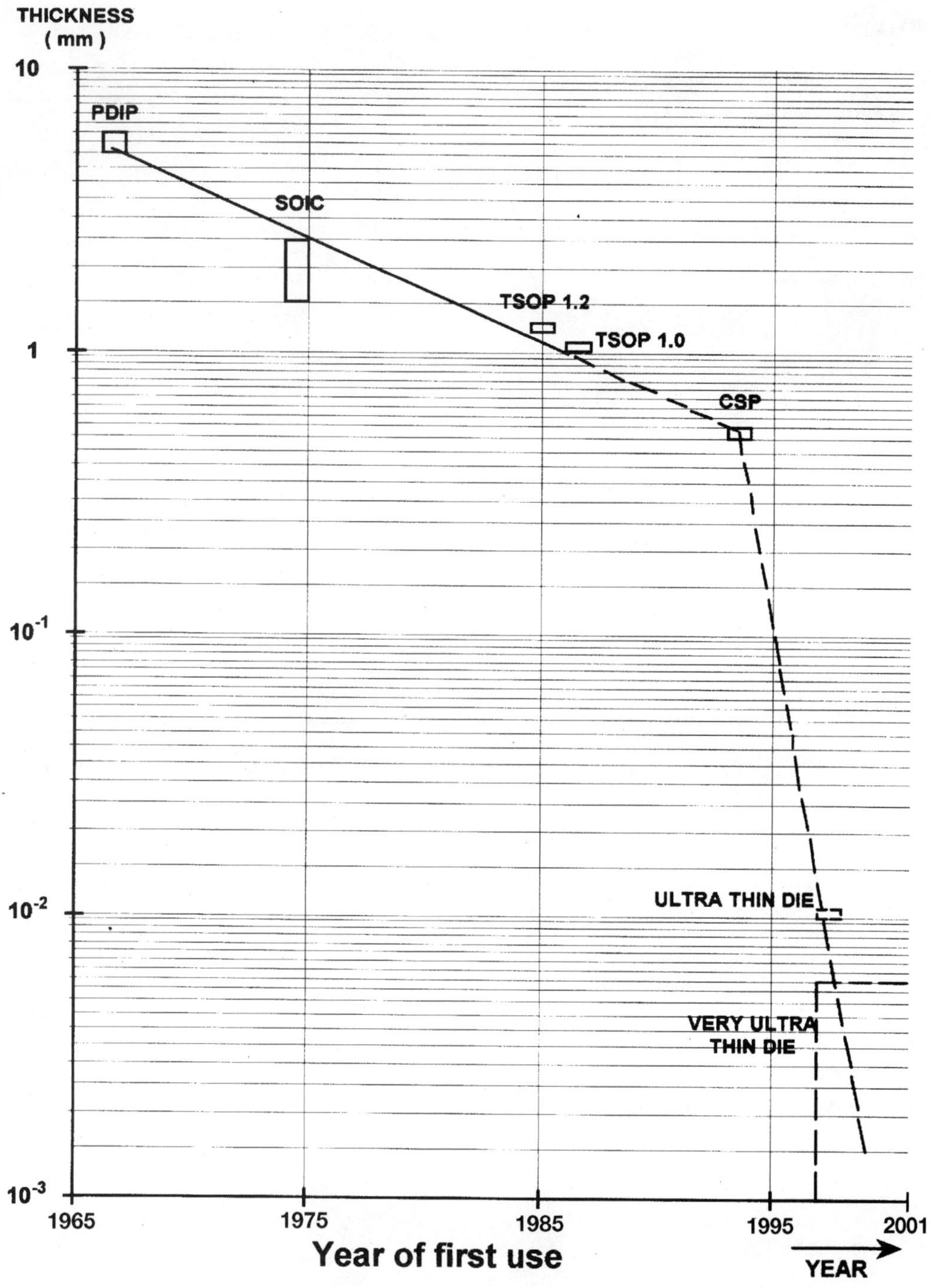

FIGURE 6.1 Evolution of thickness of the packaged die.

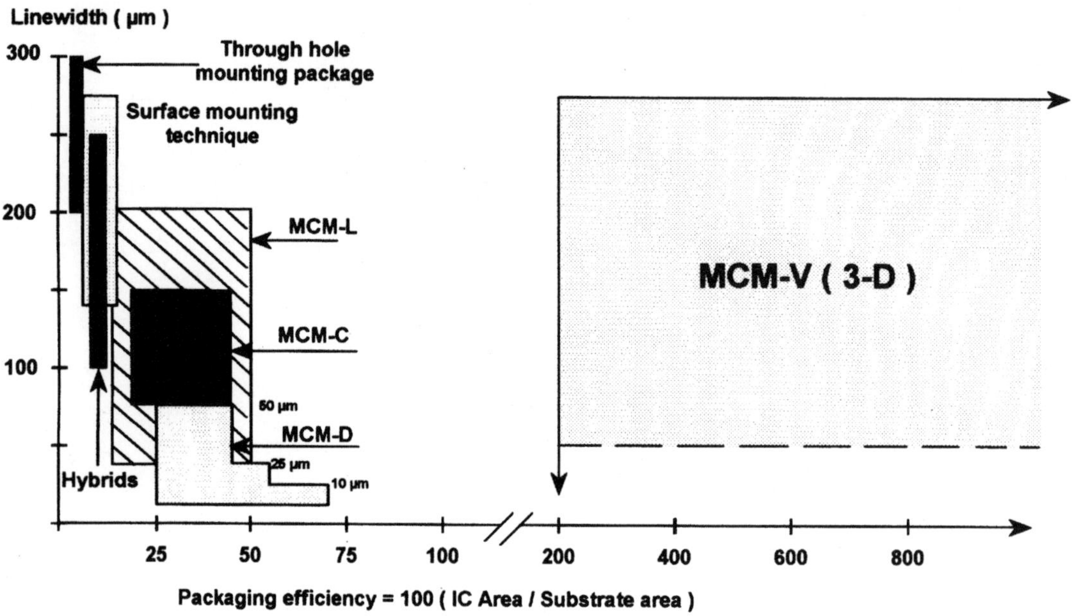

FIGURE 6.2 Packaging efficiency plot.

6.2 CLASSIFICATION

In the beginning, there is always the wafer. At this stage, two large classes of 3-D techniques can be defined:

1. Techniques utilizing dice from standard wafers
2. Techniques utilizing dice derived from modified wafers (generally rerouted)

The second class requires the availability of wafers to postprocess them. This availability exists only if the manufacturer is also a chip maker. From the level of elementary component, we can define two kinds of level of packaging (see Table 6.2):

1. Bare die
2. Packaged die [thin small outline (TSOP), and chip scale package (CSP)

6.2.1 Bare Die Assembly vs. Packaged Die Assembly

If the inherent logic of 3-D packaging determines that bare dice will be used instead of packaged dice, thus obtaining the maximum densification, it still will embrace the following disadvantages:

- Use of known good die (KGD) to get a better yield, knowing that KGDs are more expensive than standard chips
- The price of the bare dice is generally more agreeable than that of packaged dice.

At this stage, we can subdivide the bare die and packaged die into two classes (see Table 6.2).

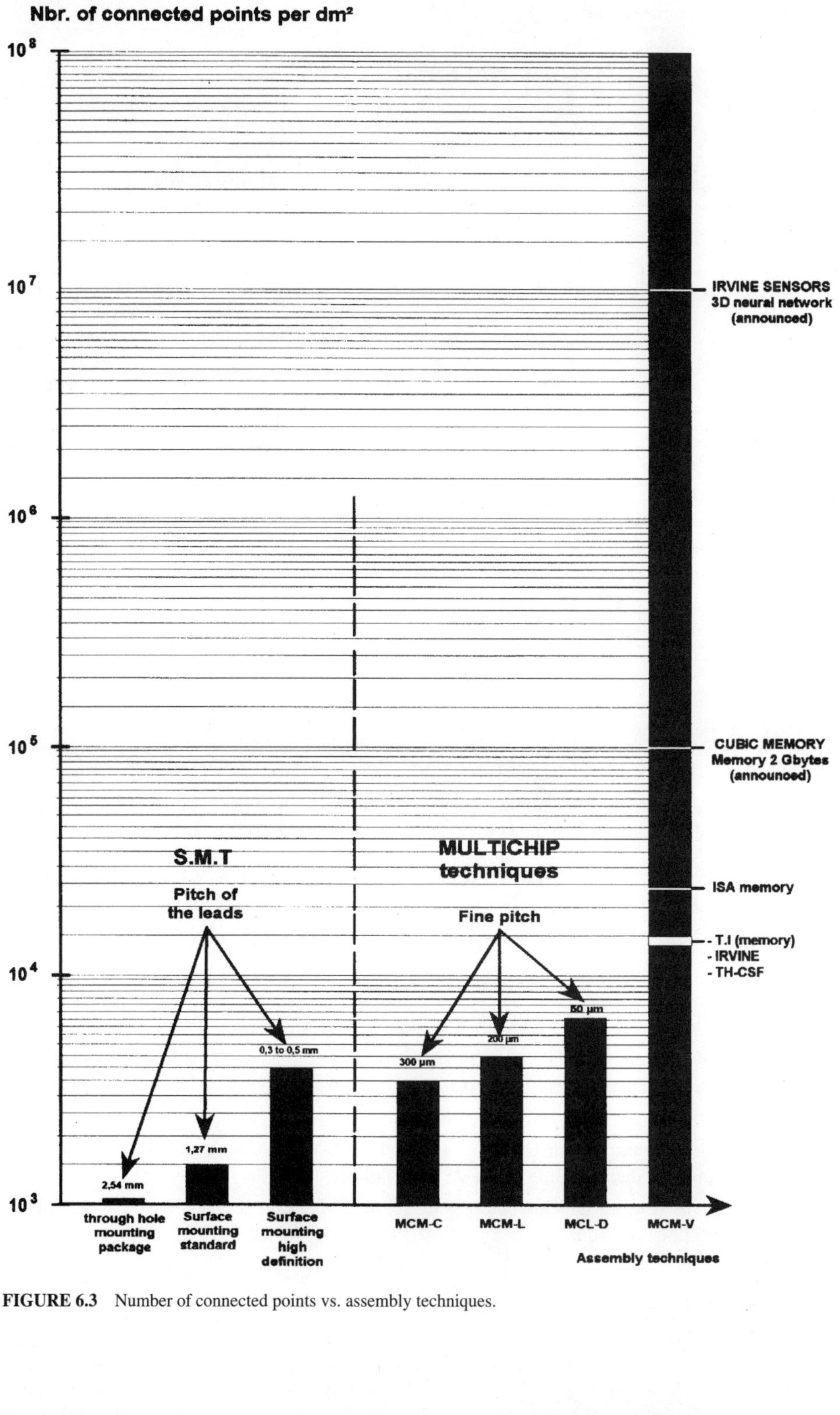

FIGURE 6.3 Number of connected points vs. assembly techniques.

Table 6.2 Review of the 3-D Techniques Worldwide

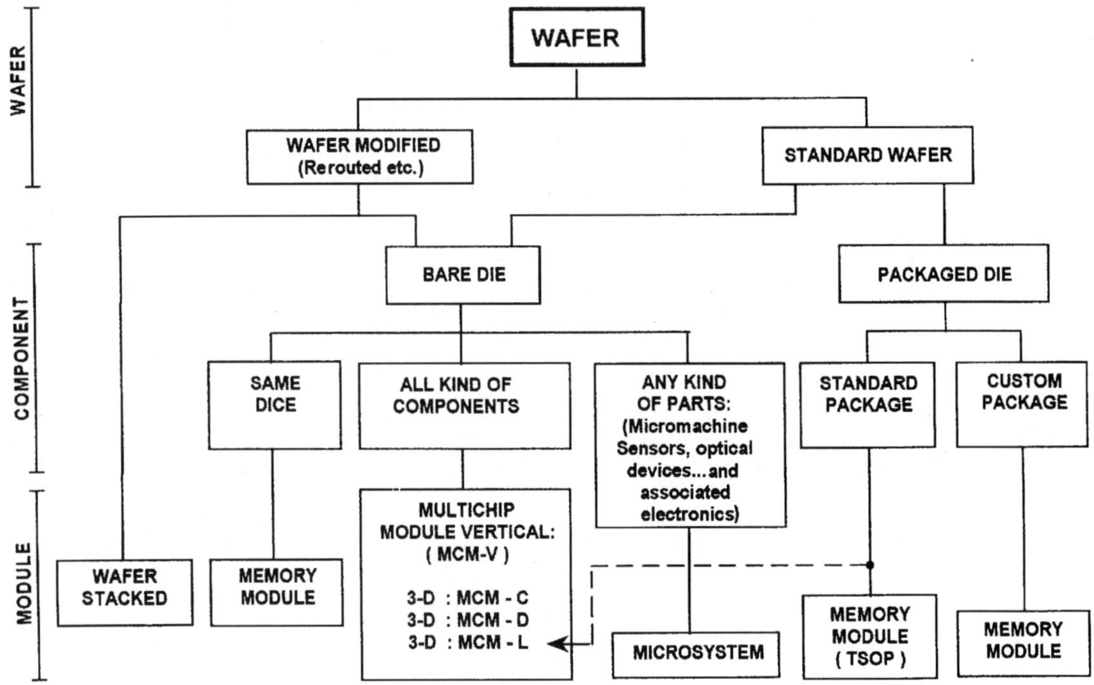

- Stacked identical chips (memory)
- Stacked mixed IC or components [MCM-V (V = vertical) and microsystem assemblies]

The stacking of memory was first implemented in the mid-1980s. Three companies developed the techniques of 3-D memory interconnection: Texas Instruments,[4] Irvine Sensors,[5] and Thomson-CSF.[6]

Within the class of stacked memory, there are four subclasses as follows:

1. Bare die assembly with standard IC chips
2. Bare die assembly with custom IC chips
3. Packaged die assembly with standard packages
4. Packaged die assembly with custom packages

To understand the strategies undertaken by these companies in contrasting directions, please refer to Table 6.2. The difference between custom and standard packaging stems from two ways of thinking:

1. The first way of thinking leads to the definition of a 3-D interconnection technique, and then to the modification of the standard chip or the standard packaged die.
2. The second way is using standard components (bare or packaged die) and then to determine a 3-D interconnection technique.

Table 6.3 brings together the companies that developed or manufactured 3-D modules. Thus, among the companies that have chosen to implement the standard components, one can see that these

Table 6.3 3-D Packaging Families by Company

Bare die assembly		Packaged die assembly		MCM assembly	Stacked wafer assembly
Standard ICs	Custom ICs	Standard packages	Custom packages	Custom modules	Custom wafers
Hitachi, Ltd. (Japan) Fujitsu Ltd. (Japan) NEC (Japan) Thomson-CSF/3D Plus (France)	Cubic Memory (USA) Irving Sensors (USA) Micron Technology (USA) Texas Instruments (USA)	Fujitsu (Japan) Mitsubishi (Japan) Thomson-CSF/3D Plus (France)	Dense-Pac (USA) Harris Corp. (USA) Motorola (USA) Samsung (Korea) Staktek Corp. (USA)	AT&T (USA) E-Systems (USA) G.E. (USA) Grumman (USA) Honeywell (USA) ISA (USA) LETI/CEA (France) Matra Marconi Space (France) Matsushita Electronic Corp. (Japan) Space Computer Corp. (USA) Trymer (USA)	Hughes Aircraft (USA) Mass Memory Technology (USA) Micron Technology (USA)
Number of large companies ÷ number of companies (%)					
100	50	100	67	70	67

are the major memory manufacturers. In two cases, they are not hindered by supply. In the case of those who have opted for the stacking of custom components, one finds tow memory manufacturers (Texas Instruments and Micron Technology) and two small firms (Irvine Sensors and Cubic Memory). In a attempt to quantify the ratio of large firms by the total number of the companies (see Table 6.3), one can see that all standard components (die or packaged die) are chosen by large firms and that custom components are mainly selected by the small firms. Another interpretation could be that the smaller firms have better performance (mainly higher densification) in comparison to cost.

6.2.2 Multichip Module Assembly

The 3-D interconnection is not exclusively dedicated to the stacking of identical chips. Many techniques have been tried (see Fig. 6.4). Generally, multiple bare dice are mounted on a 2-D multilayer substrate and interconnected using wire bonding, TAB, flip chip, or thin film metallization. The 2-D multichip modules are stacked vertically by the means of very different interconnected techniques. Table 6.4 shows some of them; the outstanding feature is the stacking of multichip module on ceramic MCM-C. For very high density, two companies use multichip modules deposited MCM-D with thin film metallization (HDI). Two other companies developed stacking of multichip module laminated MCM-L to decrease cost. The main applications are space and avionics.

6.2.3 Stacked Wafer Assembly

Wafer scale integration (WSI) has not produced the expected results. Generally, drawings showing stacked wafers do not take into consideration the need for a power supply network with decoupling capacitors, low-resistivity conductors, and so on. Only a few companies are involved in this area.

6.2.4 Stacked Heterogeneous Parts—Microsystems

Many 3-D assembly techniques are limited to one type of IC. Many companies have stacked duplicate chips such as memories, but a mixture of different ICs is more difficult to design and manufac-

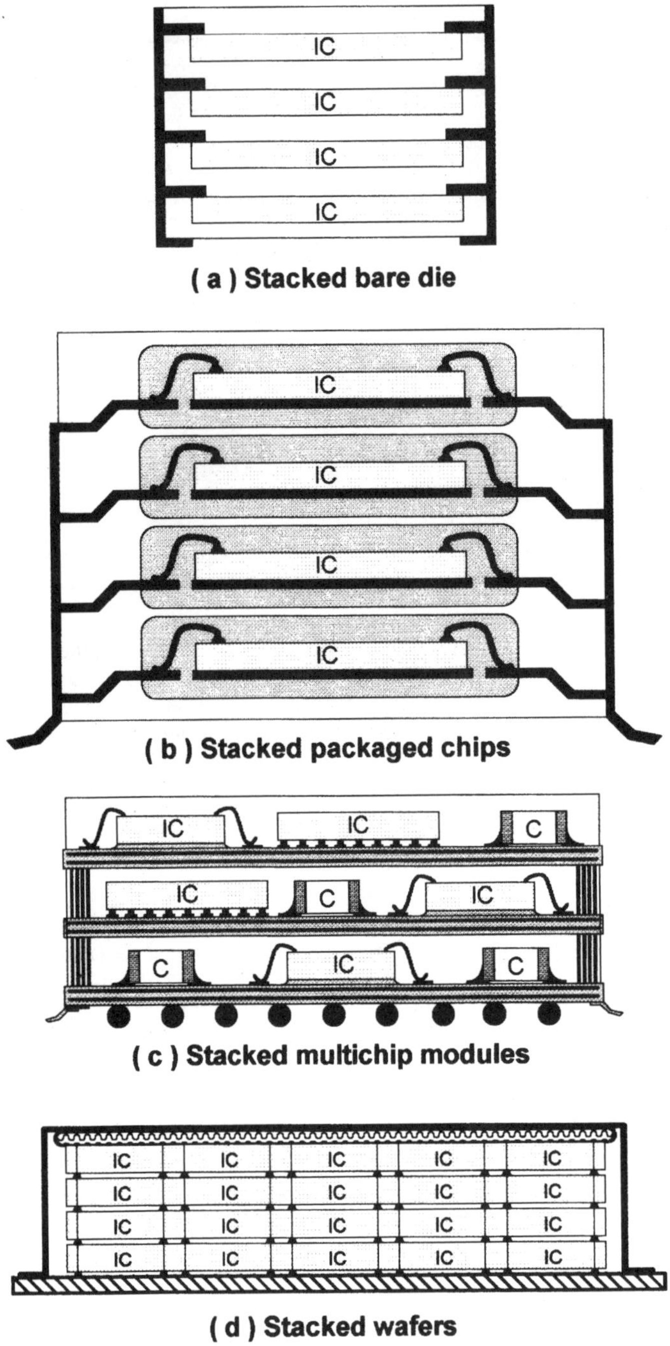

FIGURE 6.4 Simplified 3-D stacking approaches. Adapted from R. Crowley, TechSearch International, Inc.

Table 6.4 Multichip Module Assembly Techniques

MCM type	Vertical interconnection	Chip interconnection	Company
MCM-C	Fuzz buttons in plastic spacer and filled vias in substrate	Wire bond to substrate	E-Systems (USA)
MCM-C	Array connected to an IR detector with flip chip connect	Thin film metal layer from chip to substrate	Grumman (USA)
MCM-C	"Blind castellations" connecting chip carriers	TAB to LTCC carrier tubs	Harris (USA)
MCM-C	Filled vias in AlN spacers	Wire bonding	Honeywell (USA)
MCM-C	Solder-filled vias	Flip chip	Micron (USA)
MCM-C	BGA solder balls on top and bottom of substrate layers	Flip chip, TAB, or wire bond	Motorola (USA)
MCM-C	Soldered carriers to create vertical conductors on edge	Lead on chip	Trymer (USA)
MCM-D	HDI–thin film interconnect laminated to side of stack	HDI–chips first thin film connection layer	G.E., Lockheed (USA)
MCM-D	Wire bonding on MCM	Flip chip	CEA, LETI (France)
MCM-D	Multiple HDI	HDI–chips first	ISA (USA)
MCM-D	External leads butt soldered	Wire bonding	Motorola Marconi Space (France)
MCM-S*	Compliant anisotropic conductive material	Flip chip	AT&T (USA)
MCM-S	Elastomeric connector	Flip chip	Space Computer Corp. (USA)
MCM-L	Multiple chip substrate, substrate leads soldered	Flip chip, TAB, wire bond	Matsushita (Japan)
MCM-L	T-connect by plating	Wire bond, TAB, flip chip	3-D Plus, Thomson (France)

* Silicon/silicon

ture. The ultimate level is reached with the assembly of heterogeneous parts (i.e., the parts are a mixture of different ICs and mechanical components such as micromachines, microactuators, microsensors, and optical components).

6.3 MANUFACTURING TECHNIQUES

Three-dimensional packaging technique has been studied and developed for the space, military, computer, and consumer electronics industries. More than 40 companies have patented, developed, or manufactured 3-D modules. This chapter classifies and describes 30 of them. Among participating companies, the 3-D production status is at very different levels: research, development, characterization, qualification, and production. Consequently, many vertical interconnection techniques are in use. Table 6.5 brings them together and shows two classes of 3-D techniques according to the type of T-connect:

1. Intrinsic interconnection by thin film metallization
2. External interconnection using any of four methods:
   - leaded or leadless soldered edge conductors
   - glued with conductive epoxy
   - local connector
   - wiring

Many companies are concerned with external interconnections, and particularly with *soldered edge conductors.* This classification integrates all of the 3-D packaging families (bare die, packaged die, and MCM assembly) so as to present homogeneous applications. The four areas shown in Table 6.3 will be described in detail in this section.

6.3.1 Bare Die Assembly

Standard and custom chip assemblies will be considered separately.

Standard Bare Die Assembly. Bare die assemblies are produced by the following companies.

Hitachi, Ltd. (Japan). Hitachi's stacking process uses standard chips, but it requires the availability of a wafer form ball bumps on its surface.[7] Indeed, TAB was selected over wire bonding to achieve minimum thickness. After inner lead bonding, the active surface of the chip is coated and the memory chip is then tested. The chip on tape and the glass-epoxy frame will be aligned and attached with the UV-hardening epoxy (Fig. 6.5). A spacer obtained with a solder resist layer (40 μm thick) has eliminated problems due to the solder ball short circuits. This process has been mainly developed for four memories, but height chips have been assembled and have passed internal quality tests. The glass-epoxy frame was plated with a 60/40 tin/lead alloy. The Z interconnection between the PCB frame lands and the gold-plated leads of the TAB tape is made by dip soldering via semicylindrical through holes (castellation). The chip select of each chip is connected to an unique pad at the level of the PCB frame, so four lands are used for four memories. The external pads of the four-memory

Table 6.5 Vertical Interconnection Techniques by Company

Intrinsic interconnection	External interconnection				
Thin film metallization	Soldered edge conductors		Glued conductive epoxy	Local connector	Wiring
	Leadless	Leaded			
G.E. IBM Irvine Sensors ISA Thomson-CSF	Dense-Pac Grumman Micron Technology Motorola NEC T.I. Trymer	Fujitsu Hitachi Matsushita Elec. Indust. Matsushita Electr. Comp. Mitsubishi Samsung Staktek	CTS Cubic Memory Harris	AT&T E-Systems Hughes Space Computer Corp.	Matra Marconi Space CEA/LETI

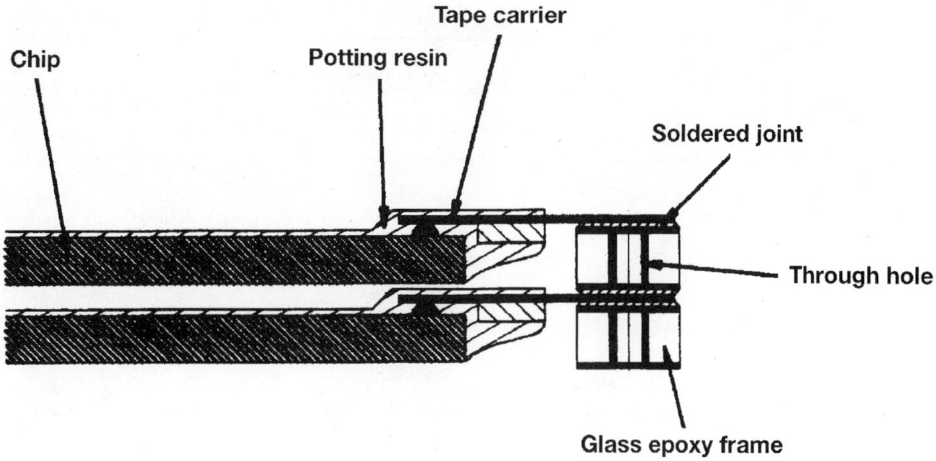

FIGURE 6.5 Hitachi's stacking process.

stack are reflow soldered to a PCB. The footprint of the stack is the same as a 400 mil small outline J-lead package (SOJ). The decrease of the thermal resistance from 35° C/W to 20° C/W is reached by adding two internal layers of copper heat conductors and an aluminum frame.

Fujitsu Ltd. (Japan). Two memories can be stacked in one standard package (SOP), or four memories in an SOJ package. This internal stacking is called *stacked chip in package* (SCP) by Fujitsu. Figure 6.6 shows these two packages; the TAB technique is used to interconnect the memories. The chips are mounted back to back, with gold bumps placed on the bond pads of each die. One end of the inner lead is bonded to the gold bumps on the upper and lower chip pads, and the other end is connected to the external leads. An end of the lead frame, generally used inside the plastic package, is sandwiched between the outer leads of the TAB tape. For instance, the two memories stored inside a 1.4 mm-thick TSOP and a 2.5 mm-thick SOJ package can store four chips.[8,9]

Matsushita Electric Industrial Co. (Japan). MEI has used a transfer-bump TAB process in which the bumps are plated on a glass substrate then transferred to the tape.[10] The bumped tape is bonded on standard memory chips; thereafter, the memory is encapsulated and electrically tested. During this step, each tape is customized to get one unique lead for chip select. The leads of each chip are fitted according to the chip's position in the stack. The stack is aligned to the printed circuit board via an insertion pin and alignment holes. The leads are soldered to the PCB using a pulsed hot bar (Fig. 6.7). This technique is limited to short stacks by the number of the extra leads used for the chip select and by the stack of the leads before soldering. Nevertheless, the height of the four-memory stack is relatively low (2.5 mm), which allows it to enter the memory card market.

NEC (Japan). NEC has presented[11] a short stack with four memories. A two-layer alumina chip carrier with a total thickness of 250 μm is used to interconnect a standard memory chip. The wafer is back-lapped up to 100 μm. Several techniques to bond the chip to the chip carrier have been tested, and the choice has been:

- Bumping the wafer by gold ball bonding (ball with a short *tail*)
- Use of high-Z retraction glue to have mechanical contact between gold wire and chip carrier

After the cure of the epoxy, the chips can be tested and burned in. A solder ball process similar to flip chip bonding is applied to the upper level of the chip carrier. The interconnection between the two sides of the chip carrier is made by metallized holes. Chip carriers are stacked and reflow soldered by four at 230° C. The total thickness for the four memories is only 1.2 mm (Fig. 6.8).

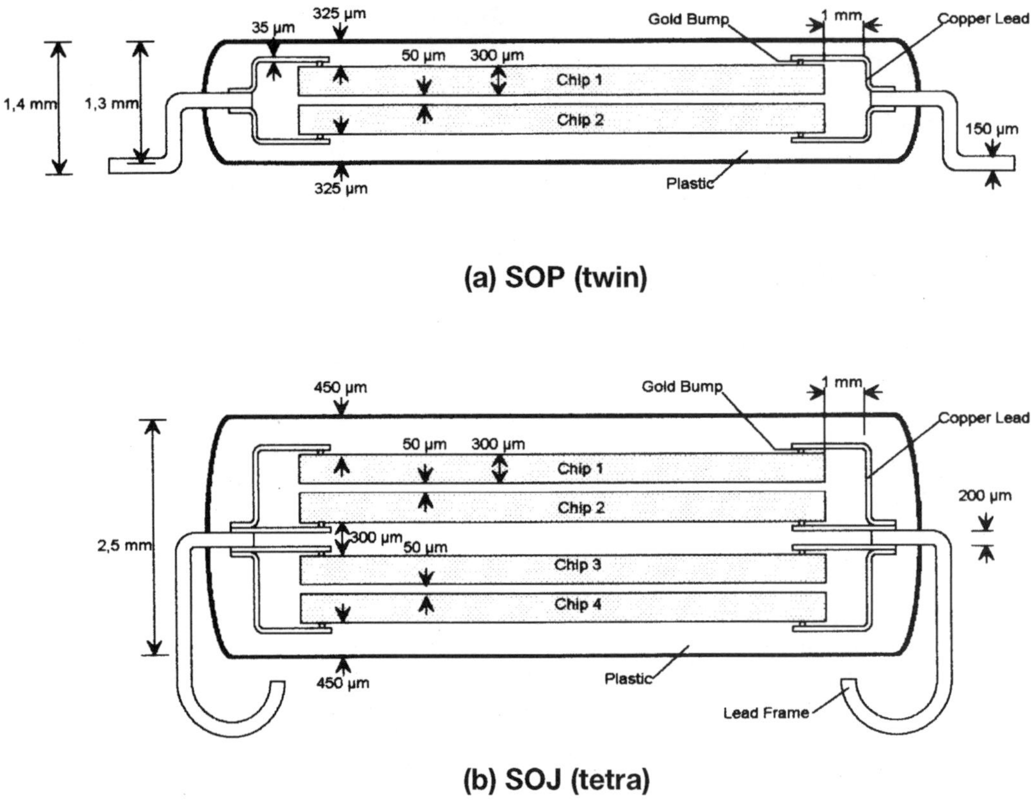

FIGURE 6.6 Fujitsu's stacked chips, cross section.

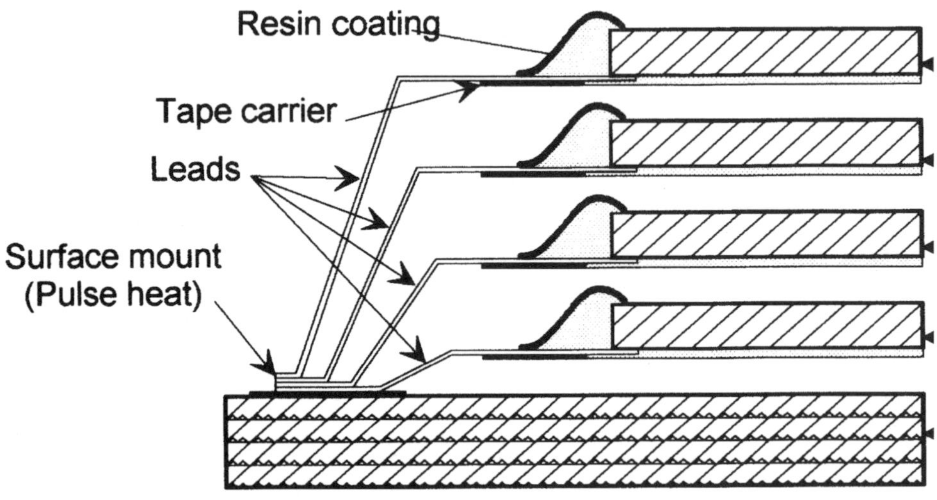

FIGURE 6.7 Matsushita's stacked TAB memory card.

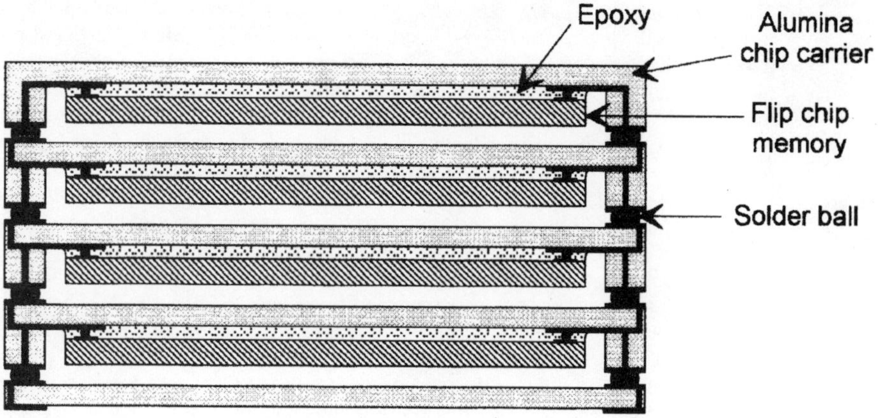

FIGURE 6.8 NEC's four-memory short stack.

3-D Plus/Thomson-CSF (France). TH-CSF has developed a 3-D technique based on chip on tape (35 mm), making it possible to use equipment dedicated to the manufacture of smart cards and the tools realized for the TAB technique (sockets and so on). After the electrical tests and eventual the burn-in, the tapes are stacked and molded with epoxy. At this step, machining by sawing (parallelepipedal block) or lathing (cylindrical block) allows a cross section of the leads to be printed on the tape (flex PCB). The entire module is metallized using a process similar to plating through holes in printed circuit boards. The surface of the module is electroless deposited, then copper, nickel and gold are electrochemically plated. The interconnect pattern on the surface of the cube or on the cylinder is generated by YAG laser trimming. A laser cuts the metal and gives a kerf that electrically isolates the conductors from the ground plane (see Fig. 6.9). Generally, a lead frame (0.5 or 0.635 mm) is embedded in the module, but the ball grid array technique can also be used. The third method is chip on board attachment.[12] In October 1995, TH-CSF launched a start-up, 3-D Plus, to produce and sell these 3-D modules.

Custom Bare Die Assembly. Custom bare die assemblies are made by the companies discussed in the following sections.

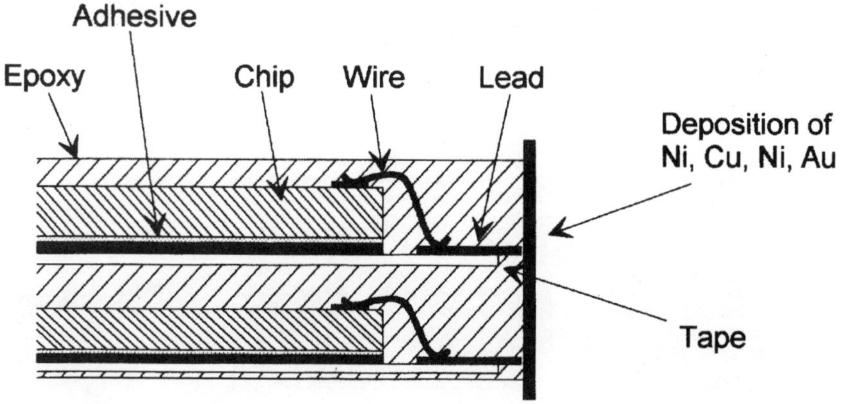

FIGURE 6.9 3-D Plus/Thomson-CSF technique. *Source:* 3-D Plus.

Cubic Memory Inc. (USA). Cubic Memory is a Santa Clara-based start-up. The primary feature of its process is the deposition of gold interconnect traces over insulating layers of polyimide. The gold traces provide a nonoxidizing surface for layer-to-layer connections and horizontal interconnection of the dice on a single layer or segment.The wafers are masked for the first layer (poly 1) of insulating polyimide, 4.8 microns thick, which protects the wafer and provides insulation for the next step. Next, comes the photoresist, wherein unwanted polyimide and the original aluminum contacts are removed. The wafer is then masked, and a 0.37 micron layer of gold (Au 1) is deposited. Another 4.8 micron layer of polyimide is added to insulate the gold traces just deposited. A second gold masking step and deposition (Au 2) are performed to provide a second level of interconnection. Then, a final 3 micron layer of polyimide is deposited for electrical and mechanical protection.

At this point, the wafer is thinned from 635 microns to 75–280 microns, depending on the application. Each wafer is sawed into segments of four dice (2 × 2), with edges beveled to approximately a 45° angle. Finally, a 1 micron layer of silicon nitride is deposited on the back of the wafer for insulation. The wafer segment is tested and stacked, and a polyimide dot is applied to act as an adhesive between each segment. Next, the gold pads, exposed at the sides of the segments as a result of the bevel cut, are connected with whiskers of conducting silver-filled epoxy. The epoxy is then cured, and the stack is physically attached to the PCB with epoxy (see Fig. 6.10). The PCB and stack are then electrically connected with horizontal whiskers of conductive epoxy, and final programming is completed. The stack is then encapsulated with a casting resin to provide protection.[13]

Irvine Sensors Corp. (USA). Irvine Sensors has developed 3-D memory stacks with from 4 to 100 memories[14]. At the level of the wafer, the I/O signals are rerouted to one edge of the chip. A polyimide multilayer with one metal layer is built. A 2.5 μm metal layer of TiW/Au is deposited by thin film sputtering and electroplating. A second layer of polyimide is applied to the wafer to cover the metal layer. The wafer is back lapped to 175–200 μm before dicing. Wafers are diced, and the physical dimensions of each die are measured and recorded to allow accurate alignment during the stacking process. The ICs are laminated into a precision aligned stack, or module. The face of the cube is lapped to expose the leads, the edge of the stack is lightly sanded to remove excess epoxy, and then the stack is plasma etched to expose the gold leads deposited on the surface of the chips. The silicon IC edges are passivated so that subsequent metallizations are not shorted to the silicon. TiW/Au processes are used to form pads and bus lines on the face of the cube. Solder bumps are deposited, and the cube can be flip chip bonded to a substrate. The memory chips are oriented perpendicular to the mounting substrate (see Fig. 6.11). For short stacks, the I/O signals are routed to one of the end caps, and solder balls are applied to the end cap. Then, the memory chips are oriented parallel to the substrate.

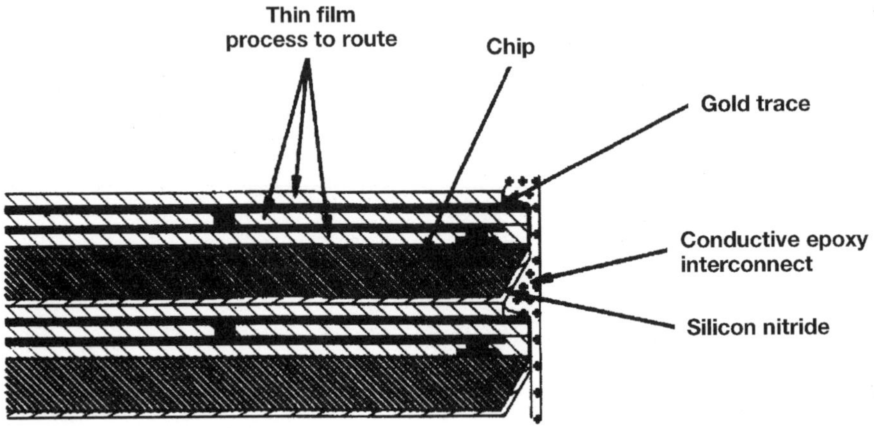

FIGURE 6.10 Cubic Memory Technology's process.

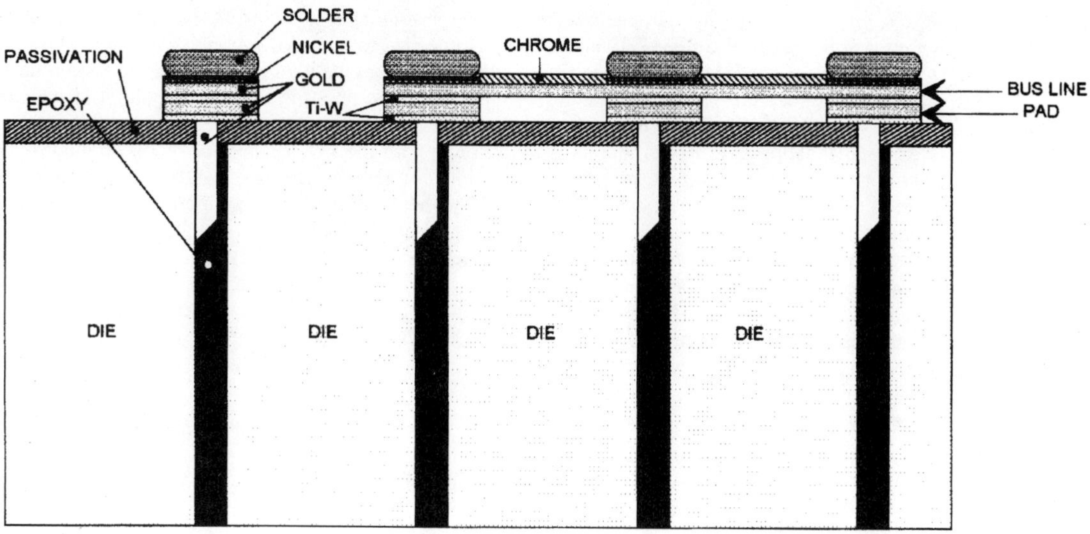

FIGURE 6.11 Irvine Sensors Corporation's stacking process (not to scale).

Micron Technology (USA). A patent issued to Micron Technology has been described, but this does not imply that Micron is pursuing the technology.[15] As shown in Fig. 6.12, a metallized substrate associated with a spacer permits the creation of a cavity. The memory chips are face bonded to the metallized substrate. A solderable package interconnection pad (PIP) on the periphery of the substrate with metallized hole is used. This spacer is adhesively attached to the top surface of the substrate during the assembly process. Multiple submodules are stacked, a nonmetallized cap is glued to the top submodule, and all the submodules are interconnected by dipping in a solder bath. A partial vacuum is applied to all the stacked vias, filling them with molten SnPb alloy.

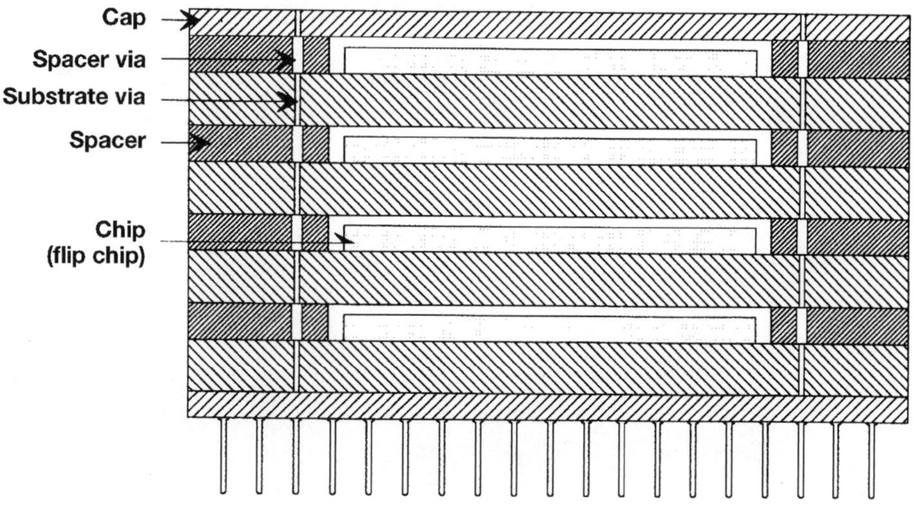

FIGURE 6.12 Micron Technology's stacked submodules. *Source:* U.S. Patent 5,128,831, July 7, 1992.

Texas Instruments (USA). TI's technique requires rerouting of the chip I/O pads to gather all the pads on one side of the chip. Pad bumping is also needed to use TAB process; therefore, additional wafer steps are required.[16] Generally, the signals issued from the memories pads are used to route to one edge of the chip. The signals can also be routed on two opposite edges of the chip to have pads for testing and burn in the chip before stacking. The rerouting of the wafer requires a thin film process, and a plating process permits to get gold bumps on the I/O pads. After dicing, the chips are inner-lead bonded to the tape by gold-gold thermocompression, and they can be electrically tested and screened. The chips are stacked and attached with an electrically insulating adhesive. Due to the fact that the leads of each memory are not connected at the level of the stack, they have to be connected to a multilayer substrate (Fig. 6.13). The leads coming from each level are cleaned and straightened. Solder bumps deposited on a silicon multilayer substrate are used to attach all the leads by reflow soldering, so the structure of this assembly is like a silicon-on-silicon MCM. The stack with vertically mounted memories can pack from 8 to 200 chips.

6.3.2 Packaged Die Assembly

Packaged die assemblies are described in detail in this section.

Standard Packaged Die Assembly. The standard packaged die assemblies are made by the companies described in the following sections.

Fujitsu Ltd. (Japan). Fujitsu has developed two techniques to stack standard thin SOP package (TSOP), but they are limited to only two single-chip packages (Fig. 6.14). The ultra-thin SOP packages (0.8 mm thick) can be mounted one top of another.[19] The external leads of each one are different; the first package has gull wing-type leads, and the second has butt-type leads. The two packages can be stacked in parallel or perpendicular.

Mitsubishi Electric Corp. (Japan). Mitsubishi has developed 3-D packaging technology based on the stacking of thin small outline package.[17] The TSOP parts have straight leads or specifically shaped leads (L or C shaped) to interconnect them with two printed circuit boards by reflow soldering (Fig. 6.15). The mounting technique requires several improvements: the solder paste must have

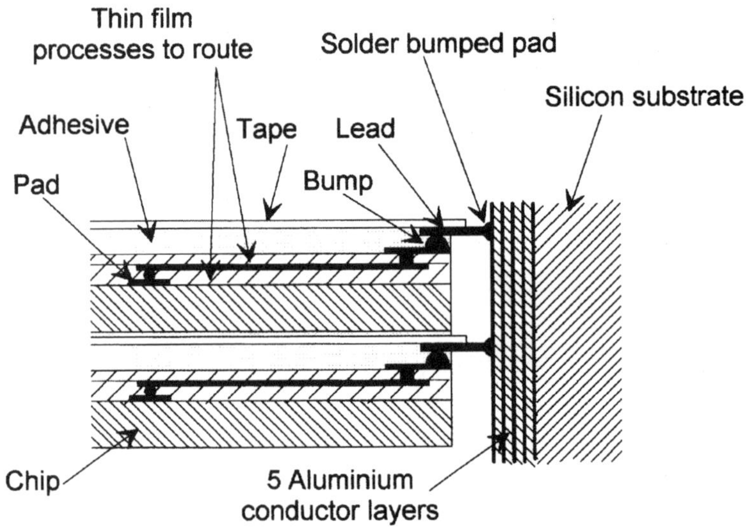

FIGURE 6.13 Texas Instruments' technique.

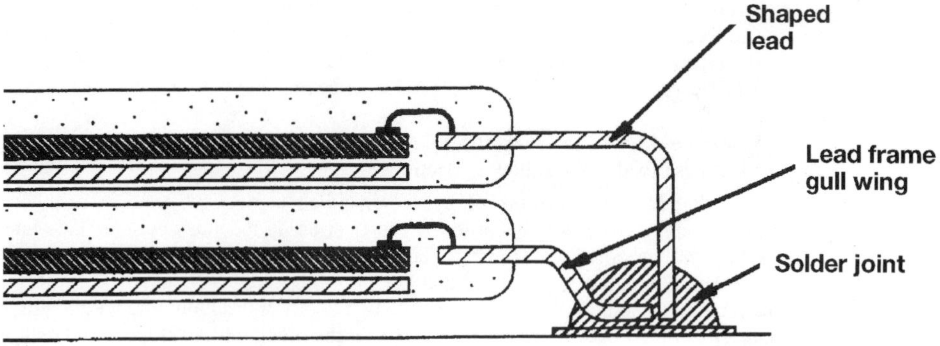

FIGURE 6.14 Fujitsu's die assembly process.

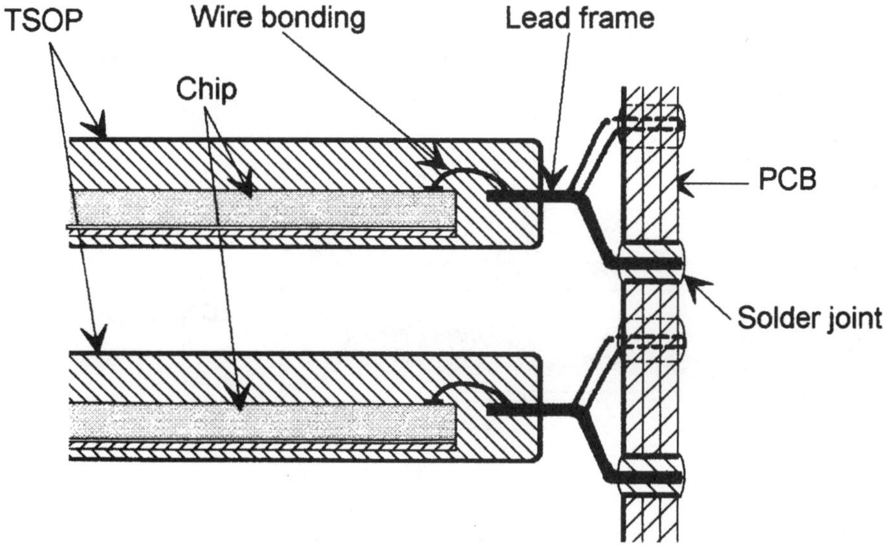

FIGURE 6.15 Mitsubishi's process.

uniform grain size to ensure a good filling into the through holes, and an automatic screen printer deposits 250 μm of solder paste on both sides of the printed circuit board with a good accuracy. In addition, special equipment has been developed to accurately align and insert all the leads coming from the TSOP into the through holes of the boards (1.27 mm for the pitch of the TSOP type II). Up to nine packages are stacked, with a pitch between each package of 2 mm (18 mm for 9 TSOP stacked).

3-D Plus/Thomson-CSF (France). The standard TSOP (type I or II) or chip-scale packages (CSP) are surface mounted on a thin printed circuit board for low-power applications or on copper lead frame. Several packages can be mounted per board (≥64). If the reverse packages are available, they are soldered on both sides. Then, the boards are stacked, and the pre-existing locating holes in the boards permit them to be centered thanks to the pins that are part of the mold. Molding with a low expansion coefficient epoxy resin gives a part having dimensions slightly larger than those of the final block. With a part with 64 blocks, a sawing will give 64 epoxy cubes on which the cross sections of the external leads of the PC boards appear on the two opposite sides of the blocks (Fig. 6.16). A collective plating of the sides of the blocks, exactly like the plating process used for the PCB, is made. The etching of the metallic layers (electroless Ni + electrochemical deposition of Cu + Ni + Au) is accomplished by means of a YAG laser beam.[18]

Custom Packaged Die Assembly. Custom packaged die assemblies are made by various companies as described in the following sections.

Dense-Pac Microsystems, Inc. (USA). Dense-Pac uses a substrate-on-chip technique. A cofired alumina chip carrier with a slot in the center or along the sides permits us to interconnect the bond pads to the top side of the substrate. The size of the ceramic carrier is only 1 mm beyond the edge of the memory die. It must be noted that die with tight-pitch peripheral bond pads on four sides cannot be interconnected. Figure 6.17 shows the two layers cofired alumina with a wire bonded chip. Each carrier can be programmed for its position in the block with a single wire bond. After that, the chip carrier is filled with epoxy, cured, and ground to 0.53 mm. An epoxy layer is screen printed on the top surface of each chip carrier, and then the carriers are stacked and cured. The vertical interconnection of the carriers is obtained by dipping in a tin/lead bath; the gap between the carriers due to the

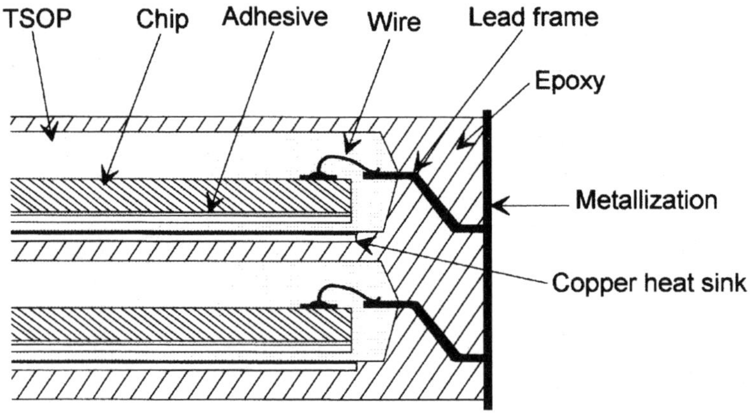

FIGURE 6.16 3-D Plus/Thomson-CSF method. *Source:* 3-D Plus.

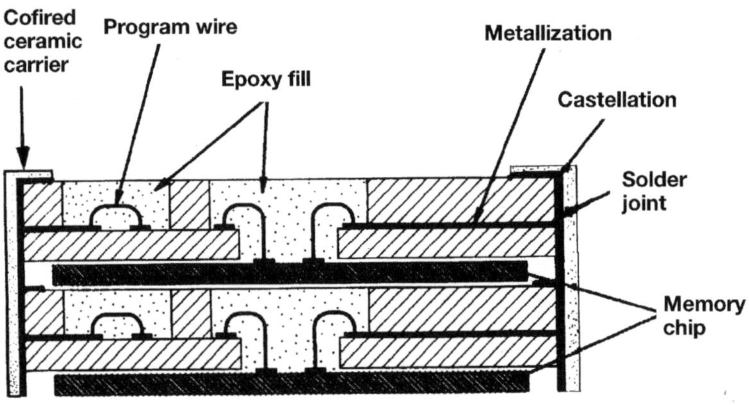

FIGURE 6.17 Dense Pac Microsystems' method.

height of the chip is jumped thanks to the capillary action of the solder bridges.[19] Several mounting techniques are proposed:

- Stacks are mounted on a lead frame and then molded.
- The stack can be directly connected to a multichip module by wire bonding to a special slab previously attached to the stack.
- The stacks can be surface mounted directly to a substrate.

Harris Corp. (USA). The basic Harris approach is stacked layers of ceramic chip carriers, or tubs, with one or two memory dice per tub. Each carrier is hermetically sealed by the next carrier in the stack. The top carrier is sealed with a ceramic lid. Silver-filled thermoplastic film material, designed to tolerate short-duration temperature exposure up to 500° C (Staystik 501 from Alpha Metals), was used to attach the die and to support the firing cycle of the glass sealing. To overcome the reliability problems met when the stacked chip carriers are connected by solder, Harris uses the *blind castellation* approach. This castellation would provide the interconnect from inside the package to the adjacent carrier above and below (Fig. 6.18). The top and bottom layers of a tub have holes that remain unfilled along the perimeter. On the adjacent layers, a field of metal is exposed through the hole. These fields of metal are interconnected internally through a solid metal via within the ceramic. After firing the ceramic, the tubs are excised from the plate with the cut line through the center of the unfilled top and bottom hole. The openings above and below the metallization are the unfilled holes that form the castellation. This is the key element of the process, because there is also a method of stacking leadless chip carrier and using solder as a means of both electrical interconnection and mechanical attachment and the inventors believe that this attachment method is unreliable. This is based on past experience with leadless chip carrier (LCC) attachment and reliability testing.[20] Finally, the cavities that are formed when tubs are stacked, aligned, and filled with a conductive material such as conductive epoxy.[21] Silver- and gold-filled thermoplastic pastes were evaluated in several environments.

Motorola (USA). The Motorola 3-D technology is based on the company's ball grid array (BGA) package technology.[22] One or more die are attached to a chip carrier, with solder balls mounted to the bottom of each carrier. The solder balls, located on the bottom of the top carrier, are soldered to pads deposited on the top of the lower carrier (Fig. 6.19). The upper substrate has solder bumps on the bottom only. The chips can be attached to each carrier using wire bonding, TAB, or flip chip attach-

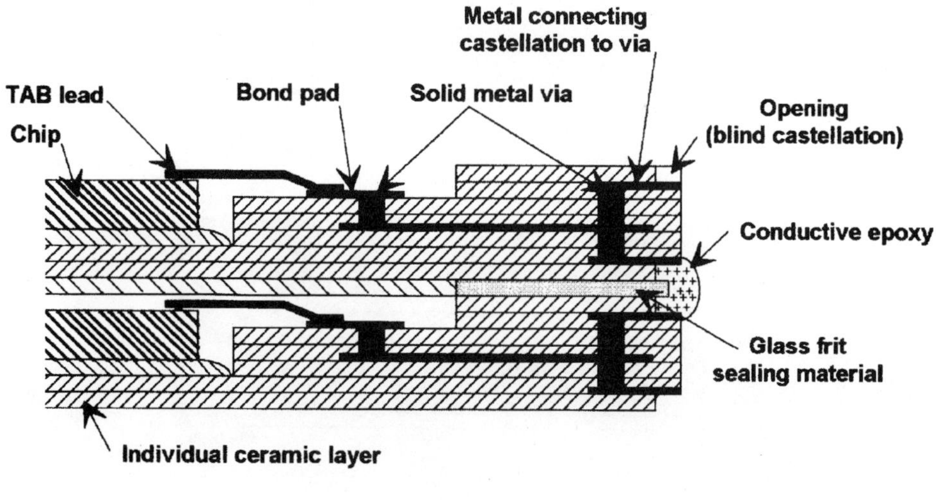

FIGURE 6.18 Harris chip-bump-TAB-CML-epoxy method.

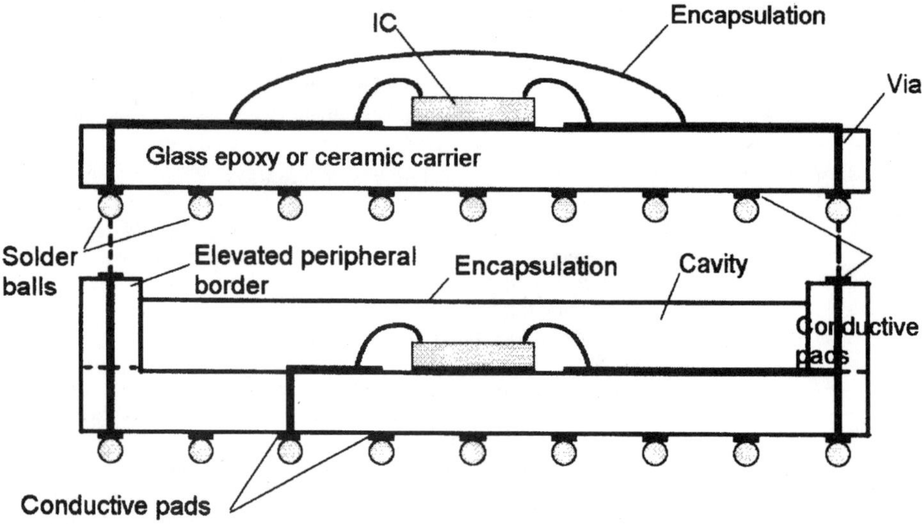

FIGURE 6.19 Motorola's stacked chip carrier approach. *Source:* Adapted from U.S. Patent 5,172,303, December 1992.

ment. Two kind of encapsulation are presented: by filling the cavity or by using a glob-top encapsulation. The height of a two-carrier stack reaches 3.2 mm.

Samsung (Korea). Each plastic package is a recessed thin small outline J-leaded (TQFP) package. As shown in Fig. 6.20, the periphery of top half of the each plastic package is removed to expose top surfaces of inner leads for interconnections between neighboring components and to create the space where top neighboring package can fit. The connection is made by reflow soldering. The stack can be attached to the substrate by surface mounting (gull-wing leads) or by socketing.[23]

Staktek Corp. (USA). Staktek has used a lead-on-chip technique and rebuilt a plastic package from bare dice. The ends of the lead-frame have to be designed with a specific shape to insert cooling fins and to conduct the signal between the different levels of the plastic packages. Of course, only two

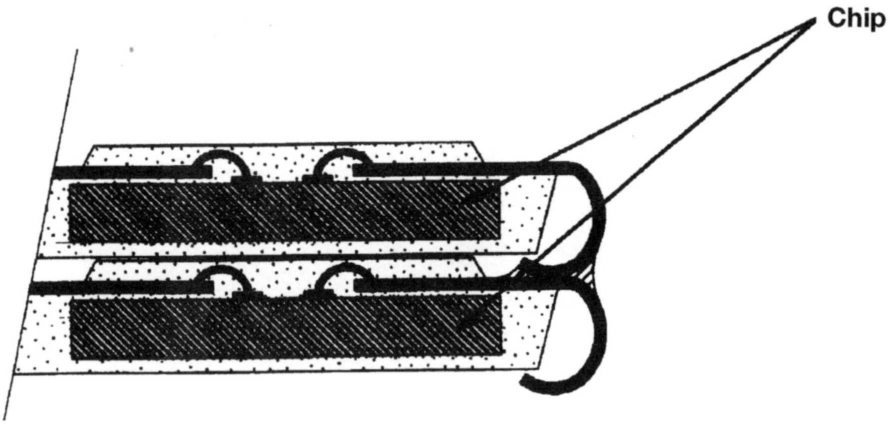

FIGURE 6.20 Samsung's packaging method.

sides are affected by this 3-D interconnection (Fig. 6.21). Thin layers of B-stage epoxy are transferred from a carrier sheet to the surface of the memory chip to act as a spacer for the lead frame. A second layer is deposited on the chip to glue the lead frame, then the chip is wire bonded to the lead frame using aluminum wedge wire bonding (due its low profile). These parts are transfer molded in plastic. After curing and deflashing, the package can be tested and burned in. The leads are formed and the cooling fins are probably attached by dipping in a solder bath.[24–26] The cooling fins improved thermal transfer in comparison with standard plastic packages (SOJ and TSOP); nevertheless, the electromagnetic interference (EMI) has to be limited.

6.3.3 Multichip Module Assembly

As seen in Table 6.4, the stacking of the multichip module can be divided in four classes:

- MCM-C stacked
- MCM-D stacked
- MCM-S stacked (silicon on silicon)
- MCM-L stacked

For each class, the companies will be presented in alphabetical order.

MCM-C Stacked. The MCM-C stacked assemblies were made by the companies listed below.

E-Systems/Norton Diamond Film (USA). E-Systems describes a 3-D packaging approach that involves the stacking of several diamond substrates. The concept for implementing diamond in 3-D MCMs is shown in Fig. 6.22. It consists of a stack of alternating layers of diamond substrates, each

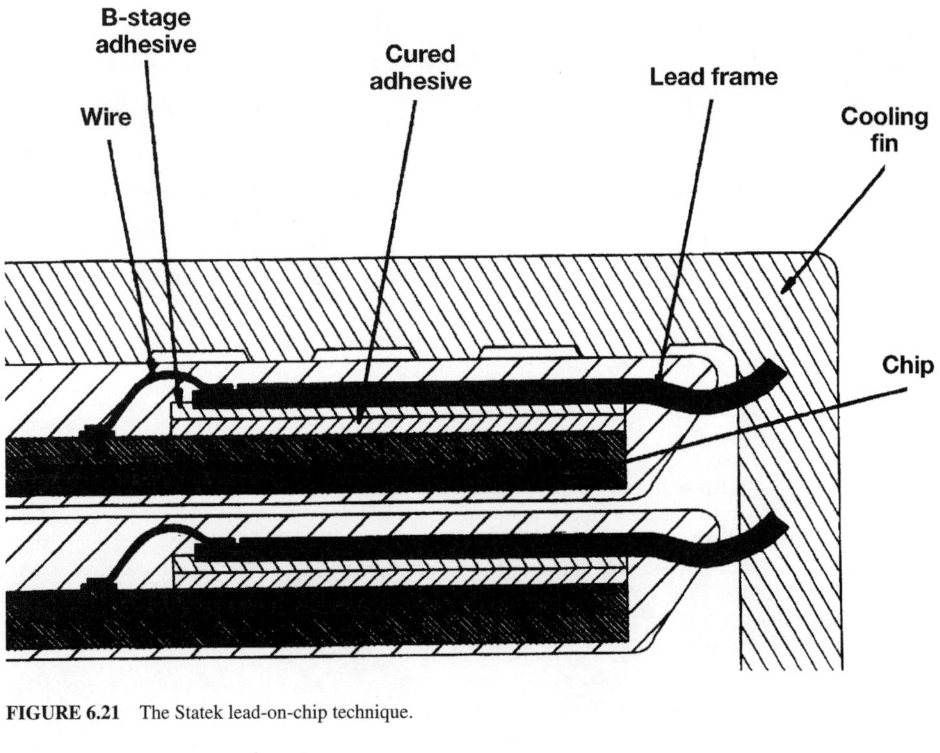

FIGURE 6.21 The Statek lead-on-chip technique.

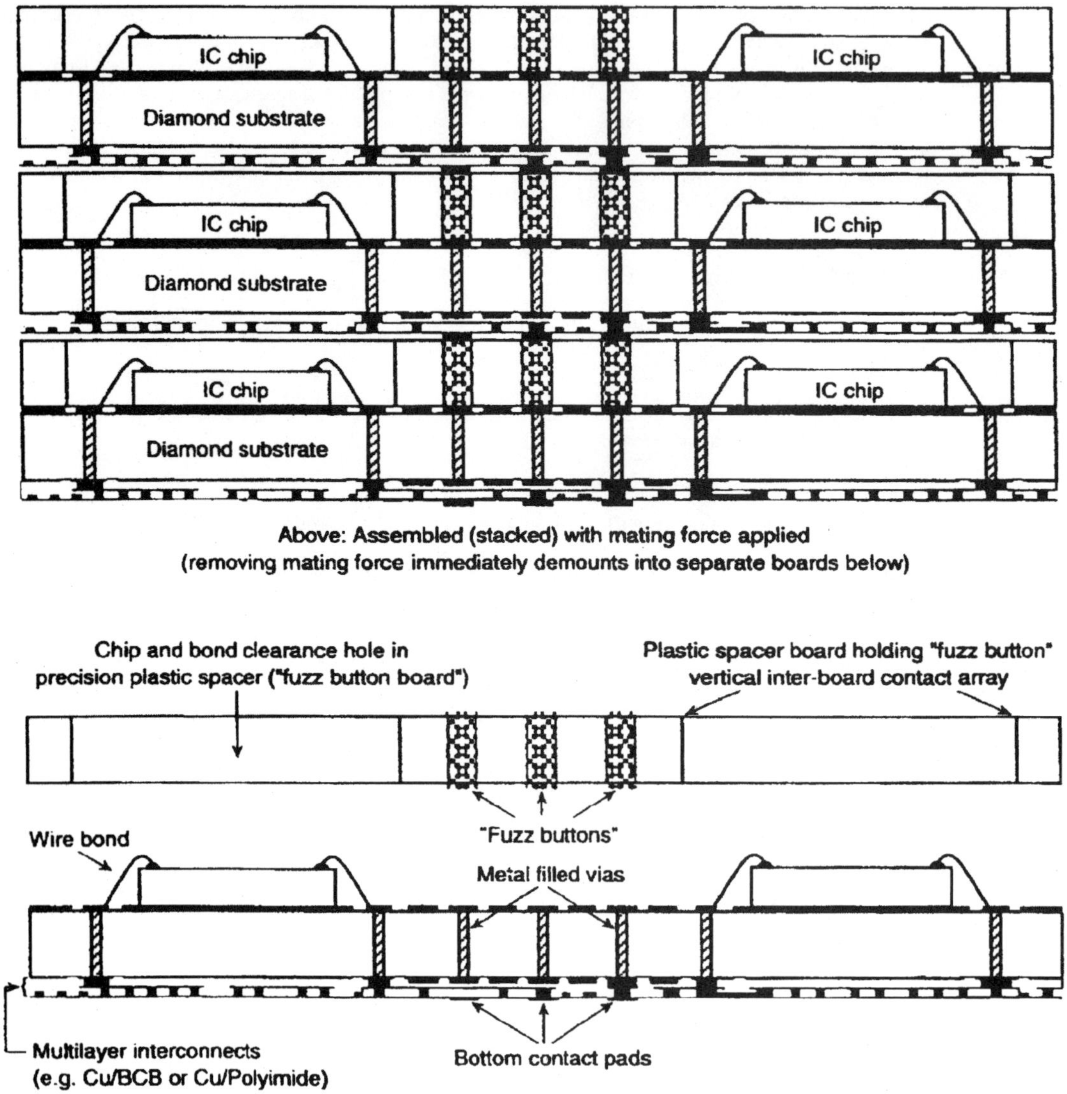

FIGURE 6.22 E-Systems' cross section of a 3-D diamond MCM assembly.

with ICs and interconnects attached to or fabricated on the diamond, and a precision plastic spacer layer. Conductive "fuzz buttons" are stuck in arrays of holes in a spacer board that is placed between the diamond boards to be vertically interconnected. The fuzz buttons, typically 0.5 mm diameter, extend beyond both sides of the spacer board by ≈0.1 mm so that they make solid contact to the pads on the boards above and below the button board or spacer board when the assembly mating force is applied. This precision plastic spacer board has clearance holes for the IC die and a large number of 0.5 mm diameter holes into which the fuzz buttons are forced.[27]

Grumman Aerospace Corp. (USA). Grumman has patented a technique of stacking alumina substrates. Contact pads are formed on the vertical faces of the stack, and the vertical interconnections are made with flip chip technology to an infrared detector array. The conductors are deposited on the alumina substrate by vacuum deposition and patterned by photolithography. Apertures are formed in the alumina substrate by laser machining. To obtain the same level for the metal conductors and the upper layer of the chips, the substrate and the individual ICs are placed face down on a flat surface; the dice have been thinned at the level of the wafer and then diced (Fig. 6.23). The groove between the aperture and the die is filled with an epoxy resin. Each multichip module is released from the flat surface, and the interconnection is made with a thin film technique.[28]

Harris Corp. (USA). The Harris (USA) MCM-C Stacked assembly was described in detail in earlier sections.

Honeywell (USA). Honeywell stacked multichip modules are built with aluminum nitride multilayer technology.[29] Each module is hermetically sealed on the periphery of its upper layer, and several rows of pads are interconnected with a ceramic spacer (also in AlN). The spacer is connected to the lower and upper multilayer by reflow soldering of the solder balls (Fig. 6.24).

Micron (USA). Described in detail in earlier sections.

Motorola (USA). Described in detail in earlier sections.

Trymer (USA). The package base consists of a one-piece, raised, solid-lead ceramic substrate shown in Fig. 6.25. The ribs are metallized by block printing on the top, bottom, and each side of the package. Nickel barrel plating is followed by gold over the printed and fired base metal. This completes the formation and processing of the package base. The one-size package uses a ceramic lid with complementary metal-plated rib leads. This allows vertical joining of individual leads after sealing the stacked package joined and interconnected by a simple solder dip, one side at a time. Any of the vertical connections can be interrupted by masking before the solder dip. Package sealing is by solder glass or gold-tin preform with a metal lid.[30]

MCM-D Stacked. The MCM-D stacked structures have been produced by the companies described below.

General Electric (USA). The basic technology used for the 3-D approach involves the GE's 2-D high-density interconnect (HDI) MCM process, also referred as a *chip first* approach. The main steps of the HDI process are:

- Fabricate ceramic substrate with milled IC cavities.

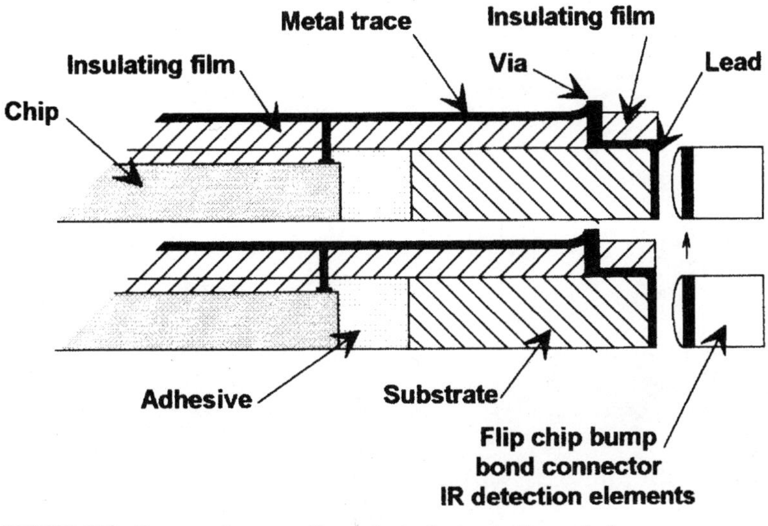

FIGURE 6.23 Grumman Aerospace Corporation's alumina stacking method.

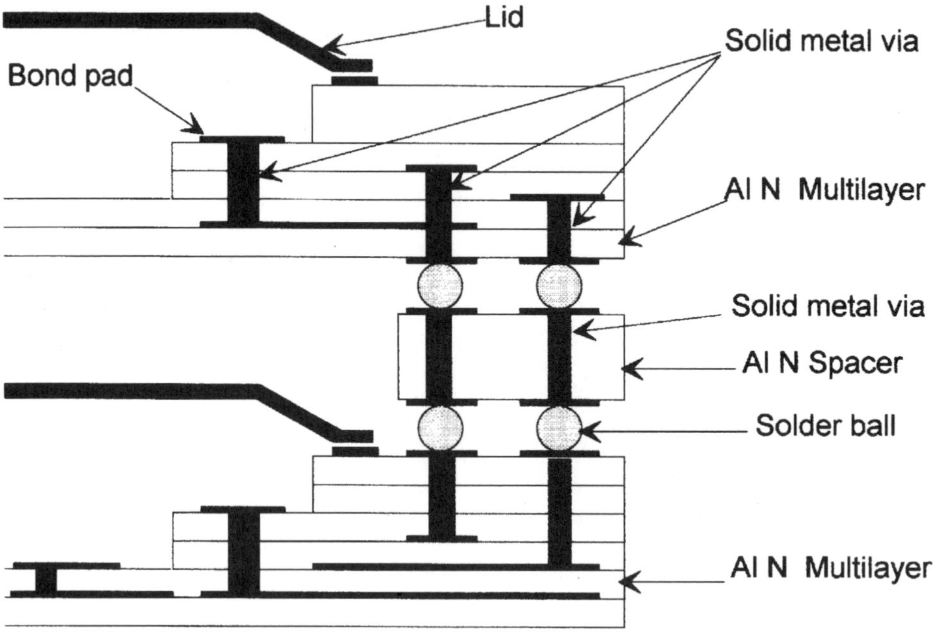

FIGURE 6.24 Honeywell/Coors MCM.

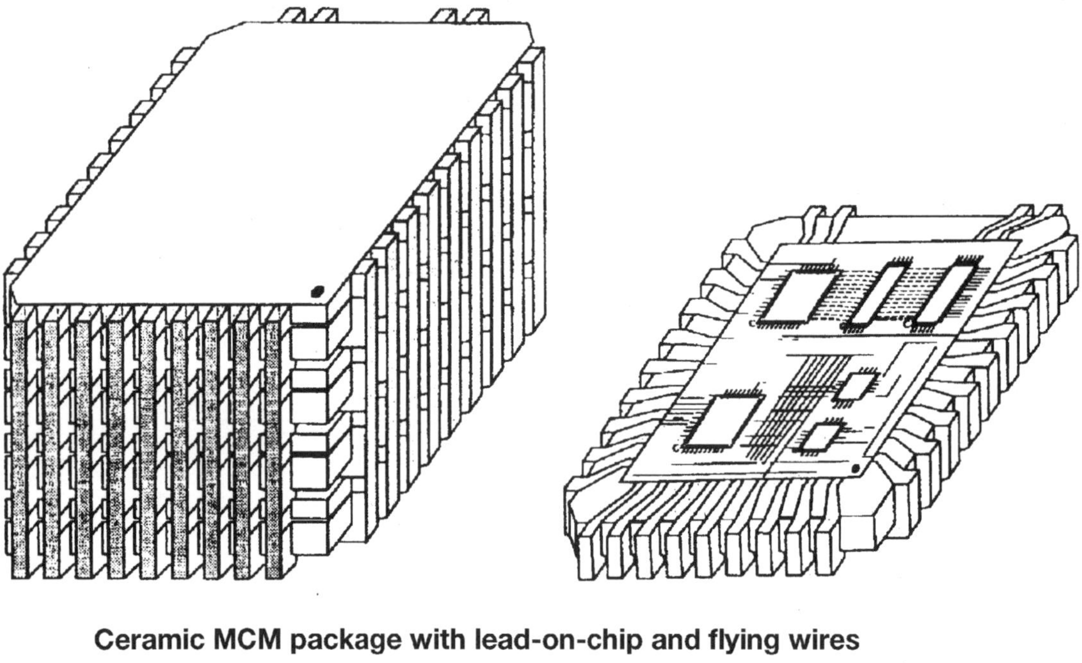

FIGURE 6.25 Trymer's stacked MCM technology with soldered castellations.

- Attach ICs to substrate.
- Laminate a 25 μm thick polyimide film over the chips.
- Laser ablate via holes to bond pads.
- Deposition by sputtering of thin film metal titanium and copper, then copper is plated, and titanium is sputtered again.
- Two to five layers can be used, depending on design requirements.

Each substrate has a planar top and bottom surface. Edge contacts are formed on one or more sides of each ceramic substrate. Several substrate are pressed together with a 38 μm thermoplastic adhesive.[31] The vertical interconnection is made with the same HDI process (Fig. 6.26). The module does not require mechanical parts to maintain z-axis conduction, and reliability is improved by avoiding the use of solder. The external interconnection between the module and the next level of packaging requires wire bonding with gold wire.

ISA (USA). The basic approach used is to fabricate *chips-first* modules on a carrier substrate. The ISA process begins with a back-lapping step to bring the IC chips to a uniform thickness (153 μm). Chips are then placed on a flat quartz carrier and encapsulated in a polymer. Via holes are drilled by laser through the polymer to the chip connection pads below. Conventional photolithography tech-

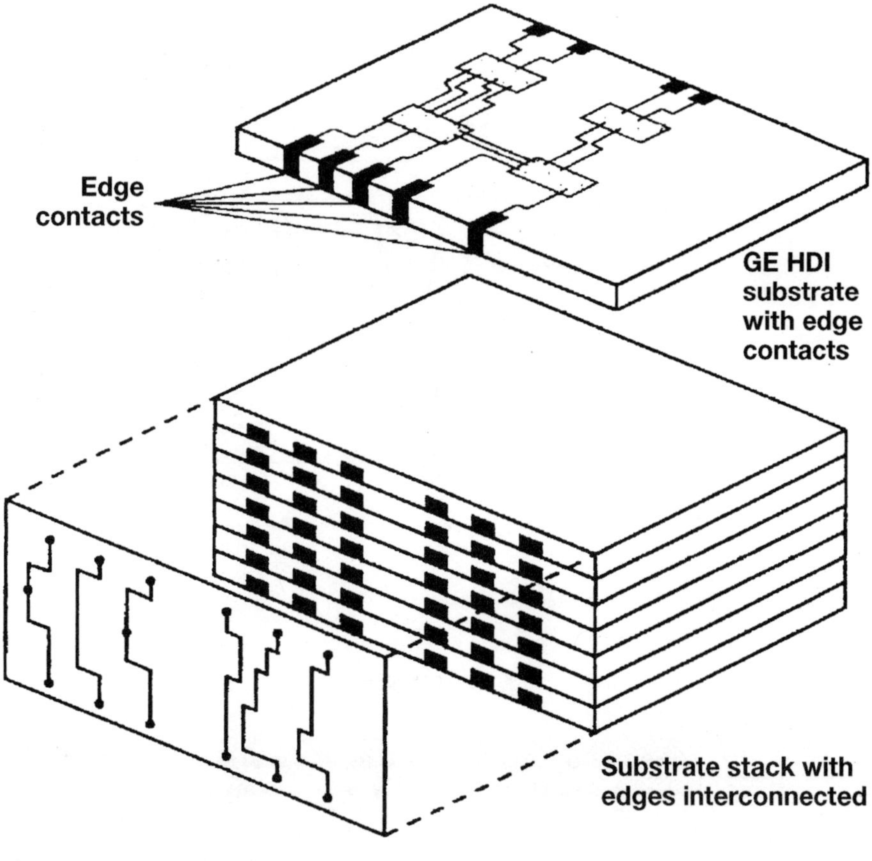

FIGURE 6.26 GE's 3-D HDI stacking approach.

niques are used to define the interconnect pattern, and the surface is metallized by sputtering. After test of the individual module (known as a *ply*), the module is released from the quartz using an excimer laser. When the polymer of the die attach material is encountered, the laser energy is absorbed. This causes a small volume of the material to vaporize and cleanly separates the ply from the quartz carrier. The ply now has no substrate and looks similar to the prepreg used to fabricate a multilayer printed circuit board. A single ply is equal to the thickness of the chips plus the interconnect (three layers = 75 μm) or 225 μm. Individual plies are laminated using standard methods. Deep vias are formed in the area between chips to provide interconnection from one ply to the next (Fig. 6.27). Metal is deposited and patterned to form the interconnect between each layer.[32]

Matra Marconi Space (France). Matra has studied three techniques to stack three or four memory dice:

1. The *crossed pairs* concept (Fig. 6.28)
2. The *tiles* concept, using four rerouted unwedged dice, with relay pads for *chip select* outputs mounted on a multilayer substrate MCM-D
3. The flip chip concept, where dice are turned upside down on a silicon MCM

The *crossed pairs* technique has been developed with four memories. Two memory chips are mounted on a two-layer ceramic substrate with adhesive, wire bonded, and tested. Two additional

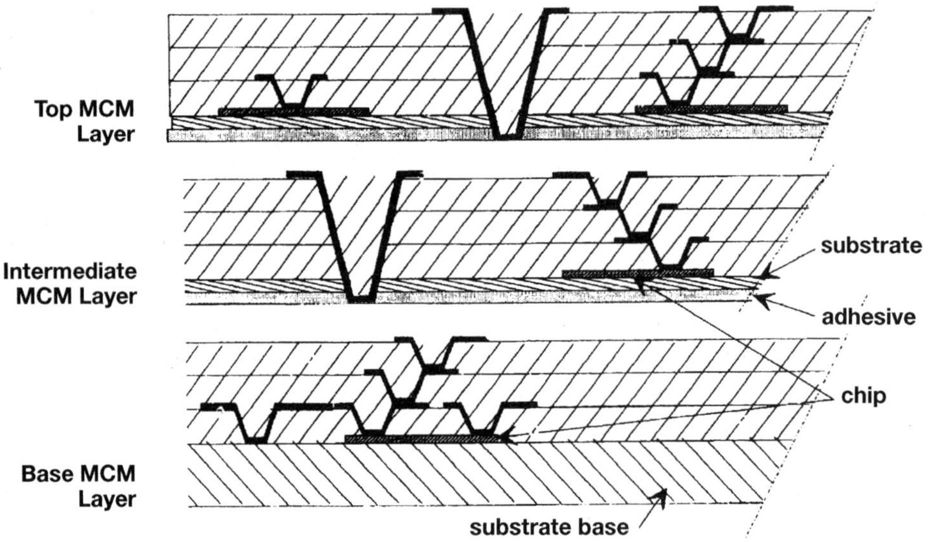

FIGURE 6.27 The ISA *chips first* approach.

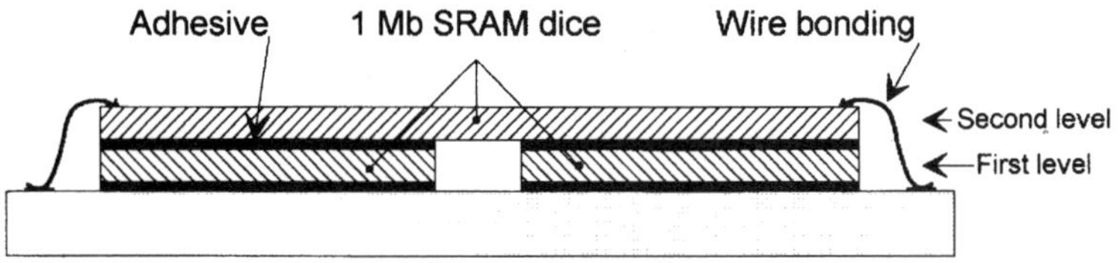

FIGURE 6.28 Matra Marconi Space's MCM module.

memory chips are mounted perpendicularly on top of the first two chips. These two chips are wire bonded to the substrate. A hierarchical packaging architecture has been used based on four stacked chips wire bonded to a thick film ceramic substrate, high-temperature cofired ceramic module MCM-C with four stacks and double-sided surface mount printed circuit boards.[33]

MCM-S Stacked. The MCM-S stacked structures are made by the companies described in the following sections.

AT&T (USA). The assembly process can be described in four basic steps. The first step consists in manufacturing a silicon MCM substrate and flip chip bonding the dice. Four MCM substrate are then mounted to one metal core multilayer board. In the third step, multiple boards are stacked vertically. The assembly of the enclosure and fans are accomplished in the last step.

The MCM substrate (25.4 × 30.5 mm) was fabricated with silicon-on-silicon process and required three thin film layers (aluminum/polyimide). Four MCM substrate were mounted to a metal core multilayer board (127 × 127 mm), and the copper core of 1 mm thick is sandwiched between two layers of interconnect (Cu and resin). The boards were nonhermetically stacked and interconnect using two kinds of anisotropic conductive polymer material (Fig. 6.29). The first material is made from silicon rubber with conductive columns made of silver-plated nickel spheres; this structure leads to a pitch of approximately 150 microns for a film thickness of 0.25 mm. The other material is composed of inserts of gold-plated nickel alloy wires magnetically aligned in a polysiloxane elastomer matrix (Fujipoly connector), and the pitch is 0.35 mm, and the thickness reaches 0.3 mm. The insertion of a layer of this material between the boards (spacing between boards of 3.8 mm) permits us to interconnect in the vertical axis, after which the entire stack of boards is clamped.[34]

Space Computer Corp. (USA). Space Computer developed a parallel processor for spaceborne applications based on a silicon multichip module. Figure 6.30 shows the configuration of a three-dimensional stack that employs elastomeric interconnect elements to provide z-axis connection. Individual substrates are bonded to aluminum nitride heat sinks and wire bonded to substrate holders which transfer the substrate I/O to the elastomer I/O regions. The substrate holder may be a separate interconnection element that has a rectangular section routed out for the substrate, or it may be incorporated as part of the heat sink. The elastomeric material is constrained by an elastomeric holder that is precisely machined to a thickness that ensures controlled deflection of the elastomeric material and maintains a prescribed substrate pitch in the stack. The entire stack assembly is held in compression

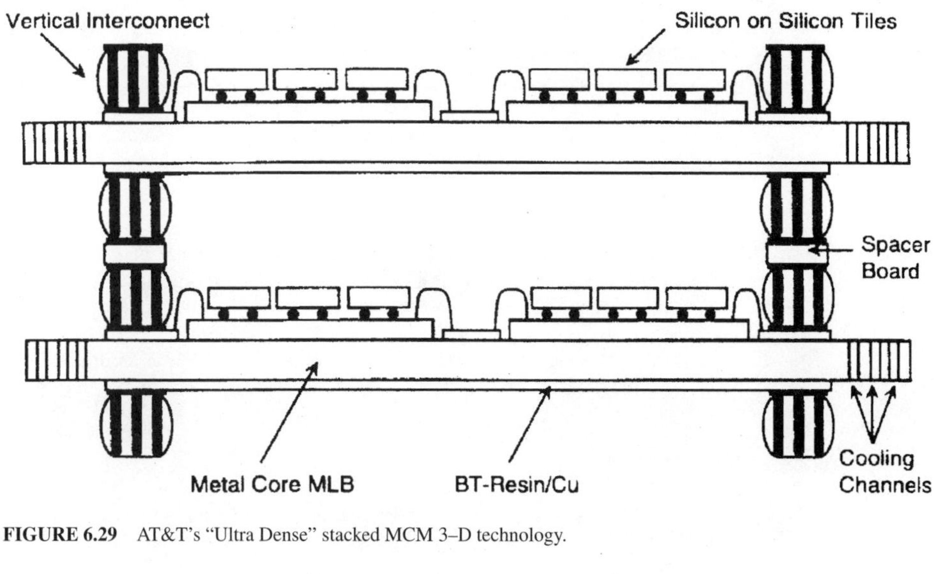

FIGURE 6.29 AT&T's "Ultra Dense" stacked MCM 3–D technology.

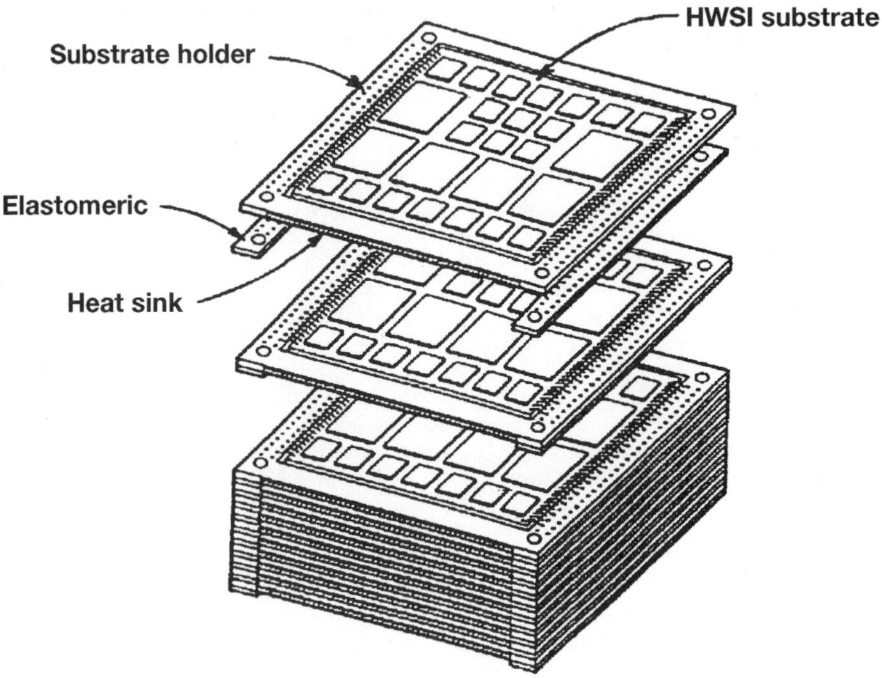

FIGURE 6.30 Space Computer Corp.'s 3-D stack.

by bolts that pass through the elastomeric holders and the substrate holders. Alignment is provided by either the bolts themselves or additional alignment pins.[35]

MCM-L Stacked. The MCM-L stacked structures are made by the companies described in the following sections.

Matsushita Electronic Components (Japan). Matsushita has mounted SRAMs and DRAMs using flip chip, TAB, or wire bonding, on a four-layer, fine-pitch, glass-epoxy board. The substrate has peripheral pads coated with solder paste. A specific C-shaped edge connector made with a liquid crystal polymer is reflow soldered to the substrate (Fig. 6. 31). All the leads are precoated with solder before the stacking up to ten layers. The stack is reflow soldered to interconnect each substrate. The lead pitch is 0.65 mm with 88 to144 pins. A ten-stack module is 23.85 mm high.[36]

3-D Plus/Thomson-CSF (France). Some of the MCM-L stacked structures are described in detail in previous sections.

LETI/CEA (France). The CEA/LETI develops a vertical interconnection technique by wire bonding the staggered silicon substrate (Fig. 6.32). The flip chip memories are soldered on one side of the substrate, and cavities are etched on the other side to constitute a cap for the chips mounted on the substrate at the underneath level. Each substrate is hermetically sealed. The vertical connection through the silicon substrate instead of the wire bonding technique is under development. Stacking of eight 2-D silicon MCM containing 8×16 Mb each gives a silicon module of 13 cm^3.

6.3.4 Stacked Wafer Assembly

More than ten years ago, a lot of colleagues had hopes of using wafer-scale integration, but technical problems (with power supplies, impedance, and so on) and manufacturing yield have limited the

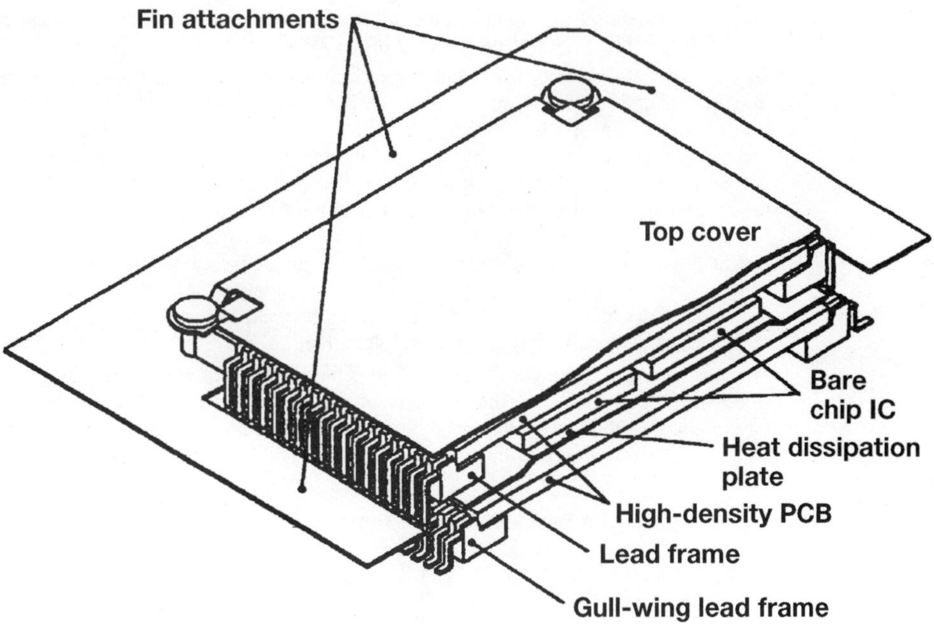

FIGURE 6.31 Matsushita Electronic Components' Mega Module.

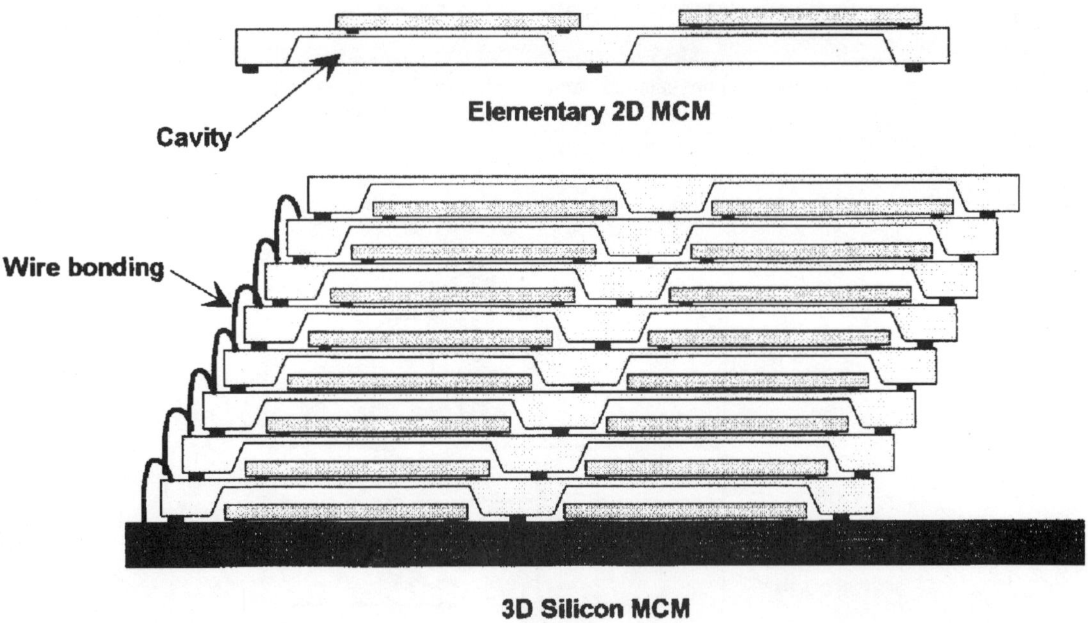

FIGURE 6.32 3-D silicon MCM.

practical applications. Fortunately, monolithic IC stacking with wafer feed-throughs can overcome a part of this problems and achieve a very high level of integration.

Hughes Aircraft (USA). The Hughes approach has the advantage of area array vertical interconnections, compared to peripheral connections. Electrical feed-throughs are produced by thermomigration of aluminum droplets through the silicon wafers at an elevated temperature (≈1000° C). A thermal gradient is applied to the wafer, which causes a concentration gradient of dissolved silicon in the aluminum droplets. Parasitic capacitance of the feed-throughs is ≈0.3 pF. A microbridge technique used to vertically interconnect the wafers has been developed (Fig. 6.33). The microbridges are fabricated by deposition a 50 μm thick spacer material on the wafer and then opening holes in the spacer to contact the feed-throughs (almost like an air bridge in the GaAs process). All the microbridges are coated with a top layer of indium-tin solder; after stacking, and the module is heated to reach the solder melting point (≈150° C) and solder each pair of microbridges together. The electrical resistance of each pair of microbridges is less than 2 Ω.[37]

Micron Technology (USA). The principle is based on using vias etched through the wafer and then filled with metal. Each hole has to be insulated by deposited insulator layer and then metallized to have a good wettability for the metal bump filled inside these holes (Fig. 6.34). Serial addressing and data access are used to minimize the total number of through vias and to increase the yield.[38]

6.3.5 Stacked Heterogeneous Parts—Microsystems

Stacked heterogeneous parts refers to interconnection in three dimensions (mechanical, optical, and electronic components) to have an *equilibrated size module.* The first silicon microactuators were developed by L.S. Fan, Y.C. Tai, and R.S. Muller in 1988,[39] and these micromachines have the potential for use in a variety of fields—in industry, where many types of industrial machinery and plants require sophisticated and delicate maintenance technology, and in medicine, where we desire advanced medical techniques that cause minimal discomfort to patients. This requires a very dense integration, and the 2-D technique cannot be used. A ten-year large scale project called *micromachine technology,* launched by MITI's Agency of Industrial Science and Technology in 1991, was a medically oriented endeavor. A project involving a smart pill, shown in Fig. 6.35, required many heterogeneous components to be 3-D assembled. A European project entitled *basic research for microsystems integration* is now dedicated to microsystems for medical applications.[40] A demonstration with micropump, chemical sensors, photovoltaic convertor, and associated electronics was made (Fig. 6.36). The structure of this microsystem illustrates the fantastic perspectives of the 3-D technique.

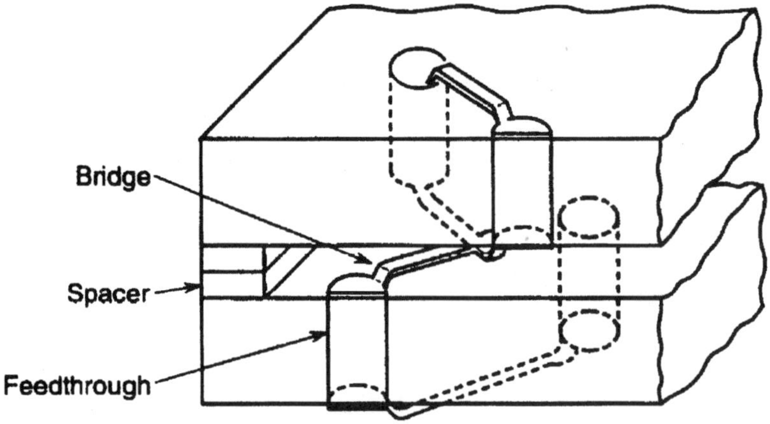

FIGURE 6.33 Hughes Aircraft microspring bridge interconnect technique.

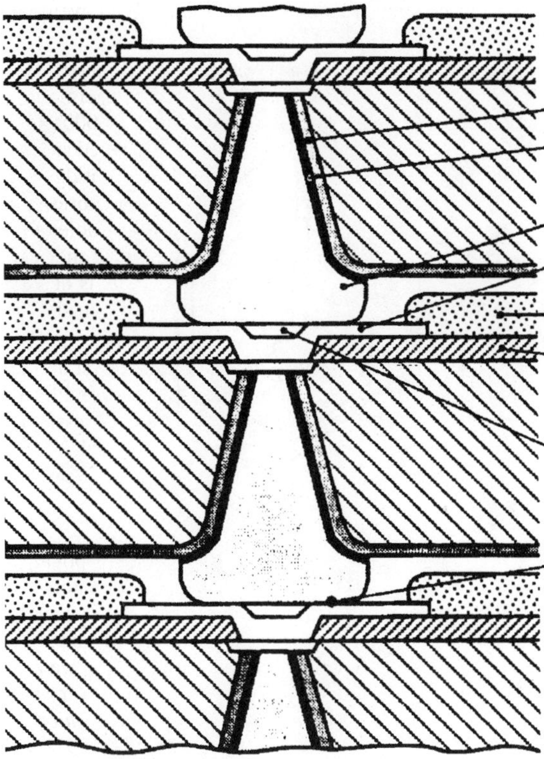

FIGURE 6.34 Micron Technology's approach.

6.4 COST

3-D packages originated in military and space applications such as infrared sensors, airborne signal processing computers, solid state recorders, and smart weapons control, where small size, light weight, and low power were required. Now, 3-D memory stacks are beginning to move from mainly custom applications into the mainstream. The commercial applications are in servers, high-end workstations, and mainframes. Use in very cost-sensitive personal computers may be two or three years off, but the perspectives for low cost are highly promising. TechSearch International Inc. plots the relative price per IC pad and ratio of IC area to package area of several single-package types, as seen in Fig. 6.37. Package types include quad flat packs (QFPs), chip-size packages (CSPs), flip chip, plastic ball-grid arrays (PBGAs), ceramic ball-grid arrays (CBGAs), tape ball-grid arrays (TBGAs), and 3-D stacked packages. The x-axis represents the price per IC pad in cents (price per I/O), and the y-axis represents the ratio of IC silicon area to the packaging area (packaging efficiency). Using 1995 prices, the figure includes only the price of the IC and package assembly, not mounting of the package to the board. The 3-D stacked packages are defined as having four to ten chips stacked in the footprint of one chip. CSP pricing reflects projected pricing for chip-size packages in full production in 1996 and 1997. Advanced packages such as flip chip, CSP, and 3-D appear price competitive with the others, and the merit factor for 3-D packages is the highest.[41]

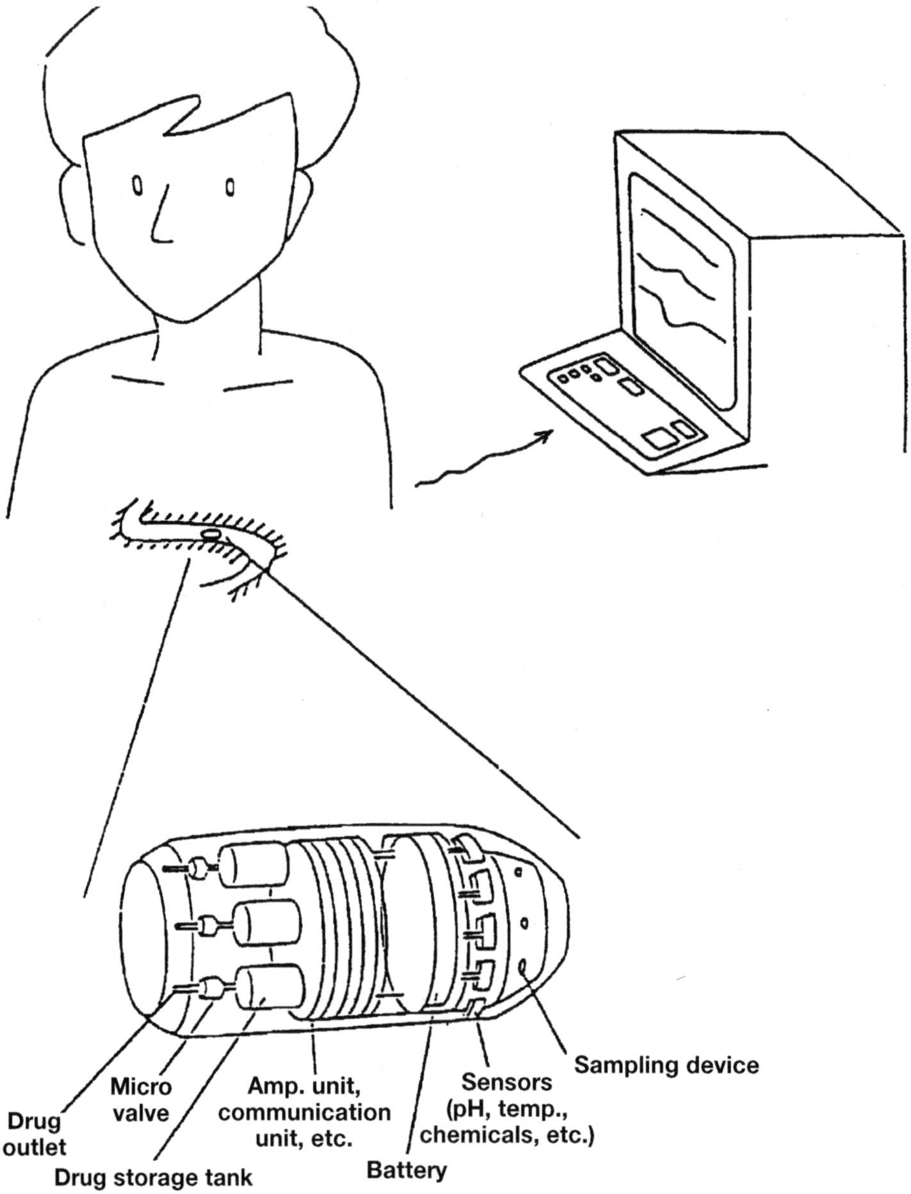

FIGURE 6.35 Smart pill.

6.5 CONCLUSION

A large proportion of 3-D techniques are memory oriented, and bare die and plastic packaged die have been stacked. The TSOP-packaged memories (1 mm thick) and in the near-future the chip-scale

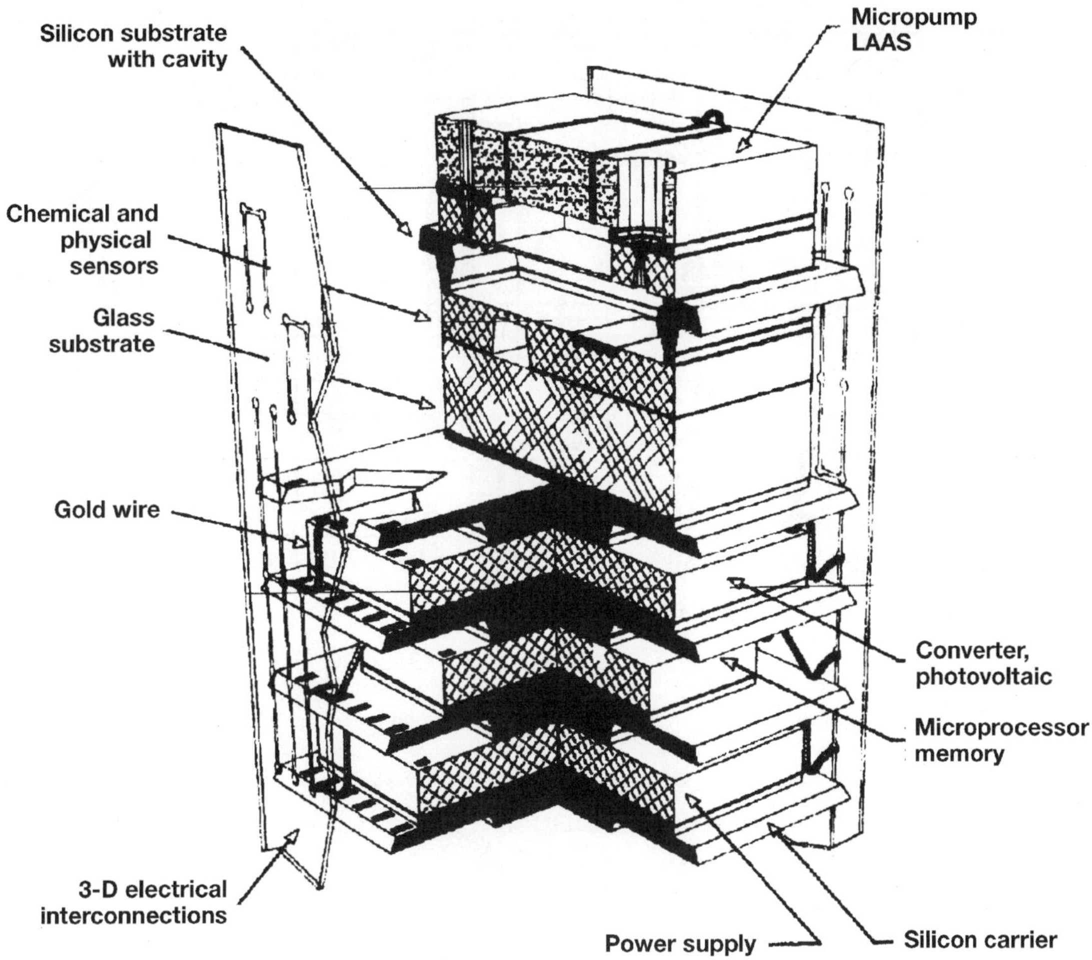

FIGURE 6.36 Cutaway view of the LAAS/CNRS/3-D Plus 3-D microsystem.

packages (0.3 to 0.5 mm thick) will be used, because they are easier to obtain, test, and handle than bare memory chips. However, the use of bare die will lead to a very dense module due to the tendency for thinning the chips (see Fig. 6.1).

Another important criterion is the decrease of capacitive and inductive loads in 3-D stacked chips so that ultra-low voltage can be used without compromising performance. In fact, when chips can be so intimately integrated that the resulting assembly can be thought of as a super chip, chip interconnect overhead (inductance, capacitance, and so on) is reduced to intrachip level. "The result is a revolutionary change in the performance-cost equation by factors of 10 to 100 in teraoperations/second (TOPS), I/O bandwidth, and bits of storage per watt-dollar," according to Carson.[42] On other hand, the thermal behavior of the 3-D interconnection could be limited but, when all the chips can communicate in a highly integrated 3-D electronics assembly, the power dissipation could be very low—less than 2 W expected for 1 TOPS.[42]

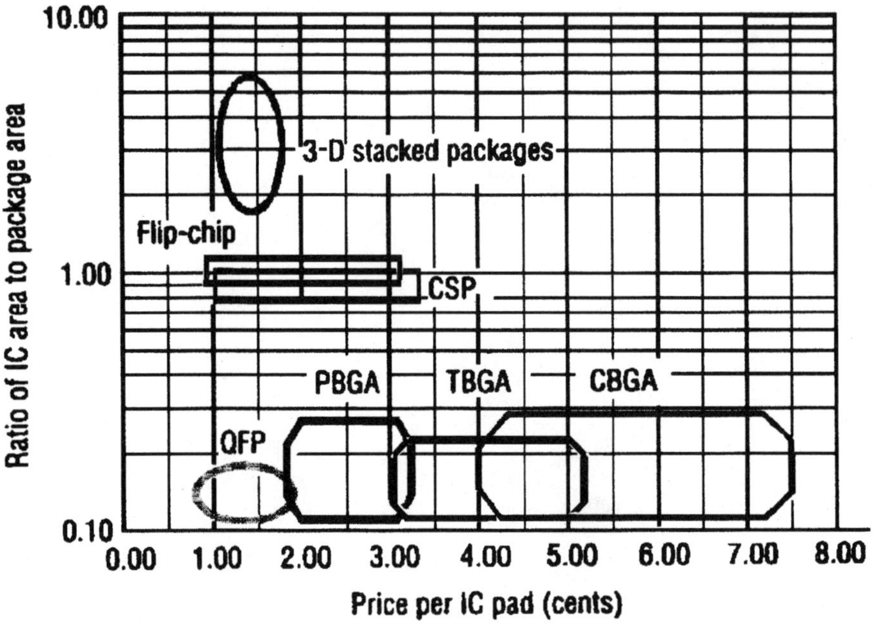

FIGURE 6.37 Package price reflects area ratios. (Only IC package and assembly are considered.)

REFERENCES

1. R. Tummala, *Georgia Tech. Magazine,* Oct 1995.
2. V.H. Ozguz, University of California, San Diego, 3-D Workshop, Los Angeles, October 1995.
3. Irvine 3-D Workshop, Los Angeles, October 1995.
4. D. Frew, "High Density Memory Packaging Technology High Speed Imaging Applications," *SPIE 1990 Proc.,* vol. 13 46, pp. 200–209.
5. S.N. Shanken, and R.J. Indin, "3-D Integrated Packaging and Interconnect Technology," *ISHM 1989 Proc.,* Baltimore, Maryland, pp. 1–7.
6. C. Val and T. Lemoine "Interconnection for Ultra Dense Multichip Module," *Proc. of 40th Electronic Components Technology Conf.,* IEEE/CHMT, Las Vegas, pp. 540–547, May 1990.
7. I. Miyano, K. Serizawa, T. Ishida, M. Yamada, and T. Kudaishi, "Fabrication and Thermal Analysis of Three Dimensionally Located LSI Packages," *9th European Hybrid Microelectronics Conference Proc.,* pp. 184–191, June 1993.
8. "Memory Package has the Same Dimensions as a TSOP or SOJ and Two to Four Times the Capacity," *Nikkei Microdevices,* April 1991, p. 80.
9. FIND, from *Fujitsu,* vol. 11, no. 2, May 1993.
10. K. Hatada, et al. "New Film Carrier Assembly Technology—Transferred Bump TAB," *IEEE CHMT,* vol. CHMT-10, no. 3, 1987.
11. 3-D Workshop, Los Angeles, October 1995.
12. C. Val and M. Leroy, "The 3-D Interconnection-Applications for Multichip Modules Vertical, MCM-V," *Proc. Int. Symposium on Microelectronics,* ISHM Florida, pp. 62–68, October 1991.
13. J.A. Ninaham, A. Pepe, R. Some, and M. Suer, "The 3-D Stack in Short Form," *Proc. 42nd Electronics Comp. and Tech. Conf.,* pp. 340–344, 1992.
14. M.F. Suer, "3-D Design Issues," *Proc. of 1993 Nepcon Conf.,* pp. 1201–1205, February 1993.
15. A.C. Fox, III and W.M. Farnworth, "High-Density Electronic Package Comprising Stacked Sub-modules which Are Electrically Interconnected by Solder-Filled Vias," U.S. Patent no. 5,128,831, July 7, 1992.
16. R. Bruns, W. Chase, and D. Frew, "Utilizing Three-Dimensional Memory Packaging and Silicon-on-Silicon Technology for Next Generation Recording Devices," *1992 ICMCM Proc.,* pp. 327–333.
17. "Mitsubishi Electric Develops Stacked Memory Module," *Denshi Zairyo,* June 1993, p. 19, from R. Crowley, TechSearch Int.
18. C.M. Val "3-D Packaging of TSOP Plastic Packages," *ISHM Proc.,* San Diego, pp. 370–375, October 1992.
19. F. Eide, "Micropackaging: An Alternative to Bare Die," *Inside ISHM,* pp. 34–35, Mar. 1993.
20. E.G. Palmer and C.M. Newton, "3-D Packaging Using Low Temperature Co-fired Ceramic (LTCC)," *Proc. Int. Conf. on MCM,* pp. 307–312, April 1993.
21. W.A. Vitriol and E. Palmer, "Qualification of 3-D Space Grade Memory Cubes Utilizing Low Temperature Co-fired Ceramic (LTCC) Technology," *IEPS Proc.,* pp. 1003–1019, September 1993.
22. P.T. Lin, "Three-Dimensional Multichip Pad Array Carrier," U.S. Patent no. 5,222,014, June 22, 1993.
23. Seung-Ho Ahn, "Development of Low Cost Three-Dimensional Packages," *MCM Proc.,* Denver 1995, pp. 269–278.
24. A.S. Conte, "MCM-L the Answer for Desktop Workstations," *ICEMM Proc.,* pp. 18–21, April 1993.
25. C. Burns, U.S. Patent 5,221,642, issued to Staktek, June 1993.
26. R.T. Crowley and E.J. Vardaman, "3-D Multichip Packaging for Memory Modules," *MCM 94 Proc.,* Denver, pp. 474–479.
27. T.J. Moravec, R.C. Eden, and D.A. Schaefer, "The Use of Diamond Substrates for Implementing 3-D MCMs," *Proc. Of 1993 Intl. Conf. On Multichip Modules,* pp. 86–91, Apr. 1993.
28. A.L. Solomon, "Framed Chip Hybrid Stacked Layer Assembly," U.S. Patent no. 5,231,304, issued July 27, 1993.
29. R.J. Jensen, W.F. Jacobsen, S.L. Palmquist, and R.K. Spielberger, "Characterisation of three-dimensional aluminum nitride multichip modules," 3-D Workshop, Los Angeles, October 1995.

30. J.M. Schroeder, "Three-Dimensional Multichip Module Packaging," *Intl. Journal of Microcircuits and Electronic Packaging,* vol. 16, no. 2, pp. 117–122, 1993.
31. G.A. Forman, et al., "GE 3-D HDI Stacked Multichip Module Technology's Impact on System Design," *Proc. 1993 Nepcon Conf.,* pp 1206–1215, Feb. 1993.
32. C.W. Eichelberger, J.L. Thomas, N. Caporale, L.A. Principe, and J.R. Doucette, "A 3-D packaging technology is applied," *Advanced Packaging,* March/April 1995, pp 39–41.
33. J.A. De Givry, "Mass Memory Packaging for Space Applications," MCM-C/ Mixed Technologies, *Proc. NATO Advanced Research Workshop,* edited by NATO ASI, series 3, High Technology, vol. 2, pp. 131–138, Florida, May 23–25, 1994.
34. T.G. Grau, R.R. Shively, L.J. Lau, J.M. Segelken, and K.L. Tai, "Ultra-Dense: An MCM Based 3-D Digital Signal Processor," *Proc. 1991 Nepcon Conf.,* pp. 495–505, Feb. 1991.
35. N.J. Teneketges, W.J. Jacobi, and L.A. Wadsworth, "A High-Performance MCM-Based Spaceborne Processor," *ICMCM Proc.,* 1992, pp. 334–341.
36. Matsushita Electronic Components, *Technical Report of IEICE,* 1993.
37. M.J. Little, R.D. Etchells, J. Grinberg, S.P. Laub, J.G. Nash, and M.W. Yung, "The 3-D Computer," *Proc. on Wafer Scale Integration,* pp. 55–64, Jan. 1989.
38. A.P. Gnadinger, "High Density Storage Using Stacked Wafers," U.S. Patent no. 5,229,647, Jul. 20, 1993.
39. L.S. Fan, Y.C. Tai, and R.S. Muller, *IEEE Transactions in Electronic Devices,* vol. 34, p. 724, 1988.
40. Esprit Project 1994–1997, "Basic Research for Microsystems Integration," BARMINT, no. 8173.
41. G. Derman, "Interconnect Packaging," *Electronic Engineering Times,* pp. 47, 74, 75, and 79, June 26, 1995.
42. J. Carson and R. Some, "Memory Stacking Drives 3-D Packaging," *Electronic Engineering Times,* pp. 58,78, June 26, 1995.

CHAPTER 7

MCM PACKAGE DESIGN

John K. Hagge
Avionics and Communications, Rockwell International

Abstract. This chapter discusses design approaches for successful packaging of multichip modules. Design success is a careful balance between meeting customers' requirements for package performance and achieving a cost that is affordable to the customer yet provides the package fabricator a reasonable profit. The customers' performance requirements include both adequate operational performance (electrical, mechanical, and thermal), and long-term reliability under various environmental stress conditions. Cost factors include materials choices (plastic, ceramic, metal), design configurations (peripheral leads, array pins, array pads, array bumps), and environmental durability (low-cost nonhermetic, high-reliability hermetic equivalent, and full hermetic).

7.1 INTRODUCTION

Multichip module packages have similar attributes and characteristics to those generally described for single-chip packages.[1–3] The MCM package

- Provides structural support for the components
- Provides protection from the environment to those components
- Either provides or houses the interconnections between the MCM components
- Provides insulation to confine electrical currents to their desired paths
- May provide shielding to confine electric fields to their desired configurations
- Provides a thermal path away from the heat generating components
- May provide distribution of power and ground currents, and/or controlled impedance at the signal input/output (I/O)
- Provides the interface for module testability

From one point of view, these are all desirable contributions of packaging. From another point of view, the package is a nuisance. As observed by a number of authors, the package

- Introduces resistance to ground and power distribution
- Introduces degradation to electrical signal lines
- Adds weight and consumes space
- Introduces thermal resistance to heat removal paths
- Introduces mechanical stresses to the integrated circuits
- May be a source of radiation particles
- Adds cost to the system

The most successful package designs will emphasize the positive attributes and minimize the negative characteristics listed above. In fact, the success of a multichip module comes from minimizing packages through the following:

- Elimination of packages between ICs, eliminating their accompanying thermal resistance, electrical resistance and delay, weight, size, and cost
- Replacement of the multiple single-chip packages with a single MCM package that occupies less total space, is lighter, and has less total lineal length and surface area to seal from environmental moisture penetration

Successful design of MCM packaging requires multidisciplinary knowledge including a host of engineering disciplines (mechanical, electrical, chemical, ceramic), along with principles of physics, chemistry, metallurgy, and materials technology.

7.2 MECHANICAL DESIGN CONSIDERATIONS FOR MCM PACKAGES

The elements of successful packaging for multichip modules include the following proactive approaches:[4,5]

- Design to ensure that applied loads will never exceed the yield strengths or fatigue strengths of the materials used.

- Design to ensure that interfaces between unlike materials do not develop significant stresses due to mismatched coefficients of thermal expansion (CTEs). Where significant stresses would be developed, design to provide compliancy at the interface.
- Design and materials selection to ensure that the heat path out of the MCM is a low thermal resistance path, never measuring more than a few degrees in IC junction temperature rise.
- Design to minimize size and weight, ensuring that inertial forces are kept low.
- Design to achieve stiff, rugged structures, ensuring that inertially induced deflections are small.
- Design to separate the electrical connections to the package from the mechanical attachments to the package.
- Design to optimize for the system level, not the module level. Design for optimized interface to the next packaging level for electrical, mechanical, and thermal interfaces.
- Design to encourage development of high-yield manufacturing and assembly processes.
- Design for reasonable package price, which will be attractive when compared to the competitive packaging choices.
- Design for good testability and producibility.

7.2.1 Package Size

As discussed elsewhere in this book, one of the chief advantages of MCM technology is the miniaturization that it provides. A great deal of importance is placed on achieving a high packaging efficiency ratio (total chip component area ÷ total MCM foot print area—see Fig. 7.1). For MCMs utilizing substrates to accomplish the chip-to-chip interconnect, much attention and effort are invested in achieving a high ratio of total component area (ICs plus passive chips) to substrate area. Significant costs may be required to improve this ratio, for example, by switching to a higher-density MCM-D substrate from a lower-density MCM-L substrate. Often overlooked in this quest are the equally important roles of miniaturization of both package size and size of the circuit board pad and fanout pattern. Total packaging efficiency depends on both circuit board pad pattern miniaturization and package miniaturization, in addition to substrate miniaturization as follows:

$$R_T = R_S \times R_P \times R_B \tag{4.1}$$

where R_T = total packaging efficiency ratio

$$= \frac{\text{Total chip component area}}{\text{Total MCM area consumed on circuit board}}$$

R_S = substrate packaging efficiency ratio

$$= \frac{\text{Total chip component area}}{\text{Substrate area}}$$

R_P = package packaging efficiency ratio

$$= \frac{\text{Substrate area}}{\text{MCM footprint area across leads}}$$

R_B = circuit board site packaging efficiency ratio

$$= \frac{\text{MCM footprint area across leads}}{\text{Total MCM area consumed on board}}$$

It is important to recognize that the next level interconnection (usually the circuit board on which the MCM is mounted) has an equally important role in packaging efficiency. The next level intercon-

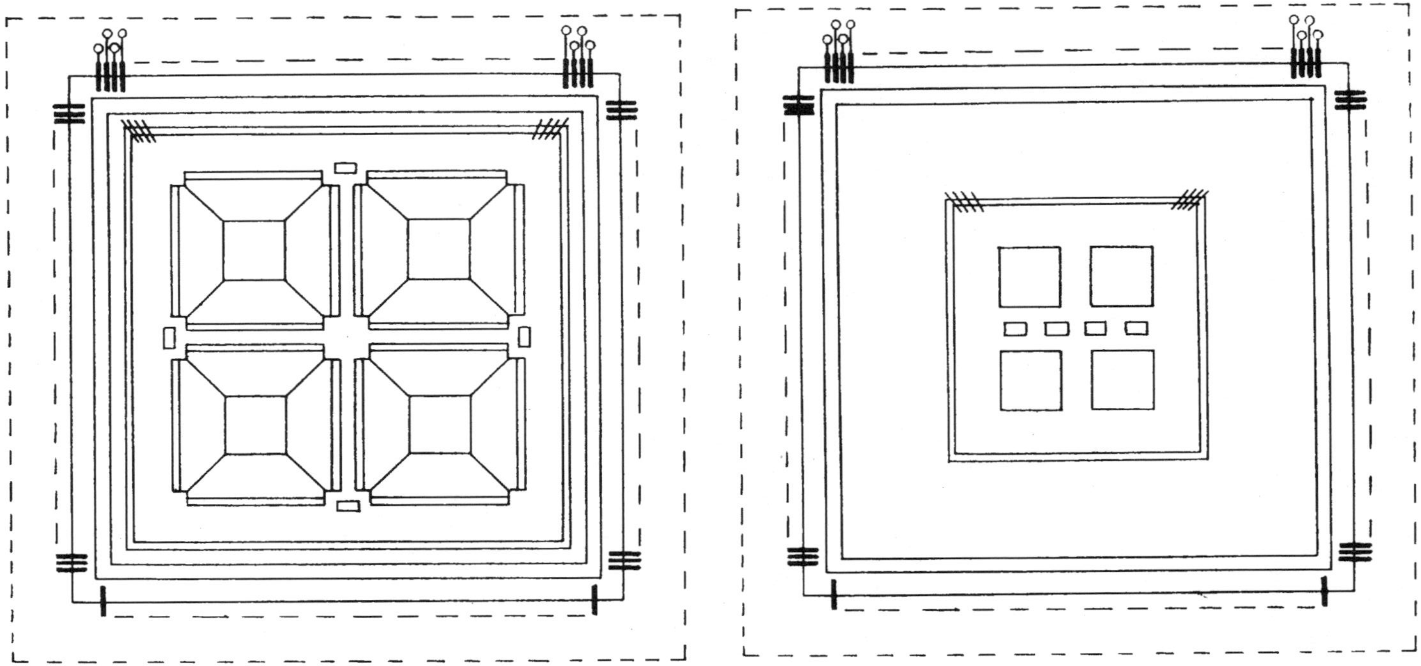

1a. TAB ICs: 5 mil ILB/12 mil OLB to ceramic substrate, 30 mil pitch wire-bonds to ceramic package, 40 mil pitch gull wing leads soldered to PWB, 100 mil pitch test pads on PWB.

- o PWB footprint = 4.4 x 4.4 inch
- o MCM Lead footprint = 3.8 x 3.8 inch
- o Substrate size = 2.8 x 2.8 inch

Limiting Constraints: package lead pitch & PWB test pads

1b. Wire-Bond ICs: 5 mil pitch to ceramic substrate, 18 mil pitch wire-bonds to ceramic package, 40 mil pitch gull wing leads soldered to PWB, 100 mil pitch test pads on PWB

- o PWB footprint = 4.4 x 4.4 inch
- o MCM Lead footprint = 3.8 x 3.8 inch
- o Substrate size = 1.6 x 1.7 inch

Limiting constraints: package lead pitch & PWB test pads

FIGURE 7.1 Comparison of MCM size vs. interconnect and package choices.

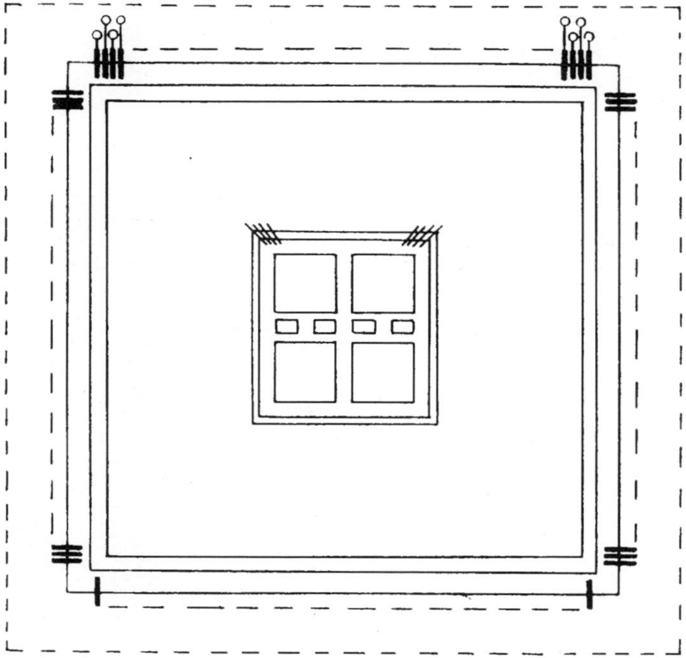

1c. Flip-Chip ICs: 10 mil pitch solder bumps to ceramic substrate, 12 mil pitch wire-bonds to ceramic package, 40 mil pitch gull wing leads soldered to PWB, 100 mil pitch test pads on PWB.

- o PWB footprint = 4.4 x 4.4 inch
- o MCM lead footprint = 3.8 x 3.8 inch
- o Substrate size = 1.1 x 1.2 inch

Limiting constraints: package lead pitch & PWB test pads

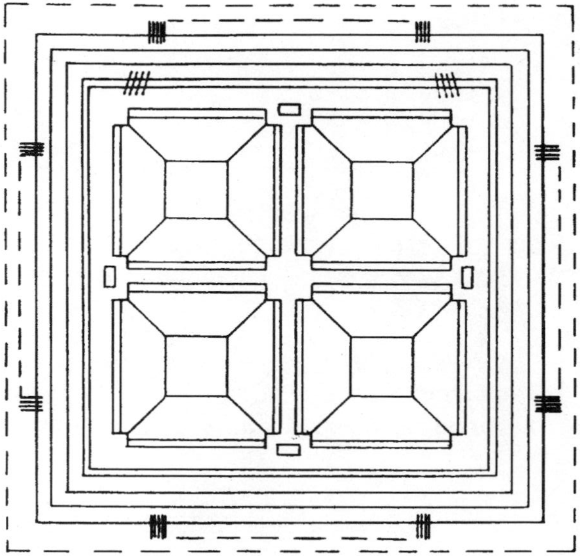

1d. TAB ICs: 5 mil ILB/12 mil OLB to ceramic substrate, 25 mil pitch wire-bonds to ceramic package, 20 mil pitch gull wing leads soldered to PWB.

- o PWB footprint = 3.7 x 3.7 inch
- o MCM lead footprint = 3.5 x 3.5 inch
- o Substrate size = 2.6 x 2.6 inch

Limiting constraint: TAB fanout area

FIGURE 7.1 *.continued*

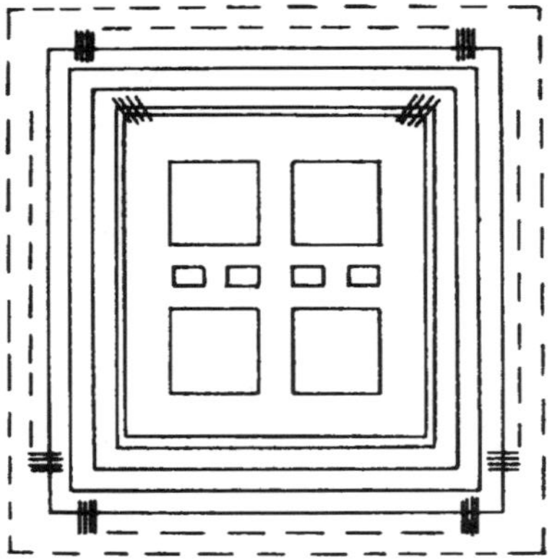

1e. Wire-Bond ICs: 5 mil pitch to ceramic substrate, 18 mil pitch wire-bonds to ceramic package, 20 mil pitch gull wing leads soldered to PWB.

- o PWB footprint = 2.45 x 2.6 inch
- o MCM lead footprint = 2.25 x 2.4 inch
- o Substrate size = 1.35 x 1.5 inch

Limiting constraint: Wire-Bond Area

1f. Flip-Chip ICs: 10 mil pitch solder bumps to ceramic substrate, 10 mil pitch wire-bonds to ceramic package, 20 mil pitch gull wing leads soldered to PWB.

- o PWB footprint = 2.45 x 2.45 inch
- o MCM lead footprint = 2.25 x 2.25 inch
- o Substrate size = 1.1 x 1.2 inch

Limiting constraint: package lead pitch

FIGURE 7.1 *.continued*

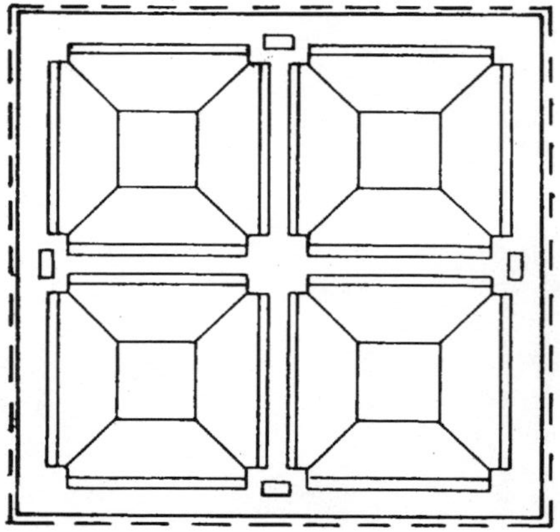

1g. TAB ICs: 5 mil ILB/12 mil OLB to ceramic or laminate substrate, 50 mil pitch array bumps soldered to PWB, glob top encapsulant.

- o PWB footprint = 2.7 x 2.7 inch
- o MCM footprint = 2.6 x 2.6 inch
- o Substrate size = 2.6 x 2.6 inch

Limiting constraint: TAB fan-out area

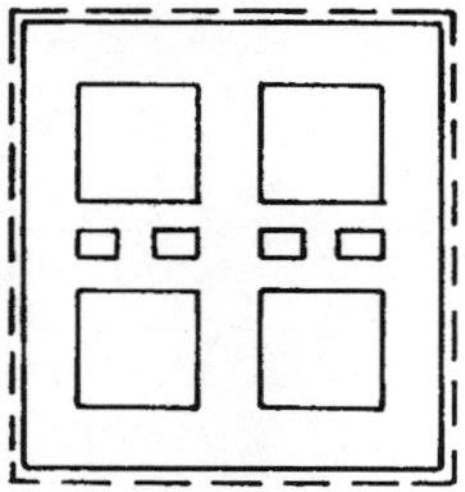

1h. Wire-Bond ICs: 5 mil pitch to ceramic or laminate substrate, 50 mil pitch array bumps soldered to PWB, glob top encapsulant.

- o PWB footprint = 1.45 x 1.6 inch
- o MCM footprint = 1.35 x 1.5
- o Substrate size = 1.35 x 1.5 inch

Limiting constraint: Wire-Bond area

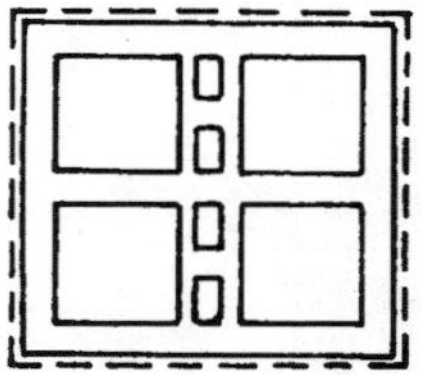

1i. Flip-Chip ICs: 10 mil pitch solder bumps to ceramic or laminate substrate, 50 mil pitch array bumps soldered tp PWB, underfill encapsulant.

- o PWB footprint = 1.2 x 1.3 inch
- o MCM footprint = 1.1 x 1.2 inch
- o Substrate size = 1.1 x 1.2 inch

Limiting constraint: IC bare die size

FIGURE 7.1 *.continued*

nection circuit board technology must be able to accomplish its interconnect without further fanout from the MCM packaging I/O (for routing or testability), or the MCM footprint area will become considerably larger to accommodate this additional fanout. Figure 7.1 and Table 7.1 illustrate these points. Each example is for a four-chip MCM composed of one processor chip, three special function ASIC chips, and four chip capacitors. All IC chips are 10 mm (0.4 in) square with 280 I/O. Total MCM I/O is 320. Figure 7.1*a* shows an example where TAB technology was selected to accomplish chip pretest. A low-cost substrate was chosen so that TAB outer-lead solder connections could be made on 0.3 mm (0.012 in) pitch with low-cost hot bar soldering. The package selected is a cofired ceramic design with fanout to a customer-imposed lead pitch of 1 mm (0.040 in). No provision was made for module-level boundary scan testing, so test pads have been added to the circuit board around the MCM pads. In this example, the substrate packaging efficiency ratio is 0.086, the package packaging efficiency ratio is 0.543, and the circuit board site packaging efficiency ratio is 0.746, giving a total packaging efficiency ratio of 0.034.

Figure 7.1*b* shows the same chip-set with similar package, board, and test constraints. This time, the chips are wire bonded to the substrate with zero fanout at the chip-to-board connection. The substrate packaging efficiency ratio has improved from 0.086 to 0.243. However, the package peripheral I/O spacing still controls the total package size, and the total packaging efficiency ratio remains at only 0.034. Figure 7.1*c* shows the same chip, package, board, and test constraints, but the chips are now all flip chip attached to the substrate. The substrate packaging efficiency ratio improves further to 0.500, but the total packaging efficiency remains unimproved.

In Figs. 7.1*d–f*, testability has been designed into the MCM so the test pads can be removed from the circuit board. The board assembly process has been improved to allow soldering the MCM leads at a pitch of 0.5 mm (0.20 in). The packaging efficiency ratios have improved dramatically by factors of 1.4, 4.1, and 4.2 times, respectively.

In Figs. 7.1*g–i,* the substrate has been changed to a ball grid array (BGA) style using 1.27 mm (0.050 in) pitch solder bumps. The circuit board technology has been improved to higher-density features, so no extra fanout area is required to accomplish routing within the board to the BGA pattern The packaging efficiency ratios have now improved further over Figs. 7.1*a–c* by factors of 2.7, 8.8, and 12.4, respectively.

Table 7.1 Effect of Die Connection, Package, Board, and Test Approaches on Packaging Efficiency

No fanout from substrate to board, boundary scan utilized	Fig. 7.1*g* $R_S = 0.098$ $R_P = 1.0$ $R_B = 0.927$ $R_T = 0.091$	Fig. 7.1*h* $R_S = 0.325$ $R_P = 1.0$ $R_B = 0.875$ $R_T = 0.284$	Fig. 7.1*i* $R_S = 0.500$ $R_P = 1.0$ $R_B = 0.846$ $R_T = 0.423$
Small fanout in package, boundary scan utilized	Fig. 7.1*d* $R_S = 0.098$ $R_P = 0.552$ $R_B = 0.895$ $R_T = 0.048$	Fig. 7.1*e* $R_S = 0.325$ $R_P = 0.376$ $R_B = 0.848$ $R_T = 0.104$	Fig. 7.1 $R_S = 0.500$ $R_P = 0.261$ $R_B = 0.843$ $R_T = 0.110$
Large fanout in package and board, no boundary scan	Fig. 7.1*a* $R_S = 0.086$ $R_P = 0.543$ $R_B = 0.746$ $R_T = 0.034$	Fig. 7.1*b* $R_S = 0.243$ $R_P = 0.188$ $R_B = 0.746$ $R_T = 0.034$	Fig. 7.1*c* $R_S = 0.500$ $R_P = 1.091$ $R_B = 0.746$ $R_T = 0.034$
Chip interconnect:	TAB with fanout	Wire bond, no fanout	Flip chip
	Increasing substrate packaging efficiency →		

It should be noticed from these examples that various choices about fanout within the package and on the circuit board can produce a variation in size of the MCM by over an order of magnitude. It is also interesting that these choices produce different limits to miniaturization. Frequent limiting constraints are

1. Fanout on PWBs
2. Peripheral package lead pitch
3. TAB fanout
4. Fanout in the MCM substrate

It may be beneficial to consider changing design approaches to eliminate these constraints. Conversely, if these constraints cannot be changed, it may be a waste of money to invest in miniaturization elsewhere in the design, as was evident in Figs. 7.1*a–c*, where investment in changing to high-density, flip chip substrates produced no reduction in overall MCM size or PWB footprint area.

A good package design will be easily scalable to various sizes to accommodate a wide variety of IC sizes, numbers, and device heights. Few-chip designs may be suitable for packaging in conventional packages originally designed for single-chip devices. These MCMs typically would be in the 1.25 × 1.25 cm (0.5 × 0.5 in) to 5.1 × 5.1 cm (2 × 2 in) sizes and would be limited in I/O count to the common size packages already tooled, in the 24 to 250 lead range. Special packages designed specifically for MCMs offer larger sizes and higher-density I/O. MCM package sizes ranging in length from 2.5 to 10 cm (1 to 4 in) are commercially available in square and rectangular configurations, some with lead counts exceeding 400 I/O. Custom packages have been built with even higher lead counts. As pointed out in Fig. 7.1, the external dimensions of the MCM footprint should be kept as small as possible relative to the substrate dimensions to maintain acceptable packaging efficiency ratios. Figure 7.2 shows this strong influence on packaging efficiency.

It should be noted that, in the previous discussion, design choices at the circuit board and package level have even more dramatic impact on total packaging efficiencies than choices about die attach and substrate technologies. Too frequently, the emphasis in packaging design is being placed on the latter two items. As is often the case, it is important to accomplish a full system design at all levels of packaging to achieve a truly optimum design.

This point is illustrated in Table 7.2. Keeping in mind that the total packaging efficiency (in percent) is in the 1 to 3 percent range for circuit board assemblies with through-hole components, and the 3 to 8 percent range with surface-mount components, all of the MCM packages listed in Table 7.2 provide improvements in packaging efficiency, compared to single-chip packages. However, only designs that have optimized both R_S and R_P achieve total packaging efficiencies above 25 percent, regardless of the selection of MCM-L, -C, or -D substrate technologies. MCMs that employ stacked chip, stacked substrate, or stacked package 3-D approaches can obviously achieve considerably higher packaging efficiencies, as discussed in a separate chapter. Four basic types of package constructions have emerged in the MCM industry, as shown in Fig. 7.3.

1. Independent substrate
2. Integral substrate
3. Postmolded substrate
4. Encapsulated substrate

Each type can have I/O versions with array pins/pads/bumps, peripheral leads, or multicontact connectors.

Independent-substrate packages house a separate substrate that provides both the interconnect between the chips and the power/ground connections to the chips. A surrounding package provides module I/O and is available in ceramic, metal, and plastic versions.

Integral-substrate packages incorporate the substrate directly as part of the package structure, usually the package base. These types of packages are usually fabricated with multilayer ceramic or multilayer laminate technologies.

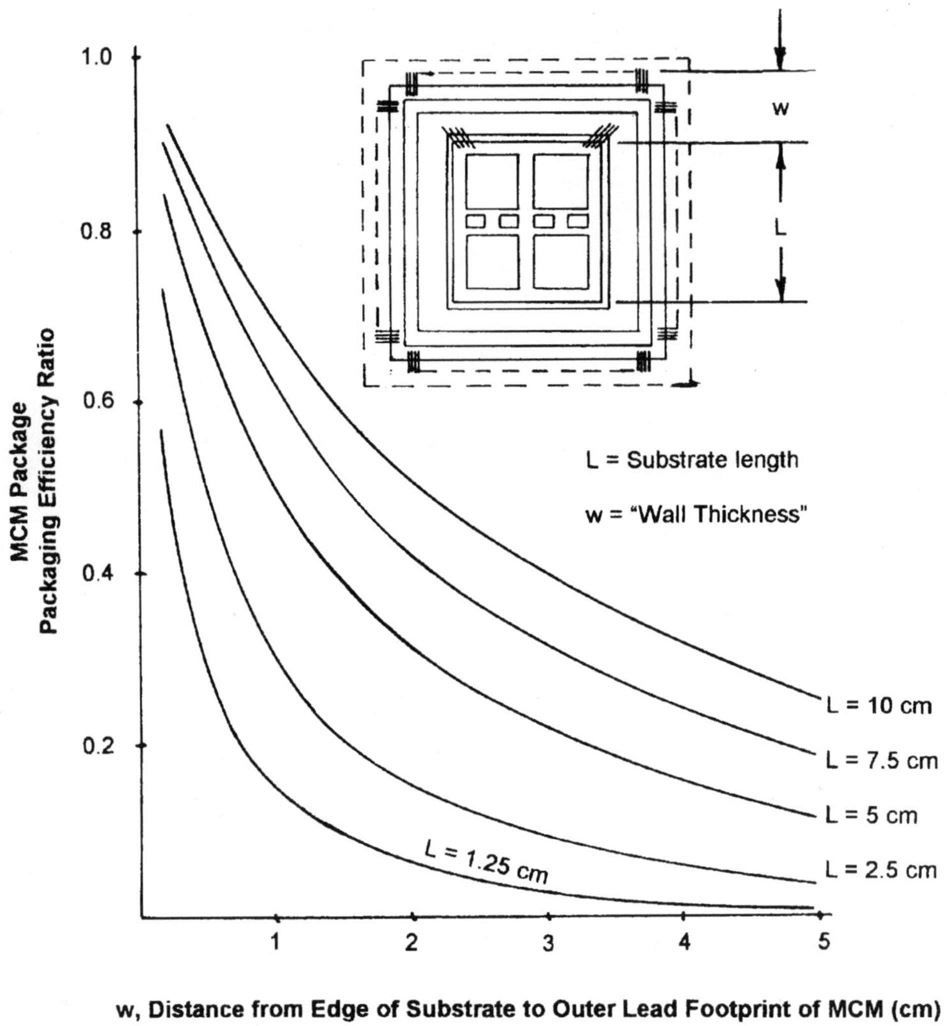

FIGURE 7.2 Effect of "wall thickness" on MCM packaging efficiency.

Postmolded substrate packages use a substrate to accomplish all chip interconnect, to fabricate or attach a leadframe or array pinout to the substrate, and to surround the IC assembly with plastic in a high-pressure molding operation.

Encapsulated-substrate packages again use a substrate to accomplish chip interconnect and then apply some type of protective coating over the ICs. The line of distinction between this type of MCM and a chip-on-board circuit board assembly is difficult to define.

Package cavity depth in the first two types will be defined by the sum of the thicknesses of the substrate, the tallest component, bond lines, interconnect height allowance (e.g., wire bond loop), plus clearance and tolerances. A clearance should be included to accommodate lid deflection. Hybrid microcircuit specifications require a minimum clearance of 0.13 mm (0.005 in) between any wire-bonds and a metal lid. It should be noted that this clearance was established when hybrid packages were fairly small.[10] MCM packages larger than 38 mm (1.5 in) probably require greater clearances

Table 7.2 Attributes of Various MCM Packages (measured, calculated, or quoted from various sources)

I/O	Substrate type	Number of ICs	Substrate size (inches)	Package type	PWB footprint (inches)	Packaging Efficiency (%) RS	RP	RT	Ref.
131	MCM-D	6	1.3 × 1.3	PPGA	1.9 × 1.9	59	47	28	6
200	MCM-L	80	6 × 9	Sealed chip on board	6 × 9	14	100	14	6
480	MCM-C	17	2 × 4	Ceramic flat pack	2.8 × 5.6	50	51	26	6
84	MCM-D	8	0.6 × 0.6	CQFP	1.4 × 1.4	52	18	9	6
168	MCM-L	4	1.0 × 1.0	Postmolded PQFP	1.2 × 1.2	38	69	26	7
325	MCM-L	4	1.4 × 1.4	Overmolded pad grid array	1.5 × 1.5	35	87	30	7
308	MCM-D	19	1.4 × 1.4	CQFP	2.3 × 2.3	77	33	25	8
386	MCM-D	33	1.8 × 3.8	Ceramic flat pack	2.5 × 4.5	45	33	30	9

than this to allow for lid deflection. Typical cavity depths might be less than 1 mm (0.040 in) for integral-substrate packages, or in the range 1.2 to 2.7 mm (0.050 to 0.110 in) for one independent-substrate package. Cavity depths can range up to 7.5 mm (0.30 in) for packaging of multiple substrates, or even more for 3-D assemblies employing stacked memory devices and/or devices mounted on edge. Postmolded plastic MCM packages tend to be as thin as half the above numbers.

7.2.2 Package I/O

Most digital MCMs are expected to require from 100 to 600 I/O, and this can generally be accomplished with package lead pitch in the 0.5 to 1.25 mm (0.020 to 0.050 in) range for perimeter leads and 0.75 to 2.5 mm (0.030 to 0.100 in) range for grid array connection packages of the anticipated package sizes. It is usually obvious that I/O in an array configuration can provide higher density interconnection than those in a perimeter lead configuration [a one-inch square package can hold 184 perimeter leads at 0.5 mm (0.020 in) pitch, 361 pads, pins, or bumps on a 1.25 mm (0.050 in) pitch array configuration, or 1,024 pads, pins, or bumps on a 0.75 mm (0.030 in) pitch array configuration). However, high density I/O on a package offers no useful purpose if its mating circuit board cannot be successfully routed away from the dense array of circuit board pads. Successful routing away from array configurations depends on having both an adequate number of layers in the circuit board and the ability to route lines between the array pads to escape from the inner rows of connections. A 0.75 mm (0.030 in) pitch array package demands at least a 0.12 mm (0.005 in) line/space circuit board technology, 0.36 mm (0.015 in) dia. pad, 0.12 mm (0.005 in) space, 0.12 mm (0.005 in) line, and 0.12 mm (0.005 in) space, for a total 0.75 mm (0.030 in) pitch.

An issue needing wider attention is the development of approaches to achieve optimum partitioning techniques and to minimize I/O requirements for multichip packages. The quantity of I/O that results from a particular partitioning decision has major impacts on packaging, testing, yield, and cost. Figure 7.4 shows the relationships between the number of I/O terminals (T) required and the number of circuits or bits (C) within the unit. At the high end, Rent's rule predicts a required number of I/O equal to:[11]

$$T = 2C^{(0.67)}$$

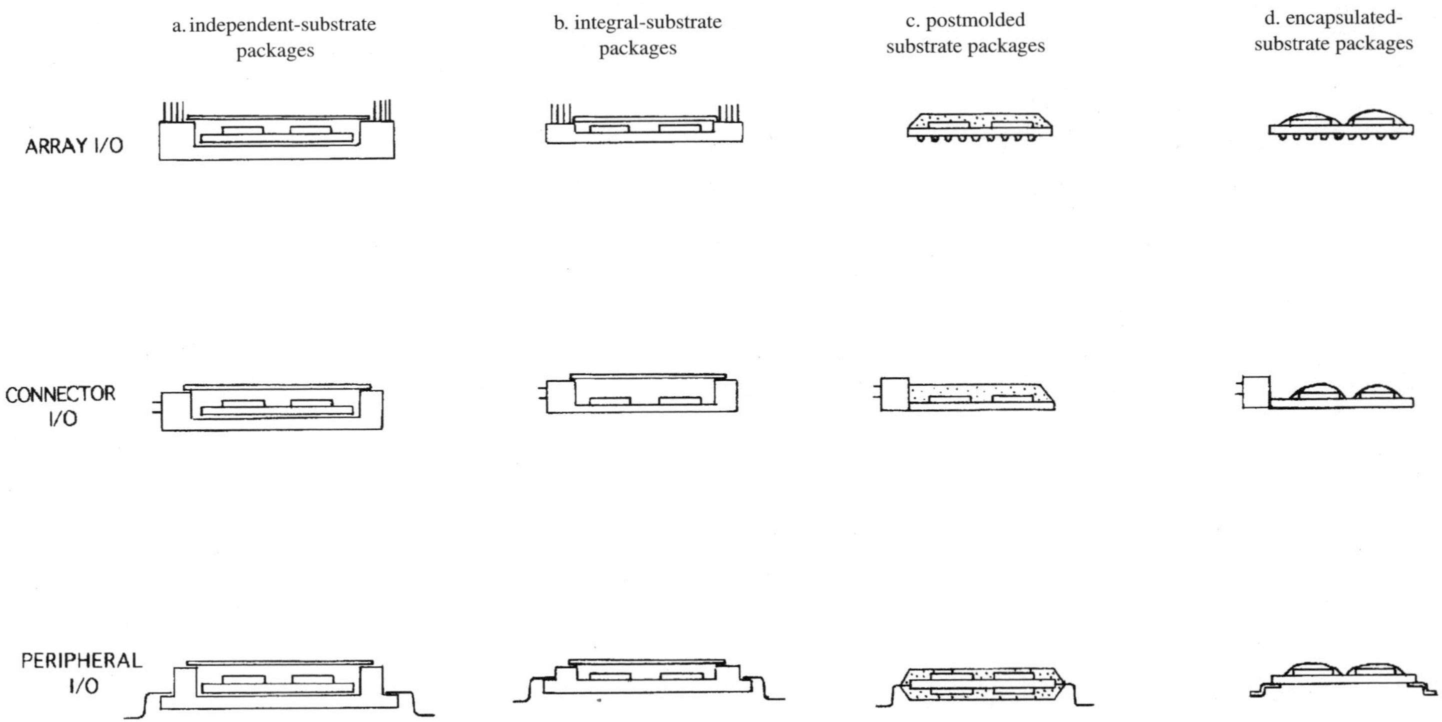

FIGURE 7.3 Basic types of MCM packages.

for randomly partitioned logic chips. This is highly relevant for mainframe computers but, in practice, VLSI logic chips are not very randomly partitioned. They have some degree of functional partitioning, which brings the required I/O down to perhaps 60 percent of the quantity predicted by Rent's rule, as shown by the line of lower slope and smaller exponent for VLSI chips[12] in Fig. 7.4. At the lower end, the relationship for memory chips is the simple binary relationship:

$$C = 2^T$$

In practice, there must also be a few additional I/O for control lines (for purposes of clock, read, write, and so on), resulting in slightly more I/O than predicted by the binary equation as shown in the two low-slope lines for SRAMs and DRAMs. Microprocessor chips fall between the logic and memory devices as shown in the figure.

An analysis of logic circuit card assemblies from a variety of applications is also shown in Fig. 7.4. Enough internal functional partitioning has been designed into typical assemblies to now require only 5 to 50 percent of the I/O predicted by Rent's rule. Also plotted in the figure are typical multichip modules. The modules containing mostly logic ICs fall very near the line for the VLSI logic circuit card assemblies. As more memory is included in the multichip modules, they move closer to the memory chip lines shown in Fig. 7.4. MCMs requiring high currents or very fast operation may need a significant number of additional I/O devoted to power or ground distribution.

At some point in the higher partitioning levels of electronics interconnection, the quantity of hard-wired I/O begins to come down; for example, the I/O count from a home computer is four (two line plus two modem), the I/O count from a TV set is four (two line plus two antenna), and the I/O count from a total aircraft is zero. However, in the near term, the functions that can reasonably be packaged in multichip modules will follow the various adaptations of Rent's rule shown in Fig. 7.4. As module circuits become more complex, they will require higher I/O counts. About the best than can be done to minimize I/O count is to apply as much functional partitioning as possible into the digital elements of the module circuit to drive it toward the lower slope curves of the figure while also moving analog signal devices into the MCM to accomplish truly mixed-signal MCMs.

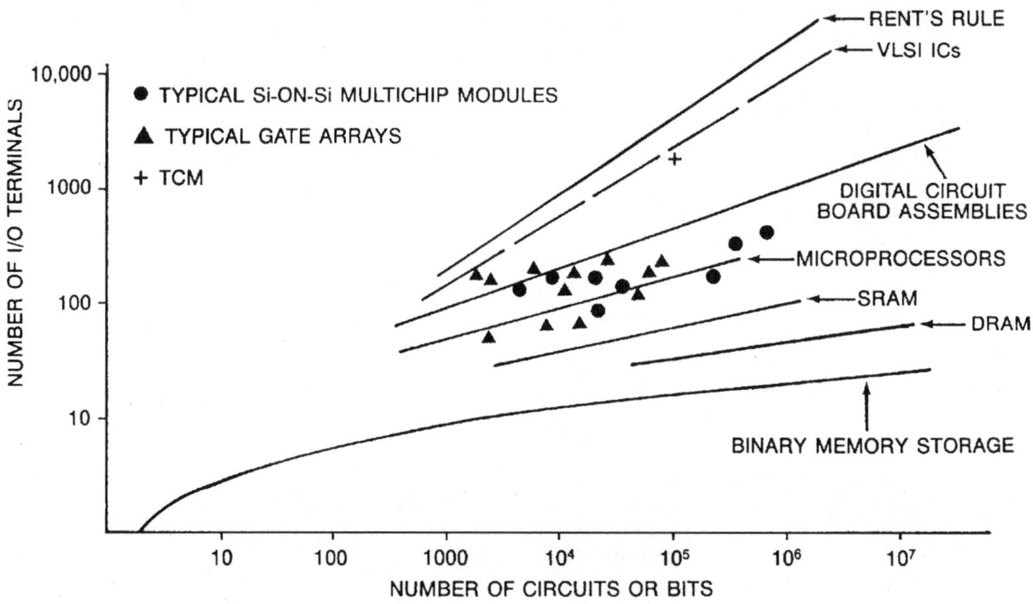

FIGURE 7.4 Module I/O count vs. circuit complexity.

Alternative approaches to avoid any unwieldy quantity of I/O would be to employ some type of multiplexed telemetry signals or to multiplex many signals over a few I/O lines. Both approaches require very high frequency (many gigahertz) multiplexing capability; this is part of the impetus behind developments toward very high frequency materials, integrated electro-optics, and vacuum microelectronics discussed elsewhere.[13]

7.2.3 Expansion Matching

In many cases, the thermal expansion behavior of the package base will need to closely match that of the substrate or the ICs, for the reasons described below. If this need for expansion matching is not attended to and the mismatch is too great, excessive straining may occur during the normal thermal excursions that accompany changes in environment and/or device status (operational or off). Silicon and gallium arsenide ICs exhibit glass-like stress-strain behavior, with a linear elastic region and virtually no plastic deformation prior to fracture. They are notch sensitive, with little propensity to arrest cracks once they begin to propagate. For these reasons, strains imparted to silicon or GaAs devices or silicon substrates by thermal expansion mismatch may cause catastrophic failure. Less dramatic but equally important is the observation that even modest straining of a silicon device may cause operational problems, because the electronic function of the doped semiconductor areas is impaired by strain-induced stresses (similar to piezoelectric effects).

In the case of plastic package or plastic encapsulation MCMs, it is important to select a plastic with a coefficient of thermal expansion (CTE) between that of the IC material and that of the circuit board material. Fortunately, newly available epoxy encapsulants and liquid crystal and epoxy molding compounds have CTEs close to circuit board substrates over normal operating temperatures, as seen in Fig. 7.5. Kovar and alumina packages are shown to be good expansion matches for alumina substrates; aluminum nitride, silicon carbide, Invar (36 percent Ni), Alloy 39 (39 percent Ni), and molybdenum-invar-molybdenum are good expansion matches for silicon devices or substrates. Experience has shown that the substrate should have thermal expansion behavior within about 1 ppm/°C of large ICs or packages to avoid damage to rigidly mounted parts during thermal cycling over a wide temperature range. Thicker bond lines, use of compliant adhesives, and/or applications involving only a narrow temperature range would relax this requirement somewhat and permit use of a wider range of materials.

7.2.4 Compliant Leads

Packages for multichip modules tend to be relatively large and, therefore, require careful analysis of the expansion mismatch between the package and the next level interconnect. This is particularly true if the package is to be expansion matched to silicon or alumina substrates for the reasons discussed above, and especially if the packaged module is to be soldered onto a (comparatively high expansion) conventional printed wiring board. Conservative design practice dictates that for extended life (several thousand cycles) under the thermal cycling conditions, the "nominal" shear stress in a solder joint should be kept below approximately 690,000 N/M^2 (100 psi). [Some authors suggest a less conservative limit of 2.7 MN/M^2 (400 psi)].[14] This "nominal" shear stress is easily computed by dividing the load on each connection by the approximate plan view area of the critical part of the solder fillet. This plan view area is the area of solder under shear stress between the lead and circuit board. This rule of thumb permits calculation of the desired spring rate for the needed compliancy in package leads, once the package size is known. Figure 7.6 provides a family of curves describing the expected mismatch between a package base and an FR-4 printed wiring board during a 180° C temperature excursion (–55 to +125° C thermal cycle). The mismatch is computed along the package diagonal, using an expansion rate of 17 ppm/°C for FR-4. From the figure it is seen that a 50 mm (2 in) square aluminum nitride package will have a total mismatch of about 0.18 mm (0.007 in) for the conditions stated above. If we assume that the mismatch is split evenly by the leads on either side of the package, then the mismatch "per side" is one-half of the total mismatch, or 0.09 mm (0.0035 in) for this example. A more conservative approach would be to presume that the

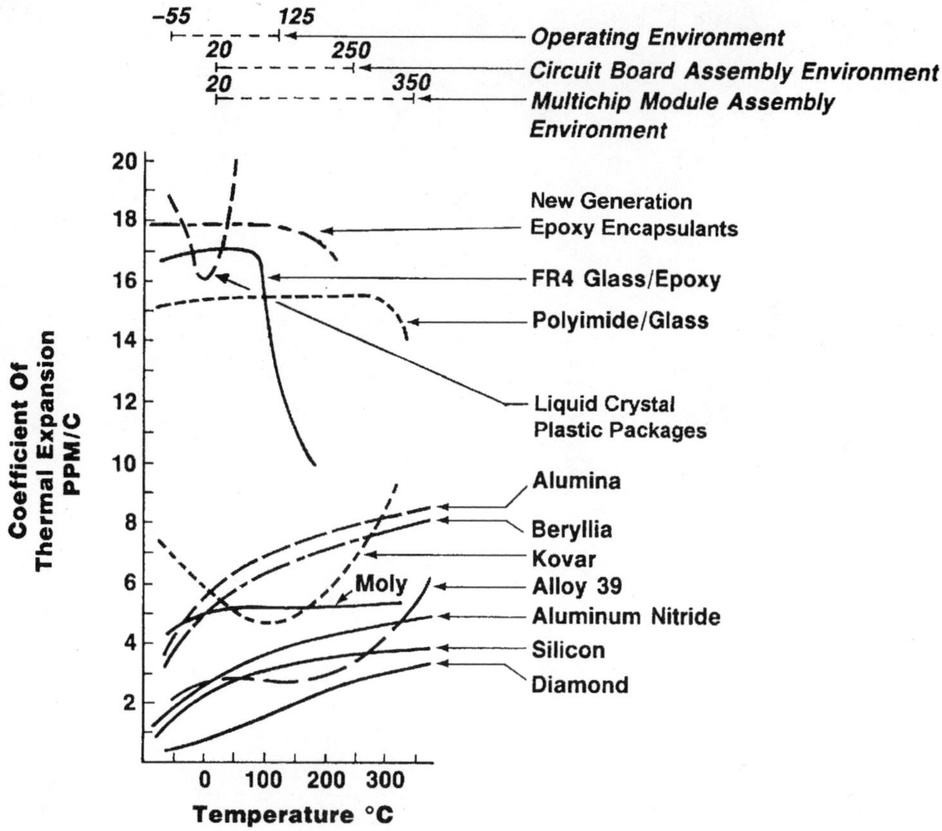

FIGURE 7.5 Thermal expansion behavior of various materials.

mechanical mounting of the part might nonuniformly restrict the movement of the package body, and that the entire expansion mismatch will need to be absorbed by the leads on one side of the part. The maximum desired spring rate for the lead may now be computed if the critical area of the solder joint can be estimated. If the width of the solder in the critical area is 0.5 mm (0.020 in) and its length is 1 mm (0.039 in), then the area is 0.5 mm^2 (7.9×10^{-4} in^2). The solder joint load transferred through the lead should therefore not exceed (690,000 N/M^2) $\times$ (0.5 mm^2) $\times$ (10^{-6} M^2/mm^2), or about 0.35 N (0.08 lb). The maximum allowable spring rate can be calculated from (0.35 N)/(0.18 mm deflection), which is 1.94 N/mm (11 lb/in).

Fine-pitch leads will tend to have smaller solder fillets, and the maximum desired spring rate may drop to 1 N/mm to (5 lb/in) or less. For comparison, measurements of lead spring rate on a limited number of currently available single chip packages indicated stiffnesses ranging from about 6 N/mm to over 76 N/mm (34 to 430 lb/in).[15] Methods to calculate spring rates for a large variety of lead forming shapes have been presented by Kotlowitz[16–19] and various other authors.[20–22]

This observation points to the compromise that is inherent in most electronics packages: the leads often provide both the electrical and the mechanical connection to the next level interconnect. This requires a compromise between lead strength for mounting integrity and lead compliance for long solder joint life under the thermal cycling conditions. For larger, MCM-type packages, the actual requirements regarding mechanical mounting and expansion accommodation become mutually untenable. Leads that are stiff enough to support the package for vibration and shock have too high a

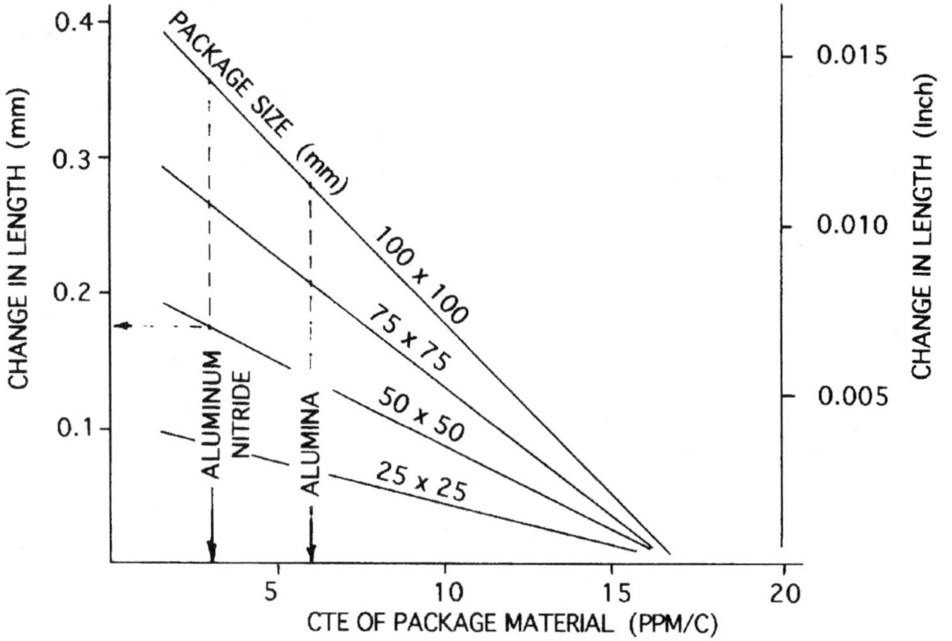

FIGURE 7.6 Predicted thermal expansion mismatch of an MCM on a circuit board.

spring rate for solder joints to survive thermal cycling. Leads that are compliant enough to pass thermal cycling will fail in vibration and shock. An alternative approach is needed to ensure component mounting integrity.

Separating the mechanical attachment from the electrical connection permits optimization of each of these aspects of component mounting. A package design providing this separation is presented in Figs. 7.7 and 7.8.[23,24] A hold-down ring provides the mechanical attachment of the package to the next-level interconnect. The ring is made from a reinforced polymer having thermal expansion behavior closely matching that of conventional printed wiring board laminate. A threaded metal ferrule in each corner of the ring receives the fastener; the ring is positioned on the package such that each corner of the ring is brought into tight contact with the PWB during component mounting. An elastomeric adhesive bonds the ring to the package and accommodates the thermal expansion mismatch between the ring and the lower-expansion package elements. The hold-down ring also ensures good contact between the package base and any heat sinking that may be provided for heat removal from the base of the package through the circuit board.

The hold-down ring and the lead retainer are staked together. This ensures the dimensional relationship between the leads and the mounting holes. In addition, because the leads are protected and supported, they may be optimized for compliance (as discussed in earlier paragraphs) as well as current carrying requirements and signal propagation characteristics, without fear of undue fragility or bending of leads during handling or assembly processes.

7.2.5 Vibration and Shock Considerations

The spectrum of typical vibration environments experienced by MCMs is shown in Fig. 7.9. The specification requirements typically might be two or three times higher g-loads than the real environments shown in Fig. 7.9, due to safety factor considerations.

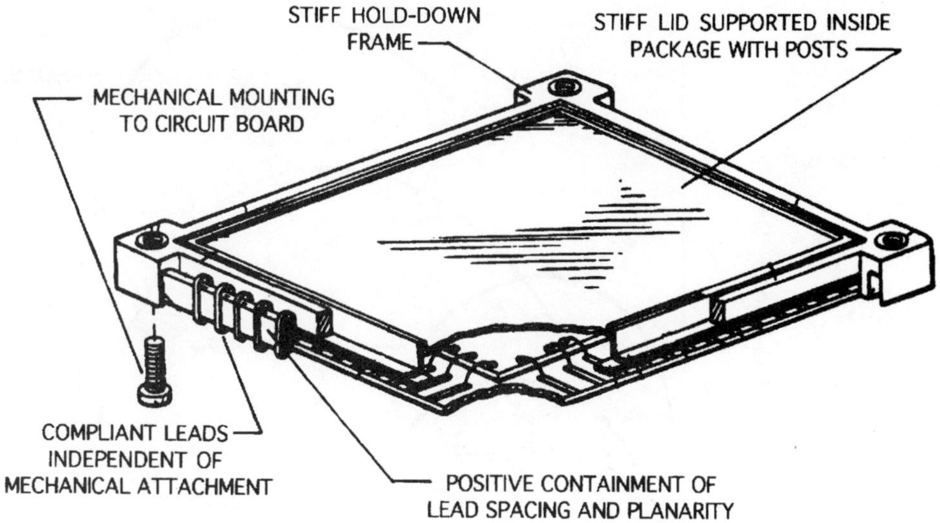

FIGURE 7.7 Features of an MCM package designed for cycling and vibration.

FIGURE 7.8 Example of package with separated mechanical/electrical interfaces.

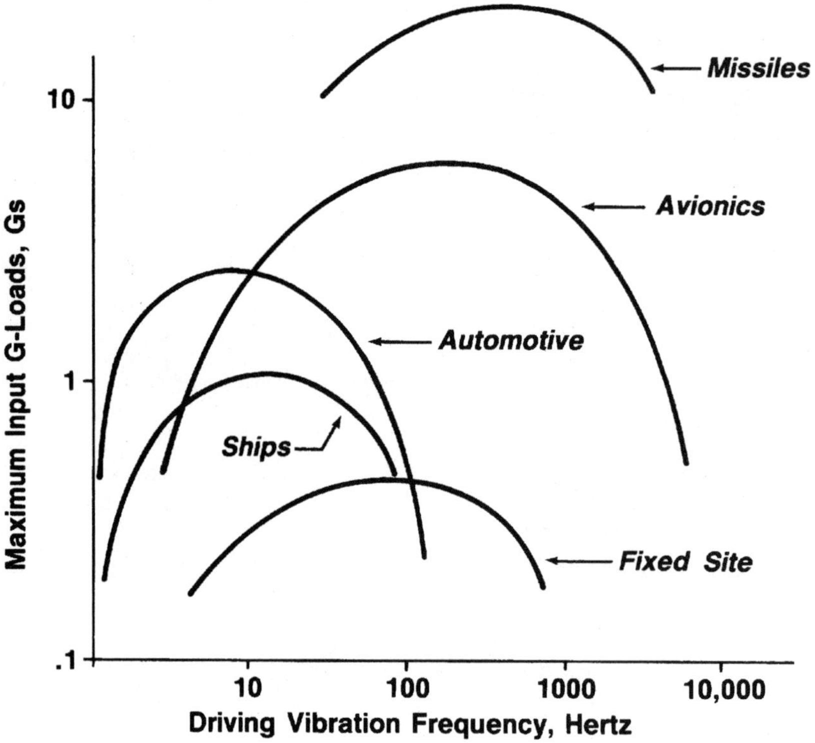

FIGURE 7.9 Features of an MCM package designed for cycling and vibration.

Packaging principles of particular importance for vibration include the following:

- Design package elements with the lowest possible mass to minimize inertial forces.
- Design very stiff elements so that (a) natural frequencies are very high (above the environmental forcing function frequencies if possible) and (b) vibration deflections (and corresponding bending stresses) are kept very low.
- Design separate electrical connection and mechanical mounting approaches.

MCMs lend themselves to very good vibration performance. All die attach and interconnect structures (wire-bond, flip chip, TAB) will have natural frequencies above 6 kHz, giving them decades of reliable life. The package cases themselves can have natural frequencies as high as 1000 Hz for large-sized ceramic packages, or even 4000 Hz for moderate sized packages. In fact, the presence of an MCM can be a stiffener on a circuit board assembly. For example, the circuit board assembly shown in the background of Fig. 7.10[24] has a natural frequency of 425 Hz. When an MCM was designed to repackage the IC components on the board into the MCM shown in the foreground of Fig. 7.10, the natural frequency of the new assembly was raised to 610 Hz due to the high stiffness of the ceramic MCM package. Under most practical conditions (assuming no resonance problems), such an increase in natural frequency will decrease vibration problems, because more energy would be required to drive higher-frequency deflections. The rear board assembly shown in Fig. 7.10 survives the normal mil-spec random vibration tests at 20 G_{rms} but fails at a more severe level of 23 G_{rms}. The MCM board assembly shown in the foreground of Fig. 7.10 survives beyond an impressive 50 G_{rms} level of severe random vibration testing.

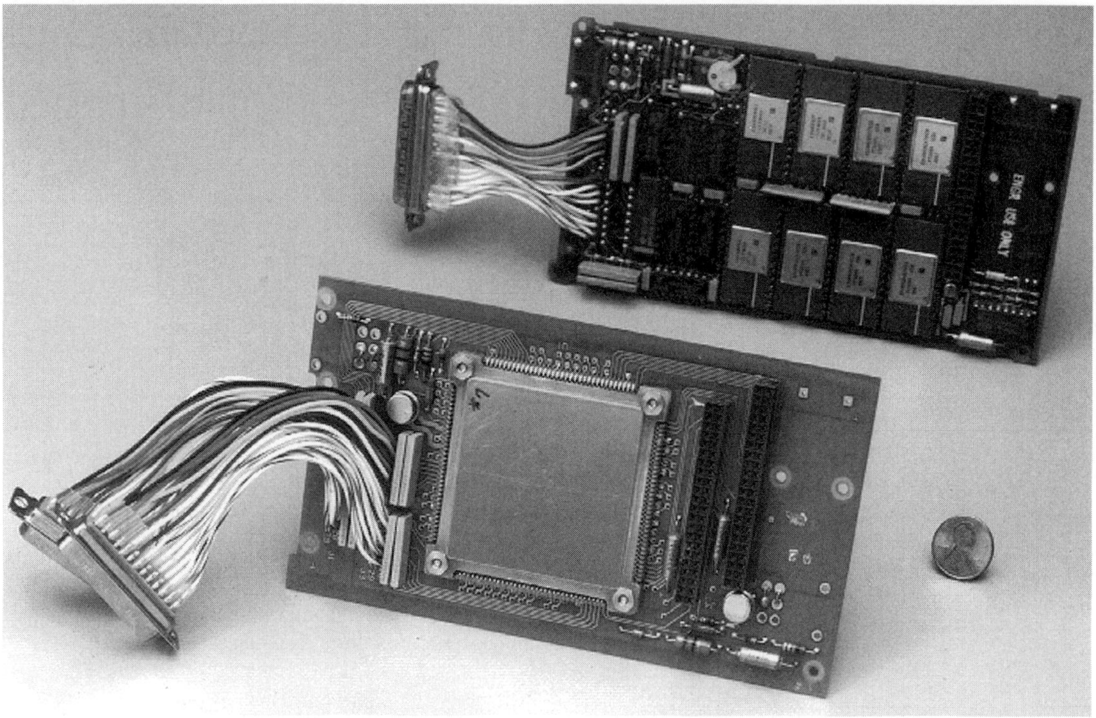

FIGURE 7.10 MCM package assembly (foreground) designed for severe vibration.

7.3 ELECTRICAL DESIGN CONSIDERATIONS FOR MCM PACKAGES

It should be remembered that a multichip module is primarily an electronic component, and that electrical performance should not be compromised in the quest for good mechanical or thermal design. The primary electrical design tasks for the packaging engineer are to construct a package configuration that provides adequate and unrestricted flow of power and ground currents, while allowing multiple signals to enter and exit the module without distortion or interference.

7.3.1 Package Lead Characteristics

The conductors used for I/O in a MCM package must provide acceptable levels of resistance, capacitance, inductance, and characteristic impedance. Resistance of leads must be kept low enough so that RC constraints of signal rise times have no adverse effect on timing delays. Most solid metals used for leads easily meet this constraint, providing only a few milliohms of resistance, as shown in Table 7.3. I/O fabricated from fired pastes have considerably higher resistivity and smaller cross section as shown below. Ceramic MCM packages using such conductors may require special design attention to avoid either excessive resistance or differential timing problems on I/O conductors of different path lengths.

Capacitance between adjacent leads must be kept low to prevent unwanted crosstalk between signal lines. Capacitance is increased between signal lines when they are closely spaced, have large areas facing the adjacent lead, have long parallel lengths, or are separated by insulators with high

Table 7.3 Lead Resistance in MCM Package

Conductor material	Resistivity (μΩ-cm)	Cross section (cm^2)	Example lead length (cm)	Package	R (mΩ)
Copper	1.7	0.0004	0.5	QFP	2
Alloy 42	70	0.0004	0.5	QFP	82
Kovar	49	0.0004	0.5	QFP	58
Tungsten paste	50–100	0.0001	2.0	BGA	1000–2000
Molybdenum paste	50–100	0.0001	2.0	BGA	1000–2000

dielectric constants. Plastic packages with solid metal leads usually have lead-to-lead capacitance in the 1–2 pF range while ceramic multilayer constructions are usually higher in the 5–10 pF range.[25] Crosstalk between sensitive signal leads can also be controlled by inserting grounded leads between and around the signal I/O or introducing grounded planes above or below the signal I/O leads. Full coaxial shielding may be needed for very sensitive high-frequency I/O lines. As is often true in interconnect, although high capacitance of signal lines is undesirable, high capacitance between power and ground I/O is generally desirable, because it tends to minimize local voltage drops during heavy current draws. Sometimes separate decoupling capacitors are also added to the package to avoid these undesirable local voltage drops.

Inductance of package I/O leads is typically not a problem until signal frequencies move above several hundred megahertz. Then I/O leads need to be designed with minimal loop areas (fairly straight lead forms and wire bond shapes) to keep inductance low. On the other hand, inductance on power leads on MCMs can create serious problems at all frequencies. The voltage noise generated here is proportional to the lead inductance multiplied by the rate of change of current flow. Simultaneous switching of several power leads can generate very high voltages due to fairly high rates of change in current flow. These induced voltages either raise the local ground voltage level or drop the local supply voltage level, which can cause switching errors. This problem can be minimized by routing the switched currents through multiple parallel paths, and proper MCM designs may require 20 to 40 percent of all package leads to be devoted to these parallel power and ground paths. Inclusion of power and ground planes within the package may also be quite helpful in controlling these lead inductance problems.

The characteristic impedance of the package I/O becomes important when signal frequency is high enough to require transmission line design. Figure 7.11[8] shows an estimation chart (the lower line) for determining when package I/O need to be designed with characteristic impedance matched to the outside interconnect. At very high frequencies or long transmission distances, optical I/O may be required, as indicated by the upper line.[26]

7.3.2 Package Body Electrical Characteristics

The advantages of including power and ground planes in package constructions to control capacitance and inductance problems were discussed in the previous section. Such planes may also be useful as shielding to confine electromagnetic fields to desired configurations. Obviously, metal packages or metal-coated plastic packages may provide similar benefits.

The important electrical properties of the insulators used in package constructions include bulk resistivity, dielectric constant, and dielectric breakdown voltage. Values of these properties for selected packaging materials are listed in Table 7.5.

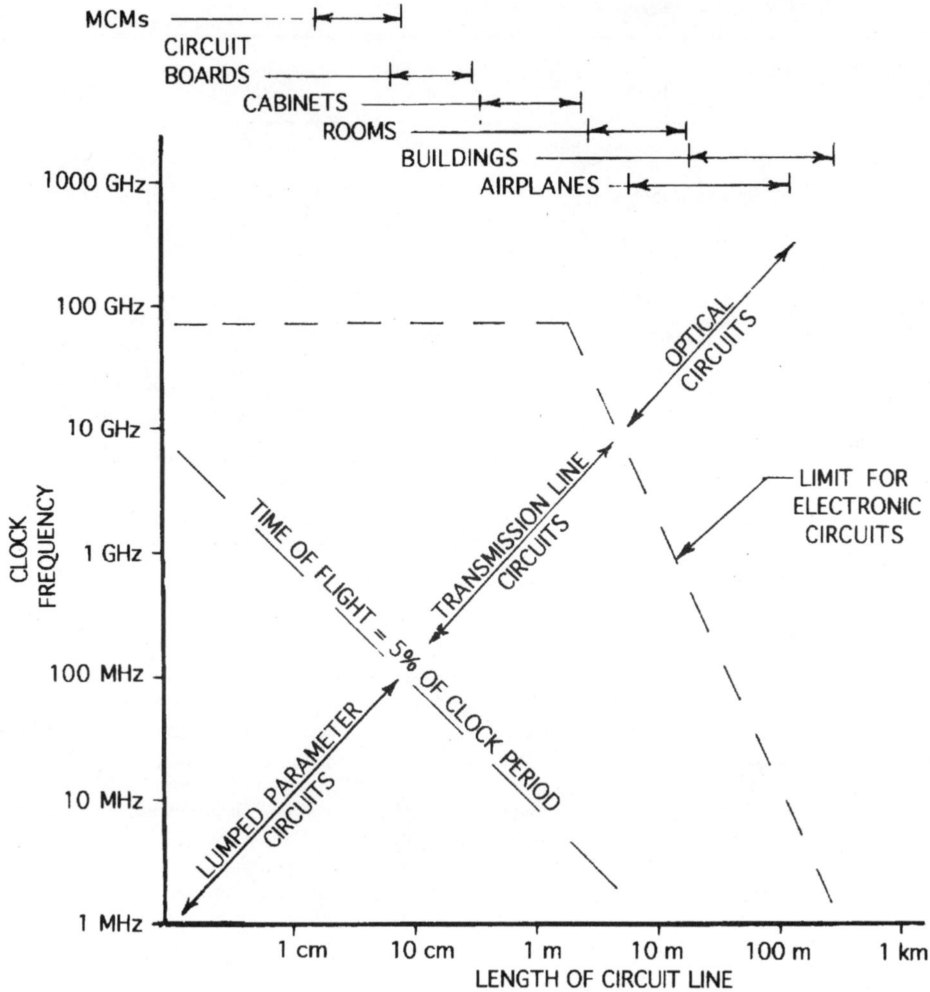

FIGURE 7.11 Estimated domains for transmission line and optical circuits.

7.4 THERMAL DESIGN CONSIDERATIONS FOR MCM PACKAGES

Two temperature related topics must be considered in the design of packages for MCMs: the operating temperature range and the internal package temperature rise due to heat dissipation.

7.4.1 Temperature Extreme Considerations

The materials used in MCM packages must be suitable for the operating temperatures of the intended applications. ICs are usually sold for operation within particular temperature ranges such as commercial (typically 0 to 70° C), industrial (typically –40 to +85° C) or military (typically –55 to +125° C, or –65 to +150° C). Temperature limitations on bare ICs are usually due to an increased risk of

Table 7.4 Electrical Properties of Packaging Materials

Material	DC volume resistivity (Ω-cm)	Dielectric constant (1 MHz)	Dielectric strength (kV/mm)
Epoxy molding compounds	10^{15}	4.3	18
Epoxy encapsulants	10^{15}	4.3	18
Alumina	10^{16}	9.7	13
Aluminum nitride	10^{14}	9.1	15
FR-4 laminate	10^{14}	4.5	16
BT laminate	10^{14}	4.3	16
Glass-to-metal seals	10^{17}	12–33	14

temporary malfunction or outright failure at extreme temperatures, whereas limitations on packaged ICs may also be related to limitations of the packaging materials used. These same limitations apply to the design of multichip module packages.

Many plastic packaging materials exhibit significant changes in modulus or expansion characteristics above or below certain temperatures, which can cause catastrophic failures in electronics modules. For example, it is fairly well known that many silicone RTV encapsulants become very rigid around –40° C, and that operation below that temperature will tear off wire bonds from an IC chip. Another well known example is that epoxies exhibit sharp increases in thermal expansion rates when raised near or above their glass transition temperatures, which can cause a variety of failures. Consequently, epoxy molding compounds used for packaging are generally limited to use somewhat below their glass transition temperatures (90 to 180° C, depending on the molding compound). It is important to select packaging materials for MCMs carefully, since they are used in a great variety of applications that range from cryogenic electronics (often below –100° C), arctic operations (around –65° C), or flight at 35,000 ft (around –55° C), to operation under desert conditions (around 100° C inside enclosures), in automotive engine compartments (around 150° C), or in down-hole oil well applications (over 200° C). Space probe applications may be even more severe.

7.4.2 Temperature Rise inside MCM Packages

As discussed elsewhere in this book, modern ICs produce heat during their operation that ranges from a few milliwatts to dozens of watts per IC. This heat must be efficiently removed through the package to prevent IC junction temperatures from exceeding the limits defined by the IC manufacturer (for functional, reliability, or materials related reasons). Detailed discussion of MCM package thermal design is discussed elsewhere in this book, but for the purposes of this packaging chapter, four primary heat removal paths should be considered as shown in Fig. 7.12:

1. Heat conducted from the ICs to the package surfaces
2. Heat convection from the package surfaces
3. Heat conducted from the MCM to the circuit board through the package I/O
4. Heat conducted to the circuit board through any underfill, spacer, or air gap between the MCM and circuit board

The internal package thermal resistance between the IC junctions and the external package surfaces is entirely a function of the materials selected and the geometries of the design. Selection of high

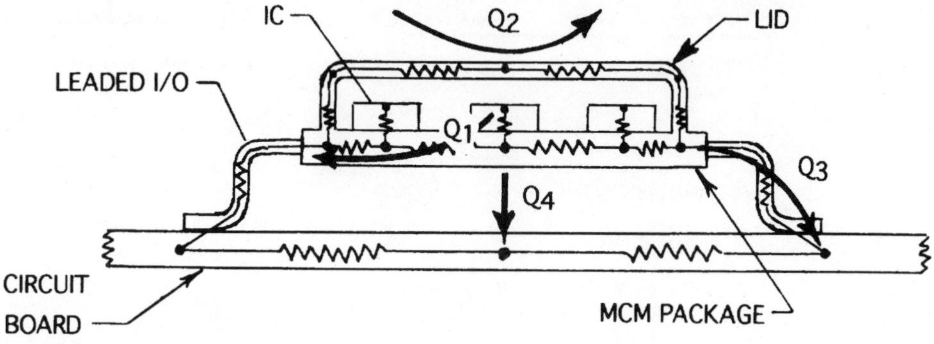

FIGURE 7.12 Heat removal paths from an MCM package.

thermal conductivity materials is extremely important, and well designed MCMs can operate with only a few degrees junction temperature rise above package temperature.[4] The conduction portions of heat removal paths fall into two types: one-dimensional heat flow (constant thermal flux), and thermal spreading (decreasing thermal flux). The three-dimensional heat flow of thermal spreading is easily solved by computer with finite element analysis. The one-dimensional flow can be quickly estimated with Fourier's conduction equation. For convenience, Table 7.5 presents some design data, and Fig. 7.13 shows anticipated temperature rises per watt of dissipation for typical constant flux thermal paths (die-bond, substrate-bond, package-bond, solder-bump I/O, copper lead I/O, iron-nickel alloy lead I/O, and air gap).

7.5 ENVIRONMENTAL DESIGN CONSIDERATIONS FOR MCM PACKAGES

MCM packages must be designed to withstand the various hazards presented by environmental conditions: cyclic changes in temperature and air pressure, high humidity, and airborne contamination.

7.5.1 Thermal Cycling

Thermal cycling results from a variety of common situations: power on-off cycles, cyclic exposures to solar heating, alternating light and heavy duty cycles, flights exposing equipment to cycles of

Table 7.5 Conduction Path Parameters

		Typical values		
Path element	Material	K (W/M°C)	Δ_χ (M)	K/Δ_χ (W/M^2°C)
Bondline	Epoxy	0.3	6×10^{-5}	5×10^{3}
Bondline	Conductive epoxy	2.0	6×10^{-5}	3×10^{4}
Bondline	Gold-tin solder	180	3×10^{-5}	6×10^{6}
Flip chip bump	Solder	30	1×10^{-4}	3×10^{5}
Package lead	Copper alloy	30	3×10^{-3}	1×10^{5}
Package lead	Fe/Ni alloy	15	3×10^{-3}	5×10^{3}
Package-board gap	Air	0.024	3×10^{-4}	8×10^{1}

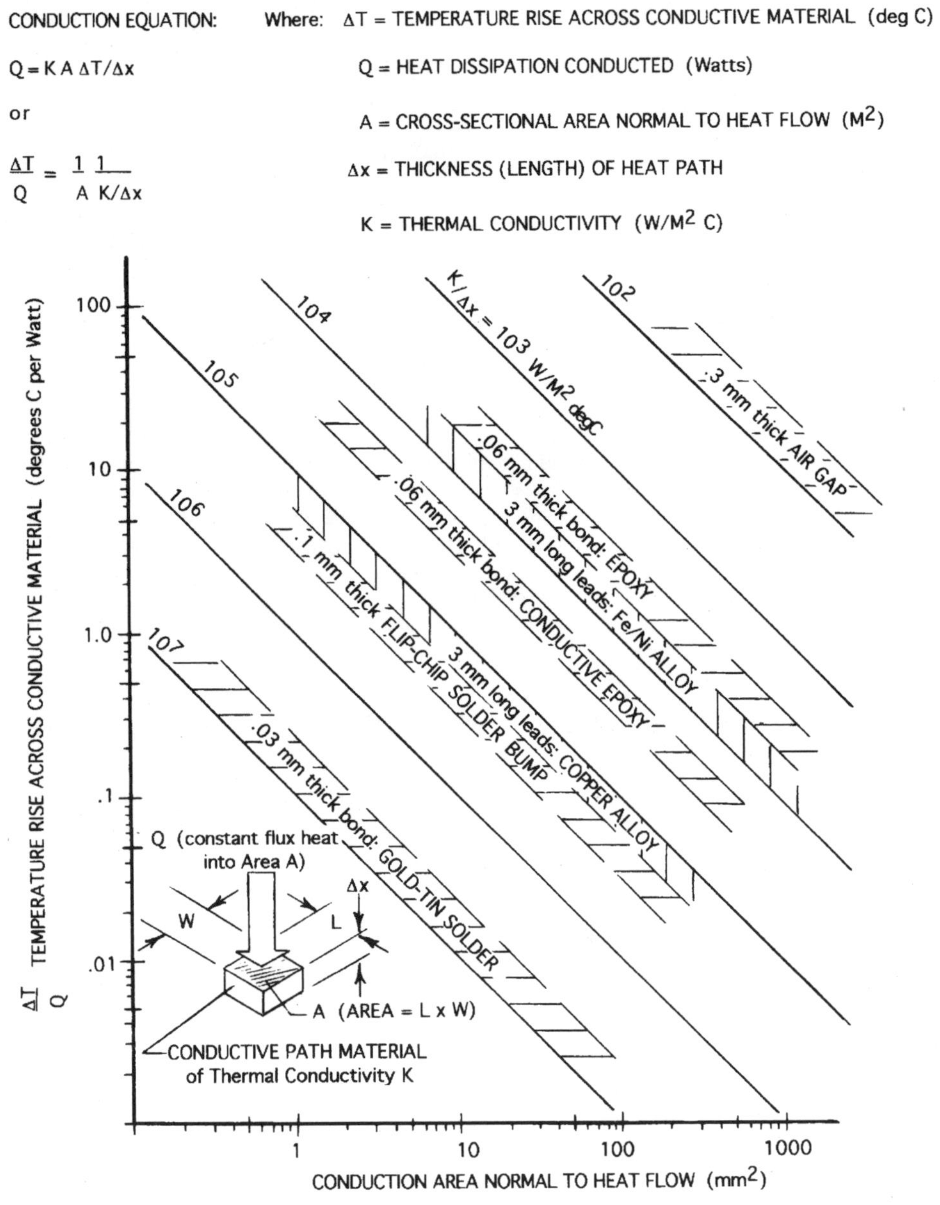

FIGURE 7.13 Temperature rise in various conduction paths.

ground temperature and high-altitude temperature, excursions between room temperature and cryogenic operation temperatures, and so on. The materials commonly used in MCM packaging can have over two orders of magnitude variation in rates of thermal expansion, as shown in Fig. 7.14. When two different materials are fastened together, the differential expansion between the materials can generate significant stresses. The challenge in MCM package design is to create designs in which

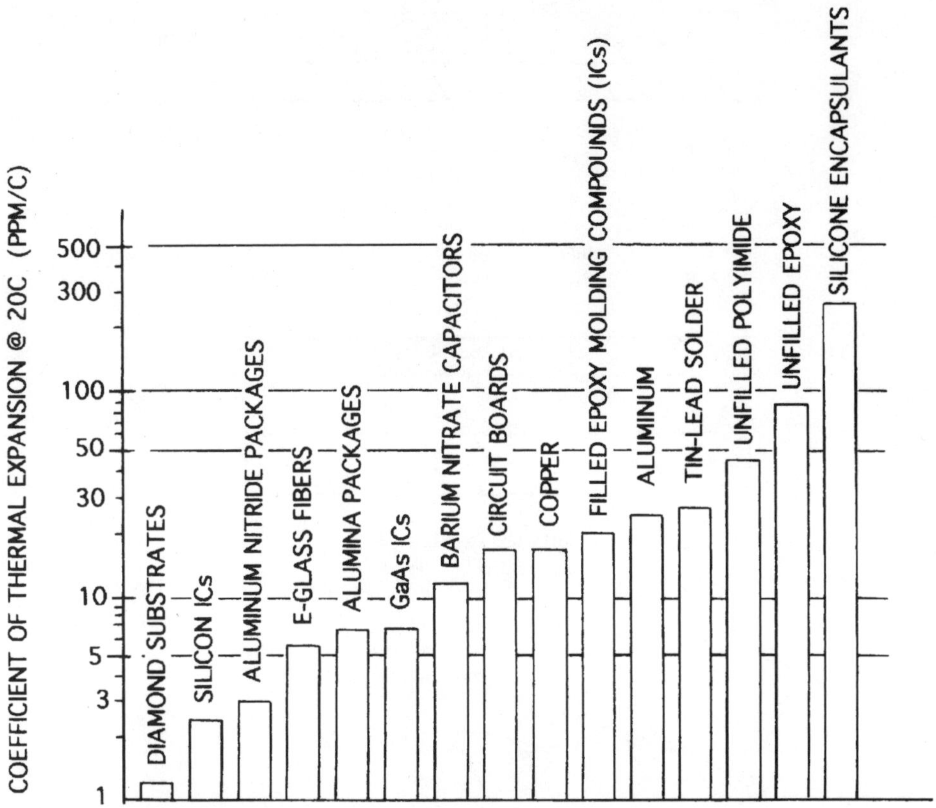

FIGURE 7.14 Wide range of CTE in packaging materials.

fatigue limit stresses or strains are never exceeded, either through the use of closely expansion-matched materials or by designing compliant elements at the interfaces. Three configurations occur frequently: (a) mismatched layers bonded together along their length (Fig. 7.15*a*), (b) two mismatched structures attached together at their ends (Fig. 7.15*b*), and (c) one structure encapsulated in a mismatched material (Fig. 7.15*c*). The figure also lists common examples of these configurations in MCM packaging.

7.5.2 Hermeticity Considerations

The term *hermetic,* by dictionary definition, means sealed to be gas tight or impervious to gas flow. In the microelectronics packaging context, hermeticity is defined as the ability of a package to prevent any significant penetration of environmental contaminants such as water vapor.[27] While many materials can be used as an effective barrier to the movement of liquid water, only glasses, ceramics, and metals are capable of forming practical barriers (hermetic seals) against the penetration of water vapor. As shown in Fig. 7.16, an epoxy cavity-style package allows a substantial amount of permeation of water vapor through its walls. Even if the package had walls as thick as 1 cm, moisture would penetrate the walls quickly, allowing equilibration inside the package with the outside environment's humidity in as little as one year. To achieve a 20-year seal against moisture penetration, the epoxy wall would need to be over 1 m thick! At the other extreme, a metal wall as thin as 0.001 inch can easily maintain hermeticity for 20 years.

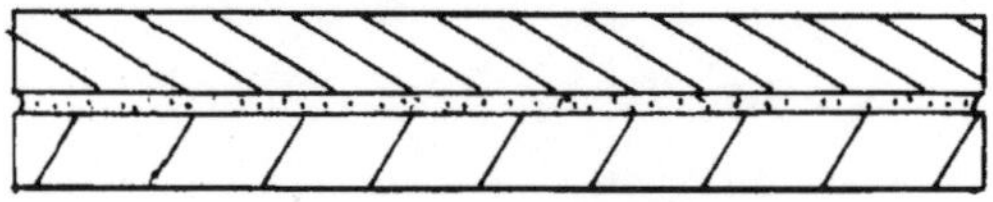

(a) bonded layers
• IC layer bonded to substrate layer
• glass cloth layers in resin layers
• conductor layers in dielectric layers
• coating layer over surface layer
• substrate layer bonded to package

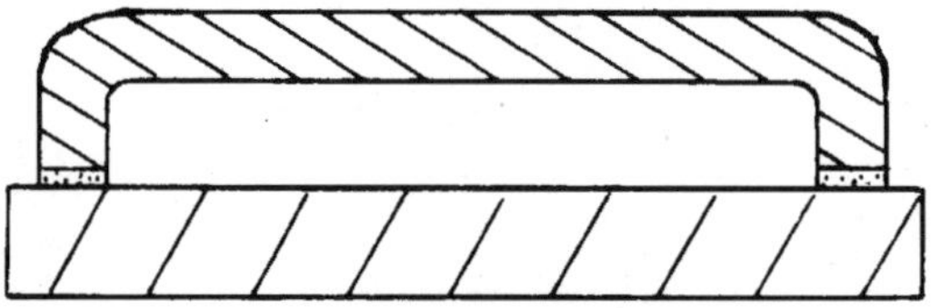

(a) attached structures
• chip R/C soldered to substrate
• component soldered into PWB holes
• SMT leads soldered to PWB
• metal lid welded to ceramic package
• transistor I/O soldered to substrate

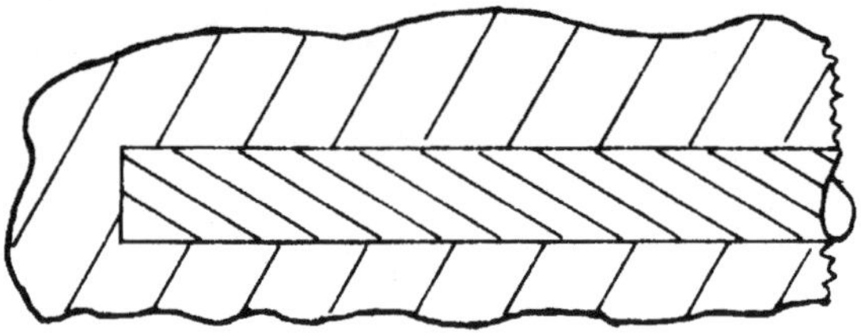

(a) encapsulated element
• FeNi alloy lead soldered in PTH
• IC encapsulated in glob-top coating
• lead frame in molding compound
• component in potted assembly
• any I/O lead in solder joint

FIGURE 7.15 Common geometries of CTE mismatch.

The presence of any liquid moisture droplets on IC surfaces can contribute to a number of failure mechanisms, including chemical corrosion, electromigration, galvanic corrosion, and damage to doped layers in the semiconductor materials. The moisture problems are most severe when liquid droplets form on vulnerable surfaces. This can happen anytime the package is cooled below the dew point of the atmosphere within the package. Under these circumstances, liquid will condense on the cold surfaces, in the same way that water condenses on a cold glass in warm humid weather. Another way liquid droplets can form on IC surfaces would be upon warming of a frosted surface. As the frost melts, it would either directly sublime to the package atmosphere or first form droplets and then

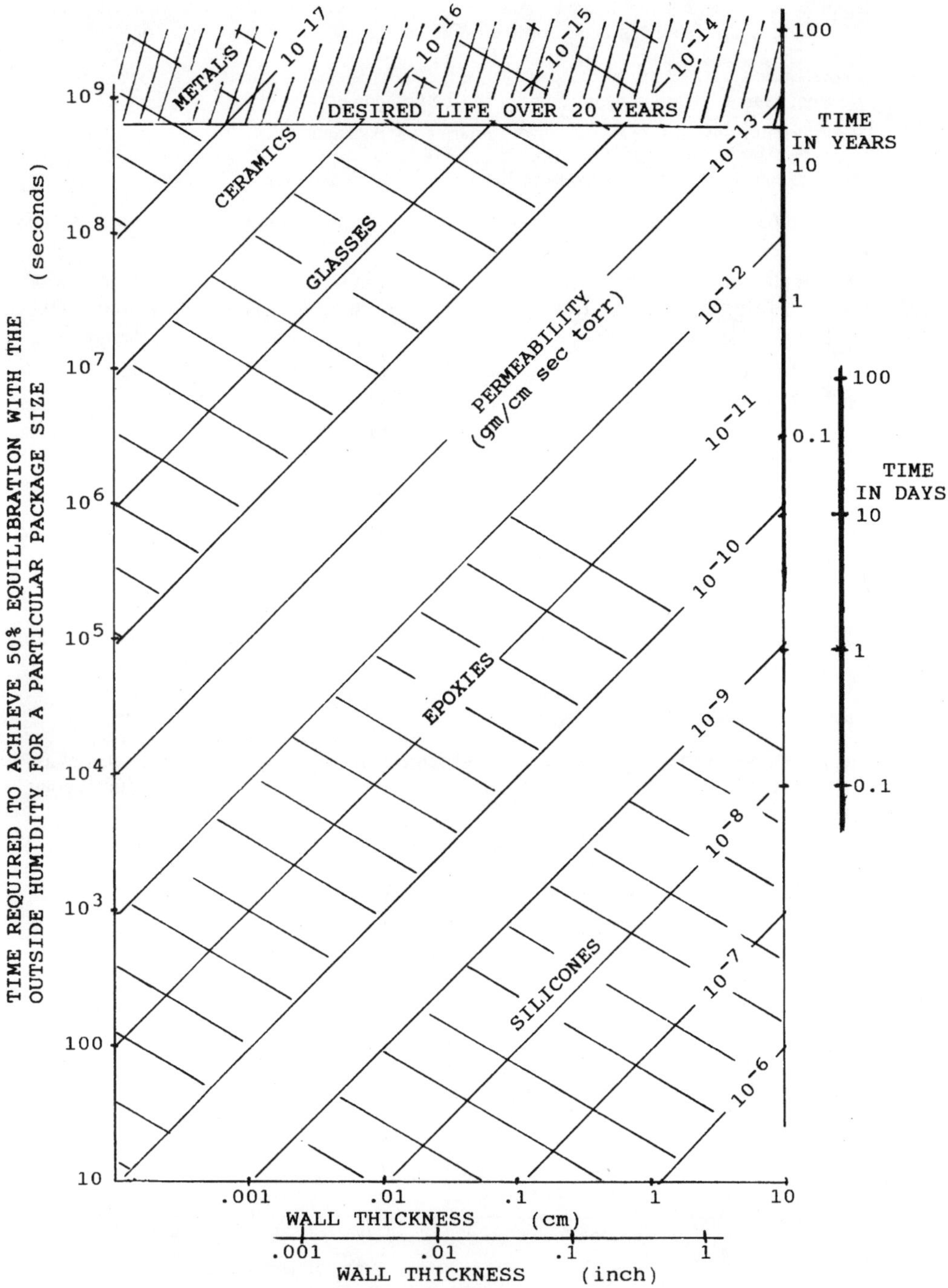

FIGURE 7.16 Moisture passage through various packaging materials.

begin to evaporate, depending upon conditions in the package. The purpose of a hermetic package is to keep water droplets from being formed on IC surfaces by keeping moisture out of the package cavity. The success of a hermetic package depends on:

1. Removing nearly all moisture from inside the package prior to sealing
2. Preventing any outside moisture from entering the package

Moisture can be present in several areas inside an electronic package:

1. Adsorbed monolayer of moisture on all package surfaces, component surfaces, and substrate surfaces.
2. Absorbed moisture within most organic materials. Epoxies (die attach, substrate attach, conductive epoxies, encapsulants) typically will absorb 0.1 to 0.4 percent by weight moisture. Polyimides (die passivations, substrate multilayers, and so on) can absorb 1 to 3 percent moisture.
3. Evolved moisture released from materials holding absorbed moisture.
4. Evolved moisture released as a reaction product of materials undergoing thermal degradation. Some die attach adhesives generate very high quantities of water in breakdown products at high temperature.
5. Moisture present in the sealing atmosphere.

The amount of moisture inside a hermetic microelectronic package can be minimized by several techniques:

1. A presealing bakeout is used to remove adsorbed, absorbed, and evolved moisture sources.
2. The bakeout atmosphere is controlled to a low moisture content. Typically, either a dry nitrogen atmosphere bake or a vacuum bake is used.

To prevent liquid droplets from forming within a package, the moisture level within the package atmosphere must be maintained low enough so that its dew point is always below freezing. Thus, as the package gets colder, no liquid water can condense on the cold surfaces until it condenses as frost. The solid frost has very little effect as an electrolyte or corrosive agent.[28] The dew (frost) point as a function of humidity ratio is plotted in Fig. 7.17, where it can be seen that any humidity ratio level

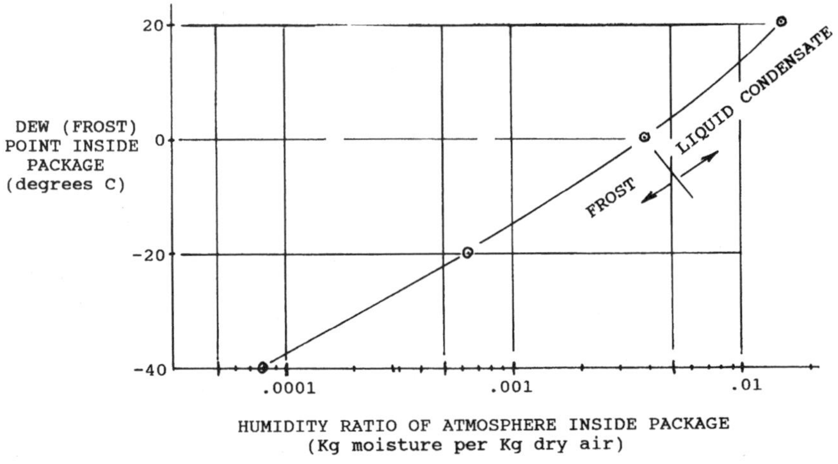

FIGURE 7.17 Moisture content inside package vs. dew point.

above 0.003788 has the possibility of condensing as a liquid upon cooling. The method of measuring moisture content within a package is established in MIL-STD-883, Method 1018. In this method, the moisture is measured as the *volume* ratio of water to air in a residual gas analyzer (RGA) test performed at 100° C. Using the densities of air and water vapor at 100° C the humidity ratios can be converted to the parts per million by volume (ppmv) that are used in the mil spec. The maximum allowable humidity ratio of 0.003788 can be converted as follows:

$$\frac{0.00378 \text{ kgH}_2\text{O}}{\text{kg air}} \times \frac{0.947 \text{ kg air/m}^3 \text{ air}}{0.598 \text{ kg H}_2\text{O/m}^3 \text{ H}_2\text{O}} \times 1{,}000{,}000 = 6{,}000 \text{ ppmv}$$

To avoid formation of liquid droplets from melted frost or premature condensation, the mil spec uses a slightly more conservative limit of 5000 ppmv (instead of 6000 ppmv). Figure 7.18 shows the RGA test ppmv values corresponding to various dew (frost) points.

As discussed above, the use of an appropriate material package which is perfectly sealed will provide more than 20 years life maintaining good protection from moisture intrusion. In practice, not all seals are perfectly made. In addition, stresses applied to the packages or seals during assembly or testing can occasionally cause tiny cracks that allow moisture intrusion. A determination is needed as to what level of tiny leak rate is large enough to cause problems. A standard test for leak rates has been established in MIL-STD-883. The most common method involves subjecting the sealed package to a pressure bomb exposure with a helium atmosphere. Helium, which has a very small molecular size, will infiltrate the package if any leaks are present. The package is then exposed to a helium detector, which will identify the rate at which helium is leaking back out of the package. Depending on package size and pressure bomb conditions used, Method 1014 of MIL-STD-883 would allow leak rates in the 10^{-8} to 10^{-7} ATM-cc/s helium range. However, it should be noted that these levels have been found by many investigators to be insufficient to protect against early moisture intrusion.[28–33] As discussed later, leak rates in the 10^{-10} to 10^{-9} range are the maximum allowable to ensure 20-year field lives. Fortunately, the great majority of packages sealed to at least the minimum spec level are in fact sealed to at least one or two orders of magnitude lower leak rates than the maximum leak rate allowed by the mil spec.[28,34]

The flow rates of helium through various size leaks in hermetic packages fall into three distinct regions, as shown in Fig. 7.19.[35] Large leak rates are governed by viscous flow equations. The flow rate is inversely proportional to the viscosity of the gas in this region. Here the diameter of the pin-

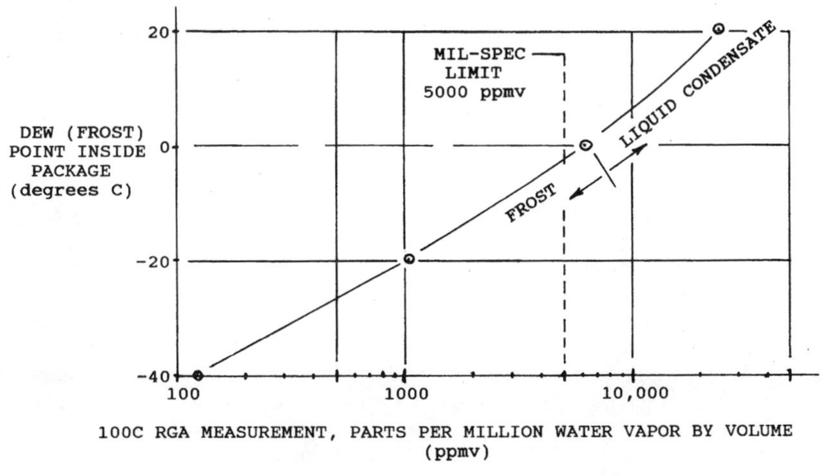

FIGURE 7.18 100° C RGA ppmv moisture inside package vs. dew point.

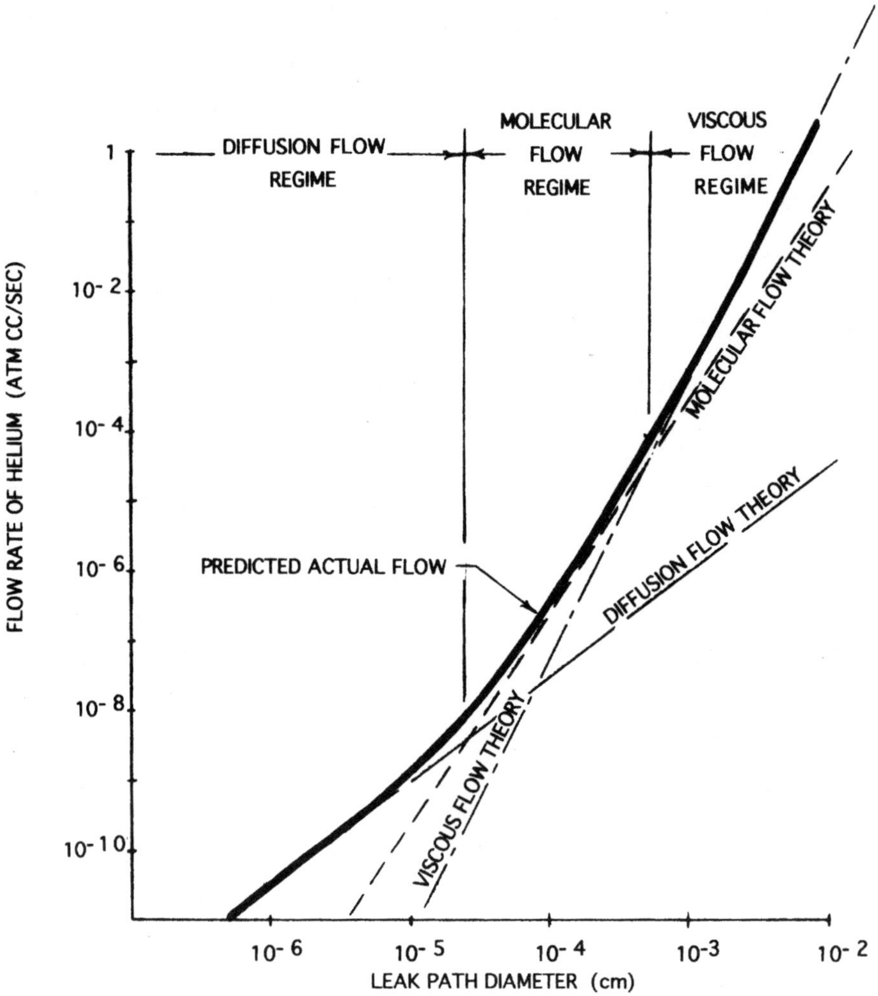

FIGURE 7.19 Leak rates of helium in various flow regimes.

hole leak is much larger than the mean free path of the gas. At lower flow rates, the flow is governed by free molecular flow equations. The flow rate here is directly proportional to the average molecular velocity of the gas. At extremely low flow rates, the flow is governed by diffusion equations and is directly proportional to the product of (mean free path) × (average molecular velocity). The heavy curve in Fig. 7.19 shows the total flow for all three types of flow.

The relationship between helium flow (as defined in the mil spec leak test) and dry air or moisture leak rates can be calculated using properties given in Table 7.6. These property values can be used to calculate the corresponding flow rates for dry air and water vapor compared to helium flow rates, resulting in the multiplication factors shown in Table 7.7. The results are plotted in Fig. 7.20. It should be noted that in the moderate leak rate region, the flow rate of helium out of a particular size crack will be about 2.7 times higher than the flow of dry air would be. In the very low leak rate region, helium flow rate is up to 10 times larger than dry air flow rates would be. This is due to helium's smaller molecule size, considerably higher average molecular velocity, and longer mean

Table 7.6 Molecular Properties: Calculation of Flow Rates, Helium, Air, Moisture[35]

		VIS	MFP	AMV
	Molecular size (Å)	Viscosity (dyne s/cm^2)	Mean free path (cm)	Average molecular velocity (cm/s)
Helium	2.3	194×10^{-6}	18.6×10^{-6}	123,000
Air	3.5	180×10^{-6}	6.4×10^{-6}	46,000
Water vapor	6.0	93×10^{-6}	4.2×10^{-6}	58,000

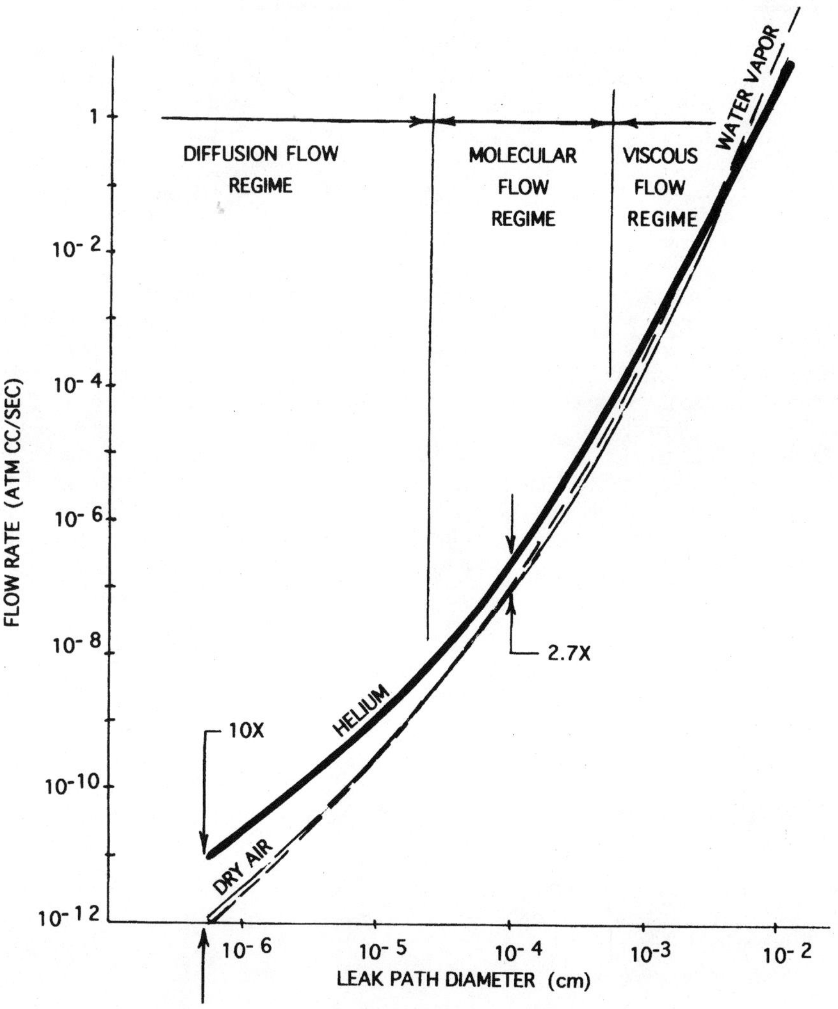

FIGURE 7.20 Flow rates of helium, dry air, and moisture through leaks of various sizes.

Table 7.7 Ratios of Flow Rates of Air-to-Helium and Moisture-to-Helium

Flow region (ATM cc/s He)	Diffusion (below 10^{-9})	Molecular (10^{-8} to 10^{-5})	Viscous (above 10^{-4})
Multiplication factor	$\frac{(MFP)_X(AMV)_X}{(MFP)_{He}(AMV)_{He}}$	$\frac{(AMV)_X}{(AMV)_{He}}$	$\frac{(VIS)_{He}}{(VIS)_X}$
$\frac{\text{Flow rate of air}}{\text{Flor rate of helium}}$	0.13	0.37	1.08
$\frac{\text{Flow rate of water vapor}}{\text{Flow rate of helium}}$	0.11	0.47	2.09

free path (see Table 7.6). Note also that the flow rate curves for dry air and water vapor are nearly identical, so the flow of any mixture of them also be nearly the same as for dry air.

Figure 7.21 shows the time to reach a damaging moisture level as a function of the leak rate measured by mil spec helium leak rate testing. There is a large variation in Fig. 7.16 because a variety of factors contribute to the rate of moisture ingress into packages of various sizes. In a fairly constant-temperature environment, moisture penetration is driven primarily by diffusion due to differences in water vapor partial pressures inside and outside the package. The local relative humidity obviously has a large effect on this rate. In an environment with varying temperature, the changing pressure inside the package due to ideal gas law temperature effects will force moisture laden air into or out of the package. The extent of temperature swing obviously has a large effect on the rate for this penetration mechanism. Figure 7.21 shows that packages with leak rates of 10^{-7} ATM-cc/s helium can be fully infiltrated by atmospheric moisture within a week to a few months, depending on conditions. It is evident from this figure that to achieve a 20-year life, helium leak rates lower than 10^{-8} ATM-cc/s helium per cc of package cavity volume are necessary. Table 7.8 lists some recommendations for maximum allowable helium leak rates to be allowed in high-reliability manufacturing operations.

Applications requiring extended reliability in the range of 20 or more years field life must therefore plan on achieving hermetic seals several orders of magnitude better than specified by conventional mil specs. Control of processes in this very low leak range will probably require advanced leak detection equipment such as has been developed by Bergquist to extend helium leak-rate detection down to 10^{-12} ATM-cc/s.[37]

Table 7.8 Leak Rate Controls Recommended for Manufacturing

Recommended maximum leak rate for hybrid/MCM packages (atm-cc/s helium)	Ref.
Below 3×10^{-9}	29
Much lower than 10^{-8}	30
1×10^{-10}	28
Substantially below 1×10^{-8}	32
Below 1×10^{-8}	33
Below 3×10^{-9} times cavity volume in cm^3	36

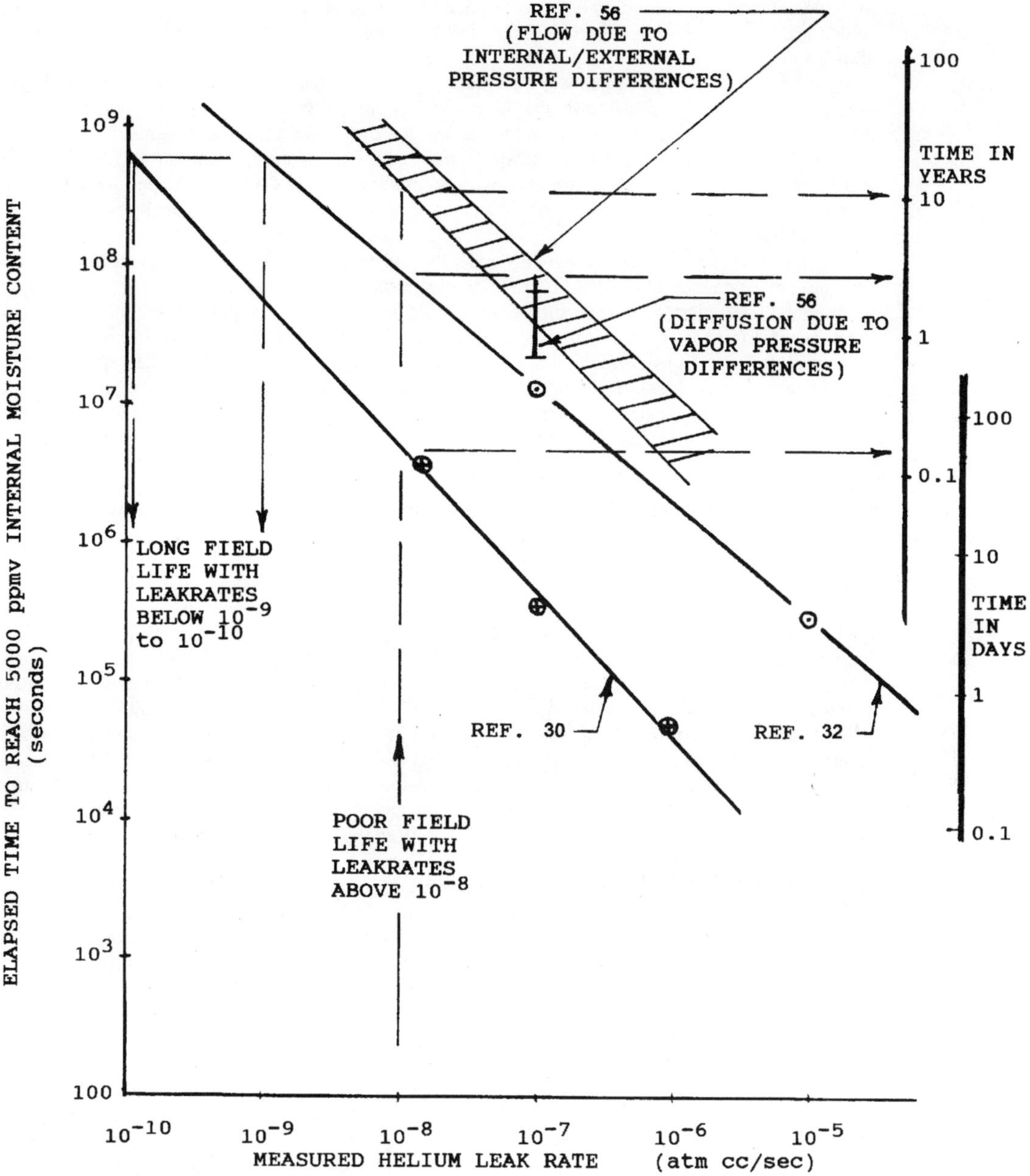

FIGURE 7.21 Elapsed time to reach 5,000 ppmv internal moisture content.

7.5.3 Pressure Cycling

Packages with sealed cavities can experience substantial stress levels as the package cavity experiences differential pressure relative to the ambient outside the package. These stresses can be driven by either processing temperatures or atmospheric pressure variations. In a typical gold-tin eutectic solder seal process operating around 327° C (600° K), the seal is formed, and the MCM package is allowed to cool to room temperature (300° K). Because the absolute temperature is now half the value it was at sealing, ideal gas law considerations dictate that the pressure inside the package is also about half, and the package lid can be easily recognized to contain a partial vacuum that pulls the lid down. Deflection must be limited to ensure that the lid does not short against components or wire bonds. Another option is to use a positive sealing pressure to compensate for ideal gas behavior, and to end up with neutral pressure at room temperature. Sealed cavity MCM packages that must operate over a range of temperatures obviously must be designed to withstand the pressure variations inside their cavities.

Another category of pressure cycling occurs in electronics subjected to repeated high- or low-pressure environments. Down-hole oil drilling electronics or submersible electronics modules must be designed for cycles of high pressure. On the other hand, airborne modules flying in unpressurized locations on airplanes above 30,000 ft altitude can easily experience pressure variations from 1 to 15 psia on each flight. Considering that a well serviced airliner might fly over 300 days per year at an average of 3–4 flights per day, avionics MCMs may accumulate >1,000 cycles per year and >20,000 cycles in their normal service life. Careful designs of lid deflection limiters and controls of lid seal stresses are essential to maintaining reliable hermetic seals under these conditions.

7.5.4 Hermetic-Equivalent Packaging, Moisture and Corrosion Resistance, Reliability without Hermeticity

Hermetic-equivalent packaging using plastic materials has been an elusive goal for many years. As was noted in Fig. 7.16, water vapor diffuses through most plastic materials very rapidly. Consequently any voids, cracks, or delaminations at IC chip or I/O surfaces will quickly achieve the same humidity levels as the surrounding ambient. Any subsequent drop in temperature below the cavity dew point will lead to condensation (as described by Figs. 7.17–7.18) and possible corrosion. Extensive testing of various encapsulants during the 1980s[38,39] identified a significant finding with silicone gel encapsulants: the presence of water molecules within the encapsulant did not produce corrosion as long as the IC surfaces were totally encapsulated and void free. Thus, prevention of formation of liquid water on surfaces that are vulnerable to corrosion is one avenue to achieving hermetic-equivalent packaging. Interest in making nonhermetic packaging as reliable as hermetic packaging grew rapidly. It was greatly accelerated by the USAF's "Reliability Without Hermeticity" (RWOH) program.[40]

Corrosion mechanisms typically require the presence of both liquid water and ionic contaminants for corrosion to occur. Reduction of ionic contamination levels below 10 ppm was demonstrated very early[41] to reduce corrosion in silicone encapsulants. The reduction in ionic content could be accomplished by either purification or by the addition of ion scavenger materials to immobilize the ions.

Modern epoxy encapsulants and molding compounds have been shown recently[42] to be similarly improved by reduction of ionic contamination, either by purification or scavengers. Figure 7.22 shows the dramatic improvements in resistance to corrosion failures with these improvements to encapsulating materials.

The physical properties of encapsulant materials are also important. An encapsulant with a very high CTE can tear wire bonds apart in thermal cycling unless its modulus of elasticity is very low. For example, even though the two materials have similarly high CTE (over 200 ppm/°C), silicone gel encapsulants (modulus below 500 psi) produce no wire bond failures, but silicone rubber encapsulants (modulus over 50,000 psi) produce extensive failures in thermal cycling.[39] On the other hand, high modulus epoxy encapsulants (over 1 million psi) perform quite reliably as long as they are expansion matched somewhere close to the circuit board materials (17 ppm/°C). Figure 7.23 displays the remarkable improvement in thermal cycling performance of encapsulant materials over the last decade.

Other approaches apply thin inorganic coatings as moisture barriers over the bare die assemblies[44,45] or over the bare die while still at the wafer level (in addition to converting the alumi-

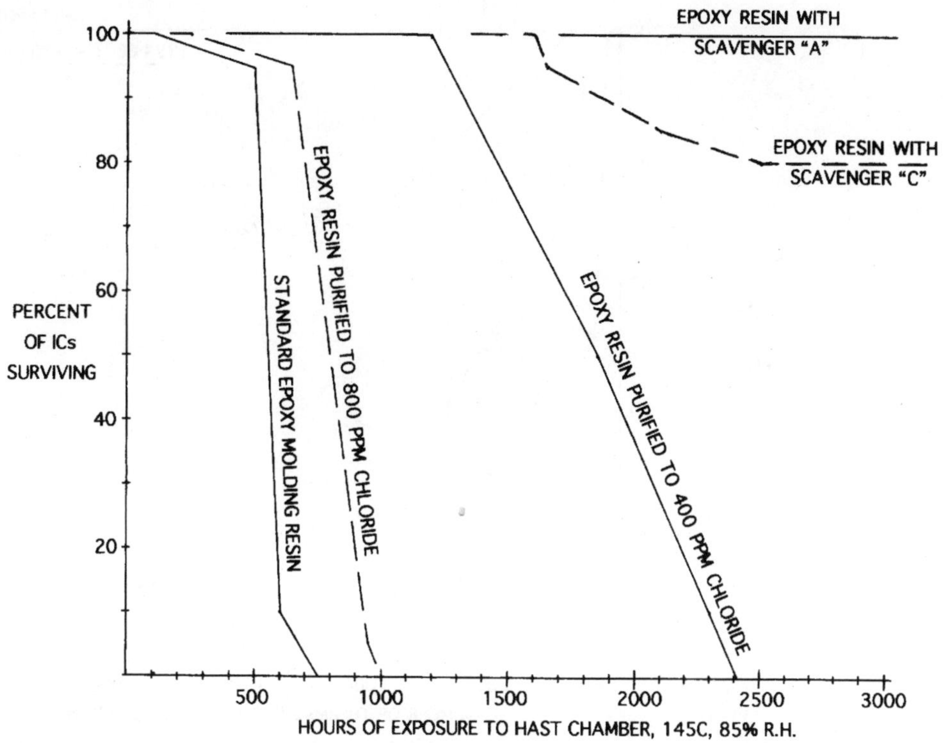

FIGURE 7.22 Recent improvements in corrosion resistance of plastic encapsulants and molding compounds (after Ref. 42).

num bond pads to corrosion-resistant gold pads). This latter approach has produced results equivalent to hermetic packaging reliability (<5 percent failures after 1000 hr at 140° C HAST testing) as shown in Fig. 7.24.[45]

7.6 EXAMPLES OF MCM PACKAGE DESIGN APPROACHES

This section discusses five MCM package design approaches representative of the technologies being used throughout industry.

7.6.1 Premolded Plastic MCM Packages

Premolded plastic cavity packages are low in cost and very convenient to assemble. They may be built simply as a frame with leads surrounding a separate substrate, a normal cavity flat pack configuration with peripheral leads, or a ball grid array cavity package. Figure 7.25*a* shows a typical cross-section, and 25*b* shows a photo of a representative MCM of this type.

7.6.2 Postmolded Plastic MCM Packages

Postmolded plastic MCM packages are particularly suited for high-volume applications, for the same reasons that molded packages are the technology of choice for high-volume single-chip devices. Fig-

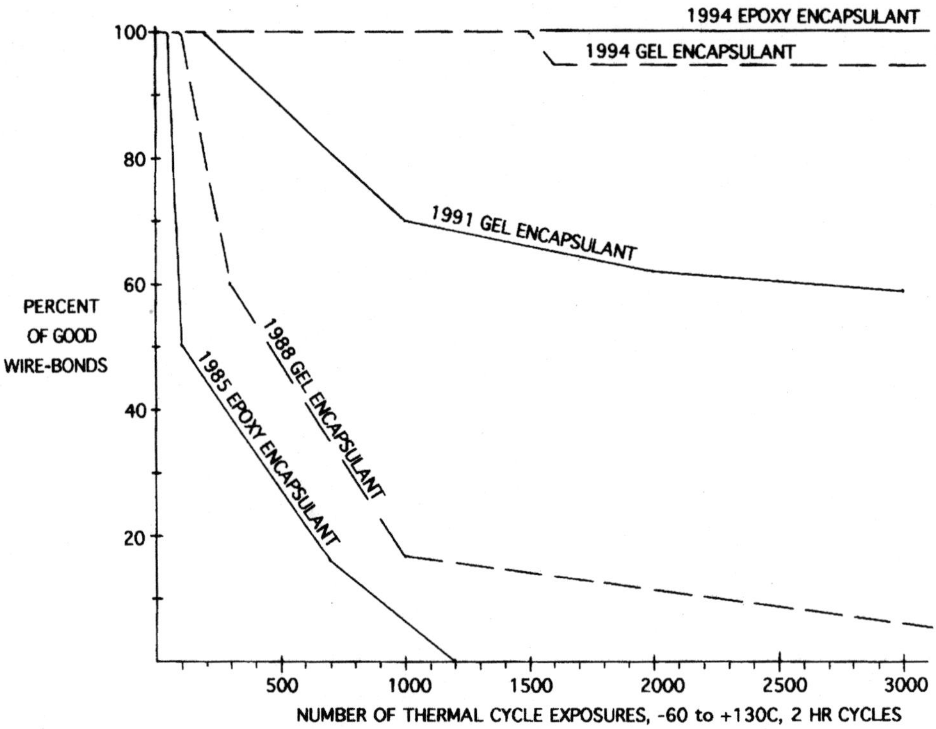

FIGURE 7.23 Recent improvements in performance of chip-on-board encapsulants.[43]

ure 7.26 shows a cross section of a postmolded plastic MCM (*a*) and a memory module (*b*). These MCMs may be identical in appearance to conventional single-chip components. Hundreds of millions of MCMs of this type have been fielded in the past few years.

7.6.3 Encapsulated Chip-on-Board MCM Packages

A convenient combination of chip-on-board die protection with an MCM-L laminate substrate is shown in Fig. 7.27. This approach is particularly suitable for rapid prototyping of new configurations and combinations of bare die and surface mount components in the same MCM.

7.6.4 Metal Hermetic MCM Packages

The metal MCM package has been the workhorse of the hybrid industry for three decades. With properly designed seals, it offers one of the most rugged and reliable forms of MCM packaging. A typical example from a space application is shown in Fig. 7.28.

7.6.5 Ceramic Hermetic MCM Packages

The ceramic MCM package has also been widely used by the hybrid industry and is available in both alumina and aluminum nitride materials. By using co-fired ceramic processes the package can be fabricated to include all interconnections between components as shown in Fig. 7.29*a*. The MCM

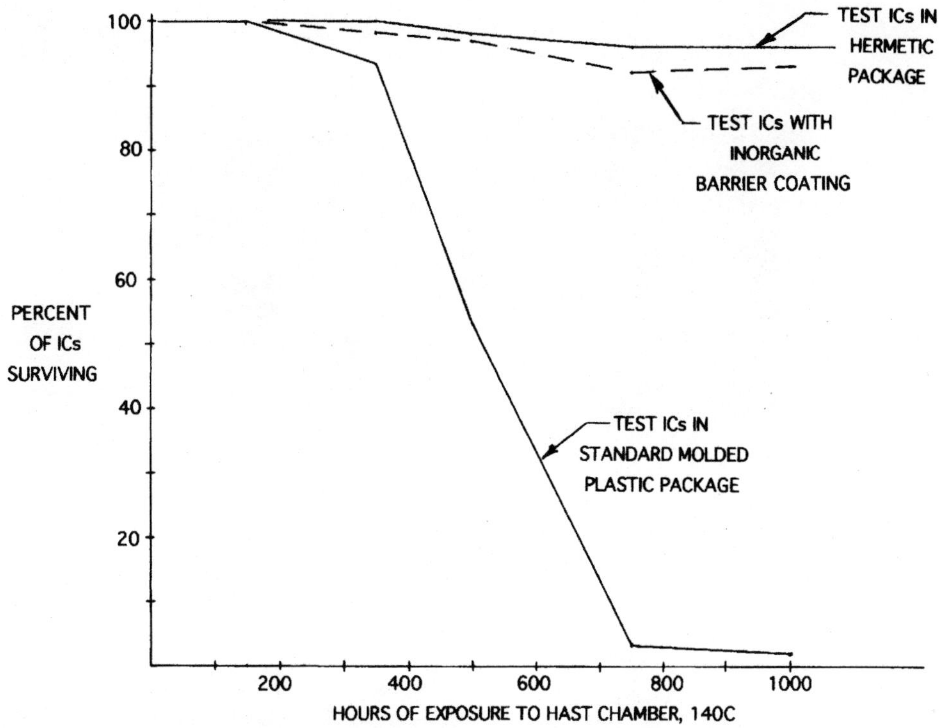

FIGURE 7.24 Hermetic-equivalent performance of nonhermetic packaging.

package shown in Fig. 7.29*b* is extensively used in sophisticated military avionics applications. This large 2 × 4 inch package houses a high density multilayer thin film substrate to interconnect the components. Figure 7.29*c* shows a similarly sized aluminum nitride package which uses silicon substrate technology to interconnect 33 ICs (the SRAMS and EPROMs are two-high stacks). Figure 7.29*d* shows the first MCM to ship in volume in a workstation application. Five ICs are interconnected in a ceramic package with 220 peripheral leads. Figure 7.29*e* shows a second-generation hyperSPARC module. Six ICs are interconnected in a ceramic package with 134 two-row pin-grid array pins.

7.7 COST CONSIDERATIONS FOR MCM PACKAGING

The great variety of MCM packaging approaches results in a wide range of cost options for the designer. The use of a package style already tooled obviously avoids substantial initial nonrecurring charges. However, just as performance optimization needs to be considered from a systems approach (as discussed in Sec. 7.2), it is important to consider cost optimization also from a systems point of view. It has been pointed out that 40 to 80 percent of a product's final cost is established by decisions made in the first 10 to 20 percent of the design cycle.[49,50] It is useful to include *design for packageability*[50–52] disciplines in making the early design decisions that so strongly influence final price. Very useful software recently has been introduced that provides trade-off visibility between various packaging technology choices.[53]

Prices of finished MCMs in production today range from under $10 to many thousands of dollars, depending upon the number of ICs, cost of ICs, and durability required of the package. At the lower

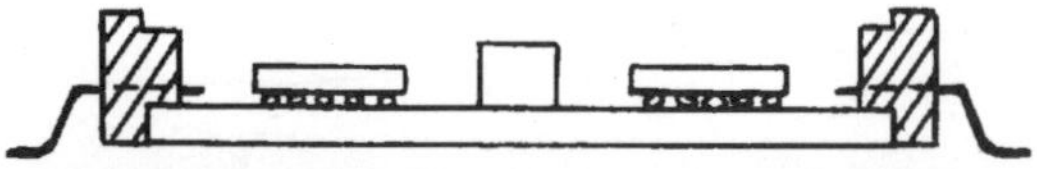

a. cross section (after Ref. 47)

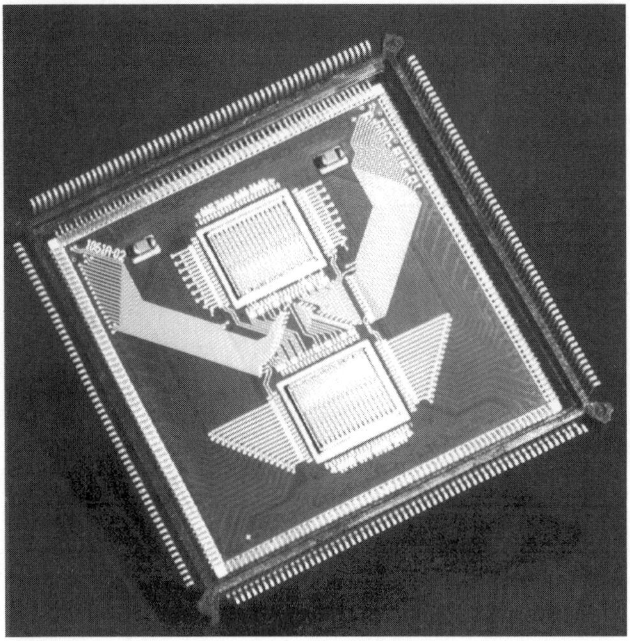

b. telecommunications MCM (from Ref. 48)

FIGURE 7.25 Typical premolded plastic package MCM.

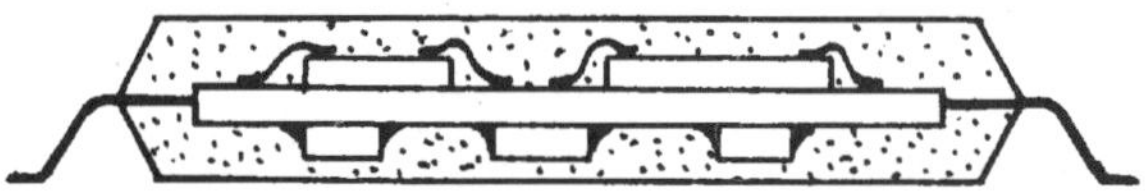

a. cross section

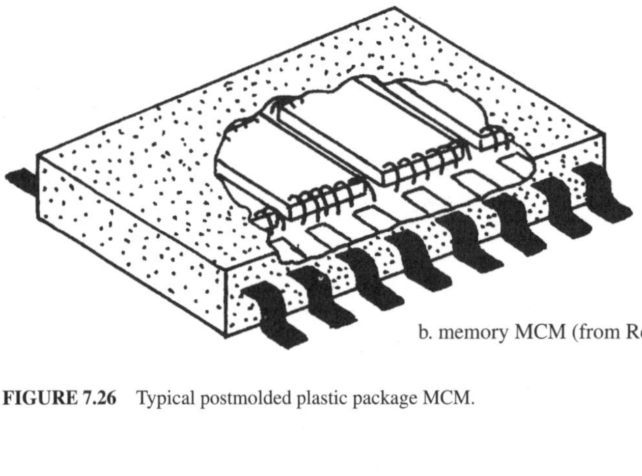

b. memory MCM (from Ref. 48)

FIGURE 7.26 Typical postmolded plastic package MCM.

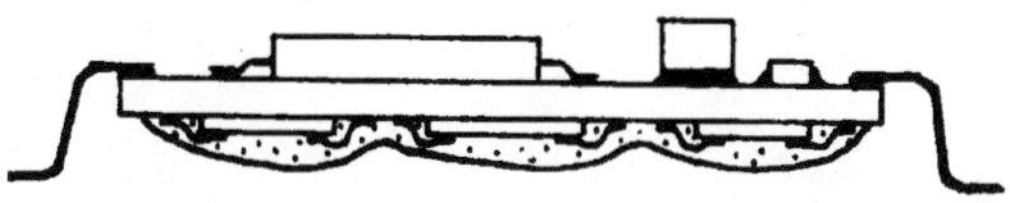

a. cross section

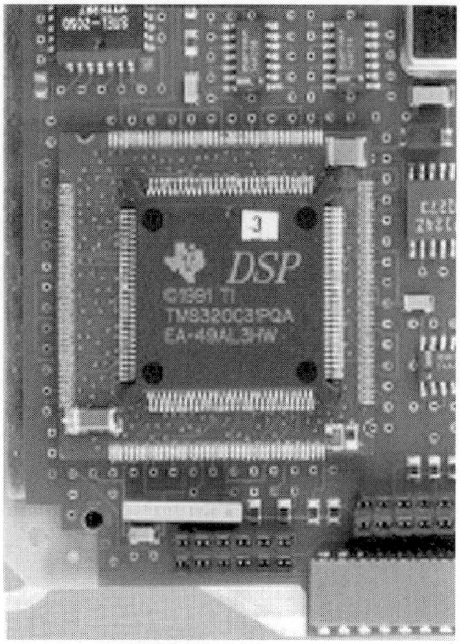

b. satellite communications MCM (courtesy of Rockwell International)

FIGURE 7.27 Typical encapsulated COB package MCM.

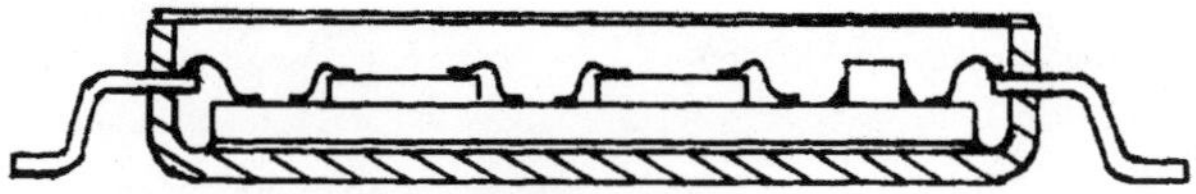

a. cross section

b. space computer MCM (from Ref. 48)

FIGURE 7.28 Typical metal package MCM.

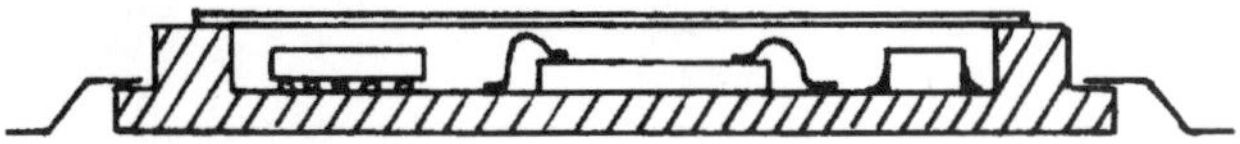

a. cross section

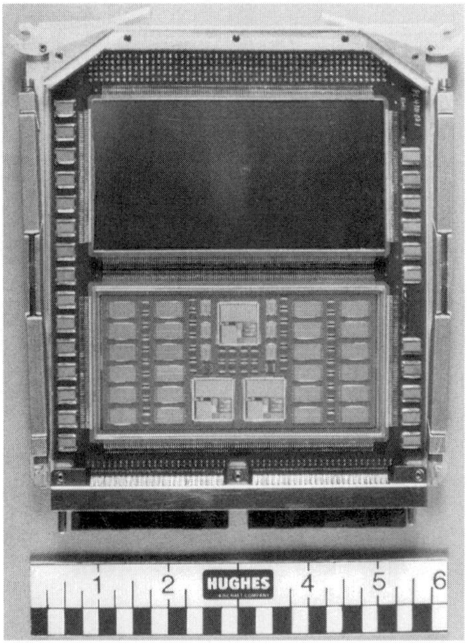

b. avionics MCM (courtesy of Hughes Aircraft)

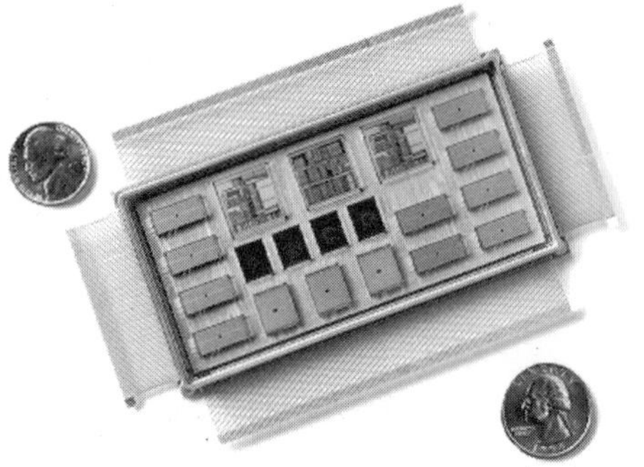

c. space flight computer
(courtesy of nChip, TRW, and JPL)

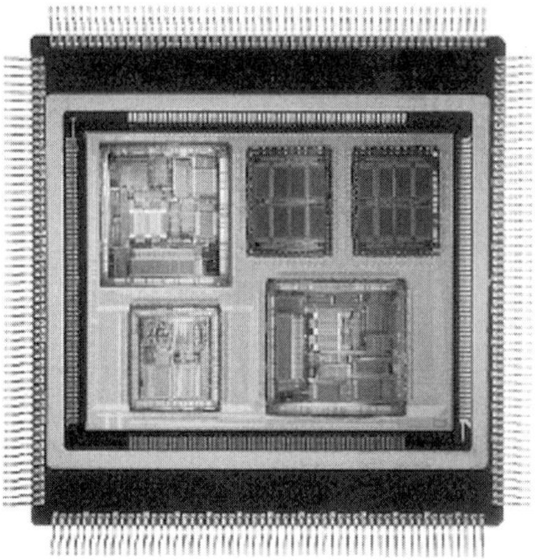

d. workstation MCM
(courtesy of nChip and Ross Technology)

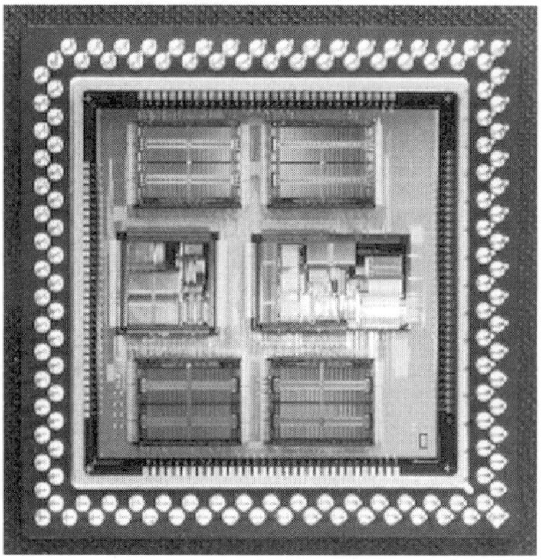

e. 150 MHz SPARC MCM
(courtesy of nChip and Ross Technology)

FIGURE 7.29 High-technology ceramic MCM packages.

end of the cost spectrum, it is not uncommon to have three approximately equal elements of total MCM cost: devices, packaging, and test. The selection of a packaging technology can affect all three of these cost areas.

The first element of total cost of a MCM is the device cost. While raw wafer costs are usually moderate (less than $\$1/in^2$ for silicon and $\$30/in^2$ for gallium arsenide), finished IC selling prices range from a few tens of dollars per square inch to hundreds of dollars per square inch, depending on feature size (submicron to several micron), process complexity (the number of layers of conductors and dielectrics, number of masks and process steps, and so forth), wafer size (75, 100, 125, 150, or 200 mm diameter), wafer yield (below 20 percent on some experimental designs to over 95 percent on mature designs), and volume of production and market demands. IC vendors have little incentive (and sometimes disincentives) to sell bare die at prices lower than their equivalent packaged parts. Special processing to add solder bumps, gold bumps, or TAB leadframes to enhance the MCM packaging design may result in cost adders of several dollars per die and tooling charges of tens of thousands of dollars per die type. The problem of known-good die costs versus assembly yields[8] must also be considered carefully.

The second element of total MCM cost is packaging cost. The packaging costs include both the cost to interconnect the ICs and the cost to surround them with a suitable protection from the environment and suitable I/O to interface to the rest of the electronic system. While it is a somewhat curious choice of cost metric, the cost per I/O pin has emerged as a popular way to describe high-volume package costs.[54,55] Plastic molded packaging typically costs around \$.01 per I/O, but tooling costs for multiple-cavity molds can cost many hundreds of thousands of dollars. Ceramic package costs range \$.03–.05 per I/O for flat pack styles and \$.05–.10 per I/O for PGA styles, with tooling costs of a few tens of thousands of dollars. Nonhermetic metal packages cost \$.03–.05 per I/O, while hermetic metal versions generally range from \$.20–.80 per I/O. These costs can be increased by several orders of magnitude if only small volumes of packages are produced, as shown by the simple amortization model shown in Fig. 7.30. There it can be seen that to approach full amortization of tooling costs requires a volume of at least ten times the ratio of tooling cost to piece part cost.

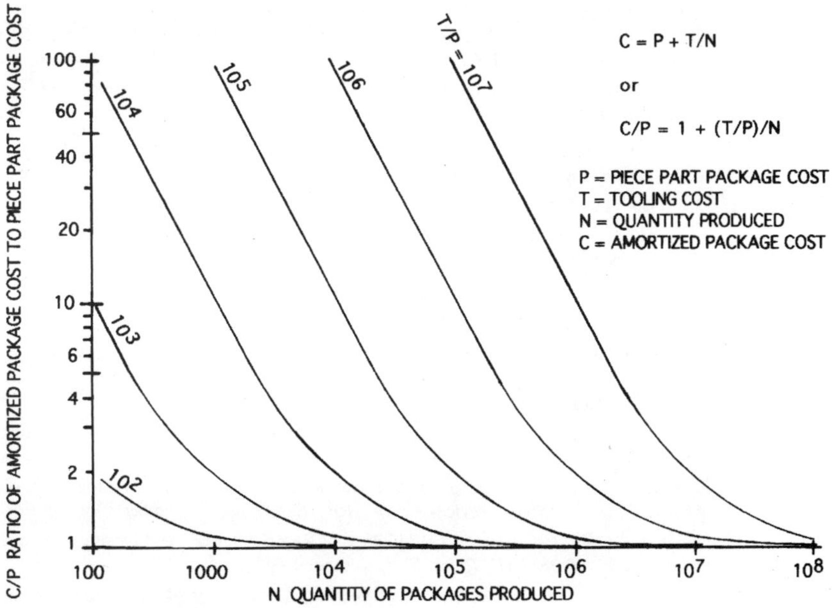

FIGURE 7.30 Amortization of MCM package tooling costs.

Packaging costs are probably most strongly dominated by the durability required for the particular application. Harsh-environment space or military applications generally require better environmental performance (hence, more expensive packaging) than has been designed into cost-sensitive, high-volume commercial applications for office or home environments. However, as manufacturers of high-volume commercial products continue to refine and improve their processes and products, there emerge selected opportunities to adapt these *commercial off-the-shelf* (COTS) approaches and products to high-reliability applications. The challenge (and responsibility) of the MCM package designer is to develop the evidence that such high-volume approaches will adequately meet the customers' needs in these very different applications.

The third element of total MCM cost is related to test. New MCMs introduced into manufacturing may require extensive levels of testing until the design has matured. As the process and design become mature, it may become advantageous to move from reliance on product testing to statistical process control approaches, and the level of MCM testing can be reduced. However, the testability of the MCM can never be eliminated—only the frequency of exercising the testability can be reduced. This requires the MCM package designer to include obvious but frequently overlooked maxims: every MCM should be designed with good testability, and every MCM package design should be accompanied by a companion low-cost test socket design (or other low-cost means of testing the MCM).

REFERENCES

1. Sampson, R.N., and Mattox, D.M., Chapter 1 in *Electronic Packaging and Interconnection Handbook*, C.A. Harper, Ed., McGraw-Hill, Inc., New York, 1991, p. 1.1.
2. Seraphim, D.P., et al, *Principles of Electronic Packaging*, McGraw-Hill, Inc., New York, 1989, pp. 1–3.
3. Rymaszewski, E.J., and Tummala, R.R., *Microelectronics Packaging Handbook*, Van Nostrand Reinhold, New York, 1989, pp. 1–4.
4. Hagge, J.K., "Ultra-Reliable Packaging for Silicon-on-Silicon WSI," *Proc. EIA/IEEE Electronic Components Conf.,* Los Angeles, May 10, 1988, pp. 282–292.
5. Hagge, J.K., "Mechanical Considerations for Reliable Interfaces in Next Generation Packaging," *Proc. NAECON,* Dayton, OH, May 1989.
6. Markstein, H.W., "Applications for MCMs," *Electronic Packaging & Production,* April 1995, pp. 48–50.
7. Trent, J., "Laminate Technology for MCMs," *Electronic Packaging & Production,* April 1995, pp. 30–31.
8. Hagge, J.K., and Wagner, R.J., "High Yield Assembly of Multichip Modules Through Known-Good IC's and Effective Test Strategies," *Proc. IEEE,* vol. 80, no.12, December 1992, pp. 1965–1994.
9. Private communication.
10. MIL-STD-883C, Method 2017, Paragraph 3.1.5.1. U.S. Government Printing Office, Washington, DC.
11. Seraphim, D.P., *Principles of Electronic Packaging*, pp. 61–67.
12. Rymaszewski, E. J., *Microelectronics Packaging Handbook*, pp. 12–14, 676–679.
13. Hagge, J.K., "State-of-the-Art Multichip Modules for Avionics," *IEEE Trans CHMT,* v. 15, no. 1, February 1992, pp. 29–42.
14. Steinberg, D.S., "Preventing Thermal Cycling and Vibration Failures in Electronic Equipment," short course, Cedar Rapids, IA, 1990.
15. Jahsman, W.E., et al, "Measurement and Modeling of In Situ Lead Stiffness of Surface Mounted Packages," *IEEE Trans CHMT,* vol. 14, no. 4, December 1991, pp. 859–869.
16. Kotlowitz, R.W., and Engelmaier, W., "Impact of Lead Compliance on the Solder Attachment Reliability of Leaded Surface Mount Devices," *Proc. IEPS,* San Diego, November 1986, pp. 841–864.
17. Kotlowitz, R.W., "Comparative Compliance of Representative Lead Designs for Surface Mounted Components," *IEEE Trans CHMT,* vol. 12, December 1989, pp. 431–448.
18. Kotlowitz, R.W., and Taylor, L.R., "Compliance Metrics for the Inclined Gull-Wing, Spider J-Bend, and Spider Gull Wing Lead Designs for Surface Mount Components," *IEEE Trans CHMT,* vol. 14, no. 4, December 1991, pp. 771–779.

19. Kotlowitz, R.W., and Gosen, G., "Compliance Metrics for the Generalized S-Bend Lead Design for Surface-Mount Components," *Proc. EIA/IEEE ECTC,* May 1992, pp. 345–353.
20. Gee, S.A., and Van Kessel, G.G.M., "Stiffness and Yielding in PLCC 'J' Lead," *Proc Int. Elect. Mfg. Symp.,* 1986, pp. 228–241.
21. Lau, J.H., "Stiffness of PLCC 'J' Lead and its Effect on Solder Joint Reliability," *ASME paper 87-WA/EEP,* 1987 Winter Annual Meeting.
22. Lau, J.H., "Stiffness of PQFP 'Gull-Wing' Lead and its Effect on Solder Joint Reliability," *IEEE Trans CHMT,* 1988, pp. 131–132.
23. Mather, J.C., "Package Structure for Multichip Modules," U.S. Patent no. 5,034,568, July 23, 1991.
24. Mather, J.C., and Minogue, G.R., "Robust Aluminum Nitride Package Technology for Multichip Modules," *Proc. Surface Mount Int.,* September 1992.
25. Pivin, D.P., "Pick the Right Package for Your Next ASIC Design," *EDN,* February 3, 1994, pp. 91–108.
26. Tsang, D.Z., "High-Speed Optical Interconnections for Digital Systems," *Lincoln Lab Journal, MIT,* vol. 4, no. 1, 1991, pp. 31–44.
27. Weiss, A., "Hermetic Packages and Sealing Techniques," *Semiconductor International,* June 1982, pp. 111–119.
28. Kumar, A., and Ozmat, B., "Package Sealing and Encapsulation," Chapter 10, *Microelectronics Packaging Handbook,* Van Nostrand Reinhold, 1989, pp. 729–736.
29. Swanson, D., and Licari, J., "Effect of Screen Tests and Burn-in on Moisture Content of Hybrid Microcircuits," *Solid State Technology,* September 1986, pp. 125–130.
30. Stroehle, D., "On the Penetration of Water Vapor into Packages with Cavities and on Maximum allowable Leak Rates," *Proc. 15th IEEE Reliability Physics Symposium,* April 1977, pp. 101–106.
31. Suloff, R.E., "A Study of Leak Rate Versus Reliability of Hybrid Packages," *Proc. Int. Microelectronics Symp.,* September 1978, pp. 121–124.
32. Der Marderosian, A., and Gionet, V., "Water Vapor Penetration Rate into Enclosures with Known Air Leak Rates," *Proc. 16th IEEE/EIA Reliability Physics Symposium,* April 1978, pp. 179–186.
33. Pinsky, D., "Hermeticity Measurements for Large Microelectronic Packages," *Proc. ISHM,* 1986, pp. 379–382.
34. Neff, G., "Hybrid Hermeticity and Failure Analysis," *Hybrid Circuit Technology,* April 1986, pp. 19–42.
35. Ganus, F., and Nenoff, L., "Microelectronic Package Hermeticity," Internal Report, 1970.
36. David, R.F.S., "Effect of Leak Rate and Package Volume on Hybrid Internal Atmosphere," *Int. Journal Hybrid Microelectronics,* vol.6, no. 1, October 1993, pp. 433–440.
37. Green, T.J., and Duncan, C.T., "Using Design of Experiments to Optimize Hermetic Package Seam Welding," *Proc. ISHM,* 1992, pp. 485–490.
38. Otsuka, K., *et al,* "A New Silicone Gel Sealing Mechanism for High Reliability Encapsulation," *IEEE Trans. CHMT,* vol. 7, no. 3, September 1984, pp. 249–256.
39. Balde, J. W., "The Effectiveness of Silicone Gels for Corrosion Prevention of Silicon Circuits...," *IEEE CHMT,* vol. 14, no. 2, June 1991, pp. 352–365.
40. Murphy, C.F., et al, "Evaluation of Non-Hermetic Coatings for MCM Applications through HAST, 85/85, and PCT," *Proc. IEPS/ISHM MCM Conf.,* Denver, April 1994, pp. 338–343.
41. Wong, C.P., "Room Temperature Vulcanized (RTV) Silicone as Integrated Circuit (IC) Coating," *Proc. ISHM,* 1981, pp. 315–319.
42. Project briefing, "Tri-Service Initiative on Plastic Package Availability," contract DLA-900-92-C-1647, Defense Logistics Agency, Dayton, OH, May 26, 1993.
43. Unpublished test results, Rockwell Int., 1994.
44. Gates, L.E., and Kubacki, R.M., "Sealed Chip-on-Board," *Electronic Packaging & Production,* September 1994, pp. 48–50.
45. Byrne, R.C., and Camilletti, R.C., "Reliability Without Hermeticity (RWOH) for Integrated Circuits: Sealed Chips for Hermetic-Like Protection," *Proc. GOMAC,* 1993, pp. 37–40.
46. Camilletti, R.C., et al, "Chip Scale Packaging with High Reliability for MCM Applications," *Proc. ISHM MCM Conf.,* Denver, April 1966.
47. Personal communication, Die-Tech, Inc., York Haven, PA, Plastic Packages for MCMs, September 1995.

48. Hagge, J.K., and Mather, J.C., "Multichip Module Package Design," Chapter 9, *Thin Film Multichip Modules,* ISHM, Reston, VA, 1992, pp. 334–337.

49. Moreno, H., and Sanborn, P., *Conceptual Design of Multichip Modules and Systems*, Klewer Academic Publishers, 1994.

50. Dehkordi, P., et al, "Impact of Packaging Technology on System Partitioning: A Case Study," *Proc. IEEE MCM Conf.,* Santa Cruz, February 1995, pp. 144–149.

51. Dehkordi, P., and Bouldin, D., "Design for Packageability: Early Consideration of Packaging from a VLSI Designer's Viewpoint," *Computer,* vol. 1, April 1993, pp. 76–81.

52. Dehkordi, P., and Bouldin, D., "Design for Packageability: The Impact of Bonding Technology on the Size and Layout of VLSI Dies," *Proc. IEEE MCM Conf.,* March 1993, pp. 153–159.

53. Savan Sys™, System Design Advisor Tool from Savantage Company, 3925 W. Braker Lane, Suite 325, Austin, TX 78759–5321.

54. Lassen, C., "Silicon Packaging: The Aluminum Nitride Domain," *Advancing Microelectronics,* Jan–Feb 1994, pp. 41–42.

55. Franzon, P.D., "MCM Package Selection: A Systems-Need Perspective," Chapter 3, *Multichip Module Technologies and Alternatives: The Basics*, D. Doane and P. Franzon, Eds., Van Nostrand Reinhold, NY, 1994, p. 113.

56. Roberts, S., "Predicting Post-Seal Atmospheric Moisture Content in Hybrid Packages," *Proc. ISHM,* 1991, pp. 212–216.

CHAPTER 8

ASSEMBLY

Dr. Wayne Johnson
Auburn University

8.1 INTRODUCTION

Multichip module assembly is the mechanical attachment and electrical interconnection of semiconductor devices and passive components to an MCM substrate. The mechanical and electrical attachment may be performed in two separate operations, as is the case of epoxy die attach and wire bonding, or may be accomplished simultaneously, as occurs in flip chip assembly. The three dominant assembly methods, wire bonding, flip chip, and tape automated bonding, are illustrated in Fig. 8.1 and will be discussed in detail in the following sections. Variations of these assembly methods are being developed and will also be examined. The *Industry Guide* published by the International Society for Hybrid Microelectronics is a helpful starting point for those new to the industry and trying to identify suppliers of materials and equipment for multichip module assembly.

Table 8.1 lists the semiconductor die-related mechanical, electrical, and material information needed to design an MCM and to define the assembly process.[1] Historically, there has not been a standard format for providing this information. Recently, the Die Information Exchange (DIE) format has been established to address this problem.[2] The chip dimensions and bond pad coordinates are needed for the physical layout of the MCM. This information, along with the physical layout of the MCM, can be loaded directly into assembly equipment such as die pick-and-place machines and wire bonders. One challenge facing the MCM industry is "die shrink." As semiconductor manufacturers continue to improve manufacturing processes, feature size decreases. For selected high-volume integrated circuits, the semiconductor manufacturer may redesign the chip using these reduced features, shrinking the size of the die. This increases the number of die per wafer, increasing yield and decreasing per-die costs. This die shrink, however, may force the redesign of the MCM, especially if flip chip or tape automated bonding is used. Careful communications must be maintained with the semiconductor foundry to ensure proper, early notification of planned die shrinks.

The bond pad dimensions and pitch may determine the type of assembly. For example, ultrasonic wire bonding can be used for finer bond pad pitch than can thermosonic wire bonding. This will be discussed in a later section.

The electrical potential of the die backside is both a physical layout and an assembly issue. For wire bond or face-up tape automated assembly, the backside of the die may be electrically connected, if necessary, to the metallized die mounting pad on the substrate with conductive epoxy. In flip chip

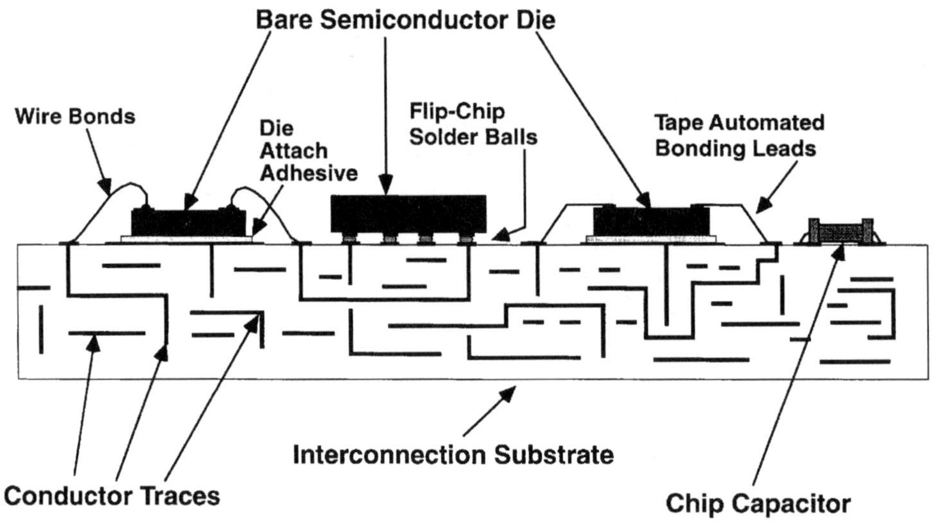

FIGURE 8.1 Cross section illustrating multichip module technology.

Table 8.1 Die-Related Information[2]

Mechanical
• chip dimensions (x, y, z) with units and tolerances
• bond pad dimensions
• bond pad x, y coordinates on chip
• recommended die attach material and limitations
Electrical
• electrical potential of die backside (floating, V_{cc}, ground, other)
• bond pad numbering identification
• power pad identification and potential
• signal pad names identification
• minimum allowable bond wire diameter
• allowable bond wire compositions and constraints
• power dissipation models or data
• I/O buffer models or data
Materials
• bond pad composition and thickness
• bump metallurgy for flip chip or tape automated bonding
• backside metallization, composition and thickness
• wafer material (Si, GaAs, InP, SiC, etc.)
• relevant chip cross-sectional description (top passivation, layer material and thickness, etc.)

assembly where the backside of the die is not in contact with the substrate metallization, backside electrical contact is more difficult. If the die backside is metallized with a wire bondable metal, then wire bonding can be used to connect the backside of a flip chip assembled die to the substrate metallization. Backside bias has been implemented in a flip chip application with GaAs dice, using a process that does not require the back of the chip to be wire bondable. The process involves aluminum wire bonding from a grounded bond pad on the substrate to a second dummy bond pad. The backside of the chip is coated with a silver-filled epoxy. The dummy end of the bond wire is loosened and the wire embedded into the wet epoxy, which is subsequently cured.[3]

Power dissipation information is important in defining substrate technology and in determining appropriate die attachment materials. Solders are commonly used for high-power dissipation applications. The thermal conductivity of solder is ten times that of a typical epoxy die attach material. Thermal modeling is used during the design of the MCM to optimize die placement and construction options to meet thermal performance requirements.

Bond pad composition and thickness are important in determining wire bondability. For solder flip chip devices, knowledge of bump metallurgy is necessary to determine assembly processes and reflow temperatures. With gold bumps, conductive adhesives may be used for attaching flip chips rather than soldering. Both solder and gold bumps are used in tape automated bonding and determine the process for inner lead bonding of the tape metallization to the bump.

Wafer material is an important factor in assembly. GaAs is more brittle than Si, and extreme care must be used in the assembly process. The thermal stability of materials for ohmic and Schottky contacts to III-VI semiconductors may limit assembly processing temperatures. High-frequency GaAs devices are often fabricated with air bridge interconnections on the chip to reduce parasitic capacitances. Additional care must be used in the handling of these die in the placement operation. A major issue in assembly design is the coefficient of thermal expansion (CTE) mismatch between the semiconductor die and the MCM substrate. For example, the CTE of Si is 3 ppm/°C, while a typical laminate MCM has a CTE of 16 ppm/°C. The attachment materials should be chosen to minimize the stress on the die. This is especially important for analog ICs since die stress can affect device characteristics. Thermal and power cycling produce cyclic stresses on the attachment material, which can result in fatigue, crack initiation, crack growth, and failure. This problem increases with increasing die size, which is the industry trend.

The design of the assembly process must be included in the initial MCM design to ensure compatibility of assembly processes, compatibility of die and substrates with assembly processing, and to address such issues as repair and rework. The facilities, materials, and processes for MCM assembly are discussed in the following sections.

8.2 FACILITIES AND HANDLING

MCM assembly requires a clean-room environment to minimize particulate contamination. Assembly operations do not involve photolithography, etching, printing, and so forth, which are used in MCM substrate fabrication. Thus, the clean-room requirements are less stringent. The two primary concerns with particulate contamination in assembly are ionically contaminated particles and electrically conductive particles. Ionically contaminated particles can lead to corrosion and long-term reliability problems. Conductive particles result in intermittent shorts. MIL-H-38534B specifies a class 100,000 clean room as defined in FED-STD-209C for the assembly of military products using bare semiconductor devices. Figure 8.2 shows the particle size limits for different clean-room designations. Most commercial assembly operations also operate in a class 100,000 or better environment.[4] A class 100,000 clean-room environment may be achieved through the construction of a room with proper air handing and filtration or through the use of modular clean-room units in which the assembly equipment is placed providing local clean environments. People are the most significant source of particulate contamination (dead skin cells and cosmetics). Clean-room gowns and gloves are used to minimize these contaminants.

Semiconductor devices should be stored in a dry nitrogen environment to prevent oxidation, corrosion, and moisture adsorption. For handling, Teflon®-tipped vacuum pickups are typically used on the top surface of the die. Metal tweezers or grippers must be used with care, because silicon, and especially gallium arsenide, are very brittle. Also, the surface passivation and metallization can be easily scratched if the die is mishandled.

Semiconductor die may be shipped in wafer form, waffle packs, gel packs, or tape-and-reel format. To separate the die in a wafer for assembly, the wafer is mounted on a plastic sheet coated with a pressure-sensitive adhesive. The wafer is then sawed with a high-speed diamond particle impregnated blade (Fig. 8.3). A water jet is sprayed over the wafer surface for cooling and to remove debris during sawing. After sawing, the plastic film is expanded (stretched) to separate the individual die, allowing them to be removed without contacting the adjacent die. To remove a die for placement, mechanical probes pierce through the expanded plastic film and push on the back of the die to facilitate vacuum pickup of the die from the adhesive coated film. Care must taken not to damage the backside of the die. Motorola has recently studied the effect of backside damage on die fracture during thermal cycling of die assembled by flip chip solder reflow.[5] An alternative is the use of adhesives that break down and release the die when exposed to UV radiation. Presentation of the die in expanded wafer form to the pick-and-place machine eliminates handling steps, which reduces cost and the potential for handling damage.

Semiconductor manufacturers have been hesitant to ship entire wafers. Die which are found to be defective during the manufacturer's wafer-level testing are marked with an ink dot. As a result, the semiconductor manufacturer's yield and cost (profit) for a particular device can be estimated. To protect die yield information, wafer sawing is often performed by the semiconductor manufacturer, and the good die are then placed in waffle packs, gel packs, or tape-and-reel for shipping. Waffle packs (Fig. 8.4) are so named because they resemble waffles with individual cells into which die are placed. The die are all oriented the same way in the cells to simplify the pick-and-place operation during MCM assembly. The disadvantage to this shipping format is the potential for the die to move within the cell and the resulting challenge (time) in aligning the die during the placement operation. Custom waffle packs with cells designed for the specific die size can reduce this problem. GEL-PAKs are an alternative (Fig. 8.5). These use a tacky gel membrane in the package to hold the die in place. There has been some concern that vapor from the gel would condense on the die I/O pads, degrading wire bondability. Current indications are that this has not been a problem. To remove the die from the gel,

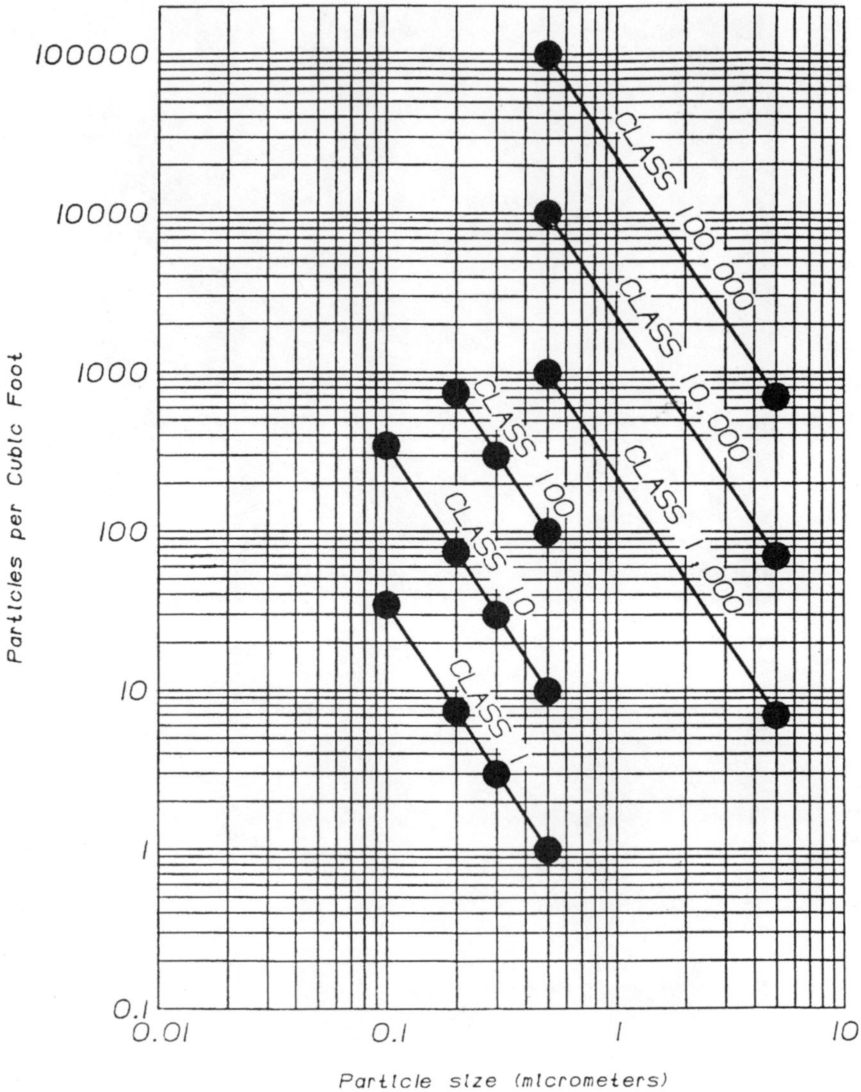

FIGURE 8.2 Class limits in particles per cubic foot of size equal to or greater than particle sizes shown. (The class limit particle concentrations are defined for class purposes only and do not necessarily represent the size distribution to be found in any particular situation.)[4]

a vacuum is applied to the bottom of the package deflecting the gel membrane over a fabric mesh reducing the contact area as illustrated in Fig. 8.6. Both waffle packs and GEL-PAKs employ a two-dimensional array of die and can present from 25 to 400 die per pack, depending on the die size, to the placement machine for assembly. For high-volume assembly where downtime for material resupply and machine feeder space are critical issues, tape-and-reel formats are preferred (Fig. 8.7). Handling is similar to that used for high-volume surface mount components. The die are arranged as a one-dimensional array in an indexed tape. The tape contains cavities, the bottom of which are coated

FIGURE 8.3 Wafer sawing. Courtesy of Kulicke and Soffa Industries, Inc. (K&S).

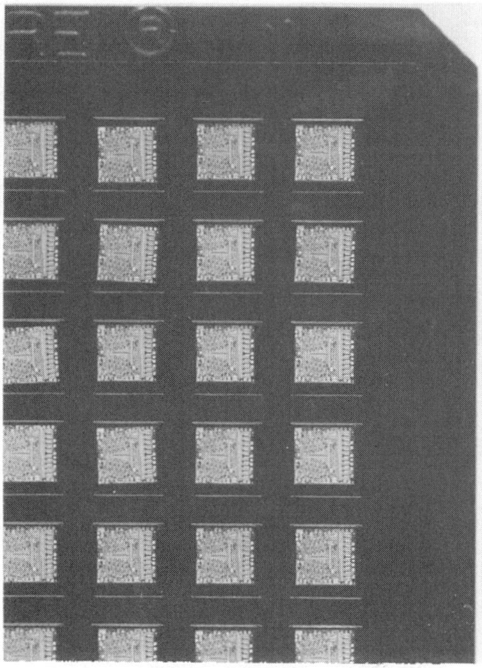

FIGURE 8.4 Die in a waffle pack. Courtesy of AVEX Electronics Inc.

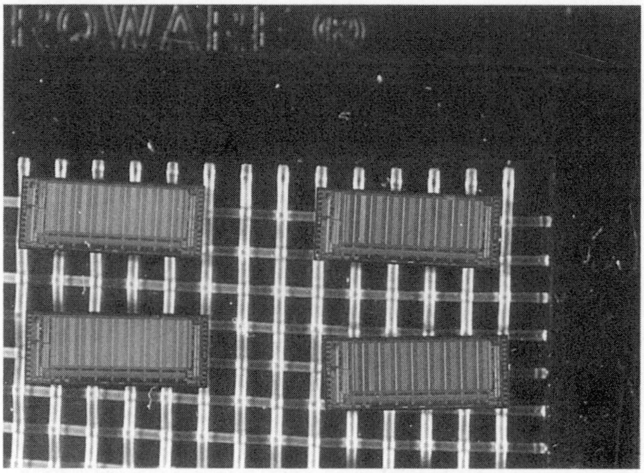

FIGURE 8.5 Die in a GEL-PAK. Courtesy of AVEX Electronics Inc.

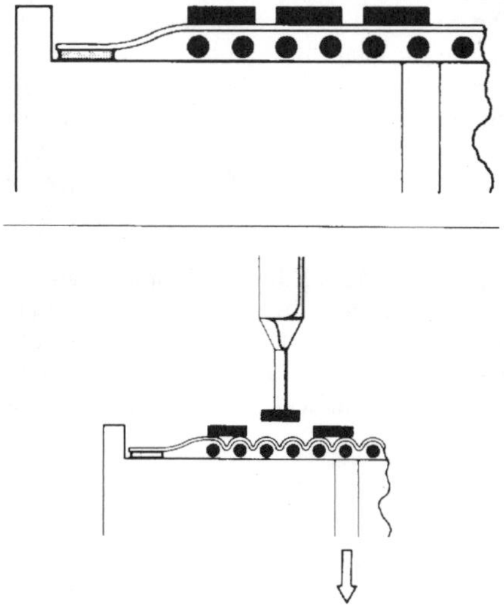

FIGURE 8.6 Illustration of vacuum release of die in GEL-PAK. Courtesy of GEL-PAK.

with a pressure sensitive adhesive. The die are placed into these cavities. The tape is wound on a reel for automatic handling and may contain 5,000 or more die. In assembly, a mechanical probe pierces through the plastic cavity and pushes on the back of the die to facilitate vacuum pickup of the from the tape. Care must exercised in this process to minimize the damage to the die backside.

8.3 ELECTROSTATIC DISCHARGE

The buildup of electrostatic charge occurs when two dissimilar materials come into contact and are then separated, transferring electrons from one material to the other. The material with the greatest electron affinity becomes negatively charged, while the other material is develops a positive charge.

FIGURE 8.7 Die in tape-and-reel format. Courtesy of AVEX Electronics Inc.

The relative magnitude and polarity of the charging effect between two materials can be estimated from the *triboelectric series* (Table 8.2).[6] For example, paper in contact with glass becomes negatively charged; while in contact with polyester, paper becomes positively charged. Common sources of electrostatic charge buildup and the resulting voltages are listed in Table.[6] In the case of metals, the charge is rapidly conducted away. However, insulators may retain static charge for hours or days. As seen in Table 8.3, high humidity and the adsorption of moisture on surfaces provide a conductive path that limits the voltage that can be generated. Semiconductor assembly plants are often constructed in humid areas of the world to take advantage of this.

The buildup of static charge can result in damage to semiconductor devices from either the electric field or by discharge of current into the device. The potential failure mechanisms are given in Table 8.4. The resulting damage can be catastrophic device failure, degraded device performance, or increased infant mortality. Table 8.5 lists the susceptibility of different device types to electrostatic discharge (ESD). Integrated circuits are usually designed with input protection to improve the device's ability to withstand ESD.

Components and modules must be protected from ESD during storage and handling. Two types of materials are used: static-free and shielding. Static-free materials conduct charge away and are classified as antistatic, static dissipative, or conductive, depending on the rate of charge dissipation. Antistatic materials dissipate charge more slowly, while conductive materials quickly dissipate charge. Waffle packs and gel packs are typically made with static-free materials. Shielding provides a

Table 8.2 Triboelectric Series[6]

Material	
Air	↑ Increasingly positive
Human skin	
Glass	
Human hair	
Wool	
For	
Paper	
Cotton	
Wood	
Hard rubber	
Acetate rayon	
Polyester	
Polyurethane	
Polyvinylchloride	
Teflon®	↓ Increasingly negative

Table 8.3 Typical Electrostatic Sources and Voltages[6]

Source	Voltage (V) 10–20% RH	Voltage (V) 65–90% RH
Walking across carpet	35,000	1,500
Walking across vinyl floor	12,000	250
Worker at bench	6,000	100
Vinyl envelope	7,000	600
Common polyethylene bag picked up from bench	20,000	1,200
Work chair padded with polyurethane foam	18,000	1,500

Table 8.4 ESD Failure Mechanisms (adapted from Ref. 7)

Failure type	Cause
Dielectric failure	An electric field-induced action ruptures oxides on a chip.
Interconnection metal failure	Melting of conductor metals due to discharge current. This can happen at a lower current level than expected, because oxides that may be underneath the metal are very poor thermal conductors and do not conduct the thermal energy away quickly.
p-n junction failure	ESD voltage/current causes melting of the silicon or metal "spiking" from one diffusion layer to the next.

Table 8.5 Susceptibility to ESD[6]

Device type	ESD susceptibility (V)
Power MOSFET	100–200
JFET	140–10,000
CMOS	250–2,000
Schottky diodes, TTL	300–2,500
Bipolar transistors	380–7,000
ECL	500
SCR	680–1,000

Faraday cage around the devices protecting them from electric fields. Shielded bags with a metal or metallized conductive layer are used to ship and store wafers, and die are stored in shielded waffle packs, gel packs, and tape-and-reel.

In the assembly area, the buildup of static charge can be eliminated by providing a conduction path to drain triboelectric generated charge to ground. First, assembly equipment should be grounded. To remove charge from operating personnel, wrist straps contact the operator's skin and provide a low-resistance path to ground. A 1 MΩ resistor is typically placed in series with the ground connection to avoid rapid discharge and to avoid the possibility of electric shock when working with "live" electrical circuitry. Antistatic gloves and gowns are also available for operating personnel. Grounded, static-free table mats are placed on work surfaces. In the assembly area, special static-free floors or grounded, static-free floor mats are used. Ionized air blowers can provide ions to neutralize buildup charge on surfaces that are difficult to connect to ground.

8.4 DIE ATTACHMENT

Attachment of the die to the substrate with epoxy adhesives and wire bonding is the most common method of bare die assembly. This approach does not require any special processing of the semiconductor die, increasing the number of devices available for MCM design. A number of die attach materials are available including epoxy, polyimide, thermoplastics, and solder.

8.4.1 Epoxy

Unfilled die attach epoxies are thermally and electrically insulating. By adding ceramic fillers such as Al_2O_3 or AlN to the epoxy, the thermal conductivity can be increased. Diamond particles have also been used to achieve higher thermal conductivity. If electrical conductivity is also required, Ag or Au particles are added instead of the ceramic fillers. A wide variety of epoxy die attach formulations are commercially available to meet varying requirements for processing (application and curing), cost, performance (electrical, thermal, mechanical), and so on. Flexible epoxies have been formulated to minimize the stress associated with the assembly of large Si die on substrates where the mismatch in CTE may range from 1:2 for alumina to 1:5 for FR-4 laminate. These low-stress, flexible epoxies typically have a low glass transition temperature, Tg. Above the Tg, the CTE of the epoxy increases by a factor of 3 to 5, but the Young's modulus decreases significantly, and the epoxy is more compliant.

Table 8.6 lists typical Ag filled epoxy properties and the military requirements per MIL-STD-883C, Method 5011 supplier requirements. MCM assemblers do not want to mix two-part epoxies. Therefore, die attach epoxies are provided as single-component systems, including the resin and the curing agent. Cross-linking occurs slowly at room temperature, increasing the viscosity and limiting the pot life of the material. Epoxies are generally stored at –40° C to improve shelf life. It is important to allow epoxy removed from the freezer to reach equilibrium with room temperature prior to dispensing. Failure to do so can result in moisture condensation on the epoxy, which will degrade the cured properties of the epoxy.

While epoxies have long been qualified for military and space applications, outgassing of epoxies remains an issue. When exposed to elevated temperatures, some polymer decomposition occurs, and low molecular weight species such as water vapor, alcohols, ammonia, carbon dioxide, methane, toluene, and xylene are released. This occurs during curing and during post-cure high-temperature exposure. Outgassing during the cure cycle is a concern if the outgassing products condense on the wire bonding pads, decreasing bondability. The outgassing tendency of an epoxy can be measured by thermogravimetric analysis (TGA). A sample of material is placed on an electronic microbalance and heated at a rate of 5 to 10° C/min, and the decrease in weight is measured. Alternatively, the sample can be rapidly ramped to a high temperature and held to measure the isothermal weight loss at the particular temperature. Earlier epoxy die attach materials contained solvents which were evaporated during the cure cycle. With large area die, removing the solvent from under the center of the die through the relatively thin bond line (0.003 in) was difficult and often resulted in trapped solvent. During subsequent TGA analysis, higher weight loss was measured due to the continued evaporation of trapped solvents. Today, most epoxy die attach materials do not contain volatile solvents. Figure 8.8 illustrates the weight loss for a typical epoxy during the cure cycle. The scanned and isothermal TGA plots for the cured epoxy are shown in Figs. 8.9 and 8.10. Higher temperature/longer cure cycles can be used to decrease weight loss during later high-temperature exposure. In hermetic packaging, a high-temperature (150° C) vacuum bake-out is used to remove moisture from the MCM prior to sealing the package. Moisture in the package after sealing combined with ionic contaminants can result in corrosion. The epoxy itself is a potential source of ionic contaminants. Chlorine chemistry is used in the synthesis of epoxy resins and must be "washed" from the resulting resin. The industry has aggressively addressed this issue, and very low ion concentrations have been achieved.

The bond strength of epoxy die attach materials decreases with increasing temperature. The materials are sufficiently strong to allow gold wire bonding at 150° C. At temperatures above 200° C, suf-

Table 8.6 Typical Ag-Filled Epoxy Properties and Military Requirements

Property	Typical results	MIL-STD-883C, Method 5011 supplier requirements
Pot life	2 weeks	1 hr minimum
Shelf life	1 year @ –40° C	For one-component systems, 1 yr @ –40° C
TGA weight loss @ 300° C (%)	0.2	≤1.0
Ionic impurities:		
conductivity of water extract	10 mS/cm	≤45 μS/cm
pH	5.5	4.0 ≤ pH ≤9.0
chloride	40 ppm	≤300 ppm
sodium	25 ppm	≤50 ppm
potassium	4 ppm	≤50 ppm
Corrosivity	passed	no visible corrosion on aluminized Mylar
Bond strength		
@25° C	5200 psi	870 psi
@150° C	890 psi	870 psi
@25° C after 1000 hr @ 150° C	6300 psi	870 psi
Volume resistivity (Ω-cm)		
@25° C	0.00015	≤0.005
@60° C	0.00014	≤0.005
@150° C	0.00016	≤0.005
@25° C after 1000 hr @150° C	0.000097	≤0.005
Thermal conductivity	1.87 W/m°K	≥1.5 W/m°C
Coefficient of linear thermal expansion		
Below Tg	55×10^{-6} in/in/°C	$\leq 65 \times 10^{-6}$ in/in/°C
Above Tg	150×10^{-6} in/in/°C	$\leq 300 \times 10^{-6}$ in/in/°C

ficient force can be applied to the die to shear it from the substrate allowing die removal for rework. Some epoxy materials are formulated with low glass transition temperatures to reduce the rework temperature.

The thermal and electrical conductivities of filled epoxies are one to two orders of magnitude lower than the materials used as fillers. A thin epoxy layer coats the particles. These interfaces degrade the thermal and electrical performance compared to the bulk filler.

Die attach epoxies are available in two forms: paste and film. The paste epoxies are most widely used. Paste epoxies can be applied from syringes manually or with fully automated dispensing systems. In some dispensing systems, the volume of material dispensed is controlled by a combination of dispense time and pressure. However, variations in the needle tip-to-substrate spacing results in epoxy volume variations. For better volume control, positive displacement systems are used. A cavity of controllable size is filled with epoxy and then the epoxy is forced from the cavity through the needle onto the substrate. For large area die, an X or ✻ pattern is dispensed to minimize voids in the die attach at the corners of the die. Stamping can be used for epoxy transfer. It is a high-speed process but does not have the volume control of syringe dispensing. Epoxy materials can also be screen printed onto the die bond pads. This is a high-speed process suitable for high-volume production. It is not used for low-volume production due to the setup time required for the printer.

Epoxy films or preforms are made by casting liquid epoxy on a release film and partially curing (B-staging) the epoxy. A woven glass fabric may be incorporated in the preform to improve handling and mechanical properties. The preforms may be stamped or cut to the desired size. Preforms are most commonly used for attaching substrates into packages. The preform exhibits minimum flow, minimizing the risk of shorting package pins while ensuring complete coverage under the substrate.

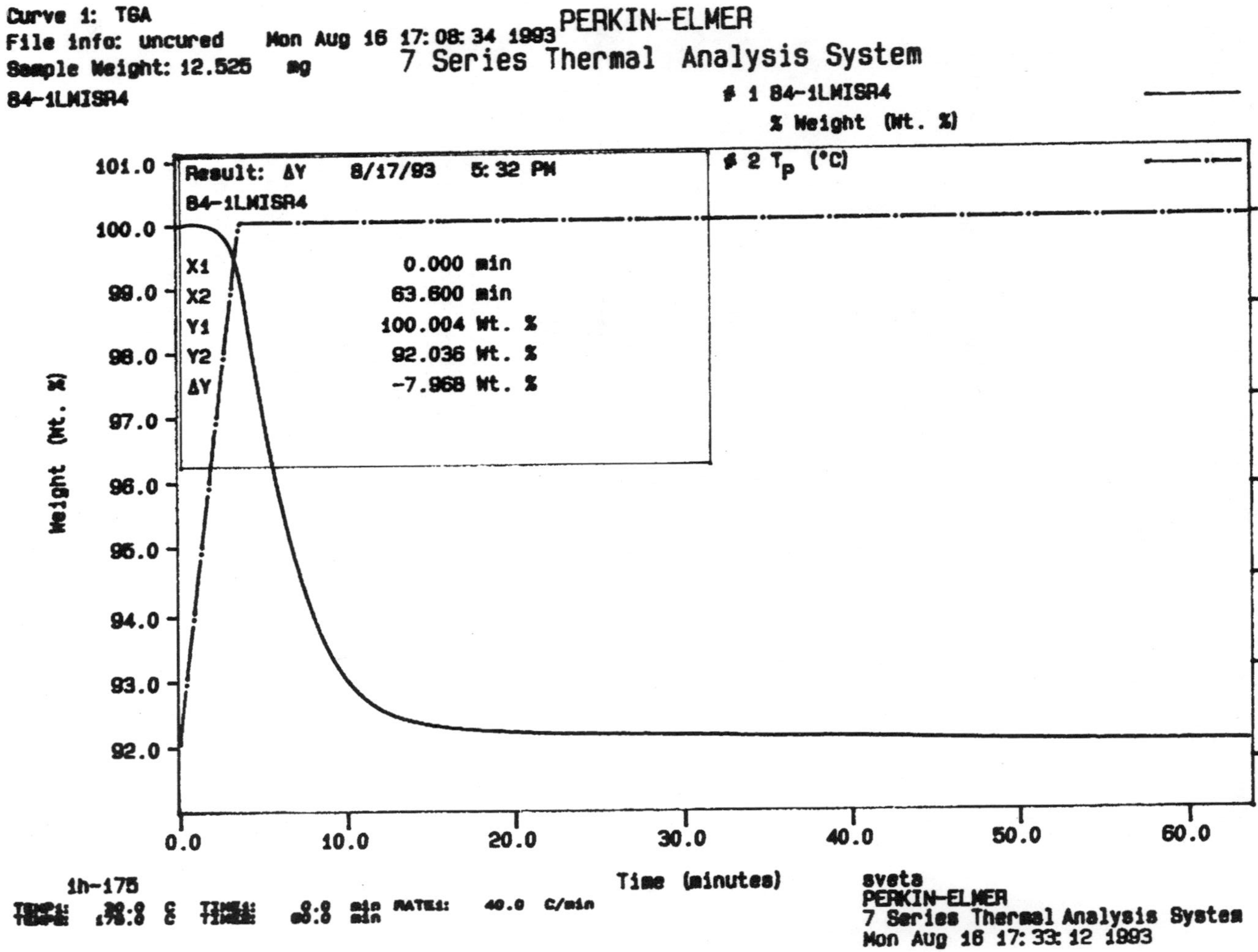

FIGURE 8.8 Thermogravimetric analysis of an uncured die attach epoxy. Courtesy of Ablestik Laboratories.

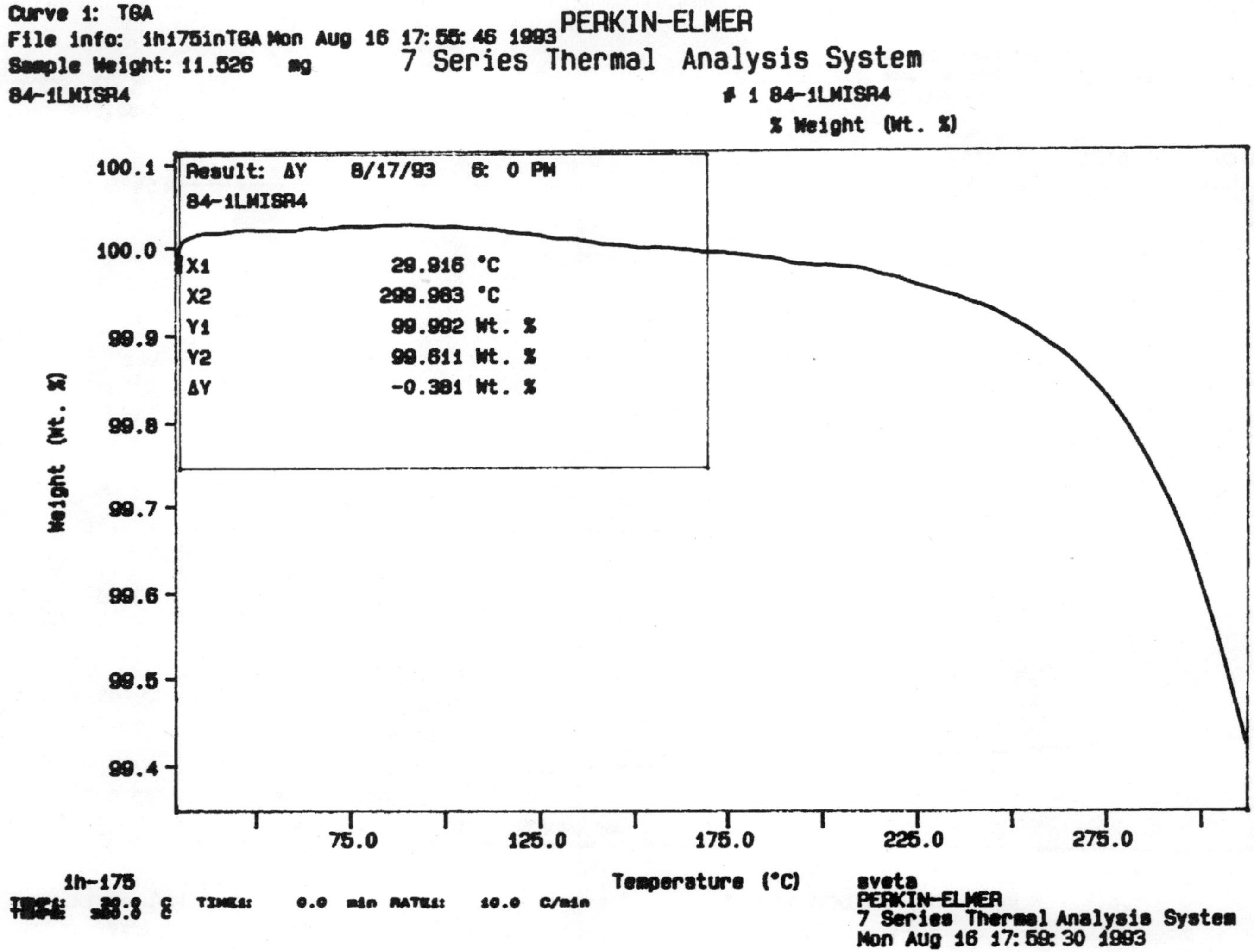

FIGURE 8.9 Thermogravimetric analysis (scan 10° C/min) of a cured die attach epoxy. Courtesy of Ablestik Laboratories.

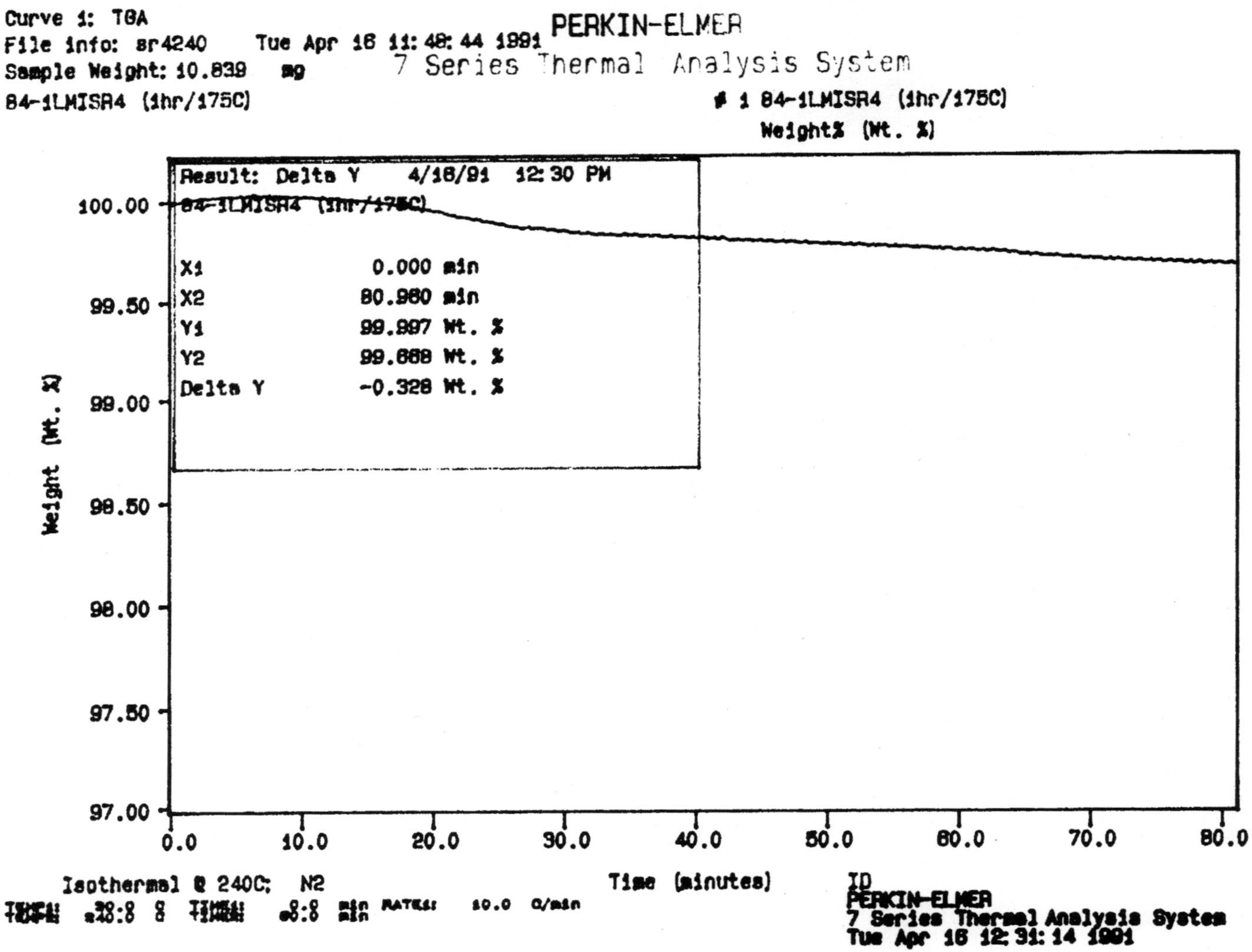

FIGURE 8.10 Thermogravimetric analysis (isothermal) of die attach epoxy. Courtesy of Ablestik Laboratories.

The preforms may be either tacky or tack-free. Tacky films have the advantage of holding the component in place but are more difficult to handle. To assemble tack-free films, the package is preheated (generally to ≈80° C), and the preform is placed in the bottom of the package. The preform softens, and the substrate is pushed into the preform under pressure. In an alternative process, the package, tack-free preform, and substrate are stacked at room temperature. A small amount of pressure is applied and maintained during the cure cycle. The same process can be used for die attach with epoxy preforms.

Die attach epoxies are thermally cured. Traditional epoxies, both pastes and preforms, require 30–60 min at 150° C. For in-line, high-volume production, shorter cure cycles are preferred. "Snap-cure" epoxies are available with cure times ranging from 1 to 5 min at temperatures ranging from 200 to 140° C.[8] As the temperature of the epoxy is increased during cure, two competing phenomena occur. The increased temperature decreases the viscosity of the polymer, while the chemical cross-linking increases the viscosity. In the case of epoxy paste, low molecular weight polymer chains may flow during this time. If this "bleed-out" flows onto wire bond pads, bondability will be degraded. Epoxy manufacturers have formulated their products to minimize bleed-out.

8.4.2 Polyimides

In general, epoxies should not be exposed to temperatures above 200° C. If higher-temperature exposure is required, polyimide die attach materials can be used. Polyimide die attach adhesives are polyamic acid dissolved in a solvent and loaded with Au or Ag fillers. Alternatively, modified thermoplastic polyimides are available and will be discussed in the following section. During cure, the solvent evaporates, and the polyamic acid imidizes through a condensation reaction, liberating water. High (250–300° C) cure temperatures are required. As previously discussed, solvent evaporation from under large chips is a processing issue. The generation of water during cure is also a concern. Polyamic esters do not release water during cure and could be considered for die attach. While polyimide is a high-temperature stable polymer (>400° C), the addition of Ag reduces the high temperature stability.[9–11] The Ag is thought to be a catalyst for the decomposition of the polyimide. Some polyimide die attach materials can be exposed to temperatures in excess of 350° C for subsequent assembly operations and can be used in continuous service at 250° C.

8.4.3 Thermoplastics

Thermoplastic materials are long-chain, linear polymers that soften on heating.[12] Thermoplastic die attach materials may be supplied as a solvent-containing paste or a solid preform. To avoid solvent entrapment under large die, the paste can be dispensed and dried prior to die placement. The substrate is then heated above the softening point of the thermoplastic, and the die are "pushed" into the thermoplastic. The amount of placement force is inversely related to the placement temperature increment above the softening point. Local heating is used for placement. It is important that the thermoplastic be uniformly heated to ensure that the die pushes into the material evenly so the die surface is parallel to the substrate for wire bonding. When cooled, the thermoplastic hardens, holding the die. There is no "cure," because the thermoplastic does not cross-link. Minimal hold time is required at high temperature, making this a high-speed assembly process. If the die is placed directly in the undried paste, longer times are required to evaporate the solvent. The primary advantage of thermoplastic die attach is the ability to reheat and remove devices for rework. Thermoplastic die attach materials typically have a low modulus resulting in lower die stress levels. Loaded (Ag, Au, Al_2O_3, AlN, and so on) thermoplastic materials available include polysiloxaneimide, polyether-amide-imide, and polysulfone. Die placement and rework temperatures depend on the specific thermoplastic used.

8.4.4 Solder

Solder die attach materials provide higher thermal conductivity than organic-based adhesives for high power dissipation applications. A number of solder alloys are available, providing a range of process temperatures. Common die attach solders are listed in Table 8.7. While available in paste form, solder preforms are most commonly used. The reflow temperatures are typically 40–60° C above the liquidus temperature. For alloys reflowed at temperatures above 350° C, forming gas (95% N_2/5% H_2) atmospheres are used. At these temperatures, the H_2 reacts with the surface oxides, removing them to allow metal-to-metal wetting. For lower temperature soldering, rosin, mildly activated (RMA) fluxes are used. Solder pastes contain RMA flux. The primary concern with flux is the potential for entrapment under the die.

Die for solder attachment require appropriate backside metallizations. For Pb/Sn-based solder alloys, a composite Ti-Ni-Ag metallization is common. The Ti provides adhesion to the Si, Ni is a solderable metal, and Ag prevents oxidation of the nickel. Ti-Ni-Au backside metallization is used with Au-Sn and Au-Ge solders. Au and Si form a eutectic alloy. In principle, silicon die with a bare silicon backside can be soldered with Au-Si eutectic solder. However, a layer of Au is usually deposited on the die backside during wafer fabrication to prevent oxidation of the Si, which inhibits wetting. Mechanical scrubbing of the die during soldering improves wetting. Process and material control is important in solder assembly to minimize voids in the solder.

Solders can fatigue and crack under the cyclic strains due to thermal or power cycling. This must be considered in designing an MCM with solder die attach.

8.4.5 Other Die Attach Materials

Loaded silicone adhesives have been developed for die attachment. Silicones can range from gels with extremely low modulus to rigid silicone rubbers. Silicone adhesives have been formulated for die attachment with excellent flexibility at temperatures as low as –60° C.[13] These materials are used when a large mismatch in CTE exists.

Low-stress cyanate ester adhesives have recently been developed for large-area die attach.[14] The modulus of the material is similar to the modified low-stress epoxies and silicone modified polyimides. Cyanate esters have higher temperature stability (300° C) than epoxies.

8.5 WIRE BONDING

Wire bonding is the dominant method for making the electrical connection between the pads on the top of the semiconductor die and the MCM substrate metallization. In the wire bonding process, the

Table 8.7 Solder Die Attach Alloys

COMPOSITION	LIQUIDUS (° C)	SOLIDUS (° C)
62.5Sn, 36.1Pb, 1.4Ag	179 eutectic	179 eutectic
95Sn, 5Pb	222	183
80Au, 20Sn	280 eutectic	280 eutectic
90Pb, 10Sn	302	275
97.5Pb, 2.5Ag	303 eutectic	303 eutectic
95Pb, 5Sn	312	308
95Pb, 5In	313	300
96.8Au, 3.2Si	363 eutectic	363 eutectic
88Au, 12Ge	356 eutectic	356 eutectic

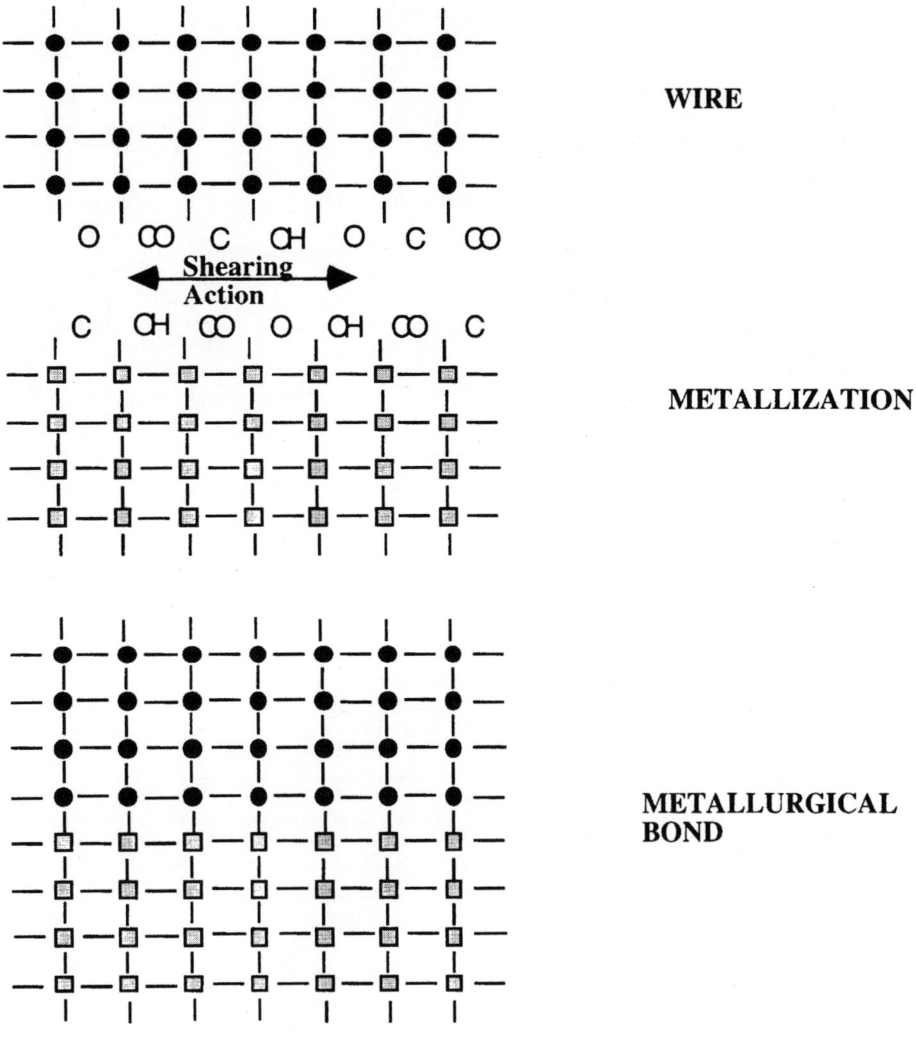

FIGURE 8.11 Atomic view of wire bonding.

wire is brought into contact with the metallized pad. The wire and pad metallization are deformed, producing a shearing action at the interface. The shearing action removes contaminants from the wire and the bond pad at the atomic level. When sufficient cleaning has occurred, the exposed surface atoms from the wire and the bond pad metallurgically bond. The concept is illustrated in Fig. 8.11. Thermosonic bonding of gold wire combines heat, pressure, and ultrasonic energy to produce the required deformation, while ultrasonic bonding of aluminum wire only requires ultrasonic energy and pressure.

8.5.1 Thermosonic Wire Bonding

The thermosonic gold wire bonding sequence is shown in Fig. 8.12. The ball is formed at the end of the wire by I^2R heating of the wire end by the discharge of a high voltage capacitor through the wire

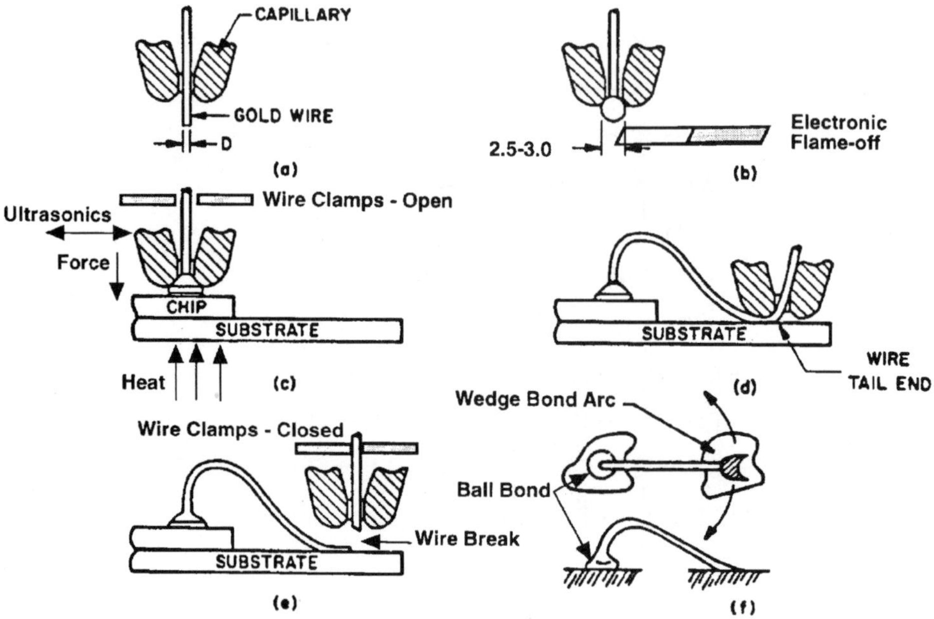

FIGURE 8.12 Thermosonic wire bonding sequence: (*a*) Gold wire is fed through a tungsten carbide capillary. (*b*) Ball is formed by capacitor discharge through wire and electrode that passes under wire. (*c*) Ball bond is made by application of force, ultrasonic energy, and heat. (*d*) With wire clamps, open capillary moves to second bond site, and wedge bond is made. (*e*) After second bond, the capillary rises, and a small amount of wire feeds out before the clamps close. When the clamps close with the capillary still rising, the wire breaks behind the second bond, and the sequence begins again. (*f*) Profile of wire bond and illustration of wire deformation during bonding.

and an electrode which passes with a small gap under the wire. The gap creates the high resistance point in the circuit, and sufficient heating occurs to melt the end of the gold wire (Au melting point = 1063° C). The size of the ball is controlled by the voltage, gap spacing, and wire tail length. Gold does not tend to oxidize and readily forms a ball to minimize surface area. Aluminum ball bonders have been developed using an argon shroud around the ball forming area to prevent oxidation. However, thermosonic aluminum ball bonding has found only limited use. In the next step in the bonding sequence, the ball is positioned over the bond pad and then the ball is brought in contact with the pad under a controlled force. Simultaneously, ultrasonic energy is applied from a transducer horn and coupled through the bonding capillary to the ball. The ultrasonic energy aids in the plastic deformation of the ball. As deformation occurs, there is a relative shearing action at the interface between the ball and the pad metallization. This shearing action cleans contaminants from the two surfaces, and metallurgical bonding occurs across the interface. Gold wire bonding requires substrate heating (125–200° C). This aids in the desorption of gases from the metallization surface. Substrate materials must be chosen that are compatible with the bonding temperature. It is important that both the wire and the pad metallization have approximately the same hardness, so the deformation and shearing occurs on both surfaces. Surface cleanliness is also important, because the amount of contamination that can be removed in the gold wire bonding process is limited. After the first bond is made, the capillary is moved to the second bonding position. During this movement, wire pulls through the capillary. Since the wire travels through the center of the capillary, the capillary can move in any direction. This is a primary advantage of thermosonic gold wire bonding in that it increases the bonding speed. Once in second bond position, ultrasonic energy and force are again applied to deform the wire forming the second bond. The capillary is designed to severely deform the wire at the back of the second bond. After the second bond is made, the capillary starts to rise. At a predetermined height, wire clamps close. The wire is no longer allowed to feed through the capillary, which is continuing to rise. As a result, the wire breaks just behind the second bond. The electronic flame-off electrode then travels under the wire tail, which protrudes from the capillary melting the end of the gold wire to form the gold ball, and the process starts all over again. Figure 8.13 shows a close-up of the bonding capillary and the electronic flame-off electrode. Thermosonic gold wire bonds are shown in Fig. 8.14.

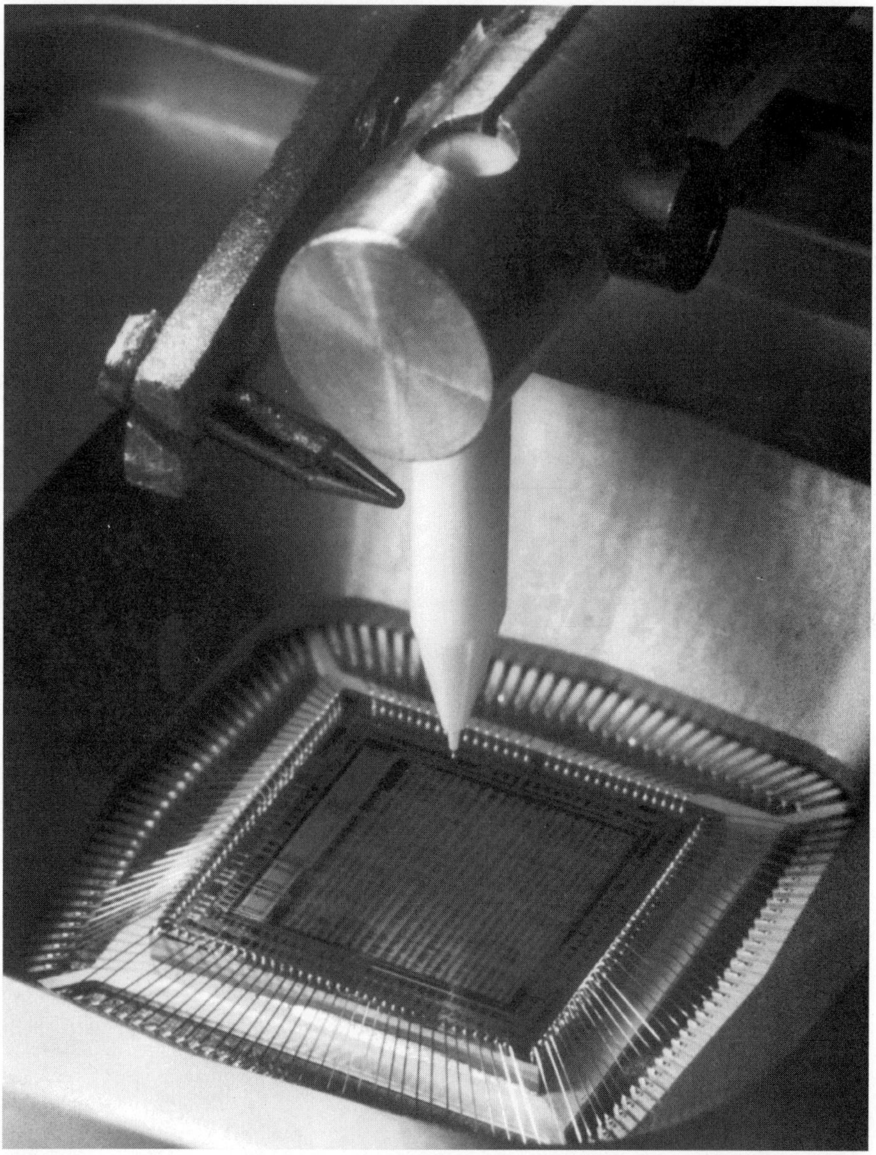

FIGURE 8.13 Close-up of thermosonic wire bonding capillary. Courtesy of Kulicke and Soffa Industries, Inc. (K&S).

FIGURE 8.14 Close-up of thermosonic gold wire bonds. Courtesy of Kulicke and Soffa Industries, Inc. (K&S).

The advantage of thermosonic gold wire bonding is speed and the fatigue resistance of gold wire. Automatic thermosonic wire bonders can bond 6–10 wires per second. The ball diameter and the geometry of the capillary limits the semiconductor bond pad pitch. As illustrated in Fig. 8.15, the pad pitch is limited by the capillary contacting a previously bonded wire.[15] Special capillary designs have been developed for fine-pitch bonding. Increased I/O density can be achieved by staggering two rows of perimeter I/O pads as illustrated in Fig. 8.16.

8.5.2 Ultrasonic Wire Bonding

Ultrasonic wire bonding is illustrated in Fig. 8.17. No ball is formed in this process which can be used with both aluminum and gold wire. The bonding sequence is similar to thermosonic wire bonding. Ultrasonic energy and force are used to provide deformation and relative shearing action between the wire and the bond pad. A significant difference is the requirement that the wire be maintained under the foot of the bonding tool. Therefore, after the first bond, the bonding tool can only move backwards to the second bond position. Since pads are located on all sides of the die, either the entire bonding head or the substrate must be rotated. The movement and stabilization of this mass reduces the overall cycle time for automatic ultrasonic wire bonding to 2–4 wires per second. In the case of aluminum wire, no substrate heat is required. Aluminum wire has a thin native oxide passivation on its surface. During the initial deformation, this oxide breaks-up and serves as an abrasive to help clean the metallization surface. Aluminum wire can be used on more contaminated surfaces than gold wire. Ultrasonic gold wire bonding requires substrate heating. While ultrasonic wire bonding is slower than thermosonic bonding, the geometry of the bonding tool allows finer die pad pitches.[16] Ultrasonic aluminum wire bonds are shown in Fig. 8.18.

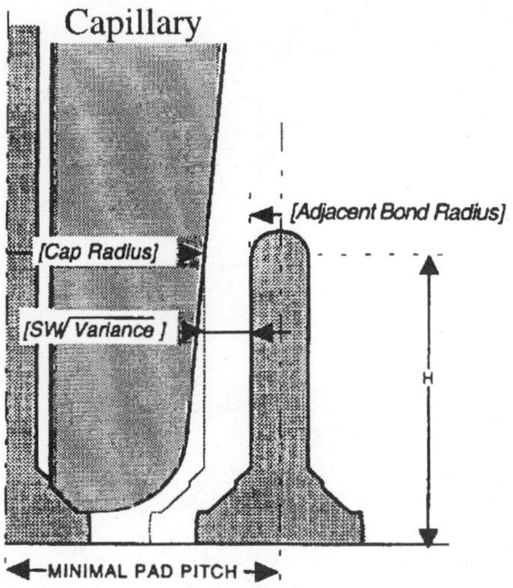

FIGURE 8.15 Minimum pitch between I/O pads is determined by wire and capillary spacing as well as tolerances on positioning accuracy.[15]

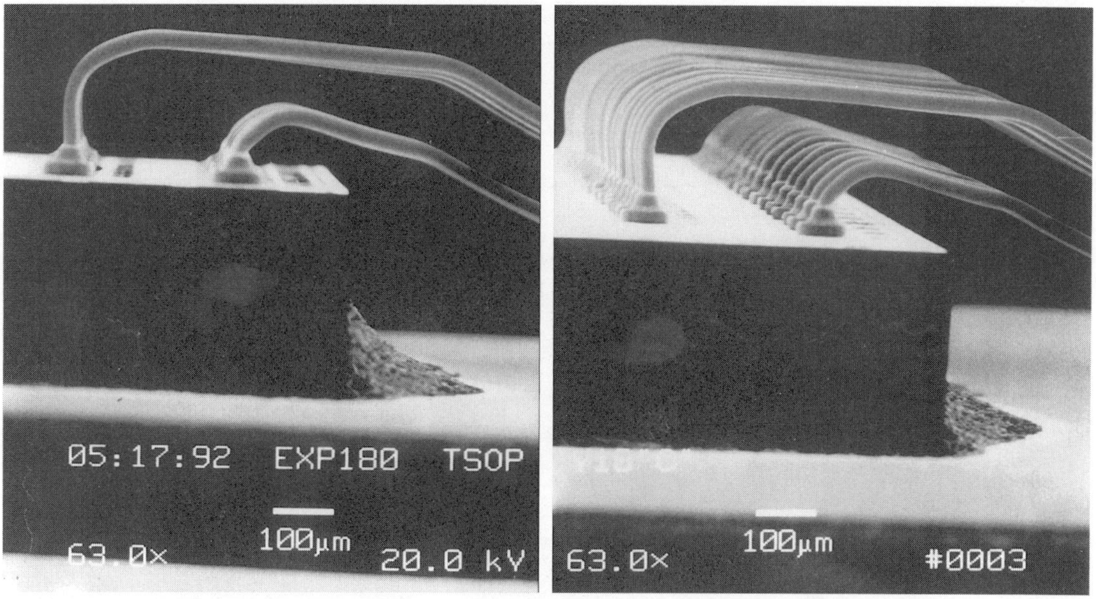

FIGURE 8.16 Photo of looping profile for two rows of perimeter I/O pads. Courtesy of Kulicke and Soffa Industries, Inc. (K&S).

8.5.3 Wire Bond Quality

To produce good wire bonds, deformation is required. Figure 8.19 illustrates the relationship between bond strength and percent deformation. If too little deformation occurs, insufficient metallurgically bonded area is formed. During pull testing, illustrated in Fig. 8.20, bond lifts (separation of the wire from the pad) occur. If too much deformation occurs, then the transition region (bond heel)

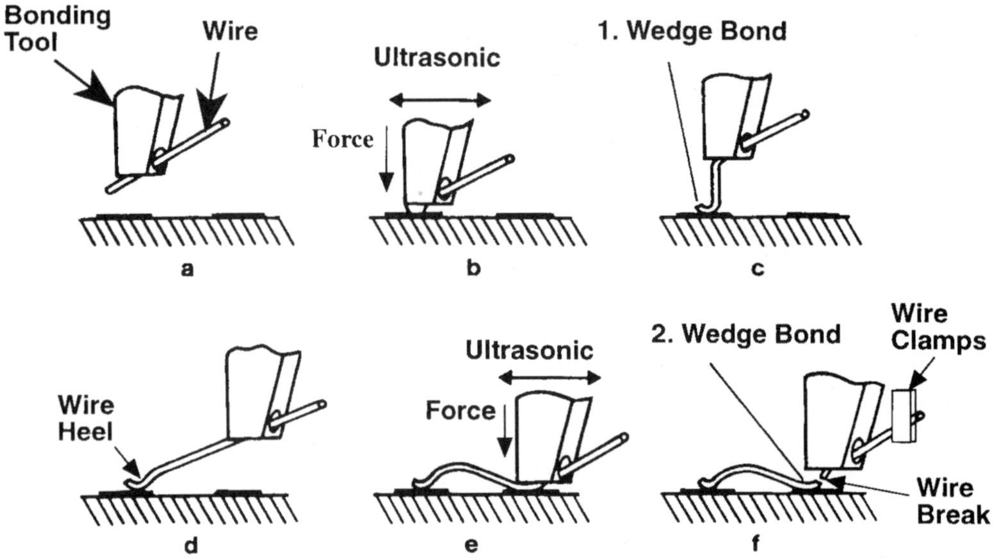

FIGURE 8.17 Ultrasonic wire bonding sequence: (a) The wire feeds through a hole in the back of the bonding tool and must be maintained under the front portion of the tool. The tool is positioned over the first bond site. (b) The first wedge bond is made using force and ultrasonic energy. (c) The bonding tool moves to the second bond. The direction of motion must maintain the wire under the front portion of the bonding tool. (d) The tool is positioned over the second bond site. (e) The second wedge bond is made. (f) The bonding tool rises, and the wire clamps close breaking the wire behind the second bond.

FIGURE 8.18 Photo of ultrasonic aluminum wire bonds. Courtesy of nChip.

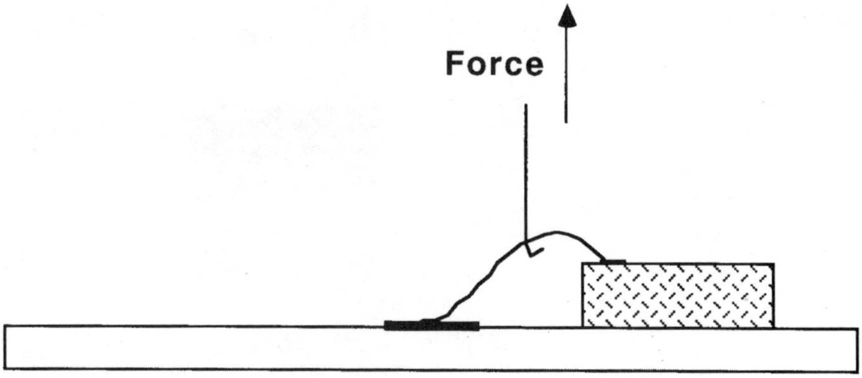

FIGURE 8.19 Illustration of wire bond pull test.

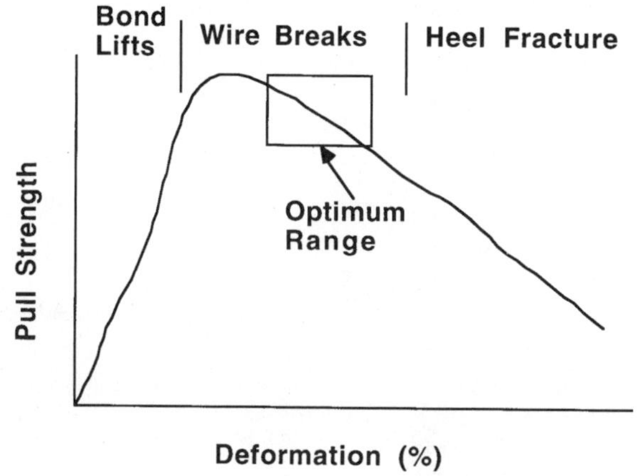

FIGURE 8.20 Wire bond pull strength vs. percent deformation.

from round wire to flattened bond is too abrupt, and cracks may develop. The heel region becomes the weak segment in the connection, and wire pull failures occur in the heel. Wire bond pull testing identifies only the weakest point in the connection. In particular, it does not adequately test the ball bond interface, because the ball bond has a relatively large area. Due to recrystallization of the wire just above the ball during the ball formation process, this is typically the weakest point in the wire, and failures occur here during the pull test. A bond shear test, illustrated in Fig. 8.21, is often used with both ball and wedge bonds to evaluate the strength of the actual bond interface. Bond pull and bond shear tests are used for initial process qualification at the beginning of a production run. These tests are used for requalification during the run if a change such as new capillary, new wire, new die wafer lot, or new substrate lot is introduced. Tests are regularly scheduled during the run for statistical process control.

8.5.4 Wire Bonding Equipment

Wire bonding equipment ranges from manual machines to fully automatic. The manual machines allow the force, time, and ultrasonic power parameters to be set independently for the first and the

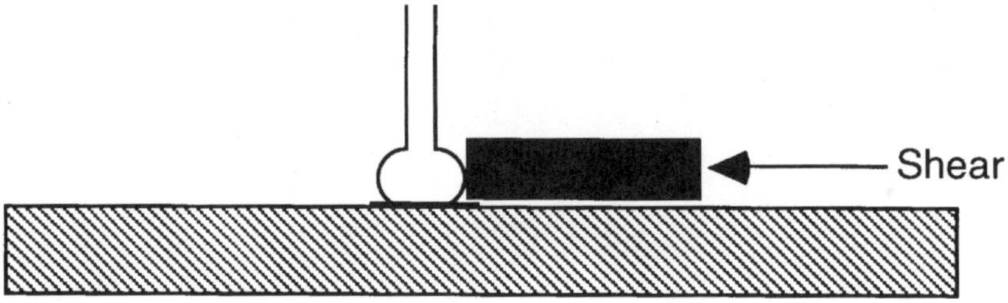

FIGURE 8.21 Illustration of wire bond shear test.

second bond. The substrate temperature (if required) is also set. The operator is responsible for substrate handling, bond positioning, and actuation of the first and second bond cycles. The looping profile, which is critical for low-profile products and for fine pitch bonding, is controlled by the operator. Manual machines are suitable for prototype production, rework, and low wire-count MCMs.

Substrate handling systems for automatic wire bonders range from operator-loaded stages to cassette-to-cassette systems for volume batch processing to rail-based handlers for in-line, high-volume production. Grayscale pattern recognition systems are used to locate reference points on the substrate and the die. X, Y, and rotational corrections are applied to the bond position coordinate data file. The initial bond coordinate file may be created by either downloading from the MCM computer-aided-design file or by manually "walking" the bonder through the bond sequence. Once programmed, the system does not require operator intervention, under normal operation. The force, time, and ultrasonic power can be individually set for each bond. This is useful, because the metallization on the die from different manufacturers may vary, and the MCM typically has die from different vendors. The bond loop profile is also programmed.

8.5.5 Bonding Wire

Bonding wire is produced by drawing the selected metal through a progress of dies of ever decreasing hole diameter. In is important that the drawing lubricants are removed from the wire after drawing and that the wire has not been nicked in the drawing process. Aluminum does not have the mechanical strength to be draw down to a diameter of 0.001 in. One percent silicon (most commonly used) or one-half percent magnesium is added to the aluminum to harden it, allowing the fabrication of 0.001-in wire. For high-current applications, 0.005–0.020-in pure aluminum wire is produced. Trace dopants are typically added to gold wire to stabilize it. Bonding wire should be used within one year of its production, as its properties change with time.

Significant work hardening occurs during the drawing process. Depending on hardness requirements, the wire can be supplied as-drawn, stress-relieved, or annealed. Heat treatments are used to stress relieve or anneal the wire. Typical wire properties are listed in Table 8.8. Gold wire is more fatigue resistant than is aluminum.

8.5.6 Substrate Metallization and Reliability Considerations

Gold is the most common substrate metallization finish for both aluminum and gold wire bonding. The thick film gold conductors used in MCM-C substrates is compatible with gold wire bonding. For aluminum wire bonding, palladium doped gold inks are used to prevent failures due to Kirkendall voiding with high temperature exposure. Differential diffusion rates in a metal couple (for example, Au and Al) result in vacancies in the rapidly diffusing material (Au). Kirkendall voiding is the coalescing of the vacancies to form voids.

Aluminum metallization is used in MCM-D substrates and is compatible with both gold and aluminum wire bonding. It is important to recognize that gold and aluminum form a series of interme-

Table 8.8 Typical Properties of Bonding Wire

	99.99% Au Hard 25 μm	99.99% Au Annealed 25 μm	Al–1%Si Hard 25 μm	Al–1%Si Annealed 25 μm
Tensile strength (g)	11–12	9–10	17–19	13–15
Elongation (%)	1–3	4–8	1–4	1–4
Burnout current (A)	0.55	0.55	0.48	0.48
Electrical resistivity (μΩ-cm)	2.4	2.4	3.1	3.1

tallic compounds: Au_5Al_2—tan, Au_2Al—tan, $AuAl_2$—purple, AuAl—white, and Au_4Al—tan. These intermetallics are electrically conductive and mechanically strong. However, they are brittle. The intermetallics are formed at elevated temperature, and care should be taken in the assembly process to limit high-temperature exposure after wire bonding. Table 8.9 presents the failure modes associated with intermetallic formation.[17] Failures are typically related to void formation. Intermetallic volumes differ from the sum volume of the two constituent elements. If sufficient intermetallic formation occurs, voids can result. By keeping metallization thin, only a limited amount of intermetallic can form. Aluminum metallization is common on MCM-D substrates.

Copper metallization is used in both MCM-L and MCM-D substrates. Copper is not directly bondable with either gold or aluminum wire. Composite Cu-Ni-Au substrate bond pads are used (Fig. 8.22). The Ni barrier layer is required since Cu rapidly diffuses through gold and will oxidize on the gold surface limiting wire bondability. For bonding with aluminum wire, the gold layer only needs to be 0.1 to 0.3 μm thick, with an absolute maximum of 1 μm to eliminate the potential of void formation related failure mechanisms. For gold wire bonding, the gold thickness is generally specified at 1–1.5 μm. However, recent results with 0.5 μm gold have been successful. The purity of the gold is critical for bondability. The gold layer should be undoped, soft, gold with trace impurities in the very low parts per million range. Impurities make the gold hard and unbondable. Vacuum-deposited gold tends to be of high purity, but the purity of plated gold varies widely and requires careful control of the plating bath.

Wire bonding to metal pads on polymer dielectrics (MCM-D and MCM-L) is more challenging than on ceramic dielectrics (MCM-C). The polymer dielectrics are relatively soft. The bonding pad may deflect under the force, heat, and ultrasonic energy associated with the bonding process. One or more steps can be taken to improve bondability on "soft substrates" by making the wire bond pad

Table 8.9 Failure Modes Associated with Intermetallic Formation[17]

Symptom	Cause	Remedy
Open metallization around bonding pad Zero pull strength, bond peeled off pad, fracture surface purple	Voiding in $AuAl_2$	Keep circuit below 400° C
Zero pull strength, bond peeled off pad, fracture surface tan	Voiding in Au_5Al_2 Thermal cycling will aggravate	Make metallization thinner or reduce time at temperature
Zero pull strength, break at heel, fracture surface tan	Intermetallic formation in heel of bond fatigues during thermal cycle	Make metallization thinner, use thicker wire, or reduce time at temperature

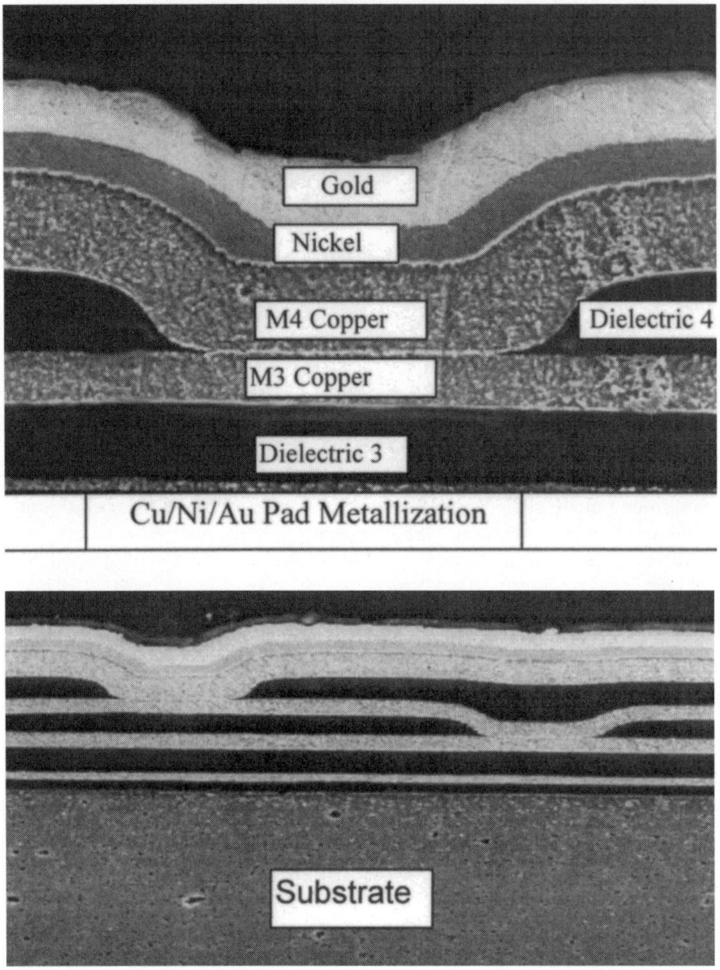

FIGURE 8.22 Typical MCM-D Cu-Ni-Au metallization for wire bond pad.

more rigid. These include increasing bond pad metal thickness and area, use of a hard metal underlayer (Ti, Cr), and/or a hard metal top layer (Ni, Pd) capped with a highly bondable metal (Au or Al).[18] Figure 8.23 illustrates the concept. Table 8.10 compares wire bonding options that have been used successfully.

If both wire bonding and Sn-Pb soldering processes are required, care must be used in the selection of the metallization finish. Gold dissolved into a Sn/Pb solder joint embrittles the joint. The volume of gold dissolved in a Sn/Pb solder joint should be maintained below 3 percent, and typically 1 to 2 percent.[19,20] If the soldering operation is performed prior to wire bonding, the nickel will diffuse into the gold layer. If the gold layer is thin to meet solder reliability requirements, the nickel may diffuse throughout the entire gold layer. The diffused nickel hardens the gold, reducing wire bondability. The formation of nickel oxide on the surface further reduces the bondability. Multiple surface finishes can be provided for the different assembly processes, but this increases substrate fabrication costs. A Ni-Co alloy diffusion layer (over thin film Cr-Cu) with a thin Au top layer has been demonstrated to be compatible with flip chip and TAB solder assembly and wire bonding.[21] Cu-Ni-Pd-thin

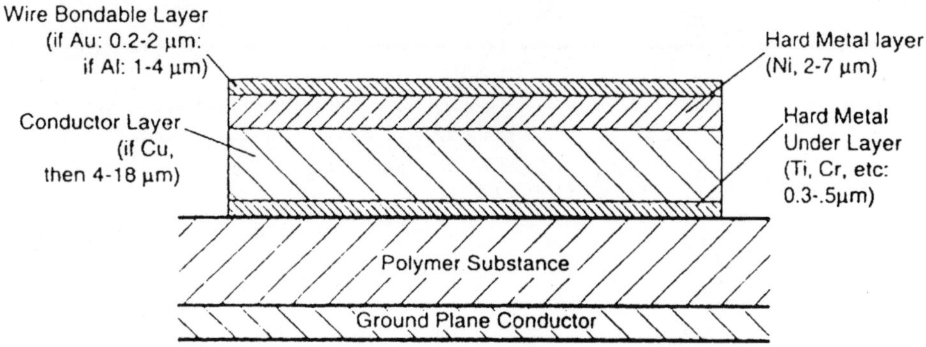

FIGURE 8.23 Structure of a generic, stiffened-metal bond pad and polymer substrate for wire bonding to MCMs.[19]

Au composite bond pads are also gold wire bondable and are compatible with soldering if mixed assembly processes are required.

8.5.7 Plasma Cleaning

As discussed, wire bonding requires a relatively clean surface. Plasma etching has been shown to improve both wire bondability and reliability.[22,23] Condensation of polymer die attach outgassing by-products is a common contaminant. Both oxygen and argon plasmas are used to clean the bonding surfaces. The contamination is generally carbon based and reacts with the monatomic oxygen created in the oxygen plasma to form volatile species (CO, CO_2, H_2O, and so forth) which are exhausted through the vacuum system. The problem with oxygen plasmas are the oxidation of the silver in the filled die attach materials. The resulting dark appearance of the silver oxide is visually unacceptable. Argon plasmas do not chemically react with the surfaces. Instead, the surfaces are sputter cleaned by bombardment from the argon ions. There has been some concern over exposing CMOS circuits to the ion bombardment. As a compromise, a mixture of oxygen and argon (98%Ar/ 2%O_2) is commonly used. In-line, batch processed plasma etching systems are available for high-volume production.

8.5.8 Rework

If it becomes necessary to remove a die after it has been wire bonded, the wires must be removed first. A manual process is to remove the wires with tweezers. IBM has developed wire bond rework equipment that directs jets of dry air or nitrogen to alternately flex the wires side to side.[24] This fatigues the wires in the heel regions and above the ball. The wire ultimately fractures and fails. All wire pieces must be removed to avoid potential shorts. Once the wires are removed, the die can be removed as discussed for the various die attach materials. If a rework option is planned, the bond pads on the substrate should be designed with sufficient space to allow placement of two bonds, because the original bond remains on the substrate pad after rework.

8.5.9 Wire Bond Summary

Wire bonding continues to be the dominant method of MCM assembly. It does not require additional wafer processing. Thus, more die are available for wire bond assembly. The process is well understood, and the reliability has been established by the single-chip packaging industry. The limitations of wire bonding are the parasitic inductance of the wire, the requirement for perimeter pads on the

Table 8.10 Comparison of Recommendations for MCM Wire Bonding[18]

Polymer, Thickness	Substrate	Bond Method	Metals, Thickness	Wire dia., Type	Power	Temp.	Time	Force
PI, 3–6 µm	Si chip	TS, high-freq.	Al/Ti, 1/0.5 µm	25 µm Au, ball		290° C		
PI, 30 µm	Si chip	TS, 60 kHz	Al/Ti 3/0.5 µm	35 µm Au, ball	120 mW	180° C		110 g
PI/13 µm + glue/5 µm	Flex PCB	US, 60 kHz	Au/Ni/Cu 2/3/17 µm	35 µm Au/Cu, wedge	1–3 W	25	360 ms	60 g
PI, 50 µm	Flex	TS, 60 kHz	Au/Ni/Cu 0.76/5/35 µm	32 µm Al, wedge/ball		200° C		
PI/50 µm + acrylic glue	Flex	US, 60 kHz	Au/Ni/Cu 1/6/18 µm	35 µm Al, wedge	130% > FR-4	25	27 ms	29 g
PI, 9 µm	Glass/ ceramic	US, 60 kHz	Au/Ni-Co/Cu 0.15/2/4/µm	32 µm Al, wedge 25 µm Au, ball		25° C 150° C		
BCB, 20 µm	Si	US, 60 kHz	Al-2%Cu 4 µm	35 µm Al, wedge		20° C		
Triazine, 50 µm rubber modified	Ceramic	TS, 60 kHz	Au/Ni/Cu/Pd/Ti 10 µm	32 [illegible] Au, wedge		120° C	50 ms	80 g
Epoxy glass, BT 0.5 mm	None	TS, 60 kHz	Au/Ni/Cu 0.5/5/17 µm	32 µm Au, B-wedge		150° C		Normal
Epoxy glass, BT 1.0 mm	None	TS, 60 kHz	Au/Ni/Cu 0.5/5/10 µm	32 µm Au, B-wedge	400 mW wedge	180	25 ms	Normal Wedge
Epoxy-aramid 100 µm	None	TS, 60 kHz	Au/Ni/Cu 0.4/3–5/18 µm	25 µm Au, ball wedge	390 mW 500 mW	165° C	55 ms 45 ms	75 g 60 g
PTFE-glass fill, 250 µm	None	TS, 60 kHz	Au/Ni/Cu 2.5/2/12 µm	10 µm Au, wedge		85° C + heated tip		28 g
PTFE-glass fill	?	TS, 60 kHz	Au/Ni/Cu 3.8/4/17 µm	18 µm Au: wedge and ball		75° C 55° C-tip		

Notes:
Most TS (ball-wedge) bonding on MCMs is ball-on-chip and wedge-on-soft substrate.
All bonder dial settings are relative. Also, (input) power given in mW is relative, due to variations in transducer efficiency.

die, and the slow, sequential nature of the process. The parasitic inductance can limit the performance of high-speed electronics. The wire inductance on power and ground pins contributes to switching or Δ-I noise. Multiple power and ground pins are used to minimize the inductance, but this consumes pads that could have be used for signal I/O. Integrated circuit designs are becoming limited by the number of perimeter I/Os. Prediction are for digital integrated circuits with over 1000 I/Os in the very near future. As the I/O count increases, wire bonder manufacturers continue to decrease the required pad pitch and have demonstrated precision loop control for staggered, dual-row I/O pads. Ultimately, the limit will be reached. In must, however, be realized that while some ICs will have over 1000 I/Os, the majority will not, and wire bonding will be used into the foreseeable future. Wire bonding speeds continue to increase to address the need for throughput in the sequential process.

8.6 TAPE AUTOMATED BONDING

The origins of tape automated bonding (TAB) can be traced to the Mini-Mod technology developed by General Electric in the mid-1960s. TAB involves first bonding the die to a polymer tape containing a copper lead pattern. A single die site is illustrated in Fig. 24. The die are bonded to the metal leads through bumps. These bumps prevent shorting of the TAB lead to the silicon surface. Once bonded, the die can be automatically handled in a tape-and-reel format. Probe pads are patterned in the copper leads to allow test and burn-in of the die prior to assembly. This is a major advantage of TAB technology. For assembly, the die are excised from the tape with copper leads protruding from the edges of the die. The leads are then formed for bonding to the substrate in either a die face up or face down configuration. The overall concept is illustrated in Fig. 8.25.

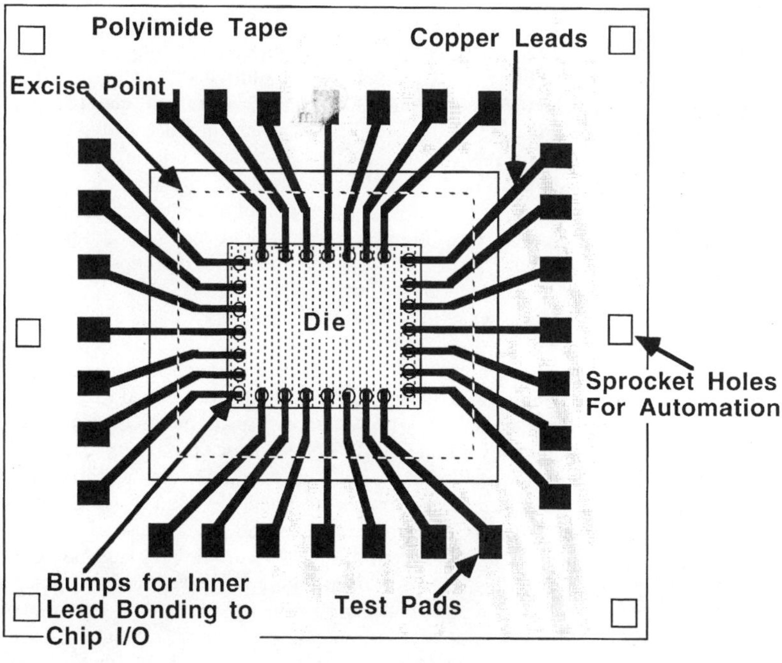

FIGURE 8.24 Illustration of die mounted to tape automated bonding (TAB) tape.

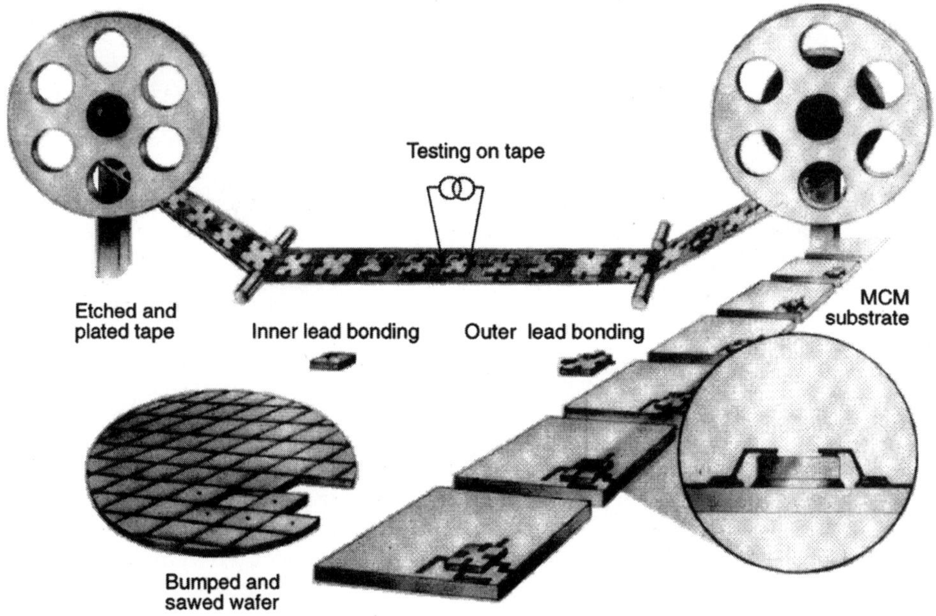

FIGURE 8.25 Illustration of TAB concept.

8.6.1 Bump Formation

Additional wafer processing steps are required to fabricate the bumps on the die. A typical process sequence is illustrated in Fig. 8.26. The interface metallurgy is first vacuum deposited onto the wafer surface. The interface metallurgy provides adhesion, serves as a diffusion barrier, and provides conductivity for the electroplating of the bump. A common interface metallurgy is titanium-tungsten/gold. The thin gold layer prevents oxidation of the Ti-W layer prior to plating. Next, a thick photoresist layer is applied and photoimaged. The bump is then electroplated, the photoresist is stripped, and the background interface layers are etched. After bumping the die, no aluminum remains exposed. Assuming no pinholes in the passivation layer, reliability concerns about potential aluminum corrosion are eliminated. Figures 8.27 and 8.28 illustrate plated gold bumps.

Gold is the most common bump metallurgy,[25] although 5%Sn-95%Pb,[26] Ni/Au,[27] and Au/Sn/Au bumps have been used. In the example of the Au/Sn/Au bump, the plated bumps are reflowed to melt the Au/Sn eutectic alloy. Figure 8.29 shows a cross section of the as-deposited Au-Sn bump, while the bump after reflow is shown in Fig. 8.30. Au/Sn eutectic and "Zeta" intermetallic phases form. A laser-assisted chemical vapor deposition process has been developed as an alternative to plating for depositing gold bumps on the wafer.[28] The system uses an argon ion laser and dimethyl gold hexafluoroacetylacetonate to deposit gold on the Al-TiW-Au device pads.

Bumped wafers are not readily available, and a number of alternative approaches have been developed. In one approach, a thermosonic wire bonder is used to bond gold and palladium-gold balls to the I/O pads.[29] After the ball bond is made, the wire clamps close, and the wire is broken above the ball. In a secondary operation, a tungsten carbide tamping tool is installed in the wire bonder, and the wire "pigtail" is tamped down. Figure 8.31 illustrates this process. Another alternative is to form the bumps on the tape. This can be done by plating the bumps onto the ends of the copper fingers on the TAB tape[30] or by a bump transfer process.[31,32] In one bump transfer process, bumps matching the pattern of the chip I/Os are electroplated plated onto a carrier substrate.[31] The carrier substrate shown in Fig. 8.32 consists of an indium tin oxide (ITO) conductive layer for elec-

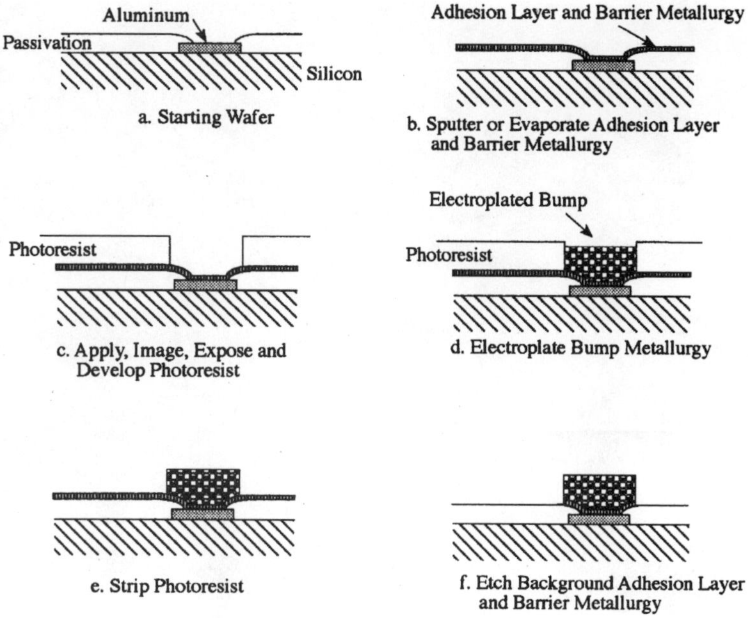

FIGURE 8.26 Bumping process for tape automated bonding.

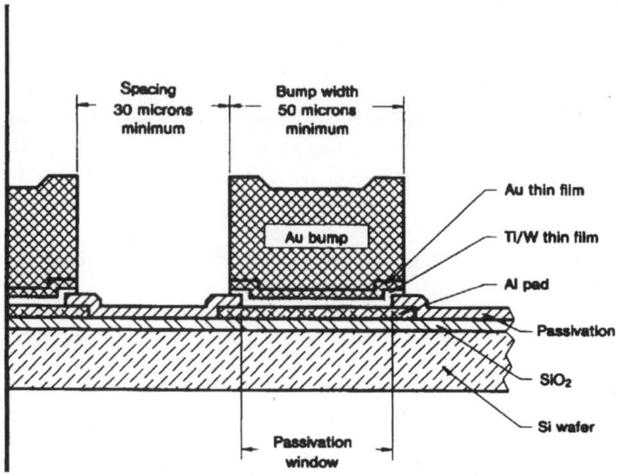

FIGURE 8.27 Illustration of bump metallurgy.

troplating and a silicon nitride/silicon dioxide mask layer on an insulating substrate. The TAB tape is then bonded to the bumps, which release from the carrier substrate, transferring from the ITO layer to the tape. In a second bump transfer process, Au balls are manufactured by melting precisely weighted gold pieces.[32] The gold balls are held in place by vacuum in a template with small holes patterned to match the die I/O pattern. The bumps then can be either bonded (transferred) to the tape or to the semiconductor die by thermocompression bonding. The process is illustrated in Fig. 8.33.

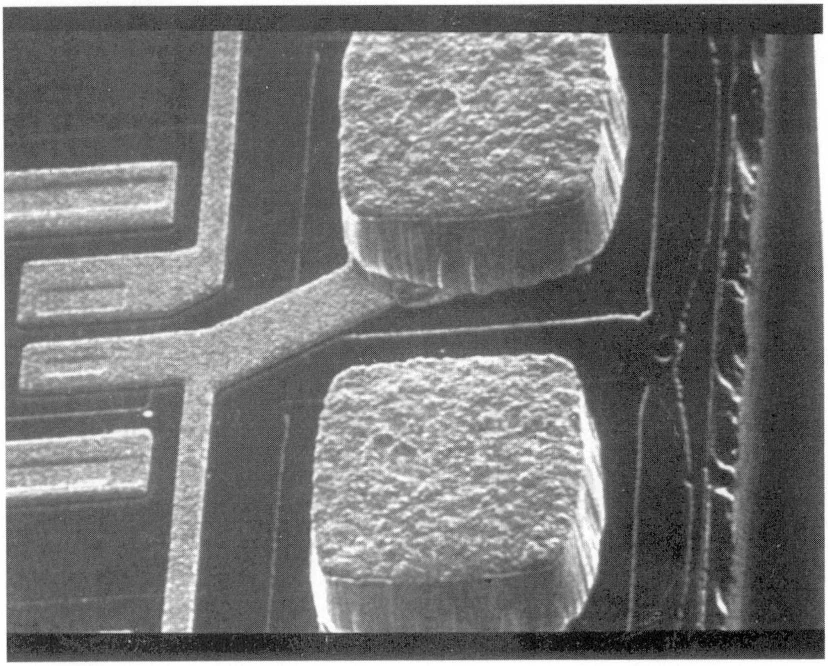

FIGURE 8.28 Photo of gold bumps on a die.

8.6.2 Inner Lead Bonding

The copper leads patterned in the tape are either Au or Sn plated, depending on the inner lead bonding (ILB) process. The original concept was to gang thermocompression (heat and force) bond all of the Au plated leads to the Au bumps simultaneously. Typical thermode temperatures are 400–500° C. As the lead count increases, the force required results in cracking of the Ti-W barrier layer and the passivation. Recent use of Ni/Au[26] bumps has been shown to eliminate the cracking. Single-point thermosonic bonding (one bond at a time) is generally used with Au plated tape and Au bumps. Thermosonic inner lead bonders are modified thermosonic wire bonders and use heat, force, and ultrasonic energy to deform the ball and the lead, producing a metallurgical bond. Single-point bonding is illustrated in Figs.8.34 and 8.35. Single-point bonding minimizes the force that must be exerted on the die and accommodates variations in bump height. Thermosonic bonding also uses lower temperatures than thermocompression. If the tape is gold bumped rather than the die, the gold bumps are either thermosonic or thermocompression bonded to the aluminum I/O pads.

Au plated tape can be gang bonded to Au/Sn/Au bumps. The reflowed eutectic Sn-Au alloy on the top surface of the bumps melts when heated by the thermode and wets to the Au plated tape. Figure 8.36 shows close-up of the lead and bump after reflow inner lead bonding. Laser soldering with a Nd:YAG laser can also be used[33] with Au plated tape and Au/Sn bumps.

Sn-plated copper tape can be gang soldered using a thermode with either 5%Sn-95%Pb bumps or Au bumps. In the case of Au bumps, the Sn melts and interdiffuses with the Au at the thermode temperature (400° C). The Au-Sn eutectic forms consuming the Sn. A single-point laser process has also been developed for heating Sn plated tape on Au bumps to form a Au-Sn eutectic bond. The process is illustrated in Figs. 8.37 and 8.38. Sn-plated tape creates some handing and reliability concerns. Sn oxidizes with time, particularly if exposed to high humidity. This limits shelf life, and flux may be required to achieve good wetting. The thickness of the Sn plating is critical. Excess Sn results in the

FIGURE 8.29 SEM photomicrograph of as-deposited Au/Sn bump cross section. Courtesy of Motorola.

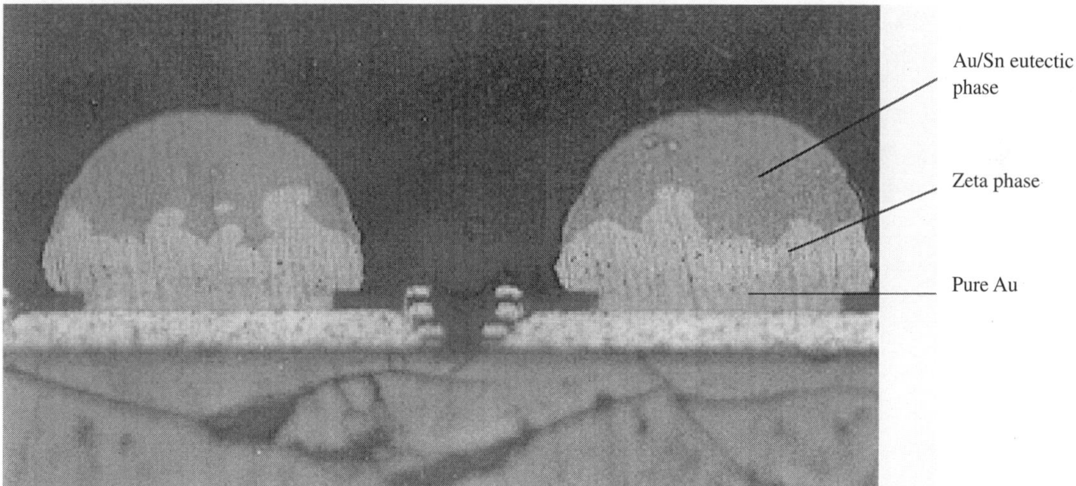

FIGURE 8.30 SEM photomicrograph of reflowed Au/Sn bump cross section showing eutectic and "zeta" phase. Courtesy of Motorola.

formation of Sn nodules, which can produce electrical shorts. Another disadvantage of Sn plating is the growth of Sn whiskers, which also can short leads. The whisker growth is a result of stresses in the plated Sn layer, and Sn whisker growth is accelerated by high temperature and humidity.[34]

Figures 8.39 and 8.40 show die inner lead bonded to tape. After inner lead bonding, a liquid epoxy is often dispensed on the die surface and then cured. This epoxy coating provides protection of the die surface and reinforces the inner lead bonds.

FIGURE 8.31 Bumps formed by adaptation of the wire bonding process. Courtesy of Kulicke and Soffa Industries, Inc. (K&S).

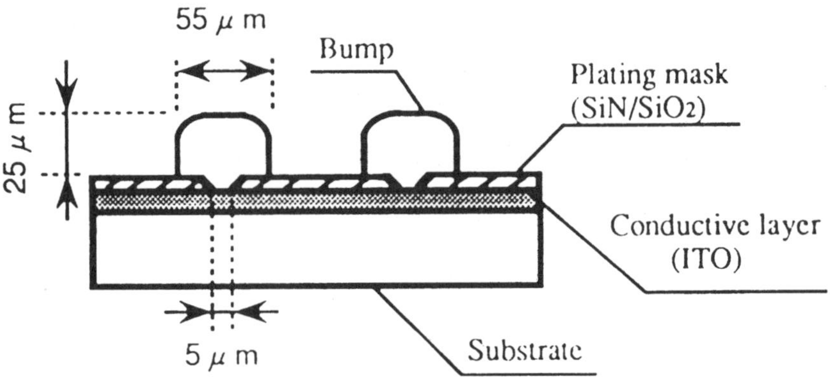

FIGURE 8.32 Structure of the substrate for transfer-bump process.[31]

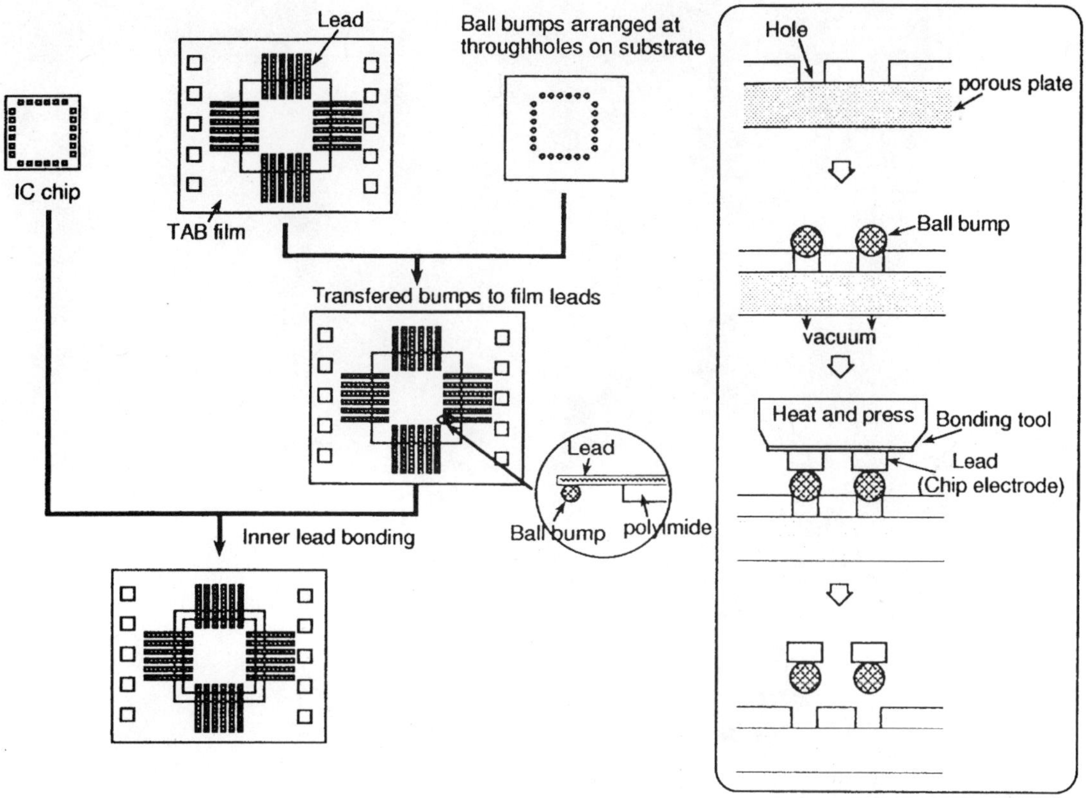

FIGURE 8.33 Tape bumping process by transferred ball bump technology.[32]

8.6.3 Testing

The ability to test and burn-in the semiconductor device prior to assembly is a major advantage of TAB. The devices can be probe tested in tape form using the relatively large test pads patterned on the tape. The devices can also be singulated and test sockets can be used for both test and burn-in.[35] Burn-in of Sn plated tape results in oxidation of the Sn, degrading solderability for outer lead bonding and Au plating is preferred.

8.6.4 Outer Lead Bonding

The first step in the outer lead bonding process is to excise the semiconductor and appropriate length of lead from the tape using a tool and die set. This is done just prior to assembly to minimize the potential for bending the fragile leads. A second tool and die set is required to form the leads for either face-up or face down bonding. The device is then ready for placement. A die attach material is often used to increase heat transfer to the substrate and to provide electrical contact (if needed) to the backside of face-up devices. Epoxy and thermoplastic die attach materials (paste and preforms) are used.

Soldering (63%Sn/37%Pb) can be used to bond both Sn and Au plated leads to the substrate metallization. The Au plating thickness is important in the reliability of soldered outer lead bonds.[36] The

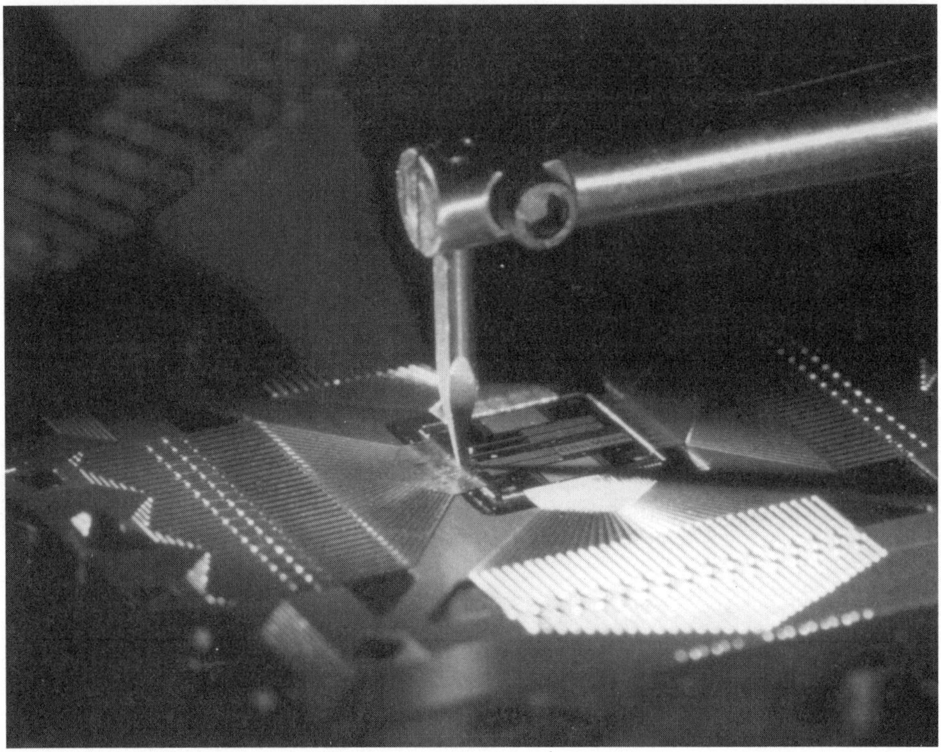

FIGURE 8.34 Single-point inner lead TAB bonding. Courtesy of Hughes Aircraft.

maximum gold plating thickness is dependent on the volume of solder in the joint. Greater than 3% gold in the solder joint has been shown to embrittle Sn/Pb solder joints.[19,20] With fine outer lead bond pitch, pad design, and solder volume are critical to avoid bridging of adjacent leads. The solder is typically pulse heated with thermodes. Four thermodes are arranged to solder all four sides of the TAB device simultaneously. The thermodes maintain the coplanarity of the thin TAB leads. After the solder melts, the power to the thermodes is cut off. The thermodes maintain force on the leads until the solder solidifies, ensuring that all leads remain in the solder. The thermodes are typically made of molybdenum, which is not wet by the solder. Flux is generally required for good solder wetting. However, IBM has developed a fluxless outer lead soldering process that uses a hot nitrogen thermode.[37] In this process, the copper tape is gold plated, and the surface oxide on the molten 63Sn/37Pb solder is pushed away by the velocity of the hot nitrogen.

Single-point thermocompression or thermosonic bonding can be used for bonding Au-plated TAB leads to Au substrate metallization. The process does not produce a liquid that can flow, bridging adjacent leads and allowing finer pitch outer lead bonding.

Figures 8.41 and 8.42 show a device tape automated bonded to a MCM-C substrate. During the formation of the TAB tape, rings of polyimide were left to maintain proper spacing between the fine-pitch TAB leads to prevent shorting.

8.6.5 Rework

TAB devices soldered to a substrate may be reworked by melting the solder and removing the device. Localized heating with thermodes or hot air allows device removal without exposing adjacent

FIGURE 8.35 Close-up of single-point inner lead TAB bonds. Courtesy of Hughes Aircraft.

devices to high temperatures. If a rework option is planned, appropriate consideration must be given to the die attach material selection, if any is used. Removal of devices assembled by thermocompression or thermosonic bonding is more difficult. The strength of the TAB lead-to-substrate metallization bond often exceeds the metallization-to-substrate adhesion. Rather than attempt to remove the entire lead, a process has been developed that leaves a portion of the TAB lead on the substrate.[38] In this process, additional single-point bonds are made along the leads in contact with the substrate metallization, as illustrated in Fig. 8.43. The leads are then cut, and the die is removed. The replacement die is attached with its leads bonded to the remnants of the previous device leads. A thermoplastic die attach material is used in this example to facilitate rework.

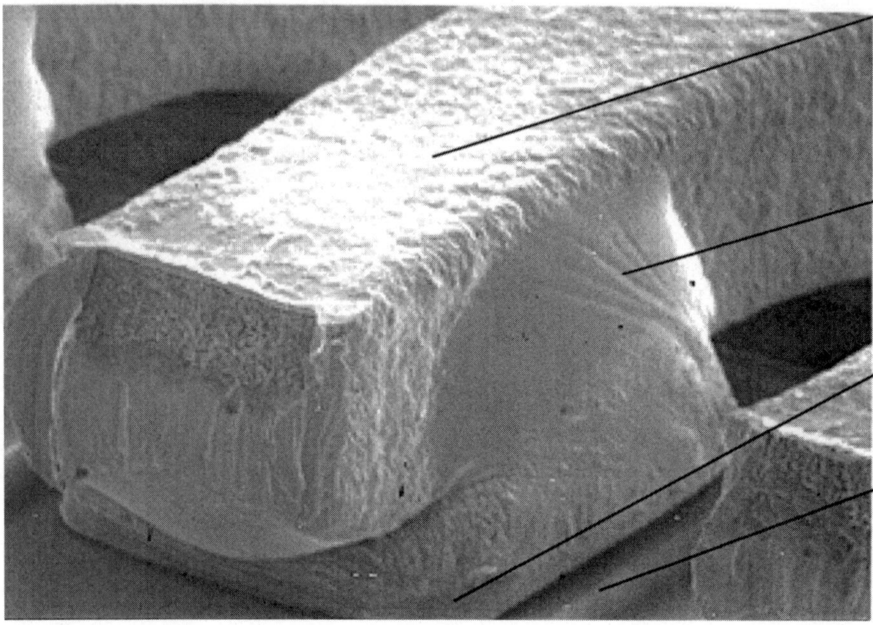

FIGURE 8.36 Close-up of the lead and bump after reflow inner lead bonding. Courtesy of Motorola.

8.6.6 Tape Automated Bonding Summary

The primary advantage of TAB is the ability to test and burn-in the semiconductor die prior to assembly. A second advantage is the sealed (no exposed aluminum) top surface of a bumped die. Multilayer TAB tape, while expensive, can be designed with a ground plane to provide controlled impedance or can be designed to access an area array of I/O die pads. The disadvantages of TAB include the need to custom design a tape for each die, the need to redesign for "die shrink," and the need for bumped die. The limited availability of bumped die has severely limited the use of TAB. Bumped tape, bump transfer, and Au wire bond ball bumps are attempts to address this limitation. Hewlett-Packard has developed a bumpless TAB process to allow the use of standard processed die.[39] In this process, the Au plated TAB inner leads are single-point thermosonically bonded directly to the aluminum pads on the die. The inner leads are formed to avoid lead shorting to the die surface and are shown in Fig. 8.44.

8.7 FLIP CHIP

Flip chip technology was first introduced by IBM in 1964 in the solid logic technology (SLT) hybrid modules in the System 360 mainframe.[40] The chips were three terminal transistors with Au/Ni plated Cu balls embedded in a Pb/Sn solder bump on the three I/O pads of the transistor. A Cr-Cu-Au interface layer was deposited between the Al transistor bond pad and the solder bump. The device was assembled to the hybrid substrate by inverting (flipping) the chip and reflow soldering the copper balls to corresponding pads on the substrate. The copper bump maintained a standoff between the die and the substrate after reflow.

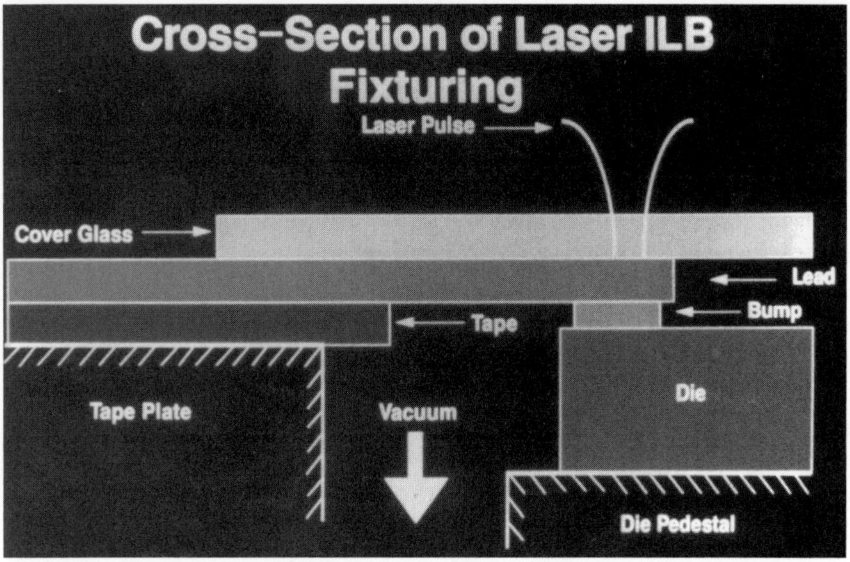

FIGURE 8.37 Setup for laser soldering of tin-plated copper lead to gold bump. Courtesy of MCC.

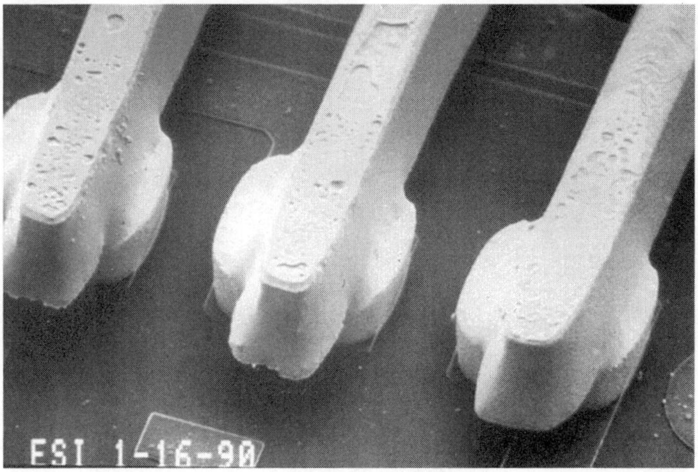

FIGURE 8.38 Tin-plated copper leads, laser soldered to gold bumps. Courtesy of MCC.

8.7.1 Bump Formation

IBM replaced the copper ball technology with a pure solder (95%Pb/5%) bump in their controlled collapse chip connection (C^4) process.[41] The flow of the solder (the collapse of the chip-to-substrate spacing) is controlled during reflow by controlling solder volume and limiting the amount of solder-wettable metal exposed on the die and the substrate. In the C^4 process, a Cr adhesion layer, a mixed Cr/Cu layer (provides resistance to multiple reflows), a Cu solderable layer, a Au flash, and 95%Pb/

FIGURE 8.39 Die after inner lead bonding. Courtesy of IMI.

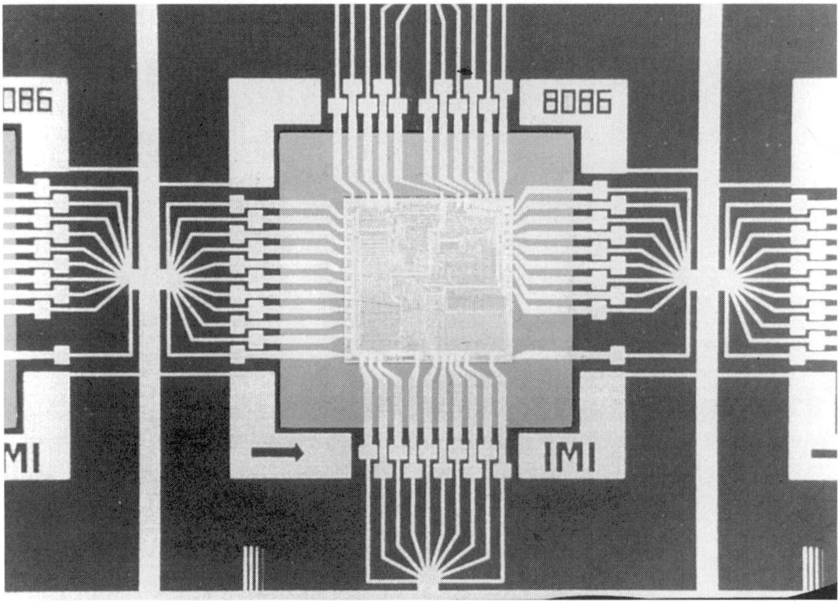

FIGURE 8.40 Die in tape after inner lead bonding. Courtesy of IMI.

5%Sn solder layer are evaporated through a shadow mask onto the semiconductor wafer. Due to the higher vapor pressure of Pb compared to Sn, the Pb deposits first, followed by the Sn resulting in a solder bump, which is segregated with Pb at the bottom and Sn on top. After evaporation, the wafer is reflowed at 350° C in an H_2 ambient to homogenize the PbSn layer. The reflow produces nearly spherical solder bumps. Die with solder bumps are shown in Figs. 8.45 and 8.46.

FIGURE 8.41 Example of TAB bonds. Courtesy of Hughes Aircraft.

FIGURE 8.42 Close-up of TAB bonds. Courtesy of Hughes Aircraft.

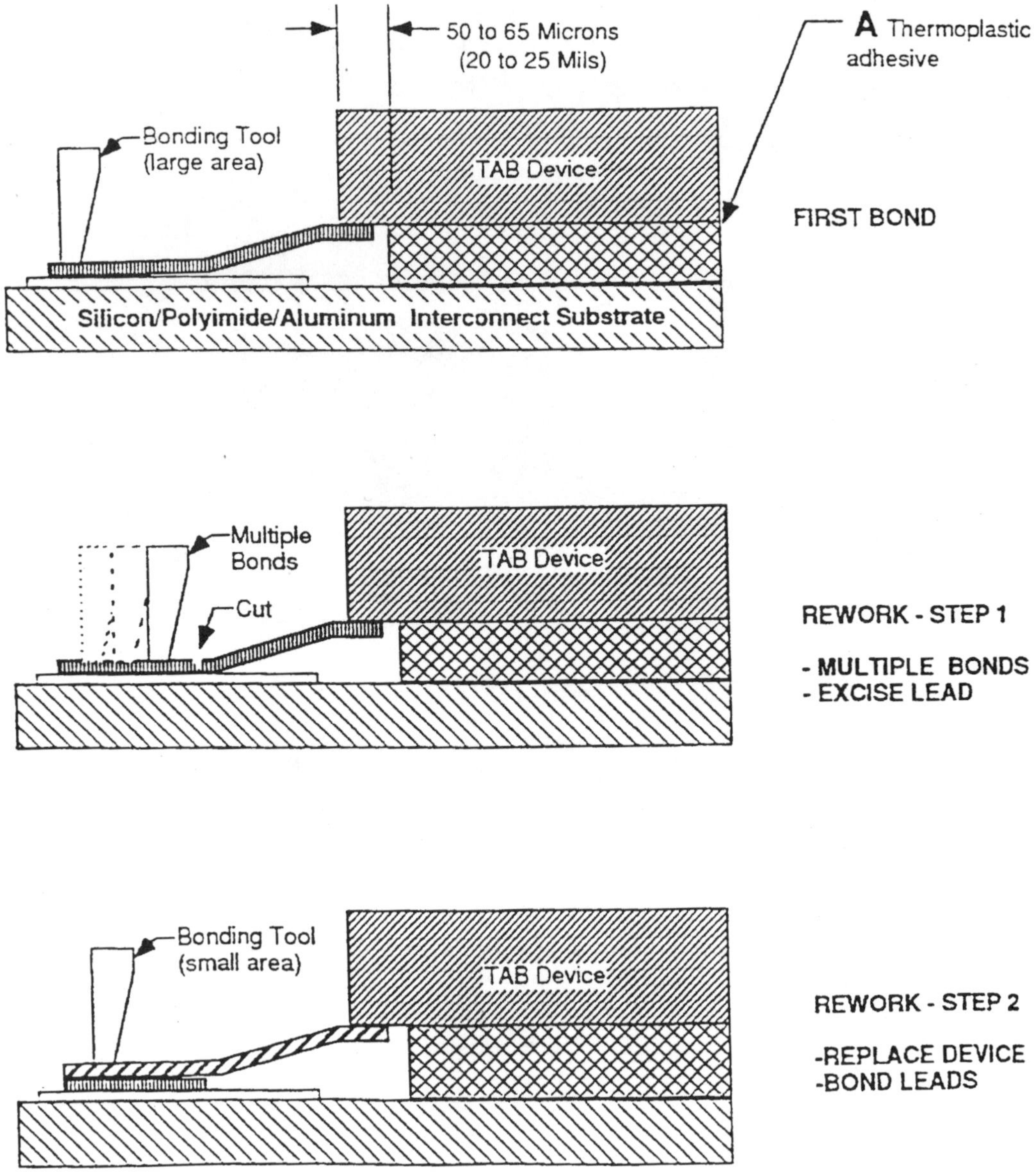

FIGURE 8.43 Repair process for TAB assembled with thermosonic single-point outer lead bonding.

Flip chip bumping processes have been developed by others that use plating to form the solder bump.[42–44] These processes are similar to the process previously described to form Au bumps for TAB (Fig. 8.26). Common adhesion/barrier layers are Cr, Ti, and TiW, whereas Cu, Ni, or Pd are used for the solderable metal layer. Au is the typical oxidation barrier to which the Pb/Sn solder is plated. After plating and etching of the background layers, the solder is reflowed to produce nearly spherical bumps. Texas Instruments has reported an electroplated copper bump technology with an

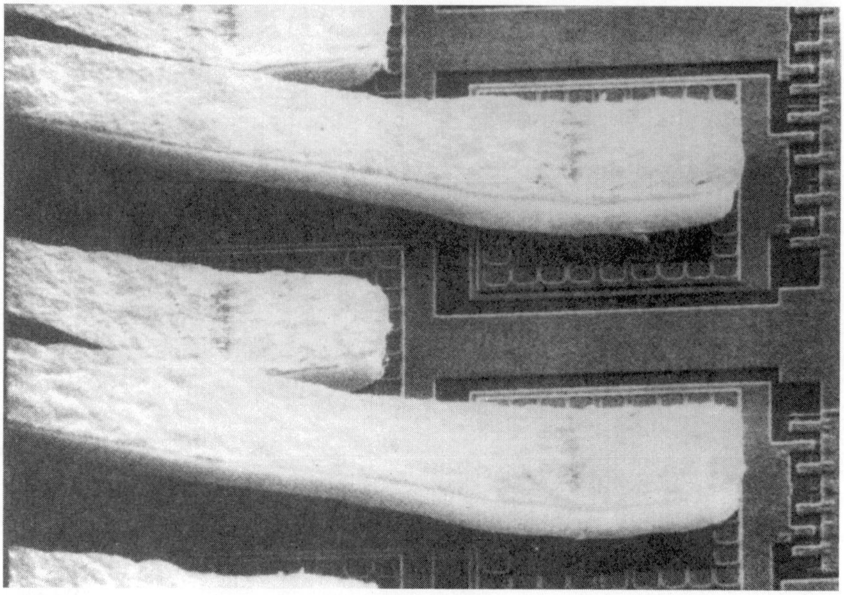

FIGURE 8.44 Bumpless TAB. Courtesy of Hewlett-Packard.

FIGURE 8.45 Low-I/O count device with solder balls along perimeter.

electroplated 10%Sn/90%Pb solder cap.[45] A sputtered Cr adhesion/barrier layer was used. Toshiba has described a copper cored solder bump formed by plating.[46,47] The bumps are made by first electroplating copper, then the solder alloy. The copper core maintains chip-to-substrate spacing during assembly. Lower melting temperature 63%Sn/37%Pb and 50%Pb/50%In solder bumps have also been used. The lower melting point Sn/Pb solders are not as fatigue resistant as the high lead content

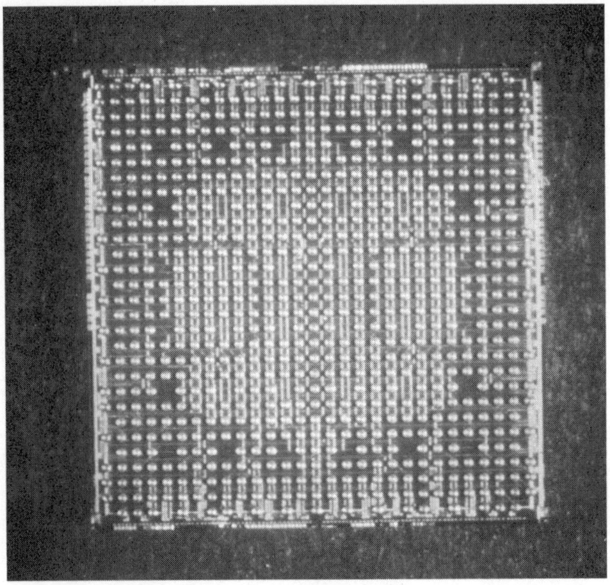

FIGURE 8.46 High I/O count device with solder balls in an area array. Courtesy of IBM.

solders. Indium based solders provide a very fatigue resistant interconnection.[48] However, indium is prone to corrosion in the presence of chlorine.

As an alternative to plating, a mechanical bumping process using solder wire in a modified thermosonic wire bonder has been demonstrated.[49] An inert atmosphere is provided around the wire during ball forming to prevent oxidation. MCNC has developed a solder bump transfer process (Fig. 8.47). In this process, a carrier substrate is fabricated with a sacrificial wettable metal layer over a nonwettable metal pad in a pattern corresponding to the I/O pads of the die. The solder bumps are then plated up on the wettable metal layer. To transfer the bump to the die, the die is positioned in contact with the solder bumps, and the solder is reflowed. During reflow, the solder wets and adheres to the under-bump metallurgy (UBM) of the die, while the sacrificial wettable metal on the substrate is dissolved into the molten solder. Once the sacrificial layer is dissolved, the solder de-wets from the substrate, and the transfer is complete. A number of organization have developed bumping processes based on a stencil printing solder paste onto the appropriate UBM and reflowing to form solder bumps.

Gold bumps either electroplated[50] or made mechanically by wire bonding[51–54] are also used for flip chips. Electroless nickel/gold bumps have also been fabricated using a zincating process to deposit a zinc layer on the aluminum bond pad.[55] The nickel bump and thin gold layer are deposited in an electroless process. No vacuum deposition or photolithography steps are required.

MCC has developed a polymer-based compliant bump.[56] In this process, a polymer is coated on the wafer and patterned to produce polymer bumps centered in the chip bond pads. The bumps are smaller in diameter than the bond pads, as shown in Fig. 8.48. The bump is then metallized so that the metal covers the top of the bump, extends down the side of the bump and then contacts the bond pad metallization. Epoxy Technology has developed a conductive polymer bumping process.[57] Silver-filled epoxy or thermoplastic polymers are screen printed onto the wafer over the I/O pads. Then either the epoxy is thermally cured or the solvent is dried from the thermoplastic bump. This approach relays on point contact to the metallization of the I/O bond pad. Aluminum has a tenacious native oxide that forms rapidly at room temperature. This insulating oxide must be penetrated or removed to form electrical contact. If initial contact is achieved, the polymers do not provide a hermetic seal, and the aluminum oxidizes over time and the contact resistance increases. Conductive

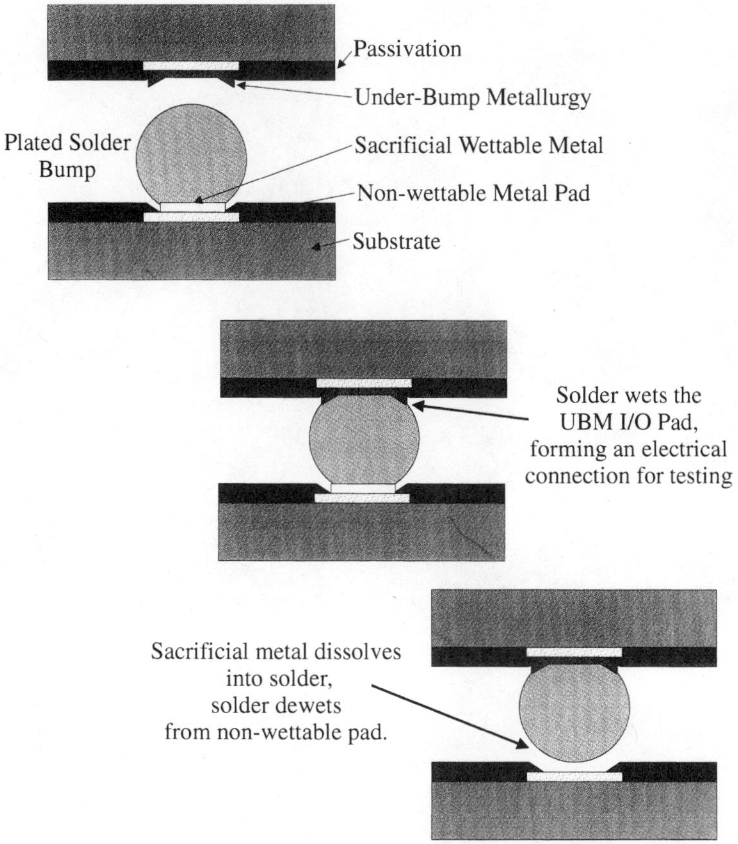

FIGURE 8.47 Illustration of MCNC solder bump transfer process. Courtesy of MCNC.

adhesive contacts have not proven reliable to aluminum metallization. In the Epoxy Technology process, the aluminum pads are first zincated to convert the surface. This is followed by electroless nickel and gold plating prior to screen printing of the conductive adhesive. Point contacts to gold do not suffer from the oxidation problems of aluminum.

Bumps may be placed in an area array over the surface of the die. This dramatically increases the number of I/O pads for a given chip size compared to peripheral pad designs. Care must be taken in the design to avoid electrical interactions between the electrical signal on the bump and the underlying circuitry. If area array connections are planned during the design of the chip, the chip can be optimized to maximize the benefits of the area array. Power, ground, and signals can be provided to the chip at the point where they are required. This improves electrical performance by eliminating the need to route all connections to the chip periphery, reducing parasitics and delays. The increased number of I/O pads available also allows the design of parallel I/O increasing data transfer rates. Serial I/O must often be used with peripheral I/O chips since the design is I/O limited. A full array of bumps may not be needed for electrical connections. However, nonelectrically connected bumps are often used to improve thermal and mechanical performance. Elimination of the periphery bond pads also reduces the chip size, increasing the number of chips per wafer and reducing per die costs.

Today, most chips are not designed for area array interconnections. Typical area arrays are 0.004-in bumps on 0.009-in or greater pitch. Peripheral designed bond pads may be on 0.004-in pitch or less. This is too fine for flip chip bumping and assembly. A redistribution layer is required to fan the

FIGURE 8.48 SEM micrograph of MCC compliant bump technology. Courtesy of MCC.

perimeter pads to a wider pitch over the die area. A polymer dielectric layer (polyimide or benzocyclobutene) is first deposited and patterned over the wafer exposing the aluminum bond pads. A thin film redistribution metallization layer is deposited and patterned. Solder bumps must then be formed. Redistribution layers adds significantly to the cost of wafer bumping. An example is shown in Fig. 8.49.

8.7.2 Assembly

Flip chip devices with 95%Pb/5%Sn bumps may be assembled by fluxing the bumps or the substrate, placing the inverted die, and reflowing at approximately 350° C. After reflow, the flux residue is cleaned. MCNC has developed a process, *plasma assisted dry soldering* (PADS), which eliminates the need for flux.[58] Prior to soldering, the assembly is exposed to atomic fluorine downstream from a fluorine-based microwave plasma. The atomic fluorine converts the surface oxides to oxyfluorides, which passivates the solder. The oxyfluorides break up when the solder melts, exposing free solder surfaces and allowing soldering without flux.

Some substrate technologies, especially MCM-L, cannot withstand the high reflow temperatures. For lower-temperature assembly, 63%Sn/37%Pb solder paste can be screened onto the substrate pads.[59] The chip with 95%Pb/5%Sn bumps is then placed into the paste, and the low melting point solder is reflowed at 230° C. In this process, the bumps do not melt, maintaining the spacing between the die and the substrate. This is illustrated in Fig. 8.50. Removal of the die with the bump intact is also possible with this process. An alternative approach is to use 63%Sn/37%Pb or 50%Pb/50%In solder bumps on the die, lowering the assembly temperature.

Gold bumped flip chips can be assembled to gold metallized pads on the substrate by thermocompression bonding.[53] Heat and pressure are used to deform the balls in contact with the Au substrate

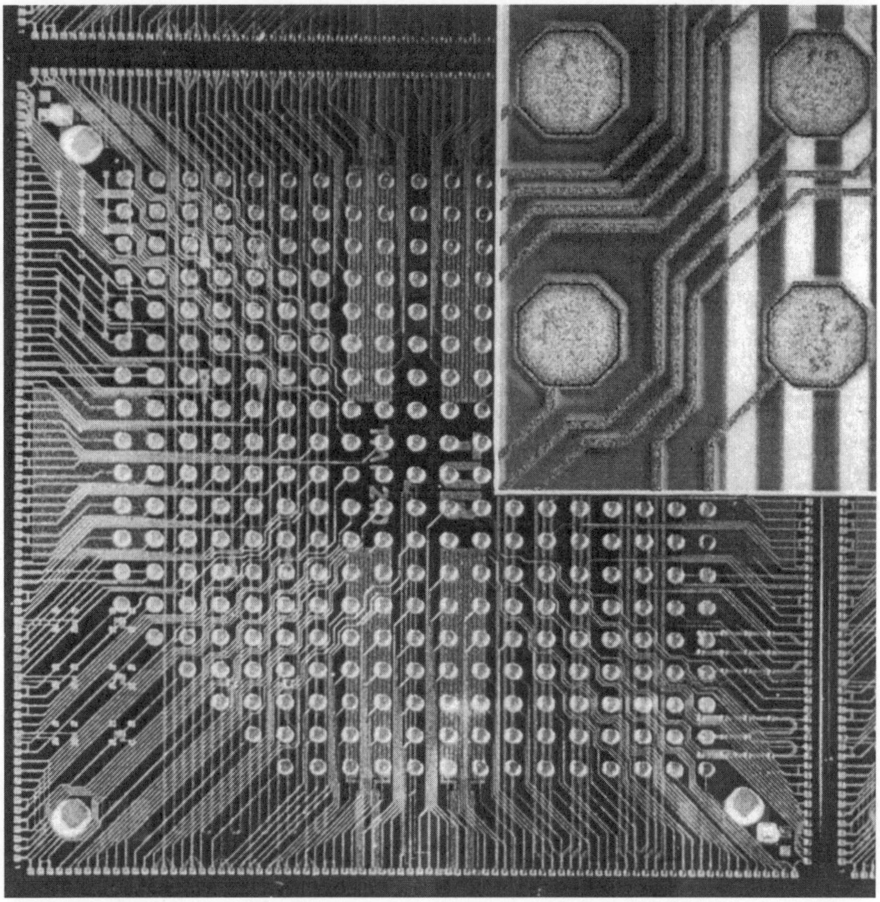

FIGURE 8.49 Die with redistribution layer for flip chip. The redistribution layers are benzocyclobutene dielectric and TiW/Au metallization. The pads are Ni:P/Au, and the pitch pad is 350 μm.

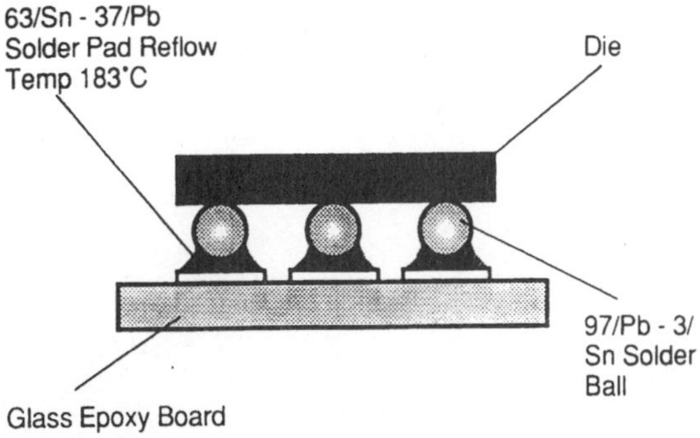

FIGURE 8.50 Illustration of eutectic solder assembly of high melting point (95Pb/5Sn) solder ball flip chip device.

pad metallization. nChip has developed a solder approach for flip chip assembly of gold wire stud bumped die.[60] The process uses a composite bump formed on the aluminum metallization of the MCM-D substrate. The bump consists of a 1000 Å sputtered Ti layer, a 1000 Å sputtered copper layer, a 6 μm plated copper layer, a 16 or 22 μm 95Pb/5Sn layer and, finally, an 8 or 2 μm indium layer. The gold wire bond stud bumped die is inverted and heated to a temperature above the melting point of indium but below the melting point of the 95Pb/5Sn solder layer. The substrate is heated to a temperature below the melting point of the indium. When the hot gold balls are brought in contact with the indium, the indium melts forming a metallurgical connection.

Anisotropic conductive adhesives have been used by Fujitsu to attach gold ball flip chip devices.[61] The epoxy resin is loaded with dielectric coated Ag particles. The die are assembled to the substrate under pressure. The pressure ruptures the dielectric coating on the particles, which are between the bump and the substrate pad, providing a gold bump-to-silver particle-to-substrate pad pressure contact electrical path. Particles between bumps are not subjected to the high pressures, and the insulating coating remains intact preventing conduction in the x-y plane. The epoxy adhesive is heat cured, maintaining the pressure contacts. The concept is illustrated in Fig. 8.51.

Isotropic conductive adhesives have also been used to connect Au bumps to the substrate metallization (Fig. 8.52).[54] The conductive adhesive is applied by screen printing on the substrate or by dipping the top of the bumps in conductive adhesive paste prior to die placement. Die with conductive epoxy bumps can be assembled in a similar manner. Figure 8.53 is a cross section of a die assembled with conductive adhesive bumps. With thermoplastic conductive adhesive bumps, the die/substrate is locally heated, softening the bump, which is then pressed onto the substrate pads. On cooling, the thermoplastic hardens and the die is attached.

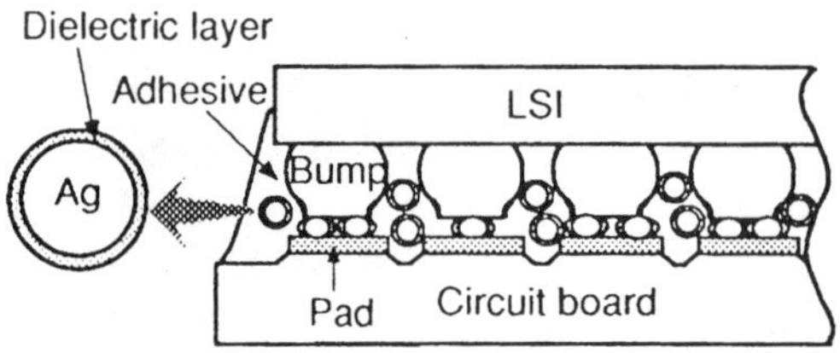

FIGURE 8.51 Assembly of gold bumped device with anisotropic adhesive.[61]

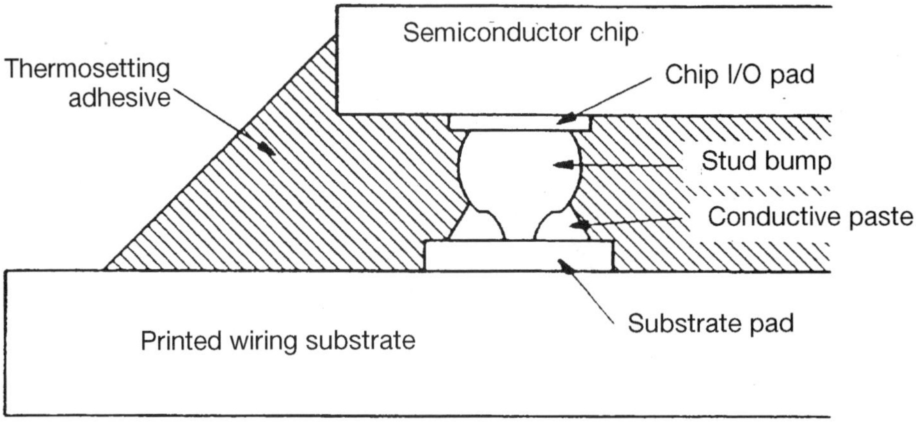

FIGURE 8.52 Assembly of device with gold stud bumps using conductive adhesive.[54]

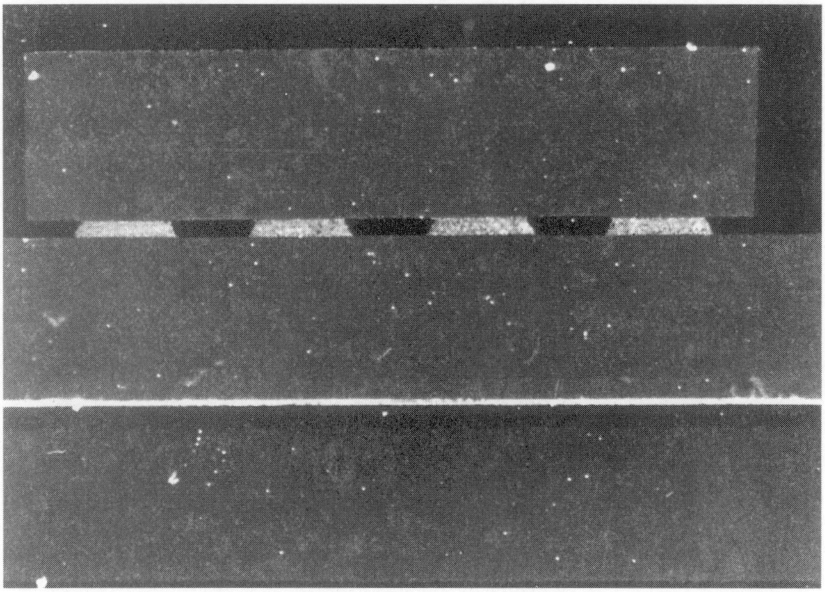

FIGURE 8.53 Cross section of a die assembled with conductive adhesive bumps. Courtesy of Epoxy Technology.

Placement accuracy is critical when using Au or conductive adhesive bumps. During assembly of solder based flip chips, the surface tension of the molten solder tends to align a die that has been misplaced.[62] However, in thermocompression or isotropic conductive adhesive attachment, there are no surface tension forces to align misplaced die.

Nonconductive adhesives that shrink on curing can be used to provide gold-to-gold pressure contacts between gold bumps and gold substrate metallization. UV curing adhesives have been used for chip-on-glass assemblies for displays. Maintaining pressure contact over time at elevated temperatures is a concern with this type of assembly. Heat curing adhesives have been used with the MCC compliant polymer bump on MCM-D and MCM-L substrates. Experimental results indicate improved elevated temperature performance with the compliant bump.[56]

The need to bump flip chip devices has limited the use of flip chip assembly. Anisotropic conductive films have been used to assemble die without bumps. The films are produced by loading polymers with conductive particles at levels below the percolation limit.[63,64] The resulting films are nonconductive, because the conductive particle do not touch to form a continuous electrical path. The adhesive film is designed to have a distributed monolayer of particles suspended in a polymer matrix. When the die are assembled to the substrate under pressure, the film is compressed, the conductive particles contact the die pad metallization, and the substrate metallization providing z-axis electrical connection. Since the film is not compressed in the x and y directions, there is no particle-to-particle contact created in the x-y plane. As previously discussed, aluminum chip bond pad metallization cannot be used. A composite pad with an outer layer of gold is required. For fine-pitch devices, the particle size must be small to avoid x-y shorts. Because only a monolayer of particles is used, the film is thin, which requires a very flat die and substrate surface.

Patterned particle z-axis adhesives have been developed by 3M to provide electrical connection only in predefined areas, as shown in Fig. 8.54.[65,66] The particles are nickel-coated polymer spheres. The die is assembled under pressure, providing pressure contact from the die pad to the nickel-plated sphere to the substrate pad. After cure and cooling, the thermosetting resin maintains the pressure contact.

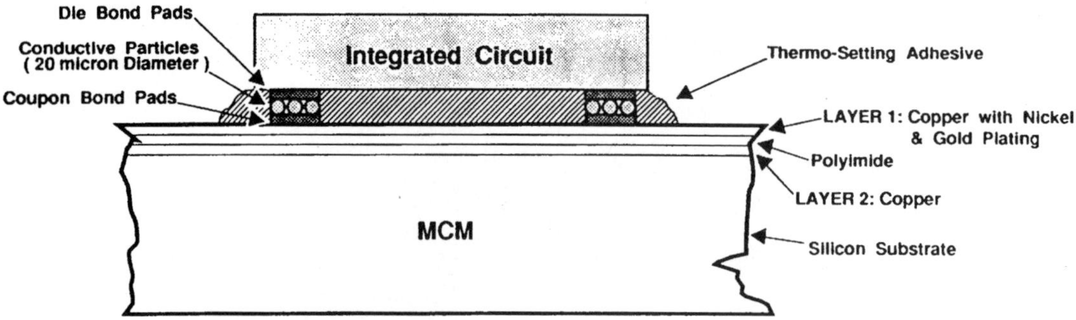

FIGURE 8.54 Cross section illustrating patterned particle Z-axis adhesives for flip chip assembly.[65]

Merix Corporation has also developed a patterned z-axis adhesive preform. The preforms are produced by screen printing two complimentary patterns: silver-filled epoxy dots and a nonconductive particle filled epoxy dielectric onto a release film. The silver dot pattern is designed to correspond to the I/O pads of the die. The epoxy is B-staged (solvent removed and partially cured). For assembly, the preform can be aligned and attached to either the die or the substrate. The die or substrate is heated to 80° C during this process. The die and substrate are then aligned and a small amount of pressure (2–4 psi) is applied. During this operation, the substrate is heated to 80° C. Once adhered, the epoxy is cured at 160° C.

8.7.3 Reliability

Coefficient of thermal expansion (CTE) mismatches between the semiconductor die and the MCM substrate result in cyclic stress on the solder joint during thermal cycling. Solder fatigue and failure results from this cyclic stress. The effect of CTE mismatch on fatigue failure is shown in Fig. 8.55.[67] Matching CTEs eliminates the stress due to thermal cycling. However, power cycling can also produce stress on the solder joint. Power dissipation in the device causes the device temperature to rise (and the device to expand) before the substrate temperature rises. The problem of fatigue failure increases with increasing die size, which is the trend in the semiconductor industry. To improve reliability, epoxy underfills are used.[68–70] The liquid epoxy is dispensed along one or two sides of the die. Capillary action pulls the epoxy under the die. The time required for complete underfill depends on the epoxy viscosity, ambient temperature, and the chip-to-substrate gap. After the epoxy has filled the gap under the die, it is thermally cured. It is important that voids not form and that good adhesion is achieved between all surfaces. Flux used during solder reflow may be removed prior to underfill. If not, the flux chemistry should be compatible with the underfull epoxy. The epoxy is loaded to improve thermal conductivity and to match the CTE of the solder. It is important that the filler remains in suspension during the capillary flow, so the underfill has a uniform CTE. The improvement in reliability is shown in Fig. 8.56.[71] The epoxy underfill material reduces the stress on the solder joints (Fig. 8.57).[70] After cure, the device cannot easily be removed for rework. This issue is being addressed by the material suppliers.

Adhesive-based flip chip approaches are still under development, and reliability data are limited. The primary concern is the loss of point-to-point contact and increasing resistance over time, especially with thermal cycling and high-temperature storage. The technology is being used in the display industry for chip-on-glass assembly. This is an ongoing research area that holds promise.

8.7.4 Rework

IBM has developed both mechanical and thermally enhanced techniques for reworking flip chip assemblies.[72,73] Figure 8.58 illustrates one mechanical method in which the die is gripped and a 1–2°

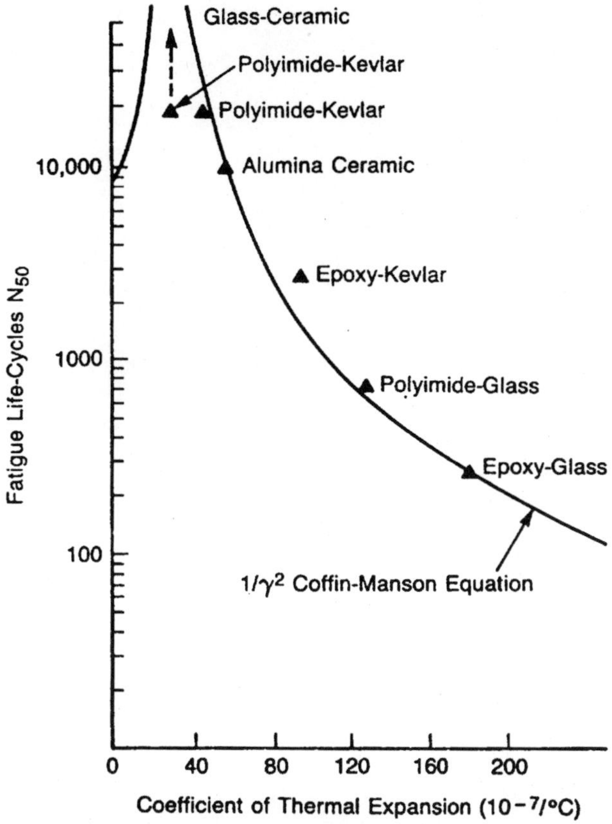

FIGURE 8.55 Relationship between substrate coefficient of thermal expansion and fatigue life for silicon flip chip die.[70]

torque is applied. This technique cannot be used for large die, because the die tend to fracture in the corners. A second mechanical method is shown in Fig. 8.59. Ultrasonic energy is used to fracture the solder joints. This method is not die size limited. Both techniques require a clearance of greater than 0.020-in around the die for the mechanical grippers. In thermal removal methods, the substrate is preheated to a temperature approximately halfway between ambient room temperature and the solder solidus temperature. Local heating of the die can be by conduction, convection, or radiation. Figure 8.60 illustrates the conduction approach. In this method, uniform contact with the die backside is necessary, which requires planarity of the heating element and parallelism. This is difficult to maintain. Figure 8.61 illustrates a hot nitrogen convection system. In this approach, planarity is not required, but shields are needed to prevent overheating of adjacent die. This limits the nearest neighbor spacing. The preferred approach is a focused IR, as illustrated in Fig. 8.62. When the solder joints melt, the die is removed with a vacuum activated mechanism.

It is necessary to remove excess solder and other debris from the bond site on the substrate prior to assembly of the replacement die. This can be accomplished using hot gas as shown in Fig. 8.63. The molten solder and debris are swept away by the velocity of the exiting gas flow. A solder wicking process using a solder wettable, porous metal slug (shown in Fig. 8.64) can also be used to remove molten solder.

Repair after the application and curing of underfill material is significantly more difficult. IBM Japan has developed a repair process for die after underfilling.[74] In this process, the defective die and

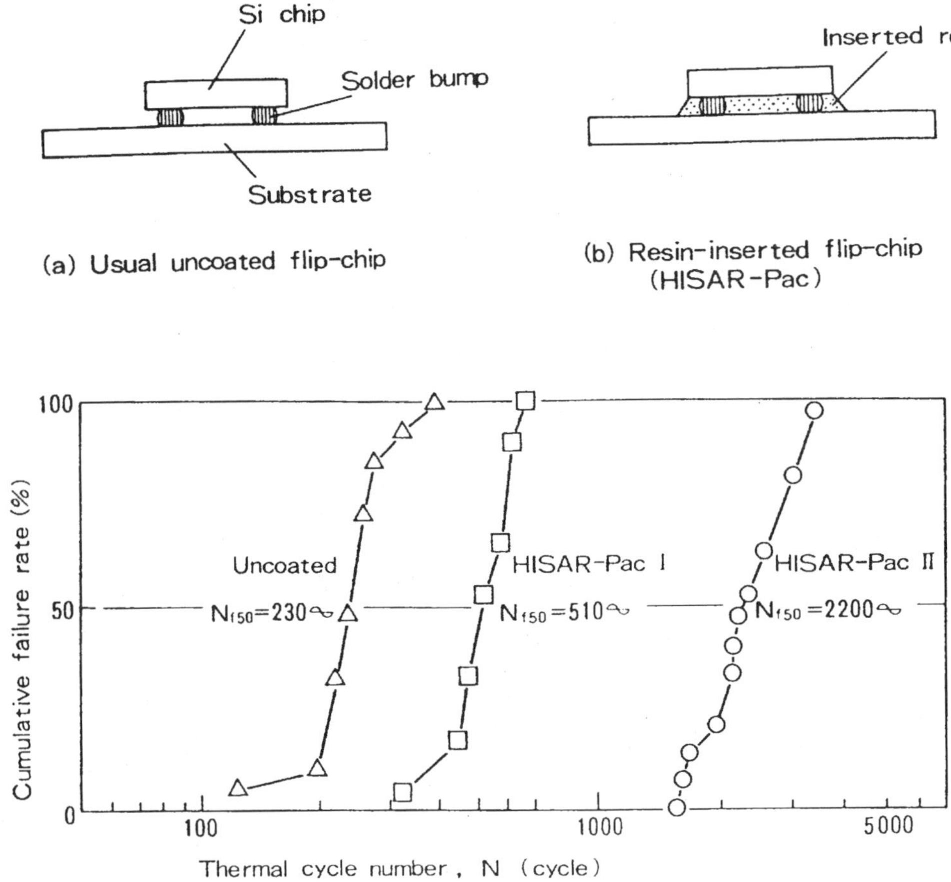

FIGURE 8.56 Cumulative device failure rate vs. number of thermal cycles with and without underfill material.[71]

approximately 50 percent of the underfill material are machined off using a milling machine. A double-blade end mill is used for machining the silicon while a single-blade is used for the encapsulant/solder layer. The remaining structure is exposed solder balls embedded in encapsulant. In this particular example, the original solder balls are high in lead content. The replacement die is eutectically reflow soldered to the half solder bump exposed in the encapsulant. After testing, the new die is underfilled.

8.7.5 Flip Chip Summary

Flip chip technology provides a high-volume, mass reflow method of assembling MCMs. The short length of the solder bump has minimal inductance, and the area array connections can improve electrical performance and meet the need for the ever increasing I/Os. The process requires bumping of the wafer. As in the case of TAB, this has limited the availability of flip chip devices to vertically integrated companies such as IBM or to high-volume applications such as automotive. As integrated circuit complexity and I/O counts increases, flip chip technology utilization will expand. Adhesive-based flip chips are being used in chip-on-glass assembly for displays but have found limited use in MCMs. Additional development is required.

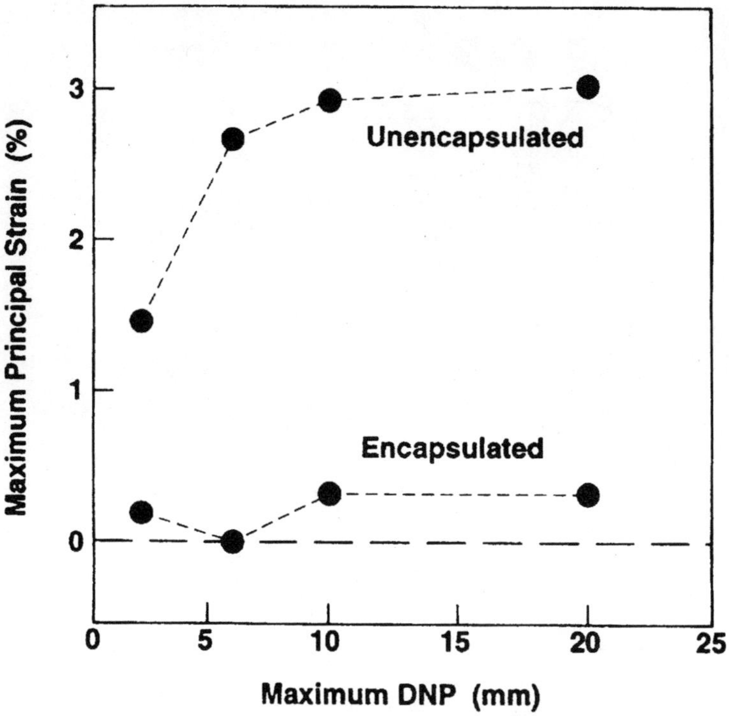

FIGURE 8.57 Maximum peripheral strain on flip chip solder joint vs. maximum distance from the neutral point.[70]

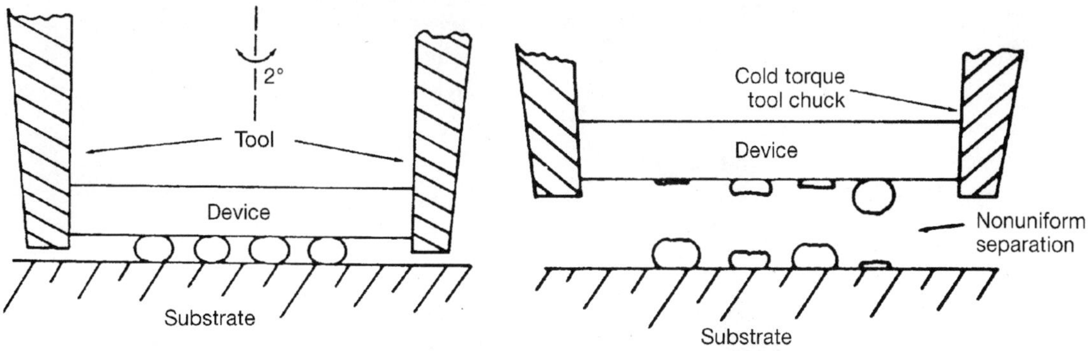

FIGURE 8.58 Illustration of mechanical flip chip removal using torque.[73]

8.8 ASSEMBLY SUMMARY

Over 90 percent of all semiconductors manufactured today are assembled using wire bonding. There is a well established infrastructure to support wire bond assembly. Equipment manufacturers con-

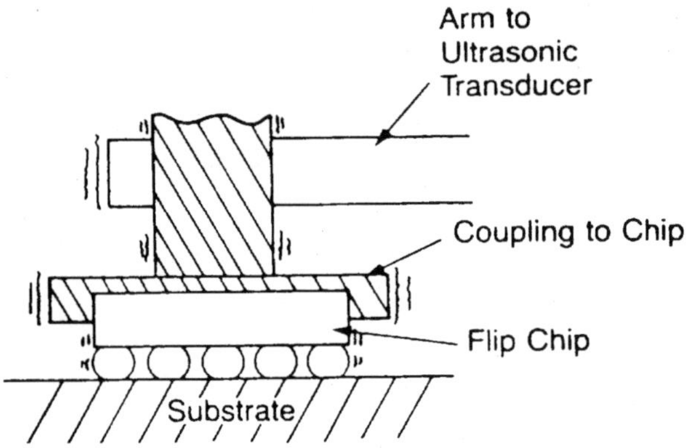

FIGURE 8.59 Illustration of ultrasonic flip chip removal process.[73]

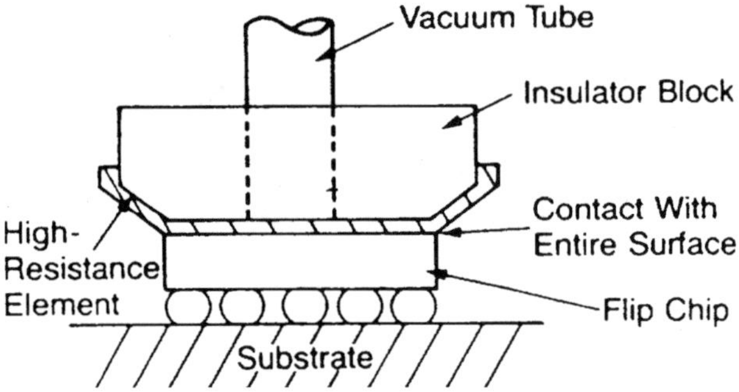

FIGURE 8.60 Illustration of conduction flip chip removal process.[73]

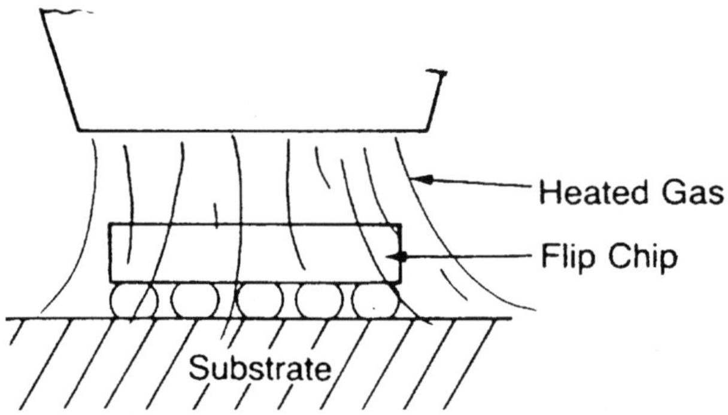

FIGURE 8.61 Illustration of hot gas process for flip chip removal.[73]

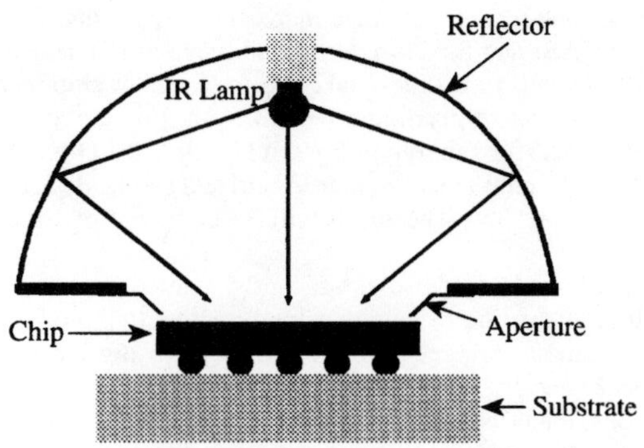

FIGURE 8.62 Illustration of focused IR removal process for flip chip removal.[73]

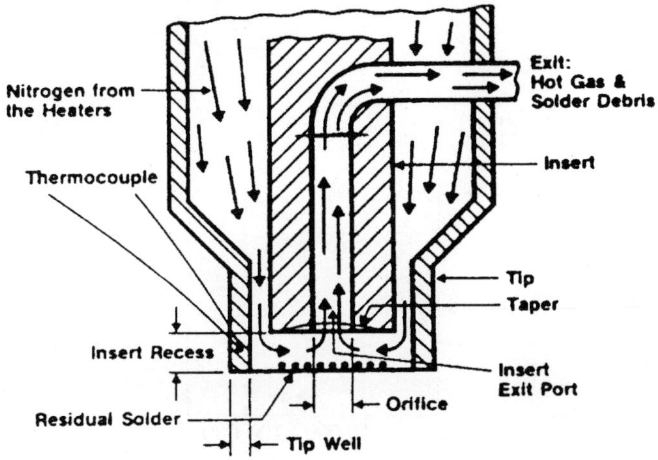

FIGURE 8.63 Illustration of bond site dressing using heated gas.[73]

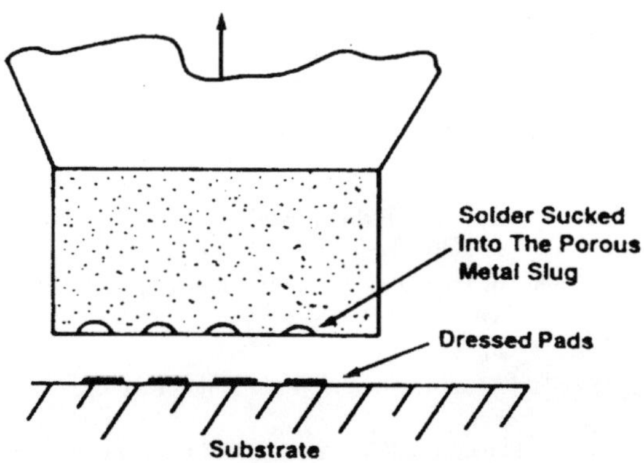

FIGURE 8.64 Illustration of bond site dressing using wicking method.[73]

tinue to increase bonding speeds and to accommodate increasing I/O counts. Figure 8.65 estimates the I/O range for wire bonding, TAB, and flip chip. For the ultimate in I/O density, area array interconnections will be required. This will most likely take the form of flip chip technology, although area array TAB has been used. To realize maximum benefit from flip chip assembly on multichip modules, the die should be designed for this type of assembly. Signal I/O and power and ground placement can be optimized over the chip area. Smaller I/O drivers can be designed for intramodule signals, because line capacitances are low. The result will be increased performance and reduced power dissipation.

For flip chip and TAB utilization to increase significantly, semiconductor manufacturers must support these technologies. It is interesting to note that Intel is providing the Pentium processor in TAB format for notebook applications, whereas Motorola/IBM are using flip chip to assemble the PowerPC into a single-chip package.

Alternative assembly options, many based on adhesives, will continue to be pursued. Increased assembly speed, reduced cost, improved reliability, and improved performance are the drivers for new assembly technologies.

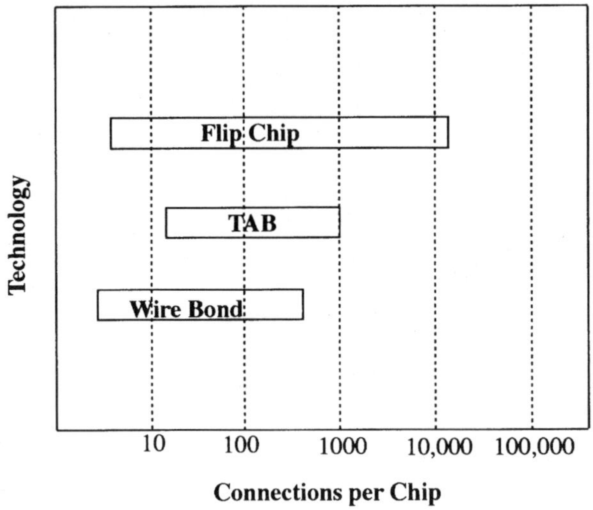

FIGURE 8.65 Estimates for the maximum number of interconnections per ship for different assembly technologies.

REFERENCES

1. Thin Film Multichip Modules, G. Messner, I. Turlik, J. Balde, and P. Garrou, Eds., International Society for Hybrid Microelectronics, ISBN 0-930815-33-5, Reston, VA, 1992, pp. 511.
2. R.E. Harr, "Easing the Availability of Bare Die: Die Information Exchange (DIE) Format," *Proc. 1994 Int. Symp. on Microelectronics,* Boston, MA, Nov. 17–19, 1994, pp. 78–83.
3. J. Goldstein, nChip, private conversation.
4. FED-STD-209C, Federal Standard Clean Room and Work Station Requirements, Controlled Environment, October 27, 1987.
5. K.X. Hu, C.-P. Yeh, B. Doot, A.F. Skipor, and K.W. Wyatt, "Die Cracking in Flip Chip-on-Board Assembly," *Proc. 45th Electronics Components and Technology Conf.,* Las Vegas, NV, May 21–24, 1995, pp. 293–299.

6. D. Hollander, "Electrostatic Discharge," *Radio Electronics,* vol. 60, no. 1, January 1990, pp. 54–57.
7. J.J. Licari and L.R. Enlow, *Hybrid Microelectronics Technology Handbook,* 1st Edition, Noyes Publications, Park Ridge, NJ (1988).
8. I.S. Adhihetty, P.W. Deal, D.Galloway, and R.P. Padmanabhan, "Evaluation of Snap-Cure Die Attach Materials," *Proc. 1992 Int. Symp. on Microelectronics,* San Francisco, CA, Oct. 19–21, 1992, pp. 128–133.
9. J.C. Bolger, "Long-Term vs. Short Term Thermal Stability of Polyimide Die Attach Adhesives," *Int. J. for Hybrid Microelectronics,* vol. 5, no. 2, Nov. 1982, pp. 496–499.
10. S.P. Anderson and H.S. Kraus, "Heat Aging Characteristics of Polyimide Chip Adhesives," *Int. J. for Hybrid Microelectronics,* vol. 4, no. 2, October 1981, pp. 190–193.
11. R. Dietz and L. Winder, "An Innovation in Materials for Attaching Silicon Devices," *Int. J. for Hybrid Microelectronics,* vol. 5, no. 2, Nov. 1982, pp. 480–486.
12. P.R. Harper, "Thermoplastic Die Attach for Hermetic Packaging," *Proc. 1994 Int. Conf. on Multichip Modules,* Denver, CO, April 13–15, 1994. pp. 56–60.
13. M.A. Lutz and R.L. Cole, "Flexible Silicone Adhesive with High Electrical Conductivity," *Proc. 39th Electronics Components Conf.,* Houston, TX, May 22–24, 1989, pp. 83–87.
14. I.Y. Chien and M.N. Nguyen, "Low Stress Polymer Die Attach Adhesive for Plastic Packages," *Proc. 44th Electronics Components Conf.,* Washington, D.C., May 1–4, 1994, pp. 580–584.
15. U. Efrat and C. Enman, "Decision Support Model for Fine Pitch Interconnection Package Design," *Proc. 1993 Int. Symp. on Microelectronics,* Dallas, TX, Nov. 9–11, 1993, pp. 419–425.
16. R. Short, "Fine Pitch Wedge Bonding for High Density Packaging," *Proc. 1994 Int. Conf. on Multichip Modules,* Denver, CO, April 13–15, 1994. pp. 50–55.
17. G.G. Harman, Reliability and Yield Problems of Wire Bonding in Microelectronics, *Int. Society for Hybrid Microelectronics,* Reston, VA, (1989).
18. G.G. Harman, "Wire Bonding to Multichip Modules and Other Soft Substrates," *Proc. 1995 Int. Conf. on Multichip Modules,* Denver, CO, April 19–21, 1995, pp. 292–301.
19. J. Glazer, P.A. Kramer and J.W. Morris, Jr., "Effect of Au on the Reliability of Fine Pitch Surface Mount Solder Joints," *Circuit World,* vol. 18, no. 4, August 1992, pp. 41–46.
20. Sherman Banks, "Reflow Soldering to Gold," *Electronic Packaging & Production,* June 1995, pp. 69–74.
21. H.M. Clearfield, M.J. Interrante, D.J. Tudyrn, and L. Economikos, "A Single Top Surface Metallurgy for Solder Flip Chip, Wirebond and TAB Interconnections," *Proc. 1994 Int. Symp. on Microelectronics,* Boston, MA, Nov. 17–19, 1994, pp. 485–490.
22. J.F. Graves, "Plasma Processing of Hybrids for Improved Bondability," *Proc. 1983 Int. Symp. on Microelectronics,* Philadelphia, PA, Oct. 31–Nov. 2, 1983, pp. 147–154.
23. M. Ohzono, "Plasma Cleaning for Fine Connection," *Proc. 1992 Int. Conf. on Multichip Modules,* Denver, CO, April 1–3, 1992, pp. 491–498.
24. C. Proietti-Bowne and P. Elenius, "Rework of Wire Bonded Devices on MCMs," *Proc. 1994 Int. Conf. on Multichip Modules,* Denver, CO, April 13–15, 1994, pp. 61–66.
25. W.C. Whitworth, R.P. Sood, L.M. Flaherty, and D.E. Brothers, "Beginning the Application of Class S TAB Module Manufacturing," *Proc. 1990 Int. Symp. on Microelectronics,* Chicago, IL, Oct. 15–17, 1990, pp. 460–467.
26. C. J. Speerschneider and J. M. Lee, "Solder Bump Reflow Tape Automated Bonding," *Proc. 2nd ASM Int. Electronic Materials and Processing Congress,* Philadelphia, PA, April 24–28, 1989, pp. 7–12.
27. A.F.J. Baggerman and J.A.H. van Gerven, "TAB Inner Lead Gang Bonding on Ni-Au Bumps," *Proc. 44th Electronics Components Conf.,* Washington, D.C., May 1–4, 1994, pp. 938–944.
28. D. Metzger, U. Beutler, J. Eldring, and H. Reichl, "Laser Bumping for Flip Chip and TAB Applications," *Proc. 44th Electronics Components Conf.,* Washington, D.C., May 1–4, 1994, pp. 910–916.
29. E.N. Larson and M.J. Brock, "Development of a Single Point Gold Bump Process for TAB Applications," *Proc. 1993 Int. Conf. on Multichip Modules,* Denver, CO, April 14–16, 1993, pp. 391–397.
30. J.S. Sallo, "Bumped Beam Tape for Automatic Gang Bonding," *Proc. 1978 Int. Microelectronics Symp.,* Minneapolis, MN, Sept. 25–27, 1978, pp. 153–156.
31. T. Kawakita and H. Hatada, "Application of Transferred Bump TAB Technology to Large Size, Multi-pin and Fine Pitch LSI Chip," *Proc. 1991 Int. Symp. on Microelectronics,* Orlando, FL, Oct. 21–23, 1991, pp. 328–332.

32. K. Tatsumi, T. Ando, Y. Ohno, M. Konda, Y. Kawakami, N. Ohikata, and T. Maruyama, "Transferred Ball Bump Technology for Tape Carrier Packages," *Proc. 1994 Int. Symp. on Microelectronics,* Boston, MA, Nov. 17–19, 1994, pp. 54–59.
33. E. Zakel, G. Azdasht, and H. Reichl, "Investigations of Laser Soldered TAB Inner Lead Contacts," *Proc. 41th Electronics Components Conf.,* Atlanta, GA, May 11–16, 1991, pp. 497–506.
34. S. Britton, "Spontaneous Growth of Whiskers on Tin Coatings: 20 Years of Observations," *Transactions of the Institute of Metal Finishing,* vol. 52, 1974, pp. 95–104.
35. L. Alton and R. Kuntz, "High Performance Test Socket System—35 mm TAB Devices," *Proc. 1991 Int. Symp. on Microelectronics,* Orlando, FL, Oct. 21–23, 1991, pp. 348–352.
36. E. Zakel, J. Villain, and H. Reichl, "Reliability and Au Concentration in OLB Solder Fillets of TAB Devices Having a 75mm Pitch," *Proc. 44th Electronics Components Conf.,* Washington, D.C., May 1–4, 1994, pp. 883–893.
37. M.J. Palmer, R.R. Horton, H.R. Bickford, and I.C. Noyan, "HAT Tool for Fluxless OLB and TAB," *Proc. 41st Electronics Components Conf.,* Atlanta, GA, May 11–16, 1991, pp. 507–510.
38. L.A. Hughes, J.W. Slemmons, P.J. Redmond, D.R. Wilson, and D.L. MacGregor, "Improved Device Attach and Rework Techniques for MCM-D Circuits using Thermoplastic Adhesives and Flip TAB Devices," *Proc. 1993 Int. Conf. on Multichip Modules,* Denver, CO, April 14–16, 1993, pp. 398–403.
39. F. Matta, B. Heflinger, R. Kaw, and V. Rastogi, "High Performance Packaging at Competitive Cost Using Demountable TAB," *Int. J. of Microcircuits and Electronic Packaging,* vol. 15, no. 2, Q. 2, 1992, pp. 61–73.
40. P.A. Totta and R.P. Sopher, "SLT Device Metallurgy and Its Monolithic Extension," *IBM Journal of Research and Development,* May 1969, pp. 226–238.
41. *Microelectronics Packaging Handbook,* R.R. Tummala and E.J. Ramaszewski, Eds., Van Nostrand Reinhold, New York, NY, 1989, pp. 376–380.
42. E.K. Yung and I. Turlik, "Electroplated Solder Joints for Flip Chip Applications," *IEEE Transactions on Components, Hybrids, and Manufacturing Technology,* vol. 14, no. 3, September 1991, pp. 549–559.
43. G. Chmiel, J. Wolf, and H. Reichl, "Lead/Tin-Bumping for Flip Chip-Applications," *Proc. 1992 Int. Symp. on Microelectronics,* 1992, pp. 336–341.
44. K.K. Yu and F. Tung, "Solder Bump Fabrication by Electroplating for Flip Chip Applications," *Proc. 15th IEEE/CHMT Int. Electronics Manufacturing Technology (IEMT) Symp.,* Santa Clara, CA, 1993, pp. 277–281.
45. K.G. Heinen, W.H. Schroen, D.R. Edwards, A.M. Wilson, R.J. Stierman, and M.A. Lamson, "Multichip Assembly with Flipped Integrated Circuits," *Proc. 39th Electronics Components Conf.,* 1989, pp. 672–680.
46. S. Kimijima, T. Miyagi, and O. Shimada, "High Density Multichip Module by Chip-on-Wafer Technology," *Proc. 1988 Int. Symp. on Microelectronics,* 1988, pp. 314–319.
47. T. Togaski, T. Uchida, Y. Ieski, M. Mori, and M. Saito, "Bump Interconnection for Chip Components and LSI Chips for High Density Modules," *Proc. 1994 Int. Symp. on Microelectronics,* Boston, MA, Nov. 17–19, 1994, pp. 266–271.
48. K. Shimizu, T. Akamatsu, T. Nakanishi, K. Karasawa, K. Hasimoto, and K. Niwa, "Solder Joint Reliability of Flip Chip Interconnection," *Proc. 1994 Int. Symp. on Microelectronics,* Boston, MA, Nov. 17–19, 1994, pp. 272–277.
49. C.R. Montgomery, "Flip Chip Assemblies Using Conventional Wire Bonding Apparatus and Commercially Available Dies," *Proc. 1993 Int. Symp. on Microelectronics,* Dallas, TX, Nov. 9–11, 1993, pp. 451–456.
50. K. Yamamura, H. Atarashi, N. Kakimoto, N. Sakota, H. Sato, M. Miyauchi, K. Naito, and T. Nukii, "Flip Chip Bonding Technology for GaAs-MMIC Power Devices," *Proc. 1994 Int. Symp. on Microelectronics,* Boston, MA, Nov. 17–19, 1994, pp. 433–438.
51. J. Kloeser, E. Zakel, A. Ostmann, J. Eldring, and H. Reichl, "Cost Effective Flip Chip Interconnections on FR-4 Boards," *Proc. 1994 Int. Symp. on Microelectronics,* Boston, MA, Nov. 17–19, 1994, pp. 491–500.
52. R.F. Cooley, "Low Cost MCM-L Flip Chip Interconnection Utilizing Gold Ball Bumping," *Proc. 1994 Int. Symp. on Microelectronics,* Boston, MA, Nov. 17–19, 1994, pp. 473–478.
53. C.E. Goodman and M.P. Metroka, "A Novel Multichip Module Assembly Approach Using Gold Ball Flip Chip Bonding," *IEEE Transactions on Components, Hybrids, and Manufacturing Technology,* vol. 15, no. 4, August 1992, pp. 457–464.

54. T. Kusagaya, H. Kira, and K. Tsunoi, "Flip Chip Mounting Using Stud Bumps and Adhesives for Encapsulation," *Proc. 1993 Int. Conf. on Multichip Modules,* Denver, CO, April 14–16, 1993. pp. 238–245.

55. J. Simon, A. Ostmann, and H. Reichl, "Electroless Bumping for TAB and Flip Chip," *Proc. 1994 Int. Symp. on Microelectronics,* Boston, MA, Nov. 17–19, 1994, pp. 439–444.

56. K. Keswick, R. German, M. Breen, and R. Nolan, "Compliant Bumps for Adhesive Flip Chip Assembly," *Proc. 44th Electronics Components Conf.,* Washington, D.C., May 1–4, 1994, pp. 7–15.

57. R.H. Estes, "Fabrication and Assembly Processes for Solderless Flip Chip Assembly," *Proc. 1992 Int. Symp. on Microelectronics,* 1992, pp. 322–335.

58. N. Koopman and S. Nangalia, "Fluxless Flip Chip Solder Joining," *Proc. NEPCON West 1995,* Anaheim, CA, February 1995.

59. K. DeHaven and J. Dietz, "Controlled Collapse Chip Connection (C^4)—An Enabling Technology," *Proc. 44th Electronics Components Conf.,* Washington, D.C., May 1–4, 1994, pp. 1–6.

60. J.L. F. Goldstein, D. Tuckerman, P. Kim, and B.S. Fernandez, "A Novel Flip Chip Process," *Proc. Surface Mount Int. Conf.,* San Jose, CA, August 1995.

61. H. Date, Y. Hozumi, H. Tokura, M. Usui, E. Horikoshi, and T. Sato, "Anisotropic Conductive Adhesive for Fine Pitch Interconnections," *Proc. 1994 Int. Symp. on Microelectronics,* Boston, MA, Nov. 17–19, 1994, pp. 57–575.

62. T. Hayashi, "An Innovative Bonding Technique for Optical Chips Using Solder Bumps that Eliminate Chip Positioning Adjustments," *IEEE Transactions on Components, Hybrids, and Manufacturing Technology,* vol. 15, no. 2, April 1992, pp. 225–230.

63. N.R. Basavanhally, D.D. Chang, B. Cranston and S.G. Segar, Jr., "Direct Chip Interconnection with Adhesive Conductor Films," *IEEE Transactions on Components, Hybrids and Manufacturing Technology,* vol. 15, no. 6, December 1992, pp. 972–976.

64. S. Jin, T.H. Tiefel, L.-H. Chen, and D.W. Dahringer, "Anisotropically Conductive Polymer Films with a Uniform Dispersion of Particles," *IEEE Transactions on Components, Hybrids and Manufacturing Technology,* vol. 16, no. 8, December 1993, pp. 972–976.

65. B.K. Gilbert, W.I. McNeff, P.J. Zabinski, W.P. Kornrumpf, S.F. Tead, R.L. Maki, K.E. Carlson, and J. Gerber, "Development of Deposited Multichip Modules with Unique Features for Applications in GaAs Signal Processors Operating Above One GHz Clock Rates," *Int. J. of Microcircuits and Electronic Packaging,* vol. 15, no. 3, Q. 3, 1992, pp. 144–159.

66. P.B. Hogerton, J.B. Hall, J.M. Pujol, and R.S. Reylek, "Investigations into the Use of Adhesives for Level-1 Microelectronic Interconnections," *Materials Research Society Symp. Proceedings,* vol. 154. 1989, pp. 415–424.

67. S.E. Greer, "Low Expansivity Organic Substrate for Flip Chip Bonding," *Proc. 28th Electronics Components Conf.,* May 1978, pp. 166–171.

68. D. Suryanarayana, R. Hsiao, T.P. Gall, and J.M. McCreart, "Enhancement of Flip Chip Fatigue Life by Encapsulation," *IEEE Transactions on Components, Hybrids, and Manufacturing Technology,* vol. 14, no. 1, March 1991, pp. 218–223.

69. D. Suryanarayana, T.Y. Wu, and J.A. Varcoe, "Encapsulants Used in Flip Chip Packages," *IEEE Transactions on Components, Hybrids, and Manufacturing Technology,* vol. 16, no. 8, December 1993, pp. 858–862.

70. J. Clementi, J. McCreary, T.M. Niu, J. Palomaki, and J. Varcoe, "Flip Chip Encapsulation on Ceramic Substrates," *Proc. 43rd Electronics Components Conf.,* Orlando, FL, June 1–4, 1993, pp. 175–181.

71. F. Nakano, T. Soga, and S. Amagi, "Resin-Insertion Effect on Thermal Cycle Resistivity of Flip Chip Mounted LSI Devices," *Proc. 1987 Int. Symp. on Microelectronics,* Boston, MA, Nov. 17–19, 1987, pp. 536–541.

72. K. Puttlitz and K. Lidestri, "Replacing Flip Chips on MLC Multichip Modules Using Focused Infrared (IR): The Basics," *Proc. 1992 Int. Symp. on Microelectronics,* San Francisco, CA, Oct. 19–21, 1992, pp. 384–390.

73. K.J. Puttlitz, Sr., "An Overview of Flip Chip Replacement Technology on MLC Multichip Modules," *Int. J. of Microcircuits and Electronic Packaging,* vol. 15, no. 3, Q. 3, 1992, pp. 113–125.

74. Y. Tsukada, Y. Mashimoto, and N. Wantanuki, "A Novel Chip Replacement Method for Encapsulated Flip Chip Bonding," *Proc. 43th Electronics Components Conf.,* Orlando, FL, June 1–4, 1993, pp. 199–204.

CHAPTER 9

MODULE-TO-BOARD CONNECTIONS

W.H. Knausenberger
RD Hikuai
THAMES, New Zealand

9.1 INTRODUCTION

Connecting an MCM to its next interconnection level (generally a printed circuit board, or PCB) entails simultaneous electrical, mechanical, and thermal connections. This chapter gives an overview of the different available options for making these connections[1] and addresses the trade-off issues associated with their use.

MCMs may, of course, be connected to the next interconnection level by solder techniques. Some designers have used long, flexible leads looping up from the MCM and then down to the PCB.[2] Others have wire bonded the edge pads of the MCM to pads on the PCB.[3] Both can have problems for all but the smallest sizes in the event of major differences in the *coefficient of thermal expansion* (CTE) between MCM and PCB. Certainly, it is not always practical to use matched CTE PCBs for the whole system. The use of the wire bonding process[4,5] has its own problems. Removing the MCM is difficult and can easily damage the interface pads, making subsequent attachment operations difficult or even impossible. Soldering is only somewhat more helpful when it comes time to remove and replace the MCM. To desolder hundreds of leads and clean their pads to permit installation of a replacement module is not easy.

Clearly, connectors are desirable and almost a necessity for the economical manufacture of complex systems. There are two basic connector approaches that a designer can utilize for the readily separable attachment of MCMs to their underlying substrate.

9.1.1 Off-the-Shelf Connector Approach

The most direct way is to use existing, commercially available socketing and connection systems developed originally for single-chip LSI packages. There are two basic variations within this approach: *surface mount* or *pin grid array* (PGA) connection hardware systems.

With surface mount, the simplest approach is to use existing JEDEC standard chip carrier socket connectors[6,7] or commercially available BGA connectors, when allowed by the application design. This can result in very simple and cost-effective solutions, particularly when there are other SM components mounted on the same board. The other variation on this approach is the use of PGA-type connector structures[8] built to utilize existing commercially available connection systems. This use of the existing high-volume technology base can lead to cost-effective solutions.

9.1.2 Customized Connector Approach

For MCM designs that cannot use off-the-shelf solutions, customized connectors generally must be utilized. There are numerous customized connection systems available. The major directions are reviewed here.

MCM connection needs can range from requiring fully socketable connections (for *field replaceable units*) to needing semipermanent and permanent connections where the MCM is an integral part of the assembly and not designed for ready removal. There are as yet no widely accepted industry-standard solutions.

9.2 REQUIREMENTS

A separable contact must meet the basic requirements of (a) being a good conductor upon being closed and (b) being a good insulator upon being opened. The quality of the connection is determined by the electrical, mechanical, thermal, and chemical properties of the materials as well as the specifics of the design and the application criteria. MCM connectors must be capable of providing a large range of I/O densities to satisfy the diversity of possible applications. These I/Os have to be provided in such a way that the I/Os meet the application requirements and successfully provide interconnection to the next level. Some of the desired characteristics of the ideal MCM connector are reviewed in the following section.

9.2.1 Electrical Requirements

Signal Interconnections. In the electrical design of the connector, one must first determine if it is necessary for the connector to have a controlled impedance. A basic question must be addressed: Is the electrical path length through the connector (expressed in time units) appreciable compared to the fastest signal rise time to be passed through the connector? If it takes half a signal rise time or longer to pass through the connector, the connector should be designed to present a uniform, controlled-impedance electrical environment to the signal. The value of the impedance should be matched to that of the interfacing interconnections on either side of the connector. The physical discontinuity of an impedance-matched connector is minimized in such a way that the connector is made all but "invisible" to the propagating signal, which can then travel on an essentially contiguous transmission line.[9–11]

Other important electrical considerations include maximizing the electrical conductivity of the path and minimizing the overall connector path length while making the length of each path through the connector to be as close to being the same as possible. This will serve to minimize signal skew and avoid adding system timing problems. Crosstalk between adjacent contacts must also be minimized and is generally helped by the addition of reference paths or planes for controlling impedance within the connector.[12] The specific requirements for each application will be a strong function of the system design trade-off choices that are made for the circuitry being interconnected.

Power/Ground Interconnections. The objective for power and ground interconnections is to provide a very low impedance (low inductance and resistance, with high power to ground capacitance) transmission path. The specific requirements of a particular system will be very dependent on the design specific trade-offs made. In the higher performance applications, separate interconnection systems are normally employed for signals and for power. However, in general, and particularly for cost-performance dominated systems, power and signal interconnections use the same interconnection medium. This is in spite of the fact that the electrical requirements for power and ground connections are quite different from those for signal interconnections.

These all purpose connections must be capable of operating for a wide range of typical electrical parameters, covering currents ranging from microamps to a few amperes and voltages from microvolts to tens of volts. In addition, it may be necessary to handle short surges (on the order of a millisecond or less) of hundreds of amperes and several thousand volts. This wide range in requirements comes about in large measure because signal interconnection needs are so different from power interconnection needs. It is for this reason that separately optimized systems are often employed for signal I/Os and for power I/Os in high end applications.

9.2.2 Thermal Capabilities

The thermal path for heat removal from the chip is normally a very critical MCM design aspect. The MCM connector is frequently a critical element in that thermal path. Chapter 13, Thermal Issues, presents a methodology for the analysis of the thermal performance of MCMs.

9.2.3 Mechanical Compatibility

The MCM connections system must be carefully designed to be mechanically compatible with the overall system. This includes careful attention to the CTE of the MCM and its compatibility with the CTE of the devices mounted on the MCM and the underlying MCM substrate. The transition from through-hole to surface-mount components has run its course in nearly all areas of electronics assembly, except for connectors. Unless additional, non-SMT compatible, mechanical attachment steps are taken, surface mount connectors have the basic drawback of having a lower level of mechanical integrity than through-hole equivalent connectors. The metallic joint, which provides the electrical continuity, must also provide the primary means of mechanical attachment. A very careful system reliability assessment[13,14] must be conducted to ensure that no undue compromises are made in this area.

9.2.4 I/O Requirements

Ranging across different applications, the I/O needs may vary from only a few to thousands per MCM. For example, the MCM I/O count in some mainframe computers is as shown below:

Computer system	MCM I/Os	Reference
IBM System ES/9000	2,772	15, 16
NEC S3600	11,540	18
Hitachi M880	528	18
Fujitsu M1800	8,640	19
Siemens H-90	1,850	20
DEC VAX9000	1,072	21
Cray J90	1,600	22

However, for the purposes of this book, we will concentrate on solutions that are most applicable to the majority of commercial applications, which range up to 500 I/Os in the mid-1990s. Because MCMs tend to be used to interconnect functionally interrelated ICs, the great majority of applications tend to have much lower I/O needs than Rent's rule[23,24] might suggest. The mainframe MCM I/O structures tend to be specialty solutions with limited application to the broader base of commercial applications.

9.2.5 MCM Connection System Reliability

The electrical, thermal, mechanical, and I/O requirements, as well as the resulting trade-offs that are made in coming up with a system design solution, affect reliability. In general, the following actions and design directions tend to reduce the risk of a poor contact in a separable contact system (and thereby improve the reliability):

- Redundant contact points
- Noble metal contact finishes
- Increased contact forces
- Increased wiping action

The CTE of silicon is about 2.8, while that of most organic dielectrics is in the range of several hundred[25] in the temperature range of interest. Most MCM constructions under development or in use today utilize metal conductors embedded in organic dielectrics (see Chaps. 3 and 4). Effectively managing the differential thermal expansion between the materials involved in the packaging system is one of the major challenges facing the physical designer. Compliant leads for interfacing to the next level of packaging are often a part of the solution.

Because of the large number of MCM to board connections possible, even a very low individual failure rate can seriously affect system reliability. Thus, it is important to use materials, techniques, and processes that will ensure a very high attachment reliability throughout the system lifetime. Several references address this subject for a variety of applications.[26–29]

To provide a high reliability in the attachment of the MCM to the next level, the following factors must be considered:

- Established design practices of the assembly factory
- Number of signal I/Os and their signal integrity requirements
- Number of P/G connections and their spatial distribution and electrical requirements
- Plugability requirements of the connection (Must it be separable, or is it sufficient to be a reworkable or permanent connection?)
- Insertion force requirements if it must be a plugable connection
- Physical height that can be allocated to the connection

- The minimum electrical properties that can be tolerated (electrical delay, resistance, inductance, and capacitance)
- The long-term reliability requirements
- The operating environment the finished assembly will experience

The answers to these questions will point the direction to the classes of connection solutions that are worth considering.

9.3 ATTACHMENT APPROACHES

The MCM-to-substrate connections can be made exclusively around the periphery [peripheral I/Os or over the area (area array I/Os) of the MCM-board interface, or as a combination of both]. This choice is one of the more important up-front decisions that the designer must make and is usually strongly influenced by the commercial availability of the parts in question and their compatibility with the assembly processes in place. The decision is made by trading off different conflicting and competing factors.

Another fundamental MCM connector decision concerns separability. If the MCM is an integral part of inexpensive, "disposable" equipment, it need not necessarily be separable. On the other hand, if the MCM is to be a field replaceable unit, it must be separable at the MCM-to-board connector level. These connections are designed to be made and unmade repeatedly and easily. Pressure is applied between the separable, mating contacts by means of mechanically applied force.

We will consider only separable connector approaches in this section. A fundamental constraint imposed by basic physics is that a separable contact requires a minimum amount of stored energy to provide a reliable contact. This means that there must be an externally applied force acting upon the mating contact surfaces. There are many types of connectors used for MCMs, but the use of gold MCM pads with gold pads on the printed wiring board is generally advisable, with a vertical force of up to 50 or more grams per contact.[1] For MCMs with I/O needs up to many hundreds of contacts, that can require very high total forces on the module. Such Z-axis connectors therefore generally require backup plates and pressure frames. Smaller MCMs can use lighter plates or perhaps even utilize the inner plate of metal core PCBs. The pressure forces must be carefully considered in using these connectors, and the pressure requirements, on a per contact basis, are essentially the same for all types.

9.3.1 Peripheral I/0 Connections

This approach is very much in tune with many present-day manufacturing processes. It has historically been the method of choice for most non-field replaceable (consumer-type) products. It is extendible to relatively high I/O numbers (within reasonable limits) by increasing the MCM size or by further decreasing the pitch of the I/O leads. It is preferred over area array connections by some manufacturing engineers today (particularly for low-cost applications) because the MCM-to-board connections will usually be more accessible and inspectable than other alternatives.

9.3.2 Array Connections

There are three general array connection methods in use today for attaching modules to PCBs. They are the *pin grid array* (PGA), *ball grid array* (BGA), and *land grid array* (LGA) approaches. The array approach is generally capable of providing more I/Os than the peripheral approach for a comparable connection pitch.

Modules with area array pins that plug into a PGA connector are the most common examples of this attachment category. However, for high-performance designs, a pinned approach is increasingly less likely to be a satisfactory solution. For example, the electrical path may be too long and have unacceptable parasitics, the size of the holes and the fact that they pass entirely through the board may affect routing, the rest of the system may be all surface mount, and so on.

BGA packaging technology has been the subject of intense development throughout the 1990s. The principal drivers for this have been the BGAs inherently higher I/O count capabilities for a given size package as well as the robustness of its assembly processes.[30,31] Instead of pins plugging into an array of holes, solder balls on the bottom of the package are soldered to an array of pads on the PCB. There have been reliability concerns[32–34] that are being carefully worked through. The BGA has become the package of choice for high I/Os (200 or greater).

In the LGA, in place of pins plugging into holes, an array of metal pads on the bottom surface of the MCM is aligned with a corresponding array on the PCB. Vertical force is applied to the module package, and the pressure on each pad is applied to some resilient contact structure. There are many types, but most require gold contact pads on the MCM and gold pads on the PCB, with an applied vertical focus on the order of 50 or more grams per contact. For MCMs with I/O needs of around 500 contacts, that imposes a force of several kilograms on the MCM. In the pressure-contact LGA, this force is generally applied externally across the whole MCM package, requiring carefully designed hardware, probably including metal pressure plates and backup plates. Examples of specific connection system solutions will be discussed in the following section.

9.4 SURVEY OF MCM ATTACHMENT OPTIONS

9.4.1 Pins

There exist numerous MCM attachment options. The common ones entail the use of pins on the MCM mating to a socket or holes in a PCB. The established nature of this connection system is its biggest advantage. The primary disadvantage of pins is their relatively large spatial requirement for plated-through holes, which can quickly lead to a requirement for large layer count, generally expensive, multilayer boards. For high-end systems, the inductance of pins can also be performance limiting.

Plugable Pins as the MCM-PCB Interface. Here, the bottom surface of the MCM typically presents a sea of pins that interface to a pin contact. The pin contacts are usually bifurcated to provide a redundant contact to the pin and enhance the reliability.

Figure 9.1 shows a representative pin connector system,[35] and Fig. 9.2 shows a plan view of a dense pin array on MCMs, as used in the IBM ES/9000 mainframe.[16]

When the I/O count becomes very large, the overall force required to open each and every pin contact simultaneously when inserting the MCM into the connector can become insignificant. *Zero-insertion force* (ZIF) connectors were developed to address this. ZIF connectors typically have bifurcated contacts with a side-actuated cam mechanism to open up the contacts for vertical insertion or removal of the pin. Figure 9.3 represents such a ZIF-openable contact system, where an independently applied sideward actuating force opens and closes the contacts.

Soldering. The same pins that can be plugged into a bifurcated pin contact also could be directly inserted into plated-through holes in the PCB and soldered in place. Wave soldering of pin grid arrays constitutes a mature and robust assembly process and provides a more permanent form of attachment. Rework and repair is generally more difficult with this method but still feasible. It can be a relatively effective solution with a small or moderate sized MCMs. Special attention must be given to a design sufficient compliance into this contact system to accommodate the differential thermal expansion that will occur over the operating temperature range. Furthermore, the presence of pins in the plated-through holes in general requires a larger diameter hole to be specified than in an SM environment. This reduces the available space for signal traces. As a consequence, more signal layers may be needed to provide the required wireability in the PCB.

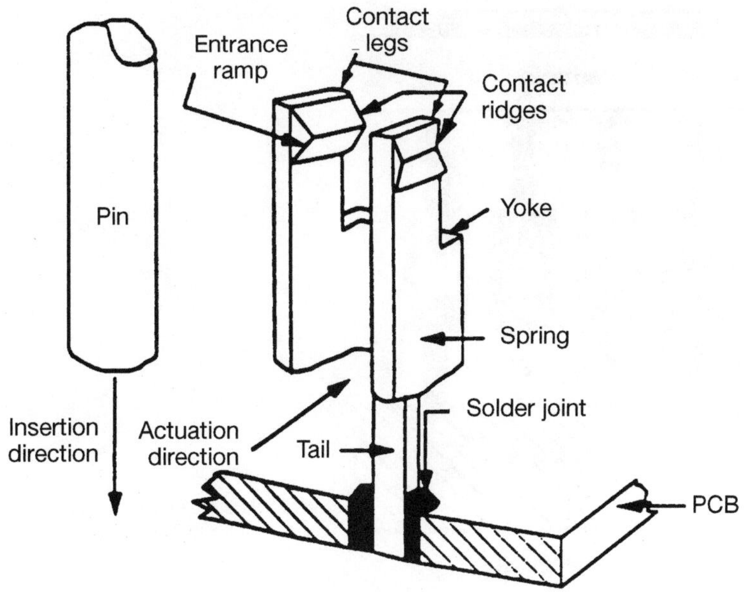

FIGURE 9.1 Bifurcated mechanical pin connector.

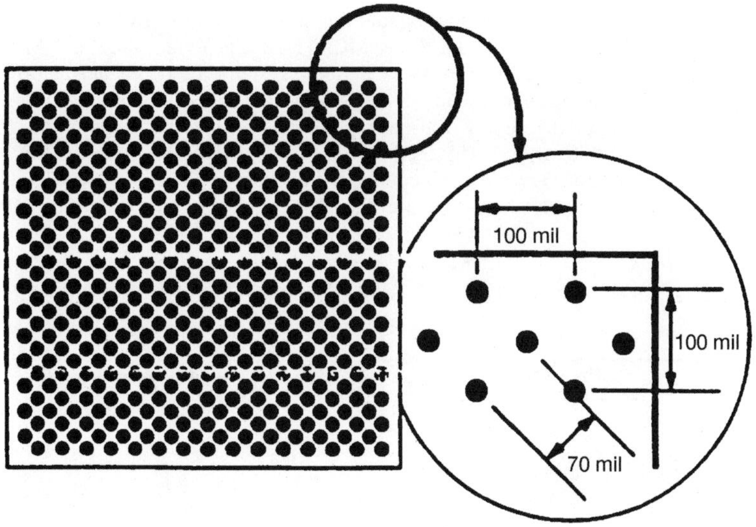

FIGURE 9.2 State-of-the-art pin density (after Ref. 35).

9.4.2 Solder Joints at the MCM–PCB Interface

Solder historically has been one of the most natural methods of attaching components to PCBs. However, the physical and mechanical features of the solder joint must be very carefully designed to

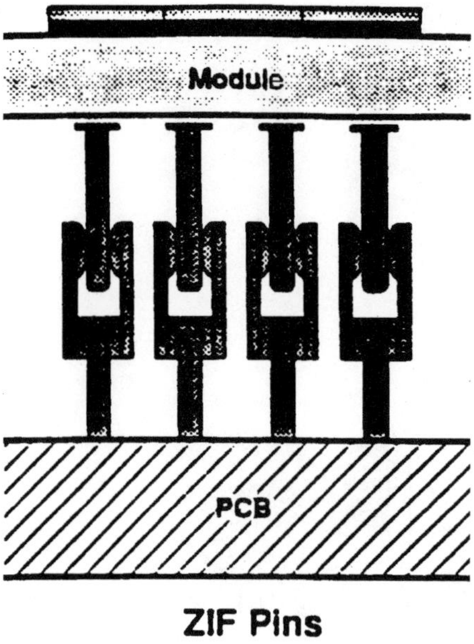

FIGURE 9.3 Zero insertion force connector concept (after Ref. 35)

achieve the desired robust characteristics. This is particularly so for MCMs, since they can be physically larger than most other electronic components, leading to heightened reliability concerns.

Solder Ball. Solder balls form the module-board interface in BGAs. This is the analog of the flip chip attachment process as applied to the MCM-PCB attachment interface. Most high-I/O (<200) component attachment needs for today's electronic systems are being satisfied by this BGA format. Particular attention must be paid to designing the proper compliance into the solder ball joints.[32,34,36,37] In practice, this will usually place size or material limits on the MCMs that can be interconnected by this method. This method can lead to very robust, high-quality attachment process with the capability for very large I/O counts.[38–41] Figure 9.4 shows a cross-sectional view of such a joint before and after reflow attachment of the MCM package to the PCB substrate.

Posts or Pillars. Post and pillar technologies evolved to provide a larger amount of compliance at the package-PCB interface than solder balls could provide. Pillars or posts are generally formed of

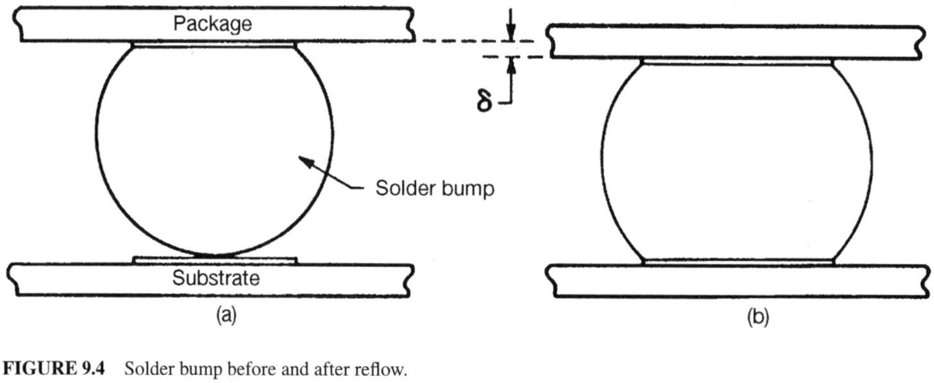

FIGURE 9.4 Solder bump before and after reflow.

solder, or solder coated wires or balls. Among the methods utilizing solder are precast pillars made of higher (than eutectic) melting temperature solder[42] and the double bump technique.[43] Cast pillar technology utilizes a 1–5 mm high pillar of 95/5 tin-lead solder which is cast (molded) in place on each MCM I/O pad. This uses the self-centering capabilities of solder bump technology as well as providing a very high level of mechanical compliance (see Fig. 9.5 for an example of self-centering during cast lead application to the MCM pads). Double bump attachment technology is another way to provide a higher solder joint with more mechanical compliance. It uses mechanical standoffs and the larger volume of solder realized by forming a bump on both each PCB I/O pad and each MCM pad to form a controlled-geometry, elongated solder joint. Figure 9.6 shows a cross section of the double bump assembly before and after reflow.

9.4.3 Compliant Springs at the MCM–PCB Interface

Compliant connectors are growing in importance, because they offer a number of capabilities that are consistent with the trend in the electronics industry toward higher-density packaging. The two principal types of compliant connectors are:

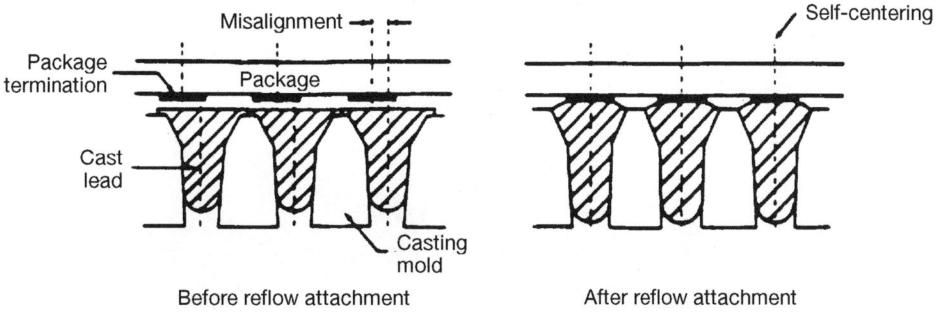

FIGURE 9.5 Attachment of cast leads to package terminations (after Ref. 42).

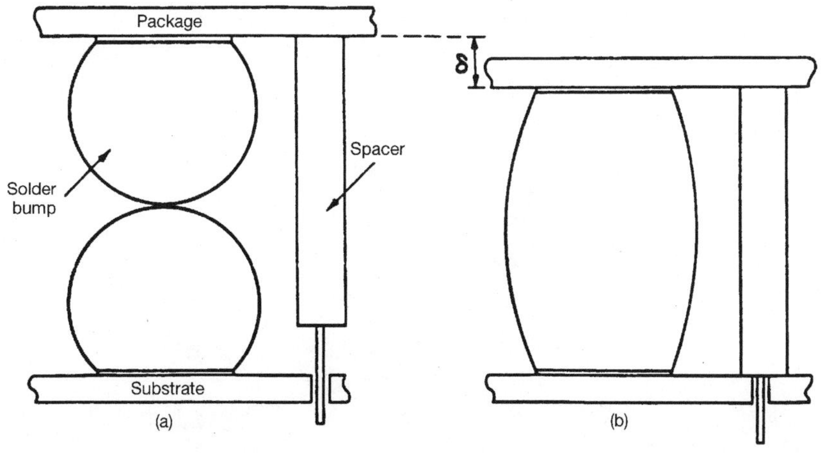

FIGURE 9.6 Double bump assembly view (*a*) before and (*b*) after reflow (after Ref. 43).

1. Elastomeric (vertically aligned, conductive balls or embedded wires in elastomer)
2. Metal spring contacts (compliance and electrical contact both provided by a spring)

Consider the interface between two planar surfaces with distributed arrays of contacts facing each other (see Fig. 9.7) where each contact is microscopically different with respect to its opposing contact. When the surfaces are mated in the general case, some contacts will mate, while others will not. Properly conductive and compliant springs interposed between the contacting surfaces will ensure that proper contact is made. There is a variety of commercial products designed to fill this need.[44] Table 9.1 provides an overview of some of them and their areas of applicability. Details about some of these connectors are described in the following sections.

Elastomer Embedded Compliant Springs. There is a variety of connectors which utilize an elastomer to provide some or al of the compliance required by their contacts. A characteristic common to these connectors is that their compliance can allow interconnections to be made over an area array. These connectors appear in two basic variations.

Electrical Contact Internal to Elastomer. In this category, an elastomer sheet which contains electrically conducting paths in the z-direction, is placed between the MCM pads and the contact pads on the PCB. The compliance and z-axis conductivity of the sheet enable a robust and reliable contact to be made between opposing, aligned contacts on the MCM and PCB. Such planar, z-axis conducting, elastomer materials come in two basic types:

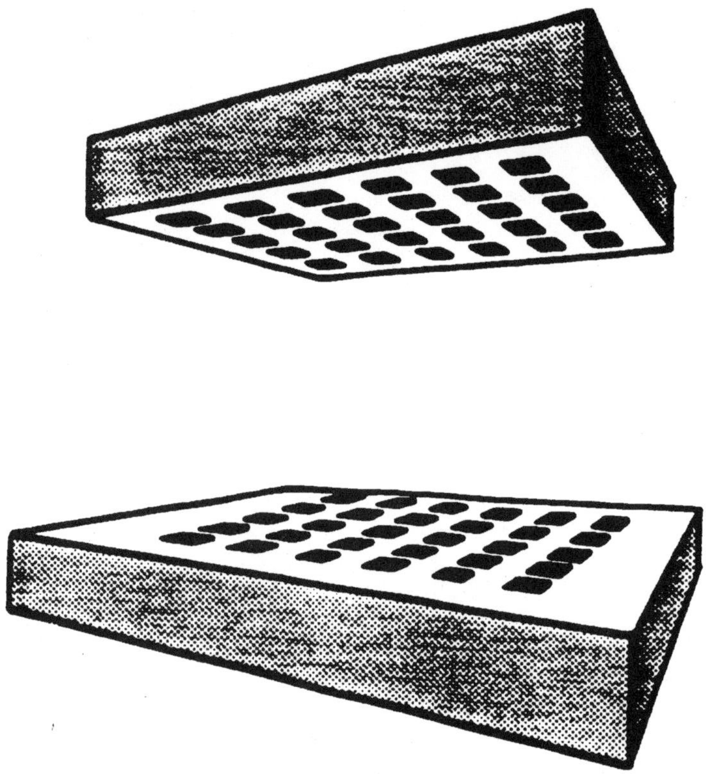

FIGURE 9.7 Planar surfaces to be electrically mated.

Table 9.1 Properties of Commercially Available Planar Connectors

Connector	Manufacturer	Description	Pitch	Compliance	Performance	Comments
Interposer	AMP	7.5–25 mil centers Ω or M shaped conductors	>10 mils 95–120 mils thick Al-epoxy housing	>10 mils	10 mΩ, wipe >100 g contact force 1.4 nH, 35 ps PD 2.5 mil wipe	1. linear contacts 2. alignment required
Isocon	Rogers	S-shaped conductors, cellular elastomer 80 mils thick	>50 mil	40 mils	<10 mΩ 10 psi min. 10,000 cycles 9 mils wipe <1 nH	1. alignment required
Flex/ elastomer	Beta Phase, Hughes, Intel	Au dots on flex circuit with elastomer backing				1. alignment required
Wire button	Tecknit, TRW, Cinch	20–50 mils dia. button	50 mils	10 mils	<10mΩ 40,000 cycles 1 oz/button 5 A, 0.5 nH 1.7 GHz 100 ps RT	1. alignment required 2. ≤15 mil pitch difficult
ECPI	AT&T-BL	10–30 mils thick aligned particles silicone polymer	>10 mils	10 mils	<10 mΩ 0.3 nH >30,000 cycles	1. redundant interconnect
W series	Fujipolymer	0.5–2 mil dia. wire 12–90 mils thick silicone polymer	>8 mils	10% thickness	<400 mΩ (25×25 mil^2) 200,000 cycles	1. redundant interconnect
PCR 301-02C	Japan Synthetic Rubber	10–40 mils thick 2 mil dia. particles in silicone polymer	>10 mils		<50 mΩ	1. redundant interconnect

1. Discrete contact, where the interconnection is formed by a single, aligned vertically conductive element (see Fig. 9.8*a*). This class of materials requires precise alignment of these conductive elements with respect to the pads being interconnected. The Rogers Isocon™ connection system[45] is an example of this type (see Fig. 9.9).
2. Redundant contact, where several conductive elements make connections per contact area (see Figs. 6.8*b* and 6.10). This class of materials requires exact alignment only between the pads being interconnected. The elastomer is inserted into the plane between the pads without any specific need for alignment. Three commercial examples of elastomeric redundant contact materials are the AT&T *elastomeric conductive polymer interconnect* (ECPI) material, the Fuji polymer W series elastomeric material, and the Japan Synthetic Rubber PCR 301-02C elastomeric material. Their cross-sectional appearance is compared in Fig. 9.11. Their characteristics are contrasted in Table 9.2. The AT&T ECPI materials[46] have vertically aligned, 1.0- or 3.0-mil diameter noble metal-plated particles which are designed to make interconnections on a 15 mil or 25 mil pitch. The Fuji polymer W series elastomeric connector material has vertically aligned, gold-plated 1-mil diameter nickel wires which protrude 0.5 mil beyond the plane of the sheet. The Japan Synthetic Rubber material is similar to the AT&T material, except that the particles are 2.0 mils in diameter and not plated. The IBM ELASTICON™ interposer structure[47] is yet another example at each conductive site. However, they are placed at an angle, which requires that the pads being interconnected be precisely aligned with a lateral offset to each other.

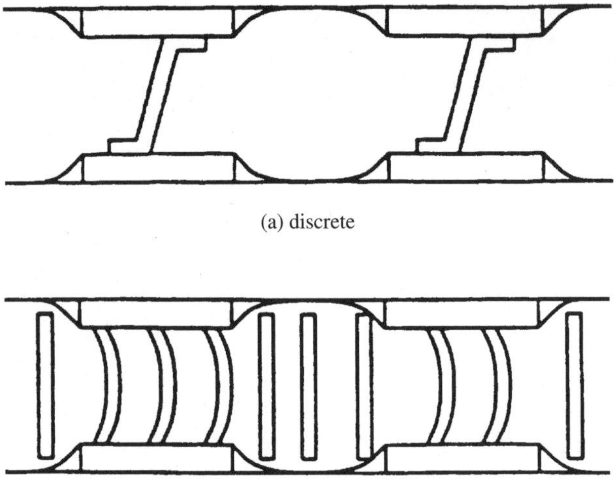

(a) discrete

(b) redundant

FIGURE 9.8 Two types of planar elastomer connectors.

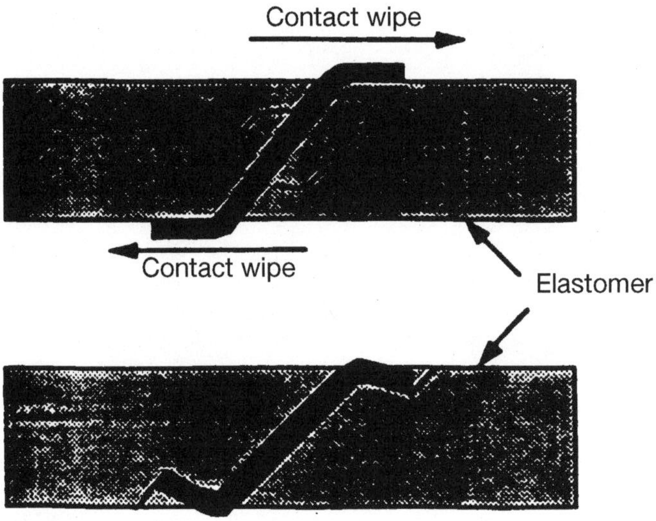

FIGURE 9.9 The Rogers ICOCON™ connector in cross section in the uncompressed and compressed states (after Ref. 45).

References 48 and 49 report on a critical comparison of some of these interconnection systems, and references 50–53 provide other useful information.

Electrical Contact External to Elastomer. Here, elastomers are generally used to help in adjusting for planar nonuniformity in the connector system and provide uniform contact pressure. The wrap MOE™[54] is an externally metallized elastomer (see Fig. 9.12) that can be used for interconnecting linear arrays of pads, as in a leadless, peripheral I/O MCM.

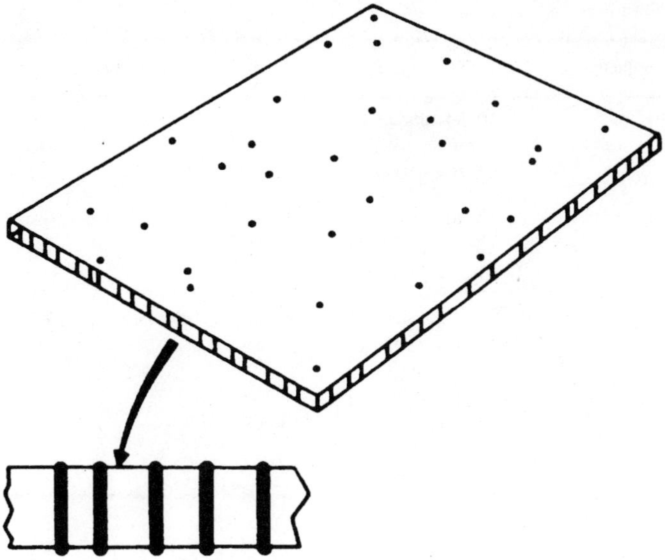

FIGURE 9.10 Elastomeric structure showing redundant conducting columns.

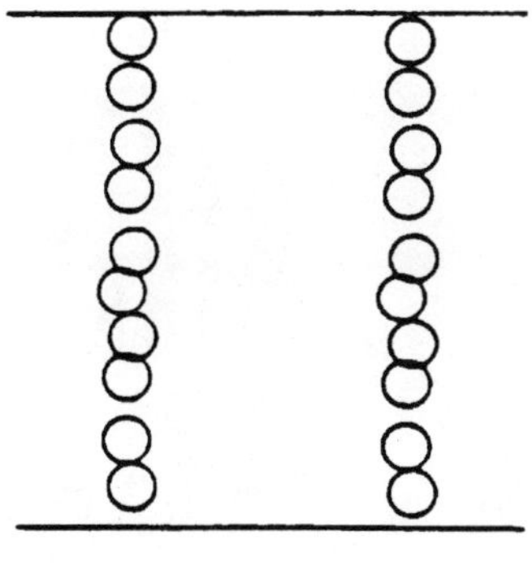

Elastomeric Conductive Polymer Interconnect, AT&T Bell Laboratories

PCR 301-02C, Japan Synthetic Rubber

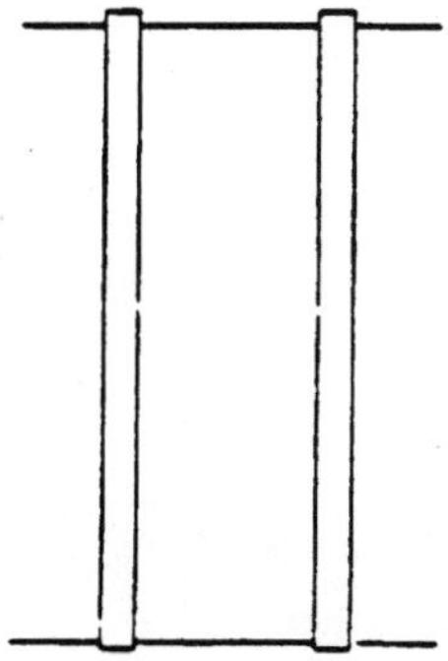

W series elastomeric connector, Fujipolymer

FIGURE 9.11 Cross section of redundant contact sheet elastomers.

Table 9.2 Examples of Planar Elastomeric Connectors

Connector	Manufacturer	Description	Pitch	Compliance	Performance
Coarse-pitch ECPI	AT&T-BL	10–30 mils thick 3-mil dia. particles silicone polymer	>10 mils	10 mils	<10 mΩ 0.3 nH >30,000 cycles
Fine-pitch ECPI	AT&T-BL	10–20 mils thick 1-mil dia. particles silicone polymer	>6 mils	5 mils	<100 mΩ 0.3 nH <1,000 cycles
W series	Fujipolymer	0.5–2 mil dia. wire 12–90 mils thick silicone polymer	>8 mils	10% thickness	<400 mΩ (25×25 mil²) 200,000 cycles
PCR 301-02C	Japan Synthetic Rubber	10–40 mils thick 2-mil dia. particles silicone polymer	>10 mils		<50 mΩ

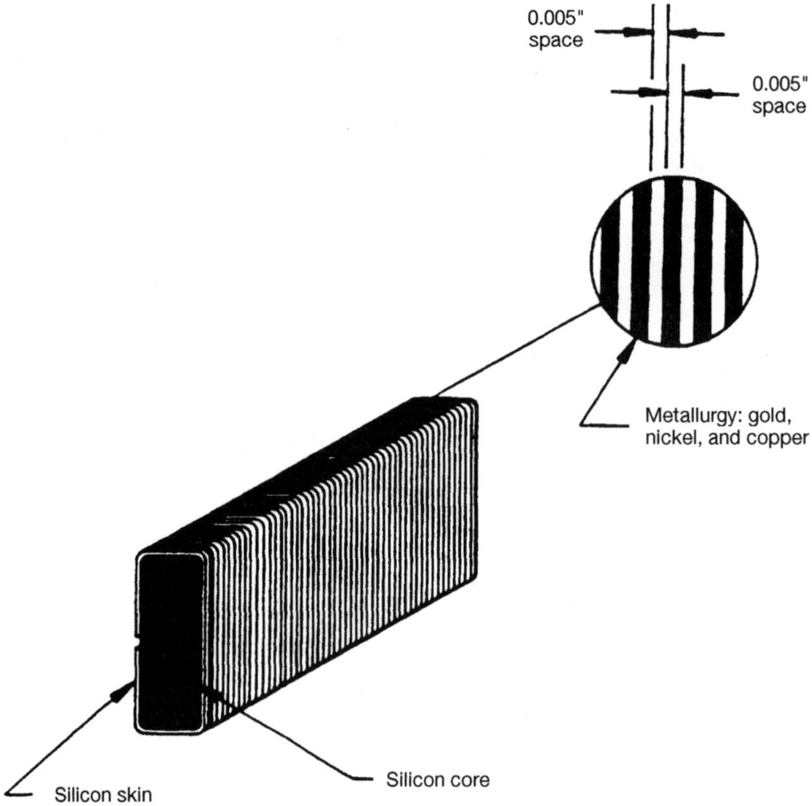

FIGURE 9.12 The wrap MOE® is available in a wide range of sizes and geometries. The simplest and most common version is shown here (after Ref. 54).

FIGURE 9.13 Button board interconnect contacts. Courtesy of TRW Electronics Systems & Technology Division.

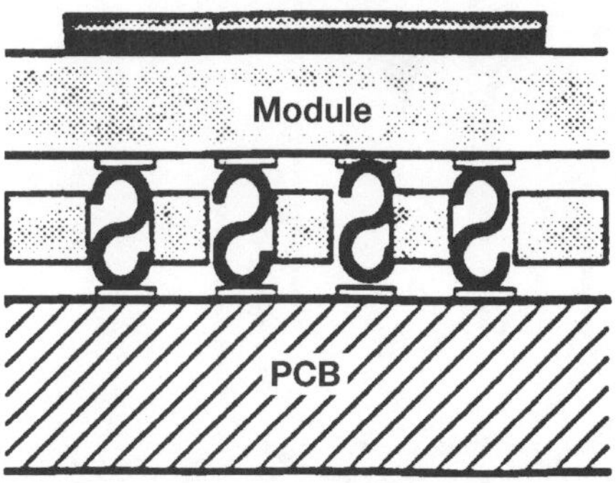

FIGURE 9.14 Rigid interposer housing with compliance provided by springs (after Ref. 35).

Rigid Housing Embedded Compliant Springs. Here, the columns of vertically conducting balls used in the elastomeric connectors are replaced by an array of metallic springs. The springs can be part of the contact system, as in the TRW "gold button" system[55,56] shown in Fig. 9.13. The two-sided button board allows direct MCM-to-PCB connections. Alternatively, the springs can be a separately applied compliant element as is represented schematically by the "S" contacts and their carrier in Fig. 9.14. Among the products of this type on the market is a family of low-inductance planar connectors from AMP, Inc.[57] Fig. 9.15 shows a cross section of the spring elements in the Interposer™, the microInterposer™, and the AMPFLAT™ products. The Interposer contacts are miniature spring contacts, while the AMPFLAT™ contacts are etched out of thin sheet stock and alternately deformed in opposing directions to provide the springs. Table 9.3 contrasts some of their characteristics.

For MCMs with perimeter leads, the *leaded multichip module connector* (LMC) can be used. Figure 9.16 illustrates the concept of the interconnection. It is a compact, low-profile ZIF connector, which is actuated by means of a plastic wedge driving the package lead into intimate contact with the

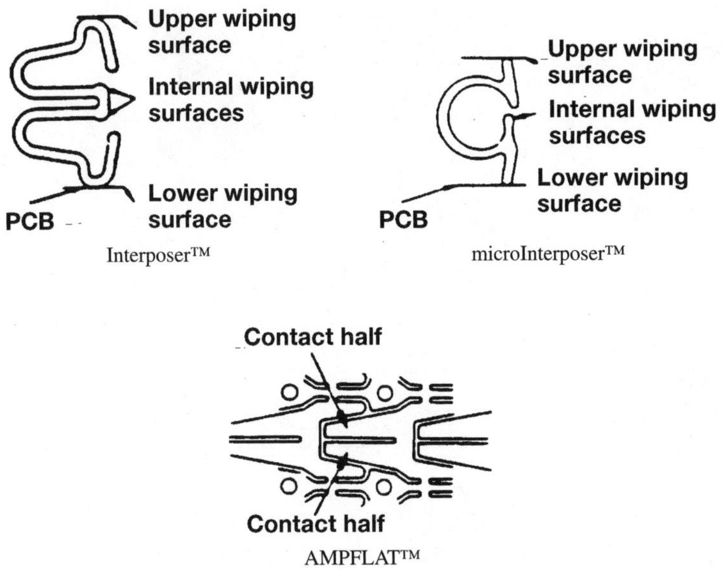

FIGURE 9.15 Contacts from a family of planar connectors for MCM-PWB connections. Courtesy of AMP, Inc.

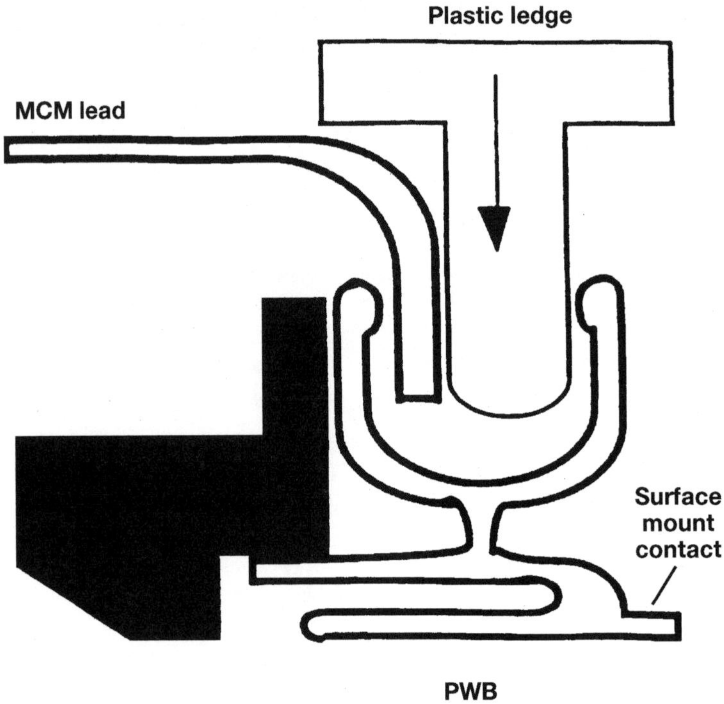

FIGURE 9.16 LMC connection concept. Courtesy of AMP, Inc.

Table 7.3 Characteristics of Family of Planar MCM-PWB Connectors Requiring Externally Applied Pressure

	Height (mils)	Minimum spacing (mils) In-line	Minimum spacing (mils) Area array	Insertion force (g)
Interposer™	120	20	60	300
microInterposer™	90	10	50	120
AMPFLAT™	9		5	

socket contact. In Fig. 9.17, this type of connector system is shown. It is surface mountable, can have a modular edge dimension (and I/O capacity), and has no need for external application of pressure.[58]

9.4.4 Conductive Adhesives

Conductive adhesives have become a viable assembly option in recent years. Adhesives of the type used for die attachment can be used but are incompatible with high-density I/O applications. Aniso-

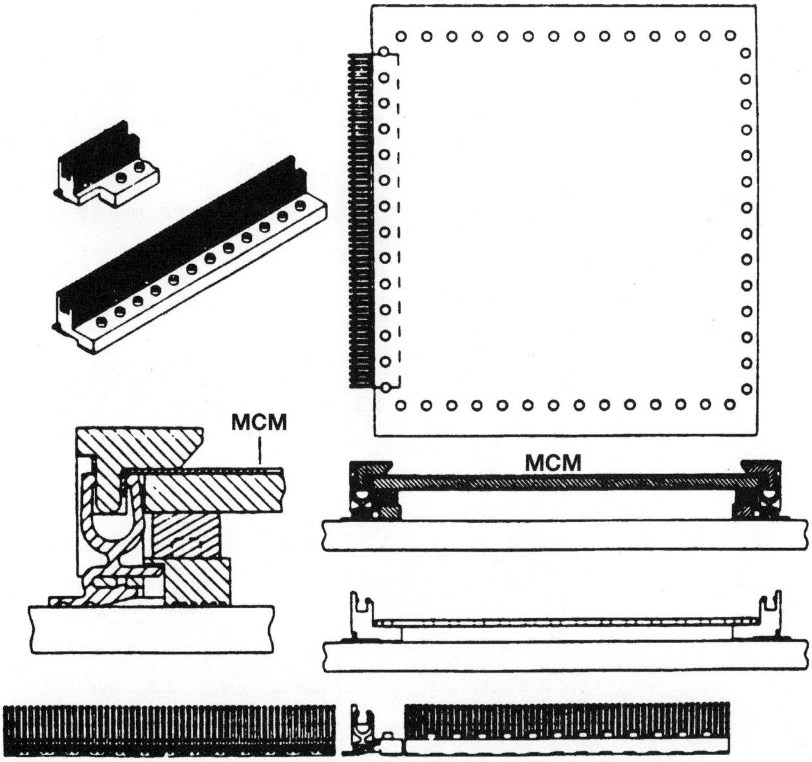

FIGURE 9.17 Leaded chip module connector (LMC) system. Courtesy of AMP, Inc.

tropic conductive adhesives, on the other hand, appear to be compatible with the highest density attachment requirements envisioned for MCMs. They are a relatively new class of adhesives that are conductive in the z-direction and insulating in the x and y directions.[59–64]

9.4.5 Flex Circuits

Figure 9.18 illustrates the concept behind a family of connectors with a novel application of shaped metal memory alloys. The shaped metal alloy allows the insertion of the MCM with zero insertion force when the connector is in the open position. The connector's actuation mechanism utilizes the heat-actuated shaped metal alloy effect.[65,66] A short burst of current causes the flex circuit contacts of the connector to close upon the MCM contact pads. Figure 9.19 shows an example of a ZIF flex circuit connector for MCM-PCB attachment by Betaflex, Inc., which utilizes shaped memory alloy metal piece parts to actuate all four sides. Figures 6.20 and 6.21 show a cross-sectional view of such a flex circuit based connector, developed by Digital Equipment Company for the multichip unit of the VAS 9000 mainframe computer.[21] Two factors are of critical importance in determining the performance of this connector design:

1. The stiffness of the flexible circuit
2. The coplanarity of the contact array

A compliant elastomer spring was used to optimize these parameters.

Flexible circuit interconnects are a way to provide very compliant MCM-PCB connections that can be designed to have excellent electrical properties.[67] The electrical properties of the interconnection are largely determined by the design of the flexible circuit, which can be tailored to the specific

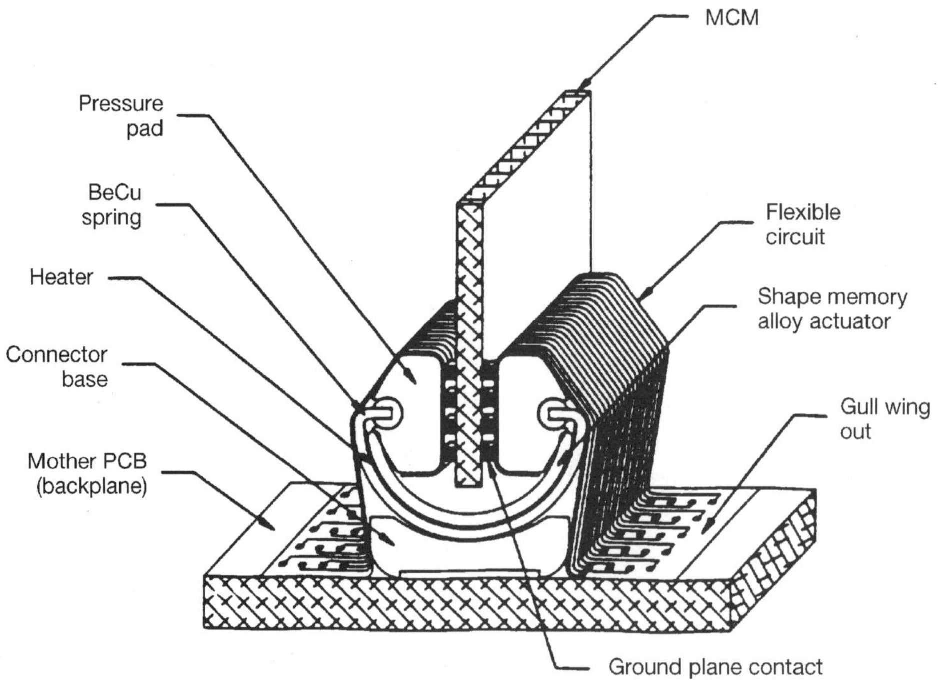

FIGURE 9.18 Flex circuit ZIF connector concept using shaped metal memory alloy actuator mechanism. Courtesy of Betaflex, Inc.

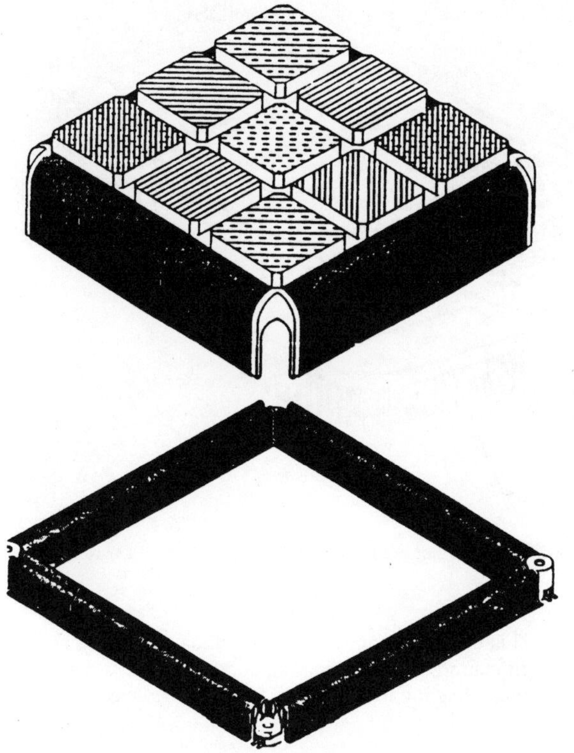

FIGURE 9.19 Cross-sectional view of flex circuit connector assembly. Courtesy of Digital Equipment Corp.

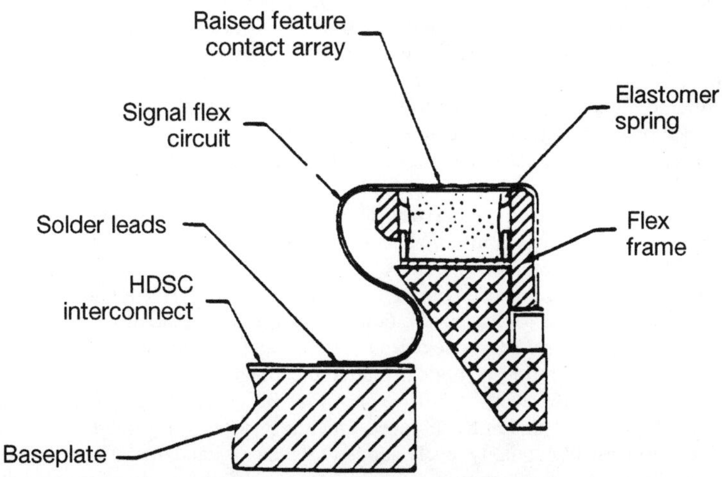

FIGURE 9.20 Cross-sectional view of flex circuit connector assembly. Courtesy of Digital Equipment Corp.

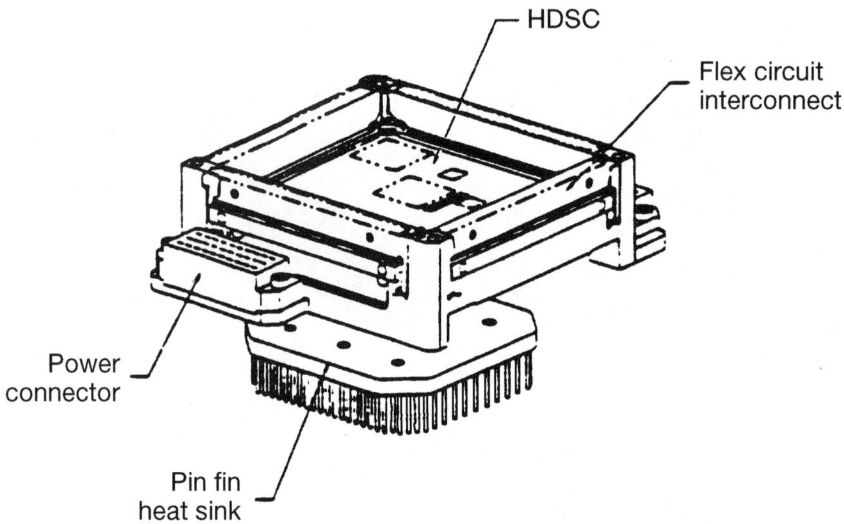

FIGURE 9.21 The VAX9000 multichip unit and its flex circuit based connector. Courtesy of Digital Equipment Corp.

needs of the application. This class of connection easily allows maintaining a very precisely controlled impedance path from mating contact to mating contact. There is considerable design activity among connector manufacturers leading to many new flex-based connector products.[53]

9.5 TRENDS AND FUTURE REQUIREMENTS

A continuing system-level design trend is the move toward ever more compact physical configurations, consuming less power and delivering more performance.[68,69] At the same time, individual components and modules in this equipment will moderately increase in size and vastly increase in capability and performance.[70] This will occur as a result of technology improvements and the payoffs associated with putting more functions into the high-speed, "chip-like" MCM environment. It is important for packaging designers to pre-anticipate each step of this and move in the same direction, delivering physical designs that support higher I/O and gate densities with ever more dense, higher-quality electrical and optical interconnections.

This will require the development of tighter peripheral I/O spacing (0.4 mm and smaller) for some applications. For many other applications, the wholesale move to area array connections and BGA packages will be the likely answer. This will be the greatest growth arena for packaging, with the technical driver being the insatiable demand for more I/Os in ever higher performance systems. These module-to-PCB area array connections will be evolved to smaller and smaller grid dimensions, allowing for larger and larger I/O counts on increasingly compact packages. As higher I/O counts are needed on the package and module level, the BGA spacings will progress from the 50 and 60 mil pitches of the early BGA packages to pitches on the order of 20 mils.

The seamless integration of optical interconnections into new generations of systems that are presently predominantly electrical will be one of the bigger challenges facing packaging designers in the future. This will be driven by the need for future systems to provide much higher signal bandwidths at extremely high quality levels, judged by today's standards. It is probable that future generations of many current electronic systems will gracefully evolve into mixed electrical/optical systems. This will present further challenges and opportunities for the connection of modules to the next interconnection level.

REFERENCES

1. *Connectors and Interconnections Handbook,* vol. 1, G. Derman, Ed., Int. Institute of Connector and Connection Technology, Deerfield, IL, March 1990.
2. J.K. Hagge, "Ultra-Reliable Packaging for Silicon-On-Silicon WSI," *IEEE Trans CHMT,* vol. 12, no. 2, June 1989, pp. 170–179.
3. *Multichip Module Compendium,* Int. Electronics Packaging Society, Wheaton, IL., December 1989.
4. G. Harmon, *Wire Bonding in Microelectronics,* ISHM Press, 1989.
5. G.G. Harman, "Wire Bonding to Multichip Modules and Other Soft Substrates," *Proc. 1995 Int. Conf. on Multichip Modules,* Denver, CO, April 1995, pp. 292–301.
6. Microelectronics Standards MS002-MS008, JEDEC Publication no. 95, Electronic Industries Association Engineering Dept., 2001 Penn. Ave., NW, Washington, DC, 20006.
7. *Connectors and Interconnections Handbook,* vol. 1, G. Ginsberg, Ed., Int. Institute of Connector and Connection Technology, Deerfield, IL, September 1979.
8. *Microelectronics Packaging Handbook,* Tummula and Rymaszewski, Eds., Van Nostrand 1989, p. 782.
9. C.J. Straghan and B.M. MacDonald, "Electrical Characterization of packages for High-Speed Integrated Circuits," *IEEE Trans. Components, Hybrids and Manuf. Technol.,* vol. CHMT-8, pp. 468–473 (1985).
10. A.J. Rainal, "Performance Limits of Electrical Interconnections to a High Speed Chip," *IEEE Transactions CHMT,* vol. CHMT-11, 1988, p. 260.
11. E.E. Davidson, "Electrical Design of a High Speed Computer Packaging System," *IEEE Trans. Components, Hybrids and Manuf. Technol.,* vol. CHMT-6, pp. 272–282, 1983.
12. G. Geurts, "Modern Connector Design Tools Assure High Speed Signal Integrity," *IEEE Proc. 45th ECTC,* Las Vegas, NV, May 1995, pp. 154–160.
13. J-P Clech, J.C. Manock, D.M. Noctor, F.E. Bader, and J.A. Augis, "A Comprehensive Surface Mount Reliability Model Covering Several Generations of Packaging and Assembly Technology," *IEEE Trans. CHMT,* vol. 16, no. 8, December 1993, pp. 949–960.
14. J-P Clech, "Solder Joint Reliability: Test Vehicles and Test Conditions Issues," *Proc. Second Workshop of the Solder Interconnect Design Team,* National Institute of Standards and Technology (NIST), Gaithersburg, MD, June 12–14, 1995.
15. E.E. Davidson et al., "The Design of ES/9000 Module," *Proc. 41st ECTC,* pp. 50–54.
16. R.R. Tummala et al., "Packaging Technology for IBM's Latest Mainframe Computers (S/390/ES9000)," *Proc. 41st ECTC,* May 1991, p. 682.
17. U. Shimada, et. al., "High Speed Computer Packaging with Multilayer Glass/Ceramic Substrate," *ISHM 1991 Proc.,* p. 176.
18. F. Kobayashi et al., "Hardware Technology for HITACHI M-880 Processor Group," *Proc. 41st ECTC,* May 1991, pp. 693–703.
19. K. Yokouchi, et. al., "Packaging Technology for High Speed Computers—Multilayer Glass/Ceramic Circuit Board," *ISHM 1991 Proc.,* p. 183.
20. H. Wessely et al., *Siemens R&D Reports,* Band 17 (1988) no. 5, Springer, p. 234.
21. L. Palmeri, "Performance Characterization of Vax 9000 MCU Flexible Circuit Interconnect," *Proc. 1991 IEPS Conf.,* p. 741.
22. D.M. Carlson, "Cray J90 Computer System: Air-Cooled Supercomputer Packaging," *Proc. 1994 Int. Electronics Packaging Society An. Conf.,* Atlanta, GA, September 1994, pp. 708–721.
23. W.R. Heller et al., "Wireability–Designing Wiring Space for Chips, Packages," *Design and Test,* August 1984, p. 43.
24. W.R. Heller et al., "Prediction for LDI," *Proc. 14th Design Automation Conf.,* June 1977, p. 32.
25. Ibid., #7, p. 278.
26. J.C. Mather and J.K. Hagge, "Material Requirements for Packaging Multichip Modules," *Proc. NEPCON West,* Anaheim, February 1990, pp. 1135–1142.
27. J.K. Hagge, "Mechanical Considerations for Reliable Interfaces in Next Generation Electronics Packaging," *Proc. NAECON,* Dayton, May 1989.
28. L. Liang et al. "High-Performance VLSI Through Package-Level Interconnects," *39th ECC,* pp. 519–523, 1989.

29. J-P Clech, D.M. Noctor, J.C. Manock, G.W. Lynott, and F.E. Bader, "Surface Mount Assembly Failure Statistics and Failure-Free Time," *IEEE Proc. 44th ECTC,* Washington DC, May 1994, pp. 555–561.
30. M.C. Loo et al., "Surface Mount Studies of Fine Pitch Micro BGA," *Proc. 1994 Int. Electronics Packaging Society An. Conf.,* Atlanta, GA, September 1994, pp. 555–561.
31. B. Freyman and R. Pennisi, "Overmolded Plastic Pad Array Carriers (OMPAC): A Low Cost, High Interconnect Density IC Packaging Solution for Consumer and Industrial Electronics," *IEEE Proc. 41st ECTC,* May 1991, pp. 176–182.
32. S. Bolton et al., "Influence of Plastic BGA Design/Materials Upon Solder Joint Reliability," *Proc. 1994 Int. Electronics Packaging Society An. Conf.,* Atlanta, GA, September 1994, pp. 725–739.
33. T.R. Lindley, "BGA Solder Joint Reliability Study for Automotive Electronics," *Proc. 1995 Int. Conf. on Multichip Modules,* Denver, CO, April 1995, pp. 308–313.
34.
35. N.G. Koopman, "I/O Options for MCNC Multichip Package," *Proc. 41st ECTC.,* pp. 234–244.
36. W. Engelmaier, "Functional Cycles and Surface Mounting Attachment Reliability," *Surface Mount Technology,* ISHM Monograph Series 6984-002, p. 87.
37. P.M. Hall, "Forces, Moments, and Displacements During Thermal Chamber Cycling of Leadless Ceramic Chip Carriers Soldered to Printed Boards," *IEEE Trans. on CHMT,* vol. CHMT-7, no. 4, Dec. 1984, p. 314.
38. F.F. Cappo et. al., "Highly Manufacturable Multi-Layered Ceramic Surface Mounted Package," *1991 IEEE/IEMT Proc.,* p. 176.
39. B. Freyman et al., "Overmolded Plastic Pad Array Carriers (OMPAC): A Low Cost, High Interconnect Density IC Packaging Solution for Consumer and Industrial Electronics," *1991 ECTC Proc.,* p. 176.
40. G. Coors et. al., "Pad Array Carriers—The Surface Mount Alternative," *Proc. 1991 IEPS Conf.,* p. 373.
41. R. Satoh, et. al., "Development of a New Micro-Solder Bonding Method for VLSI," *1983 IEPS Proc.,* p. 455.
42. J. Fisher, "Cast Leads for Surface Attachment," *IEEE Trans. on CHMT,* vol. CHMT-3, 34th ECC, p. 219, Dec. 1984, p. 306.
43. F. Lovasco, "Double-Bump Surface Mount Technology for Very High I/O Interconnections," *IPC Fall 1989 Meeting,* IPC-TP-835.
44. K.Wright et. al., "A New Contact/Connector Design for High-Density High-Performance Connectors," *IEEE Proc. 45th ECTC,* Las Vegas, NV, May 1995, pp. 134–139.
45. S.S. Simpson and M.E. St. Lawrence, "A High Density Land Grid Array (LGA) Connector With Wipe and High Compliance," *IEPS 1990 Proc.,* September 1990, p. 951–971.
46. J.A. Fulton, R.C. Moore, and W.R. Lambert, "Applications and Reliability of the AT&T Elastomeric Conductive Polymer Interconnection (ECPI) System," *IEPS 1990 Proc.,* September 1990, p. 930–943.
47. D-Y. Shih et al., "A Novel Elastomeric Connector for Packaging Interconnections, Testing and Burn-in Applications," *IEEE Proc. 45th ECTC,* Las Vegas, NV, May 1995, pp. 126–133.
48. W.R. Lambert and W.H. Knausenberger, "Elastomeric Connectors—Attributes, Comparisons and Potential," *Proc. of 1991 NEPCON West,* p. 1512.
49. G. Rosen, P.T. Robinson, V. Florescu, and M.T. Singer, "A Comparison of Metal-in-Elastomer Connectors—The Influence of Structure on Mechanical and Electrical Performance," *IICIT 1990 Proc.,* Montreal.
50. C.A. Haque, "Characterization of the Metal-in Elastomer Contact Material," *Proc. 35th Holm Conf. on Electrical Contacts,* 1989, p. 117–128.
51. H. Yonekura, "Oriented Wire Through Connectors for High Density Contacts," *Proc. NEPCON West 1990,* February 1990, p. 57–71.
52. R.C. Moore, J.A. Fulton, and W.R. Lambert, "Reliability of Anisotropically Conductive Elastomeric Interconnects," *EXPO SMT 1990 Proc.,* July 1990, San Jose.
53. H.W. Markstein, "Connectors for High Speed Signals," *Electronic Packaging & Production,* Sept. 1991, p. 44.
54. J. Kunihiro and L. Buchof, "The Versatility and Reliability of Elastomeric Connectors," *1990 IEPS Proc.*, Marlborough, MA, p. 915.
55. C.W. Pike and R. Hasan, "Wire Button Contacts as a Connection Alternative: Design Opportunities and Challenges," *IEPS 1990 Proc.,* September 1990, p. 944–950.

56. P. Hurt, "Advanced High Speed Packaging and Interconnect Developments," *IEPS 1991 Proc.,* p. 1527.
57. M. Freedman, "MCM Interconnection Options," *1991 NEPCON West Proc.,* p. 1527.
58. D. Grabbe, "ModuPak: A New, Low Cost, High Speed MCM Socketing System," *1991 NEPCON West Proc.,* p. 1537.
59. S. Etemad, X. Quan, and N.A. Sanders, *Appl. Phys. Let.* 48, p. 607 (1986).
60. X. Quan, *J. Polym. Chem., Poly. Phys. Ed.,* 25, p. 1557 (1987).
61. S. Crum, "Conductive Adhesive Use on Upswing," *Electronic Packaging & Production,* August 1991, p. 26.
62. S. Spitz, "Conductive Polymers Come Out From Labs," *Electronic Packaging & Production,* Feb. 1991, p. 64.
63. S. O'Brien, "Conductive Adhesive Specifier Guide," *Connector Specifier Magazine,* May 1995, pp. 13–15.
64. C.H. Lee and K.I. Loh, "Fine Pitch COG Interconnection Using Anisotropically Conductive Adhesives," *IEEE Proc. 45th ECTC,* Las Vegas, NV, May 1995, pp. 121–125.
65. H.Kent, "Multi-Chip Module Connectors and Sockets: High Density and High Speed," *1991 IEPS Proc.,* p. 726.
66. J. Krumme, "LIF and ZIF High Density and High Signal Integrity Flexible Circuitry Connectors," *1990 NEPCON East Proc.,* p. 231.
67. D.M. Bruner, "Z-Axis Adhesive Film Interconnect Using 150 Micron & 75 Micron Pitch Flex Circuits," *Proc. 1994 Int. Electronics Packaging Society An. Conf.,* Atlanta, GA, September 1994, pp. 364–375.
68. M.R. Pinnel, et. al., "Interconnection System Requirements and Modeling," *AT&T Technical Journal,* vol. 66, Issue 4, July 1987, p. 31.
69. W.H. Knausenberger, "Design and Packaging Trends in Electronic Systems," *Proc. IPC Electronic Packaging Conf.,* Los Angeles, CA, October 27, 1994.
70. J.B. Gillett, "Trends in High Performance Connector Requirements," *Proc. 1991 NEPCON East,* p. 255.

CHAPTER 10

MCM DESIGN

James R. Trent
Motorola Semiconductor Products Sector

10.1 INTRODUCTION

In today's ever changing design environments, customers desire more and more system-level support and shorter and shorter cycle times. Product cycle times are becoming shorter which, in turn, requires design process times to shrink. To reduce product cycle times, engineers are focusing on front-end modeling and analysis so that the emphasis will be on concurrent engineering during the design process instead of the traditionally slow, iterative product development process that is graphically illustrated in Fig. 10.1.

Therefore, design is no longer synonymous with computer aided design (CAD) but combines the requirements of CAD with the need for computer aided engineering (CAE). As CAD and CAE become better integrated, computer aided manufacturing (CAM) will follow directly behind so that all the engineering tools from conception through product delivery are tightly integrated.

10.2 CHAPTER PURPOSE

The purposes of this chapter are to present the process by which MCMs are designed today using tools provided by CAD/CAE companies, to discuss the future direction of the CAD/CAE companies, to outline a commonly used design process from design conception through final test, and to provide

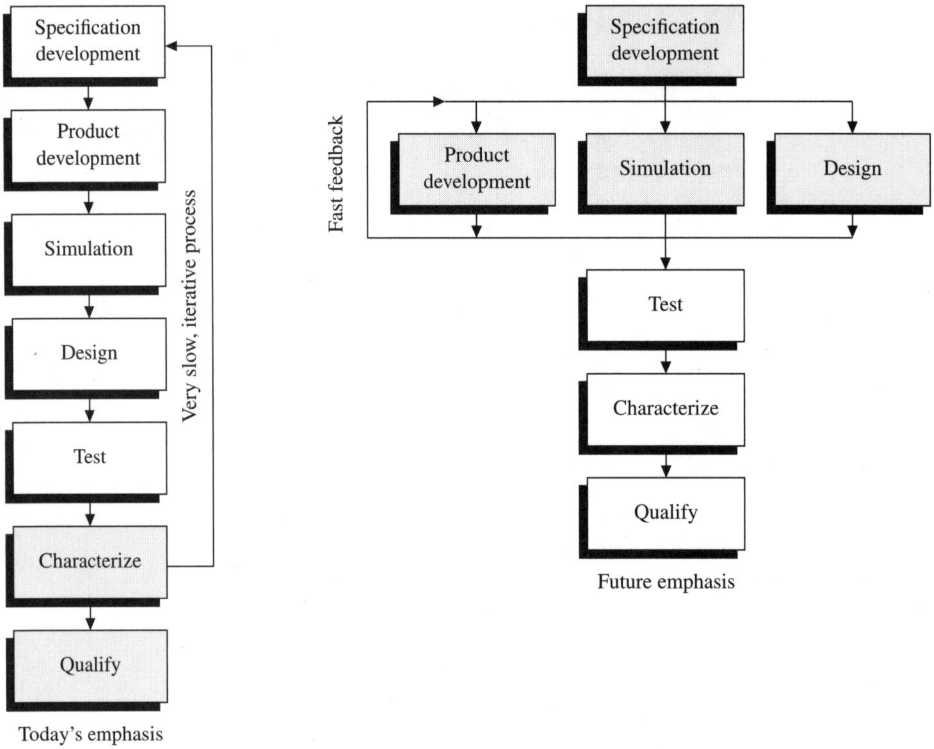

FIGURE 10.1 Improving the product development cycle: (left) traditional and (right) increased emphasis on concurrent engineering, in development.

helpful hints for ensuring first-pass success. The process will show how CAD/CAE tools play an important part in the design, development, and test of MCM products.

10.3 REASONS FOR A DESIGN INFRASTRUCTURE

Design infrastructures have been put in place over the years to ensure that the intent of the design is captured during the design phase. Before computer aided design systems became available, the product development cycle times were rather long. Many hours were spent prototyping, evaluating, and testing the system. This system was very prone to error because of the many tedious, manual operations put in place to mediate errors.

To ensure that designs meet the timing and signal integrity requirements, engineers would rely on experience or perform some "back of the envelope" calculations. Most design cycles included an electrical breadboard design, followed by an engineering board design, and ended with the final product. This process took months, many engineering hours, and deep pockets to validate that the design performed to the requirements.

Today, product development cycle times are becoming shorter and shorter. The faster a product can be designed, implemented, and placed in the market, the more revenue the company will receive. Being late on product introductions is not normally a good business strategy.

To support the reduction in product development cycle times, more sophisticated design system and tools are becoming available. The faster the machine and the more accurate the tools, the faster the product can be designed and placed into the customer's hand as product. Breadboard designs and engineering models are being eliminated. Simulation and test of modules are being accurately performed on computers today that can provide a high level of confidence in the development of a functional IC, module, and/or system.

Simulation and modeling begins at the IC level. With the increase of edge rates and higher clock cycles, more simulation and analysis is being performed at the system level. In the near future, semiconductor companies will no longer be able to provide only the silicon. System houses want semiconductor suppliers to provide more system-level support or solutions as well as the IC, thus moving greater system integration away from the customer. This can be observed by the need for MCMs in several high-end applications or increased functionality added into the integrated circuits.

10.4 MCM CAD PROCESS

An MCM design process is more than just a package design process or a circuit board design process. MCMs require package design expertise, manufacturing expertise, a collection of silicon data for design, manufacturing and test, an integrated CAD system, a module test philosophy, and system-level design expertise. The goals of an MCM design process should be as follows:

- First-pass success
- On-time execution
- Design for test
- Design for manufacturing
- Meet or exceed electrical and thermal requirements
- Use of lab data to validate modeling
- Use of a seamless interfaces between CAD/CAE tools and systems

To accomplish these goals, a good design infrastructure needs to be established. The first step is to develop a process that will work for MCM technologies. Figure 10.2 represents a generic MCM design process. The CAD/CAE system must support all aspects of a package design from schematic capture, simulation, and modeling to the physical design and production of the manufacturing data through test program generation. Figure 10.2 shows a serial design methodology when in fact many of the design tasks can be performed in parallel or concurrently.

Several companies offer very comparable CAD/CAE systems for designing MCM substrates and producing the manufacturing data for the substrate supplier. Each company has positive and negative aspects. No CAD/CAE system far exceeds the others. The companies that offer good interfaces to third-party tools and highly integrated electrical, mechanical, and thermal modeling tools are a step ahead.

Not all MCM designs require the use of the entire design process as shown in Fig. 10.2. If enough analysis has been performed on similar type packages, much of the CAE modeling can be eliminated.

For example, assume a 35 mm ball grid array (BGA) package has been designed and qualified using the identical material set planned for the next design. The silicon passivation material is identical, but the chip size is smaller, and the power dissipation less. Since the chip size can affect the power dissipation of the package, a cursory thermal analysis should be performed. However, the models used on the previous design should have been parameterized to make the analysis simple and fast. If the same packaging material set is planned, any mechanical stress analysis can be bypassed, because the size of the chip is smaller and the passivation material the same; thus, the stresses in the package would be presumed less. With the chip size reduced, any coefficient of thermal expansion (CTE) mismatch problem with the package-to-board interconnects and delamination of the die from the die flag will also be reduced.

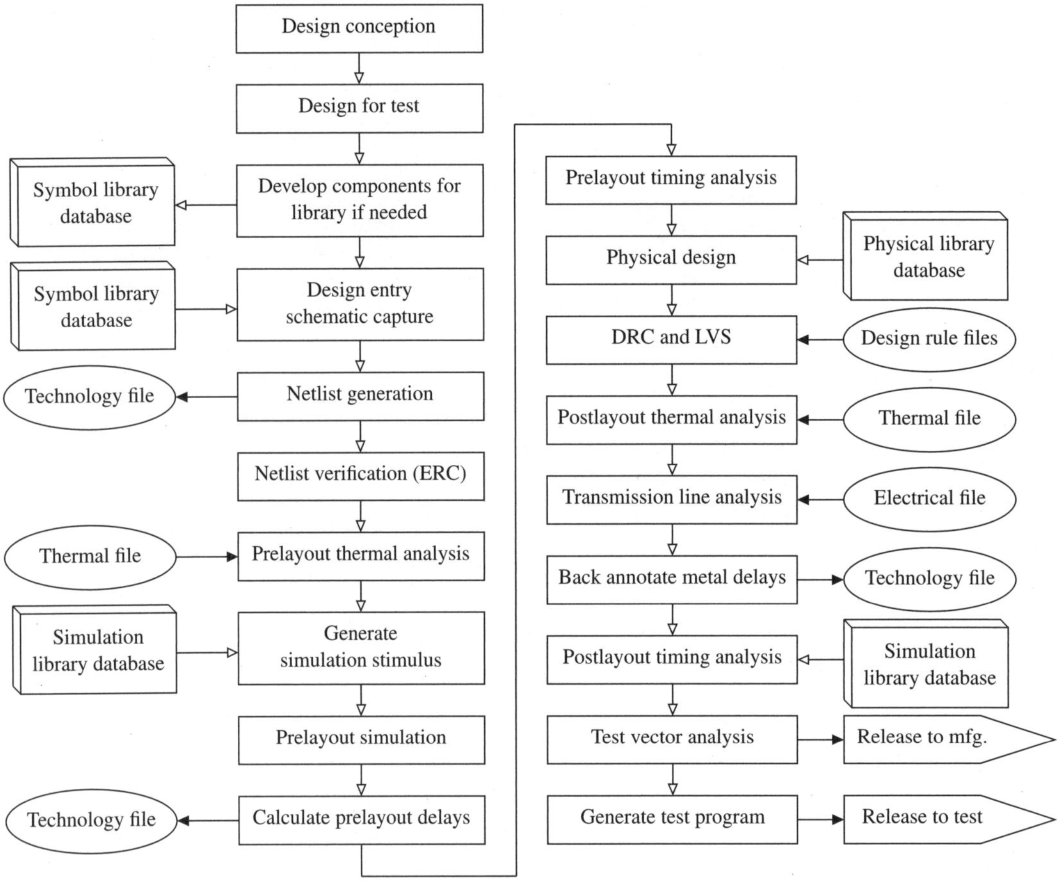

FIGURE 10.2 MCM design flow.

Short cuts or elimination of process steps should not be taken lightly. A clear understanding of all the parameters, risks, resources, cost, and schedule should be taken into account before making short cut decisions.

Most of the available CAD/CAE systems available today do not offer the full solution from finite element analysis to behavior model simulation. The semiautomated CAD/CAE systems available today focus on design cycle reduction instead of precise modeling accuracy. To reduce cycle time for modeling, behavioral level models are required, whether it is for logic, electrical, mechanical, or thermal analysis. Keep in mind that behavioral level models do not always offer the detail required to determine if a problem exists.

To enable the integration of different point tools from several suppliers, thus providing flexibility, CAD/CAE systems should be built around a framework. This tends to be a problem today in the CAD industry, because interfacing different tools to each other is a very complex problem. Large CAD/CAE companies allow partners to access their interface information so that tools can be written to interface periphery or support tools. These tools can complement their current tool suite but rarely provide a major function. Autorouting software, however, seems to be an exception. Also, CAD vendors realize they cannot support the requirements of all the unique needs from the user community, so CAD/CAE companies encourage smaller companies to develop these periphery tools. However, with all the different design processes CAD/CAE tools need and the different interface protocols, having clean interfaces to several existing tools is a challenge.

A framework developed around the CAD/CAE system can help the user develop standards and processes for a given design flow. The design framework should provide an electronic interface to all the CAD point tools, which begins with schematic entry and ends with the design files required for manufacturing and test. To go from a schematic to a completed module ready for test, a comprehensive CAD infrastructure must be in place to ensure that good, functional modules are produced on the first spin.

Because each MCM design requires a unique substrate design, the most complex and evolutionary infrastructure required for MCM development is providing a design framework to enable module simulation, physical design, and analysis. The framework is a difficult challenge, because integrating several choice point tools to develop a stand-alone CAD system requires multiple interfaces and several talented engineers to understand and analyze the operation of each point tool.

10.4.1 Design Conception

To begin the MCM design process, the need for a module must first be identified. Once the decision has been made to use a module, the circuit design must be partitioned into small functional blocks to extract a good MCM chipset. The chipset should provide good use of the MCM by reducing the number of total I/O while adding more functionality into a reduced footprint, thus increasing the performance per square inch of the mother board.

Several MCM substrate technologies exist throughout the industry:

- Thin film copper polyimide deposited on an aluminum, ceramic, or silicon substrate material, referred to as MCM-D
- Ceramic substrate material, referred to as MCM-C
- Laminated printed circuit board substrate material, referred to as MCM-L
- A derivative or combination of the previous three methods

All the substrate technologies available have their advantages and disadvantages, so the selection of the appropriate substrate technology is specific to the application, manufacturing capability, and cost targets. The MCM substrate technology also requires specific design rules and features during the physical design.

The substrate technology selected for a given application influences several aspects of an MCM:

- Electrical performance
- Thermal dissipation capability
- Die attach methodology
- Packaging technology
- Manufacturing processes
- Cost

The die attach method and packaging technology also influence the electrical and thermal performance of the module along with cost. Therefore, the MCM supplier must choose their targeted marketplace and standardize on the substrate and die attach technology. This method puts constraints on the MCM chipset, but in turn, reduces design cycle time, nonrecurring engineering cost, and piece part cost.

10.4.2 Starting the Design Phase

The design phase begins with the collection of data. Many questions must be asked at the onset of the design to ensure that the customer requirements will be met.

Information required for an MCM design. Many of the questions and the data required for collection are listed in the generic design checklist shown below. The checklist outlines typical items that must be considered before starting an MCM design.

DESIGN CHECKLIST

Specific die information
- ___ Diced die size
- ___ Die thickness
- ___ x/y pad coordinates (soft and hard copy)
- ___ High powered die photograph or good drawing of die
- ___ Die diagram with die pads numbered and no-connects shown
- ___ Back side die biasing potential
- ___ Back side material
- ___ Bond pad size
- ___ Bond pad metallurgy
- ___ Passivation opening
- ___ Passivation material
- ___ Expected power dissipation per die
- ___ Acceptable die operating junction temperature (Tj)

Specific package information
- ___ Package size
- ___ List of fixed signal I/O
- ___ List of nonstandard power and ground I/O
- ___ List of I/O test points
- ___ List of substrate test points
- ___ Simultaneous switching noise and crosstalk considerations

Design information
- ___ Netlist (identify format)
- ___ Schematic
- ___ Desired die placement
- ___ Identify critical and priority signals
- ___ Crosstalk minimizing considerations
- ___ Wire bond diagrams which includes wires in the accurate pad location
- ___ Flip chip I/O configuration

Analysis
- ___ Crosstalk analysis
- ___ Propagation delay results
- ___ Thermal analysis
- ___ Expected ambient temperature
- ___ Free or forced air convection (air flow velocity)
- ___ System board size anticipated
- ___ Heat sink used

Design for testability (DFT)
- ___ Define special test consideration (JTAG, BIST, test points, etc.)
- ___ Define desired test point I/O locations
- ___ Determine specific die testing desired (standard or AC tested)
- ___ Determine die burn-in preference (die or module level)

Test program generation
- ___ Determine if simulation files are compatible
- ___ Determine use of hardware, behavioral level, or gate level models
- ___ Determine acceptable fault coverage

Environmental conditions
- ___ Operating temperature range anticipated
- ___ Expected ambient temperature
- ___ Expected reliability

Design for Test. Design for test should be addressed at the beginning of the design phase, because the method for testing the MCM may influence the design. There are several ways to increase the testability of a module, such as the use of boundary scan for ASICs, built-in self-test (BIST) for memories and large functional blocks, pinning out all the internal I/O, and adding test points on the substrate surface. Each method allows easier isolation of failures and improved fault coverage.

To increase the testability and internal node validation of an MCM, access to all the internal nets is helpful. However, this is normally only possible when MCMs have a small number of die. The act of adding test points can greatly influence the physical design. These test points should be included in the schematic so that the information is passed to the layout designer. This influences the I/O pin assignments. However, one of the reasons for using an MCM is to decrease the I/O on the system board, which conflicts with having access to all the I/O. Even if all the internal signals are brought out to the I/O of the package, they do not have to be routed on the system board. If extra I/O exists, this option should be considered. Before implementing this test feature, be sure to evaluate all the consequences. Added I/O test points are likely to decrease the quality of the signal and increase the routing complexity.

Joint Test Action Group (JTAG) is another test option to consider. JTAG has developed a specification to introduce standard test interfaces and boundary scan architecture. A group of companies participated in JTAG to promote the standardization of boundary scan I/O cells for implementation into integrated circuits. The boundary scan standardization will help semiconductor companies begin to support the test needs required for circuit board designs and MCMs.

In MCM applications, boundary scan increases the fault coverage of internal interconnects that cannot be distributed throughout the I/O connector. The IEEE 1149.1-1990 (JTAG) standard partitions scan channel rings at each I/O signal buffer to provide control and access through a scan-based tester.

Some gate array families offer boundary scan cells located near the input and output signal buffers. During the design of the gate arrays and/or MCMs, the designer must determine the test strategy for the device so that the gate arrays can be designed with or without the boundary scan cell enabled. Testability of an MCM is very important to increase the fault coverage during module test with a limited number of test vectors; therefore, whenever boundary scan can be a practical solution, it should be used.

During normal operation of an IC, the boundary scan cells will be configured so that signals will not be disrupted. However, when the JTAG cells are enabled during test mode, the test stimulus will be applied to the JTAG cells and the test response captured in a boundary scan latch. Once the test response is captured, the results can be output through the boundary scan I/O port for inspection.

If memory devices are included on a MCM, a BIST system can be implemented that would provide a self-test of the memory devices with only a few test vectors. Exhaustive memory tests can be quite complex and time consuming for a production environment. BIST circuitry can perform memory device testing while the tester is applying stimulus to other I/O cells within the MCM. This allows testing to be performed in parallel, thus decreasing cycle time of the production test. Also, excellent fault coverage is achieve through the use of a BIST system.

BIST should be incorporated into test strategies for the near future to decrease the test time required for production while increasing the fault coverage.

10.4.3 Library Generation

Well maintained libraries are essential to producing error-free designs. Many different libraries are required for MCM designs. The libraries contain the following information:

- Die libraries that contain specific information relating to the die such as x-y coordinates, I/O names, I/O numbers, I/O buffer type (input, output, bidirectional), test information, and electrical specific information
- Schematic symbols
- Physical design symbols

- Package symbols
- Logic models that enable simulation, which include behavioral, gate, or hardware models
- Material properties for thermal and mechanical analysis

These libraries are stored in separate databases used in conjunction with a specific CAD/CAE tool to increase productivity through automation, ensure that the information is consistent from design to design, and reduce errors due to incorrect data.

However, libraries are only as good as the library manager. If the library manager does *not* have a good configuration management procedure that monitors the check-out and check-in of data, errors can easily be introduced into the system. If precautions are used up front during the library generation, errors can be eliminated, and cycle time eventually will be reduced.

If the libraries are being developed manually, strict review policies should be in place to ensure that the data manually developed is accurate. A two-person independent checking procedure is recommended. Once the libraries are reviewed and are known good, only the library manager should allow data to be entered into the "golden" database.

Library management can be a time-consuming process that will not show all the benefits immediately. Producing one design that fails for a customer on a critical time line will provide all the reasoning to the importance of a good library management system.

10.4.4 Design Capture

Design capture is the entry point into the design system. Entering the circuit schematic and related material properties (such as assembly data, processing data, circuit board stack-up, fabrication material, and so on) into machine-readable form are referred to a *design capture.* Creation of the design can be performed online, which is typically the case today with designs starting from scratch, or annotated from other analysis or design tools.

Once the chipset is selected, the design begins by importing the package and die symbols into the schematic capture point tool. The symbols can be manually generated in the tool or generated automatically through scripts and database files. As mentioned earlier, an automated procedure reduces the opportunities for errors being induced into the system.

The die information used to generate the schematic is extremely critical in an MCM design. The die pad location, pad number, and pad name must be checked to ensure that they accurately match the die information. Any unconnected pads or pads not highlighted on the die supplier's data must be identified in the MCM die database. This will prevent errors during design and manufacturing.

Many times, for wire bonded parts, the information received is a wire bond diagram that does not highlight multiple power, ground, or signal pins that reside next to one another. In these cases, the number of single-chip package I/O does not match the number of die I/O. Normally, an MCM customer does not have access to the die information or has partitioned a design from a system board that uses all single-chip packages. Therefore, a customer who receives a schematic for review is concerned only with checking that the MCM design matches the system board design; thus, the mismatch in I/O between the die and the single-chip package is confusing. A methodology for communicating with the customer and documenting the difference in the I/O is imperative to customer satisfaction and a good review that will prevent errors. In most cases, designers prefer to control only one design instead of two, which shows the package I/O and the die I/O. However, methods exist for developing one design that can toggle between both views—one for the customer that shows package I/O, and one for the MCM design that shows the die I/O.

Once the die symbols are produced, the user can then logically enter the design's interconnectivity into the schematic capture point tool. After entering the schematic into the point tool, an electronic rules check (ERC) should be performed to check for errors in the schematic, such as duplicate pin, unconnected pins, shorts, opens, and so forth. Once the schematic has passed the online ERC and visual checks, a netlist is generated for use in the logic simulator and the physical design tool.

10.4.5 Netlist Generation

As mentioned in the previous section, the package I/O schematic and the die I/O schematic will be different in appearance. They will also produce a different netlist because of the different I/O numbers and names. Therefore, special care must be taken when generating the netlist to ensure that the die I/O netlist is produced for the physical design, and the package I/O is produced for customer review. To prevent errors, automating this process would be beneficial to ensure that only a die I/O netlist is used for physical design.

Different netlists may be necessary, depending on the tools used for physical design and simulation. In many cases, the netlist format used for the physical design will be different from that used for logic simulation.

Physical design tools today are moving toward online signal integrity analysis to help optimize and better automate routing; therefore, parameters are placed in the netlist for the physical design tool to use. For instance, the required delay on a net can be produced automatically by adding additional trace length during the routing phase if the information is attached to a specific net. This option will automatically increase the line length to match the specified delay.

10.4.6 Prelayout Design Evaluation

Before beginning the actual routing phase, prelayout logic simulation, thermal analysis, and electrical analysis is recommended to ensure that the design will match the requirements.

To help the layout designer meet the electrical requirements of the design, a prelayout design evaluation is recommended to place constraints on those nets that are critical or require special attention. This evaluation may include specific placement of the die along with the rotation so that the critical net routing is predetermined.

Prelayout Logic Simulation. To ensure that the MCM chipset functions per the specification, device-specific test stimuli can be written to simulate the functionality of the module. To enable simulation, the netlist generated from the MCM schematic and a functional model of each device in the design must be accessible within the design framework.

One or more of three different types of functional chip models can be used:

1. Gate-level model
2. Behavioral-level model
3. Hardware model.

All three models can be used together in a single MCM design. Acquiring, producing, and compiling these models can be an enormous effort.

The gate-level model represents the chip's raw gate design. For example, gate-level models are generated during an ASIC design cycle.

Behavioral level models are a high-level software model used to emulate the chip as a "black box" function, (i.e., output response to input stimulus). This software model is written using a hardware description language.

Hardware models can also be used to emulate chips in an MCM design. Hardware models are generally used when the gate-level and/or behavioral-level model becomes very complex, decreases the simulator's efficiency, or is unavailable. However, to use a hardware model of a chip, test hardware, referred to as a *hardware modeler,* is required. The hardware modeler also requires a single-chip package of the device under test along with the correct interface software and hardware.

By using the module's netlist, the module's test stimulus, and one or a combination of the chip modeling techniques, the MCM can be logically simulated to ensure that a good design and an accurate netlist exist. Figure 10.3 shows a typical output from logic simulation.

Prelayout Electrical Analysis. The prelayout electrical analysis is key to ensuring good signal integrity and electrical performance of the MCM. The prelayout electrical analysis is used to estab-

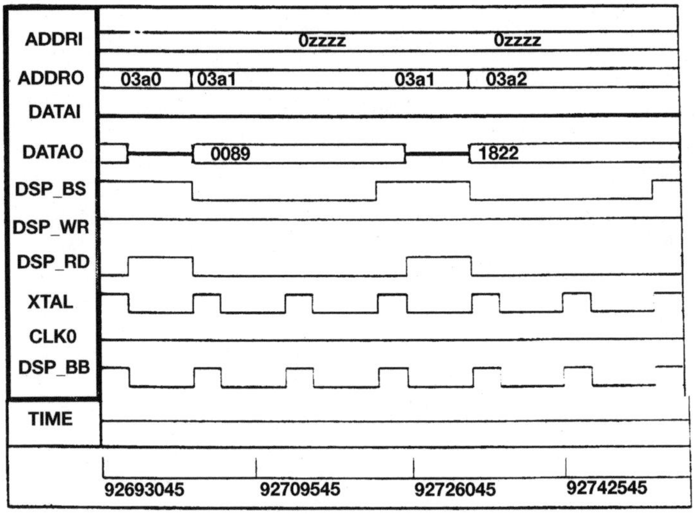

FIGURE 10.3 Logic simulation output.

lish design guidelines for the layout designer. The design guidelines communicate requirements so that the physical design will provide good signal integrity for all nets. Each net will be given different amounts of attention, depending on the criticality. For instance, a latched data signal that has 20 ns to settle from transmission line effects before being clocked will receive less attention than the clock signal latching the data. In fact, special attention to the propagation delay, overshoot, undershoot, and ringing of the signal should be analyzed. Figure 10.4 shows a typical transmission line analysis output.

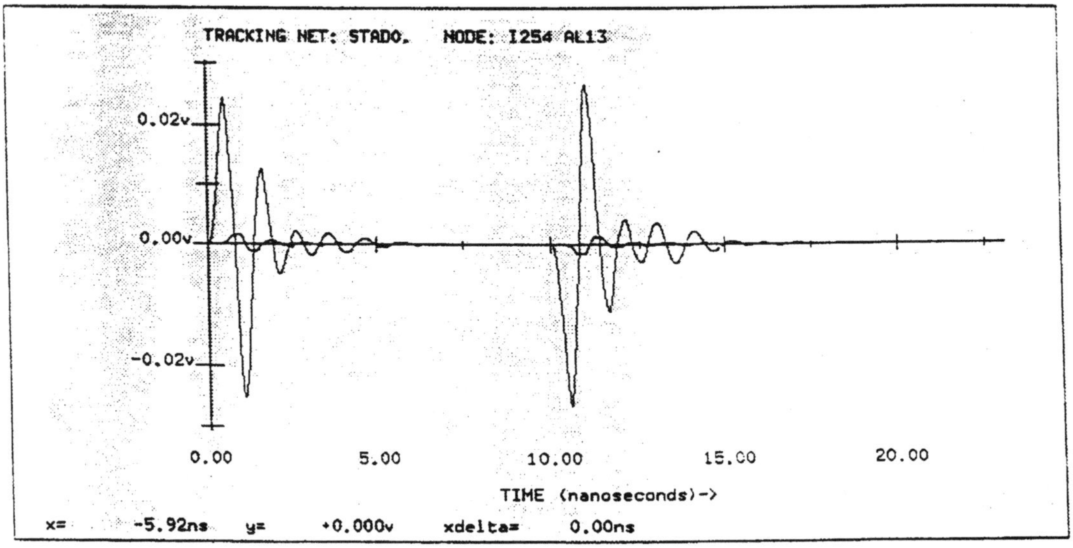

FIGURE 10.4 Transmission line analysis output.

The table below lists a series of items that must be examined during the prelayout electrical analysis phase:

PRELAYOUT ELECTRICAL ANALYSIS CHECKLIST

Layer stackup
- ___ Layer thickness
- ___ Layer assignment (power, ground, signal)

Approximate placement of die
- ___ Include die rotation
- ___ Include placement with pin #1 package and die identified
- ___ Identify die and package I/O configuration

Identify critical nets and priority nets
- ___ Approximate signal location layer by layer
- ___ Indicate physical design rules to follow
- ___ Define package I/O assignments (exact location)

Die I/O fanout
- ___ Perform die I/O fanout pattern for ease of routing

Identify bus routing required
- ___ Identify routing philosophy
- ___ Indicate desired physical design rules to follow
- ___ Provide a general package I/O assignment

All other nets
- ___ Indicate desired physical design rules to follow

Design for test
- ___ Work with the test engineer to identify desired test features
- ___ Perform calculation to highlight potential signal quality problems
- ___ Indicate desired physical design rules to follow

Rule of thumb requirements
- ___ Indicate general physical design rules to follow
- ___ Highlight routing keep-outs

CAD/CAE companies are just beginning to introduce applications that can help define many of the items listed above within the framework. These applications allow the engineer to perform electrical analysis and topology engineering for high-speed board designs. Any netlist changes—such as T-junctions, H-junctions, necessary delays, resistors, and capacitors—can be backannotated into the netlist used for physical design. For example, a T-junction would break one net into two and attach any delay information onto the net. If a resistor is necessary to produce a monotonic waveform, the resistor and required net changes would be incorporated into the netlist after backannotation. The drawback to using this type of application up front in the prelayout design phase is that I/V curves for all the I/O buffers are required to accurately model the transmission line effects.

Note that a standard has been introduced into the market called *I/O buffer information specification* (IBIS). IBIS is an emerging standard based on a behavioral-level device description supported by most transmission line simulators offered today. IBIS started because of the reluctance of die manufacturers to share buffer design information or SPICE models with their customers. Also, many SPICE modeling tools are customized which make the system houses prefer one standard as opposed to using several multiple SPICE programs.

IBIS specifies a standard format for die manufacturers to communicate buffer response to their customers. IBIS allows die suppliers to provide buffer I-V curves in a standard format. The format is being used by most CAD/CAE software companies to import buffer information into their transmission line analysis tools. This IBIS standard allows the user to receive the necessary information for performing a transmission line analysis without access to company-proprietary SPICE models or buffer designs.

SPICE, the most widely used IC circuit simulation software, is typically the basis of third-party circuit simulators. In many cases, commercial IC circuit simulators are based on a version of SPICE. Simulation tools generally work best in a particular frequency range, and the models used in SPICE are most accurate below 1 GHz. Microwave simulation packages are typically used above 1 GHz.

Prelayout electrical analysis also requires the use of a behavioral-level electrical analysis application to run in conjunction with the application. Note that behavioral-level model applications are not as accurate as SPICE. However, the behavioral level models normally suffice for most applications. Also, behavioral-level models, such as IBIS would provide, run several times faster than transistor-level SPICE models. Thus, if the required information can be gathered up front, the application can reduce up-front engineering time that now requires several SPICE runs to evaluate specific nets. "What if" scenarios can also be performed. This application is also good for redesign efforts because the recipe is not lost from one design to the next. All the necessary information is stored in the netlist.

Prelayout Thermal Analysis. Before beginning the physical design, an initial thermal analysis should be performed using thermal dissipation values for each chip, system environment parameters, and specific board geometries to determine the physical die location that thermally optimizes the package. This could be very important if the heat dissipation causes the die junction temperature to approach the maximum temperature allowed before reducing reliability and functionality of the die. Figure 10.5 shows a typical thermal analysis output.

The two hottest devices should be placed as far from one another as possible to optimize the heat spreading area. If the thermal gradient impinges another chip, the ability to dissipate power will be reduced, and the junction temperatures will be higher.

Placement of the devices can be very dependent on the routing requirements, but the thermal dissipation of the chips should also be considered. Ideally, a prelayout thermal analysis is performed to optimize placement based on thermal considerations. Because the reliability of chips is highly dependent on the junction temperature, maintaining the junction temperature as low as possible is recommended. Therefore, the prelayout thermal analysis is used to approximate the thermally optimized die location prior to placement.

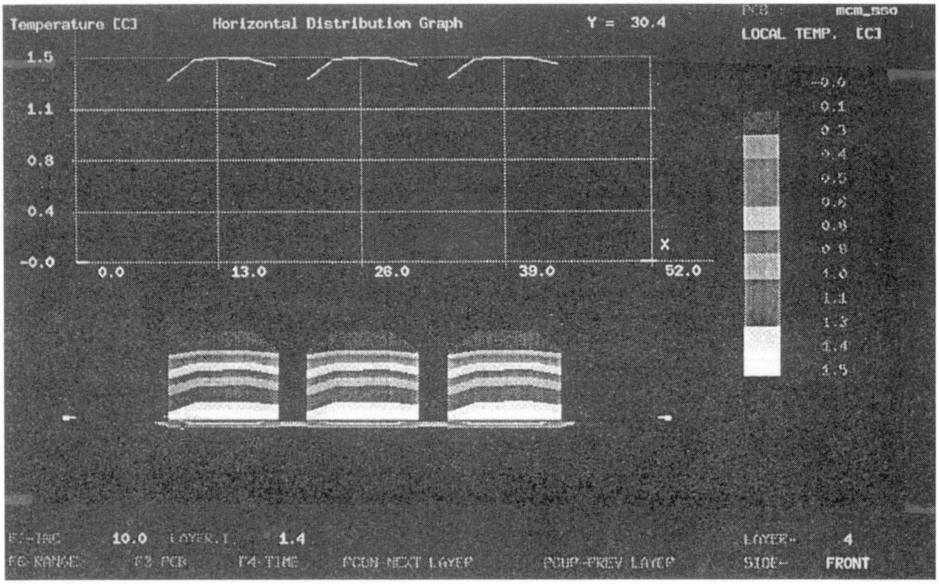

FIGURE 10.5 Thermal analysis output.

10.4.7 Physical Design

Physical design refers to the implementation of an electrical or system schematic in its tangible form. The development of a substrate with attached components (die, capacitors, resistors, and so on) and the associated interconnection is the result of this process. See Fig. 10.6 for a graphical top layer representation.

Physical design is not just the place and route of components. Physical design comprises several additional tasks, which will be addressed in this section.

Because the internal connections and signal I/O for MCM packages are unique for each new design, a good, cohesive physical design system must be established. Physical design is the heart of the design system. All pre- and post-analysis of package design influences the physical design output. To develop quick turnaround design specific substrates, substrate companies must provide comprehensive design guidelines that are specific to their fabrication processes. The substrate-specific guidelines should provide high-volume and high-yielding fabrication processes.

Once the chipset has been successfully simulated, a netlist is generated from the schematic for downloading into the physical design point tool. The netlist provides the interconnects between com-

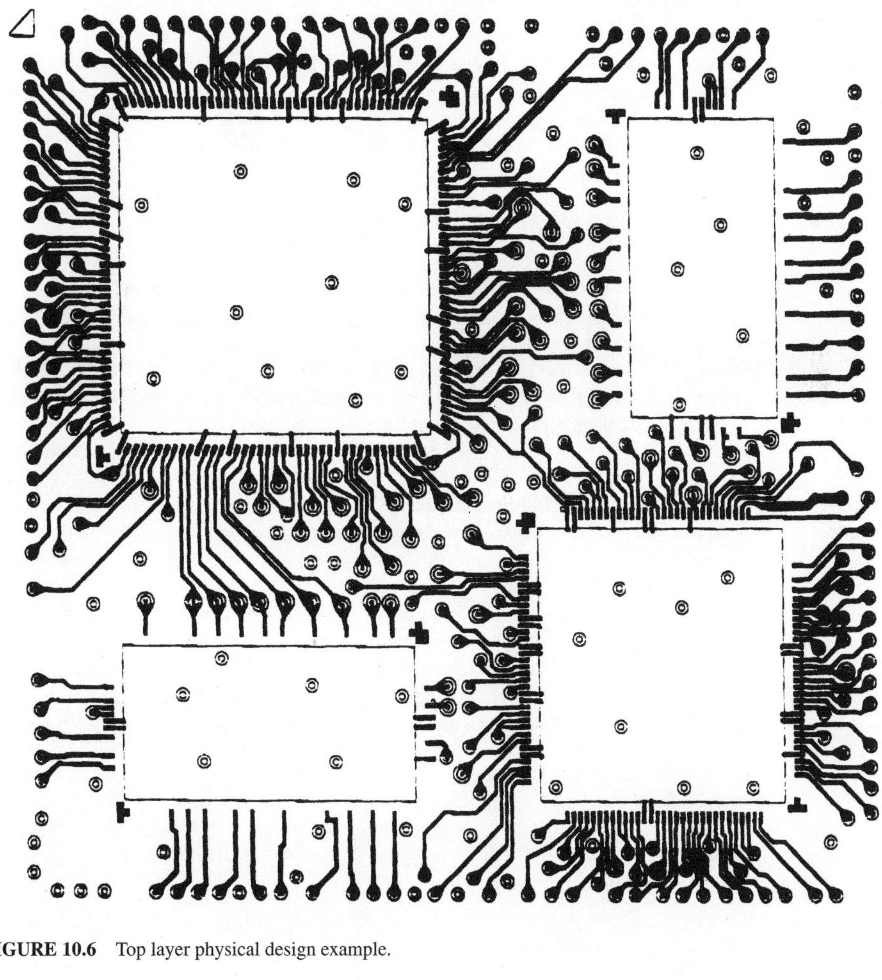

FIGURE 10.6 Top layer physical design example.

ponents and the package I/O locations. Each component I/O and package I/O location is assigned a name and/or number during the schematic entry process.

Two features are extremely important when selecting a physical design tool:

1. *An online design rule checking (DRC) capability.* This feature allows the user to enter the substrate and assembly manufacturing design rules into the tool. During the design process, the tool will ensure that the established rules are followed. If a design rule is violated, a DRC report or a graphical feature on the screen will highlight the violation.
2. *Netlist driven design capability.* This feature allows the tool to follow an established design-specific substrate interconnectivity list. This netlist is used for communicating the interconnectivity to an automatic routing tool or for highlighting nets for manual routing.

Getting Started. The physical design does not start without detailed planning. As mentioned previously, a prelayout design review should be performed to establish the layer stackup, design rules, critical nets, routing priority, and so forth. The layer stackup and substrate manufacturing design rules must be entered into the tool before proceeding to the physical design. The package outline and any relevant routing or placement keep-out should also be established.

Symbol Generation. Once the design rules, layer stackup, and general packaging rules have been entered, the next process is symbol generation. Symbols must be generated in the physical design tool to represent each component and the selected package type. Each symbol, whether it is a die, resistor, capacitor, or the package (namely, for identifying the package I/O locations) must be produced. A name and/or number is assigned to the symbol I/O that corresponds directly to the name and/or number entered into the schematic. Tools exist for automatic generation of the symbols but, in many cases, the symbols are manually drawn with the physical design tool.

Component Placement. Physical design begins with component placement. This process can be automatic or manual. For placement of wire bonded devices, the component system should include the via fanout pattern so that placement takes into account the via fanout area required. If an automatic algorithm is used, it should be sophisticated enough not to simply locate components next to another component with only consideration of interconnectivity. This approach ignores critical net routing. Placement routings must rank nets hierarchically based on critical nets. In many cases, manual routing is preferred because of the poor placement algorithm in existence today. The designer should not only use critical net timing and signal quality for component placement but should also consider using the initial thermal analysis.

Manufacturing and test considerations should be considered as well. Pick-and-place and TAB machines may impose certain constraints on the placement of the components. Certain types of die attach methods can support higher density placement, such as flip chip versus wire bond. Such constraints can be accommodated by establishing a manufacturing design rule database for a given application (die attach method, manufacturing equipment, etc.).

When placing components manually, a "rat's nest" that shows the interconnectivity of all the nets to the package I/O and device being placed will appear. By observing the rat's nest, the best rotation and placement of the device can be approximated. When placing components manually, the most critical components should be selected first. Then, during the placement process, the component should be rotated in all directions to observe wire crossovers. The designer should have a good idea of the recommended rotation if a good prelayout design review had been performed. Note that this process may be required to observe interconnectivity of all chips in a system during the prelayout design phase.

Routing. Many different automatic routing algorithms exist today. The standard printed circuit board routers have not been successful on MCM designs because of the difficult requirements. The via density on a typical MCM design is at least ten times that of printed circuit board design for single-chip package.

This can be envisioned easily by comparing the footprint of a single-chip package with that of the die located in that package. The typical single-chip QFP package I/O pitch is 25 mils, whereas the I/O die pitch is 5 mils.

For an MCM-L device (laminate substrate material), the MCM design requires fanout to relatively large via land pads from 5 mil pitch devices. The top layer of a design is normally congested with die fanout and vias connecting the die I/O to internal MCM layers. To fan out the die I/O in a small area, all angle routing is a necessity. Most routers today support only orthogonal routing or some 45° angle routing. Object- or shaped-based routers are required to meet the all angle routing requirements.

Automatic placement tools now offered for use with MCM designs have problems successfully routing 100 percent of certain substrate technologies. The angle routing and large via technology, in conjunction with dense substrate designs, prevent the algorithms from completing. If an automatic router completes only 90 percent of the routing, the remaining 10 percent can be extremely difficult to route manually. The automatic router must complete 95 percent before the automatic routing tool can be considered successful. If only 90 percent of the routing has been completed by the automatic router, manually routing the entire design may be less time consuming than routing the remaining 10 percent. However, keep in mind that if the router completes 95 percent of the routing, this does not guarantee that the design will be completely routed manually. The percentages provided are simply rules of thumb. Several parameters such as substrate technology, interconnect technology, density, and so on must be considered.

Manual routing is accomplished by using the routing tools in an interactive mode. From the netlist, a rat's nest is generated, which shows the connectivity from one component I/O location to another. By following the rat's nest, the designer simply draws a line per the established design rules from point to point. When a connection is complete, the rat's nest for that particular net disappears. Once all the rat's nests disappear and no online design rule errors exist, the initial physical design is complete. During the manual routing process, the designer must adhere to the design rules. The physical design tool should inform the designer if a design rule is being violated or simply not allow the designer to violate the rule.

The interactive design capability (manual routing) is a very important feature. Most of the ground and power planes are manually produced by making shapes. The editing features of the interactive portion allow the designer to make custom shapes, depending on the need and density of the design. Also, most typical MCM designs require manual routing on portions of the design. For example, no commercially available tool exists that can automatically fan out all chips available in today's market from the die I/O (whether flip chip, wire bond, TAB, or other) to the substrate pads. This process is typically performed manually.

Once the fanout is completed and the symbols have been placed on to the substrate, the critical net routing is performed. Designers perform critical net routing in several ways. Manually routing the critical nets provides the designer the most control. The entire board is available for routing and any necessary shielding. Some designers prefer to perform several automatic routing passes. The first pass would be performed on critical nets. Subsequent routing passes include busses or other priority nets. Also, automatic routing software exists for manufacturing glossing or clean-up passes for optimizing the design for substrate manufacturing. Manual review of the design for manufacturing optimization is recommended, followed by a manual cleanup.

Several physical design software companies are offering transmission line analysis tools that are tightly integrated with the routing software. Electrical design rules can be entered into the tools so that the routing software follows the specified rules similar to the online physical design rule checking performed. Integrating transmission line tools with router software requires electrical models of the components I/O buffers to be developed prior to routing. Collection of this data is difficult and is sometime impossible to collect prior to the routing phase. Estimating the I/O buffer performance is not recommended unless a good understanding of the design trade-offs has been acquired.

Design Rule Checking. Design rule checking (DRC) is a critical part of the design process. After completing the physical layout, the design can be checked using an online or independent design rule-checking tool. This provides a sanity check for assurance that no potential problems were overlooked during the physical design.

The substrate manufacturing and assembly design rules are entered into the physical design tool to control the routing, via, and die placement on the substrate to drive the online DRC tool. If the designer violates any rules, the design rules incorporated into the tool flag the designer of the violation. Using an online design rule driven place-and-route tool is a must for complex module design. DRC reports can also be generated that highlight any design rule violations.

Design rule checking is also performed by use of independent tools. In this case, a DRC deck, which contains all the pertinent design specific rules written to perform off line checking, is used to independently check the completed physical design. The DRC deck may discover additional problems not found during the place and route tool's online DRC. Several iterations on the physical design may occur until the DRC results received are satisfactory for production.

Design organizations have also begun to incorporate design rule checking on the manufacturing data provided to the substrate supplier. In some cases, errors have occurred in the design because the manufacturing data did not accurately represent the design data. This design rule checking is being performed in two different ways.

One method reads the manufacturing artwork into the physical design tool. Once the artwork has been successfully read into the tool, a netlist is generated from the design or attached to the manufacturing artwork. To perform this task, the CAD/CAE company must provide the capability or custom software must be written to perform the task. Once the artwork is associated with a netlist, the design should appear like a normal design. The DRC report can be generated and compared to an earlier DRC report (prior to making the manufacturing data) to check for discrepancies. The offline DRC checking mentioned above could also be run on the imported design.

The second method is very similar, except an independent CAM tool is used. Using a tool other than the one used to make the design can be very beneficial. The CAM tool takes the manufacturing output along with the rules specified by the substrate supplier. The design is then checked for violations of those design rules.

Performing a DRC is highly recommended, since physical design tools allow design rule overrides and de-highlighting of design rule violations. Just to ensure that a mistake has not occurred, run a DRC.

Layout Versus Schematic Checking. As a sanity check for the program manager and the designer, a layout versus schematic (LVS) check can be performed to verify that the physical design truly matches the schematic and/or netlist. The LVS check can be performed manually or automatically.

The manual process is very tedious and time consuming. The designer plots each layer and simply reviews the plots with the netlist line by line and via by via.

Several automatic versions can be performed. If the physical design tool can generate a netlist independently from the input netlist, whether it is generated from the manufacturing data or just the finished design, the input netlist and the generated netlist can be compared for differences. To check if the physical design tool can generate a netlist independently, route only a few signals and then generate the netlist. If the generated netlist only consists of the few routed nets and not the entire netlist, the tool works properly. Reading the manufacturing data back into the physical design tool and attaching the input netlist can be a second method for performing LVS. All the nets should be connected or a problem exists.

Physical design tools have become sophisticated enough to ensure that the design matches the netlist >99 percent of the time. However, if the goal is to ensure first-pass success, an independent LVS is recommended.

Manufacturing Optimization. Physical design tools are adding glossing or clean-up routines that remove stubs, center traces between objects (via, shapes, lines), evenly spaces bus lines, and so on. These tools are extremely important for optimizing the design for manufacturability. Once the physical design is complete, it is highly recommended to manually review the plots of the designs to ensure that no obvious manufacturing problems exist. Observing the plots on large D- or E-size paper may highlight manufacturing problems unnoticed from a computer screen.

10.4.8 Backannotation

Backannotation is a process of transferring data received during a specific design cycle and incorporating the known or actual data back into previously used tools, thus updating the tool's database. Many of the analysis tools used in a design process can benefit from actual information, because many assumptions are made during the prelayout phase that enable cursory analyses to be performed.

Backannotation can occur prior to the completion of the physical design. As mentioned in the prelayout electrical section, engineering tools are becoming available that allow the engineer to place constraints on the designer by performing "what if" scenarios. This is performed by analyzing certain nets in the design and determining the constraints required. The information is then backannotated into the netlist and schematic when required by placing line length constraints on a net (specified as delay information) and adding components (such as resistors).

Also, during the physical design, modifications, or changes to the design may force a change to the netlist. If the system I/O signals are not fixed prior to the physical design, the layout designer may optimize the routing by manipulating the I/O locations of a select number of signals. To ensure that the databases are accurate, this information must then be backannotated into the schematic and netlist.

10.4.9 Manufacturing Data

The manufacturing data is used to communicated the physical design information to the substrate companies, assembly contractors, and in many cases the mechanical drawing departments. Several different formats for interfacing between the different environments exist such as Gerber, Calma GDSII, DXF, IGES, and so forth. Custom interfaces are written, in many cases, to enable a clean transition into the desired tool.

Many laminate substrate manufacturers perform Gerber or DXF, while silicon-based substrate manufacturers prefer Calma GDSII. Whichever format is used, the designer must ensure that the manufacturing data matches the real design. This can be checked by reading the manufacturing data back into the design tool to check the manufacturing design rules or use a CAM tool to verify the that design meets the manufacturer's design rules. This is discusses in the design rule checking section.

Improving interfaces is extremely important in the design and analysis arena. Assembly subcontractors are beginning to see the advantages of tighter interfaces with the CAD tools. As die become larger and more complex, data entry errors become more commonplace. For instance, teaching a wire bonder the process to wire bond a complex 4-chip, 1000+ wire bond device is very error prone and labor intensive. Developing an interface between the physical design information and the assembly equipment, thus automating the process and eliminating a labor intensive process, can increase automation, reduce cycle time, eliminate manual entry errors, and improve accuracy.

10.4.10 Postlayout Analysis

Once the physical design is complete and the placement is fixed along with the routing, specific information can be backannotated into a specific tool, thus allowing more accurate results. Electrical analysis tools will use the routing and via geometries to calculate crosstalk, signal integrity, and simultaneous switching noise. The thermal analysis tools will use the accurate die placement (and, if used, thermal via locations and quantity) to better estimate the junction temperature. A more accurate timing analysis can now be performed with propagation information received from the actual routing between chips.

Logic Simulation. The delay information received from the physical design information can be used to transform the logic simulation into an accurate timing simulation. Without this information, manhattan distances (the x and y point to point distance) or one unit length for distance is used, thus enabling only a logical simulation to be performed.

Electrical Analysis. Electrical analysis can be performed on all signals to ensure that simultaneous switching noise and noise coupling is minimized. If certain nets have problems, they can be fixed by rerouting the signals using the physical design tool.

The propagation delays can be backannotated into the chip models to enable postlayout simulation. Once simulation and timing analysis are performed, problems induced by the physical design can be determined. If problems exist, changing the line length by rerouting the signal can significantly improve the results. Each physical design change requires new crosstalk analysis, simulation, and timing analyses to be performed to ensure that no additional problems were created.

Thermal Analysis. Once the physical design is finalized, thermal analysis can be performed again to accurately estimate the junction temperature of the die in the final package configuration. Insight into the reliability of the module can be obtained, because a lower die junction temperature usually translates into a more reliable product.

10.4.11 MCM Test Strategy

Developing a comprehensive, realistic test strategy is key to delivering high-quality, cost-effective MCM units to the user. Test discussions should begin during initial module design discussions. To achieve high module fault coverage, MCM testing should not only encompass final module test, but all the required ingredients must be tested prior to integration. This means not only *known good die* but also *known good substrates* are required.

To minimize the effect of substrate component failure on assembly yield, special testing prior to assembly should include *automatic optical inspection* (AOI) on each layer, along with a substrate-only opens and shorts test.

Following assembly and prior to final test, the units should be tested for opens and shorts to determine the assembly yield and to reduce the number of bad units reaching final test. This should simply be a DC continuity test that confirms proper electrical interconnect of the entire package. The units surviving the opens and shorts test proceed to the final test. Increased substrate and die level testing performed prior to integration reduces the dependence on the final test. This means that the final test fault coverage can be reduced while still providing high confidence in the functionality of the module.

To provide the MCM user with known-good modules, several issues should be considered prior to the final test step. The goal of final test is to provide high fault coverage, which could entail the use of more than one type of test methodology: (1) functional level, (2) system level, and/or (3) in-circuit test.

As mentioned earlier, design for test prior to module design must occur to ensure that all the test issues are considered that may affect the physical design. Therefore, a well defined test strategy is essential to the introduction of a successful MCM product. However, there are many issues that can prevent a clear test strategy from being developed for a module. Among the issues involved are:

- Whether the dice have been fully tested before assembly
- Whether a test access exists to all nodes of interest
- Whether repair will be performed
- Whether design for testability (DFT) features have been included on the ICs on the module
- Whether simulation for the module can be performed

In a sense, these issues can be summarized by asking whether the module will be tested as a chip or as a board.

When testing chips, IC manufacturers typically have gate-level models to which stimuli are written for test vector generation. Very high internal test coverage is desired. Expensive high-speed testers apply test vectors and perform parametric measurements. Since repair is not possible, normal testing does not include fault isolation.

Board manufacturers typically use a two-level test approach. In the first test level, in-circuit testing (ICT) is performed to verify that all the parts on the board were assembled properly. In this method, each device is tested separately on the module. Test patterns are not designed to thoroughly test the semiconductor chip's functionality but rather to test that the circuit board is assembled correctly and to isolate manufacturing defects for repair. In-circuit testers have full support for board fault diagnosis as well as software for design-for-testability analysis, automatic test generation, and routines to help turn off the rest of the board while one device is tested. The second level of test is a system test to verify the total board's functionality. The finished board is placed in a special test fixture or in an actual system and run through a diagnostic pattern. Again, diagnostics is a key requirement that provides a path to efficient circuit board repair.

Multichip modules have characteristics that require new approaches to test. Of primary concern is cost: test cost must be on the order of IC test cost for modules to be competitive. But modules can be much more complex than a single chip. Writing high fault coverage test vectors for a module can be an order of magnitude more complex than the sum of the efforts to write vectors for the ICs on the module. Another major concern is the level of die test before MCM assembly. If the dice on a module have been pretested, then the assembled module can be tested less stringently than a module with untested dice. A module with design for testability features will, of course, be tested more completely at a cost lower than that of testing a module with no DFT features. Another key part of an MCM test strategy is repair. Modules are sometimes repaired and sometimes not; the repair decision has to be made before any test strategies can be developed. Depending on the above-mentioned and many other parameters, the best strategy for testing an MCM can differ greatly from module to module.

MCMs can be tested in one of four ways: (1) the IC approach, (2) assembly testing using JTAG boundary scan, (3) assembly testing using in-circuit test, and (4) system testing. Strengths and weaknesses of each approach are listed in Table 10.1. In the IC test strategy, test patterns developed through module simulation are verified through test vector analysis and used to test finished modules. The system offers a high degree of automation and minimizes the occurrence of problems in which module outputs differ from expected test results. This test strategy is well suited to high-volume, low- to medium-cost modules, especially when repair is not part of the manufacturing process. In many cases, however, the IC strategy is inadequate for modules. Reasons for inadequacy include the fact that developing vectors to exercise an entire module with high fault coverage can be extremely difficult. Additionally, the proper simulation models and tools are not always available. A final reason to seek another strategy is that the IC strategy does not lend itself to good fault diagnostics on the test floor. For these reasons, an assembly test strategy should be developed for MCMs.

Table 10.1 Final Test Strategy Strengths and Weaknesses

Strategy	Strengths	Weaknesses
IC	Full functional test	Complex test generation Models required No fault diagnostics
Assembly JTAG	Simple vectors Good diagnostics	May not exist on dice Module function untested
Assembly ICT	Vectors not complex Good internal test Fair diagnostics	Requires 100 percent access and extensive DFT Module function untested
System	Module tested in real system Good coverage	Diagnostics can be complex New system on test floor

Assembly testing is designed to determine that a given module is put together correctly with the correct parts. The intent is to test for manufacturing defects rather than testing for full module functionality. There are two ways in which this can be done. The recommended method is to have known-good dice that support the IEEE 1149.1 boundary scan standard (also known as JTAG boundary scan). When JTAG-compatible parts are used, tests to verify that correct parts are properly interconnected on the substrate are simple to generate. Boundary scan vectors also provide the capability of excellent fault diagnosis to support a repair strategy. Using known-good die is recommended, because JTAG is less than ideal for extensive chip internal test (because vectors are applied serially).

The second assembly test method is in-circuit testing. This method tests each IC in isolation from its neighbors. The advantage of this method is that the manufacturer's vectors are likely to provide full internal testing of the die (whereas boundary scan is best suited to external testing). Additionally, in-circuit testing provides relatively straightforward fault diagnosis. The disadvantage to in-circuit testing is that certain DFT criteria must be met. The most important criteria are that all module nodes are accessible externally, all internal nodes have tristatable outputs, no feedback loops exist, and no nodes are tied to the power rails. The primary goal of ICT is to catch assembly faults. Because of tester memory limitations, test coverage on all dice on a module may be reduced compared with stand-alone die vector coverage. It becomes a secondary goal to test the die. But, compared with JTAG, ICT offers much better internal test.

Clear advantages of the assembly test approach are that simulation models are not needed for all components, and test vectors need not be developed to functionally test the entire module. A possible disadvantage is that the finished module is not fully tested for functionality and performance. A known-good die strategy is key to not needing functional testing for each module.

In cases where previous methods cannot be performed or provide incomplete coverage, system test can be used. System testing requires a special hardware board or system that accurately reflects the module's end use in a system. Untested multichip modules are inserted into the system board, and diagnostics are run to thoroughly test the module in system usage. Many advantages exist to this approach in that the module is functionally and performance tested to operate in the actual system environment. Diagnostic capability can be good, depending on the software used to do module test and diagnostics. Weaknesses center on the availability of the dedicated system test boards and the logistics of using many different system board designs and software.

In all four methods discussed, design for test (DFT) should be gleamed as the overall theme. DFT discussions must occur at the beginning of the MCM design phase.

10.5 SUMMARY

This chapter has presented a process that can be used to design multichip modules using CAD/CAE tools. Several helpful hints for designing MCMs were provided during the chapter that can help the reader establish a CAD system that meets the first pass success criteria. To meet the 10× cycle time program goals being established throughout the country, first-pass success is imperative. Without it, 10× cycle time reduction will never be achieved.

This chapter also highlighted the important role CAD/CAE tools play in the design, development, and test of MCM products. To ensure first-pass success, the tools must be tightly integrated, and the designers must be very meticulous. As IC manufacturers begin to offer a larger portfolio of known good die, the manual checking processes discussed in this chapter will be reduced, because the die data will be provided in soft copy form, thus not requiring manual or scanned entry into the MCM CAD/CAE system.

The CAD/CAE tools that are selected for the design of MCMs must have certain criteria today that will eliminate the possibility of errors and a road map for improved integration between design and analysis tools in the future.

With today's ever changing design environments and the reduction in product cycle times, more emphasis is being placed on the up-front engineering phase of the design process to ensure first-pass success. As tools become better integrated in which seamless interfaces are established between thermal, mechanical, and electrical analysis tools, shorter design cycles times can be achieved.

ACKNOWLEDGMENT

Special thanks to Andrew Flint for writing the MCM Test Strategy section.

SUGGESTED READINGS

Blood, Jr., W., and J. Kellar, "MultiChip Modules Present New Challenges to Place and Route Tool," *Proc. WESCON '92.*

Flint, A., and W. Blood, Jr., "MCM Test Strategy: Board Test in an IC Environment," *Proc. International Conference on MCM, ISHM and IEPS,* pp. 429–434 (1993).

Posse, K.E., "A Design for Testability Architecture for MultiChip Modules," *Proc. International Test Conference, IEEE,* pp. 113–121 (1991).

Ritter, T., Chapter 11, "CAD/CAE," *ISHM Thin Film,* pp. 375–433.

Trent, J.R., "Test Philosophy for MultiChip Modules," *Proc. International Conference on MCM, ISHM and IEPS,* pp. 444–451 (1992).

Trent, J.R., and A. Flint, "MCM DESIGN, TEST, and KGD STRATEGIES," *Proc. SEMICON 94 International Multi-Chip Module Conference,* January 13, 1994, pp. 65–75.

CHAPTER 11

MCM ELECTRICAL PERFORMANCE ANALYSIS

Robert C. Frye
Bell Laboratories, Lucent Technologies

11.1 INTRODUCTION

This chapter discusses analysis of the electrical performance of multichip modules and the role of technology alternatives in that performance. It presents the basic interrelationships between signal delays and design, physical structure, and choice of materials. It is intended to give technology developers an appreciation for the impact that their directions can have on electrical performance and a basis upon which to tailor their technology, and to give designers a basis upon which to select from a variety of technology alternatives.

Section 2 discusses the basic building blocks of electronic interconnections viewed from the perspective of MCM technology. Section 3 explores some of the unique consequences of physical scale and the role of structure in electrical performance, and Sec. 4 explores the role of materials. The remaining sections build on these concepts to examine signal delays in specific cases of interest and to provide some insight and perspective on the importance of factors of design and technology in MCM performance.

Technology choices are generally made at a high level in the design process. Therefore, the methods that are generally most useful focus on estimation and rules of thumb, rather than on detailed calculations. This chapter emphasizes these techniques.

11.2 BASIC BUILDING BLOCKS OF ELECTRICAL INTERCONNECTIONS

This section describes the basic electronic elements—conducting lines, transmitter and receiver circuits—that are used in electronic interconnections.

11.2.1 Transmission Lines

Even relatively unsophisticated electronic systems today operate at speeds that demand some consideration of transmission line effects. Electromagnetic signals in the form of voltage and current travel along interconnections at roughly the speed of light reduced by a factor of the square root of the relative dielectric constant of the medium between the conductors. If signals change slowly enough that all parts of a conductor are at more or less the same voltage, then the system is in the *quasistatic* regime. In such a case, the conductor can be considered to have a single voltage over its entire length and obeys Kirchoff's laws of voltage and current. This is the familiar concept of a "wire."

If, on the other hand, the signal varies so rapidly that the voltage at any location on the conductor differs significantly from the voltage at another location because of the finite speed of transmission, then the system is in an *electrodynamic,* or *transmission line* regime. Figure 11.1 shows the relationship of these regimes of interconnection operation to the signal path length and maximum operating

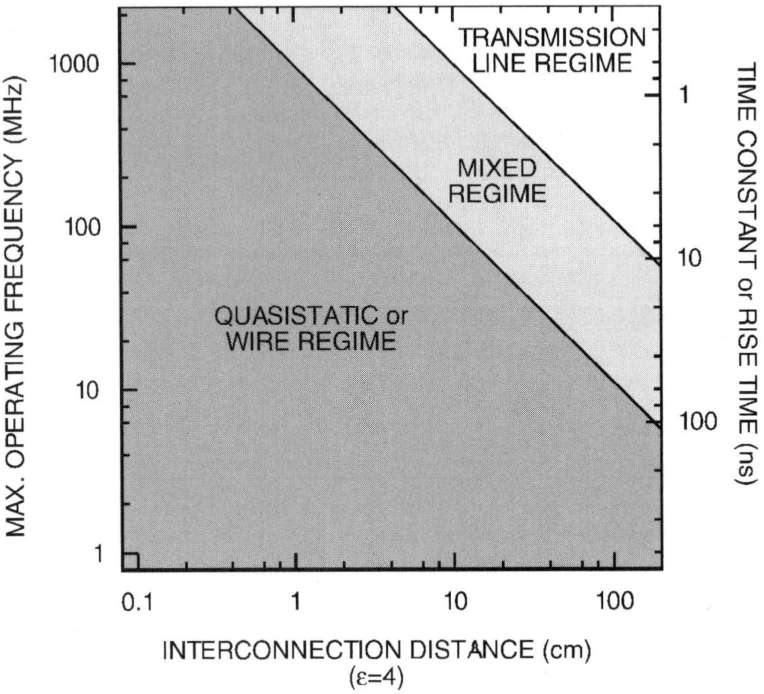

FIGURE 11.1 Signal operating regimes as a function of system size and speed.

frequency of the interconnection. Often, signal speeds are described by a time constant or rise time, τ, that is related to the maximum frequency of operation, f, by

$$\tau = 1/(2\pi f) \tag{11.1}$$

When the time required for a signal to propagate the length of the interconnection is equal to the signal rise time, then transmission line effects clearly are very important. In this case, a signal input at one end of the line has assumed a new value before anything begins to happen at the other. This defines the upper right-hand "transmission line regime" of the plot in Fig. 11.1. On the other hand, if the signal varies ten times more slowly, or if the line is ten time shorter, then to a reasonable approximation the voltage is the same along the entire line at any given time and quasistatically follows the signal. Between these extremes, transmission line effects vary in importance, depending on the systems details and requirements.

The circuit behavior of transmission lines is analyzed by treating the line as a series cascade of many short segments.[1] Each of these segments is chosen to be short enough to be in the quasistatic limit, where it can be modeled as simple lumped circuit elements that obey the basic laws of network theory. The conventional model of a transmission line is shown in Fig. 11.2. Here, L, R, C and G are, respectively, the inductance, resistance, capacitance, and conductance per unit length of the line.

If the length increment, dx, is chosen to be arbitrarily small, then the voltage along the line, $V(x)$, and the current, $I(x)$, are given by the coupled transmission line equations

$$\frac{dV}{dx} = -(i\omega L + R)I \tag{11.2}$$

and

$$\frac{dI}{dx} = -(i\omega C + G)V \tag{11.3}$$

These equations describe the propagation of voltage and current waves along the line. Here, ω is the angular frequency ($\omega = 2\pi f$). Two important parameters describing this wave propagation are the propagation constant, K, and characteristic impedance, Z_0. Voltage waves, for example, are described by

$$V = V_A e^{i(\omega t - Kx)} + V_B e^{i(\omega t + Kx)} \tag{11.4}$$

where V_A is the amplitude of a wave traveling in the positive x-direction, and V_B is the amplitude of a wave traveling in the opposite direction. Similarly, for the current,

$$I = I_A e^{i(\omega t - Kx)} + I_B e^{i(\omega t + Kx)} \tag{11.5}$$

The propagation constant, K, is given by

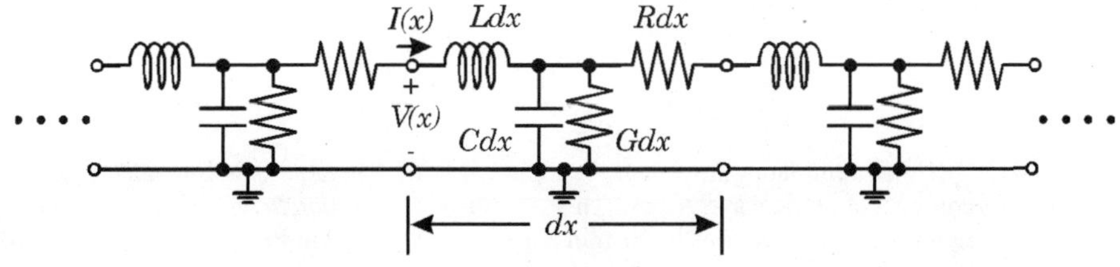

FIGURE 11.2 Conventional model of a transmission line.

$$K = \sqrt{-(i\omega L + R)(i\omega C + G)} \tag{11.6}$$

Current along the transmission line obeys a similar equation. The magnitude of the current waves is related to the magnitude of the voltage wave by the characteristic impedance, given by

$$Z_0 \equiv \frac{V_A}{I_A} = \sqrt{(i\omega L + R)/(i\omega C + G)} = -\frac{V_B}{I_B} \tag{11.7}$$

The characteristic impedance is not a real impedance, in the sense of a resistor, for example. Instead, it describes a relationship that exists between signal components on the line. A general signal that is present along a transmission line may consist of many components at many different frequencies traveling in both directions. All of these individual components are superimposed to make up the overall signal.

Propagation velocity of a signal along the transmission line is given by

$$v_p = \omega/(\mathrm{Re}[K]) \tag{11.8}$$

which for an ideal line ($R = G = 0$) is

$$v_p = 1/\sqrt{LC} \text{ (ideal line)} \tag{11.9}$$

Also, the characteristic impedance for an ideal line is given by

$$Z_0 = \sqrt{L/C} \text{ (ideal line)} \tag{11.10}$$

Many of the problems of signal transmission in MCMs come about because, in general, both the line resistance per unit length, R, and the conductance per unit length, G, are nonzero. In this case, as Eqs. 11.5 and 11.6 show, both the propagation constant and the characteristic impedance become complex and frequency dependent. An imaginary component in the propagation constant leads to exponentially decaying waves (Eq. 11.4). This simply means that signals can be transmitted for only a limited distance before their energy is dissipated. Furthermore, a frequency-dependent propagation velocity means that signals suffer from dispersion; different Fourier components of the signal travel at different speeds and don't arrive at their destination with the proper synchronization. The result is distorted, and sometimes unusable, signals.

These equations demonstrate some of the basic links between signal transmission and the materials that are used to make electronic interconnections. The line resistance per unit length is mainly a function of the metal used for the line and its cross-sectional dimensions. The conductance per unit length is mainly determined by the insulator that isolates the line.

The characteristic impedance of transmission lines becomes especially important at high frequencies and long transmission distances, as indicated in Fig. 11.1. In Eqs. 11.4 and 11.5, the values of V_A, V_B, I_A, and I_B are determined by the consideration of appropriate boundary conditions. A common example is a line that is terminated by a general impedance, Z_T. In this case, a forward traveling wave of amplitude V_A will give rise to a reflected wave of amplitude V_B. The ratio of these two components, known as the *reflection coefficient,* is given by

$$\Gamma \equiv \frac{V_B}{V_A} = \frac{Z_T - Z_0}{Z_T + Z_0} \tag{11.11}$$

If the terminating impedance is identical to the characteristic impedance of the line, the reflected component of the wave is zero. This is referred to as a *matched termination* and is a necessary condition for interconnections in the transmission line regime in Fig. 11.1. Problems in MCM interconnections often arise in cases where either the metal line or the dielectric are lossy (i.e., R or G are nonzero). In such cases, as Eq. 11.7 indicates, Z_0 becomes a complex, frequency-dependent parameter. More importantly, its frequency dependence is such that a simple network of discrete terminating components can at best match the impedance over only a limited frequency range. In general, it is not possible to perfectly terminate a lossy line.

For low-loss lines, termination is a simple matter. In such cases, Z_0 is real and independent of frequency, and the terminating impedance, Z_T, is a simple resistor. Characteristic impedances of interconnection structures with realistic geometries typically range from 20 to 100 Ω, with 50 and 75 Ω being commonly used.

11.2.2 I/O Buffers

The analytical description of signals on transmission lines is an example of the general class of boundary value problems. Solutions of Eqs. 11.2 and 11.3 are dictated by what happens at the boundaries, i.e., the ends of the lines. These are connected to input or output (I/O) buffers on the chips that are being interconnected. Generally, these can vary widely, depending on the application. However, most MCMs are currently used in digital applications, and the chips that they interconnect are usually either ECL or CMOS circuits. These two broad classes of circuits operate in a different manner and place different demands on the interconnections. The following sections discuss typical I/O buffers of each kind, in addition to TTL and various new I/O buffer families.

Low-Impedance Current Drivers. The earliest multichip module technologies were developed to address the needs of high-performance digital computers. These systems generally used bipolar, emitter-coupled logic (ECL) circuitry. Although ECL originated as a high performance bipolar circuit design, ECL drivers are currently also used in BICMOS,[2] a technology that combines high-density, low-power CMOS logic with the high-speed drive capabilities of ECL I/O buffers. The basic elements of an ECL interconnection are shown in Fig. 11.3.

The output from an ECL driver chip is the open emitter of an npn transistor. This transistor drives current into a termination, typically 50 or 75 Ω. The collector voltage, V_{CC}, is generally ground (0V) in ECL circuits, and the termination voltage, V_{TT}, is –2V. The signal at the termination, V_{SIG}, is –1.75 V when the driver transistor is in a logic LOW state, and –0.9 V in a logic HIGH state.

ECL circuits are ideally suited to driving resistively terminated transmission lines. They were originally designed for this purpose and take advantage of the bipolar transistor's relatively low output resistance compared with the load. Their low voltages keep overall power dissipation small, and the low voltage swing allows for fast switching. The important feature of ECL circuit operation from an interconnection point of view is that at all times, in either a logic LOW or logic HIGH state, electrical current flows along the line. Operation of ECL circuits in the transmission line regime, with matched terminations, requires that the total line resistance must be only a small fraction (e.g. 10 percent) of the terminating resistance. Otherwise, in addition to the signal reflections and dispersion resulting from a lossy line, the added series resistance of the line will cause V_{SIG} to be reduced. This can interfere with the correct operation of the receiver circuitry.

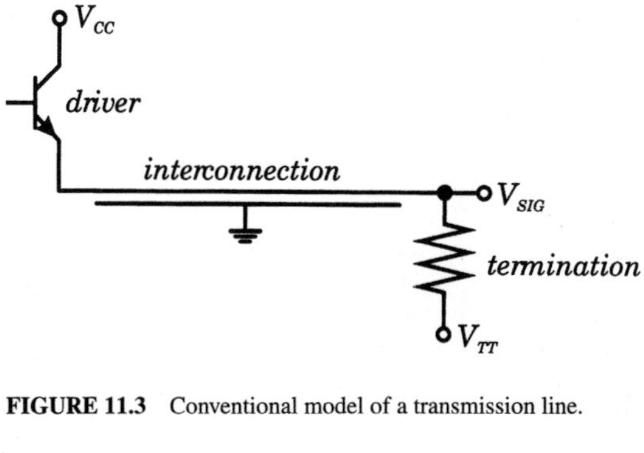

FIGURE 11.3 Conventional model of a transmission line.

Early generations of *metal oxide semiconductor* (MOS) and *complementary metal oxide semiconductor* (CMOS) technologies, based on insulated gate field-effect transistors, did not have enough drive current to be used like this with matched terminations. To drive a resistive load, the on-state resistance of the driver transistors has to be, at most, a small fraction of the terminating resistance. This would have required unreasonably large transistor sizes. Recent reductions in CMOS gate lengths, however, have made it feasible to make CMOS circuits that mimic ECL, sometimes called positive-ECL or PECL drivers.[3,4] In addition, a host of new bus standards are now being proposed that operate in substantially this same fashion.[5,6] Most of these new standards are based on CMOS technology operating in a low-impedance, current drive mode.

High-Impedance Voltage Drivers. Although they are not frequently encountered in high-performance applications, transistor-transistor logic (TTL) I/O buffers are sometimes used in MCMs. TTL is a bipolar transistor-based family of logic circuits that predates ECL. They are widely used in many electronic applications, especially in low-cost consumer electronics. They operate by switching their output voltage between two voltage levels, and they are designed to drive this voltage signal into a high-impedance load. They are generally not suited for operation in the transmission line regime.

One special feature of TTL that may be an important consideration for MCM packaging technology is that some TTL circuits may require the addition of pull-up resistors. Like ECL, TTL chips are being replaced in some instances by CMOS circuits designed to mimic TTL operation.

Early generations of CMOS circuits were too slow to require transmission lines and were operated in a TTL-like mode. The advantages of CMOS were its low cost, high levels of integration, and low static power dissipation. A typical CMOS interconnection circuit is shown in Fig. 11.4. In contrast to the ECL circuit, the termination in a CMOS circuit is usually a high impedance—most often a small capacitor (several pF). In operation, either the p-channel pull-up transistor or the n-channel pull-down transistor is "on" (i.e., conductive) and the other is "off." The interconnection is thus either connected to the drain voltage, V_{DD}, or the source voltage, V_{SS}. Typically, V_{DD} is either 5 or 3.3 V, and V_{SS} is at ground potential (0V).

The important distinction between the interconnections for CMOS shown here and those for ECL discussed above is that CMOS operates by charging and discharging a capacitive load. Current flows along the interconnection only during the charge or discharge. Because there is no static current flow, as there is in ECL, CMOS circuits are more tolerant of line resistance. Line resistance may effect the overall speed of the circuit but, given enough charging or discharging time, the output signal will eventually assume the correct value. The trade-off for this type of operation is that it will not work in the transmission line regime. It is only suitable for use in the quasistatic regime.

11.3 ROLE OF PHYSICAL SCALE AND STRUCTURE IN TRANSMISSION LINES

One of the primary motivations for MCMs is to improve packaging density. Shortening the distance between chips, and thereby reducing propagation times and parasitic loading, is the main contribut-

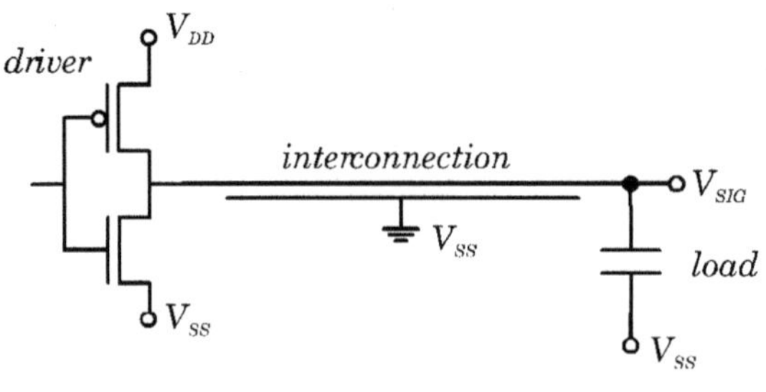

FIGURE 11.4 CMOS interconnection.

ing factor in the performance improvements that result from MCM technology. Improved packaging density, however, is generally accompanied by changes in the physical dimensions of the interconnecting lines. These changes also have an impact on the performance.

11.3.1 Transmission Line Properties

Figure 11.5 shows a generalized cross section of a typical interconnecting line on an MCM. This structure is often referred to as an *embedded microstripline.* It consists of a line of width w and thickness, t_m, separated from a conducting ground plane by an insulator of thickness t_i. the line is embedded within a dielectric medium, having an overlying layer of thickness, t_o, between it and the air. Other common structures found in MCMs are the stripline, in which the interconnecting line is sandwiched between two conducting planes, and the microstripline, in which the overlying insulator is absent. The electrical differences in these structures are generally insignificant.

Because the embedded microstripline is a geometry commonly encountered in conventional printed wiring boards, it has been extensively analyzed and modeled to derive the line's inductance, capacitance, resistance, and conductance per-unit length based on its geometry. Approximate analytical models for this have been largely superseded in more recent years by the ready availability of accurate numerical field-solvers. Rather than focus too closely on the details of these models, it is instead useful to consider some general properties relating the line's electrical properties to its geometry.

As a first-order approximation, the velocity of signal propagation on a line is just the reduced speed of light,

$$v_p \approx \frac{c}{\sqrt{\varepsilon_{rel}}} \approx \frac{1}{\sqrt{LC}} \tag{11.12}$$

where c is the speed of light in a vacuum. This relationship holds for lines that are not excessively lossy. This means that the product of the line's inductance and capacitance per unit length is largely invariant to its geometry. Increasing the line's width generally increases its capacitance per unit length and reduces its inductance per unit length.

Consider what happens when the conductor is raised to some voltage with respect to the ground plane. Charges on the line give rise to electric field lines that terminate on an equal and opposite charge distributed over the ground plane. Capacitance is a measure of the net charge on the line divided by the voltage. Charge is related to the field in the insulator by Gauss's law; it is simply proportional to the total flux. The field, in turn, is simply the gradient of the potential, ϕ. The potential obeys Laplace's equation

$$\nabla^2\phi = 0 \tag{11.13}$$

The significance of this is that Laplace's equation has no inherent scale. Its solutions remain valid for any arbitrary isotropic scaling of its dimensions. This implies that the inductance and capacitance

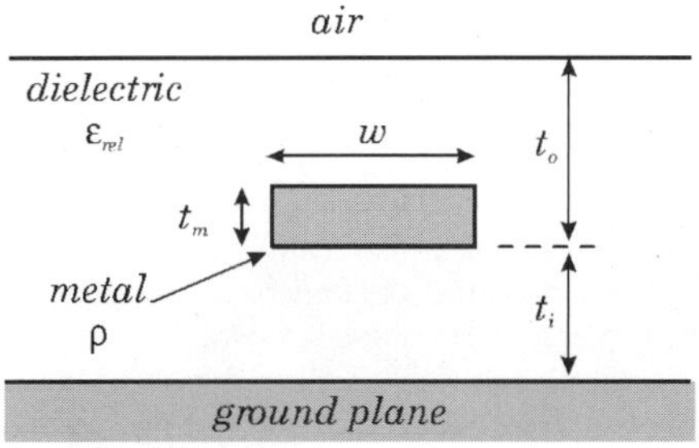

FIGURE 11.5 Cross section of a typical embedded microstripline interconnection.

per unit length of the line are determined not by the absolute dimensions, but rather by their relative proportions.

The dominant factor in determining a line's capacitance or inductance per unit length is the ratio of the line's width, w, to its separation from the ground plane, t_i. A value significantly less than unity (narrow lines far from the ground plane) results in high-inductance, low-capacitance lines. Such lines have a high characteristic impedance, usually in the 100 Ω range. For values significantly greater than unity (wide lines close to the ground plane), the lines have a high capacitance and low inductance, and consequently a low characteristic impedance, usually in the 20 Ω range. Wide geometric variations give rise to surprisingly small variations in characteristic impedance.

In the limit of very wide lines, the line's capacitance per unit length can be estimated by a simple parallel-plate formula

$$C \approx \frac{\varepsilon_0 \varepsilon_{rel} w}{t_i}, \quad w \gg t_i \tag{11.14}$$

In this same limit, inductance per unit length is approximately

$$L \approx \frac{\mu_0 t_i}{w}, \quad w \gg t_i \tag{11.15}$$

For narrower lines, account must be taken of the additional sidewall capacitance. Equations 11.14 and 11.15 underestimate capacitance per unit length and overestimate inductance per unit length. Most practical line structures fall somewhat below the wide-line limit, but it is still useful to understand the basic geometric dependences illustrated in these two equations. Analytical approximations for some common structures have been developed and are commonly used.[7,8]

The parallel conductance per unit length, which will be discussed in more detail in a subsequent section, can be modeled by a complex dielectric constant and scales exactly like the capacitance per unit length. Resistance per unit length, in sharp contrast, is strongly geometry dependent. We will also discuss this in more detail, in conjunction with the metal's material properties.

11.3.2 When Is Controlled Impedance Important?

As discussed above, whenever a signal changes so quickly that the ends of the interconnecting line are not at the same voltage, the line behaves as a transmission line rather than as a simple wire. A low-loss, controlled-impedance transmission line with a matched resistive termination is one method commonly used to allow high-frequency signals to be sent over large distances. Often, especially in high-performance electronic applications, the chips in an MCM may have I/O buffers that are designed to operate in this fashion, such as the ECL and PECL circuits discussed above.

Figure 11.1 illustrates size and speed limits for a simple point-to-point interconnection. Often, however, overall system-level interconnections take place on more than one level in the packaging hierarchy. A common example is illustrated in Fig. 11.6. In this example, an MCM is designed to function in a larger, external system in which the lines have some characteristic impedance, Z_{EXT}. Resistive terminations for these lines are provided on the MCM, preferably on the chips themselves. Clearly, the system-level interconnections, which may be very long, need to be controlled impedance. The question is, do the internal lines on the MCM also need to be the same impedance?

Obviously, making the MCM lines have the same characteristic impedance as the system-level interconnections will result in the best performance. However, overemphasizing the need for controlled-impedance connections in some instances has driven MCM technology developers toward unnecessarily expensive structures. It is important, then, to understand the effect of using line geometries in the MCM that result in nonideal characteristic impedances.

In all of the cases shown in Fig. 11.6, the signals flowing along the MCM lines are driving into an impedance of Z_{EXT}. One condition that has to be satisfied is that the line resistance has to be low

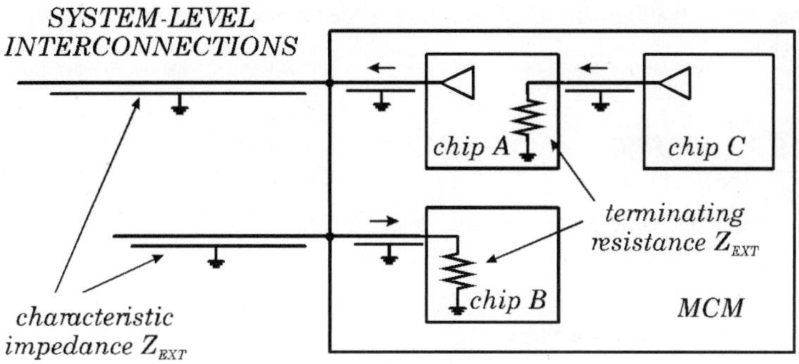

FIGURE 11.6 Controlled-impedance interconnection in a packaging hierarchy.

enough that it does not form a resistive divider with the terminating resistor. For a line of length l, this condition is

$$Rl < 0.1Z_{EXT} \tag{11.16}$$

This condition ensures that the signal is not overly attenuated by the series resistance of the line. If we consider a worst-case estimate of the line's effect on the signal, we can view its total lumped inductance as an unwanted series impedance that will also attenuate the signal. The lines inductance is insignificant provided that

$$\tau_{min} > \frac{3Ll}{Z_{EXT}} \tag{11.17}$$

or

$$Ll < \frac{Z_{EXT}\tau_{min}}{3} \tag{11.18}$$

In Eqs. 11.17 and 11.18, τ_{min} is the minimum signal rise time or time constant. The factor of three ensures that the fastest signals fall well below the bandwidth roll-off of the circuit.

Similarly, the MCM line's capacitance forms a conductive shunt in parallel with the terminating resistance, forming an RC low-pass filter. Signals will pass through provided

$$\tau_{min} > 3Z_{EXT}Cl \tag{11.19}$$

or

$$Cl < \frac{\tau_{min}}{3Z_{EXT}} \tag{11.20}$$

If the conditions in Eqs. 11.16, 11.18, and 11.20 are satisfied, then it is not necessary to make the internal characteristic impedance of the MCM equal to the external system. Note that all three of these conditions are dependent on the line length. In smaller MCMs, these conditions are more readily satisfied. Also, if Eqs. 11.18 and 11.20 are satisfied, we find by taking their product that

$$LCl^2 = \frac{l^2}{v_p^2} < \frac{\tau_{min}^2}{9} \tag{11.21}$$

In other words, the lines are short compared to the time of flight length of the fastest signal edges in the system.

11.4 MATERIALS PROPERTIES AND TRANSMISSION LINES

Both material and structural parameters determine the electrical properties of interconnecting transmission lines. These, in turn, impact performance.

11.4.1 Metal Resistivity and Permeability

Nearly all board-level interconnections in electronic systems are copper. This is also true for MCM-L and some versions of MCM-D, which have extended many aspects of printed wiring board technology to smeller features. Silicon technology, on the other hand, has relied predominantly on aluminum conductors. Cofired ceramic MCM-C often incorporates refractory metals such as tungsten or molybdenum to withstand their high firing temperatures. In all of these technologies, gold is sometimes used to provide a corrosion-resistant, solder-wettable finish. It seldom, however, plays a significant role in overall conduction of current, because of its high cost.

Metals contribute to transmission line properties through their influence on the line's resistance and inductance per unit length. Figure 11.7 shows a model of an incremental length of transmission similar to the one in Fig. 11.2. In this model, however, the series impedance elements of the line have been separated into the external inductance, L_{EXT}, and the surface impedance, Z_S.

The total inductance per unit length in a transmission line arises from two sources. The external inductance accounts for energy stored in the magnetic field outside the conductors. This component of inductance, as discussed above, is scale invariant. Inductance also arises from the penetration of magnetic field into the conductor. This internal inductance, along with the resistance, make up the surface impedance of the conductor.

Resistance per unit length at low frequencies, where skin effects are unimportant, is

$$R = \rho/(t_m w_m) \tag{2.22}$$

The internal inductance per unit length is more difficult to compute. For rectangular line cross-sections commonly found in MCMs, it can be approximated by

$$L_{INT} \approx \mu t_m/(3w_m)\,,\ w_m > t_m \tag{2.23}$$

Note that resistance and internal inductance do not have the same geometric dependence. Internal inductance is proportional to the metal's permeability. Most of the metals commonly used in electronic interconnections are diamagnetic, and have the permeability of free space,

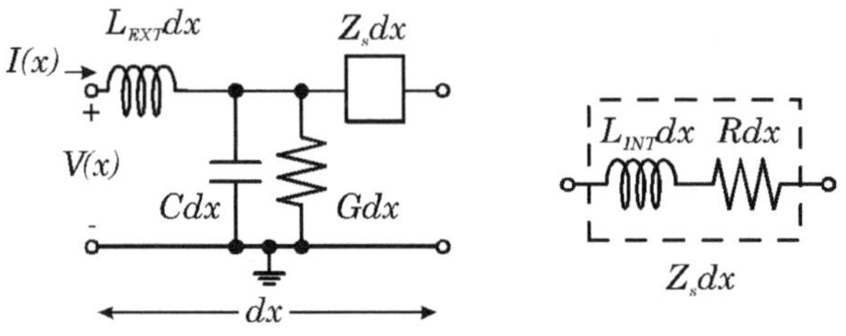

FIGURE 11.7 Transmission line model.

$$\mu_0 = 12.566\text{nH/cm} \tag{2.24}$$

The use of permeable metals such as nickel or iron in electronic interconnections can significantly increase internal inductance, because relative permeabilities can be large. For this reason, it is usually avoided.

For typical line dimensions, which tend to have a flat aspect ratio, internal inductance is small compared to the external inductance. It can be significant, however, for lines that approach a square aspect ratio. One consequence of internal inductance is to slightly reduce the propagation velocity below the speed of light in the dielectric.

11.4.2 Resistance and Skin Effects

The scale invariance of transmission line properties clearly does not apply to surface impedance losses in the metal. Resistance per unit length of the line scales proportionally to its cross-sectional area. Scale changes have a profound effect on transmission line resistive loss. Highly miniaturized MCM technologies, particularly thin-film MCM-D, must contend with this. Internal inductance is determined by the aspect ratio of the line's cross section and is at first glance scale invariant. However, at high frequencies, the current flow through the conductor is nonuniform, leading to frequency-dependent changes in the internal inductance.

At high frequencies, the skin effect limits current flow to the surface skin of the metal. For a metal of resistivity, ρ, and permeability, μ, the skin depth, δ, is given by

$$\delta = \sqrt{\frac{\rho}{\pi f \mu}} \tag{2.25}$$

where f is the signal frequency. Figure 11.8 shows the skin depth as a function of signal frequency for two commonly used interconnection metals, copper and aluminum.

As Fig. 11.8 shows, for frequencies above 100 MHz, conduction in these metals begins to be limited to a depth of several microns. This is accompanied by a rise in the apparent resistance per unit length of the line when the skin depth becomes comparable to or less than the metal line's thickness, and a corresponding decrease in the internal inductance. In thin-film MCM-D structures, for which metal thicknesses are typically only a few microns, this usually occurs at frequencies around 1 GHz. MCM-L or MCM-C structures usually have much thicker metal lines, so it may be important to consider losses from skin-effect at frequencies as low as 100 MHz.

Resistance and internal inductance per unit length can be estimated at high frequency by replacing the metal thickness, t_m, in Eqs. 11.20 and 11.21 with δ, when $\delta < t_m$. External inductance and capacitance per unit length are independent of skin depth.

Note that, because of its lower resistivity, copper has a smaller skin depth at a given frequency than aluminum. In cases where skin effect dominates the line resistance, the 58 percent difference in resistivities for these metals (2.65 μΩcm for aluminum, 1.68 μΩcm for copper) is effectively reduced to only 26 percent.

11.4.3 Dielectric Constant

A dielectric has two important effects on signal transmission. It directly effects the capacitance per unit length of the line and thereby the propagation velocity. It can also exhibit losses, usually frequency dependent, that lead to signal attenuation.

Referring to the circuit model in Fig. 11.7, the transmission line has a parallel shunt admittance per unit length given by

$$Y_p = i\omega C + G \tag{2.26}$$

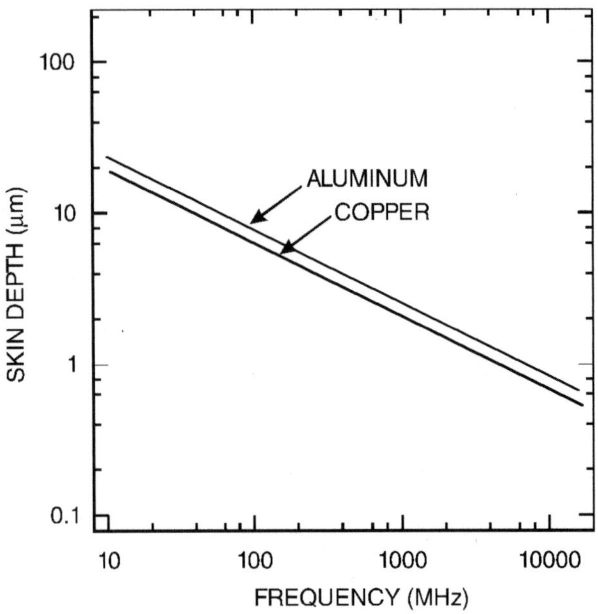

FIGURE 11.8 Skin depth vs. signal frequency for copper and aluminum.

where the circuit element C is directly related to the relative dielectric constant, and the circuit element G is related to the dielectric's conductivity. Often, these two properties of the dielectric are expressed as a single, complex dielectric constant

$$\varepsilon = \varepsilon' + i\varepsilon'' \tag{11.27}$$

The imaginary component of the dielectric constant, ε'', is related to the dielectric's conductivity, σ, by

$$\varepsilon'' = \sigma/\omega \tag{11.28}$$

Often, conductivity in dielectrics is frequency dependent. Insulators that have a very high DC resistance may show appreciable conduction at higher frequencies. Sometimes, this is expressed as a loss tangent*

$$\tan(\delta) \equiv \frac{\varepsilon''}{\varepsilon'} = \frac{Re[Y_p]}{Im[Y_p]} \tag{11.29}$$

Typically, the loss tangent is much smaller than unity for useful dielectrics. For a particular line geometry with a known capacitance per unit length, the conductance per unit length can be computed from the loss tangent by

$$G = \omega C \tan(\delta) \tag{11.30}$$

Because loss tangent is very frequency dependent, care should be taken to ensure that appropriate values are used in computing losses.

* The symbol "δ" in this equation should not be confused with skin depth. However, it is almost universally used in both cases.

Note that conductance per unit length scales in exactly the same manner as capacitance per unit length. Thus, isotropically scaling the cross-sectional dimensions of interconnections have no effect on the line's conductance per unit length. A practical consequence of this is that dielectric loss in MCMs is seldom a concern. A given magnitude of dielectric loss translates to a corresponding level of attenuation per unit distance as a signal propagates along the line. The attenuation per unit length attributable to dielectric loss is roughly the same for the very small transmission lines in MCMs as it is for very large ones used in conventional interconnections. However, because the transmission distance in MCMs is so short, losses from the dielectric are almost always insignificant.

11.5 ESTIMATING DELAYS

The performance characteristics of multichip module technologies can be analyzed using a variety of methods, each of which has its own advantages and disadvantages. Analytical methods have been used, for example, usually to optimize transmission line structures.[9–11] A distinct advantage of analytical methods is that the structural and material parameters influencing behaviors of interest, such as delay and bandwidth, are explicitly present in the equations describing the behavior. As long as these equations are not too complex, the roles of those technological parameters can be identified.

There are two serious drawbacks to closed-form analytical evaluation of interconnection technologies. For the most part, the methods of analytical techniques are only well developed for linear systems, often in the frequency domain. However, most of the electronic circuits of interest in MCM applications are digital and operate in a substantially nonlinear fashion. Even if we are willing to model their behavior with linear approximations, a second drawback of analytical methods is that they are only practical for very simple circuits. A common system configuration like several chips connected to a bus—a problem of great practical interest—poses too formidable a task for closed-form analysis. While it is possible in principle, any resulting equations would be too large and complex to be comprehensible and useful.

Simple time-constant estimation can be used to gain insight into the various elements contributing to interconnection delays.[12–14] This circumvents some of the problems of complexity, providing an overview of the problem that is often very useful in developing a technology. Most detailed calculations of interconnection delays use powerful methods of numerical simulation. While these provide the most accuracy, they are not as useful for technology development and provide little insight into the basic physical processes involved or their relationship to technological parameters. For this, it is more instructive to consider some simple analytical estimates of signal delays.

11.5.1 Time-Constant Estimates for Point-to-Point Interconnections

Figure 11.9 shows the basic elements of signal transmission in a simple, point-to-point interconnection. A driver is connected to a load through a transmission line of length l. In this simple model, the driver is represented by an ideal voltage source having an output resistance R_s, and the load is a general impedance Z_L.

In high-performance electronic systems, designers generally seek to reduce the delay between the signal that is being generated at the driver and the signal that eventually appears at the load. Most of these systems are digital. The signals consist of binary voltage levels representing logic states of the circuit. As Fig. 11.9 shows, however, an input representing a perfect step function transition from a low to a high state will not result in a perfect step function at the output. Instead, the output will be a somewhat delayed and rounded, degraded version of the input.

Two different mechanisms contribute to this behavior. One is simple delay resulting from the finite propagation velocity of the signal along the transmission line. The delay associated with this is indicated by τ_p in Fig. 11.9. Propagation delay is generally associated with the reactive elements of the line, L and C. Propagation delay does not limit the maximum frequency of signal transmission. The second element of delay, indicated by τ_r in Fig. 11.9, is a result of the finite bandwidth of the

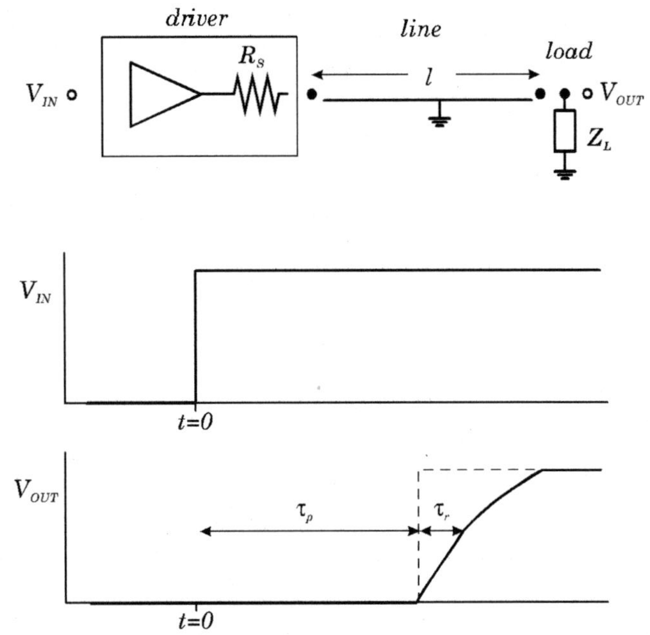

FIGURE 11.9 Signal transmission elements in a point-to-point interconnection.

interconnection. This is generally associated with the dissipative elements of the line, R and G, as well as the driver resistance and the load. It usually manifests itself as an increase in the signal rise time. In contrast to the propagation delay, this second source of delay does limit the maximum frequency of operation.

Much of the robustness of digital circuitry is a result of its insensitivity to this kind of signal degradation. The receiver circuit that senses the output voltage at the load is usually designed to regard all signals below a certain voltage as being one logic state, and those above as being another. For convenience, total signal delay is often measured from the midpoint of the input signal's rising (or falling) edge to the midpoint of the corresponding output signal's rising (or falling) edge. In general, since real driver circuits are nonlinear, rising delays are not necessarily equal to falling delays.

In static circuits, signals are asserted by some set of drivers, and then, at some later time, all of the receivers are interrogated to determine their state. The time between assertion and interrogation is chosen to be long enough to allow all of the signals of interest enough time to achieve their appropriate state. In such systems, it doesn't matter whether the delay resulted from propagation time or from bandwidth degradation—both have the same effect on the performance.

Let us consider delays for a common case like the one shown in Fig. 11.4. Here, the load impedance is a capacitor. The driver circuit is nonlinear, and as such does not have a well defined resistance. For purposes of delay estimation, however, the buffer's source or sink current, which are measured at a defined output voltage, can be used to calculate a reasonable value. A typical high-performance CMOS buffer, for example, will deliver 24 mA of current at 0.5 V, giving a 20.8 Ω output resistance.

There are several sources of delay for such a circuit. The first, obviously, is the propagation delay

$$\tau_p = l\sqrt{LC} \tag{2.31}$$

At the load end of the circuit, the signal's behavior will be approximately of the form

$$V(t) \sim V_0(1 - e^{-(t-\tau_p)/\tau}),\ t > \tau_p \tag{11.32}$$

The midpoint delay shown in Fig. 11.9, τ_r, is related to the time constant for the signal in Eq. 11.32, τ, by

$$\tau_r = ln(2)\tau \tag{11.33}$$

Several components may contribute to the overall signal time constant. One, for example, is the distributed *RC* time constant of the line itself:

$$\tau_1 = \frac{RCl^2}{2} \tag{11.34}$$

This is the time needed for the transmission line to charge its own capacitance through its own resistance. Another is the *RC* time constant for the driver to charge the line's capacitance:

$$\tau_2 = R_S Cl \tag{11.35}$$

Also, the driver has to charge the load capacitance through both its own resistance and the line's resistance:

$$\tau_3 = (R_S + Rl)C_L \tag{11.36}$$

With a capacitive load, the line's inductance may also limit bandwidth. The time constant for this is

$$\tau_4 = \sqrt{LlC_L} \tag{11.37}$$

The total signal delay for the interconnection includes both the propagation delay and the rise time. Rise time degradation in these interconnections, as in a series of cascaded filters, can be approximated by the square root of the sum of the squares of the individual time constants,

$$\tau_d \approx \tau_p + ln(2)\sqrt{(\tau_1^2 + \tau_2^2 + \tau_3^2 + \tau_4^2)} \tag{11.38}$$

This result does not include dielectric losses, since these are usually not important in MCMs.

A useful figure of merit for any particular MCM technology can be derived by considering this expression in the limit of very wide lines driven by an ideal driver ($R_S = 0$). In such a case, we can see from Eqs. 11.14, 11.15, 11.22, and 11.23 that τ_2, τ_3, and τ_4 become vanishingly small, leaving

$$\tau_{limit} = l\sqrt{\mu_0 \varepsilon_0 \varepsilon_{rel}} + \frac{ln(2)\rho\varepsilon_0\varepsilon_{rel}l^2}{2t_m t_i} \tag{11.39}$$

Equation 11.39 represents the minimum delay obtainable within the constraints of a particular MCM technology for a connection of length l. The first term, which is linear in length, simply represents the speed-of-light limitation, and clearly illustrates the role that dielectric constant plays. The second term, which is quadratic in length, is more interesting. It is determined by the materials properties of both the dielectric and the metal as well as their thicknesses.

For sufficiently short lines, the first term in Eq. 11.39 dominates the delay. In this regime, delays are dictated by the speed of light. for longer lines, however, material and structural parameters dominate the delays, which grow quadratically with length. Clearly, this is a poor regime in which to operate. The break point between the two regimes occurs at a critical length, l_{crit}, for which the two terms are equal:

$$l_{crit} = \frac{2t_m t_i \sqrt{\mu_0/(\varepsilon_0 \varepsilon_{rel})}}{ln(2)\rho} = \frac{2\sqrt{\mu_0 \varepsilon_0 \varepsilon_{rel}}}{ln(2)R_m C_i} \tag{11.40}$$

where R_m is the sheet resistance of the metal and C_i is the insulator capacitance per unit area. The critical length is an indication of the maximum interconnection distance for which a particular substrate structure is well suited. The fundamental conclusion is simply that larger distances call for thicker films. Values of the critical length for example technologies will be presented in a subsequent section.

Time-constant delay estimation is a very useful technique for optimizing the performance of a particular MCM technology and for understanding the important sources of delay. Figure 11.10 shows an example of a breakdown of delay sources as a function of metal linewidth for a typical point-to-point interconnection in a thin-film, MCM-D module. The two components shown in the lower part of the figure, the propagation delay and the distributed RC delay of the line itself, can be seen for very wide lines to approach the technology limit described in Eq. 11.39. The result shown for a loaded line includes the additional delay that results from a capacitive load, and the top-most curve includes a finite driver resistance as well. One thing that becomes very apparent in this kind of analysis is that the propagation delay, which has often been the focus of a great deal of technology effort in MCMs, constitutes only a fraction of the overall delay. A second observation is that the total delay is significantly greater than the technology limit. The remaining sources of delay are attributable either to the finite source resistance or load capacitance.

This particular example shows the sensitivity of the delay to the linewidth, holding other parameters of the technology (such as film thicknesses) constant. It can be seen here that a minimum in delay occurs for a particular linewidth. Similar delay breakdowns can be performed to optimize other parameters. Analyses of this kind have led us to the view that minimizing delays in MCMs is as much a problem in I/O buffer design as it is a problem in packaging and interconnections.

11.5.2 Time-Constant Estimates for Bus Interconnections

A bus, as opposed to a simple point-to-point interconnection, is a line with multiple drivers and receivers connected along it. It also poses a more difficult problem for analysis, because there are

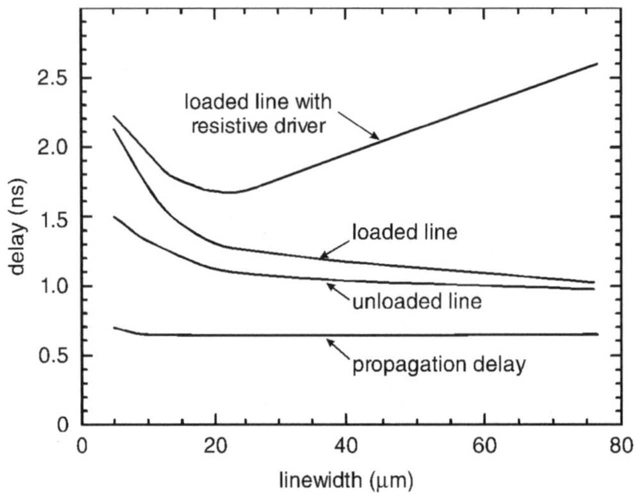

FIGURE 11.10 Time constant estimates of delay in an example MCM-D.

many more possible configurations, most of which are too complex for the simple methods used above. Buses are such common architectural elements in MCMs that it is worthwhile and instructive to consider a model for a very simple bus, shown in Fig. 11.11.

In this model, n capacitive loads of value C_L are connected along a line at regular intervals of length l_0. The transmission line for each of these intervals is modeled by its lumped equivalents. Dielectric conductance has been assumed to be negligible in this example. This lumped model approximation for the loaded line can be seen to be very similar to the model in Fig. 11.2. In this case, however, the added loads along the line change the effective capacitance per unit length. The effective series impedance per unit length of the line,

$$Z_S = (i\omega L l_0 + R l_0)/l_0 = i\omega L + R \tag{11.41}$$

which is the same as for an ordinary line. The effective parallel admittance per unit length, however is given by

$$Y_P = (i\omega C l_0 + i\omega C_L)/l_0 = i\omega(C + C_L/l_0) \tag{11.42}$$

In other words, the effective capacitance per unit length of the line has been increased by the periodic attachment of the additional load capacitance at intervals l_0.

This has two important effects. One is that the effective characteristic impedance of the line (neglecting resistive loss) is lowered:

$$Z_{EFF} = \sqrt{L/(C + C_L/l_0)} \tag{11.43}$$

Furthermore, the propagation velocity is also lowered:

$$v_{EFF} = 1/\sqrt{L(C + C_L/l_0)} \tag{11.44}$$

For a total bus with n stages, the total propagation delay becomes

$$\tau_P = n l_0 \sqrt{L(C + C_L/l_0)} \tag{11.45}$$

By analogy with Eq. 11.34, the distributed RC time-constant of the bus is

$$\tau_1 = \frac{R(C + C_L/l_0)(n l_0)^2}{2} \tag{11.46}$$

and, by analogy with Eq. 11.35, the time constant for the driver to charge the total capacitance of the bus is

$$\tau_2 = R_S(C + C_L/l_0) n l_0 \tag{11.47}$$

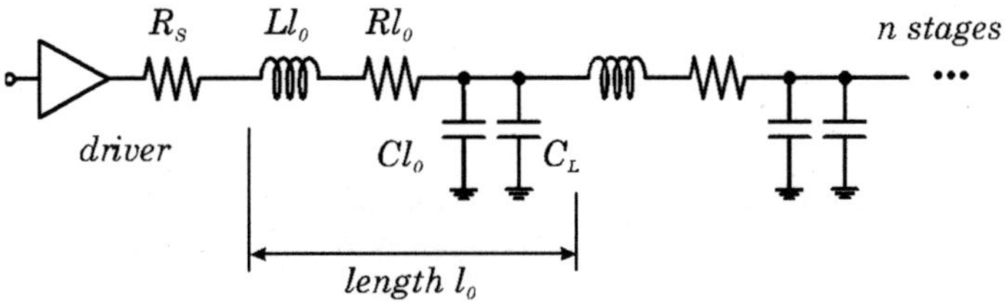

FIGURE 11.11 Model of a simple bus.

Because the load capacitance is already included as part of the line, there is no counterpart of Eqs. 11.36 and 11.37 in this case. The overall time constant estimate for the delay on the bus is

$$\tau_d = \tau_p + ln(2)\sqrt{(\tau_1^2 + \tau_2^2)} \tag{11.48}$$

Just as we defined a critical length for a simple point-to-point interconnection, it is possible to define a critical length for the bus. In this case, however, because the length between loads is fixed, it is equivalent to a critical number of loads. It is computed as before by assuming a perfect driver and then finding the value of n for which the linear and quadratic terms in the delay are equal. The critical number of loads is found to be

$$n_{crit} = \frac{2}{ln(2)Rl_0}\sqrt{\frac{L}{C + C_L/l_0}} = \frac{2}{ln(2)}\frac{Z_{EFF}}{Rl_0} \tag{11.49}$$

As the number of loads attached to the bus begins to exceed this number, delays begin to increase quadratically with the number of loads. It is apparent from Eq. 11.49 that this is a consequence of the line resistance. If the total lumped resistance in a bus begins to exceed the effective characteristic impedance of the loaded line, then the delays become excessive. Equation 11.49 is useful figure of merit for evaluating the suitability of a particular MCM structure for use in a bus. Values of the critical number of loads for example technologies will be presented in the next section.

11.6 DELAY SIMULATIONS AND EXAMPLES

Figure 11.12 compares the results of the time-constant estimation analysis discussed above with a simulation of the same circuit.[15] The time-constant analysis gives results that are surprisingly accurate and permits identification of the various elements contributing to the delay, as well as their relative importance. The only drawback to time-constant analysis is that its usefulness in more

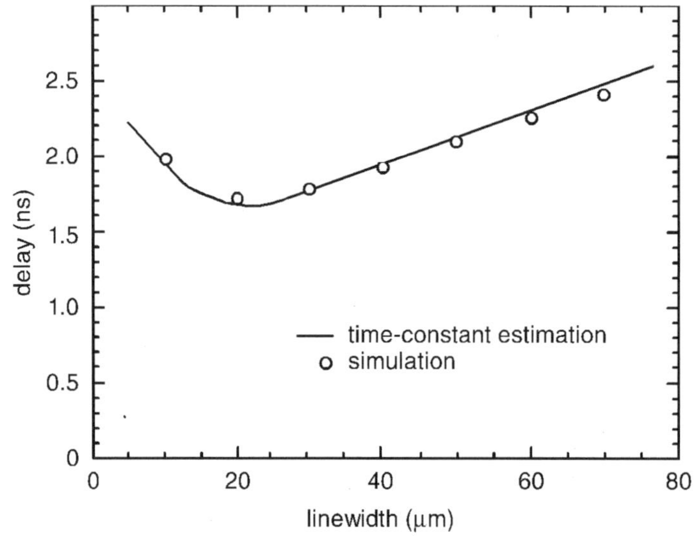

FIGURE 11.12 Comparison of time-constant analysis[6] with simulation.

complicated circuits is limited. Real I/O buffers are nonlinear, and most bus interconnections do not conform to the simple, ideal structure discussed above. Electronic circuit and system designers often use simulation methods to accurately model their circuit and circumvent these problems of nonlinearity and complexity. There have been significant advances in the ability of simulation tools to model distributed circuit elements like transmission lines,[16,17] extending their usefulness to problems of interconnection delays.

To illustrate the use of simulation methods, we have chosen several example MCM structures. Various elements of the example structures are listed in Table 1.1.[15] Values of the line parameters, defined in Fig. 11.2, are listed in the first part of the table. Parameters for MCM-C are typical of modules made with a cofired ceramic technology[18,19] and MCM-L are typical of advanced laminate technology with fairly aggressive linewidth rules.[20,21] MCM-D structures have been made by a variety of approaches. We have chosen two example structures, one typical of a high-performance technology,[22–24] and the other representative of lower-cost versions of MCM-D that are becoming more popular for less performance-driven applications.[25–27] Table 11.1 also shows, for comparison, typical values of these same parameters for conventional packaging, assuming leaded packages on a printed wiring board.

In addition to structural parameters of the lines, Table 11.1 shows the derived transmission line parameters for these example structures. These are the parameter values that are used in the circuit simulations. Several other important circuit elements are shown beneath. The method of chip attachment plays a key role in determining some of these. The average distance between chips, l_0, is greater for wire-bond attachment than for flip chip because of the extra area required around the periphery of each chip to accommodate wire bonds. Furthermore, in both MCM-C and MCM-L technologies, the vias are typically much larger, and the additional area needed for them results in a greater average chip-to-chip spacing and a correspondingly lower packaging density. For conventional packaging, with the added requirement for lead frames associated with the chip packages, the distance is greater still.

The remaining variables in Table 11.1 are part of the circuit models illustrated in Fig. 11.13. The packaging capacitance, C_{PKG}, and inductance, L_{PKG}, are each typical of their respective attachment method. On the chip side of the package, C_{PAD} and C_{ESD} represent the chips' I/O pad capacitance and electrostatic discharge protection circuitry, respectively, and are the same for all technologies. An advantage of simulations over the time-constant estimation methods discussed above is that more detailed modeling of this kind can be easily incorporated into the analysis without overly complicating it.

Because the focus of this analysis is on the MCM technologies, not the drivers, we have used a linear voltage driver with an output resistance, R_S, of 25 Ω in the simulations. This value of output impedance approximates a high-performance, high drive-current CMOS buffer. It is important to keep in mind, however, that the same variational method can be applied using more realistic, nonlinear I/O buffer circuit models in the simulations. This is one of the principal advantages of this approach over alternative methods of analysis.

Figure 11.13 shows several common interconnection configurations that we evaluated for each of the example technologies. The first consists of a simple point-to-point connection between a driver chip and a receiver chip. We chose the distance of the interconnection to be ten times the average chip-to-chip spacing for each technology to simulate a worst-case connection.

In most designs, however, the performance of the system is not limited by point-to-point connections, but rather by the more heavily loaded bus interconnections. We have examined simple bus structures consisting of four and eight loads connected serially at a spacing equal to the average chip-to-chip spacing, and another consisting of eight loads connected as two four-load serial buses driven in parallel. In all cases, the buses were unterminated.

The interconnection delays computed by circuit simulation for each of the example MCM technologies in each circuit configuration are compared along with the results for conventional packaging in Fig. 11.4. Circuit simulations were performed using ADVICE,[28] which is built upon a SPICE[29] compliant simulator with added features, including a lossy transmission line model that we utilized. Computed delays are measured from the midpoint of the rising edge at the circuit input to the midpoint of the corresponding rising edge at the load. For the bus structures, the delay is measured at the load farthest from the driver.

Table 11.1 Example MCM technology parameters.

	conventional packaging	MCM-C	MCM-L	MCM-D high-performance	MCM-D low-cost
Line structure and material parameters					
dielectric	glass/epoxy	alumina	glass/epoxy	polyimide	polyimide
ε_{rel}	4.2	9.4	4.2	3.0	3.0
metal	Cu	sintered Mo	Cu	Cu	Al
ρ ($\mu\Omega$–cm)	1.6	10	1.6	1.6	2.8
t_i (μm)	500	200	120	15	5
t_o (μm)	500	200	120	10	6
t_m (μm)	13	20	13	2.5	1
w (μm)	250	125	50	25	20
Line characteristics					
Z_0 (Ω)	89	55	97	60	35
R (Ω/cm)	0.05	0.40	0.24	2.5	14
L (nH/cm)	6.1	5.9	7.2	3.7	2.1
C (pF/cm)	0.78	1.94	0.76	1.01	1.75
Interconnection circuit parameters					
attachment	conventional	wire-bond	wire-bond	flip-chip	flip-chip
l_0 (cm)	3.0	1.5	1.5	1.0	1.0
C_{PKG} (pF)	5	0.3	0.3	0.3	0.3
L_{PKG} (nH)	12	0.5	0.5	0.025	0.025
C_{PAD} (pF)	0.6	0.6	0.6	0.6	0.6
C_{ESD} (pF)	2.0	2.0	2.0	2.0	2.0
R_S (Ω)	25	25	25	25	25
Figures of merit					
l_{crit} (cm)	21,661	1,426	5,221	148	11
n_{crit}	1036	188	416	36	4

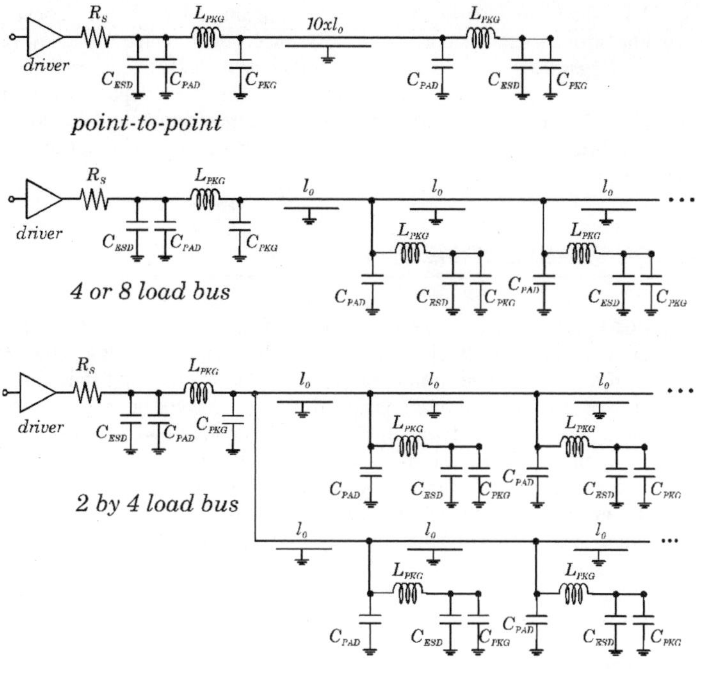

FIGURE 11.13 Example interconnection configurations.

The delays shown in Fig. 11.14 account for only the portion that arises directly from the interconnections. They do not include internal delays in the I/O buffers themselves. Stage-up delays in CMOS buffer circuits are often comparable to these interconnection delays, and in MCM systems, buffer r-design can result in significant additional decreases in total delay.[30,31] An important factor to keep in mind when considering changes in technology is not only the direct impact that they may have on interconnection delay but also their possible consequences for buffer delays.

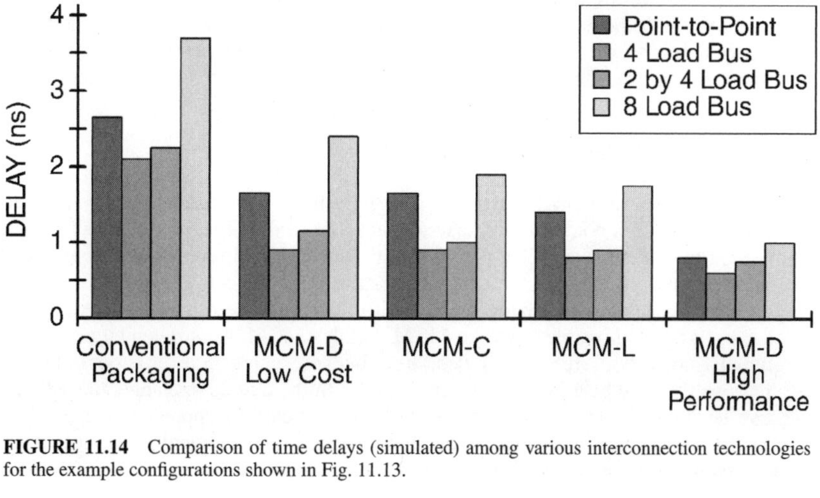

FIGURE 11.14 Comparison of time delays (simulated) among various interconnection technologies for the example configurations shown in Fig. 11.13.

As Fig. 11.14 shows, delays in MCMs are significantly less than in conventional packaging. Interestingly, the strongest factor correlating to reduced interconnection delay is the length of the interconnection. Regardless of the substrate technology or chip attachment method, shorter wires translate into reduced interconnection delays. The only significant departure from this general rule is in low-cost MCM-D technologies where the interconnection length and bus loading are close to or exceed the critical values listed in Table 11.1. Even in this case, performance is considerably better than that of conventional packaging.

11.7 A SYSTEM DESIGN PERSPECTIVE ON DELAYS

Analyses of MCM performance for various choices of interconnection structure and materials often focus on reduction of delays. It should be recognized in making technology choices for MCMs, however, that reduction in interconnection delay does not necessarily translate directly into improved system performance. Systems are often designed using a concept of "delay budget." A certain fraction of the system's clock cycle, typically 10 to 20 percent, is budgeted to interconnection delays. Implementing the same system in an MCM, even if this eliminates delays entirely, will only result in a 10 to 20 percent performance improvement. The rest of the delay is in the chips' internal logic operations and is unaffected by the interconnections. Many MCM applications are redesigns of existing systems and, as such, may not benefit greatly from reduced interconnection delay.

11.7.1 Delay vs. Interconnection Length

Figure 11.15 shows a computation of minimum interconnection delay as a function of distance [32]. These calculations are based on time-constant estimates, but also include the effects of stage-up delays in CMOS I/O buffers. As shown above, interconnection delays are strongly influenced by driver resistance. Lowering the output resistance of a CMOS buffer, for example, will generally result in reduced interconnection delay. However, this is obtained at the expense of increased stage-up delay in the buffer itself. The calculations in Fig. 11.15 include these additional delays and have optimized linewidth and buffer sizes (using 0.9 μm gate technology, in this example) to obtain minimum overall delay for each option. The figure compares on-chip delays with chip-to-chip delays using conventional packages on a printed wiring board, and chip-to-chip delays on a low cost silicon-on-silicon MCM.

Calculations like this are commonly used in partitioning studies to find optimum ways to package various parts of a system. As distances increase, on-chip delays grow quadratically. MCM interconnections, with their larger dimensions, can offer delay advantages despite the penalty imposed by buffer delays to get from one chip to another. At a still greater distance, these delays also begin to turn upward, exceeding those of conventional interconnections on a board.

11.7.2 System Size and Delay Budget

At first glance, Fig. 11.15 suggests that there is little performance advantage to be gained by an MCM. This does not, however, take any account of the differences in packaging density offered by the various alternatives. The way to cast these results into a proper perspective is to consider their implications in light of a given interconnection delay budget. This is illustrated in Fig. 11.16. Within an interconnection delay budget of 1 ns, for example, there is no choice but to remain on-chip. At 1.5 ns, it is also possible to use an MCM, but this MCM is not much larger than a chip and is consequently not very useful. For a much larger budget, such as 3 ns, it is possible to put the system in a reasonably sized board or a fairly large MCM. In the 2 to 2.5 ns range, the MCM becomes an attractive technology, since it is not possible to use a board of appreciable size. Moreover, because the MCM offers much higher packaging density than a conventional board, it is possible in some cases to package a lot more circuitry in an MCM than on a board.

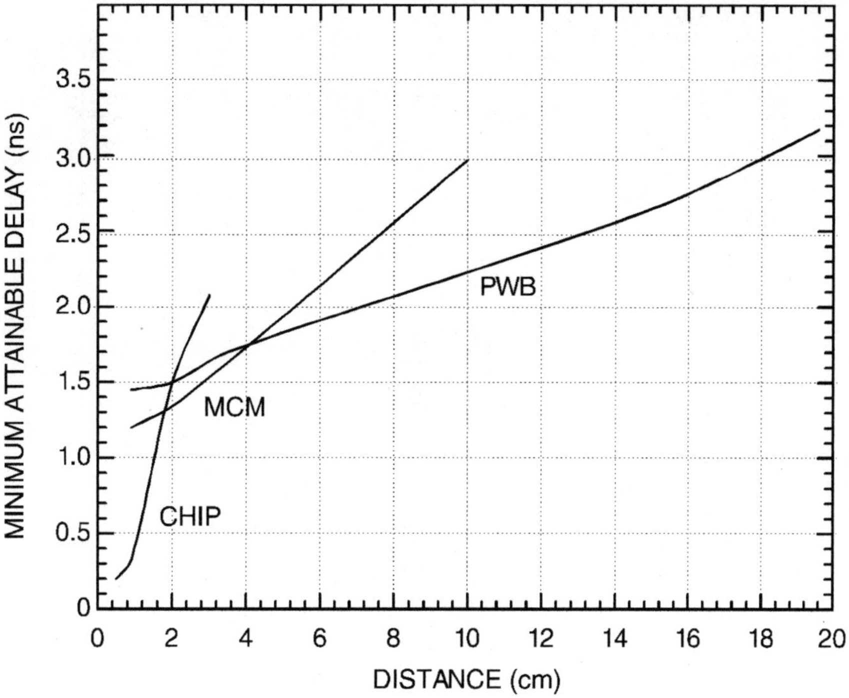

FIGURE 11.15 Calculated interconnection delays, including CMOS buffer delays, for various interconnection options as a function of distance.

Analyses like this demonstrate a perspective from which to view the role of delays in electronic packaging of systems. The true benefit of decreased interconnection delays is that they allow systems designers to put more circuitry (and consequently more functionality) within the confines dictated by their delay budget. Decreased delays combined with higher packaging densities make it possible to use MCMs to build more complex, higher-speed systems. for CMOS-based systems, these benefits are realizable within a performance window. The highest-speed systems still demand single-chip solutions, and lower-speed systems can be made with conventional packaging.

11.8 CONCLUSIONS

The earliest applications of MCMs were in high-end mainframe computers, aimed primarily at bipolar circuits driving matched impedance transmission lines. Interconnection lengths in such systems were sometimes on the order of one meter. The improvements in packaging density afforded by MCMs had a significant impact on the performance of such systems.

High-performance digital systems have been steadily migrating toward CMOS technology. The very high levels of integration offered by CMOS, as compared to earlier generations of bipolar technology, accomplishes the same reduction in interconnection length that MCMs once offered. This has narrowed the application window for performance-driven MCMs. Many digital systems fit onto a single chip or can comfortably meet their performance requirements with conventional packaging. Thus, while MCM technologies can cut interconnection delays in half, this may not have enough impact on overall system performance to justify the use of MCMs in many applications.

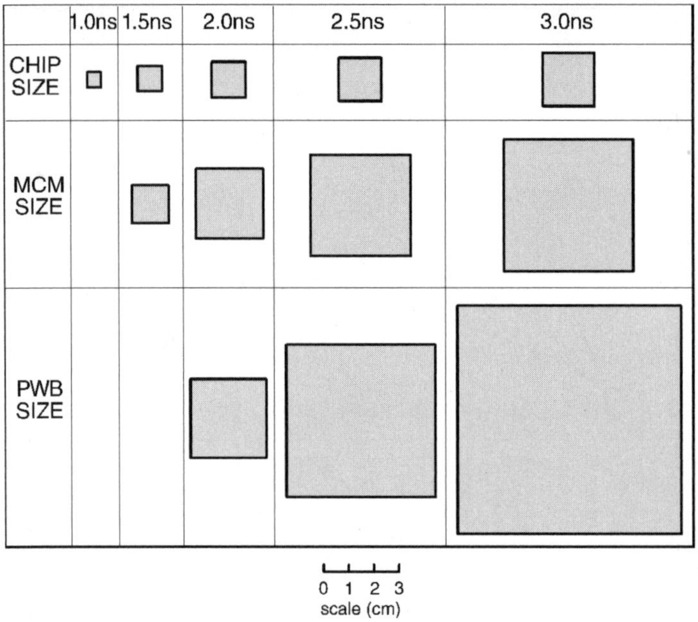

FIGURE 11.16 System packaging options for various delay budgets.

The application window shown in Fig. 11.16 for MCMs may prove to be a significant one. System designer have traditionally depended on the steady improvement in CMOS speed and density to meet ever growing demands for higher-performance, lower-cost digital systems. As these demands begin to strain the limits of CMOS technology, other avenues of improvement will be needed. Until then, however, it is likely that other considerations, such as cost, system size or system weight, will play a much larger role in MCM technology.

REFERENCES

1. L.N. Dworsky, *Modern Transmission Line Theory and Applications*, John Wiley & Sons, NY, ISBN 0-471-04086-X (1979).
2. N.S. Pickles and M.C. Lefebvre, "ECL I/O buffers for BiCMOS integrated systems: a tutorial overview," *Proc. Sixth Annual IEEE Int. ASIC Conference and Exhibit,* 436–443 Rochester, NY, Sept. 27–Oct. 1 (1993).
3. M. Pedersen and P. Metz, "A CMOS to 100k ECL interface circuit," *ISSCC89,* pp 226–227, Feb. (1989).
4. T.J. Gabara, "A 600 MHz 100k ECL output buffer fabricated in 0.9 micron CMOS," *Proc. TENCON,* Australia, Nov. (1992).
5. B. Gunning, L. Yuan, T. Nguyen, and T. Wong, "A CMOS low-voltage-swing transmission-line transceiver," *Proc. ISSCC* (1992).
6. J.H. Quigly, J.S. Caravella, and W.J. Niel, "Current mode transceiver logic, (CMTL) for reduced swing CMOS, chip to chip communication," *Proc. sixth Annual IEEE Int. ASIC Conference and Exhibit,* 452–457, Rochester, NY, Sept. 27–Oct. 1 (1993).
7. H. Howe, *Stripline Circuit Design*, Artech, Dedham, MA (1974).
8. H.R. Kaupp, "Characteristics of microstrip transmission lines," *IEEE Trans, Electronic Computers,* EC-16, 185–193, (1967).

9. J.R. Brews, "Overshoot-controlled RLC interconnections," *IEEE Trans. Electron Devices*, 38, 76 (1991).
10. R.C. Frye, "Optimal self-damped lossy transmission line interconnections for multichip modules," *IEEE Trans. Circuits and Systems–II*, 39, 765 (1992).
11. R.C. Frye, "Balancing performance and cost in CMOS-based thin film multichip modules," *Proc. IEEE Multi-Chip Module Conf.*, 6, Santa Cruz, CA (1993).
12. E.E. Davidson, "Electrical design of a high speed computer packaging system," *IEEE Trans. Components, Hybrids, and Manuf. Tech.*, CHMT-6, 272 (1983).
13. A.J. Rainal, "Performance limits of electrical interconnections to a high-speed chip," *IEEE Trans. Components, Hybrids, and Manuf. Tech.*, 11, 260 (1988).
14. R.C. Frye, "Physical scaling and interconnection delays in multichip modules," *IEEE Trans. Components, Packaging and Manuf. Tech. B*, 17, 30 (1994).
15. R.C. Frye and R.A. Lorenz, "Identifying elements of delay in multichip module technologies," AT&T Bell Laboratories internal memorandum, (1994).
16. F.Y. Chang, "Waveform relaxation analysis of RLGC transmission lines," *IEEE Trans. Circuits and Systems*, CAS-37, 1394 (1990).
17. F. Romeo and M. Santomauro, "Time-domain simulation of *n* coupled transmission lines," *IEEE Trans. Microwave Theory Tech.*, MTT5, 131 (1987).
18. A.J. Blodgett and D.R. Barbour, "Thermal conduction module: a high-performance multilayer ceramic package," *IBM J. Res. Develop.* 26, 30 (1982).
19. Y. Baba, S. Segawa, S. Fukunaga, J. Shigemi, and H. Ochi, "High-density multilayer ceramic substrate," *Int. J. Microcircuits and Electronic Packaging*, 16, 344 (1993).
20. L.M. Higgins III, "Perspectives on multi-chip modules: substrate alternatives," *Proc. IEEE Multi-Chip Module Conf.*, 12, Santa Cruz, CA (1992).
21. C. Reynolds, M. Freda, and L. DiFrancesco, "MCM-L structures, case technologies and design guidelines," *Proc. Int. Conf. Multichip Modules*, 575, Denver, CO (1993).
22. C.J. Bartlett, J. M. Segelken, and N.A. Teneketges, "Multi-chip packaging design for VSLI-based systems," *IEEE Trans. Components, Hybrids and Manuf. Tech.*, CHMT-10, 647 (1987).
23. R.R. Johnson, "Multichip modules: next generation packages," *IEEE Spectrum*, 34, March (1990).
24. K. Prasad and E. Perfecto, "Multilevel thin film packaging: applications and processes for high performance systems," *IEEE Trans. Components, Packaging and Manuf. Tech. B*, 17, 38 (1994).
25. T. Rucker, V. Murali, and R. Shukla, "A wet etch process for multi-chip modules," *Proc. Int. Conf. Multichip Modules*, 251, Denver, CO (1992).
26. W.E. Pence and J. Mosely, "Design of a silicon-on-silicon multichip module for a high-performance PS/2 workstation," *Proc. IEEE Multi-Chip Module Conf.*, 110, Santa Cruz, CA (1992).
27. M.Y. Lau, K.L. Tai, R.C. Frye, M. Saito, and D.D. Bacon, "A versatile IC fabrication compatible MCM-D for high performance and low cost applications," *Proc. Int. Conf. Multichip Modules*, 107, Denver, CO (1993).
28. L.W. Nagel, "ADVICE for circuit simulation," *Proc. 1980 Int. Symp. on Circuits and Systems*, Houston, TX, April 28, 1980.
29. L.W. Nagel and D.O. Pederson, "Simulation program with integrated circuit emphasis," *Proc. Sixteenth Midwest Symposium on Circuit Theory*, Waterloo, Canada, April 12, 1973.
30. T. Gabara, W. Fischer, S. Knauer, R. Frye, K. Tai, and M. Lau, "An I/O CMOS buffer set for silicon multichip modules," *Proc. IEEE Multi-Chip Module Conf.*, 147, Santa Cruz, CA (1993).
31. R. Frye, T. Gabara, K. Tai, W. Fischer, and S. Knauer, "Performance evaluation of MCM chip-to-chip interconnections using custom I/O buffer designs," *Proc. Sixth IEEE Int. ASIC Conf.*, 464, Rochester, NY (1993).
32. R.C. Frye and K.L. Tai, "Silicon multichip modules: evolving a high-performance technology from low-end applications," *Proc. IUMRS Int. Conf. Electronic Materials,* Hsinchu, Taiwan, Dec. 19–22, (1994).

CHAPTER 12

ELECTRONIC PACKAGING FOR HIGH-PERFORMANCE DIGITAL ICS

Barry K. Gilbert
Special Processor Development Group, Mayo Foundation

12.1 INTRODUCTION

Since the early 1980s, digital integrated circuits have become sharply divided into two categories: the mainstream silicon-based CMOS devices with very high transistor counts and clock rates in the 20–200 MHz range, and a small class of low gate count (never greater than 100,000 gates, and usually much fewer) very high clock rate (200–2000 MHz) silicon emitter coupled logic (ECL) and gallium arsenide (GaAs) components. These chips have been employed in supercomputers and in radar and communications systems where the high-speed performance was more important to system performance than the gate count.[1–4]

These two nearly completely separate performance regimes have been accompanied by completely different approaches to the packaging of these components. Until recently, the CMOS devices have been packaged satisfactorily in single-chip packages (frequently with very high pin counts). However, the parasitic reactances of the package leads have caused significant degradation in the maximum performance available from the "fast" chips since the mid-1980s. As early as 1985, designers of the highest-performance components began considering the use of multichip modules (MCMs) as a possible solution to the performance loss they were observing when using the single-chip packages. In this chapter, we will explore the evolution in the use of MCMs for these high-speed components, indicate where the remaining technological problems lie, and indicate the probable course of evolution of electronic packaging for the highest speed components (both digital *and* analog components) over the next five to seven years.

The differences in packaging requirements for the slower CMOS chips and for the faster ECL and GaAs chips are instructive: while the new MCM technologies represent perhaps a *better* way to package the CMOS chips in a reduced space, MCMs are not the *only* way to accomplish CMOS chip encapsulation and interconnect. Conversely, *for the fastest chips,* the MCM technologies appeared in 1986, and appear now, to be *the enabling packaging technology* that will allow the full speed potential of these devices to be exploited.[5–8]

In the following section, we will demonstrate by means of several recently implemented examples that MCMs can yield the high packaging performance that single-chip packaging and PC board interconnect can no longer provide. In the remainder of this chapter, we will explore the electrical, electromagnetic, and thermomechanical features required by the highest-speed chips and the ways in which the "conventional" MCM technologies can be modified to achieve these performance goals. It is our purpose in this chapter to provide an understanding of the advantages and constraints of the MCM packaging approach for high-performance systems, not the rigorous mathematics of transmission line theory or thermal modeling. Thus, the number of equations has been held to a minimum, and equations are only introduced when they can improve the reader's intuitive understanding of the design concepts. However, references are provided to articles in the technical literature where the more detailed mathematics may be found.

12.2 MCMS OPERATING IN SUPPORT OF HIGH CLOCK RATE DIGITAL ICS

While the use of MCMs to support very high clock rate digital (and analog) integrated circuits is in its infancy, the original projections that MCMs would serve well in this role are being validated. A few examples will be presented to demonstrate the performance levels that can be achieved with the MCM packaging approaches.

Figure 12.1 is a photomicrograph of an organic-based MCM-L designed and assembled in our laboratory and containing seven GaAs integrated circuits. The chips were attached to the MCM with silver-loaded epoxy (70–80 percent by weight, 20–25 percent by volume) and electrically attached with gold ball bonds. In perhaps the most extreme example of the performance of digital integrated circuits on an MCM, these GaAs circuits operated at clock rates up to 9.4 GHz, with signal rise and fall times in the range of 20–30 ps (Fig. 12.2).

FIGURE 12.1 Photo of experimental digital circuit built with seven small GaAs MESFET chips on a six-layer copper/Teflon™ MCM-L. All connections are with wire bonds. The meandered lines serve to delay specific signals to account for setup and hold time requirements of several of the chips. Triangular structures are 50-Ω passive coplanar line microwave probes.

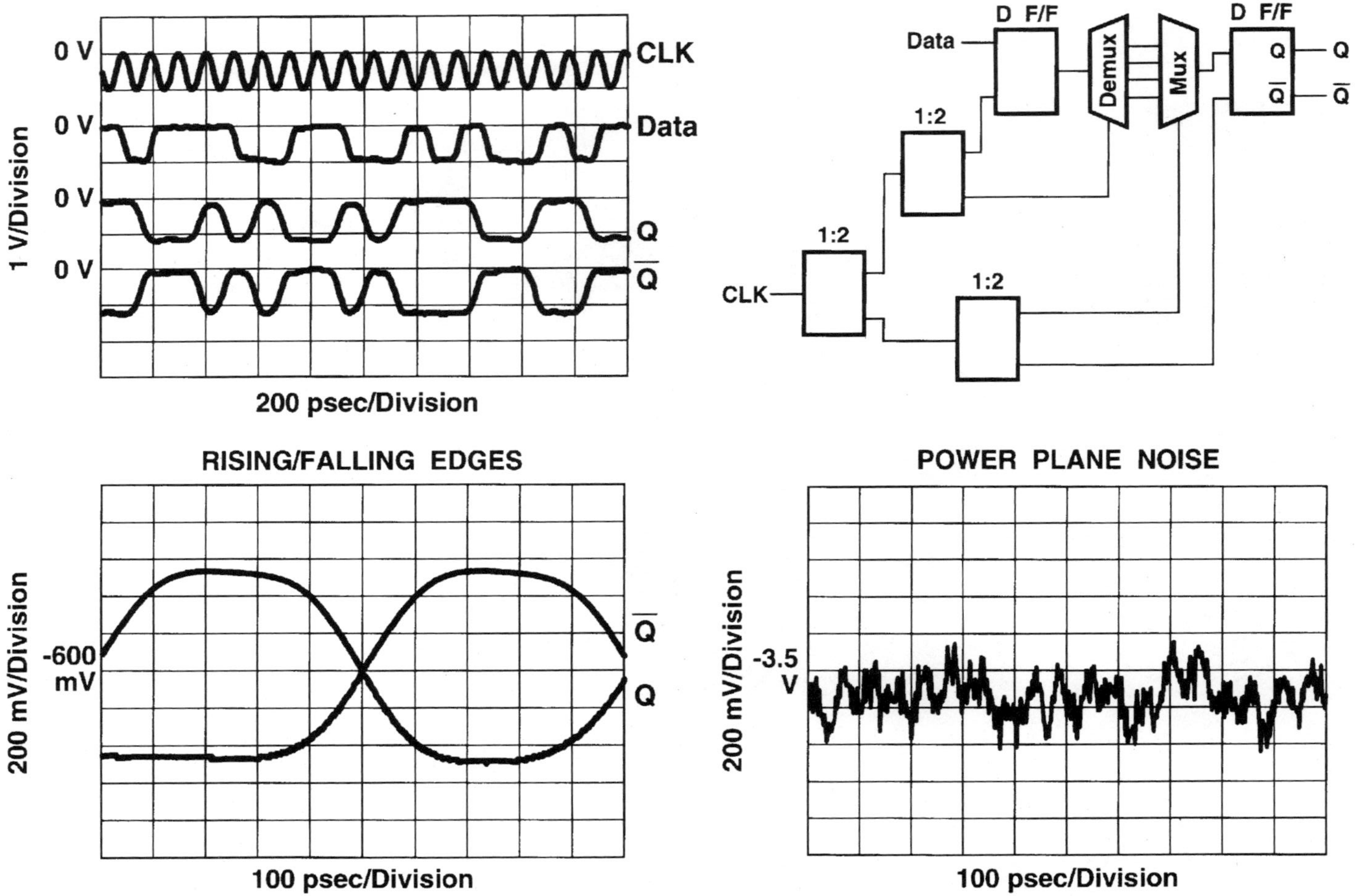

FIGURE 12.2 Measurement results from the seven-chip circuit of Fig. 12.1. The circuit performed correctly up to a 9.4 GHz clock rate, demonstrating that MCMs in general and MCM-Ls in particular are capable of supporting extremely high clock rates. Waveform conformation is excellent, although the power plane noise is higher than would be desirable for full production systems.

As a second example of a different type of MCM-D (referred to as a *chips-first* MCM-D; see later discussion), a six-chip circuit was fabricated using an organic-based MCM-D referred to by its developer as *high density interconnect* (HDI). In this type of MCM (see later discussion), the chips are actually buried beneath the layers of interconnect and power distribution metalization, and hence are invisible from the surface (Fig. 12.3). As will be described, this type of MCM can deliver extremely high system performance. Figure 12.4 depicts oscilloscope measurements of the performance of one of the two-chip GaAs circuits incorporated into this test vehicle. The 2.5 GHz clock rate performance exceeded the manufacturer's specifications for the chips by 800 MHz, demonstrating that, in a properly designed electronic packaging and interconnect environment, the chips may actually work faster on the MCM than predicted from computer simulations or even from measurements in single-chip packages in the vendor's test facility, due to the reduction in interconnect parasitics.

Figure 12.5 is a photograph of an MCM fabricated as alternating layers of aluminum and silicon dioxide dielectric on a silicon wafer substrate, containing 16 GaAs integrated circuits. The maximum performance of the bare chips, based on computer simulations and measurements of chip performance in single-chip packages, is in the range of 300–700 MHz, with one small subcircuit indicated to

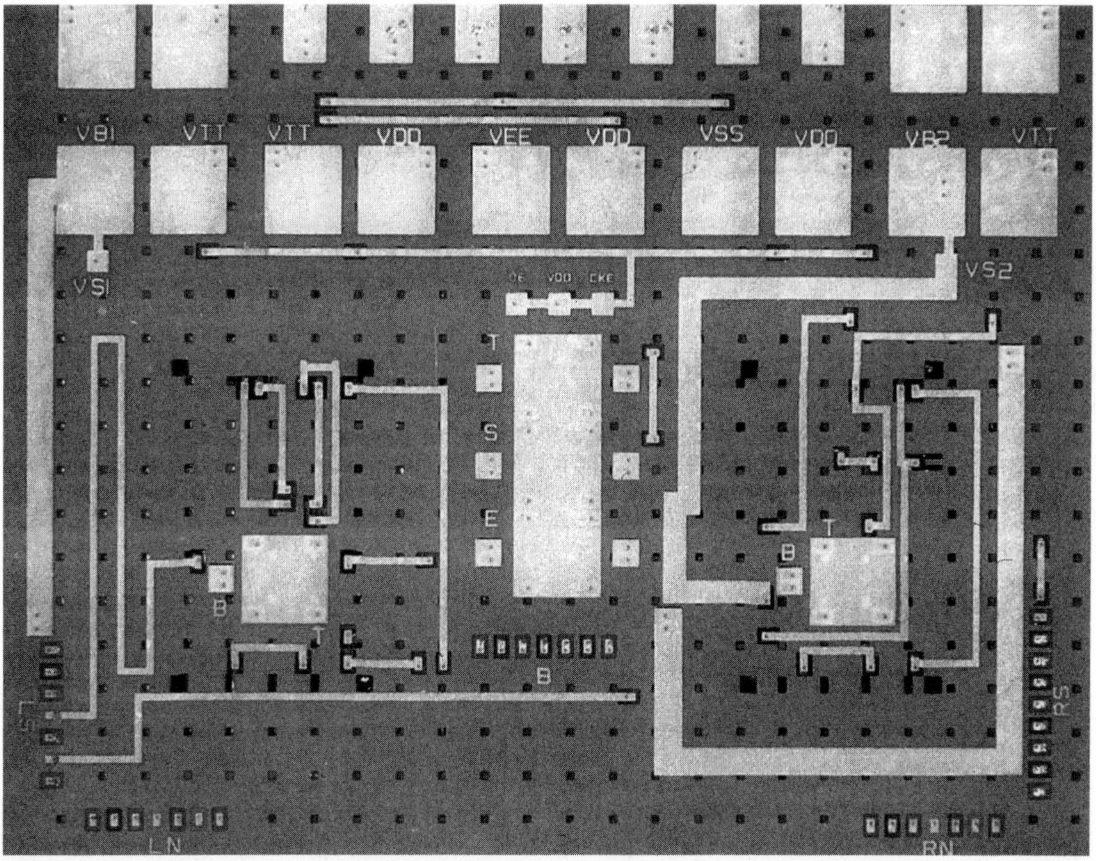

FIGURE 12.3 Micrograph of a portion of a chips-first, organic-based MCM-D. The GaAs chips are buried beneath the metal ground planes. Microstrip interconnects are visible on the surface above the top ground plane. The two large squares and rectangular metal pad are capacitor sites, and the row of small rectangular pads along the upper edge are probe card contact sites. *Source:* fabricated by GE CR&D Center.

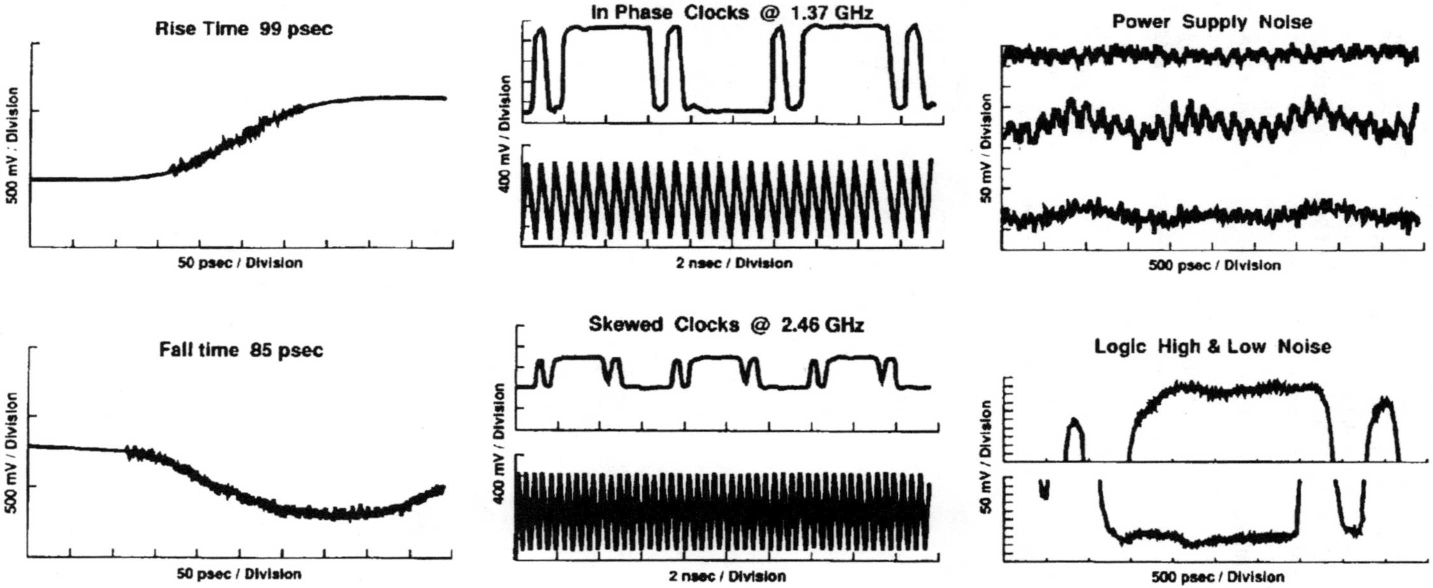

FIGURE 12.4 Clock and data waveforms taken from a two-chip circuit fabricated on the six-layer test coupon of Fig. 12.3. Note the excellent waveform conformation at the 1.37 GHz and 2.46 GHz clock rates, demonstrating that chips-first MCM-Ds can support high system clock rates.

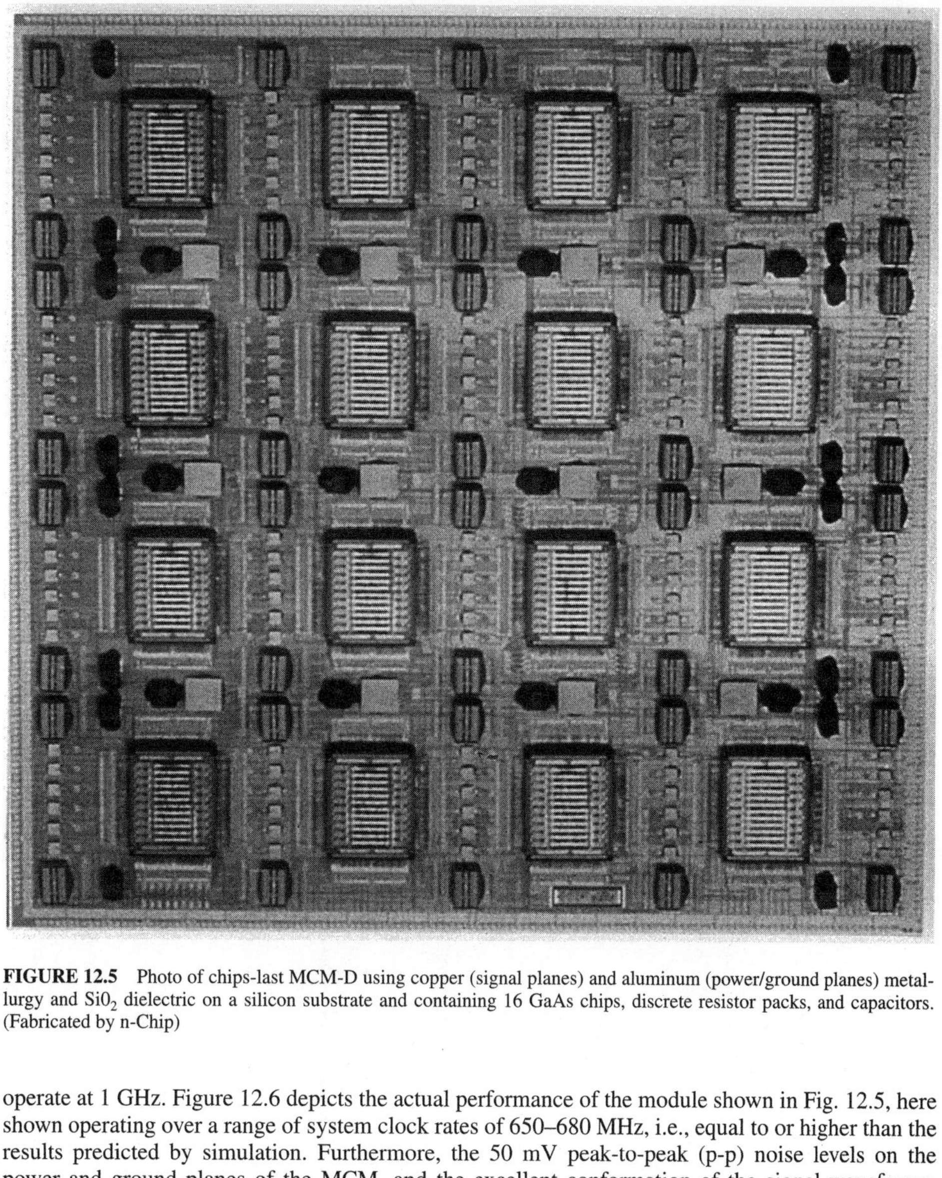

FIGURE 12.5 Photo of chips-last MCM-D using copper (signal planes) and aluminum (power/ground planes) metallurgy and SiO_2 dielectric on a silicon substrate and containing 16 GaAs chips, discrete resistor packs, and capacitors. (Fabricated by n-Chip)

operate at 1 GHz. Figure 12.6 depicts the actual performance of the module shown in Fig. 12.5, here shown operating over a range of system clock rates of 650–680 MHz, i.e., equal to or higher than the results predicted by simulation. Furthermore, the 50 mV peak-to-peak (p-p) noise levels on the power and ground planes of the MCM, and the excellent conformation of the signal waveforms, underscore the high quality of the electrical environment provided to the chips by the MCMs. Very similar results were obtained when the same 16-chip circuit was assembled using a "chips-first" MCM technology very similar to that described above.

These examples demonstrate the prediction made years ago that MCMs, almost regardless of the fabrication approach or materials employed in the MCM stackup, can consistently yield electrical performance better than the best achievable from conventional single-chip packaging and interconnects based on conventional PC board technology. In a number of tests of MCMs fabricated with different technologies, the clock rate performance of the MCMs has been equal to or greater than that predicted by the chip vendors for the bare chips alone.

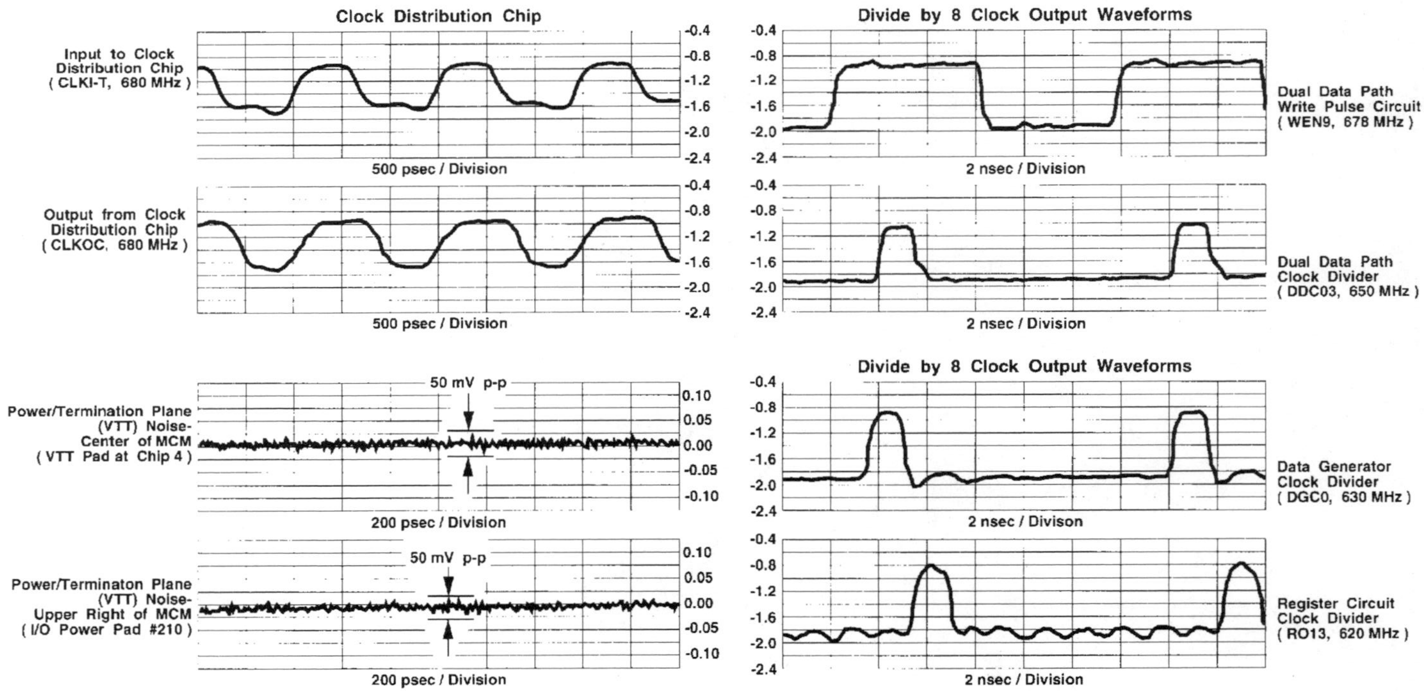

FIGURE 12.6 Clock, data, and power plane measurements taken at selected internal locations on silicon-SiO_2-aluminum MCM-D of Fig. 12.7. Note excellent waveform conformation and low levels of power and ground plane noise.

Why do the chips perform better in MCMs than the vendors' predictions for the bare devices? The fundamental measurements that the chip vendors make on their transistors, which are in turn incorporated into their simulation models for chip performance, are made with measurement probes that are contaminated by reactive parasitics. Thus, the transistor models contain "hidden" parasitics based on the device measurements with less-than-ideal probes, which makes the transistors and the chips themselves appear to exhibit less performance than is actually the case. When the real chips, which thus perform better than the models predict, are encapsulated in MCMs, which have fewer parasitics than the single-chip package, the "real" performance of the chips manifests itself, and the entire circuit performs much better than the models predict. This apparently impossible behavior has been observed many times since the late 1980s.

How are these high performance results achieved? Can MCMs fabricated for use with conventional CMOS chips, and manufactured without any special attempts to optimize or "tune" the MCM materials composition or structure, be used to achieve this type of extremely high performance? What, if any, special design, fabrication, or materials processing steps must be undertaken to ensure the type of performance that MCMs are capable of delivering? The brief answer is that conventional MCMs, using standard MCM design rules, will *not* necessarily yield these high performance levels. The remainder of this chapter will describe in detail the additional steps and precautions which must be undertaken to allow the maximum performance of high clock rate chips (whether silicon ECL, GaAs, or even CMOS) to be exploited.

12.3 MCMS AND HIGH-SPEED IC DESIGN

MCMs are attractive to the high-speed design community for two principal reasons: they can minimize the interconnect propagation delay and can provide a better thermal environment for the chips. The first advantage of MCM packaging for high-speed integrated circuits is that the interconnects between chips can be constrained to extremely short lengths when compared with single-chip packages and circuit boards, because the chips can be packed tightly together. The removal of the chips from the single-chip packages removes the additional propagation times required for the signals to traverse the distance from the outer edge of the package to the die well, which can amount to 100–250 ps out of the "source" package, and the same duration into the "destination" package. Package delays of 200–500 ps are unacceptable if the system clock rate is 1 GHz; a delay through only two packages that is equivalent to 20–50 percent of the entire clock cycle is unacceptable. The MCM packaging approach presents an opportunity to minimize the interchip propagation delay.

Second, the fact that the chips are more tightly packed creates higher power densities than in circuit boards, up to 10–50 W/cm^2 or even more. The second principal advantage of MCMs is that they can provide an improved thermal conduction path to the heat sink in comparison to singly packaged chips. Thus, the higher power densities can be accounted for by the improved thermal paths that can be designed into the MCMs.

While shorter interconnect delays and improved thermal performance are the two paramount features of MCMs that interest high-speed designers, these two issues *do not by themselves* guarantee that the MCM will work well at high system clock rates. A lengthy list of second order design and fabrication features of the MCM must be addressed if the MCMs are to perform properly at high clock rates and wide signal bandwidths. If these issues are not accounted for, the MCMs will not function at the intended system clock rates. These requisite features will be treated in detail shortly. First, however, we will comment on several of the primary strengths and limitations of each of the MCM technologies from the viewpoint of their application to high-speed systems. Because the details of these technologies are described elsewhere in this book, we will concentrate only on those features in each type of MCM that are either strongly supportive of high-speed system performance or create problems for high-speed devices. To the present time, no single MCM technology incorporates all the features that would be ideal for high-speed system performance.

12.4 REVIEW OF THE DIFFERENT MCM TYPES

12.4.1 MCM-Ls

Perhaps the most obvious MCM technology represents a "shrinking" of conventional multilayer printed circuit (PC) boards. As a result, the attributes and weaknesses of MCM-Ls are derive directly from the features of printed wiring boards (PWBs). For example, MCM-L line widths are typically 2-4 mils for 50-Ω stripline or 4–8 mils for 50-Ω microstrip, nearly comparable to the 8 mils of conventional PC boards. In addition, since MCM-L lines are typically fabricated by etching "quarter ounce" copper impregnated onto a laminate dielectric layer, the lines are fairly thick (approximately 8 μm) in comparison to those of other types of MCMs. This thickness translates into lower DC and AC line losses for propagating high-frequency signals.

Conversely, MCM-Ls have interconnect density limitations imposed by their approach to the fabrication of vias, the metal columns used to connect signals in one layer to signals in another layer. In most MCM-L fabrication approaches, the vias are *through;* that is, they are mechanically drilled or laser drilled completely through the layer stackup. These through holes decrease the signal routing density that can be achieved throughout the MCM-L, since each through via "disables" a routing row and column on all signal layers in its vicinity. Furthermore, in many MCM-L technologies, the via diameters are much larger than the line widths, e.g., 4-mil or 8-mil lines but 6-mil diameter vias and 12-mil cover pads over the via holes. In such a case, routing densities are further degraded because even if lines could be routed in theory on 8- or 16-mil pitch (i.e., line center to line center), the vias force the effective line pitch to be much greater, e.g., 20–25 mils. It must be noted that it is not primarily the minimum line pitch that causes the problem, but rather that the large via/cover pad structure has been observed to cause significant signal reflections on the interconnects that degrade signal integrity and system noise margin. Although at 200 MHz clock rates these reflections are not apparent, at higher clock rates and commensurately faster signal rise times, the reflections become more apparent and more important. Recognizing the line density and signal reflection problems, several MCM-L fabricators have recently developed and have begun to introduce fabrication processes that will yield vias no larger than the line widths, and line widths as small as 1 mil for use at 2-mil pitch. However, such small vias do not yet represent the norm for the MCM-L industry.

Individual metal layers of an MCM-L are defined with the same type of film artwork typically used in printed circuit boards, and the technology is stated by its proponents to support a very large number of metal layers if required (nearly equivalent to the 30 or more layers found in the most sophisticated PC boards). Chips are attached to the MCMs with epoxy, and electrically connected by wire bonding, tape automated bonding (TAB), or a process referred to as "flip-TAB," in which the TABed chip is placed face down on the MCM, the TAB leads are attached to the board, and the excess TAB lead material is trimmed off close to the edges of the chips. MCM-L fabricators are also employing "flip chip" attachment approaches, and in fact the so-called PCMCIA cards (now being referred to as simply "PC cards") for laptop computers are being manufactured largely with MCM-L substrates and flip chip attachment methods. The MCM-L technology, relying as it does on extensions of a decades-old, mature PC board technology base, is thus able to produce very inexpensive MCMs that are proving to be ideal for many low-cost applications, which do not require maximum routing densities. Figure 12.7 is a micrograph of a small portion of the top surface of a small MCM-L containing eight metal layers. The striplines are 4 mils wide with 6-mil diameter drilled via holes and 12-mil contact pads around the perimeter of the vias, while the microstrip lines (visible on the exposed surface of the MCM-L in this figure) are 8 mils in width, with the same size vias and cover pads as for the stripline.

Although MCM-Ls appear to represent merely a straightforward evolution of conventional printed circuit board technology to finer geometries and the attachment of bare chips, this area may yet yield breakthrough innovations. For example, at least three companies are developing a true "roll-to-roll" fabrication approach for MCM-Ls, in which the individual laminated layers will be separately patterned, and all layers will then be aligned and pressed together in a continuous process. At least one of the companies working on this technology is actively bringing "on line" a capability

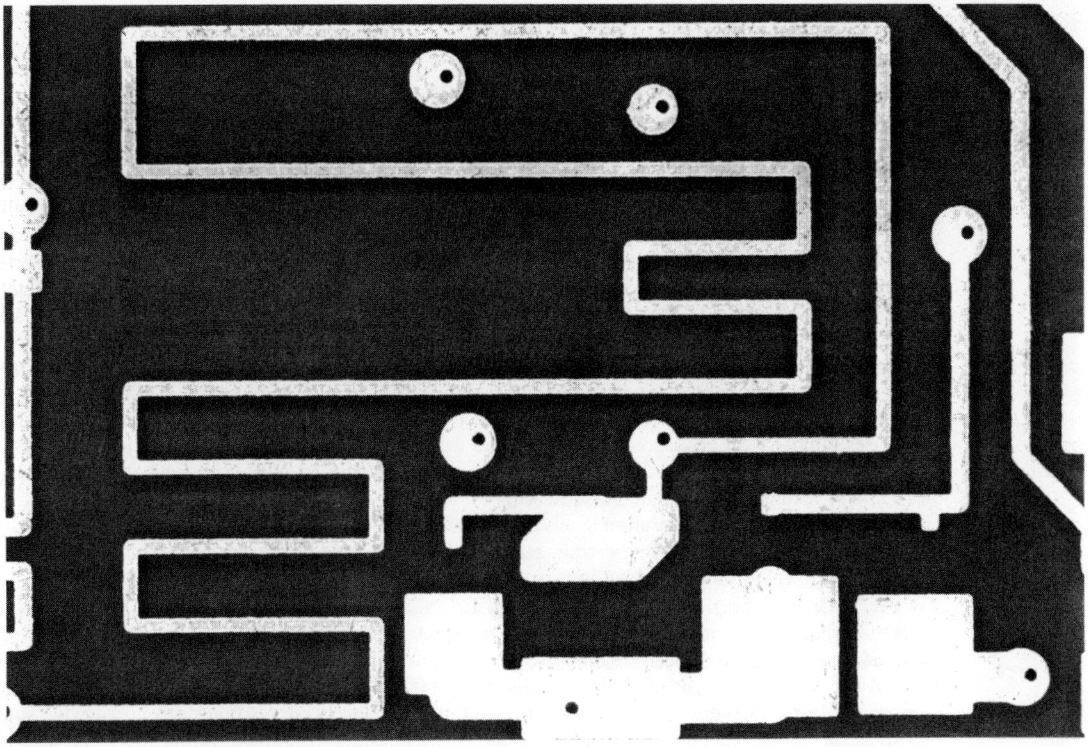

FIGURE 12.7 Surface of MCM-L showing 8-mil wide microstrip interconnects, 6-mil drilled-through vias, and 12-mil cover pads surrounding the via holes. (Fabricated by Acsist)

for the manufacture of a continuous strip of MCM-L material, 1–2 ft in width, at the rate of up to five *miles* per day. As fabrication methodologies such as this continue to mature, the price of low-performance MCM-Ls should decrease rapidly. However, there is still a problem of achieving tight lateral-dimension registration between layers as they are brought together for lamination, which results in a requirement at present to employ large diameter vias and capture pads to ensure contact between the interconnects in opposing layers. Although not a problem for low-speed subsystems, as noted above the large vias presently do pose a significant problem for high-speed systems.

12.4.2 MCM-Cs

Multilayer, cofired ceramic circuit boards have been known and exploited in the microwave industry for at least three decades, where they have typically been referred to as "hybrids." In this technology, individual sheets 5–25 mils in thickness, consisting of alumina powder in an organic binder and having the initial appearance of damp cardboard (referred to as the "green tape" form) are each patterned with a paste containing a powdered, high melting point metal such as molybdenum or tungsten, using silk screening techniques. The individual layers are then aligned and brought into contact, pressed together, and fired in a kiln at 1500–1650° C, burning out the binder and hardening and fusing the layers together. The tungsten or molybdenum particles fuse to create the signal lines and metallized planes; however, because the resistivity of these metals is relatively high, the lines have a relatively high resistance. To minimize the expense inherent in complex multilayer structures and to constrain interconnect line losses for high-frequency signals, the hybrids have typically been designed with microstrip interconnects on the top surface only (microstrip lines can be wider and exhibit less DC and AC resistance than buried lines for a given line impedance), with the buried layers usually reserved for power and ground planes.

During the past decade, the line resistance problem in ceramic multilayer structures has been somewhat alleviated. A *glass ceramic* material has been developed which fuses at approximately 600–900° C. At these reduced firing temperatures, lower-resistivity metals such as silver can survive the firing process, and thus can replace the tungsten. By shrinking the line widths of these ceramic circuit boards down to 4-6 mils in much the same way as has been done for the MCM-Ls from the PC board technology, ceramic MCMs, or MCM-Cs, have evolved. These improved hybrids or MCM-Cs have been used over the past decade in a number of microwave systems for military applications.

While the MCM-Cs typically suffer from greater than optimum line widths and pitches (for the same reasons: the via diameters may be larger than the line widths), they have a significant advantage for the fastest integrated circuits: the thermal conductivity of the ceramic, and its thermal mass, typically result in a good or even excellent (if, for example, Be0 or aluminum nitride are employed) thermal environment. MCM-Cs with a large number of metal layers have been demonstrated, but the cost increases rapidly with layer counts much greater than six to ten. Chips are attached to the MCM-C either with epoxy or solder, and the electrical connections are made with wire bond or flip chip attachment techniques (see later discussion). Finally, the MCM-Cs are mechanically very robust (they can even tolerate acceleration levels in the tens of thousands of Gs) and, because of their ability to be plated and to tolerate local heating, are naturally well suited to soldered, sintered, or welded lid attachment processes, and the resulting structures are inherently hermetically sealed.

MCM-Cs have been used as the substrates for microwave and millimeter wave subsystems in very demanding mechanical and thermal environments, e.g., in proximity fuses for artillery shells and radar seekers in high-performance missiles (i.e., the Advanced Medium Range Air-to-Air Missile, AMRAAM). Because of their cost and long fabrication cycles, MCM-Cs are, however, no longer as clearly the substrates of choice for high-performance systems that need not survive such harsh environmental conditions; this role is increasingly being subsumed by the MCM-Ls and, as indicated below, the MCM-Ds.

12.4.3 MCM-Ds

New processes have been developed to produce the last of the MCM types to be discussed here, the so-called *deposited MCMs,* or MCM-Ds. This additive process, with photolithographic definition of metal interconnects, vias, and other structures, is very similar to the one used to fabricate integrated circuits. As a result, the MCMs produced in this manner share many of the same strengths and weaknesses of integrated circuit fabrication. First, the dielectric layers can be either organic, such as polyimide or benzocyclobutane (BCB) built up layer by layer on a ceramic or silicon support substrate, or (in several cases) may be silicon dioxide built up layer by layer on a silicon substrate. The details of the fabrication processes used for organic and inorganic MCM-Ds are presented elsewhere in this book.

A principal advantage of the MCM-D process is its ability to create very precisely defined features through photolithographic processes similar to those employed in the fabrication of integrated circuits. Although MCM-D line widths are typically 25–75 μm, 10–15 μm lines have been achieved, and the control of line width, line thickness, and line cross section is good and continually improving. Since the vias are defined photolithographically as well, they can typically be the same widths as the lines, allowing effective line pitches of, typically, 50–125 μm; pitches as small as 25–50 μm (using 10–15 μm lines) have been demonstrated in an organic process and are actually used in the silicon-SiO_2-based production processes.

As depicted in Fig. 12.8, vias can be either solid plated posts (*solid post vias*), or can be created by etching (or burning) holes in the organic layers, then plating the sides of the vias and backfilling the plated holes with organic material (*open vias*). The vias thus need not penetrate all layers of the MCM (which increases routing densities) and are referred to as blind or buried vias. A via that must

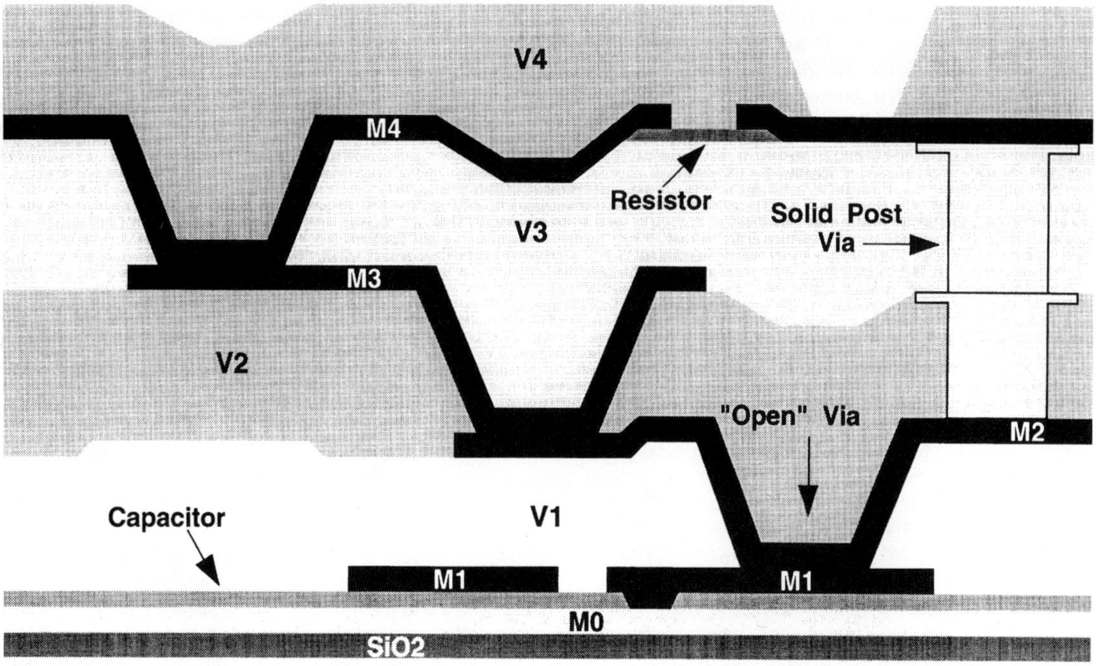

FIGURE 12.8 Schematic cross section of two different idealized organic-based MCM-Ds showing the "open via" approach (left) and the "solid post" style of vias (right). Any given MCM-D fabrication process uses one or the other via style, but not both.

penetrate multiple dielectric layers can be fairly easily fabricated as a continuous vertical pillar if it is a solid via. If the process uses open vias, each layer penetration must be offset from the single-layer vias above and below, creating a *spiral staircase* structure (although a few companies regularly employ a *stacked coffee cup* open via process).

Thermally conductive vias (i.e., generally with no electrical function) are installed under the chips in the organic MCM-Ds to remove heat to the substrate. They are not needed in the silicon-SiO_2 MCM-Ds because of the high thermal conductivity of the substrate itself. The solid post vias appear to have better electrical and thermal properties than the open vias and, at high frequencies, are far superior to the MCM-L or MCM-C vias because they are the same widths as the lines themselves, thus causing no signal reflections.

The number of metal layers available in most MCM-D processes is limited to four to six, which (for reasons to be discussed in detail below) may be insufficient for high-speed digital systems. However, some MCM-D vendors have demonstrated the ability to fabricate up to eight metal layers (all of the designs that emerge from our laboratory employ a minimum of five metal layers). Finally, chips are attached to the MCM-Ds with epoxy and electrically attached with wire bond, TAB, flip-TAB, or flip chip. At the time of this writing, the various forms of TAB are being used less frequently, whereas wire bond and flip chip attachment techniques are widely employed. Simulations and carefully conducted laboratory measurements indicate that wire bond connections are typically able to carry high-speed signals satisfactorily up to clock rates of 3 GHz and rise times down to 50–100 ps, if their lengths are held below 75 mils. Similar simulations and measurements indicate that flip chip attachment can support pulse rates in excess of 10 GHz, perhaps to pulse rates in the several tens of GHz range.

While the strong suit of the MCM-D for high routing density, slow-speed CMOS designs is its fine line and high routing density capability (up to 1000 cm routing per cm^2 of area per layer has been reported), its weakness for the highest-speed designs is the fact that the narrower the line, the greater the DC and AC losses. To minimize these losses, designers of high-performance MCMs will always trade routing density for wider lines, provided that the desired line impedance is maintained.

The MCM-D processes are at present more expensive than, e.g., the MCM-L process, because of higher nonrecurring engineering and manufacturing equipment costs. These costs are decreasing somewhat over the years, although it is unclear whether MCM-Ds can ever achieve the potential cost efficiency of MCM-Ls.

12.4.4 Chips-First MCMs

Finally, the terms *chips-last* and *chips-first* are used increasingly in the MCM community. MCM-L, MCM-D, and MCM-C technologies are usually implemented as chips-last structures; that is, the MCMs are fabricated complete with their signal and power/ground layers, then the chips are attached as the *last* part of the substrate assembly task, exactly as in a PC board (Fig. 12.5). However, several companies are producing MCMs that use an entirely different approach. The chips are *first* attached to a bare support substrate or, in some cases, a substrate that contains ground and power plane layers (the substrate may be fabricated either as an MCM-D or an MCM-L) with their active surfaces up, then layers of organic dielectric and metal (both power/ground and/or signal) are built up *above the chips*.[9–12] Electrical contact to the chips is made by using a laser to create the via openings, then plating down the open sides of the vias to make electrical contact between the interconnect lines and the chip pads.

This approach completely obviates the need for wire bonds (the cross section of one such structure appears in Fig. 12.9) and has electrical characteristics as good as flip chip attachment methods.[9–11] The chips-first MCM tends to be extremely robust when exposed to thermal and mechanical shock because (a) the chips are completely encapsulated by the support substrate and the covering layers of metal and plastic, and (b) there are no bond wires or other unsupported metal structures to be damaged by thermal or mechanical assault. For example, it has been demonstrated that a chips-first MCM can be dipped in liquid nitrogen and then dropped onto a hot plate literally hundreds of times over a duration of several years and still be completely operational. The same studies have demonstrated that these modules can withstand acceleration levels to many thousands of Gs without any damage whatsoever.[11–12]

12.4.5 Chip-On-Board

One final packaging technology has been described in the literature; however, because it is not a true "fine-line" technology, it is mentioned here only for completeness. The "chip-on-board" technology is exactly as its name implies: bare chips are epoxy attached to conventional circuit boards, and are wire bonded or flip chip attached to make electrical contact. The chips are then protected by completely covering them with a glob of epoxy or other conformal organic material. The chip-on-board approach is presently being used in a few consumer products where subsystem clock rates in the 100–250 MHz range are required for unusual applications (video controllers, wireless PC cards, and so on), but the wiring density requirements are not extreme, and the signal rise times are not below 500 ps. However, it remains to be seen whether chip-on-board approaches will continue to be employed, or whether true MCM-Ls will take their place for high-speed consumer applications as the MCM-L fabrication costs continue to decrease.

12.5 MCM FEATURES REQUIRED FOR HIGH CLOCK RATE DIGITAL SYSTEMS

What specific features must be incorporated into MCM modules to allow them to support the fastest digital or analog microwave integrated circuits? Most or all desired features will be obvious, but the ramifications of these requirements on the MCM fabrication process can be profound. In this section, the desired features will be enumerated. In the next section, the impact of these requirements on the design and detailed fabrication of the MCMs will be reviewed.

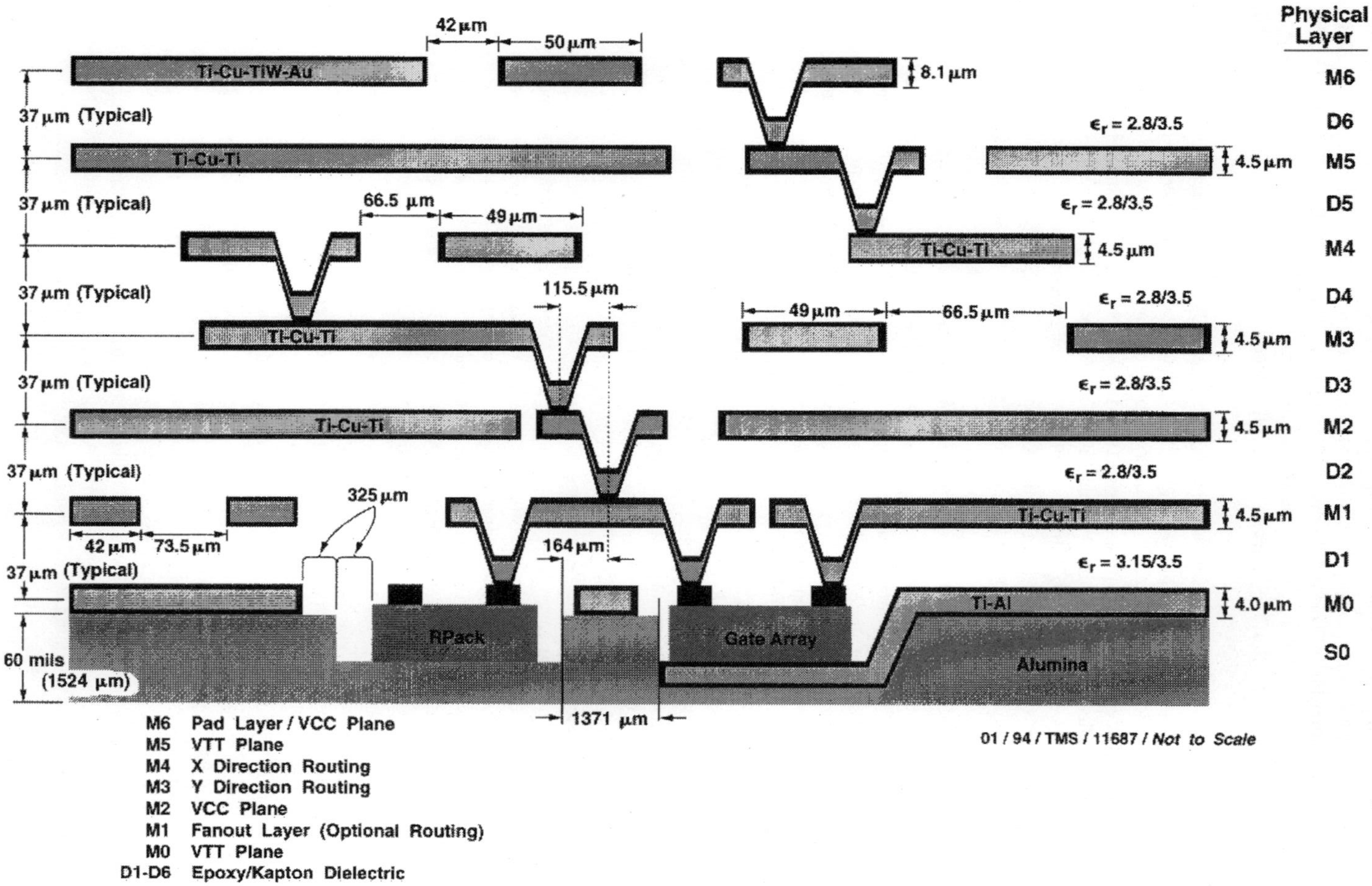

FIGURE 12.9 Cross-sectional view of a typical chips-first organic-based MCM-D technology. The chips and terminator resistor packs are set into wells milled in the ceramic base. Layers of Kapton and metal are deposited and patterned above the chips to create power, ground, and interconnect layers. Wire bond or TAB connections are not required.

1. Many of the highest-speed chips require a terminated transmission line environment to function properly, and sometimes to function at all. For example, ECL and most GaAs families employ open emitter line drivers, which must have a resistor installed between the output and the termination reference voltage, V_{tt}, for the outputs to switch states. Thus, the MCM technology must be able to support all of the hardware capabilities necessary to establish a terminated transmission line environment.
2. In a terminated transmission line environment, the MCM interconnects must exhibit low signal loss at all frequencies of interest. Signal losses much greater than 10 percent of the full-scale signal amplitude are considered to be unacceptable from the standpoint of maintaining signal integrity and adequate system noise margin. Thus, line losses across all frequencies of interest, and all line lengths of interest, must be maintained at less than ~1 dB (a very stringent requirement, sometimes downgraded to 2 dB in the interests of engineering expediency and cost control). In CMOS systems not employing terminated transmission lines, higher DC resistive losses in the lines result in two to three times longer settling times for the signal voltages to achieve their final values. The lengthened settling times in turn require that the system clock rates be decreased, thus depriving the system of some of its inherent performance and throughput.
3. The MCM technology must be able to support multiple metal layers in a variety of stackup configurations. The requirement for many metal layers is not driven by the need for routing tracks. Most designs, even quite complex ones, usually can be completed with two, or at most three, layers. (On occasion, an entire layer is set aside for clock trees in the highest speed designs.) The additional metal layers are used for the distribution of power voltages and ground; some designs, using GaAs or silicon ECL, require two or even three DC power voltages and ground, and on occasion there is a need for multiple ground planes. Low-speed designs can often partition a single metal plane layer or even rely on wide metal traces to deliver multiple DC voltages to the chips. As the clock rates increase, and the edge rates of the signal rise times decrease, experience has demonstrated that the best performance is obtained by committing an entire metal plane to a single DC voltage or to ground. Power and ground must be delivered on full area metallized planes, not on traces, to minimize resistive losses and inductive reactance problems for the return path currents that flow in these planes.

   Furthermore, to ensure optimum high-speed performance, the signal lines must be routed as microstrip or stripline transmission lines, which means that the routing planes must be sandwiched between power and/or ground planes. For example, two layers of stripline and a layer of microstrip would directly generate the requirement for (from the deepest layer to the uppermost layer) an AC ground plane (which can be assigned to be ground or any DC voltage), a signal plane (lowermost stripline), an AC ground plane, a signal plane (uppermost stripline), an AC ground plane, and a top signal plane (microstrip plane), for a total of six layers. In some designs, the AC ground plane separating the two stripline layers is omitted, decreasing the number of required metal planes to five, but somewhat increasing crosstalk between signals on the two different layers. In designs with special power supply features or shielding requirements, the number of required layers can increase to seven or even eight.
4. Large amounts of decoupling capacitance between power and ground rails must be provided in close proximity to the power leads of each chip to provide the instantaneous demands for current that occur during state switching. Otherwise, large amplitude voltage transients can occur on the power and ground planes, directly degrading the noise margin of the multichip system. The capacitance must exhibit low series inductance to be effective at the high switching frequencies typical of the fastest integrated circuits. This requirement is particularly stringent for large CMOS chips operating at clock rates greater than 100 MHz, since a disadvantageous characteristic of CMOS is that its generation of current and voltage transients becomes much worse as clock rates rise and signal rise times (both internal and external) decrease.

5. The MCM must be capable of removing large amounts of heat from very constrained areas while exhibiting very low temperature rise above ambient, since the fastest chips tend to exhibit large power densities. For example, the highest-speed GaAs chips are now designed to dissipate 50–60 W/cm^2, equivalent to nearly 150–180 W/cm^2 in silicon, because the thermal conductivity of the GaAs chips is roughly one-third that of silicon.

The requirement that an MCM be able to support very high clock rate system operation (whether for silicon ECL, GaAs, or high clock rate CMOS) in turn mandates physical, material, or metallurgical features that fall into two subcategories:

- Those that are similar in kind to (but may be different in degree from) those found in MCMs intended for 20–100 MHz CMOS chips (e.g., a good heat removal environment, good thermal expansion match between the chips and the substrate, and the ability to replace defective die without destroying the module)
- Those that are not required for and do not appear at all in MCMs intended for lower-speed systems (e.g., low-loss lines, transmission line interconnects, multiple metal layers devoted to power and ground delivery, and integral decoupling capacitors between power and ground planes)

In most cases, even those additional features that are not required for slower-speed MCMs are merely extensions or refinements of existing design rules or fabrication steps. Because of the nature of these differences, when working with a new MCM vendor, the author's research team strongly advocates to the vendor the inclusion of these additional features from the outset of the process development step, when they are easy to incorporate.

12.6 TRANSMISSION LINE EFFECT ON MCM DESIGN AND FABRICATION

The most profound impact on the design and fabrication of MCMs for high clock rate systems is the requirement for a transmission line interconnect environment. At low clock rates and, more particularly, relatively slow signal rise times, the interconnects can be treated merely as wires, with the assumption that any signal applied to one end of an interconnect appears instantly at the other end (one or more sources connected to one or more destinations is defined as a *signal net*), without any degradation in its wave shape.

An alternate approach to the analysis of interconnect behavior, and the one typically employed in the design of interconnects for CMOS systems, has been to assume that each output transistor driving a signal into an interconnect is really charging a load capacitance through a load resistance (i.e., that the signal net has an exponential charging behavior and an associated RC time constant). The capacitance is assumed to be the sum of the capacitance of the signal trace and the input capacitances of any receiving devices, i.e., the input capacitances of the receiving transistors of downstream chips on the same signal net. The resistive portion of the RC time constant, in turn, is assumed to be approximately the sum of the output resistance of the drive transistor and the series resistance of the line. The input resistances of the receiving devices are assumed to be so large as not to affect the RC time constant of the circuit. Provided that the line voltage has risen above the threshold level by the time of the next clock pulse and is not severely affected by crosstalk, the signal behavior is generally assumed to be acceptable.

There exists a third approach to the treatment of interconnects. As has been known for decades,[1,13] whenever signal rise times become in some sense *fast*, the interconnects begin to behave as transmission lines, whether they are designed as such or not. If (a) the line behaves as a transmission line of unknown characteristics, (b) the signal rise times are fast, and (c) no resistive damping is incorporated into the signal net, then the incident signal will *reflect* from the end of the signal net, causing disturbances that propagate back and forth along the entire length of the net. This *reflected wave* will distort the intended conformation of the signal at every receiving point along the net, for a duration that may be quite long compared to the rise time duration of the incident wavefront. Until the reflection decays and the signal amplitude stabilizes, the receiving gates may make an incorrect decision as to whether the wavefront represents a *zero* or a *one*, thereby requiring the system clock rate to be decreased or degrading the noise robustness of the signal net.

If the designer wishes to control the behavior of the interconnects when they are exposed to "fast" signal edges, the interconnect must be designed so that it will propagate the signal without distortion; i.e., a properly designed transmission line structure must be created. A useful rule of thumb is that a transmission line signal propagation environment must be intentionally constructed for the interconnects whenever the round trip propagation delay of the signal from the drive transistor to the farthest destination on a net is greater than one-quarter to one-half the signal rise time. For example, if the signal rise time is 200 ps (approximately that of modern silicon ECL and GaAs), and if the propagation delay in a given dielectric environment is 80 ps/cm (200 ps/in), then, assuming the more relaxed 50 percent of rise time condition, any line longer than 1 cm would have to be designed as a transmission line, with a specific line impedance and terminated with a resistor whose value "matches" the impedance of the line.[1,13] Note that it is the dielectric constant, ε_r, of the dielectric that determines the propagation delay.[1,13]

The constraint specified above is extremely stringent; almost all lines would have to be terminated in such a case. Only signal nets shorter than 1.67 cm could be allowed to settle, or *ring out* without some sort of resistive damping. (In such a case the designer actually relies on the source resistance of the driving transistor, and only that resistance, to provide the damping. This approach requires that the reflections make several round trips on the line before the transistor source resistance will have fully absorbed the reflections, and thus it only works for very short lines with brief round trip propagation delays.) In practice, the high-speed logic families typically are open collector/drain or open source/emitter logic, which *requires* that a resistor be attached from the output pin to the reference or termination voltage, V_{tt}, for a signal to appear at all. In most but not all cases, a resistor or group of resistors is employed for this purpose, allowing the output transistors to switch and also terminating the signal net, either at the source or final destination end of the net.

The need to create the terminated line environment is one of two major issues from which arise nearly all of the differences in the design of an MCM for high clock rate applications. A comment regarding the optimum impedance for transmission line interconnects in high clock rate digital systems may be helpful. Throughout this chapter, we assume that the line impedances will be 50 Ω, even though line impedances used in actual high-performance signal processors and supercomputers have varied from approximately 40 Ω to as high as 92 Ω. However, many pieces of high-performance test equipment have been designed for the analog microwave community, who have long employed 50-Ω impedances in their designs. The ability to employ commercial test equipment developed for the large analog microwave market is attractive for high-performance signal processor designers; to make best use of this equipment, a 50-Ω line environment is desirable.

As a second part of the line impedance trade-off, as line impedance decreases, the power dissipated in the line drive and termination circuitry of each signal net rises quadratically with linear decreases in line impedance (assuming a constant logic swing), and the chips operate at higher temperature. For example, the silicon ECL 100K family can drive 25-Ω transmission lines but then operate extremely "hot." As another negative aspect of low line impedances, the return path currents flowing in the power and ground planes are higher if the lines are low impedance. Thus, if the power planes have not been well designed, the higher currents can cause larger power and ground plane disturbances than if less return path current were flowing. Conversely, the lower-impedance lines are more robust electrically, because magnetically induced crosstalk voltages will be lower than in a higher-impedance line, and the larger amount of current flowing in the interconnect means that the propagating voltage wavefront is less likely to be disturbed by the uncompensated parasitic capacitances which are always present in any transmission line interconnect. One way of conceptualizing the robustness of a lower impedance line is that the parasitic capacitances can be charged to the line voltage more rapidly by high currents than by low currents; alternatively, a given parasitic capacitance causes a smaller line impedance discontinuity on a low impedance line. When all the boundary conditions cited above have been weighed and considered, the selection of a 50-Ω line impedance appears quite reasonable in most cases.

The most common transmission line structures employed in MCMs are microstrip, stripline, and offset stripline, with coplanar lines employed in special situations. A signal line separated from a single ground plane by a layer of dielectric creates a microstrip, whereas a signal line sandwiched between two opposing ground planes (again, separated from the metal planes by dielectric spacers) constitutes a stripline[1,13] (see Fig. 12.10). Offset stripline sandwiches two or even three stripline layers between a pair of ground planes, also as depicted in Fig. 12.10. Offset stripline suffers from several unique problems, because the two or three signal planes are not shielded from one another, resulting in a somewhat higher level of crosstalk than in the other two types of transmission line. Furthermore, designing and especially fabricating both routing layers to achieve a particular impedance value, and also the same impedance value, is more complicated than for stripline and microstrip. If there are two offset stripline layers, widths and thicknesses of the lines in both layers must be identical, the dielectric thicknesses above the upper stripline and below the lower stripline must be identical, the dielectric separation between the two stripline layers must also be tightly controlled, and all dielectric sublayers must have identical dielectric constants. If three stripline layers are included, it is nearly impossible to design the central stripline layer to have the same impedance as the upper and lower striplines. Furthermore, the number of fabrication variables to be controlled is even greater—so great, in fact, that the triple structure is rarely even attempted.

A coplanar transmission line (not illustrated in Fig. 12.10) is typically a signal line placed on a dielectric layer, usually without nearby opposing ground planes, flanked on one or both sides by parallel-running lines that are connected (usually at numerous locations along the run length) to ground. Such a structure is referred to as a *single-ended coplanar line.* Alternatively, two tightly coupled parallel traces can be driven with signals of equal magnitude but 180° out of phase, creating a differential coplanar line. In a differential or odd-mode signal pair, a virtual ground is induced between the traces due to the resultant orientation of the electromagnetic fields.

The coplanar lines have advantages and disadvantages as compared to the other line types. They can be designed over a very wide range of impedances by controlling the spacing between the lines and the individual line widths. Typically, the lines can be designed to be wider for a given impedance than either the microstrip or the stripline, which will result in lower DC losses. AC losses, however, may or may not be reduced when compared to the equivalent microstrip or stripline transmission line. As the signal frequency increases, the current density will tend to be the greatest at the edges of the conductors due to the strong localized electric fields between the signal line and the ground line. As a result, thin lines exhibit much higher losses as compared to conventional transmission lines such as a microstrip or a stripline. Thick conductors, however, tend to have less inherent loss (see Fig. 12.11). In addition, the current distribution in the center conductor does not change appreciably with frequency for small w/d ratios (where w is the width of the center conductor and d is the spacing between the coplanar grounds) but does change for larger w/d ratios.[13a] This finding suggests that smaller w/d ratios are beneficial to minimize dispersive effects and AC losses. Furthermore, discontinuities caused by vias to deeper signal layers to make contact with passive components such as resistors can be avoided by simply installing the components so that they bridge the space between the signal trace and the coplanar ground.

However, because there usually is no opposing ground plane, or the ground plane is too far away from the lines to intersect the radiated electromagnetic fringing fields surrounding these structures, for a given set of dimensions, coplanar lines tend to be more susceptible to intersignal crosstalk than microstrip or stripline if many sets of them are placed side by side. Though the crosstalk can be decreased by moving the coplanar line groups farther apart, routing density is then severely compromised. While these lines can easily be fabricated (usually on the uppermost dielectric layer of an MCM), their disadvantages usually outweigh their advantages except in very special circumstances, e.g., when very high or very low impedances are required in a design or when complete microstrip or stripline structures cannot be fabricated for cost or mechanical reasons. Thus, they are rarely used, particularly in digital signal processors, and will not be discussed further here.

The following discussions concentrate primarily on pure microstrip and stripline. However, specific designs must sometimes employ offset stripline if there are insufficient metal layers available from the particular MCM process.

- **Stripline**
  - **5 mil Wide Line with Ground / Power Planes 8 mils Above and Below**
- **Microstrip**
  - **8 mil Wide Line with Ground / Power Plane 8 mils Away**
- **Offset Stripline**
  - **3 Signal Layers Illustrated**
  - **8 mil Wide Lines with Ground / Power Planes 15 mils Above and Below**

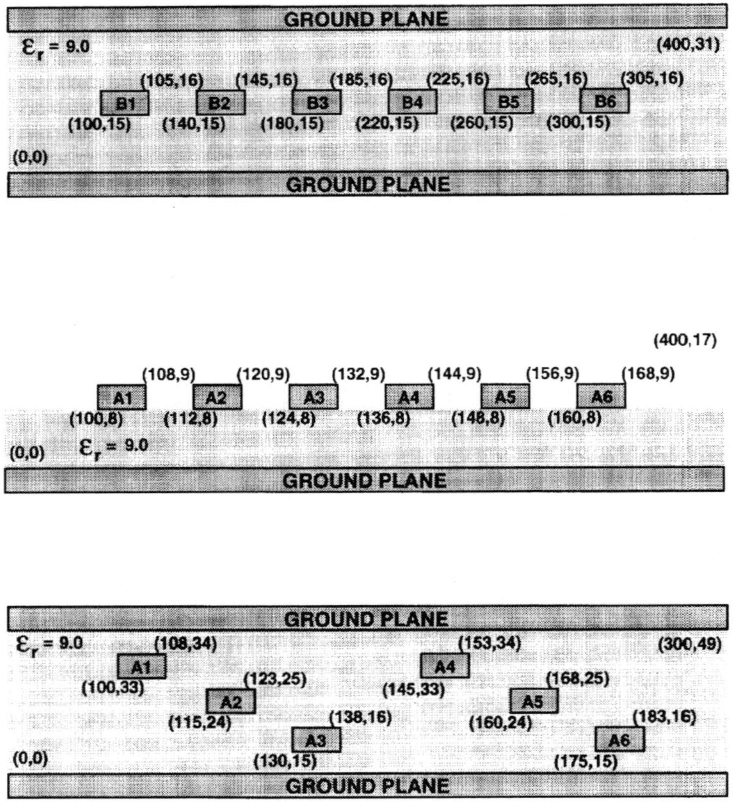

FIGURE 12.10 Illustration of stripline, offset stripline, and microstrip transmission line interconnects as employed in MCMs to achieve 50-Ω line impedance. (All coordinates and dimensions in mils; typical MCM-C dielectric constants and dimensions assumed.

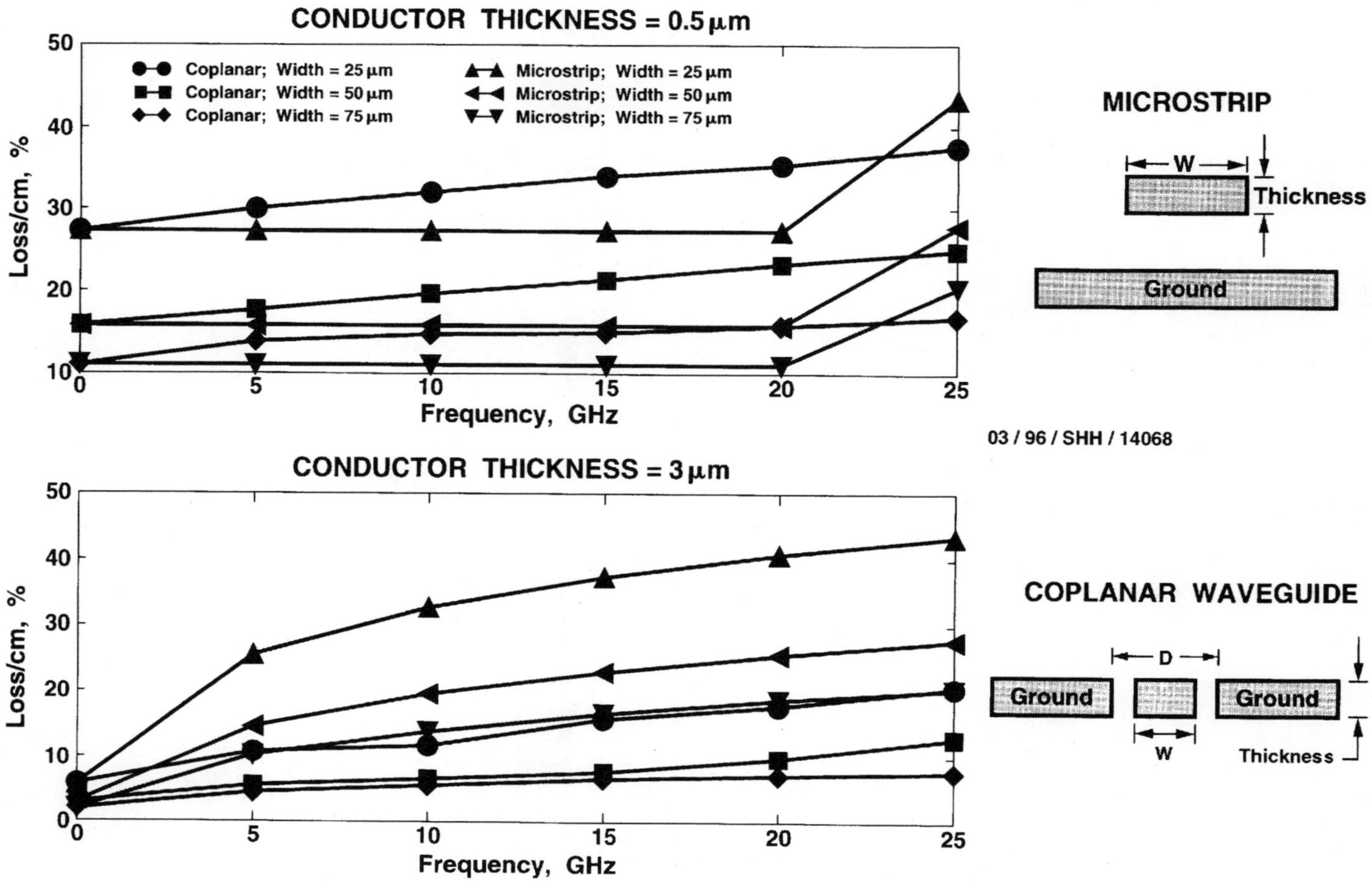

FIGURE 12.11 Comparison of skin effect losses of 1 cm coplanar waveguide (measured) and microstrip (simulated) as a function of frequency, line width, and line thickness. The relationships are complex: increases in line thickness are beneficial up to 5–10 GHz and decrease thereafter in importance. Under certain conditions, coplanar line losses can be lower than microstrip losses for the thicker conductors. (Figure is for gold lines with only skin effect losses considered; length = 1 cm.)

Closed-form equations have been available for several decades that give approximate values for the impedance of microstrip line and for stripline. These *approximate* formulae[1,13] are as follows, first for microstrip and then for stripline:

$$Z_{0ms} = 87\left(\ln\frac{5.98h}{0.8w+t}\right)\sqrt{\varepsilon_r + 1.41} \tag{12.1}$$

in which Z_{0ms} is the impedance of the microstrip line, w is the width and t is the thickness of the microstrip conductor, h is the thickness of the dielectric above the ground plane on which the conductor is sitting, ε_r is the relative dielectric constant of the dielectric, and

$$Z_{0sl} = \frac{60\left[\ln\frac{4b}{0.67}\prod w\left(0.8+\frac{t}{w}\right)\right]}{\sqrt{\varepsilon_r}} \tag{12.2}$$

provided that $w/(b-t) < 0.35$ and $t/b < 0.25$.

In this equation, Z_{0sl} is the impedance of the stripline, w is the width, t is the thickness of the stripline conductor, b is the total thickness of the dielectric between the upper and lower ground planes, and ε_r is the relative dielectric constant of the dielectric. However, note that even the original developers of the approximate formula for the impedance of a stripline considered it to be sufficiently accurate only if both dimensional boundary conditions are simultaneously satisfied. For the actual dimensions of present or evolving MCMs, either or both of these boundary conditions may come close to violation, in which case the accuracy of the stripline impedance equation may degrade. As will be discussed in greater detail below, for the dimensions typical of modern MCMs and particularly for frequencies in the range of a few to a few hundred megahertz, no completely correct formula has yet been derived to replace those above, in spite of more than a decade of research.

Furthermore, the accuracy of either or both of the equations presented above may be weakened at the small dimensions of MCMs for reasons not envisioned by their original developers. In the case of such small dimensions, the actual capacitance of these tiny microstrip or stripline structures may be higher than assumed from simple parallel-plate capacitance models, because the electrostatic fringing fields, either to AC ground or to closely spaced adjacent signal lines, may account for a high percentage of the total electrostatic component of the impedance. In such a case, with the capacitive component underestimated, the actual line impedance may be less than predicted by the equations above. Furthermore, the actual inductances of the interconnects may also be miscalculated, again because of the small sizes of the signal interconnects. Thus, in such cases, computationally costly but more accurate numerical *field solvers,* as discussed later, may be required to develop sufficiently accurate estimates of the line impedance of very small MCM transmission line interconnects.

As an inspection of the equations given above will reveal, the impedance of a transmission line interconnect is proportional to the reciprocal of the square root of the dielectric constant of the insulator, proportional to the natural logarithm of the dielectric thickness, and inversely proportional to the natural logarithm of the width and thickness of the metal lines themselves. Thus, the different independent variables of dielectric constant and thickness, metal dimension, and so forth, will have different relative effects on the ability of a fabricator to control line impedance.

However, many MCM fabricators still do not recognize how tightly these parameters must be controlled to ensure the highest performance (many fabricators have stated that their business model does not even include high-performance MCMs). Even when they do recognize the importance of these issues, there may be few mechanisms in their fabrication process to allow rigorous control of the as-deposited thickness of the dielectric layers, the ε_r of the dielectrics, or the dimensions or sheet resistance of the metal layers. For example, if the dielectric material is spun or sprayed on and then cured, the fabricator may not know the ε_r to closer than 0.2, and may not be able to control the layer thickness to better than ±10 percent. These levels of variation will not affect lower clock rate systems but can result in unacceptable high-frequency performance. The system designer should request this information from the fabricator and, if it is not forthcoming, request that cross sections be made of existing modules and measurements made to determine as many of the as-fabricated physical parameters as possible. As will be described later, the cross sections can yield other valuable information that will assist in the design process.

The signal traces themselves must be of uniform width and cross section along their entire lengths. Scallops, nicks, and "peel-ups" along the edges of the lines, and even holes within the lines, indicate poor process control and should be considered unacceptable. Figure 12.12 depicts an extreme example. The cross sections of the lines should be, in the best case, rectangular but are often trapezoidal due to the metal etching process. While trapezoidal lines are acceptable, they *do* change the line impedance *and* exhibit different DC and AC resistances in comparison to rectangular lines (see below). The impedance difference can be several ohms and create unexpected performance problems.

If the MCM fabricator does not have definitive data describing the dielectric, line impedance, and loss characteristics of his process (and most do not), the design team should seriously consider working with the MCM fabricator to design, construct, and conduct electrical measurements on a small MCM containing a variety of test structures (referred to through the remainder of this chapter as *passive test coupons*), including transmission lines of varying lengths, which are specifically intended to be tested with high-frequency equipment such as the Hewlett-Packard Model 8510 Vector Network Analyzer. Test coupons of this type, described in greater detail below, yield a wealth of data valuable to all participants in the effort.

Although a number of transmission line termination schemes have been described,[1,13] the two most typical implementations are series source and shunt termination (Fig. 12.13). Series source termination (panel D) is the least consumptive of power but also the "slowest," since a half-amplitude signal enters the transmission line at the source end and does not become full amplitude until a full reflection occurs. This scheme also requires *two* resistors. Shunt termination (panel C) is "fast," because a full-amplitude signal propagates along the interconnect. It is also the most robust (because larger currents flow in the line, more effectively "swamping" any impedance discontinuities and maintaining good signal noise margins) but the most consumptive of power (since there is always DC current flow from the source transistor, through the termination resistor, and into V_{tt}). Because shunt termination minimizes the number of resistors that must be implemented, it is presently the termination scheme of choice.

How are the termination resistors implemented on the MCM? The resistors can be discrete devices or resistor packs fabricated on small silicon or alumina slabs, which in turn are mounted onto the MCM along with the chips and connected to each signal net with wire bonds and traces. Discrete resistors surrounding the integrated circuit are depicted in Fig. 12.14, and resistor packs were used in

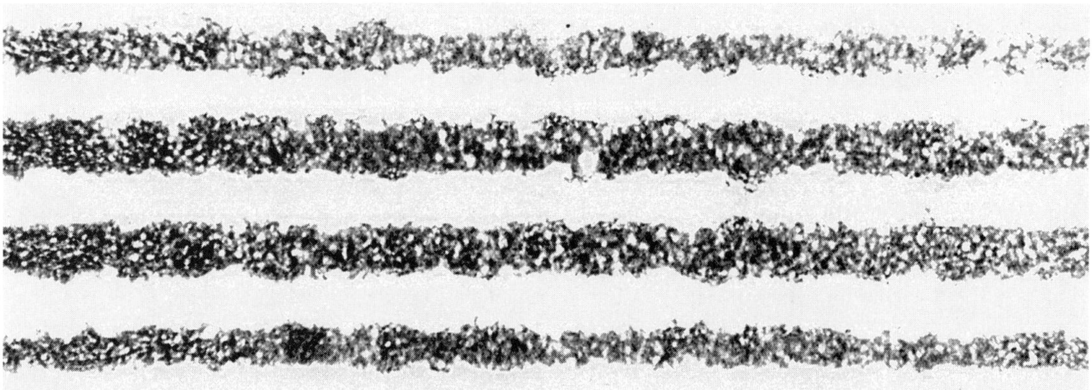

FIGURE 12.12 Example of patterned 1-mil wide lines and 1-mil spaces on MCM-C with insufficient control of patterning process. Note scallops and whiskers along edges of lines and apparent voids within the line metallization. These increase line resistance.

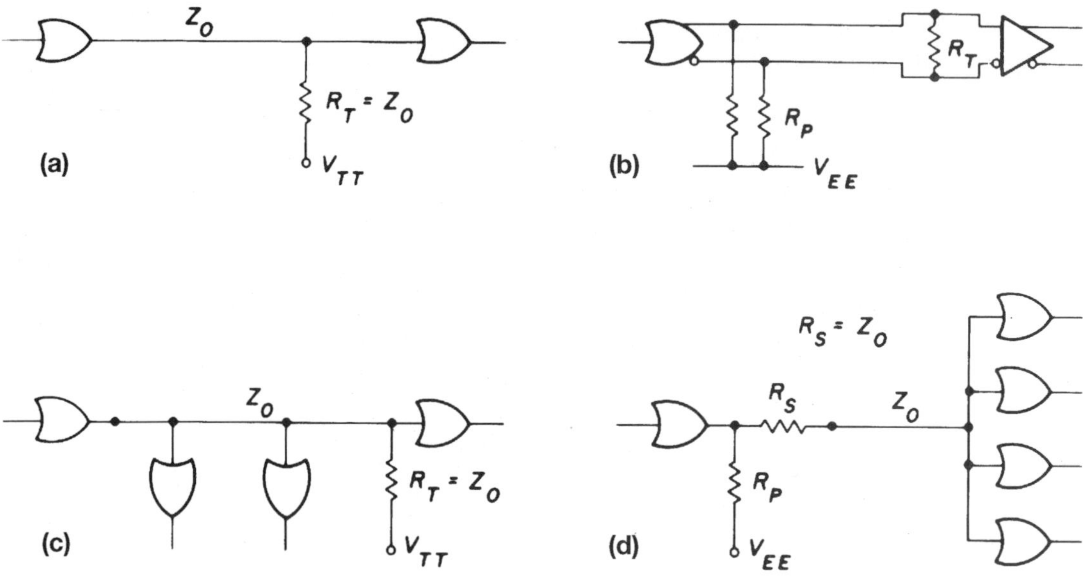

FIGURE 12.13 Different transmission line termination protocols used in MCMs intended for high clock rate operation: (*a*), single-ended transmission, (*b*) differential transmission, (*c*) shunt termination, and (*d*) series termination. The most commonly employed protocol is depicted by (*c*), and the least power consumptive by (*d*). The protocol of schematic (*b*) is the most noise immune but works only for single-source, single-destination signal nets.

the module depicted in Fig. 12.5. However, the use of discrete components, or even resistor packs, increases the amount of final assembly required, increases the need for rework, and may decrease final yield (and hence increase module cost). Although discrete resistor packs are now being widely employed in those designs that require resistors for termination, biasing, or other purposes, the optimum approach would be to deposit integral resistors directly onto the ceramic or silicon substrate, i.e., to incorporate them directly into the MCM fabrication process itself. At present, a few MCM fabricators have incorporated integral resistors into their process. The resistors may be either nichrome or tantalum nitride, depending on the features of the rest of the process. The resistors need to be within ±10 percent, and preferably ±5 percent, of the nominal values, where *nominal* is typically assumed to be the same as the line impedance. However, line impedances still vary from fabrication lot to lot, and yield and fabrication cost issues are still not totally resolved. A slightly different approach to resistor value selection may yield better results, as described below.

Clearly the optimal situation would be if the termination resistance were the same value as the line impedance, in which case the transmission line appears to the propagating signal as a pure delay, with no other effects on the wavefront.[1,13] However, if process variations dictate that neither the impedance of the line nor the value of the termination resistor can be guaranteed, then these values must be selected to avoid the worst of all possible conditions. Such a condition occurs when the line impedance is *higher* than the value of the termination resistance. In that situation, a partial-amplitude wavefront reflection will be generated at the site of the terminator resistor and propagate back along the line toward the driving gate. The polarity of the reflection will be *negative*, i.e., *in the direction of the logic threshold* (i.e., a low-to-high transition will reflect a partial pulse with a lower voltage than the logic high level, while a high-to-low transition will reflect a partial pulse with a higher voltage than the logic low level). These reflected partial pulses will have the direct effect of degrading the ability of the line to tolerate additional noise contamination caused by other phenomena. That is, the ability of the receiving gates on the signal net to resist false triggering situations as the instantaneous signal voltage approaches the logic threshold will be compromised.

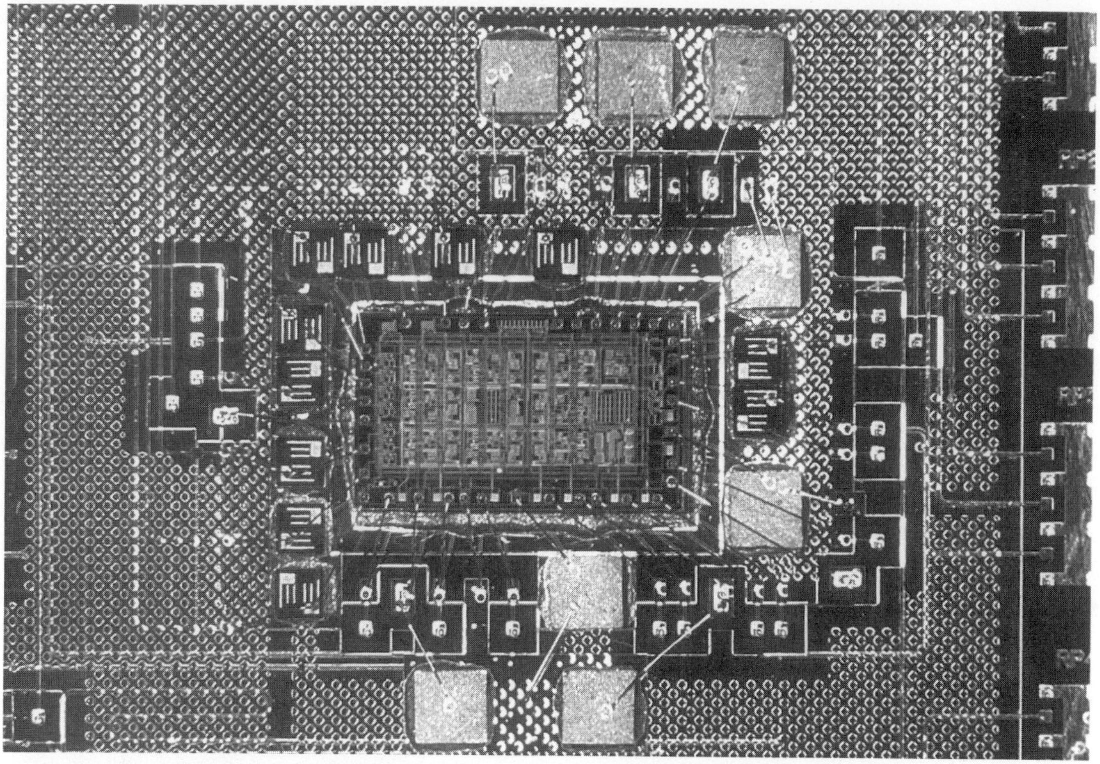

FIGURE 12.14 Micrograph of a high-speed GaAs chip wire bonded on a mid-1980s organic MCM-D. Note the multiple discrete resistors (small, dark squares), capacitors (large, white squares), and long wire bonds. Integration of discrete components into the substrate would simplify assembly, test, and rework, and result in lower costs. Unit fabricated by Honeywell.

Typically, the signal degradation *from all sources* should not exceed 20 percent of the total logic swing, with not more than half of that, or 10 percent, caused by negative voltage reflections. Thus, it is useful to set the value of the termination resistor such that, even with fabrication variations of both the resistor and the line impedance, the lowest value of termination resistance expected from the fabrication process is slightly higher than the highest line impedance to be expected. In this case, any reflections will be *positive*, i.e., away from threshold, which (although they will degrade the switching speed of some logic families such as ECL) will not degrade the noise margins of the signal nets. Note that the goal is to set the values of the integral terminator resistors, accounting for their statistical distribution of manufacturing tolerances, to be just slightly higher than the impedance of the interconnects, accounting for all of their statistical variability in manufacturing tolerances.

As will be noted later, the line impedances are subject to the vagaries of many more processing variables than the resistors, many of which will be difficult to control; thus, improving resistor tolerance is the best way to ensure that the resistors in all cases are just slightly higher in value than the line impedances. However, the MCM vendors have as yet not been able to fabricate these resistors with as-deposited tolerances better than ±20 percent, whereas ±10 percent or even ±5 percent is desirable. Only one or two processes appear to have achieved ±5 percent tolerances. Laser trimming of the resistors after fabrication has been demonstrated successfully with both nichrome and tantalum nitride resistors. In the former case, the material is simply vaporized, whereas, in the latter case, the TaN is converted to TaO. At the time of this writing, no company has announced a fully automated method of laser trimming large numbers of resistors, so this remains an unsolved problem.

Until automated trimming of in situ resistors is reduced to practice, discrete resistor packs will continue to be the implementation of choice, because the resistor packs actually installed on an MCM can be pretested and selected to have the desired values (which, in this case as well, should be slightly greater than the greatest line impedance expected as a result of process variations). Finally, it should be noted that the "pull-down" resistors, labeled R_p in Fig. 12.13, typically are assigned values in the 420–560 Ω range. Their tolerances need not be held to very exact levels (±20 percent is satisfactory), since they exert little influence on the propagation behavior of the interconnect.

Whether integral resistors or resistor packs are employed, the designers and fabricators need to ensure that the resistors demonstrate minimum capacitive and inductive parasitic reactances. Although the common view is that integral resistors will always exhibit lower levels of parasitic reactance, such need not be the case. Very well designed discrete resistors can have very low reactances (Fig. 12.15). In either case, the best approach is to minimize the physical size of the resistors, provided that the power dissipation of the terminators (up to 15–40 mW) does not create a local heating problem on the MCM substrate. The optimal resistor size to minimize reactive parasitics would be one "square" of resistive material; however, to minimize excessive local heating problems, terminator resistors of approximately four or five "squares" appear to be the best compromise.

12.7 ACHIEVING LOW-LOSS INTERCONNECTS ON MCMS

The use of terminated transmission line protocols (whether series, shunt, or some other protocols are employed) forces the designers to focus on the "lossiness" of the signal interconnects. If the lines themselves are not completely lossless, the line resistance in effect forms a series resistive circuit with the terminator resistor; some of the input signal is dissipated across the line resistance rather than the full logic swing appearing only across the terminator resistor. For example, if the line exhibits 5 Ω of resistance, and the terminator is 50 Ω, the signal appearing at the terminator (and the destination gates) will be only 91 percent of the maximum signal swing—a loss of nearly 1 dB of signal noise margin. (The issue is actually more complex, since the output resistance of the drive transistor must also be considered.)

Figure 12.16 illustrates the effects of line resistance on signal amplitude along a 26.3 mm long shunt terminated signal net with nonnegligible resistive losses. In this particular example, approximately 180 mV of signal amplitude, out of a total maximum signal swing of 1 V, or 18 percent (slightly less than 2 dB) of the full-scale swing, is lost at the final destination. Note also the significant lengthening of the rise time of the propagating signal (from approximately 400 to 800 ps) between the source and the final destination, indicating that the higher harmonic components of the rising edge are the ones most significantly attenuated by the dispersive losses along the line. The problem is similar, though less serious, if series source termination is employed, because the voltage on the line will eventually stabilize at almost exactly the open circuit voltage of the drive transistor. The signal loss problem is not only a DC problem but also an AC problem, because, as will be described in greater detail later, the resistive losses of the interconnects become substantially greater at high frequencies than they are at DC. These frequency dependent losses are often referred to as *skin effect* losses.

As a result of these types of loss phenomena, a significant effort must be made to minimize line losses in multichip modules. In fact, excessive line resistances represent one of the principal drawbacks of MCMs if not thoroughly accounted for (this particular problem does not appear in conventional PC boards). Signal lines that are made as wide and as thick as possible exhibit a larger line cross-sectional area, which decreases DC line losses according to the formula $R = \rho L/A$, where ρ is the resistivity of the metal, L, is the length of the line, and A is its cross-sectional area. Thus, the signal lines should be designed to be as wide as possible, consistent with maintenance of the desired line impedance, which in turn depends on the thickness and ε_r of the dielectric being employed. All other variables held constant, the dielectric with the lowest possible ε_r should be employed, since the signal lines can be widened as the ε_r is decreased while holding line impedance constant, according to the approximate formulae for stripline and microstrip presented earlier.

Measurement bandwidth = 45 MHz to 18 GHz

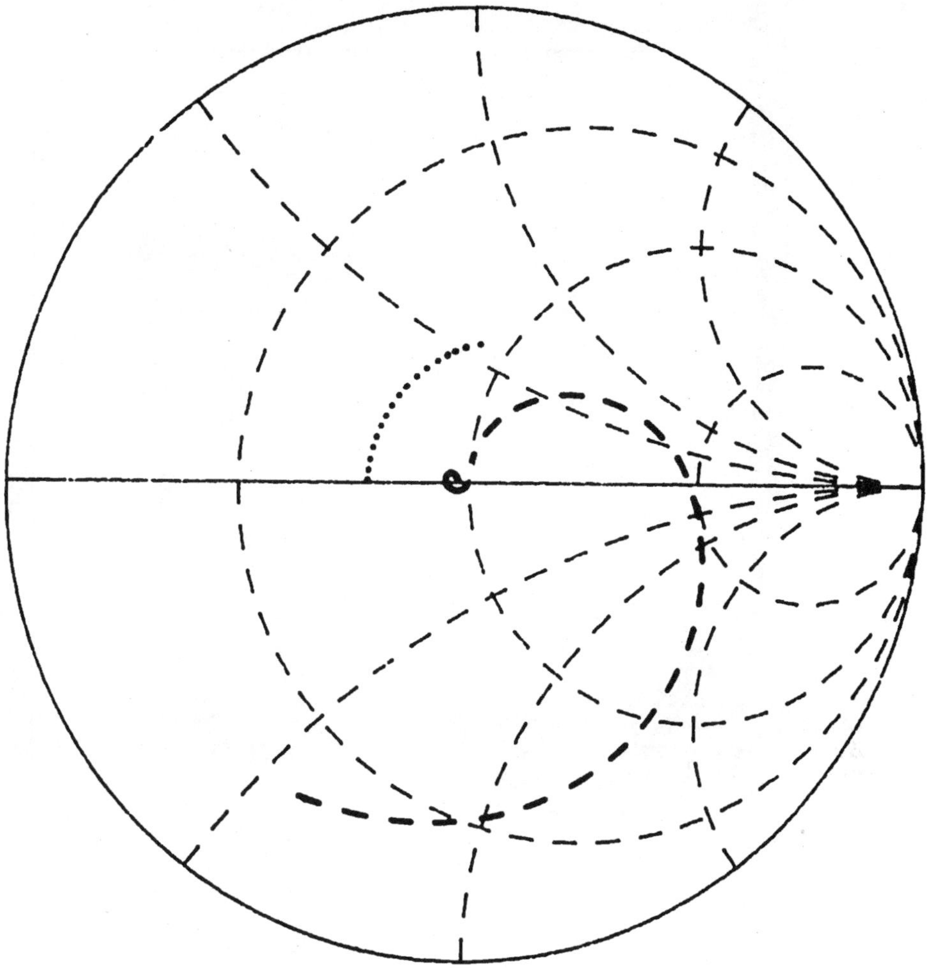

——— 50 Ω discrete wire bondable chip resistor installed on GE HDI coupon (Semi-Films WBCR-000-5000 BG)

· · · · · · · 35 Ω deposited nichrome "Meandor" resistor on GE HDI coupon

— — — 50 Ω discrete SMT chip resistor epoxy bonded to 3M Corp. two-layer MCM-D (Mini-Systems MSR20G-50R0F-U)

FIGURE 12.15 Smith chart display of impedance characteristics (freq. range of 45 MHz to 18 GHz) of discrete chip resistors and deposited thin film nichrome resistors on two types of organic-based MCM-Ds. The discrete resistor (small, closed curve) had a lower reactance than the integral resistors.

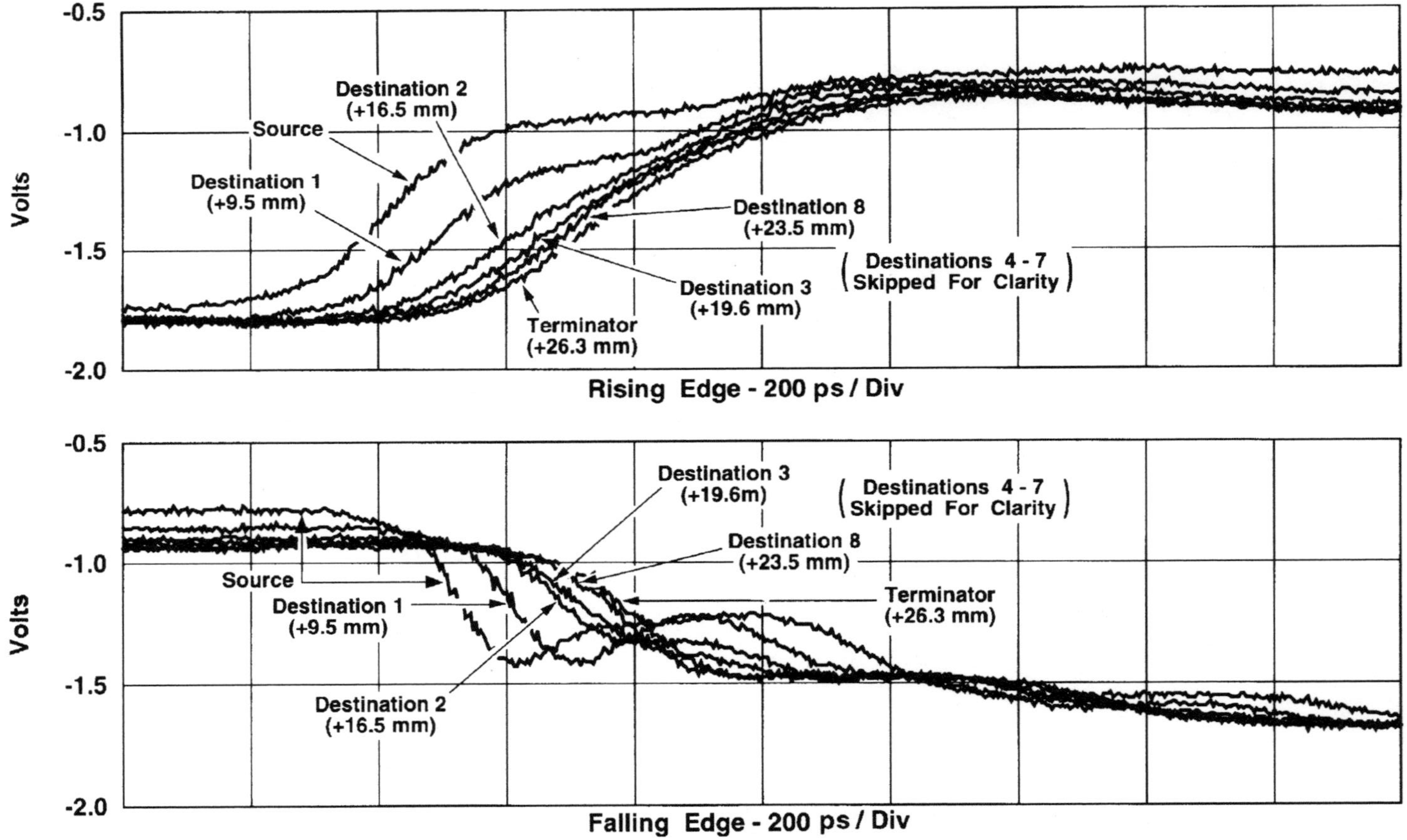

FIGURE 12.16 Waveforms taken at selected locations along a 26-mm long, shunt-terminated, 50-Ω transmission line signal net from a copper-aluminum-SiO_2-silicon MCM-D, showing a loss of signal amplitude at successive destination nodes (Test coupon fabricated by n-Chip).

The effect of line width on line losses can be observed in Fig. 12.17, a swept frequency measurement from 45 MHz to 26 GHz, of signal loss in three stripline structures and three microstrip structures from the same fabricator using three different line lengths. The impedances of all six lines were nominally 50 Ω. Note that the transmission losses (S_{21}) in the upper panels are progressively greater as the line lengths increase, for either the stripline or the microstrip. Note also that the losses are greater for the striplines, which are 34 μm wide, in comparison to the microstrip lines, which are 75 μm wide, on an equivalent length comparison basis. Figure 12.17 also introduces the concept of S-parameters, which is discussed briefly in the Appendix at the conclusion of this chapter. The two lower panels in this figure illustrate the concept of return loss, the amount of energy reflected back to the source. As described later, these curves should have values as small as possible, indicating that very little energy is reflected back to the source (i.e., most energy should be propagated toward the far end or destination end of the line).

The MCM fabricators should also be encouraged to make their lines *thicker*, also to increase cross-sectional areas and the total perimeter of the line. As will be described later, it has been a widely held view for many years that decreases in line losses could be achieved primarily by increases in the perimeter dimension of the line. However, for MCM line dimensions (as contrasted with printed wiring board dimensions), it will be noted that, for frequency components up to nearly 5 GHz, there is a real benefit from the actual increase in the line thickness as well. Most present organic-based MCM-D processes fabricate lines 5 μm thick. More recently, lines of even 8 μm thickness have been demonstrated. However, thicker lines result in increased stress buildup in the MCMs and also in layer-to-layer degradation in planarity, which in turn result in yield loss and an inability to fabricate large numbers of metal layers.

The metal deposition processes must be carefully reviewed. At present, both aluminum and copper metallurgy is employed in MCM-Ds, with silver employed in low-temperature cofired MCM-Cs. The process should strive for the lowest possible line resistance, but voids or inclusions incorporated during the metal deposition process can seriously degrade (i.e., increase) the *as-deposited* resistivity of the metal to values well above those found in handbooks for the bulk resistivity of the same material. It is the as-deposited resistivity that really determines the DC losses in the interconnects. For that reason, the resistance formula presented earlier really should be written as $R = \rho' L/A$, where ρ' is the as-deposited resistivity of the metal (i.e., greater than the handbook value for its bulk resistivity), L is the length of the line, and A is its cross-sectional area.

Copper is typically plated, but slight differences in the plating recipe can result in as-deposited resistivities 25–50 percent greater than the bulk value. Sputter deposition of copper is also being employed and, in at least one MCM recipe, has demonstrated much more uniform results than electroplating. Several research groups have argued for the use of aluminum, on the premise that the as-deposited resistivity of aluminum may be demonstrably lower than the as-deposited resistivity of copper in some cases. Silver lines overplated with a thin layer of gold have also been suggested. Based merely on the differences in bulk resistivity between copper (2.57 Ω/sq) and silver (2.51 Ω/sq), the silver lines should result in losses only a few percent lower than for copper. However, here again, the issue is the as-deposited characteristics of the metals, not their bulk values. It is possible that silver might exhibit better behavior (i.e., lower resistivity) during the deposition processes than copper. There have been some experimental studies to ascertain the differences between the as-deposited line resistances of copper, silver, and gold interconnects (Fig. 12.18). The deposition process also needs to strive for rectangular line cross section, which yields the highest perimeter value for a given nominal line width. Trapezoidal lines, with a somewhat diminished perimeter and cross section that must be correctly accounted for when performing line impedance calculations, are the rule at present (Fig. 12.19), although at least one vendor of organic MCM-Ds has demonstrated nearly perfectly rectangular lines in a full production process.

The need in copper-based MCM-D processes to protect the metal lines from corrosion, and to ensure good adhesion of the lines to the underlying and overlying organic dielectric, has in the past created some problems for the high-frequency community. Typically, an adhesion promotion layer such as titanium or nickel is deposited directly on the organic layer, with the copper, aluminum, or silver deposited thereafter. The titanium oxidizes slightly, and the TiO then forms oxygen-carbon bonds with the organic molecules in the dielectric layer. Similarly, overcoating of the deposited cop-

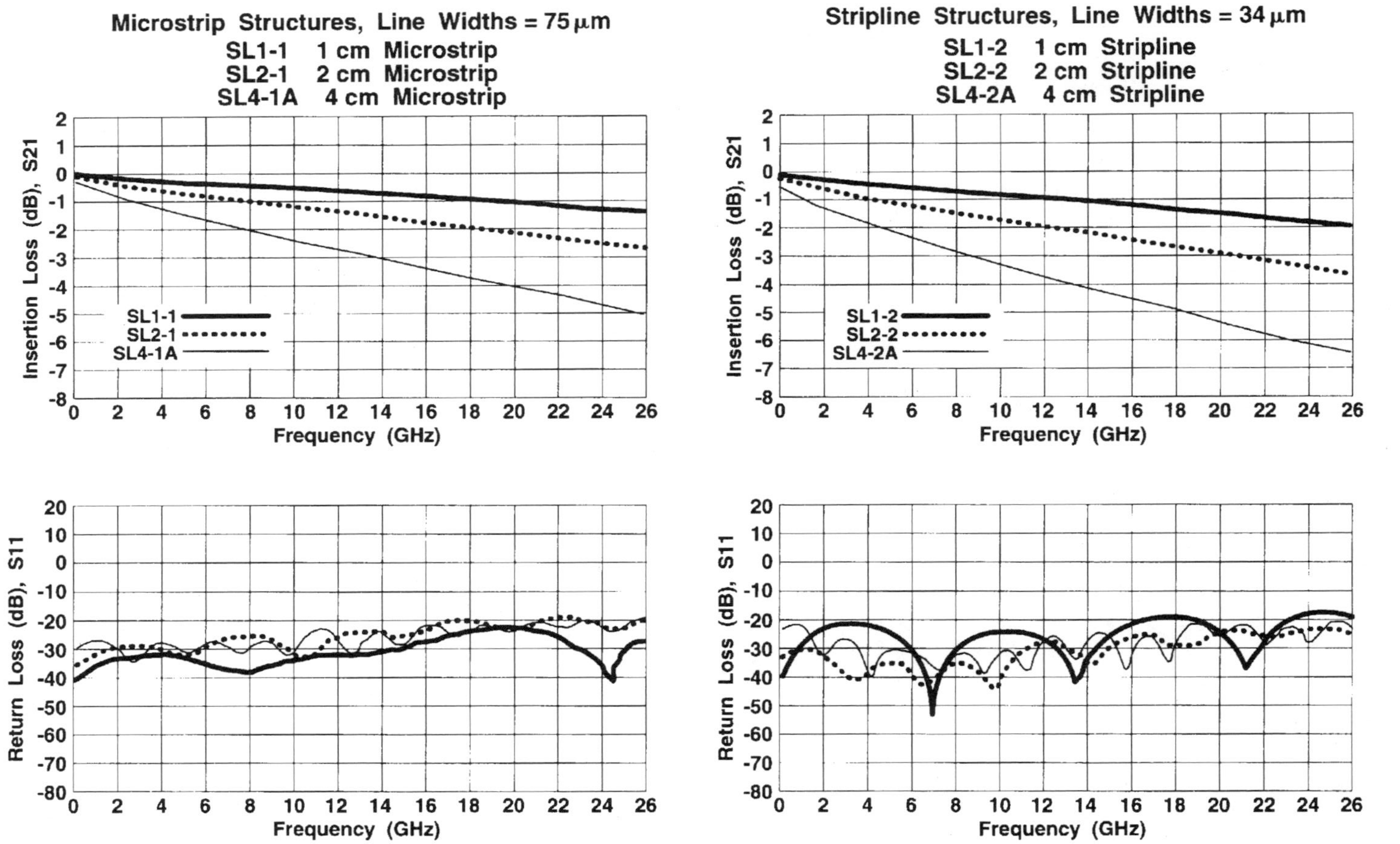

FIGURE 12.17 Frequency-dependent transmission loss (insertion loss) and reflection loss (return loss) on striplines and microstrip lines of different lengths. Losses are progressively worse for longer lines and worse for the narrower stripline than for the microstrip lines. (Fabricated by Texas Instruments; measurements made with HP 8510C network analyzer and Cascade Corp. microwave probes.)

Mayo standard passive coupon, microstrip version. Silver conductors, photodefinable polyimide, dielectric constant = 2.9. Gold conductors, conventional polyimide, dielectric constant = 3.4. Length, 4 cm; width, 12 μm, thickness, 3 μm.

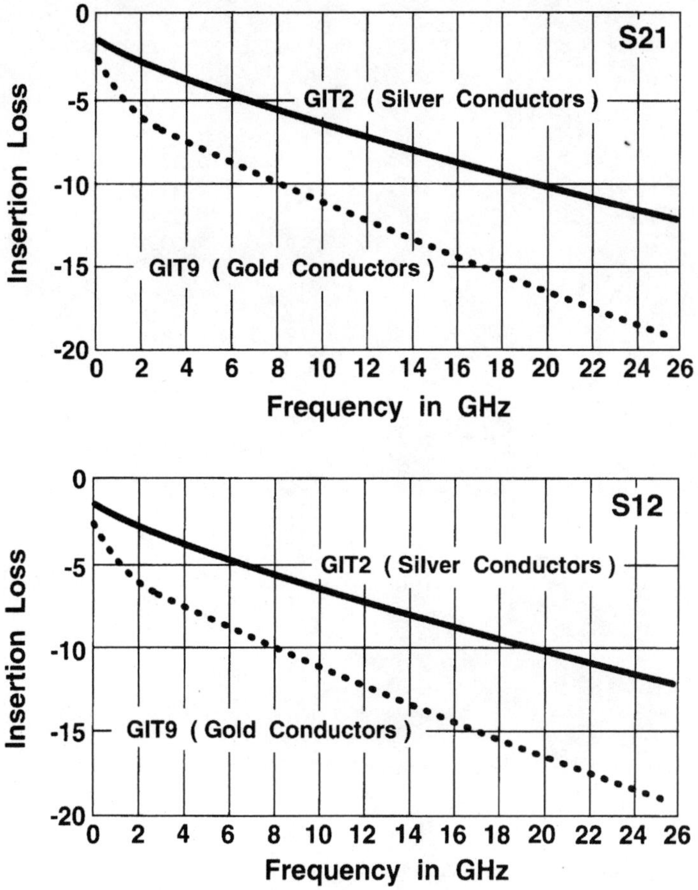

FIGURE 12.18 Insertion loss comparison measurements of gold and silver microstrip transmission line structures. The silver lines show insertion loss levels 7 dB lower at 26 GHz than do the gold lines. Coupon fabricated by Georgia Institute of Technology. (Measurements with HP 8510C network analyzer and Cascade Corp. microwave probes.)

per also promotes adhesion with the subsequent organic layer. Finally, the titanium or nickel protects the copper lines from corrosion caused by the polyamic acid that evolves during the curing process in some MCM-D recipes. However, it is now generally agreed that only a very thin layer of the Ni is required to achieve corrosion protection, and only a thin layer of Ti or Cr is needed for adhesion promotion—on the order of 500–3,000 Å.

If the protective metal barrier layer is not applied very carefully, a line cross section such as that depicted in Fig. 12.19 may result. Note the deposited adhesion-promoting metal layer beneath this line and the cap-shaped structure on top of the line, which is a heavy deposit of barrier metal (in this particular example, gold was employed). As in Fig. 12.19, when a high-resistance barrier metal such as nickel or titanium is applied too liberally, the actual shape of the conductor may be altered, which

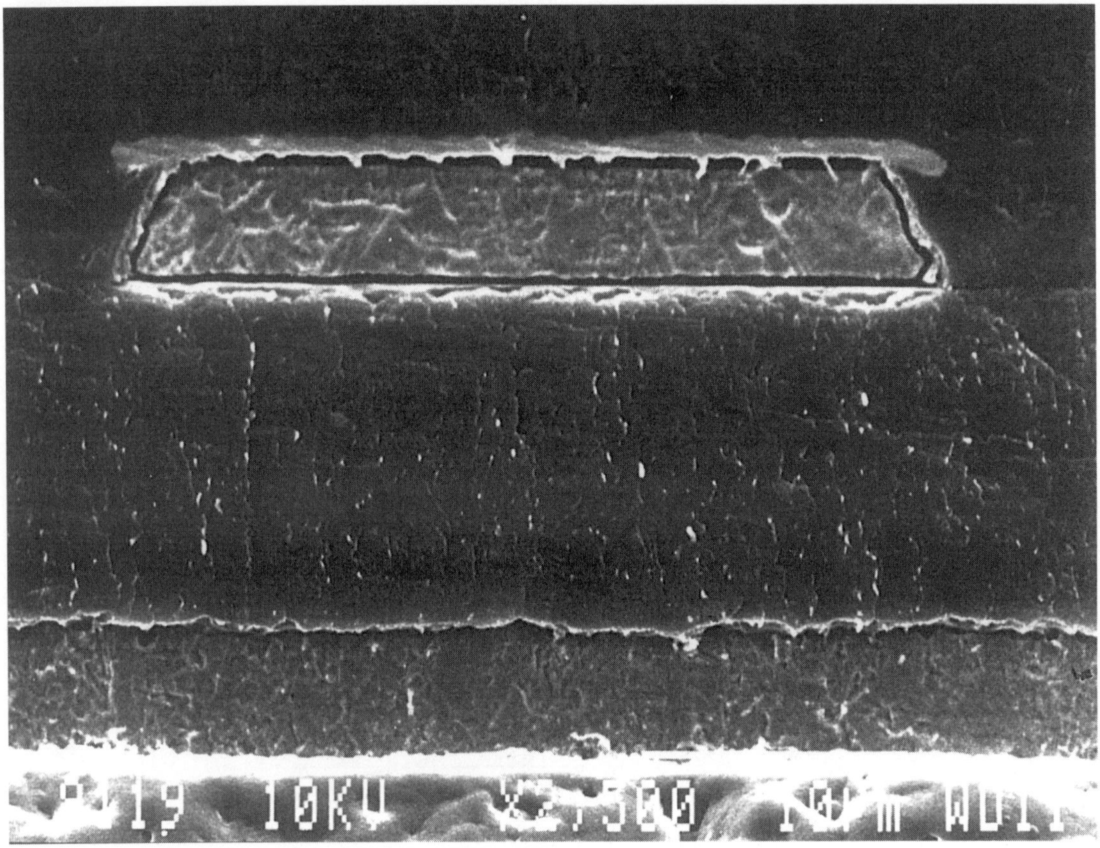

FIGURE 12.19 Cross section of a copper interconnect in an organic MCM-D. Note trapezoidal line cross section. Lighter shaded region below the line and the metal "cap" above the line are layers of adhesion promotion and barrier metal deposited before and after the copper line is deposited.

not only changes its line impedance but, in addition, the DC and AC resistances of the line. The fact that the AC current tends to flow preferentially in the upper and lower surfaces of the line (the phenomenon referred to as skin effect) may result in the current flowing largely in the high-resistance metal rather than in the copper. This issue can be resolved by controlling the fabrication process to ensure that only a few thousand angstroms of the protective metals are deposited.

12.8 MULTIPLE POWER AND GROUND PLANES IN HIGH CLOCK RATE MCMS

In the design of MCMs intended for CMOS chips, power and ground can be (and usually have been) delivered on traces similar to the signal traces and, in fact, typically share the same layers. However, MCMs intended for use with silicon ECL or GaAs devices have always required the dedication of complete metal planes to the distribution of power and ground for two reasons (and modern CMOS chips, operating at 300–500 MHz clock rates, are turning to this type of implementation as well). First, the most commonly used transmission line structures, i.e., microstrip lines, striplines, or offset striplines (defined by the diagrams of Fig. 12.10), require that the signal layers be underlaid by (for a microstrip structure) or sandwiched between (for the stripline or offset stripline structure) continuous planes that are, at the least, *AC grounded.* (That is, they can be either true ground planes or else DC power delivery planes decoupled to the ground planes through capacitors.)

Second, the power/ground distribution system must employ continuous planes to provide a minimum-impedance *return current path* back to the point of origin, i.e., back to the source transistor, since, in transmission line interconnects, the current in each signal net in effect flows out of the driver transistor, then into the signal net itself, through the terminator resistor to the V_{tt} plane, and back to the vicinity of the power and ground pins on the source chip (as illustrated at three different frequencies in the simulations of Fig. 12.20). If the effective series resistance and inductance of the

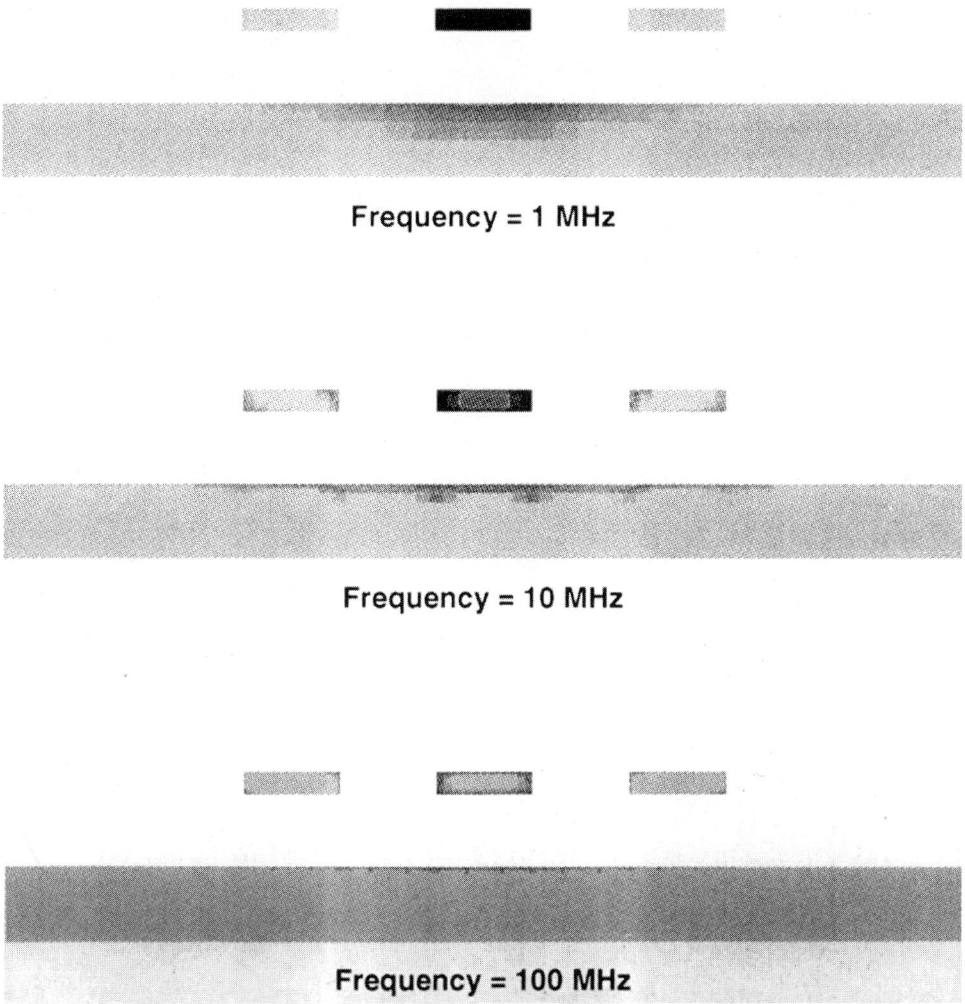

FIGURE 12.20 Computer simulation of current flow through the cross section of a rectangular conductor on an MCM. Uniform current flow at very low frequencies (upper panel), current flow on only the surface (skin effect limited current flow) at high frequencies (lower panel), and nonuniform current flow at intermediate frequencies and conductor dimensions (middle panel). The middle case can occur in high clock rate MCMs. Note patterns of return path current flow in ground plane. (100 × 30 μm lines, 100 μm above a 1,000 × 100 μm ground plane, air dielectric.)

return path through the AC ground planes are too great, the fluctuating components of these currents will cause Ldi/dt voltage spikes on the power and ground planes. These spikes are apparent in Fig. 12.21, in which differences in the layout of the ground planes of a small printed circuit board (on which several 24-pin, 250-mil square leaded chip carriers were installed) resulted in large differences in instantaneous voltage spikes on these planes. Furthermore, these exhibited different amplitudes at different locations on the planes (points A through D, separated by successive 125-mil increments from a full ground plane just beyond point A). The clock rate of the subsystem depicted in this figure was 1.92 GHz, with signal rise times in the 100–200 ps range.

Integrated circuits depend on stable power and ground voltages to determine correct voltage thresholds and logic low and high values. Any instantaneous voltage variations on these planes will directly degrade the system noise margins by an equivalent amount. In fact, in extreme cases, they may render the digital chips on different portions of a PC board or MCM incapable of communicating with one another at all, because their thresholds and logic low and high values may be completely out of alignment with one another.[1,13] Thus, the ability to minimize return path inductances and resistances is paramount. The resistive losses in the planes can be minimized by making the planes as thick as possible. However, since most MCM-D fabrication processes limit metal thicknesses to 5–8 μm at most, this solution has its limits. MCM-L processes, with their typically thicker metal layers, fare somewhat better in this regard.

The resistive *and* the inductive losses can be further minimized by employing the maximum number of power and ground planes, one plane for each voltage and ground. (This concept can be extended by distributing the same DC voltage on more than one physical plane). The use of multiple planes, all decoupled to one another with high-quality capacitors, in effect connects the plane inductances and resistances in parallel with one another, shunting their effective values down by the parallel combined values of the individual plane inductances and resistances. In a very large PC board designed in our laboratory in the early 1980s for a high-performance multichip GaAs circuit operating in the GHz range, a total of 12 power and ground planes were used in each board to minimize inductive and resistive parasitics.[2,3] The technique was so successful that it was used repeatedly thereafter, most recently in the early 1990s for a mixed-technology (TTL, GaAs, and silicon ECL) system operating at even higher multigigahertz clock rates. (In this much smaller board, eight power and ground layers were employed.) It is for these reasons that the distribution of power and ground on narrow metal traces will almost certainly create significant problems: the traces exhibit extremely high values of resistance and, in particular, inductance. As a result, the instability of the instantaneous voltages appearing at the power and ground pins of the chips can be so extreme as to render the circuits nonfunctional.

The MCM-D processes are incapable of providing large numbers of metal planes; many MCM-D processes are presently limited to a total of 5–6 metal planes (this restriction is often relaxed for the MCM-L processes). For silicon ECL and most families of GaAs chips, the minimum number of planes is six, as follows: two signal (two stripline), one ground plane, one chip voltage plane (–4.5 V or –5.2 V), one terminator voltage plane (–2 V), and one pad layer for chip attachment pads, electrical bond pads, and pads for discrete capacitors and resistors. (One GaAs logic family uses a single voltage, –2 V, for both the power and termination voltages.) The pad layer can often support microstrip interconnects as well, thereby creating a partial third signal layer (partial because the pads cause blockages which prevent unrestricted routing of the microstrip interconnects).

For processes that can support fewer than six layers, five layers can be made to work by combining two layers (typically by sharing two power rails on the same physical layer), but this approach is definitely *not* recommended for systems with clock rates much greater than 200 MHz. The availability of one, two, or three additional layers (for a total of seven, eight, or nine) would be ideal, yielding additional design flexibility for any one or more of the following:

- Full stripline implementation for both stripline interconnect layers (as opposed to offset stripline)
- Separation of the microstrip layer from the pad layer (thus creating an "embedded microstrip" structure and somewhat diminishing the number of blockages for the microstrip routes)
- Dedicating a layer to buried terminator resistors (see Fig. 12.22) or isolated capacitors (not area array decoupling capacitors, discussed later)

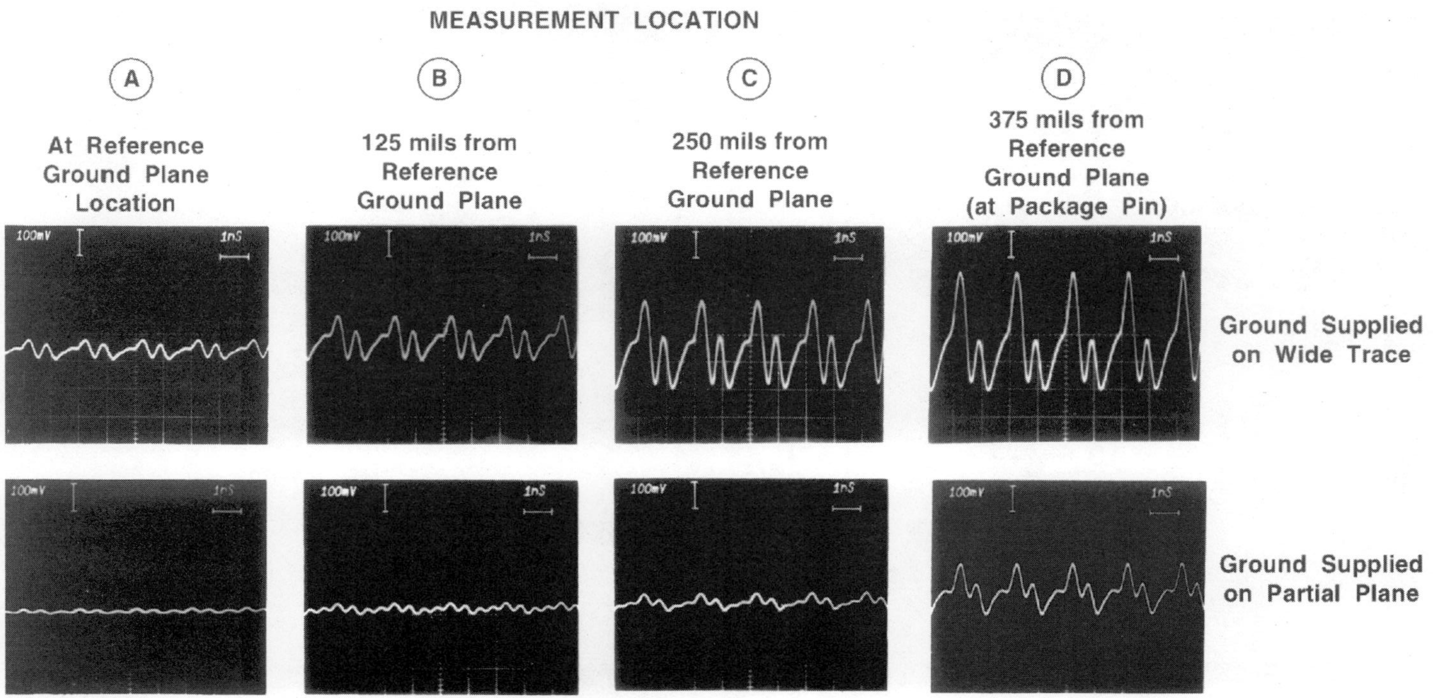

IGURE 12.21 Effect of delivery of ground on traces and full planes in a small circuit board at high system clock rates and fast signal rise times. Points A through D are separated quentially to 125 mils. Noise levels are higher on the ground supply implemented as a trace, compared with the ground supply implemented as a plane.

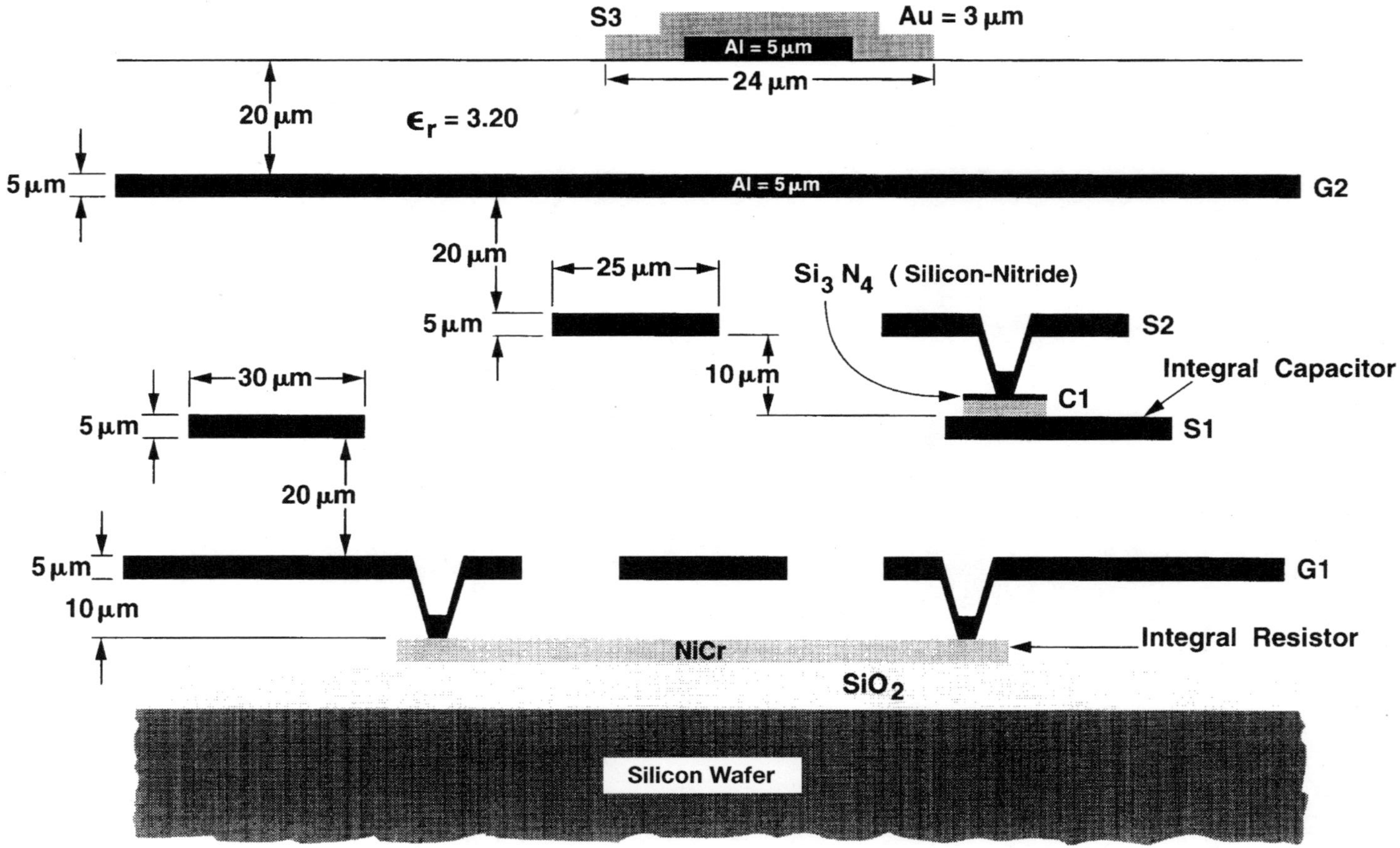

FIGURE 12.22 Cross section of an ideal MCM-D for high clock rate digital signal processing applications. Most present organic MCM-D processes support five to six metal layers—fewer than the ideal eight to ten layers. (Aluminum/polyimide MCM fabricated by Hughes Aircraft.)

- Reserving an additional signal layer for very densely interconnected designs or to isolate the clock nets and/or special signal or clock nets on a restricted-use routing layer

However, due to planarization problems at each metal and dielectric layer, MCM-D fabrication yield typically decreases with each additional layer, the principal factor that has restricted most MCM-D fabricators to the four to six layers noted above. At least one MCM-D fabricator has demonstrated eight metal layers, but widespread production use of so many metal layers must await improved planarization methods. The limitations and restrictions on the high clock rate systems designs that can actually be accomplished with differing numbers of available metal layers are roughly summarized in Table 12.1.

Another problem that must be accounted for in some MCM-D processes has been the inability to fabricate adjacent metal layers (separated by a dielectric layer, of course) in which both metal layers contain a high percentage of metal coverage. The limitation on this type of stackup, which is very desirable to create adjacent ground and power planes, is that the intervening dielectric layer must be absolutely free of pinholes and/or metal dendrite growth through the dielectric layer to prevent the occurrence of power-to-ground short circuits. Most MCM-D processes have by now gained excellent control over pinholes and dendrite growth through the dielectric.

In the case of MCM-L (and also printed wiring board) designs, which do not have an underlying silicon or alumina support structure as do the MCM-Ds nor the rigidity of the MCM-Cs, bowing of

Table 12.1 Impact of Total Number of Metal Layers in MCM Substrates on Design of Subsystems (Illustrative Implementations Only)

Number of metal layers	Design restrictions	Suggested layer assignments
1	Not feasible for any but simplest designs	
2	Digital systems up to 20–50 MHz Analog microwave designs using microstrip interconnects with no crossovers	One routing/attach, one shared power/ground plane
3	Digital systems up to ~50 MHz with split-plane power delivery Analog microwave designs using microstrip interconnects	Two routing, one shared power/ground plane
4	Digital systems up to ~1 GHz Analog microwave designs requiring microstrip and stripline interconnect	Two routing, one power plane, one ground plane
5	Digital systems up to ~2.5 GHz Most analog microwave systems	Two routing, one power plane, two ground planes
6	Digital systems up to 10–12 GHz Analog microwave systems up to 12–15 GHz	Two routing, one power plane, two ground planes
7	All digital and analog or mixed-signal designs are feasible	Two signal, one pad/signal, one power plane, two ground planes, integrated small capacitor or embedded resistor layer
8	All digital and analog or mixed-signal designs are feasible	Two signal, one pad/signal, one power plane, two ground planes, integrated small capacitor or embedded resistor layer

the substrate can occur. For example, if several mostly metal planes, such as power/ground planes, are all juxtaposed to one another (separated, of course, by dielectric layers), the structure may bow because of tension caused by the metal layers in the stackup. To prevent the bowing, particularly when a large number of metal layers are incorporated into the design, the metal layer stackup must be carefully arranged so that a symmetry of metal planes is created about an imaginary plane running through the exact center of the stackup, with equal numbers of high-coverage metal planes on each side of the imaginary central plane through the MCM.

Layer-by-layer evenness of metal coverage is an issue in low firing temperature MCM-Cs, though the effect is proportionately less because the silver is not stiff compared to the ceramic. However, differential shrinkage between the layers in the stackup (due to vagaries of the mechanical stacking process, differences in density of the ceramic particles in the "green" layers, or uneven firing) can cause a type of warpage referred to as *potato-chipping*. An MCM substrate curled in this manner can make component attachment and other assembly steps difficult or impossible, since the substrate does not "sit flat" in the assembly and bonding jigs. Probing and testing of the high-speed designs then becomes difficult because the lack of planarity of the MCM prevents repeatable touchdown of the planar probes. Finally, if the MCM-C is to be mounted in an hermetic package, there is once again the problem of bonding the curled substrate into the bottom of the package. Designers of such MCM-Cs must work closely with the fabricator and follow all design rules to minimize the occurrence of this problem.

In spite of the above warnings about stacking too many solid metal planes adjacent to one another, the value of a metal layer stackup in which adjacent planes are dedicated to power and ground is the creation of a classical parallel plate capacitor between the two planes. The need for large amounts of capacitance between the power and ground planes, whether created by the parallel plate capacitance of the adjacent planes or by the installation of multiple discrete capacitors, serves a critical function in support of high-speed integrated circuit: the capacitors serve as a local *battery* to supply charge to operate the chips during their rapid state switching. Because silicon ECL and GaAs gates and, more recently, the highest-speed CMOS typically switch states in «1 ns, the power supplies are typically too far removed from the MCMs and their chips to respond to the rapid changes in current requirements imposed on the power and ground rails.

If large amounts of local charge storage are *not* provided, the power planes experience partial *rail collapse*, i.e., the supply voltage momentarily sags toward ground potential (Fig. 12.23, upper graphs), while the ground plane, which receives a huge influx of current during the switching cycle, *bounces* upward toward the supply voltage value (Fig. 12.23, lower graphs). It is often argued that this behavior is actually caused by the series inductance in the power and ground planes which resist the instantaneous change in current. In fact, the distributed capacitance and inductance across the entire MCM or circuit board interact with one another as a distributed three dimensional LC circuit with persistent resonant behavior and a high "Q" value, as is apparent from the intermediate and long duration simulations of Fig. 12.23. (The small amount of damping in this model was provided only by the sheet resistance of the power and ground planes themselves.) The result is that, without extensive decoupling for each chip and resistive damping between the power and ground planes (often provided by the terminator resistors themselves), the voltage on the ground and power planes can oscillate for a considerable period of time after each clock-induced state switch. Finally, careful readers will note that the voltages on the vertical axes of this figure are much too high for any presently used logic families. In fact, this simulation was for a single power FET, in one location on a 6.5 × 6.5-in printed wiring board, driven at a 5 MHz clock rate. Note the complexity of the resonant behavior of the board. Simulations and measurements of more complex circuits operating at much higher clock rates have demonstrated similar, though even more complex, behavior of the power and ground planes.

To be effective in minimizing rail collapse, the capacitance, whether *integral* (i.e., formed by the adjacent planes themselves) or discrete (i.e., added as separate capacitors at the time of chip assembly) must be of high quality across all of the frequencies that appear in the switching transients (i.e., often up to extremely high frequencies). The achievement of this high quality is more difficult than at first might be expected, as will be noted below. Second, to overcome the deleterious effects of the inductances in the planes, bond wires, and so forth, a decades-old rule of thumb states that the capac-

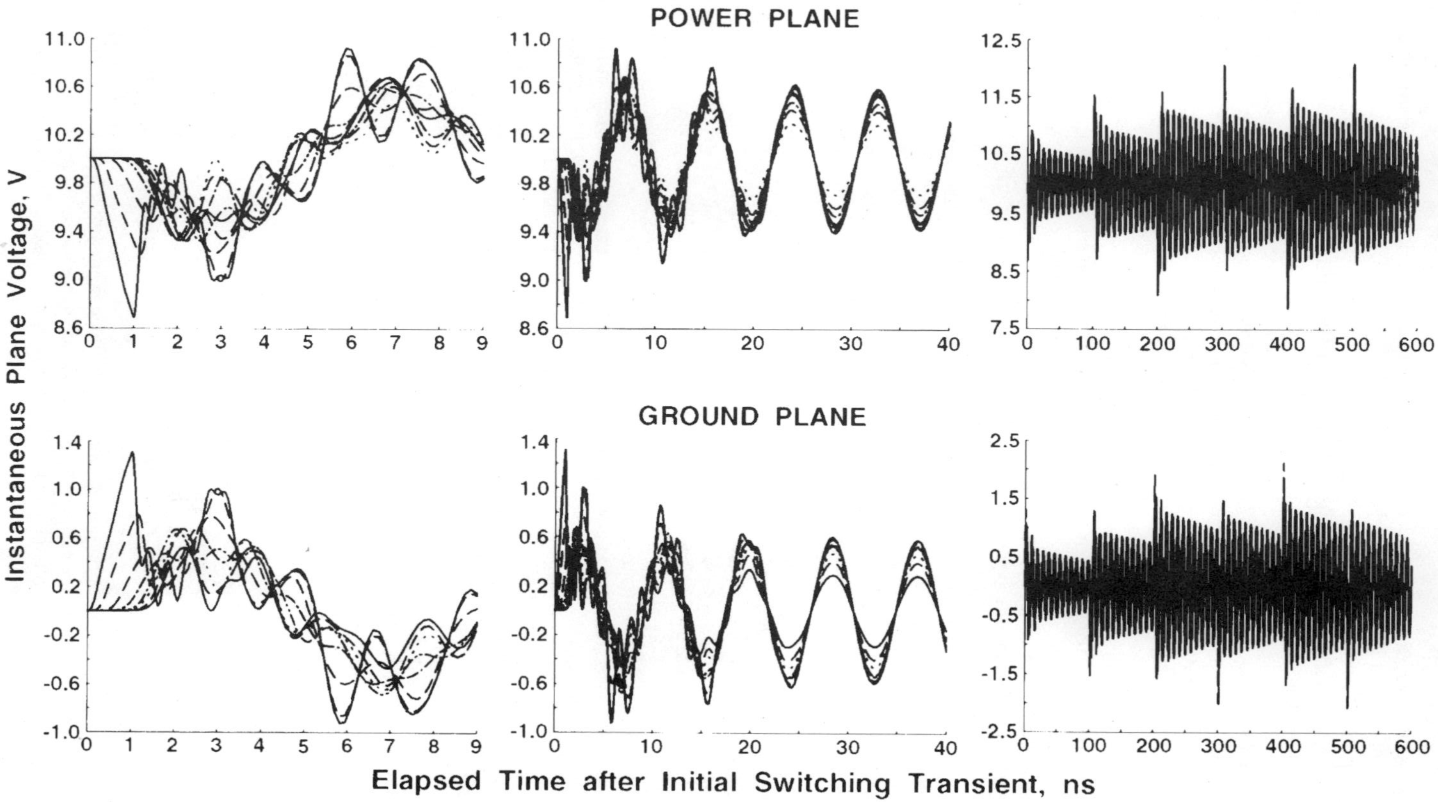

FIGURE 12.23 Computer simulation of power plane (upper graphs) and ground plane (lower graphs) behavior of a 6.5 × 6.5-in PC board when disturbed by switching transients of a single high-current transistor. Shown are voltages at eight "listening" locations across the board. Early disturbances on upper and lower left-hand graphs are reflections from far edges and corners of the board. The center column of graphs shows buildup of oscillations at resonant frequency of the board, caused by a single switching transient. The right-hand column of graphs shows gradual decay of oscillation amplitude, which is then reboosted by the next switching transient. Note the 180° phase difference between power and ground plane voltages. (0.375 A, 1 ns current pulse applied at 5 MHz rate, ε_r 4.2, dielectric thickness 0.062 in, 0.0007-in copper planes, sheet resistance 1 mΩ/sq FET at location 0.25 in., 0.25 in and power/ground plane feeds at 6.25 in, 6.25 in from board origin.)

itance must be physically very close to all the power and ground pads of each chip (a design rule that has been recently reconfirmed in our laboratory with new modeling tools that simulate in detail the behavior of power and ground planes in high clock rate systems[13b]).

The numerical value of the decoupling capacitance formed between adjacent power and ground planes is directly related to the ε_r of the dielectric separating them, the dielectric thickness, and the areas of the planes, by the standard equation $C = \varepsilon_0 \varepsilon_r A/d$, where ε_0 is the permittivity of free space, ε_r is the relative dielectric constant of the insulator, A is the capacitor area, and d is the thickness of the insulator. Organic-based MCM-Ds typically employ dielectric layers approximately 10–25 μm thick, with values of approximately 3–3.5, and thus demonstrate only approximately 193 pF/cm^2 for 10 μm of dielectric and 79 pF/cm^2 for 25 μm of dielectric. Neither is sufficient to decouple the planes without the use of auxiliary capacitors. MCM-Ls and MCM-Cs do not even approach the capacitance/unit areas of the organic MCM-Ds.

In an alternate implementation, one vendor of SiO_2-based MCM-Ds has developed the ability to deposit alumina ($\varepsilon_r \approx 10$) in very thin (approximately 0.3 μm) layers through the anodization of the aluminum metal in the ground plane, resulting in capacitances of approximately 50 nF/cm^2.[14] Numerous tests in our laboratory during the past several years demonstrate that the SiO_2-based MCMs do have sufficient integral capacitance to prevent rail collapse even in very fast circuits. Figure 12.5 is a photograph of a 16-chip MCM-D fabricated with the SiO_2 process using GaAs integrated circuits. The module was designed to employ additional discrete capacitors, as pictured in Fig. 12.5. Modules were assembled and operated both with and without the auxiliary capacitors to verify that, in fact, the capacitance formed by the adjacent power and ground planes was sufficient to decouple the planes from one another. Figure 12.6 depicts measured data from one of these modules. There was no significant difference in the amplitude of the noise on the power and ground planes (lower left graphs) between the modules with and without the additional discrete capacitors installed (i.e., 50 mV p-p). This result indicates that, at least at this chip density, the integral decoupling capacitance was sufficient to suppress the noise on the planes. Larger chips, up to 20 mm square (compared with the chips in Fig. 12.5, which were approximately 4 × 6 mm in size), with higher gate counts (100,000–250,000, compared to a few thousand gates), packed more densely (minimum transistor feature sizes of 0.3–0.5 μm compared to the 0.7 μm feature sizes), and operating at the same clock rates (i.e., 300–700 MHz) would almost certainly render the integral decoupling capacitance available from this process insufficient.

Because MCM-Ls, MCM-Cs, and organic-based MCM-Ds cannot establish sufficiently large parallel plate capacitance between the power and ground planes as a result of the relatively thick and/or low ε_r dielectric layers, discrete decoupling capacitors must be installed adjacent to each chip. For the highest-performance chips, each power pin may require its own decoupling capacitor. To make matters worse, not all discrete capacitors are equally good at supplying charge, with weaknesses either at high frequencies (the most typical situation) or at low frequencies. Poor capacitor performance at high frequencies is typically caused by the physical design and construction of the capacitor, by the quality of its dielectric, or both. These devices are often manufactured with metal end caps that contribute a significant amount of series inductance to the device—to such an extent that many become nearly pure inductors above frequencies of 100–200 MHz and, hence, become useless for the intended decoupling function. Figure 12.24 depicts swept frequency measurements of four discrete capacitors. Three become primarily inductive at a few hundred megahertz, while only one functions well up to frequencies of 5–6 GHz. (But note its resonant spike at approximately 3 GHz.) Conversely, small "split plate" microwave capacitors are commercially available that work well at extremely high frequencies but, because of their low capacitance values (10–200 pF), are insufficient for the lower-frequency components. These problems often require the use of two or even three types of discrete capacitors in MCM-Ls, MCM-Cs, and organic-based MCMs, one capacitor type for extremely low frequency decoupling, one for the mid-range frequencies, and a third type for the highest frequencies.

The task of selecting appropriate capacitors is made even more difficult by the fact that many vendors of small capacitors do not publish detailed performance specifications for their devices. Such information as the resonant frequency and Q of the capacitor (a capacitor can have a Q value because of the inductance of its metal end caps and the dissipative losses of its dielectric), its loss and trans-

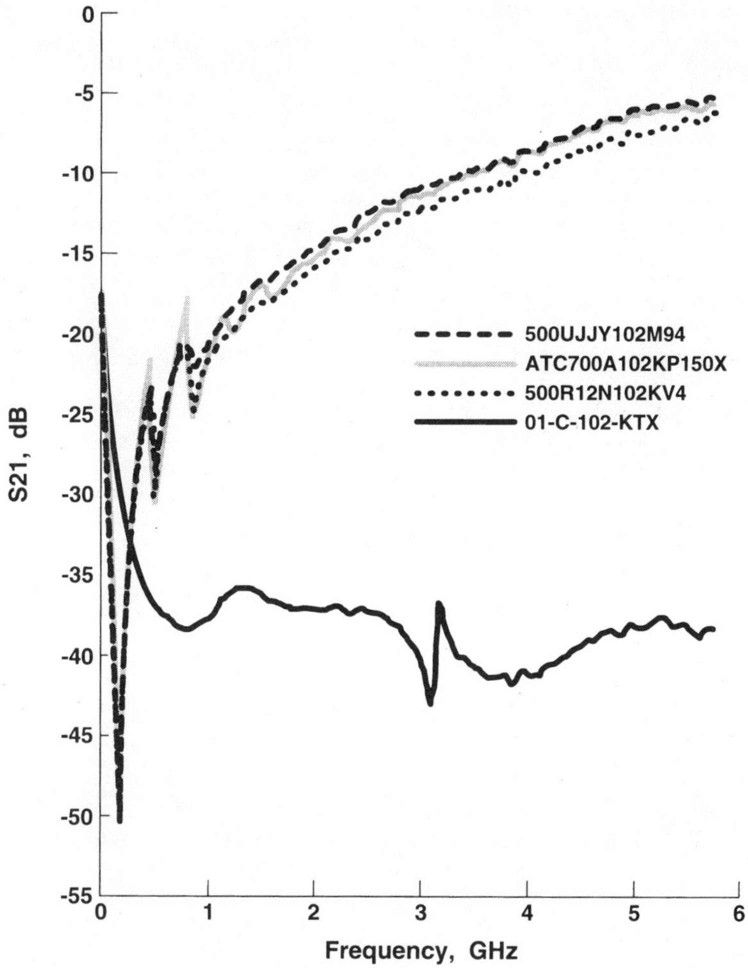

FIGURE 12.24 High-frequency performance of four discrete decoupling chip capacitors. Three capacitors become nearly inductive above 200 MHz. Note the resonance of the best capacitor at 3.3 GHz.

mission behavior in series and shunt configurations, and the conditions under which these values were measured is largely unavailable from the vendors. All of these would parameters assist the design team in the selection process. In fact, in many instances, the vendors do not possess this data themselves. More subtle types of information, such as the fact that the small parallel plate capacitors exhibit markedly different electrical behavior if they are mounted onto the MCM in a "plates-vertical" or "plates-horizontal" configuration, is almost never even described.

What type of information should the designer request from the capacitor manufacturers? At the very least, the resonant frequency of the capacitor and its Q at that frequency should be requested. Assuming that such data were available, the capacitor with the lowest Q and the highest resonant frequency above the proposed system clock rate should be selected, since the highest impedance of the capacitor is typically observed at frequencies approaching the resonant frequency, and, for good decoupling, a capacitor with *low* impedance is desired. The capacitor will function best in this role if its resonant frequency is three to four times greater than the system clock rate, or even more. Note, however, that the configuration in which this measurement is made is important. The capacitor should be measured with one of its ends tied to the test signal source and the other to ground, in a T-section structure (i.e., the capacitor forms the vertical portion of the T; the signal is fed into the top left lead of the T, and the top right arm of the T is left open or terminated in 50 Ω to ground).

An even better set of measurement data to request from a capacitor vendor, assuming the same type of T-section test configuration just described, is a complete S_{21} swept-frequency measurement between a few tens of megahertz and a frequency three to four times higher than the intended clock rate of the system. The data would then resemble that of Fig. 12.24. Although the resonant frequency and Q of capacitors is *sometimes* available from vendors who specialize in the marketing of components for the microwave community, in our experience the S_{21} measurements are *never* available. This dearth of information often requires the end user to test capacitors with swept frequency equipment such as the Hewlett Packard Model 8510 network analyzer. The author's laboratory conducted just such a study on more than 400 commercially available capacitors in an attempt to address this problem.

As depicted in Figs. 12.5 and 12.14, the requirement to install many capacitors in such a small space on an organic MCM-D creates major assembly and rework problems and hence greatly increases module costs and decreases yield. (The gold-surfaced capacitors appear as bright squares in Fig. 12.14.) It is for this reason that methods have been developed for the inorganic MCM-D processes, and are being developed for the organic MCM-D and MCM-L processes, to allow the incorporation of integral decoupling capacitance directly into the substrate fabrication process. One supplier of organic dielectrics in sheet form is developing a very thin sandwich structure (consisting of a polyimide support layer, followed by a thin layer of metal, followed by an organic binder filled with high-ε_r particles, followed by a final layer of metal) that would allow fabricators of MCM-Ls and organic MCM-Ds to incorporate effective area decoupling layers into their substrates. The sandwich structure would be available in roll format, allowing the dielectric sandwich to be installed as an appliqué directly into MCM substrates during their layer buildup. The possibility thus exists of greatly improving the performance of the MCM substrates while simultaneously decreasing the fabrication, assembly, and rework costs of the finished modules.

One possible approach to increasing the amount of decoupling capacitance to values much higher even than the 50 nF/cm^2 achieved by at least one inorganic MCM-D supplier is to substitute inorganic dielectrics with extremely high ε_r values ($\varepsilon_r > 100$), such as barium titanate, and the so-called "ferroelectrics" such as lead zirconate titanate,[15–17] or lead lanthanum zirconium titanate, for some of the normal dielectric layers of the MCM substrate. These exotic materials can exhibit very high ε_r values and acceptable dissipation factors from frequencies in the megahertz range up to several tens of gigahertz. For example, several of the ferroelectric materials exhibit both extremely high ε_r values and very rapid depolarization times (i.e., the ability to release energy stored in the electric field very rapidly, within subnanosecond time periods). Some of these formulations have been successfully deposited in robust pinhole-free layers as thin as a few thousand Angstroms. Tests in the author's laboratory indicate that small capacitors fabricated with such materials behave as excellent capacitors from a few tens of megahertz up to at least 26 GHz (Fig. 12.25). An active area of research at present is the investigation of methods of incorporating these materials in the form of very thin films, no greater than 1 μm, as the dielectric layers that separate the power and ground planes in MCM substrates. Unfortunately, because these materials must at present be deposited at temperatures of 600–700° C and solidified at 850° C, these high-ε_r materials can be considered at present only for incorporation into MCM-Cs or MCM-Ds fabricated on silicon or ceramic substrates. The development of lower-temperature deposition processes could thus greatly extend the usefulness of the high-ε_r materials.

12.9 ELECTRICAL INTERCONNECT OF CHIPS TO THE MCM

MCMs employed for high clock rate applications must support the propagation of signals with fast edges and wide signal bandwidths—not only through the signal traces on the MCM itself but also

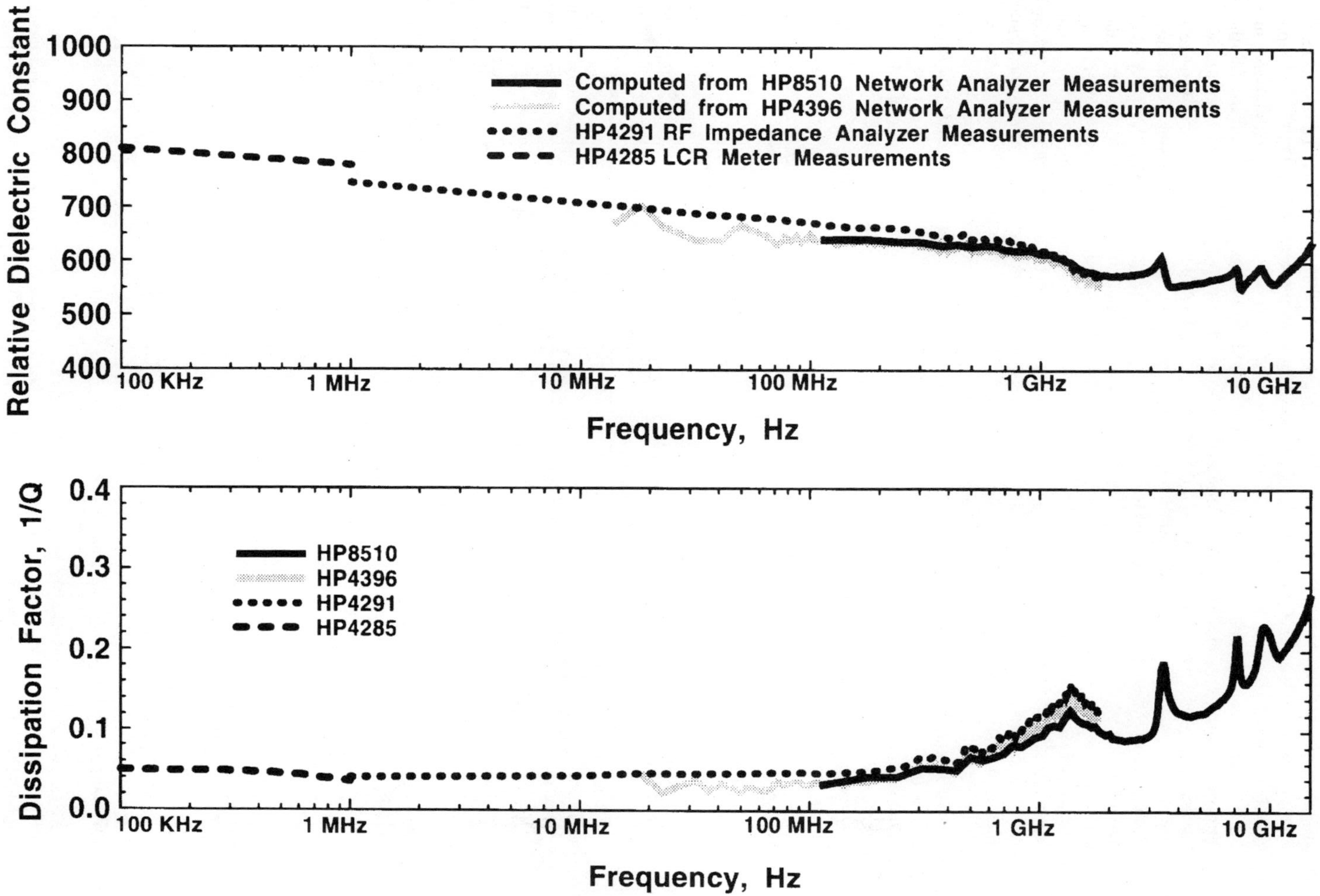

FIGURE 12.25 Frequency-dependent electrical properties of 0.48 μm thin films of the ferroelectric material lead zirconate titanate. Four different instruments were used to study this material over a range of frequencies. The material retains ε_r values above 500 up to frequencies in excess of 10 GHz. The spikes at high frequencies are not properties of the material but resonances of the small capacitor structures used to measure the properties of the films. (Top electric = 1.2 μm gold, bottom electrode = 0.1 μm Ti-Pt. Dielectric = PZT (52/48), thickness 0.482 μm, 10 × 10 μm shunt capacitor on 50 Ω coplanar waveguide [CPW].)

through the bond wires or whatever connections are made between the MCM signal pads and the chip signal pads themselves. Not all methods for connecting the chip signal pads to the MCM traces are equally capable of serving in this role without degrading the signals. Wire bonding, either gold ball bonding for GaAs chips that employ gold metallurgy or aluminum wedge bonding for the silicon and increasing numbers of GaAs chips that use aluminum metallurgy, is still the most common attachment method. Gold ball bonders can be adjusted to create wire bonds of only 30–40 mils in length plus the thickness of the die itself, and modern aluminum wedge bonders can also be programmed to shape or form the wire after the first bond has been made. The fine pitch bonding capabilities of aluminum wedge bonders, as low as 3 mils, are a good match for modern chips, which are now employing pad pitches as small as 3–3.5 mils (somewhat below the capabilities of ball bonders). In either case, the total curved length of the wire bond is typically on the order of 50–60 mils minimum, which represents a nontrivial amount of additional series inductance in the circuit path. It is possible to shorten the wire bonds to 20–30 mils if the chips are placed into wells in the MCM substrate, which allows a very flat, loop-free bond wire to be installed; however, because the cost of placing the die in wells is prohibitive in all but the most demanding designs, this approach is almost never employed.

For the highest-speed chips, i.e, those operating at clock rates above, e.g., 1–2 GHz, wire bonds present an added inductance in the signal path on the order of 1 nH/100 mils. Ribbon bonds (in the shape of flat tape with, e.g., 1 × 3 mil dimensions) exhibit slightly lower inductances per unit length (on the order of 1 nH/mm). The plots presented in Fig. 12.26 depict a set of measurements of a spe-

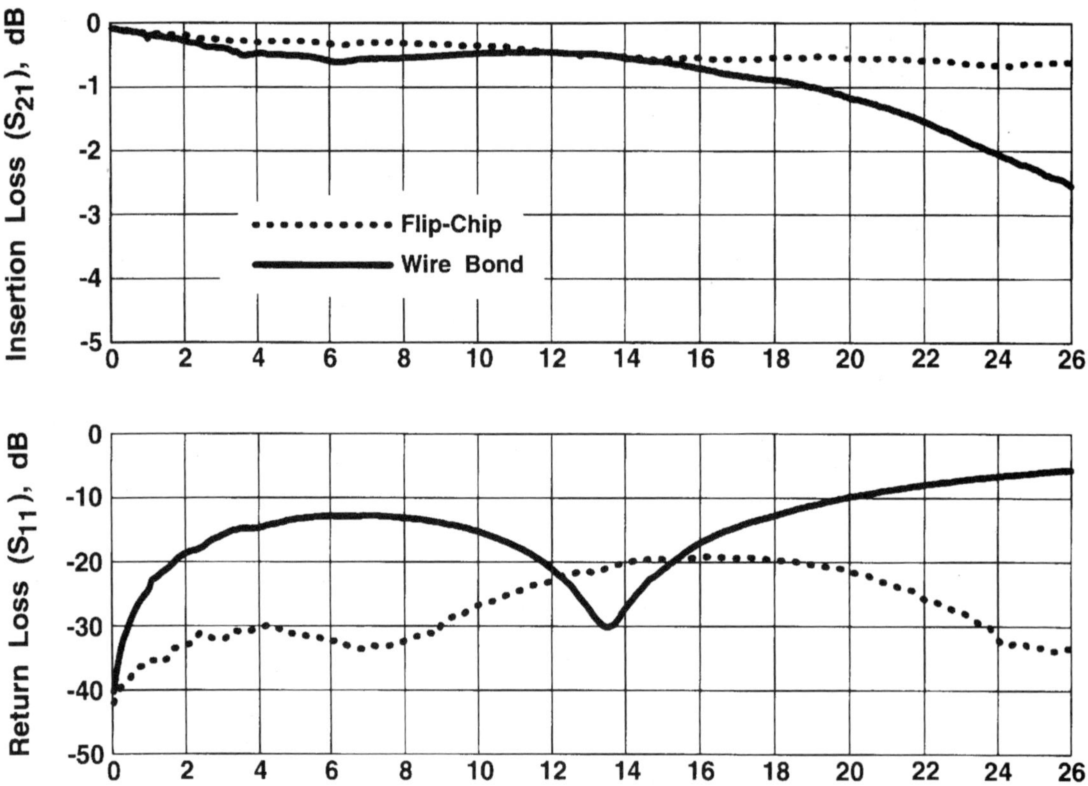

FIGURE 12.26 Measurement-based comparisons of the high-frequency characteristics of two in-series, 15-mil wire bonds connecting three sections of 65-mil coplanar lines, with two in-series flip chip attachments using 1-mil gold balls. Even at these nearly unachievably short (in production environments) lengths, wire bond loss characteristics are noticeably worse than those of flip chip attachments.

cial test structure containing three very short (65-mil) coplanar transmission line segments, representing a circuit board trace, an on-chip interconnect, and a final circuit board trace. The connections between the circuit board traces and the chip traces were made both with 15-mil wire bonds (which in actual production are so short as to be essentially unachievable) and with small flip chip gold balls. Even such short wire bonds increase the insertion losses and the return losses at frequencies above 1 GHz. The differences in loss behavior in this figure may appear small. However, the *best* assembly practice known to the author could not achieve wire bonds shorter than 20–25 mils using recessed die well cavities, making this set of data the best ever likely to be observed for wire bonds. Even in this case, the wire bonds yield noticeably poorer transmission performance than the flip chip connections. While the effects of wire bonds may in fact may be negligible from an operational standpoint for digital components operating up to 1–3 GHz clock rates, the degradations introduced by wire bonds of even 20–25 mil lengths are definitely *not* negligible for certain types of microwave systems such as A/D converter modules, where disturbances of these magnitudes have disproportionately large deleterious effects. Worse still, in daily production practice, wire bond lengths of 50–100 mils and even longer are the norm.

The loss of control of signal interconnect voltage standing wave ratio (VSWR) in a high clock rate digital system can be significant (and even more serious in an analog microwave or mixed analog-digital system). As the VSWR increases to values beyond 1.2–1.3, the amplitude of a logic signal can experience a waxing and waning of its full-scale amplitude along the length of a signal net due to the signal reflections and resultant standing waves that can appear on the traces. Depending on the clock rate of the system and the lengths of the interconnects, the resulting decreases in signal amplitude at different points along the signal net can result in significant decreases in effective noise margin on the signal nets; in the most blatant cases, the signal amplitude can fall to nearly zero at some destinations while achieving more than full amplitude at other destinations on the same net.

Unfortunately, there are no useful rules of thumb to assess the relative importance of issues such as bond wire type and length which, if they existed, could serve as a guide to the design team. The only way to ascertain whether a given interconnect approach, including the effects of bond wires, will achieve the desired operational behavior in a system under design is to simulate representative interconnects. At the very least, the longest or most frequency-critical interconnects planned for the system implementation should be simulated in both the time and frequency domains. Typically, such simulations must be conducted with a variety of circuit simulation and electromagnetic modeling tools. No single tool will at present give a complete view of interconnect performance. A good example of this type of simulation approach can be found in Refs. 17*a* and 17*b*.

As a result of the VSWR considerations described above, and for space reasons, wire bonding is already being replaced by other electrical chip connection methods as chip clock rates increase and rise times decrease. By far the most attractive method of making electrical contact with high-speed digital chips is the flip chip approach, of which two types have been described in the literature and are summarized here. IBM's C4 flip chip solder attachment process, introduced in the 1970s, deposits solder balls on the chip pads, then attaches the face-down chips to the MCM by reflowing the solder balls onto matching solder-coated attachment pads on the MCM. The chips thus are elevated above the MCM by a few mils, i.e., by the height of the reflowed solder balls.

Thermomechanically, the original IBM approach has well documented limitations in regard to maximum chip size, power dissipation, and operating temperature for acceptable levels of stress.[18] Since there was an air gap between the bottom of the chips and the top of the MCM substrate, heat dissipation into the MCM was so minimal that the original IBM implementation removed heat from the chips by making thermal contact with the back surfaces of each chip with a piston-like structure. Furthermore, limitations had to be placed on the maximum edge dimensions of the chips because, for a given temperature difference between the chips and the underlying MCM substrate, stresses increased with increasing edge dimensions until, at very large sizes, the solder bonds would fatigue from thermal cycling,[18] and the chips would actually pop loose from the substrate. However, as will be described below, more recently an even simpler and much better solution to the flip chip heat transfer problem has been implemented that has improved the thermomechanical environment of chips attached in this manner. Notwithstanding any thermal problems introduced by flip chip techniques, from an electrical perspective, flip chip attachment is one of the two or three best methods of making power and logic connections to the chips. In fact, it is worth considerable effort to overcome the thermomechanical problems of flip chip attachment methods because, as noted above in the discussion of Fig. 12.26, the inductive and capacitive parasitics associated with the solder balls are quite minimal.

A second type of flip chip attachment under development in several laboratories relies on adhesives to attach the chips, active surface down, to the MCMs. As an example of one of these approaches (of which at least five variations have been described so far), a thin plastic "decal" is fabricated to exactly the edge dimensions of the chip to be attached. Small electrically conductive "Z-axis" regions are incorporated into the plastic decal which are in the same locations as, and only slightly smaller than, the chip pads themselves. (The term "Z-axis" implies that the electrical conduction is perpendicular to the x-y plane of the MCM surface, i.e., in the "Z" direction.) In one such method,[9] the electrical conductivity is provided by small, metal-coated plastic microspheres embedded in the decal. The decal is applied to the MCM in alignment with its contact pads, and the chip is inverted and pressed into the top surface of the decal. A two-step thermal curing cycle is then initiated, with the chip still under pressure. In the first step, the decal softens and adheres firmly to the MCM and to the chip surface; in the second step, the plastic hardens. Thereafter, as the material cools, it shrinks, pulling the chip firmly toward the MCM and allowing the metal coated plastic microspheres to make contact with the chip and MCM contact pads.

These adhesive attachment approaches, although still under development, offer considerable promise for electrical attachment of high-performance integrated circuits. For example, Fig. 12.27 depicts a high-performance GaAs chip attached to a small MCM-D using this technique. Although

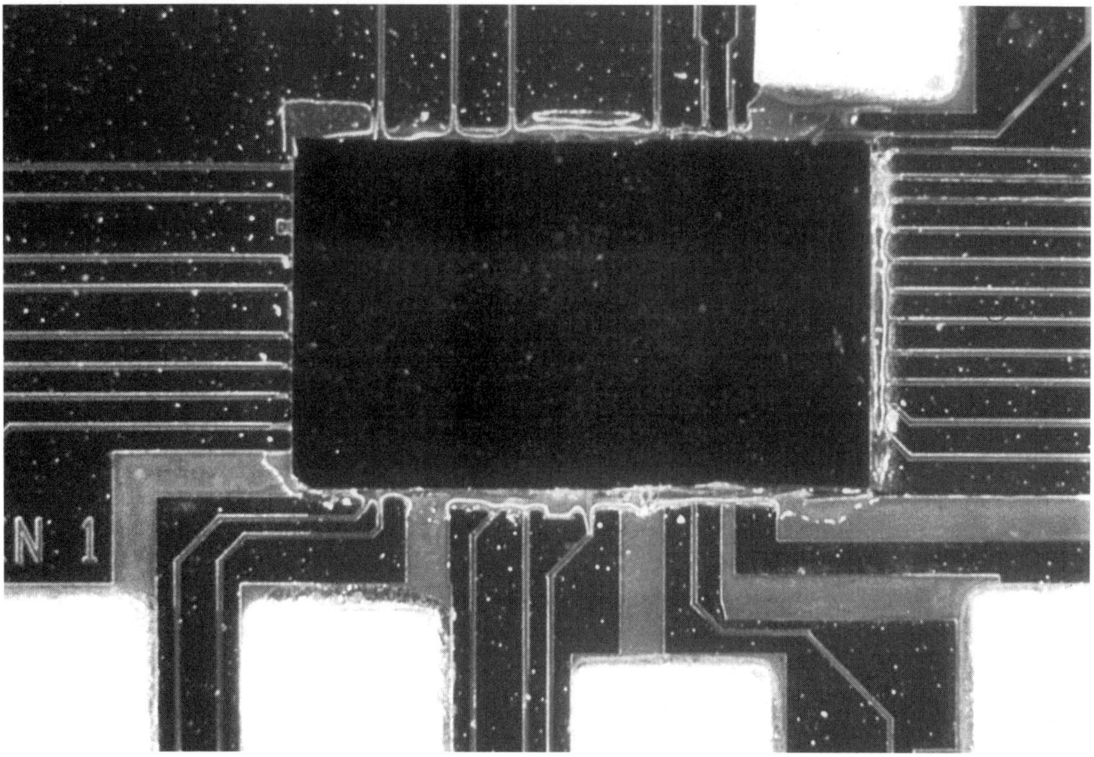

FIGURE 12.27 Micrograph of a GaAs chip attached to small MCM-D using flip adhesive z-axis conductive organic adhesive. A small amount of adhesive is visible around the edges of the chip. (Assembled by 3M Corp.)

the integrated circuit vendor stated that this GaAs chip should operate at clock rates only to 1.7 GHz, the device operated at 2.3–2.5 GHz when attached with the Z-axis conductive material. (As noted earlier, the vendor's lower performance estimates for his own chips came about because the effects of parasitic loads in his test probes were in effect incorporated into his models for chip performance.) Figure 12.28 depicts output waveforms measured from this test structure. Note the 110 ps rise times and 90 ps fall times, the fastest ever observed in our laboratory for any first generation GaAs MESFET part type. (The edge speeds of this chip, when delivered in single-chip packages, were rated by the chip vendor at 200–250 ps, for similar reasons as described in the parenthetic comment above.) The fast edge rates have been found to be a faithful measure of the level of parasitic inductance and capacitance; the lower the parasitics, the faster the edge speeds observed.

As an added benefit, the integrated circuit in Fig. 12.27 was found to operate at 37–39° C, in comparison to the 55° C typical of this device when operating in the conventional "face up" configuration. It appears that even though the adhesive material has the low thermal conductivity typical of organics, the decals are only a few mils thick and cover the entire active surface area of the chip, thereby giving them the ability to transfer large quantities of heat if the MCM has the ability to remove it.

The solder ball flip chip attachment process has also matured along a path somewhat parallel to the flip adhesive approach. Although the solder ball deposition and reflow steps have been retained, specially developed curable organic materials such as epoxies have appeared which are flowed into the air gap between the chips and the substrate. After hardening, the epoxy *underfill* material serves to relieve stress from the solder balls during thermal cycling, thereby remarkably improving the tolerance of the conducting joints to repeated thermal cycling. The underfill also serves as a thermal path that performs better than the air gap itself, significantly decreasing the operating temperature of the chips. Experiments are even under way to load the epoxy with highly thermally conductive particles such as aluminum nitride and diamond, but so far this approach has not been successful, because viscosity increases markedly decrease the volume-filling behavior of the particle-loaded epoxy. In the present embodiments of the solder ball flip chip attachment method, pads may be virtually anywhere on the surface of the chip (the original "area array" approach used by IBM) or may be confined to the periphery of the chip. In the latter case, the pads may be of different size or shape or have different sized openings in the chip passivation from those to be used for conventional wire bond attachment, but the advantage of peripheral flip chip attachment pads is that the chips can be employed in either a wire bonded face-up configuration, or in a flip chip face-down configuration.

It has been pointed out that the performance of very high-speed flip-attached integrated circuits may degrade when the air gap is replaced by organic material, because the underfill has a higher dielectric constant than air. According to this argument, crosstalk may increase, and signal propagation speed may decrease, in the on-chip interconnects. Although this type of degradation may indeed occur, it has not yet been observed in any subsystems of which the author is aware.

Finally, the chips-first type of MCM offers the potential for equivalent or slightly better electrical performance to the various flip chip approaches, since even the solder balls or metal coated microspheres are absent from the electrical interconnects. We have observed effectively the same excellent electrical performance from the chips-first process as from the flip attachment methods. (Note the 100 ps rise times and 85 ps fall times in Fig. 12.4, recorded from a chips-first MCM-D.)

12.10 THERMAL ENVIRONMENT OF HIGH CLOCK RATE DIGITAL MCMS

The thermal load placed on an MCM by its integrated circuits (and terminator resistors, if any) is typically substantially greater than in a circuit board environment, especially if the chips are of the high clock rate, high power/gate variety. The high thermal load derives from the high heat dissipation of many of the highest performance chips, and also from the high packing density of the bare chips on the MCM. Thermal loads above 50 W/cm^2 are already being experienced in GaAs chips. This is in some sense equivalent to heat loads two to three times greater in silicon, since the thermal conductivity of GaAs (k = 0.46 W/cm/° C) is roughly one-third that of silicon (k = 1.3–1.5 W/cm/° C).

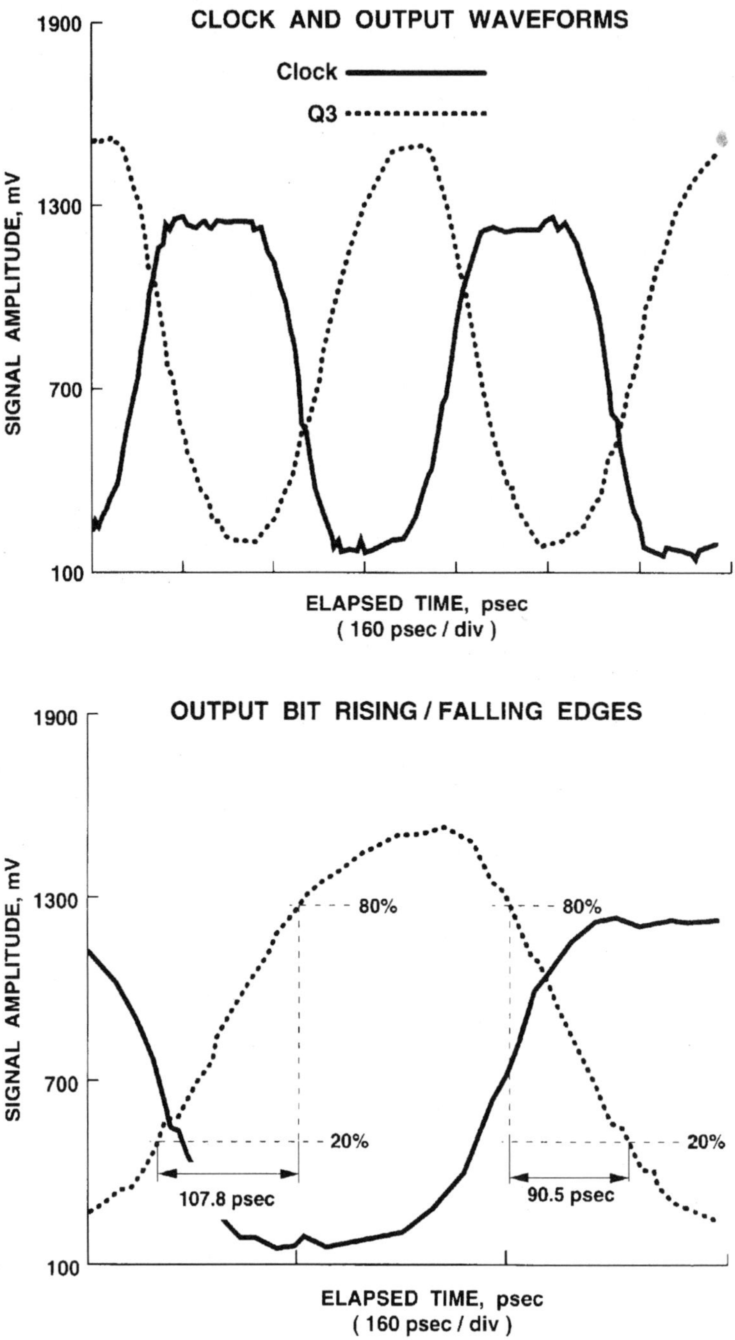

FIGURE 12.28 Performance of a small GaAs chip attached to organic chips-last MCM-D using z-axis conductive adhesive. Note 1.8 GHz clock rate performance, excellent waveform conformation, 108 ps rise times, and 90 ps fall times.

Because an entire chapter of this book is devoted to the thermal design of MCMs, only a few examples will be presented here that are particularly relevant to the design of very high-speed systems. Some of the fastest integrated circuits are those fabricated with GaAs heterojunction bipolar transistors (HBTs). These devices at present dissipate as much as 3–6 mW/gate, and (unlike silicon ECL) presently have only a limited capability for automatic temperature compensation. In fact, self-heating of GaAs transistors is so pronounced that it can be detected in the current/voltage (I/V) curves of the individual transistors as a degradation in transistor gain at the highest operating powers. Digital HBT chips thus experience serious heating problems that are manifested in shifts in logic high, low, and threshold voltages as a function of system clock rate. The chip temperature tends to increase when the devices are on and decreases when the devices are off, even during switching cycles. Careful circuit design can minimize, but not eliminate, these thermal effects for digital components.

For analog-to-digital converters (ADCs) fabricated with GaAs HBTs, the thermal problem is much worse. Slight changes in instantaneous operating temperature at various locations on the chip can actually degrade the number of useful bits, or sample resolution, which an ADC can produce. A seven-bit ADC can easily lose an entire significant bit or more due to thermally induced performance degradation of the transistors. Very high-performance ADCs [e.g., 8 bits of resolution at 3 gigasamples/second (Gsamp/s), and 6 bits at 6 Gsamp/s] are emerging from the research laboratory into practical applications (e.g., all-digital receivers for radar, electronic warfare, and communications applications) at the time of this writing, and the thermal problems are being observed directly in the test laboratory. For such chips, the MCMs must be designed to account for the extreme thermal sensitivity of the chips.

Figure 12.29 is a photo of a small MCM-L approximately 2.5 cm square, containing a 3 Gsamp/s 8-bit ADC and an 8:64 demultiplexer, both fabricated using GaAs HBTs. The ADC die dissipates 60 W/cm^2 (equivalent to 180 W/cm^2 in silicon), and the demultiplexer slightly less. Each GaAs die sits in its own cavity, at the bottom of which is a copper pedestal which is in turn an integral part of a copper cold plate covering the entire back surface of the MCM. Furthermore, the chips are thinned from 25 mils during fabrication to 7 mils before installation into the cavities to improve their heat transfer properties. Either silver-filled epoxy or reflowed solder can be used to attach the chips to the copper sinks. The final package structure is illustrated in schematic cross section in Fig. 12.30. Figure 12.31 is a graphic representation of the thermal environment of the sensitive ADC chip, calculated with the fluid dynamics modeling tool ANSYS. (Symmetry considerations made it necessary to model only one quadrant of the chip.) As shown in Fig. 12.31, by using these extreme measures to control heat flow, it is possible to constrain the temperature rise of the ADC chip to approximately 7° C above ambient.[19,20] A similar set of simulations for a flip chip attachment approach with epoxy underfill (depicted in schematic cross section in Fig. 12.32) indicates that it may be possible to constrain the temperature rise of the ADC chip to 12–15° C above ambient.

In the future, designers of high-speed analog, digital, and mixed-signal systems will be confronted with these types of extreme situations, almost certainly to a much greater extent even than those teams trying to package the largest, fastest CMOS chips. The design teams must thus be very keenly aware not only of the thermal loads generated by the chips but also of the detailed effects these loads have on chip electrical performance, and all of the ways that the packaging technology can alleviate these thermal loads and mitigate degradation in electrical performance.

12.11 TESTING OF HIGH-PERFORMANCE MCMS

12.11.1 Test Fixtures for MCMs

Because the problem of testing MCMs is discussed elsewhere in this text, we will concentrate here on the special issues that must be dealt with in the testing of very high-performance MCMs. Suffice it to say that for commercial products, many of which are encapsulated with an epoxy overcoat

FIGURE 12.29 Micrograph of 2.5 cm sq six-layer MCM-L used to package a 3 Gsamp/s 8-bit GaAs A/D converter (chip on left) and a 9:72 GaAs demultiplexer (the 9th bit is an out-of-range flag bit). Chips are mounted in cavities in MCM-L directly to copper heat spreader base. The small gray squares are power plane decoupling capacitors. A lead frame with 25-mil pitch (15-mil leads and 10-mil spaces) was brazed directly onto the MCM-L substrate. Drill holes and plated ring match with metal sidewall and removable lid structure, obviating the need to insert the MCM-L substrate into a separate package. Output data rate from the MCM is 24 Gb/s—375 Mb/s on each of 64 output lines. (MCM-L substrate fabricated by W.L. Gore & Associates.)

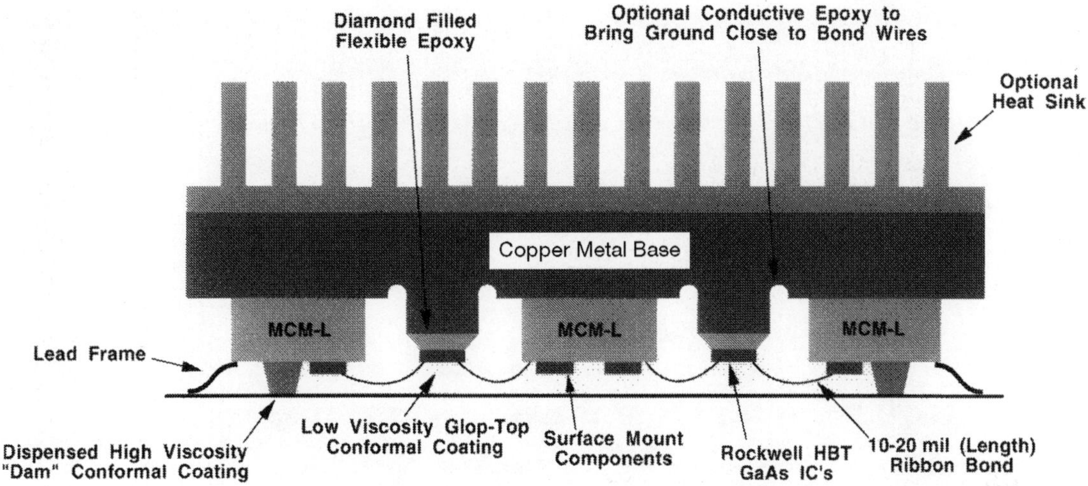

FIGURE 12.30 Concept for a complete package for a digital radar receiver using MCM-L technology and chips-last mounting. Note that the copper base protrudes into openings in the MCM-L substrate to form bases upon which chips are mounted directly for removal of high heat loads. Flush mounting of fastest, hottest chips with substrate surface allows very short (25-mil) wire bond lengths. Cooler and slower chips and passive components are mounted directly on the substrate surface.

(glob-top) for nonhermetic physical protection, rework is not feasible. If the assembled MCM is tested and found nonfunctional, it is simply discarded. In such an environment, a premium is placed on the knowledge that all installed die will be fully functional (the *known-good die* issue). Conversely, if the MCM contains high-performance, high-value components, there is a much higher likelihood that the rework of nonfunctional components will be attempted.

Unlike the low-cost MCMs with glob-top overcoats, high-performance MCMs are more typically installed into some type of ceramic or metal package and then sealed in, e.g., with an epoxied, welded, or soldered metal lid. These MCMs must thus be tested and reworked *before* they are sealed into the package, either during the prototype stage or in production. Their small physical size and fragility prior to encapsulation generally require the design and fabrication of specialized test fixtures tailored to each MCM footprint. These fixtures are capable of supporting the MCM in a variety of ways:

- Delivery of power and ground
- Removal of heat
- Provision of a base upon which microprobes can be mounted to aid in troubleshooting problems inside the MCM
- Provision of a physical housing for the MCM which is movable between different types of test stations or between test and rework stations

The test fixture and test station must provide the ability to place needle probes or integrated circuit-type probe cards, or even high-performance microwave probes, into internal test points or onto the external pads around the edges of the modules. A critical part of the test fixture problem for high-performance modules is that dozens, scores, or even hundreds of very high-speed analog and/or digital signals often must be cleanly injected into or extracted from very closely spaced peripheral pads on the MCM, often with pad pitches of 10 mils or less.

Figure 12.33 shows a typical probe station setup for a high-speed digital MCM. Note the bladed probe card at the top of the MCM, which delivers the power and ground and low-speed control signals to pads on one edge of the MCM. Note also the three specialty probes on the left, right, and

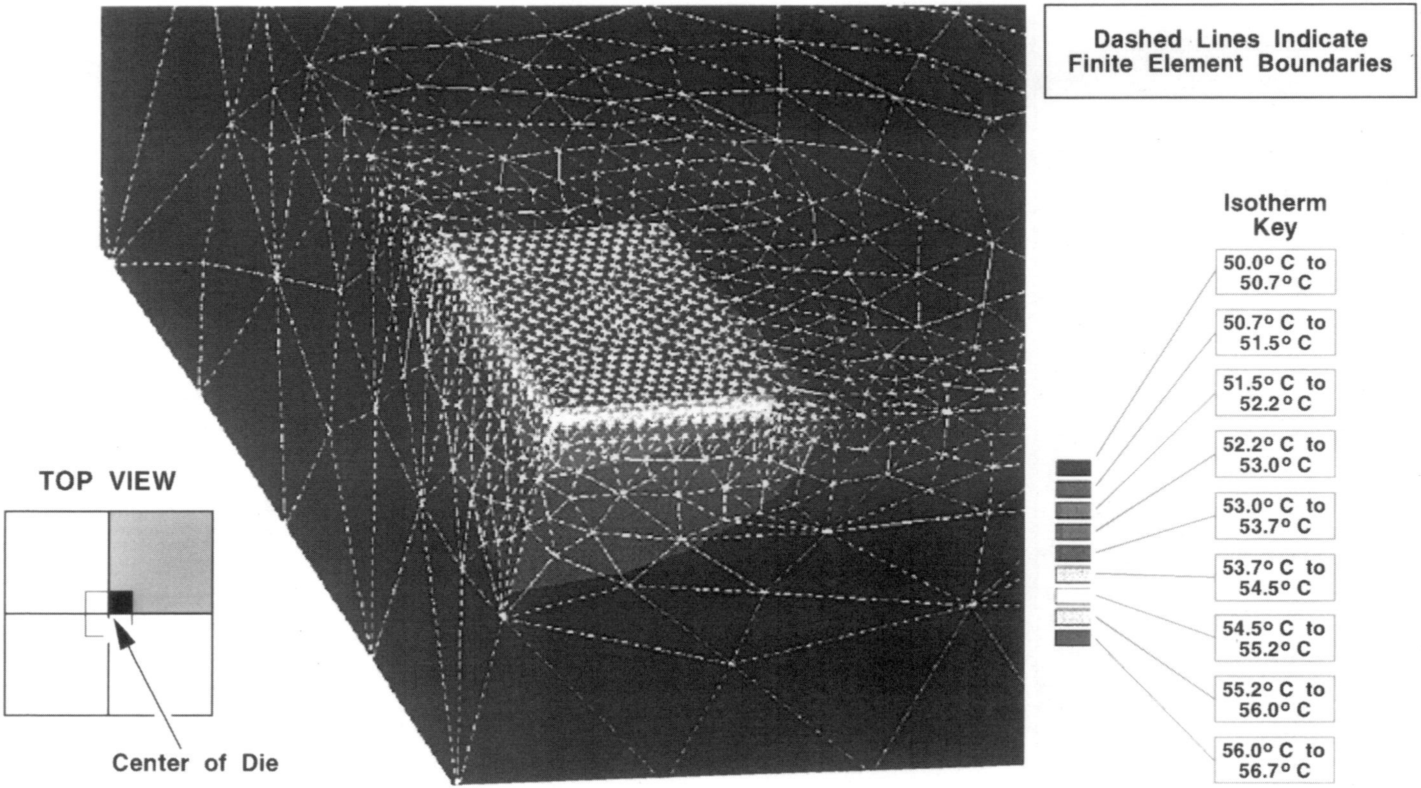

FIGURE 12.31 Thermal modeling results, using ANSYS fluid dynamics modeling tool, from MCM-L package structure of Fig. 12.30. The hottest GaAs chips in this actual system (depicted in Fig. 12.29) dissipate 59 W/cm^2, equivalent to a 180 W/cm^2 silicon chip. Only one quadrant of chip need be modeled. Maximum thermal rise is less than 7° C. (Highest chip surface temperature = 56.7° C for 7-mil thick chip in thermal-well configuration; chip attached to copper heat sink using 1-mil silver-filled epoxy.)

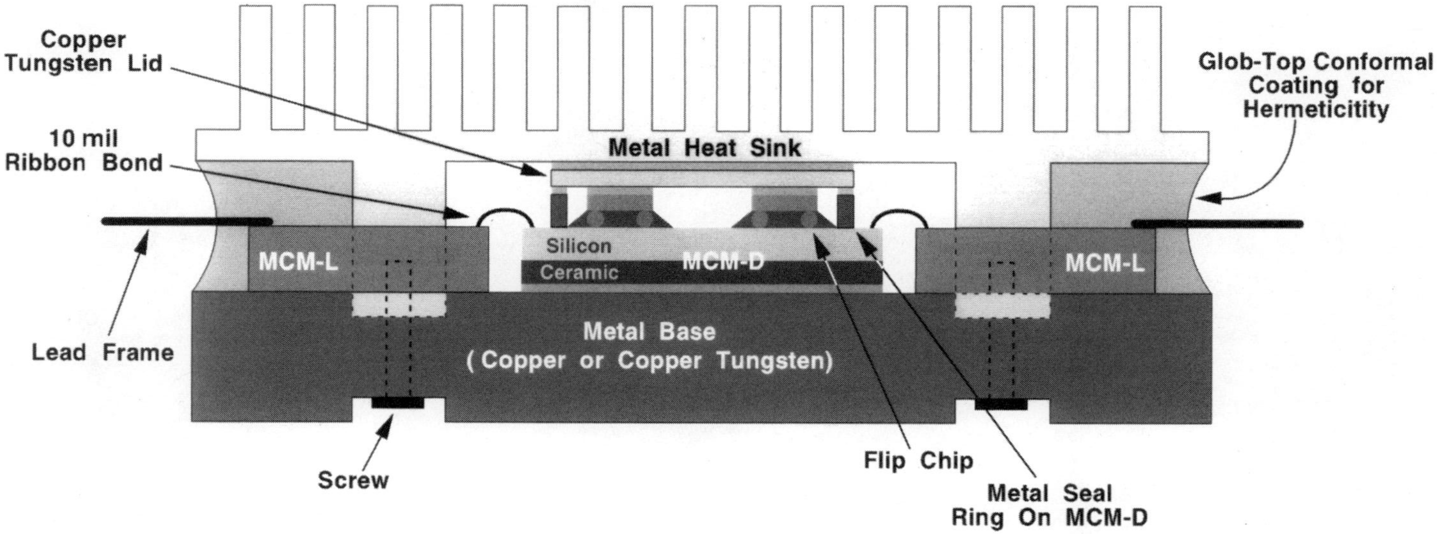

FEATURES:

- MCM-D Provides Integral Resistors and Capacitors
- Flip Chip Technology Provides "Best" VSWR Performance for Analog Signals
- All Signal Paths Provide Controlled Impedance and Minimized Discontinuities
- MCM-L for Future Growth; MCM-D Will not Require Redesign
- Metal Base and Lid Provide Assembly Solutions, Thermal Management, Shielding
- Removable Lid for Easy Access and Probing

FIGURE 12.32 Concept of a complete package for a digital radar receiver using MCM-D technology and flip chip mounting, with high thermal conductivity underfill material under the chips. Simulations indicate that flip chip attachment greatly improves electrical performance by obviating the need for wire bonds. Thermal rise can be held to approx. 12° C, which is not as good as shown in Fig. 12.31, but acceptable.

FIGURE 12.33 Probe station of high clock rate MCMs, showing low-frequency probe card pins (upper edge) and microwave coplanar transmission line probes (left, right, and lower edges.)

lower edges of the photo. The white triangular ceramic blades carry coplanar transmission lines on their under surfaces (not visible in this photo), which end in gold balls that make contact with specially positioned pads on the MCM. If the MCM were a passive test coupon only, the photo would be the same but with the upper bladed probe set removed. The coplanar probes exhibit bandwidths as high as 60 GHz and are excellent for injecting and retrieving high clock rate or wideband signals. Unfortunately, each probe can carry a maximum of only 4–8 such signals, limiting the total number of high-speed signals that can be tested at the edge of the module to 16 to 32.

Special bladed probe cards, much like an integrated circuit probe card but employing ceramic blades each carrying a coplanar line ending in a short needle, are also available, but they must be assembled uniquely for each MCM, are extremely costly, and have a frequency response limited to perhaps 400–1000 MHz. (The inductance in the needle at the end of each coplanar line may be the frequency limiting factor in these probes.) The purchase of a bladed probe card may, however, be justifiable in a manufacturing environment if, for example, a large number of MCMs must be probed prior to lid sealing. Finally, tiny coaxial probes have 40 GHz bandwidths and can apply signals to or extract signals from very small structures, e.g., signal lines down to 10 μm in width.

An alternative approach to testing of high-performance, high pin count MCMs in prototype quantities has been used in the author's laboratory. A small printed circuit board (actually composed of four "quadrant" window frame boards) is attached to an aluminum block into which liquid-filled cooling channels have been machined (Fig. 12.34). The circuit board accepts power and ground feed from an external power supply, delivering this power through buried power planes to contact pads on the inner edge of the "window" in its center. Controlled-impedance transmission line signal traces lead from the outer edges of the circuit board to its inner edges at the edge of the "window." A set of terminator resistor packs is mounted onto the circuit board such that any signal traces can be terminated at the discretion of the test engineer.

Into the "window" in the center of the circuit board is inserted a "daughter board," which can be fabricated as a PC board, as a very simple MCM-L, or even as a simple MCM-C. The daughter board is in direct thermal contact with the metal cold plate, and the controlled impedance transmission line signal leads that extend to its outer edges match the positioning of the signal leads on the inner edges of the four window frame circuit boards. Physical attachment and electrical contact is made between the daughter board and the window frame members with four clamp-type elastomeric strip connectors, which exhibit excellent high-frequency electrical performance but are easily released to allow removal of the daughter board from the test fixture (e.g., for transfer to a rework station). The bare (i.e., unpackaged) MCM is attached to the cold plate through the hole in the center of the daughter board with a tiny amount of conductive grease and held in place by a vacuum channel in the cold plate. The transmission line leads on the edges of the MCM are then wire bonded to the inner signal lead contacts on the daughter board. It has thus been possible to provide a transmission line signal environment from the outer edges of the tiny MCM to large 100-mil pitch "pin headers" or microwave connectors of as many as 400–500 leads without compromising the ability to test and rework the MCM.

Unfortunately, the approach just described also has severe limitations. Note that the MCM under test must be wire or ribbon bonded into the test fixture, which is a costly process and complicates its removal and reinstallation into a package. This approach thus only works for very small quantities of prototype modules. Although a variety of test approaches based on elastomeric connections or ball grid arrays have been proposed to allow testing of very high-performance modules while still permitting rework, in the view of the author this issue remains an unsolved problem. There is still not a good mechanism to supply power and ground, plus 500 or more very high-speed signals, to an unpackaged MCM.

The approach taken most recently in our own laboratory to circumvent this limitation is to design the MCM and package such that testing and rework of the MCM is possible *after* it has been installed in a leaded package but before the lid has been sealed into place. Alternatively, the MCM substrate itself can be designed to accept a solder-attached lead frame and a metal lid (see Fig. 12.29). Once the leads on the package are available, it is possible to design test fixtures (see Fig. 12.35) to provide high-speed signals to the package leads even if very tight lead pitches are required. The MCM pictured in Fig. 12.29 accepts one differential analog signal with a 1.5 GHz bandwidth and one differential 3 GHz clock signal, and transmits 72 single-ended signals, each toggling at 375 Mb/s.

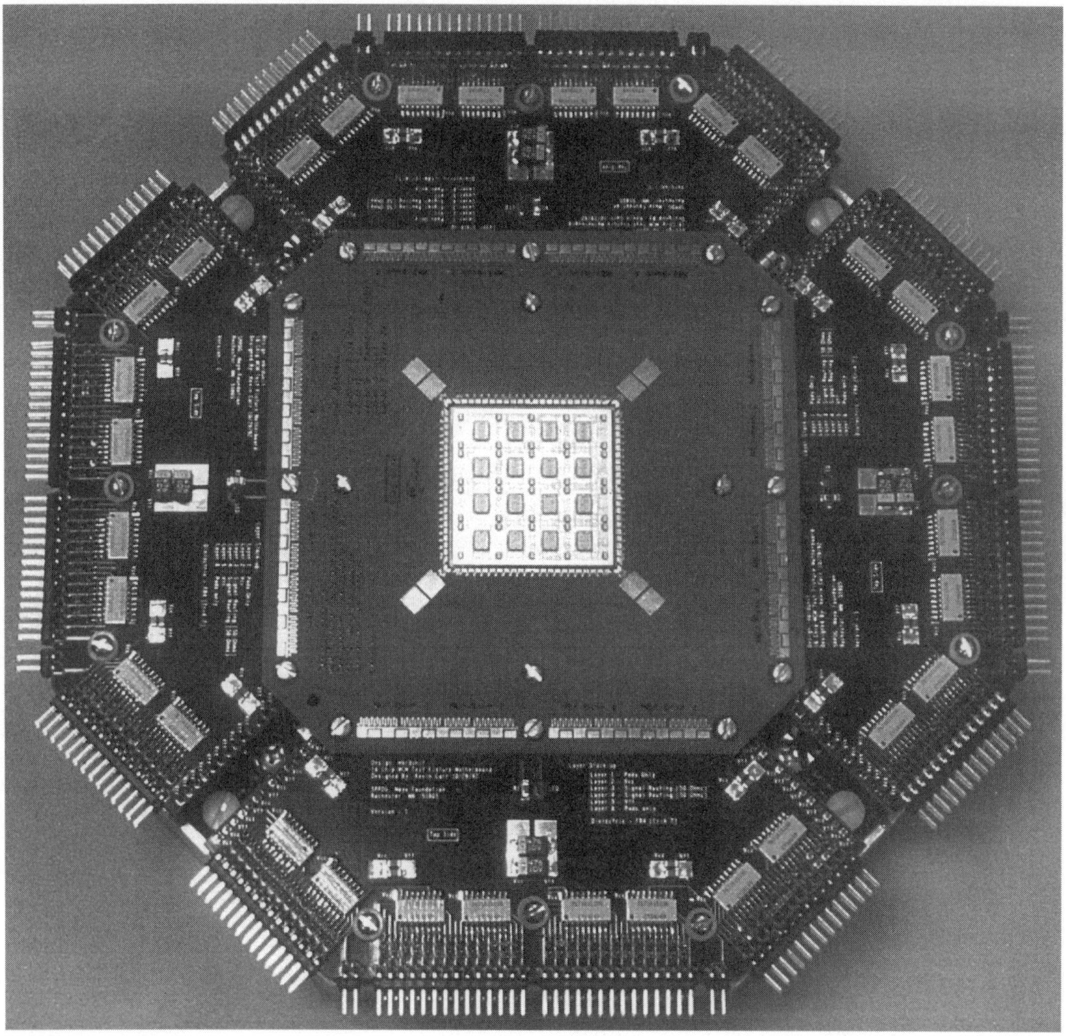

FIGURE 12.34 Test fixtures for high clock rate MCMs. Four outer mother board segments supply power, ground, controlled-impedance interconnects, probe contact sites, and signal line terminations. The inner removable daughter board contacts the MCM with wire bonds. The daughter board, MCM, and underlying cold plate chuck are removable as a single unit to allow rework of the MCM and reinstallation or swapping of MCMs without disturbing the test setup.

12.11.2 Passive Testing of MCMs

To employ the best features and circumvent the limitations of a given MCM fabrication process, the designer of high-performance subsystems must understand the characteristics of the process in detail. The designer needs to understand in detail the impedance behavior of the interconnects, their AC and DC resistive losses, the magnitudes of reflections caused by vias, crosstalk characteristics for various line widths and spacings, and so on. If such information were available from the fabricators, it could be used to develop design guidelines for maximum signal line lengths for a given off-chip clock rate and signal rise time, the best selection of transmission line type (i.e., stripline, microstrip,

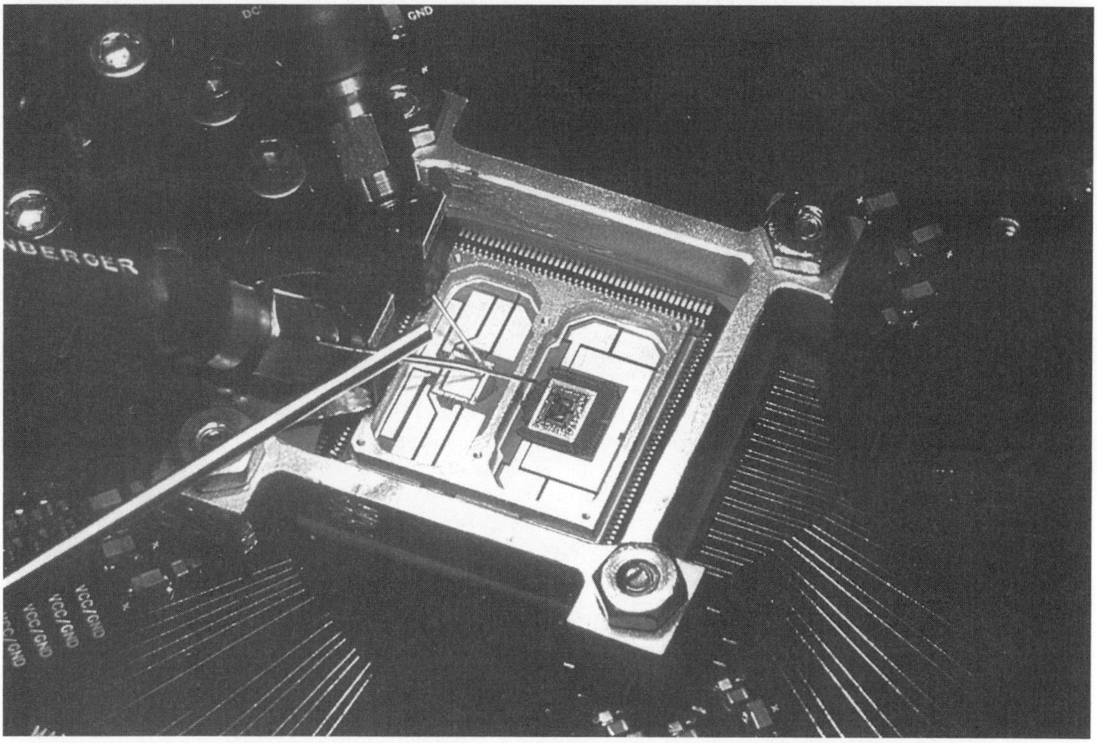

FIGURE 12.35 Photo of MCM-L of Fig. 12.29 in a specially developed high-frequency test socket. Demultiplexer die is installed first and tested (as depicted here) and, if found functional, is returned to the assembly lab for installation of A/D converter chip. Design of the test socket allows for multiple insertions and removals of the MCM under test while providing multigigahertz signal propagation characteristics at very low VSWR values. (MCM-L substrate fabricated by W.L. Gore & Associates.)

coplanar lines, and so on), termination schemes, and the like. However, in most cases, the fabricators do not have such detailed information regarding their processes, especially if the processes are evolving over time. The author's laboratory employs high-frequency test structures, referred to throughout this chapter as *passive test coupons,* to extract this type of information. The test coupons commonly include striplines, microstrips, and coplanar lines of varying lengths, chains of vias, bond pads, attachment sites for discrete resistors and capacitors, as well as integral thin film resistors and capacitors (if available in the specific MCM fabrication process). Some processes can support integral inductors as well (see later discussion).

From these passive structures, the test team can ascertain and monitor many of the electrical features of the MCMs produced by a given fabricator. This data includes the impedance of stripline, microstrip, and coplanar lines, and their high-frequency transmission and reflection behavior (often referred to as their "S parameter representations"; S-parameter data includes information regarding the AC losses in the lines), the amount of crosstalk as a function of frequency between lines of various widths and spacings, and the effects of via chains on signal integrity. If the cross-sectional shapes of the lines and the interlayer dielectric thicknesses can be determined from cross sections of the test coupons, then the dielectric constant of the materials can also be ascertained, as can any differences between the stated dimensions of lines, planes, and vias, and their actual as-fabricated dimensions.

If the MCM process supports integral resistors, careful measurements can ascertain such critical characteristics as the as-deposited resistance and their statistical variation, the values and statistical variations of the resistances after "trimming" with a laser (as-deposited values can be ±20 percent, while a few vendors have demonstrated a post-trim statistical variation of ±2 percent[20a]), and the self-inductance of the resistors (a critical parameter for designers of gigahertz clock rate systems to know, and directly related to the length of the resistor). An assortment of integral inductors placed on the passive test coupon will allow the design team to ascertain the range of inductance, quality factor (Q) values, and resonant frequencies that can be attained for these passive elements. A wide range of inductor characteristics can often be achieved, which may greatly assist the design teams for specific types of designs. (See the later discussion on mixed signal MCMs.) If all of these structures are very carefully measured and also simulated (see later discussions), the design team will gain a much better understanding of how much performance a given MCM technology can be expected to provide. Passive test coupons also play a very important role in the verification of electromagnetic modeling (EM) algorithms and *EM computer aided design tools,* as will be described below.

Perhaps the most important feature which needs to be designed into the test structures, and which is omitted from virtually all of them, is a set of "verification sites" for the high-frequency probes themselves (presented schematically in Fig. 12.36). If the probe and cabling artifacts and parasitics are not calibrated out of the test setup with verification sites physically located on the MCMs themselves, the test results can be ambiguous or incorrect. Until a particular MCM fabrication process is thoroughly understood, verification sites should be included whenever possible on the MCMs so that they are present if needed. For a complete description of the design of such modules, the reader is referred to Refs. 7, 9, 21. Figure 12.37 is the top-layer artwork from one of these test coupons. Portions of the single and dual microstrip lines, as well as several verification sites for the test probes along the edges (the groups of apparently isolated pads), are visible in this figure.

12.11.3 External Testing of Active-Circuit MCMs

Completely external testing of high pin count, high clock rate digital MCMs represents a partially solved problem, although the costs of the commercial test equipment are high. While production test

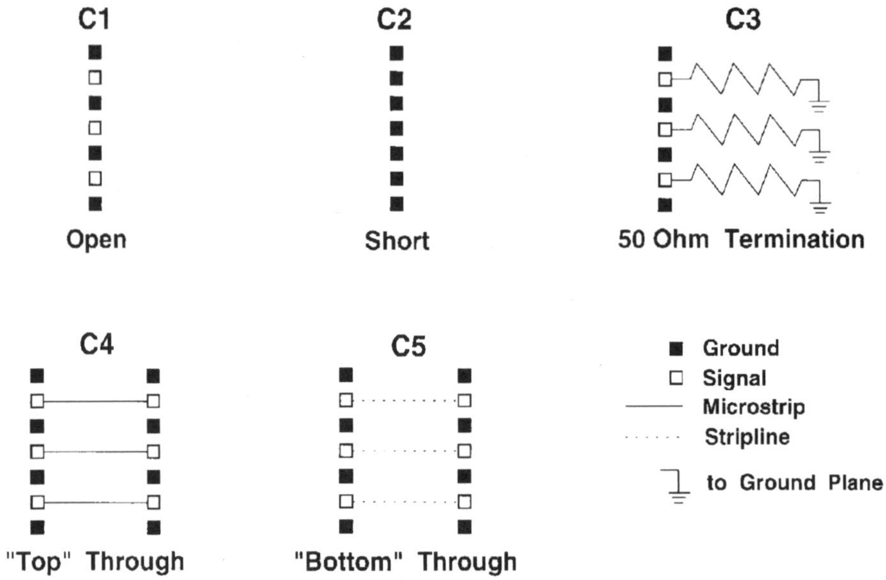

FIGURE 12.36 Schematic diagram of high-frequency test probe calibration sites designed for inclusion on MCMs during fabrication. "Top" and "bottom" through sites represent "in and out" connections for S_{21} parameter measurements for microstrip and stripline lines, respectively.

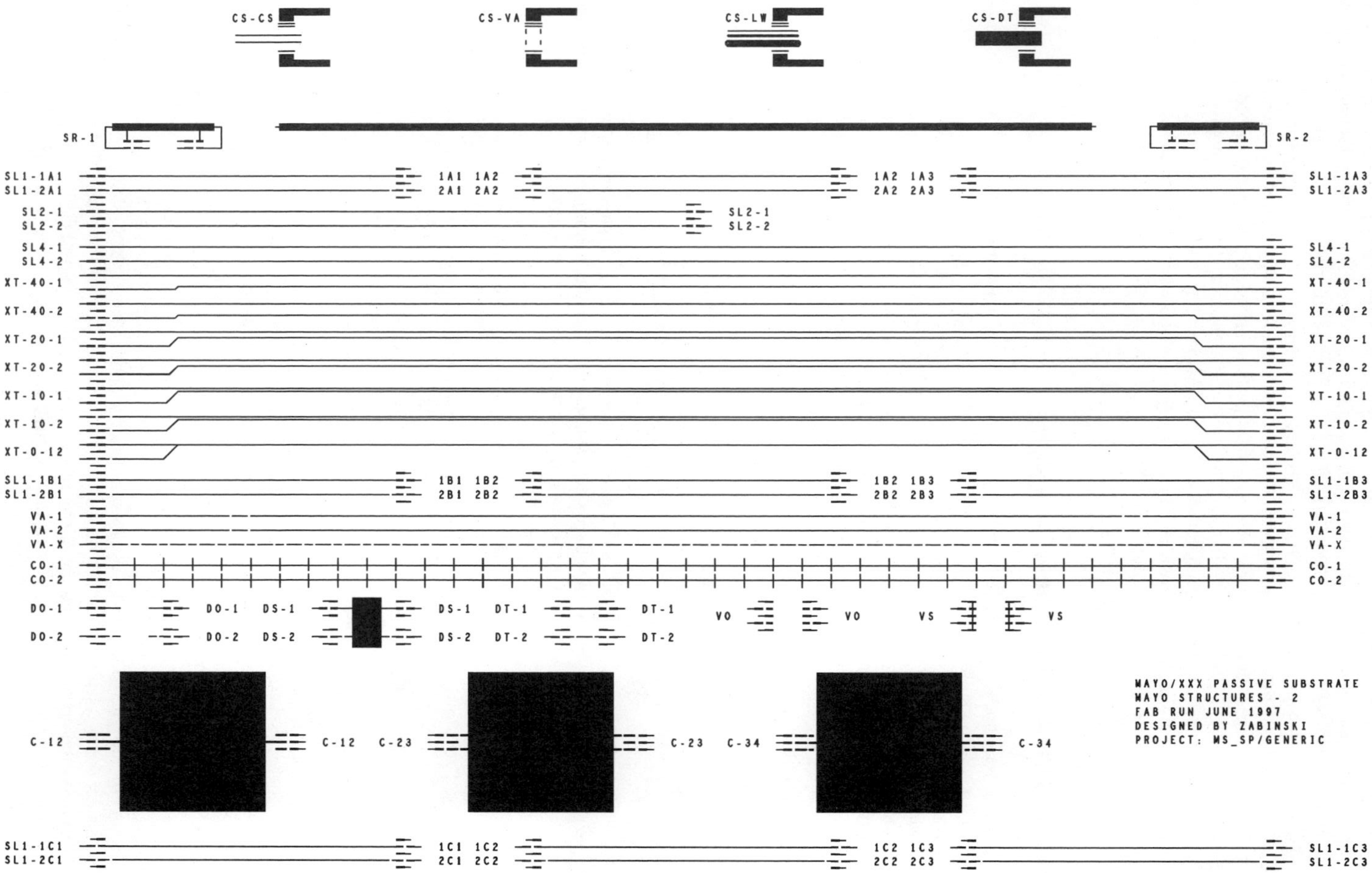

FIGURE 12.37 Artwork from a "passive test coupon" used to collect data on high-frequency interconnect performance on MCMs. Note the single and double line segments with contact pads designed for microwave probes, and the probe verification sites (rows of seemingly unconnected pads) along the edges of the coupon.

equipment with hundreds of high-speed pins does exist, these testers are extremely expensive and are still not fast enough to test the highest-speed MCMs. Furthermore, even though the data bit rates on each tester pin are in the 200–700 MHz range, the accuracy of signal edge placement, i.e., the edge skew resolution, between pins is not always as good as desired. Alternatively, relatively new test equipment encompassing both data pattern generators and data samplers is available for the testing of prototype MCMs up to 1 GHz data rates per pin, although the price of the test suite per pin is also very high. Both the production testers and the prototype testers are modular and can be expanded in capability to match the needs and the budget of the test environment. External testing of an MCM operating much above 1 GHz clock rates must be worked out on a case-by-case basis and always represents a tour de force, since very little test equipment is commercially available at these speeds.

12.11.4 Built-In Self-Test of High-Performance MCMs

It has frequently been stated within the MCM design community that, while MCMs containing CMOS chips can be completely tested at full operating speed (at least in theory) with external test equipment, on-chip self-test hardware exploiting a variety of test approaches (see below) is the best way to ensure complete functionality of the modules. As illustrated earlier, for the high-speed design community, it is even more difficult to verify correct module operation at full operating speed with external test equipment only; the test equipment is either very expensive to deploy in production settings or does not exist at all. As a result, designers of a high clock rate system should consider incorporating built-in test features in the chips themselves.

The two approaches that are most likely to yield high confidence of correct full-speed operation are the boundary scan approach, as exemplified by the IEEE 1149.1 Boundary Scan Protocol, or signature analysis, or both. In the boundary scan approach, each input and output pin on every chip passes through a flip flop, all of which can be chained together into a long shift register encompassing every I/O bit on the chip. The input flip flops in the shift register can be serially loaded with an input test vector at low (even kilohertz) clock rates. Each chip can then have one high-speed clock pulse applied, which will allow each chip to progress through one step of operation at the full system clock rate. The output data from that one step can then be clocked into the output flip flops and read out of the chip serially at low clock rates. While this approach will not necessarily guarantee complete testing of the internal functions of every chip, it will at least be possible to verify whether the chips are operating properly at their I/O pins.

Perhaps a better implementation of built-in self test for the fastest chips in an MCM environment is to configure the flip flops on the input pins so that, while they remain part of the boundary scan register structure, they also implement a *linear feedback shift register* (LFSR). An initial test vector value can either be hard wired into the chip or can be preloaded through the serial data input test port on the chip. Typically, LFSRs can be clocked as fast as the rest of the chip, so they are capable of internally generating a new test vector at every clock pulse. By simulating the performance of the LFSR prior to chip fabrication, the expected output test value can be predicted for each chip at every clock pulse if the initial test vector is known. Each output test vector can be clocked out of the chip and off of the MCM through a parallel test bus if sufficient I/O pins are available on the MCM (and if a sufficiently fast, high I/O count logic analyzer is available), or through the serial test port at slow speed, or sampled with a similar LFSR and "wound up" to create a final signature for each set of test patterns.

There is often a reluctance on the part of chip designers to devote 5 to 15 percent of their chip real estate to on-chip test features. However, because it has become clear that self-test in some cases may be the only way to verify full-speed operation of the highest-speed integrated circuits and subsequently the MCMs onto which they are installed, high-speed system designers are more willing to accept the area penalty and the one- or two-gate delay penalty in their fastest paths. The author's laboratory attempts to incorporate some amount of boundary scan or LFSR circuitry into every new chip if at all possible, even for very low-complexity chips operating in the 10 GHz range, in which real estate is at an extreme premium.

12.12 REWORK AND REPAIR OF HIGH CLOCK RATE MCMS

The best approach to the rework and repair of high-speed MCMs (or in fact any MCMs) is no rework at all. Rework raises the risk of damaging an MCM beyond recovery, or at the very least increases the cost of the final modules. It is recognized by the fabricators that the avoidance of rework depends on the availability of known-good die, the cleanliness of the starting substrates and chips, and the perfection of high-yield techniques for chip attachment and electrical connection. All of these topics are being developed rapidly. However, until these issues have completely obviated MCM failure during manufacture, some rework will be necessary. Techniques developed decades ago for the rework and repair of so-called hybrid microcircuits (i.e., microwave circuits consisting of individual transistors on single-layer ceramic substrates) have been extended somewhat and applied successfully to the rework of chips-last MCMs. Rework techniques have also been developed and demonstrated for the chips-first MCMs by "lifting off" the overlayers of metal and dielectric, replacing the failed components, and replacing the metal and dielectric layers.

Rework of MCMs containing CMOS chips in an unterminated line environment requires:

- The breaking of the wire bonds to the bad chip
- The removal of the chip by local heating to soften the epoxy or solder (the same local heating approach is used if the chip is flip attached)
- The "plucking up" of the chip
- Cleanup of the chip site and removal of wire bond fragments from the MCM bond pads (or more often, abandoning the fragments in place)
- Installation of a replacement die
- Reconnection of the electrical pads on the chip

Rework of high-performance MCMs is arguably more difficult *at present* than the rework of CMOS-based MCMs because of the large number of additional components on the faster MCMs, the somewhat fragile structure of the organic based MCM-Ls or MCM-Ds, and the way that these components are interconnected. When the point in the evolution of MCM technology is reached when all passive components, i.e., terminator resistors and decoupling capacitors, on a high-performance MCM are integral to the MCM, rework for the faster modules should be identical to that for the CMOS-based MCMs. However, as long as discrete passive components are employed, additional electrical bond connections typically must be made between the chips and the discrete components, between the chips and the MCM, and between the discrete components and the MCM, all of which may require removal and replacement. An extreme example of this problem appears in Fig. 12.14, a high clock rate, organic-based, chips-last MCM-D designed in the mid 1980s. Note the multiple wire bonds (many quite long) between different components as well as between the components and the contact pads on the MCM.

Rework is thus made more complicated because not only must chips be replaced at times but so must the terminator resistors. Furthermore, if one component is damaged, its replacement may in turn damage other components connected to the original one being removed, and so on until the module must be scrapped. Figure 12.38 depicts a number of approaches to interconnect chip bond pads and discrete terminator resistors into a transmission line signal interconnect on an MCM. Configurations C and F, while not necessarily the best approach electrically, at least have the advantage of isolating the potential damage caused by replacing either a faulty integrated circuit or a faulty resistor.

If *resistor packs* rather than discrete resistors are employed, the rework problem is further complicated, because one resistor on the pack may be damaged during assembly while the others remain intact throughout the wire bonding operation. The rework technician is then faced with the unenviable task of removing an entire pack and all its functioning wire bonds and resistors, bonding to an unused resistor on the same pack, or installing a single discrete resistor to replace the failed one (if there is even room to do so on the MCM). The permutations of problems which can arise when large

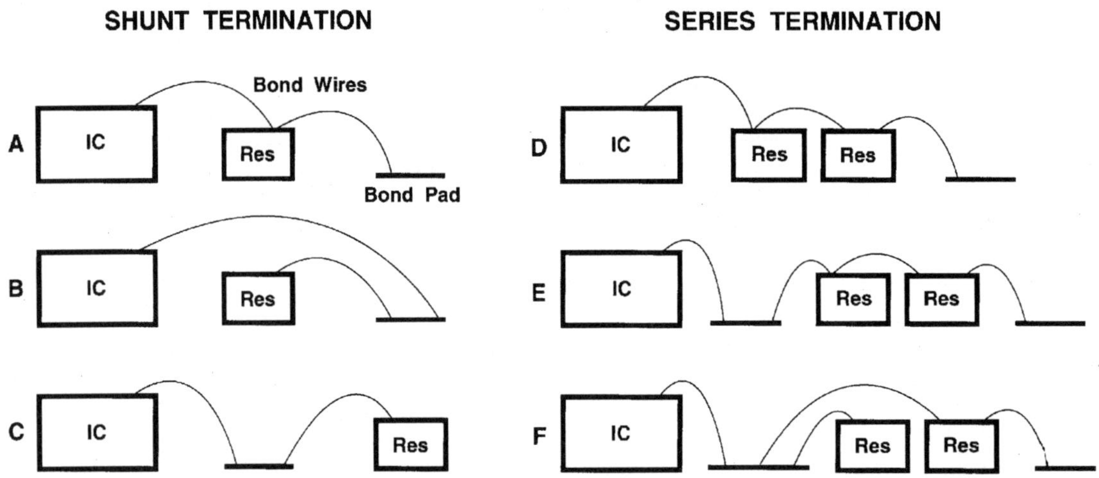

FIGURE 12.38 Six different methods of wire bonding chips and resistors into a circuit on an MCM. Methods C and F are the easiest to rework but not electrically optimal.

numbers of components must be interconnected are almost endless. As a result, the only practical long-term solution to the high-performance MCM rework problem is to integrate the passive components directly into the MCM substrate itself. The rework issue is yet another reason why fabricators of MCMs must be pressed to develop their MCM processes to support integral termination resistors, decoupling capacitors, and integral inductors.

12.13 PACKAGING OF MCMS

The completed, tested MCMs must be placed into a protective housing that delivers power and ground, removes heat, and is physically robust. Until recently, the most common (and almost the only) embodiment of the external package has typically resembled a large single-chip carrier, either metal with metal leads and metal-glass lead seals, or a large ceramic package similar to a leaded chip carrier with wire bond attachment sites around the inner edges of the cavity of the carrier. With a brazed or welded lid, these packages can be hermetic as well.

In an alternative approach to a high-speed package, presently under development in several laboratories, the ceramic support that serves as the base of the MCM is made slightly oversized, with a specialized multilayered seal ring incorporated around the extreme outer edges of the MCM. A ceramic or metal *single layer and metalization* (SLAM) lid, with the appearance of an inverted open sardine can, is then placed over the top of the module and brazed to the metal seal ring. The signal leads exit under the edges of the SLAM lid, because the seal ring itself is actually built up in the following sequence:

1. A layer of metal deposited directly on the ceramic support
2. A layer of inorganic dielectric
3. A layer of interconnect traces
4. A layer of dielectric
5. A layer of metal that serves as the actual metal attachment for the SLAM lid

Because the signal traces are sandwiched between grounded metal layers, they can be controlled impedance transmission lines. Connection between the metal layers in the MCM-D and the metal traces on the ceramic substrate can be made simply with vias from the signal lines (and the power/ground planes) down to the metal lines on the ceramic.

An encapsulation approach that preserves good, though not excellent, signal integrity employs an MCM-L substrate with a ball grid array fabricated on its underside. The solder reflow balls, typically on 50–100-mil area array pitch, are contacted with through vias from the top side of the substrate and, because of their small size, do not cause too much distortion of the propagating signal wavefronts, provided that the signal rise times are not much less than 200 ps. The components on the top side of the package are then protected with an epoxy overmolding, and the entire assembly is reflow soldered to a conventional circuit board. The MCMs are not reworkable, but they are inexpensive. This type of packaging has been used in the author's laboratory up to 100 MHz. It will probably serve satisfactorily for systems whose clock rates are less than ~500 MHz, although the upper limit of this range has yet to be explored in detail.

Perhaps the most intriguing approach to package development has been reported by the chips-first MCM fabricators.[12] Because the chips are in effect completely conformally coated and covered by the overcoating layers of plastic and metal, these fabricators reasoned that an additional layer of metal deposited over the entire upper surface of the MCM and down along its sides would provide an impervious outer package that would reliably encapsulate and protect the MCM without need for an additional outer package (Fig. 12.39). While it was felt that such an encapsulation scheme might not be truly hermetic, it might yield the requisite reliability without the need for hermeticity. In practice, all tests appear to indicate that this packaging approach is very reliable and perhaps even "hermetic," although the definition of hermeticity in such a case is unclear (i.e., because there is no existing internal void volume into or from which gasses and moisture can leak). Packages have been subjected to 1,000 hours of exposure to 135° C temperatures and 85 percent relative humidity, thermally shocked 100 times from –65° C to +150° C, and sprayed with salt water, with no ill effects. At the outer periphery of the carriers described above, a brazed lead frame can be installed to allow the large carrier to be surface mounted to a supporting PC board.

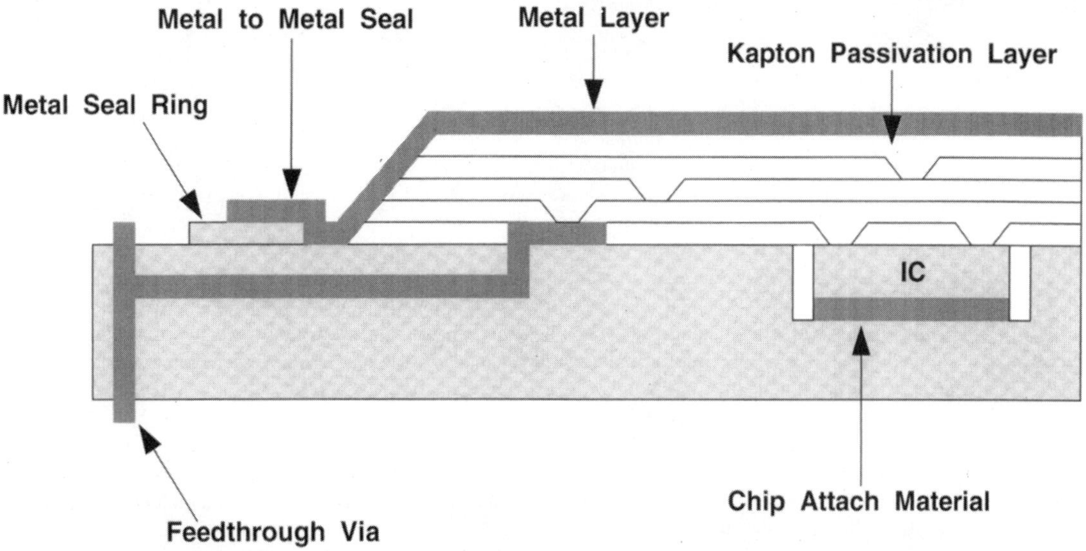

FIGURE 12.39 Technique for incorporating near-hermetic metal lid on chips-first MCM-D. The metal layer is sputter deposited directly onto the uppermost Kapton layer and made continuous with the metal seal ring on the ceramic substrate. (Courtesy of CR&D Center.)

Several companies have recently described sockets for MCM packages; at least one of these sockets purportedly has been designed for high clock rate/fast signal rise time environments. Very recently, various combinations of sockets for MCMs with ball grid arrays on their undersides have been described. Versions of these ball grid sockets with and without intervening pin headers are now available commercially.

12.14 ADDITIONAL REQUIREMENTS FOR FIRST-PASS FUNCTIONALITY: ELECTROMAGNETIC MODELING TOOLS

The approach used for years in the development of PC boards, i.e., "design, fabricate, test, modify, test, redesign, refabricate" does not work well for MCMs of any type, because the nonrecurring engineering charges are sufficiently high to make this approach unattractive. Furthermore, it is typically impossible to install engineering change wires on an MCM, because of the small dimensions of the interconnects and the fragility of the components. In addition, as noted earlier, rules of thumb and even empirical impedance formulae may lose accuracy at very small geometries, requiring the use of direct numerical techniques and CAD tools to derive even such fundamental values as line impedances. Thus, the methods based on the increasingly automated *electronic computer aided design* (ECAD) tools first developed for PC boards must be used for the design of MCMs as well.

The inability to employ ad hoc design methods for high clock rate MCMs is even more severe than for CMOS-based MCMs. The electrical behavior of the interconnects, power and ground planes, and passive and active components at high clock rates and wide signal bandwidths is too difficult to predict by intuition or rules of thumb. Ad hoc methods thus may never allow a design to converge to a fully functional module, no matter how many design and fabrication iterations are attempted. The design team must employ computer software-based modeling tools capable of predicting the electromagnetic behavior of the MCMs before they are committed to fabrication.

12.14.1 Electromagnetic Parameter Extraction

The electromagnetic environment is often counterintuitive and must be treated with great care. The discussion of the electromagnetic environment will be subdivided into a subset of systems that must transmit signals with bandwidths less than 2–5 GHz (referred to as the *quasistatic region*) and those systems that must transmit signals with bandwidths greater than 2–5 GHz (referred to as the *full-wave* case). The mathematics and computer programs used in modeling structures in the quasistatic regime are considerably simpler than those for the full-wave case, and typically run to completion on powerful computer workstations in seconds, minutes, or tens of minutes. Fortunately, most digital MCMs, even those employing present day high-performance commercial silicon and GaAs chips, can be correctly modeled under the assumption of quasistatic behavior only, because many of the *higher-order signal modes* do not propagate in the lower frequency ranges and hence can be omitted from the simulations. As a result, the following paragraphs will discuss only the quasistatic environment. Full-wave modeling tools, which are both comprehensive (i.e., which account for most or all possible higher-order propagating wavefront modes) *and* computationally efficient are currently the subject of very active research.[21a–e]

The quasistatic EM tools can be further subdivided into *parameter extractors* (also referred to as *field solvers*) and *wavefront simulators*. The task of the parameter extractors is to analyze the two-dimensional or three-dimensional structure of a complex metal-dielectric sandwich containing multiple signal nets sharing two or more signal layers, with the signal layers either shielded or not shielded from one another by intervening ground planes. The input to the parameter extractor is typically a cross-sectional diagram of the metal/dielectric sandwich (Fig. 12.40) and a plan view of the interconnect structures (or the equivalent CAD layout file).

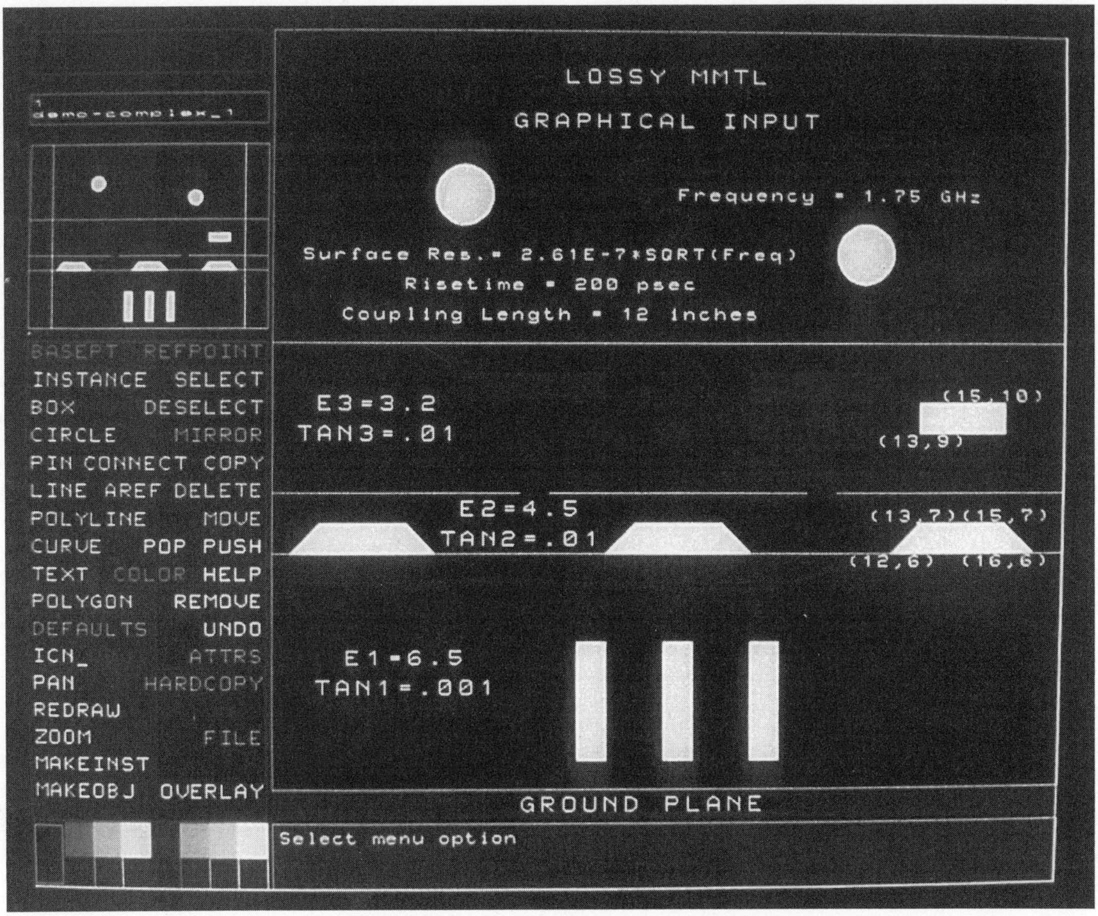

FIGURE 12.40 Photo of workstation screen, a drawing of an example MCM cross section. Geometric data from the diagram is extracted by a special compiler that generates an input data set for an electromagnetic parameter extraction (field solver) modeling tool. Conductors can be immersed in different dielectrics.

The output of the parameter extractor is a numerical characterization, in matrix representation, of the electromagnetic interactions between all of the various interconnects and ground and power planes. Included in this list are:

1. The inductance and capacitance (per unit length) between all signal lines and the AC ground planes (i.e., the *self*-inductance and capacitance values, which appear as the diagonal element values in the inductance and capacitance matrices for the structure)
2. The inductance and capacitance per unit length between all signal lines and all other signal lines (the *mutual* capacitances and inductances, represented as the off-diagonal element values in the inductance and capacitance matrices)
3. The behavior of the individual lines, including their DC and AC resistive losses (referred to as the *resistance matrix* of the interconnect system)
4. The signal propagation velocities of each interconnect

5. The effective relative dielectric constant (the effective ε_r value) observed by each interconnect
6. The energy leakage (if any) from each line through the dielectric to the ground planes (referred to as the *conductance matrix* of the structure)
7. The capacitance that occurs between signal lines on different layers when they cross over or under one another (the crossover capacitance)[21d]

If some of these values are dependent functions of frequency (e.g., the resistances of the signal lines or the dielectric losses) then the elements in the matrices become functions of frequency rather than constant values (again, stated per unit length).

While the static parameter data does not by itself allow an estimate of the conformation (shape) of the propagating wavefronts, it can be extremely useful in its own right. For example, by the mid 1980s, reasonably accurate parameter extractors were available, but no really accurate wavefront modelers were yet in existence. The waveform modelers of the time were based on the so-called *lattice diagram,* which employs a concept of the reflection coefficient as a measure of the amplitude of a wavefront reflection at an impedance discontinuity on a transmission line.[1] Although better than nothing, lattice diagrams were found not to be accurate in many instances.[3] However, using the parameter extraction data, it was possible to estimate line impedances, reflections, and, most important, interline crosstalk (Fig. 12.41), which in turn made it possible to design ceramic single-chip carriers[4] that performed essentially as predicted by the simulation results.

These extractors have continued to undergo refinement, and the majority are now equal to the task of analyzing the behavior of PC boards up to system clock rates of at least several hundred megahertz. However, these tools may or may not work correctly when applied to the dimensions and structural details typical of multichip modules. For example, the numerical techniques employed by the extractors must be able to account correctly for the very small absolute dimensions and small cross-sectional radii of curvature of the interconnects and vias of MCMs, which are much smaller than those typically found in circuit boards. Slight differences in the way that the numerical solutions are implemented can make a large difference in accuracy.[22–24] For example, as depicted in Fig. 12.19, metal line cross sections in MCM structures are typically not rectangular but trapezoidal in shape or distorted in other ways. The impedance of a line with a trapezoidal cross section may be different by 10–20 percent from that of a line with a rectangular cross section. To achieve maximum fidelity in the simulations, the design team must account for as many structural irregularities as possible. The design team must thus assure themselves that the field solver that they select is sufficiently flexible and powerful to account correctly for these types of irregularly shaped structures and very small dimensions.

The extractors should also be able to compensate for the fact that the dielectric layers in MCMs cannot always be assumed to be perfectly planar, because the dielectric layers, whether applied in liquid or sheet form, typically conform to the metal features beneath them. Figure 12.42, taken from the parameter extractor developed in the laboratory of the author, demonstrates that, if the parameter extractor is equal to the task, nonplanar interfaces can be correctly accounted for by building up slabs and layers of dielectric and metal with any shapes necessary to approximate the nonplanar dielectric-metal-dielectric interfaces, or with tools that can directly analyze the shapes of the dielectric layers as they flow over the conductors.

Even in the quasistatic regime, which has been studied intensively for many years, some issues still issues are not correctly handled by *any* of the field solvers, because satisfactory theoretical solutions have not been achieved. Perhaps the most well known of these is the problem of AC resistive losses in the small metal interconnects on the MCMs. The high-frequency and high clock rate design community has assumed for years that the frequency-dependent resistive losses in a metal interconnect were adequately treated through the definition of the so-called skin effect[25–26] in which, at high frequencies, the current in a conductor is assumed to flow only near the surface or "skin" of the conductor, with a current density that decreases exponentially with depth into the conductor. The skin depth, δ, is defined as the distance into the conductor at which the current density has decreased to $1/e$, or roughly 37 percent, of its density at the surface of the conductor. The tacit assumption is that, at three or more skin depths into the conductor, current density has decreased to essentially zero. The skin depth δ is a function of several independent variables, including frequency, as defined by:

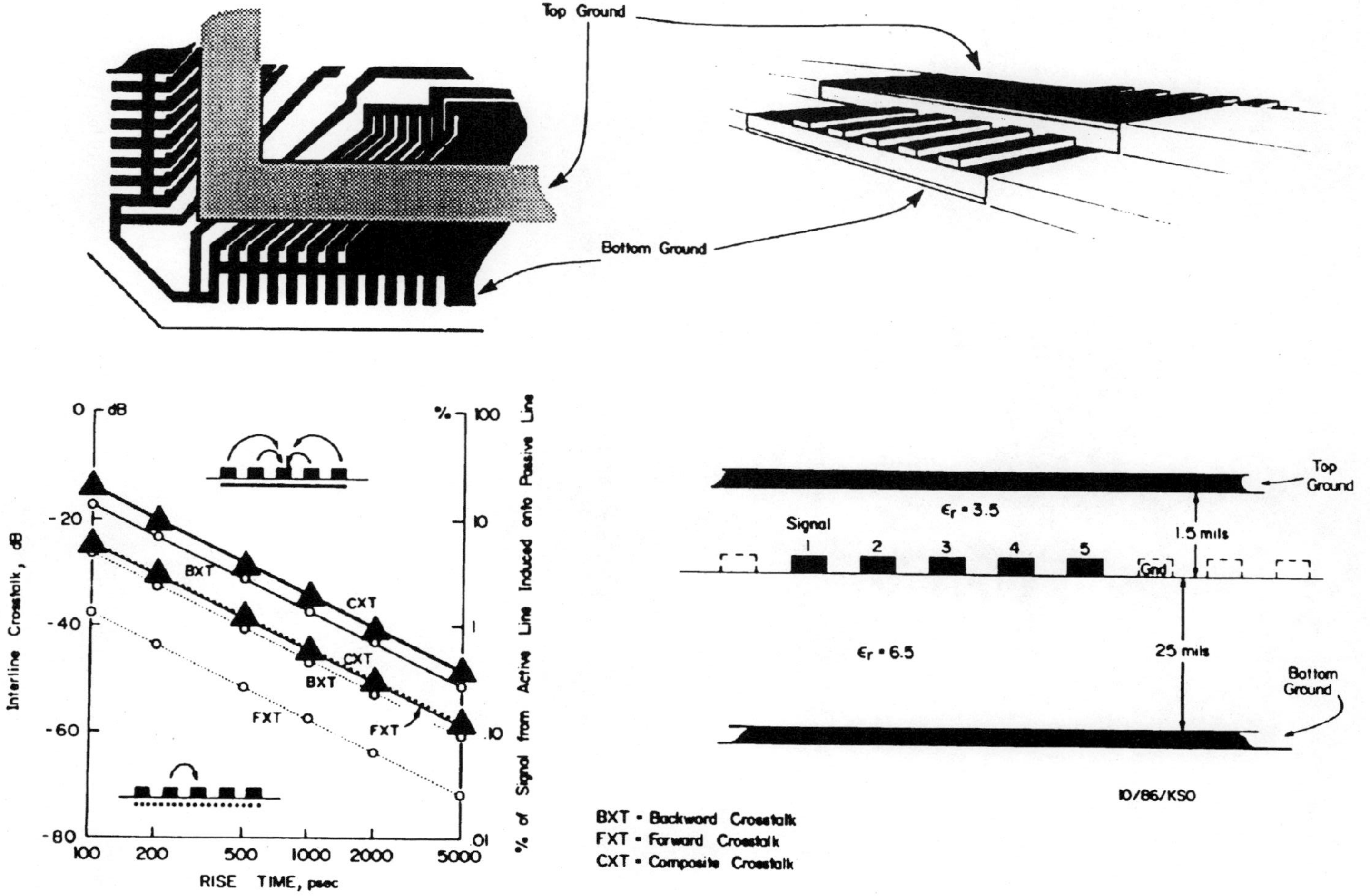

FIGURE 12.41 Illustration that output data from a field solver EM tool generates useful results, even without waveform simulation. Line impedance, propagation speed, and crosstalk (forward, backward, and composite) as a function of signal rise time were used to design a high-performance single-chip ceramic carrier as early as the mid-1980s.

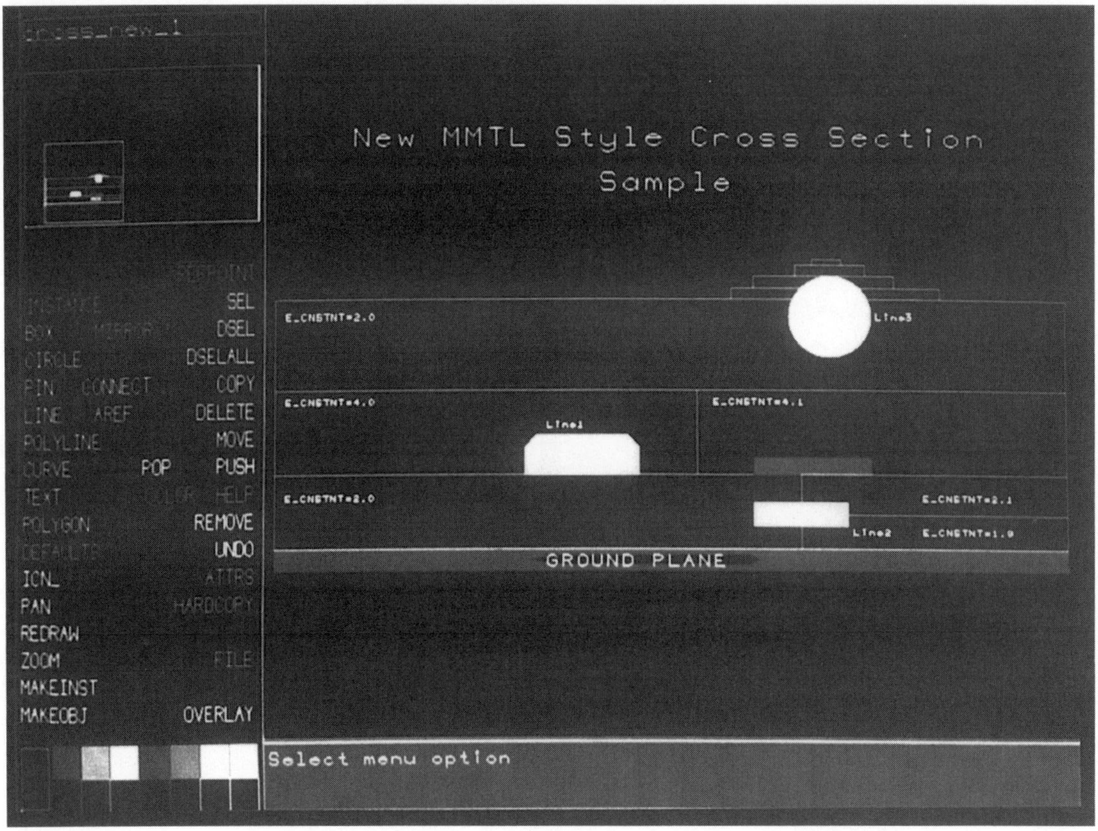

FIGURE 12.42 Photo of workstation screen, a diagram of an example MCM cross section used as input data for an advanced field solver that can approximate the real-world nonplanar flow of dielectrics over metal conductors and dielectrics of different geometrics.

$$\delta = \frac{1}{-(\pi\mu_r\mu_0 sf)} \tag{12.3}$$

where μ_o = permeability of free space, equal to $4\pi \times 10^{-7}$ Henrys/meter
μ_r = relative permeability of the conductor (the product of these two variables is simply referred to as the permeability of the metal, μ)
s = the conductivity of the conductor
f = the frequency of interest in hertz

With the current flowing only near the surface, another dependent variable, the surface resistance, R_s, can also be defined as the equivalent DC resistance of a line whose thickness is equal to the skin depth. In equation form, the expression for surface resistance is:

$$R_s = \sqrt{\frac{\pi f\mu}{s}} \tag{12.4}$$

Equations 12.3 and 12.4 were derived decades ago under the assumption that current flow is in a pair of opposing solid metal slabs of semi-infinite dimensions.[26] As pointed out in Ref. 25, the assumptions on which Eqs. 12.3 and 12.4 were derived hold only as long as the depth (thickness) of the conductor and any radii of curvature of the conductor are infinitely large in comparison to the skin depth. Table 12.2 presents the skin depths for a variety of metals at several different frequencies of interest.

While the skin effect assumptions are valid for signal lines of dimensions typical of PC boards, *they are frequently not true for signal lines found on MCMs*, because the thicknesses and/or the radii of curvature of the conductor are not much greater than the skin depth of that particular metal at the frequency of interest. For example, the corners of MCM conductors have very small radii of curvature, and the thickness of the conductors (~5 μm) is not considerably greater than the skin depth (0.5–2.5 μm for frequencies up to 20 GHz). The general situation is illustrated by the simulation of Fig. 12.20 (which, the careful reader will note, actually assumed interconnect dimensions and frequencies typical of circuit boards, not MCMs), which depicts the current flow in the following cases:

1. The very low frequency case, in which the current flow is uniform across the entire cross section of the interconnect (upper panel)
2. The skin effect limited case, in which current flow is confined to a thin layer on the surface of the conductor (lower panel)
3. The intermediate case, typical of many MCM-based systems, in which the current flows throughout the cross section of the conductor, but not uniformly so (middle panel of Fig. 12.20)

Thus at "intermediate" frequencies, the total resistance of the conductor will be lower than that predicted by Eq. 12.4, and the inductance of the conductor will be higher than predicted by assuming that all magnetic fields are completely external to the conductor.

An heuristic explanation for the decrease in total conductor inductance and increase in total conductor resistance with increasing frequency is embodied in Fig. 12.43, which shows two parallel conductors (as well as the underlying ground plane, which serves as a return path for the current flow), each subdivided into "deep" and "surface" filaments. At low frequencies, all filaments carry current, and each filament contributes to the total inductance of the conductor. At extremely high frequencies dominated by skin effect, no current flows in the deep filaments. As a result, these deep filaments do not contribute to the total inductance of the conductor. At intermediate frequencies, the deep filaments carry some current and thus contribute intermediate amounts of inductance to the total for that conductor. Conversely, when current flows in the deeper filaments of the conductor, the AC resistance of the conductor decreases, because the effective cross-sectional area of the conductor increases.

The total inductance of a conductor for a particular metal and cross section typical of MCM dimensions, over a range of frequencies, is depicted in Fig. 12.44. At low frequencies, the inductance of the conductor is predicted well by assuming that current flows uniformly through the entire cross

Table 12.2 Skin Depth in Various Metals at Selected Frequencies

			Skin depth					
Metal	ρ (μΩ-cm)	R_s (Ω/sq) $1{\times}10^{-7}\sqrt{F}$	δ (1 GHz)	δ (5 GHz)	δ (10 G Hz)	δ (15 G Hz)	δ (20 GHz)	δ (25 GHz)
Aluminum	2.65	3.23	2.59	1.16	0.819	0.669	0.579	0.518
Chromium	12.9	7.14	5.71	2.56	1.81	1.48	1.28	1.14
Copper	1.67	2.57	2.05	0.920	0.650	0.531	0.460	0.411
Gold	2.35	3.05	2.43	1.09	0.772	0.630	0.546	0.488
Nichrome	100	19.9	15.9	7.12	5.03	4.11	3.56	3.18
Nickel	6.84	5.20	4.16	1.86	1.32	1.07	0.931	0.832
Silver	1.59	2.51	2.01	0.897	0.635	0.518	0.449	0.401
Titanium	42.9	12.9	10.3	4.61	3.26	2.66	2.31	2.06

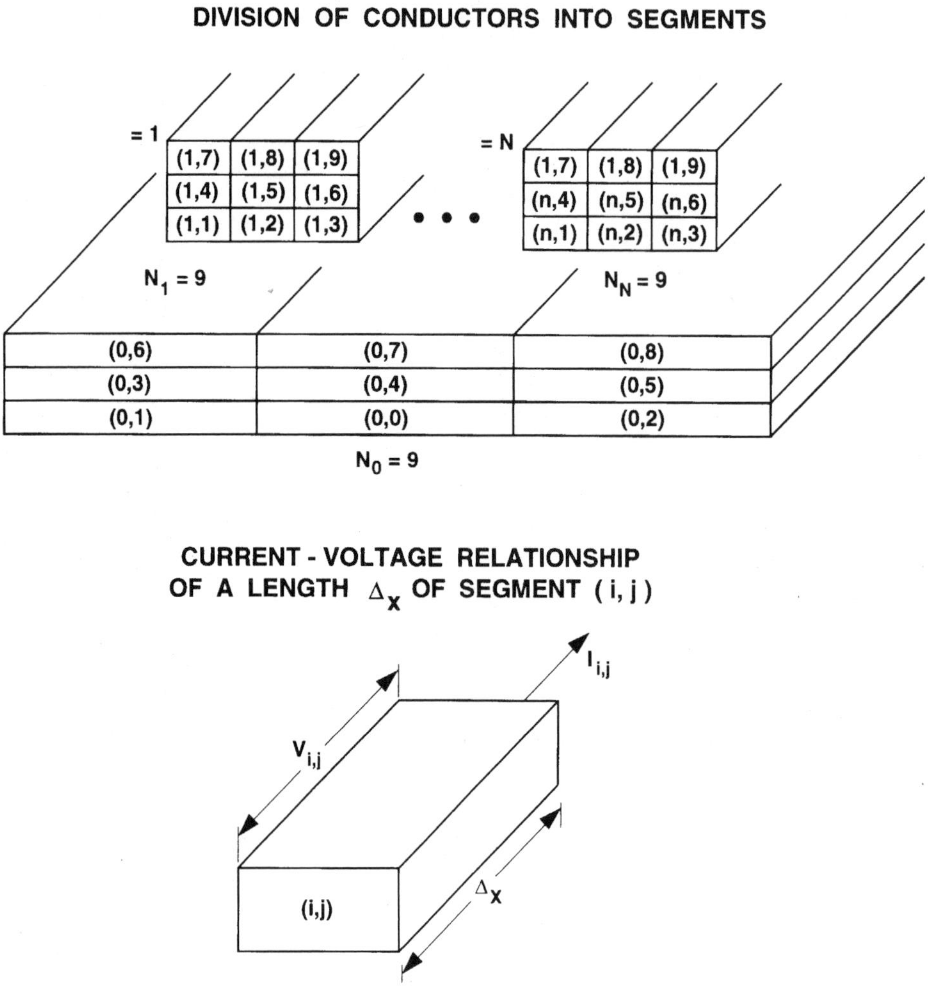

FIGURE 12.43 Diagrammatic basis for heuristic explanation of internal inductance. At low frequencies, current flows uniformly through all filaments in conductors, and wire inductance is greatest. At highest frequencies, because current flows only in surface filaments (skin effect), wire inductance is lowest. At intermediate frequencies, because some current flows in internal filaments, wire inductance value is intermediate.

section of the conductor. At the highest frequencies, the inductance is determined by the current flowing only in the outer filaments of the conductor and hence is at its lowest value; in the intermediate frequencies, skin effect is only partial, since some current does flow in the deepest filaments. Because of the small dimensions of MCM interconnects, we often find that the circuit is operating in the intermediate range of the curve in Fig. 12.44, not in the frequency range in which the skin effect dominates. In a circuit board at the same frequency, the interconnects would probably (because of their larger dimensions) be operating in a range in which the skin effect assumptions are valid.

As a result of the above considerations, the calculated impedance of the interconnect as a function of frequency, when compared to actual measurements from test coupons, has been found to be in error by as much as 10–20 percent, i.e., sufficiently large to cause signal reflections on the interconnect and degradations in system noise margin. Some commercially available field solvers do try to

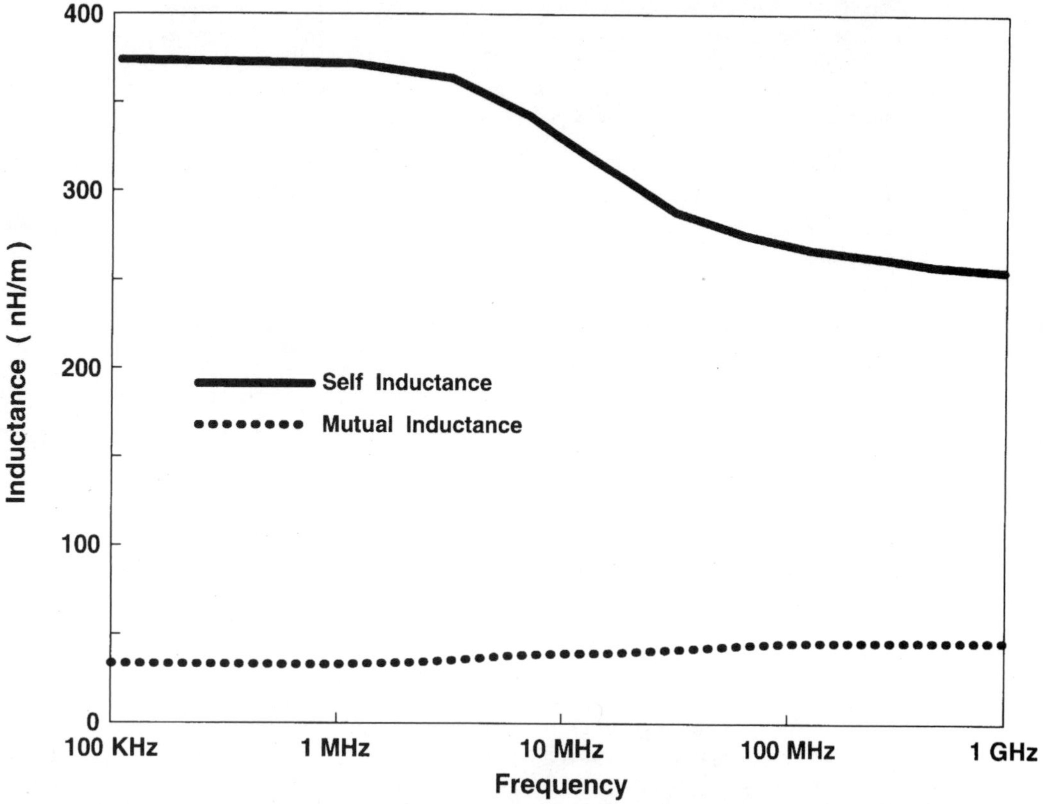

FIGURE 12.44 Computer simulation of frequency-dependent inductance of an MCM copper microstrip. Highest inductance is at low frequencies, and lowest inductance (skin effect case) is at highest frequencies, with a smooth transition at intermediate frequencies. Simulation was conducted with the modified Weeks' method, which is now known to provide erroneous results except in the single-line case illustrated here. Most electromagnetic modeling tools estimate total line inductance in the transition region (sloped center portion of upper curve), but no completely accurate equation exists for this region in the case of multiple adjacent signal lines.

deal adequately with some of the underlying phenomena, such as the effects of sharp-radius corners on small conductors. Several packages also employ an approximation of the line resistance for the intermediate-frequency region between the purely DC resistive case, and the high-frequency regime where the line resistance can be assumed to be dominated by the skin effect (as represented by Eq. 12.4). However, the internal inductance problem is still an open issue. Several research groups are attempting to develop a comprehensive method of calculating correctly the total inductance as a function of frequency, and some progress has been made.[27] However, at the time this text was proofread, those few solvers that do appear to account correctly for the inductive and resistive effects in conductors that violate the classical skin effect assumptions are very computation intensive and may execute slowly. Design teams should be aware that their field solvers may very well be giving incorrect values for some of these variables.

12.14.2 Simulation of Propagating Signal Wavefronts

The second capability in the EM modeling armamentarium that is really required to perform high-frequency or high clock rate MCM designs is wavefront simulation. Ultimately, it is the absolute magnitude and shape of the propagating signal wavefront sensed by each line receiver on a signal net (and the stability of the DC supply voltages on the power and ground planes, as described below) that determines the line receiver's ability to make a correct decision as to whether a logic high or low signal is being received. An accurate wavefront simulator must thus be able to account correctly and simultaneously for the most complex conditions to be expected in a real-world MCM:

1. Signal nets with multiple destinations and multiple signal sources (i.e., wired-OR or wired-AND logic)
2. Stubs, branches, or vias on the line or improperly terminated lines which, in turn, cause signal reflections
3. Crosstalk between adjacent, tightly packed signal nets
4. Reflections caused by crosstalk from adjacent lines and crosstalk caused by reflections on adjacent lines

Wavefront simulators that do in fact account for all of these issues using closed form solutions (typically in the frequency domain), at least within the quasistatic regime, have been developed by nonprofit organizations[24] and are gradually becoming available. For example, Fig. 12.45 shows results from a wavefront simulation study using the tool described in Ref. 28 of a PC board and a unique type of single-chip package intended for use at signal rates approaching 1 GHz. The study, which yielded results that were initially counterintuitive but turned out to be correct, demonstrated that the single-chip package simply was not up to the task of handling these fast wavefronts correctly. As a result, the packaging was redesigned to accommodate the fast wavefronts.

12.14.3 Modeling Noise Disturbances in Power and Ground Planes

As noted briefly above, the logic gates of digital chips rely not only on the amplitude and conformation of the digital signals on their interconnects to make correct decisions regarding the logic level of an incoming signal; the gates also rely on the reference voltages provided by the power and ground planes. If these voltages are instantaneously unstable because of noise injected into the planes by the pulsatile current drain caused by the chips themselves as they switch internal state following the leading edge of each clock pulse, the gates may make incorrect decisions regarding the state of their incoming signals even if the signals themselves are "clean." Thus, the ability to model the power and ground plane behavior of an MCM before it is fabricated has been a priority among high clock rate systems designers since the late 1960s. However, because the complexity of this modeling problem is extremely high (for reasons beyond the scope of this chapter), little progress was made in the development of such an algorithm for more than two decades, and this impasse continued until recently.

Figure 12.23 depicts results from a recently developed power and ground plane modeling tool that serves as a beginning for this type of capability. The figure depicts the behavior of a small PC board of dimensions 6.5 × 6.5 in, containing one power plane and one ground plane, when subjected to a periodic sequence of current transients each of 375 mA and 1 ns rise/fall times, injected across the power and ground planes at a single location on the board. The upper panels depict the voltage instability at eight different locations (and different from the current injection point) on the power plane, and the lower panels depicts the instantaneous voltage behavior at the eight matching locations on the ground plane.

Note that the ground plane does not remain at actual "ground" potential, an effect referred to earlier as *ground bounce.* In fact, the voltage on the power plane decreases at the same instant that the voltage at the matching location on the ground plane increases, a combination of instantaneous events referred to earlier as *rail collapse.* From left to right, the three columns of panels depict the short (9 ns), intermediate (40 ns), and long period (600 ns) behavior of the planes at these eight power/ground locations. Even though only a single disturbance location is modeled here, the panels show the complexity of the behavior of the planes: they behave as three dimensional high-Q circuits,

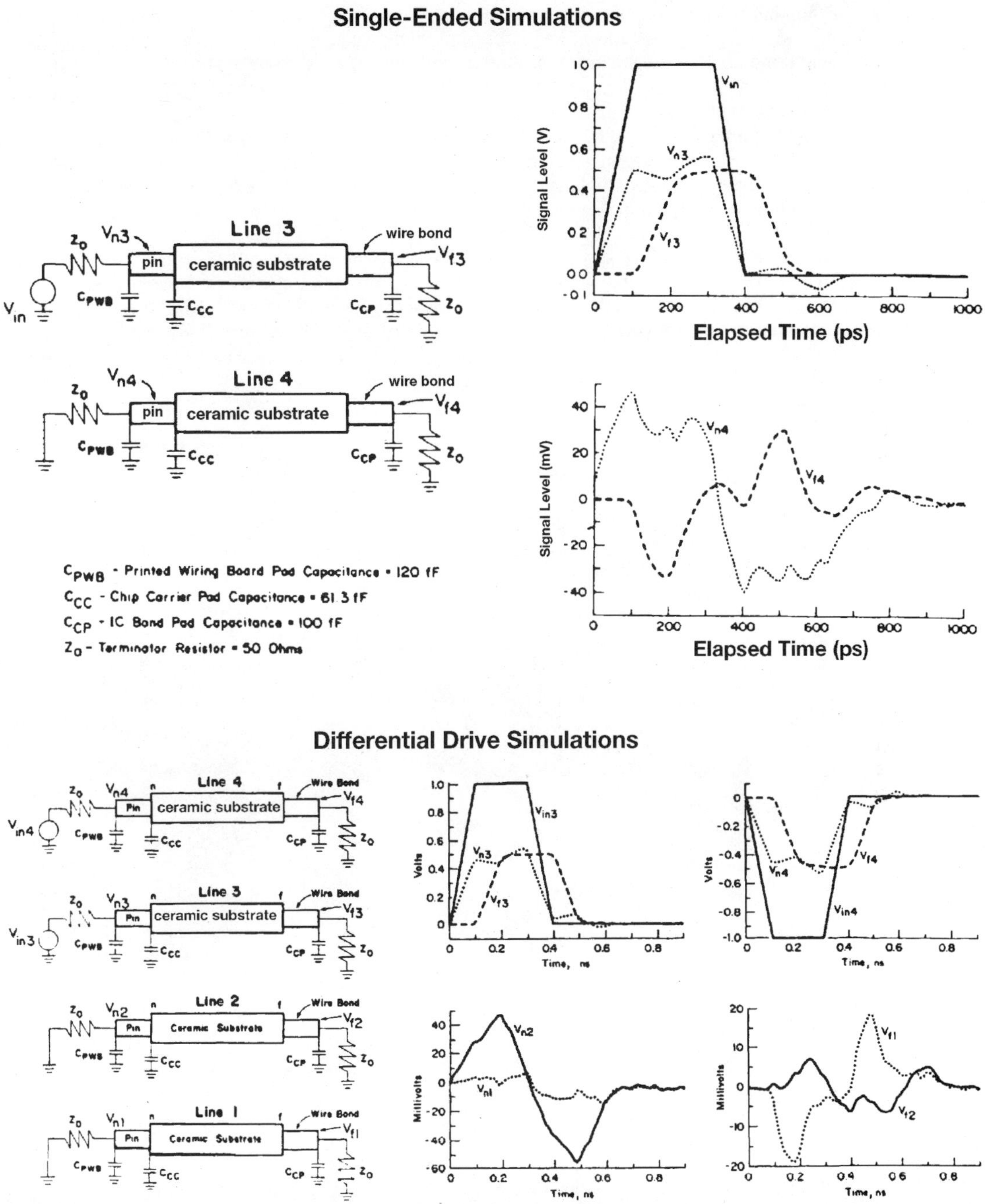

FIGURE 12.45 Electromagnetic simulation results of waveform propagation in a complex PC board/chip carrier structure for fast rise time signals, including reflection and crosstalk effects. The ability to simulate wavefront propagation in dense wiring meshes allows designers to understand the EM environment of high clock rate systems.

exhibiting voltage oscillations in space and time. The initial disturbances within the first 5–10 ns are the propagating voltage transients as they reflect from the edges of the planes at the edge of the board. In the intermediate and long term, the voltage behavior becomes more regular as the energy pumped into the planes excites their resonant behavior, with oscillations continuing for many tens of nanoseconds or longer.

In this simulation, the duration of the ringing is exaggerated, because the only damping was provided by the series resistance of the current source. In an actual MCM, termination resistors and other resistive parasitics would suppress the ringing in a much shorter time. However, also in the real world, every power and ground contact, from every chip location, represents a disturbance location, of which there are likely to be dozens on a given MCM. Some are in temporal phase with one another, and some are out of phase (because of the distances between the various excitation points, slight differences in excitation event times, and the propagation speed of the wavefronts). Thus, the disturbance amplitudes and phases will vary at each location on the MCM and over time. It is important for the designer to be able to study the *voltage amplitude noise surface* of a given MCM (an early version of such a noise surface appears in Fig. 12.46) to make sure that voltage disturbance

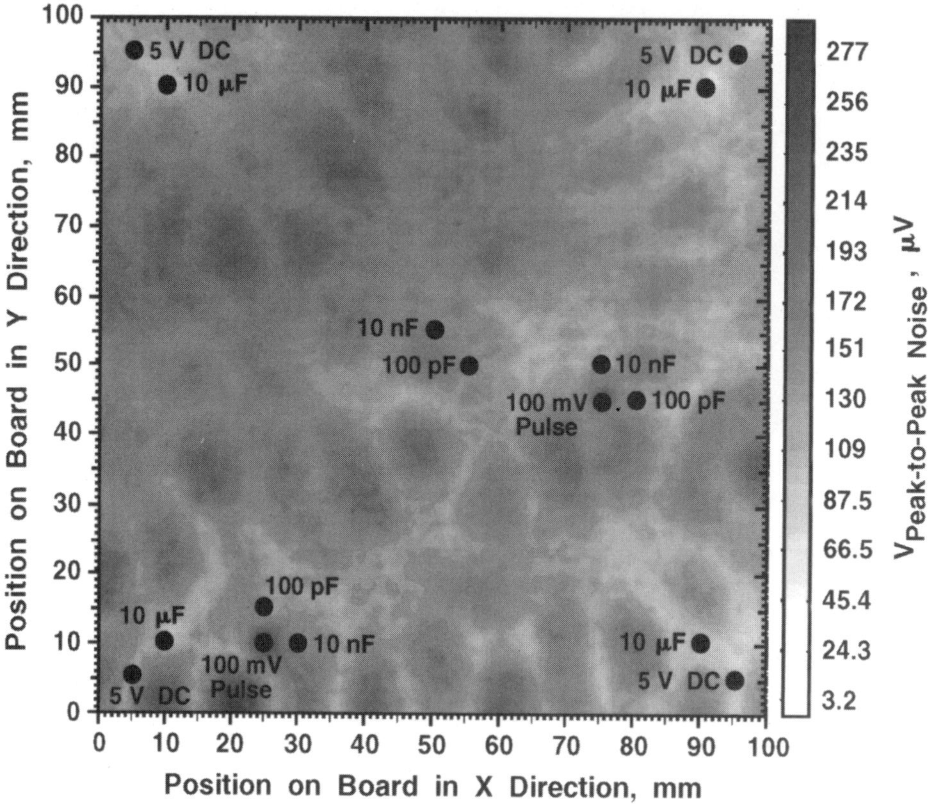

FIGURE 12.46 "Noise surface" simulation of 100 × 100 mm multilayer circuit board containing several power supplies, decoupling capacitors, and 100 mV pulsed noise sources (representing switching chips). In this simulation, using new tools developed in the author's laboratory, the worst-case peak-to-peak disturbance regions on the power plane are a very low 0.2 mV. Note the very "cool" regions surrounding decoupling capacitors. Typical circuit board operating parameters, with many more chips and much larger switching transient amplitudes caused by high gate count CMOS chips, would exhibit much worse peak-to-peak disturbances.

"hot spots" do not occur, and if they do, to be able to change the layout or design to suppress them. The complexity of the power/ground plane structure to be modeled, and the large number of disparate events that result in the final noise behavior of the MCM, significantly complicates this particular modeling problem. Clearly, the modeling of power and ground plane behavior is in its infancy, but the urgent need for such a tool will drive the EM modeling community to create better methods of carrying out the simulations. With faster, larger, higher-current, noisier chips being deployed on an increasingly frequent basis, this type of modeling tool is becoming available to designers just in time.

12.14.4 Has Your EM Modeling Tool Been Validated?

One final cautionary note is in order regarding EM modeling tools. Although many EM modeling tool suites have become commercially available over the past few years, it is not clear that all of these tools work as advertised. This statement is based on 17 years of development of EM modeling tools in the author's laboratory. The problem may best be summarized by "Murphy's Law," namely, "Anything that can possibly go wrong, will!" An EM modeling tool must be verified for correctness at every stage of its development, a complex task involving many facets:

1. As in the case of the skin effect and internal inductance for small cross section conductors, the fundamental mathematics may not be available to model a particular phenomenon correctly.
2. The mathematical modeling formulae must be verified for correct derivation. Countless examples exist in the literature of incorrect mathematical derivations that, when implemented in computer code, gave incorrect answers for structures with known analytic solutions.[29]
3. The computer code that implements the mathematics must be validated against actual measured results from passive and active test coupons (yet another reason why test coupons are so valuable to the advancement of the MCM technology). Such coding errors can be subtle and may not reveal themselves until the code has been validated against many types of interconnect structures.
4. The test coupons used to validate the EM mathematics and software may be incorrectly designed, or processed differently from that expected by the test coupon designers, so that the modeled structure is not the same as the fabricated, tested structure. For example, if the interconnect lines are fabricated with a very thick layer of barrier metal (Fig. 12.19), but the simulations assume a thin layer of barrier metal, the measured and simulated results will not agree, because the skin effect losses will be higher in the measured results.

In the types of cases described above, months of diligent work are often required to identify and correct the disparities. The problem becomes even more difficult to unravel when two or more anomalies occur simultaneously, such as a test coupon with abnormally high-resistance metal and a simultaneous error in the parameter extraction code resulting in incorrect frequency-dependent line resistance values.

A painful lesson that has been learned from numerous episodes of identifying and correcting these types of anomalies in EM modeling algorithms and CAD tools is that the design team should be *absolutely certain* that the EM modeling tools they select *have been validated against real-world structures under controlled laboratory measurement conditions.* EM modeling tools are particularly difficult to validate because, as noted above, unlike a chip router tool whose incorrect behavior is immediately evident (i.e., the signal net routes are obviously connected incorrectly or not at all), errors in EM modeling tools are usually subtle and difficult to diagnose. Nonetheless, if the EM modeling tools have not been validated in this way, and unless the proof of such validation is forthcoming, the design team should either select a different tool suite or work with the EM tool vendor to undertake the validation process.

12.15 THE FUTURE: MIXED-SIGNAL MULTICHIP MODULES

With the high performance of MCMs clearly demonstrated for digital integrated circuits, and with the steps for their high clock rate performance becoming better understood,[30] the next expected step in development is a requirement to package high clock rate digital chips on the same MCM with microwave or millimeter wave analog integrated circuits. The reasons for interest in this type of structure derive from the fact that, in an increasing number of systems (both military and civilian), the data of interest may initially be transmitted and received as analog signals (e.g., portable phones, cellular telephones, high-definition television, military radio, aircraft- and ground-based radar systems, global positioning system (GPS) receivers, and so on). Then, after reception (or before transmission, or both), the incoming data needs to be amplified by, e.g., a low-noise amplifier from the very low levels received by an antenna, followed by analog or digital demodulation, followed by some type of digital processing. Examples of this type of subsystem include an A/D converter and demultiplexer operating at gigahertz sampling rates, direct digital synthesizers (DDSs), GPS receivers, entire digital wireless radio subsystems, digital wireless modems such as embodied in the new *cellular digital packet data* (CDPD) standard, and transceivers for *synchronous optical network* (SONET) fiber links. In such cases, an attractive packaging approach is the incorporation of both analog and digital chips on the same MCM to decrease transmission losses and decrease package size.

This "mixed signal" packaging approach now appears much more feasible since the demonstrations during the past few years that MCMs can support wide bandwidth (digital) signals. However, just as was necessary to achieve high-speed digital operation, MCMs will have to demonstrate several extended or additional features before the mixed signal capability will be completely validated. Two principal issues that must be resolved are extremely low signal amplitude loss and extremely good signal shielding and power and ground plane isolation. A third, although second-order issue, is the need to provide integral inductors for the analog circuits.

First, because analog signals entering an MCM may be at levels of –60 to –130 dBm prior to amplification, losses in the interconnects must be extremely low, and VSWR values must be extremely well controlled, although perhaps only for centimeter-length interconnect runs. Second, low-noise analog amplifiers need quiet electrical environments in which to operate. However, digital chips tend to generate high-level signals that generate crosstalk in adjacent interconnect lines and also tend to inject noise transients into the power and ground delivery systems. Thus, MCMs will have to be extended to provide very low loss conductors and really excellent shielding between different chips and different signals. Isolation levels greater than 60–70 dB will almost certainly be required.

Third, it will be very advantageous if integral inductors can be incorporated directly into the MCMs as depicted in Fig. 12.47. Present discrete inductors require considerable surface real estate on those few mixed-mode MCMs that have been fabricated so far. Several MCM fabricators are developing small families of spiral inductors, often spanning more than one metal layer, that can be used directly from the design libraries in the CAD system used to lay out the MCMs. The ultimate challenge will be to incorporate high-permeability materials into the spiral structures to increase the values of inductance and Q that can be achieved in a small volume within the MCM layers, ultimately yielding integral inductors with performance characteristics factors even more favorable than those which have recently been demonstrated (Fig. 12.48).

As an early demonstration that mixed mode MCMs will become feasible, Fig. 12.49 depicts a microminiaturized six-channel GPS receiver implemented on one double-sided MCM-L with dimensions 1.37 × 1.43 inches. The analog and digital circuits were not constrained to be on opposite sides of the MCM; they were simply placed where needed, with careful shielding techniques used to isolate critical lines and power/ground planes. Connections were made between components on the two sides, employing the "through" vias typical of the MCM-L process. The receivers operate correctly, yielding position data from the 1.6 GHz L1 channels of all GPS satellites visible at the time of any given test.[31] It is of interest to note that of the total of approximately 150 components installed on the MCM, only 11 were active devices, and all the rest were passive components. The ability to integrate

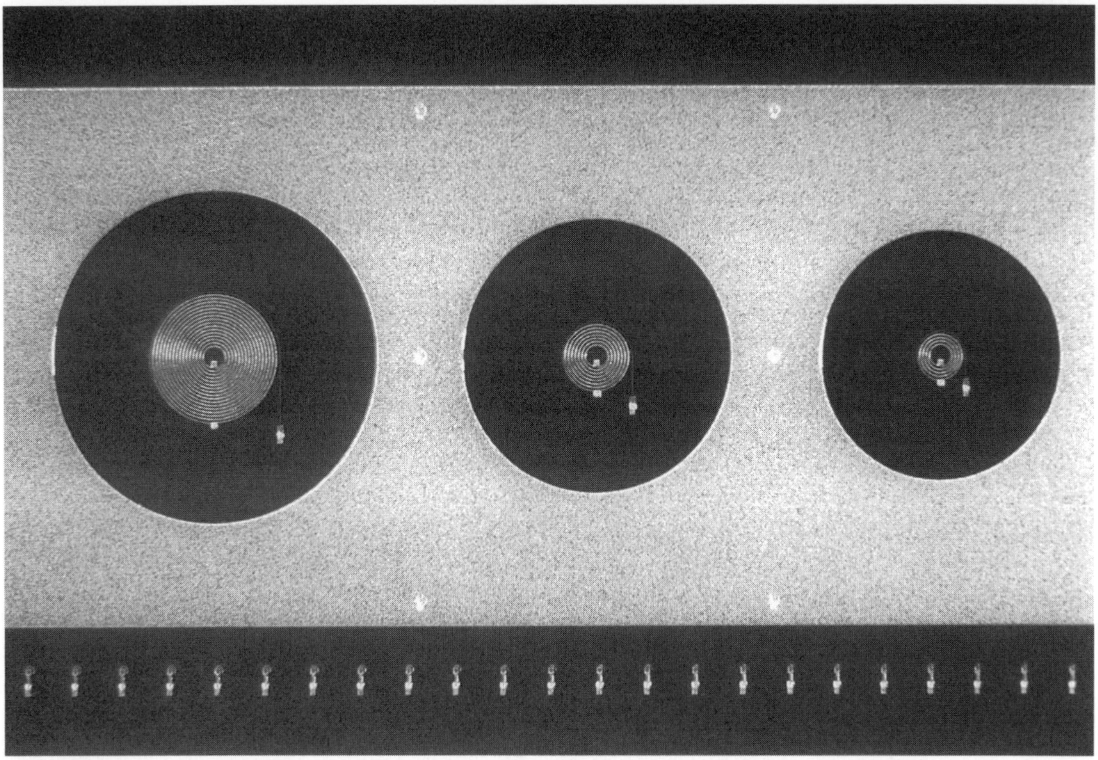

FIGURE 12.47 Photo of small section of an experimental organic MCM-D, showing three spiral inductors incorporated into microstrip layer. (Fabricated by Hughes Aircraft.)

a variety of passive components into an MCM substrate would considerably simplify the assembly and hence the cost of such a module and would have allowed the module size to shrink to less than one inch square.

As a second example of mixed-signal MCMs requiring very high levels of attention to design detail, Fig. 12.29 is a photomicrograph of an MCM-L containing a 3 Gsamp/s, 8-bit (plus one out-of-range bit) GaAs A/D converter, and a matching 9:72 demultiplexer. The package was required to provide a stable electrical environment for the A/D converter chip without allowing noise generated by the demultiplexer to disturb the converter. Because the best of these A/D converter chips is capable of providing approximately 42 dB of dynamic range, the MCM must not add contamination to the incoming analog signals to any greater degree. At the time of this writing, the first chips tested in these modules have been demonstrating front-to-back (i.e., from all chip and package effects) dynamic range of 36 dB. This early result does indicate that MCMs can provide good packaging for these types of chip sets.

A few successful proof-of-concept projects such as the two described here will raise the confidence of the design community in their ability to design and test high-performance mixed-signal subsystems. However, only with further experience will it become clear if present day MCMs can provide the requisite high-quality electrical environment. As the technology of the chips continues to improve, mixed-signal systems will provide design and fabrication challenges for MCMs for years to come. Mixed-mode MCMs will surely result in the creation of entirely new electronic systems, in turn revolutionizing the power and sophistication of electronic communications for the military and civilian worlds alike in the late 1990s and beyond.

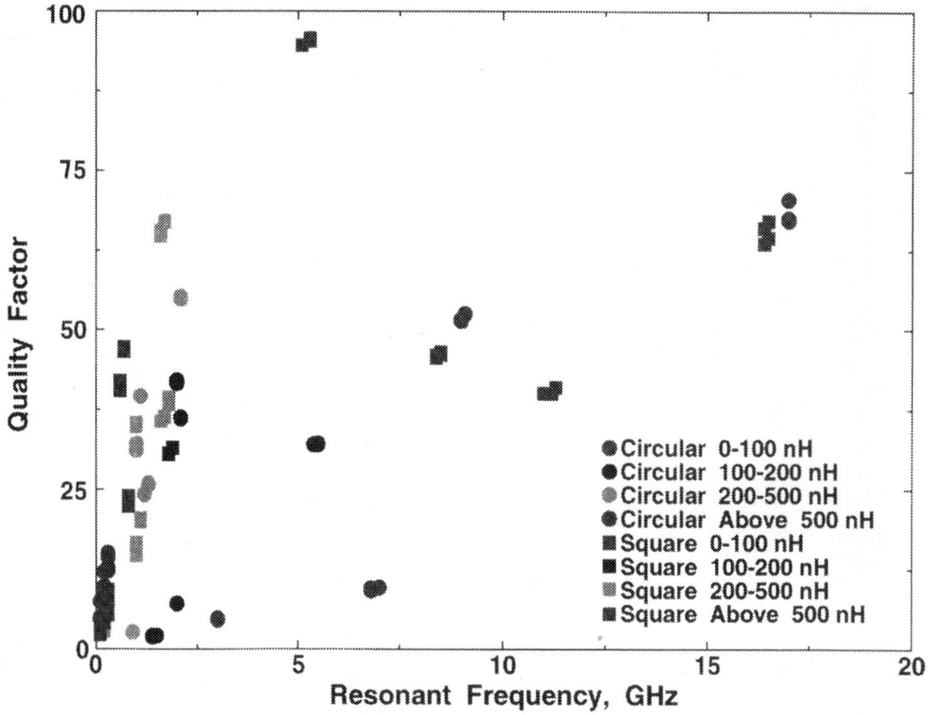

FIGURE 12.48 Inductance values and Q factors measured from a variety of single-layer and multilayer spiral inductors fabricated in a typical organic MCM-D process. Note that many inductors exhibit Q values above 25 and resonant frequencies from 100 MHz to beyond 15 GHz. The high inductance and Q values are often achieved by stacking up four layers of spirals. (Hughes HDA3 five metal layer process, aluminum metallization, polyimide dielectric, $\varepsilon_r = 3.2$, inductor layer count 1–4, 4–96 total turns, 10–25 μm line widths, 20–40 μm spacing. Measured with HP4285A LCR meter equipped with Micromanipulator 44D probes.)

ACKNOWLEDGMENTS

The author wishes to thank the staff of the Special Purpose Processor Development Group, Mayo Foundation, for development of the MCMs and test results presented in this chapter; the Mayo team and Dr. G.-W. Pan and coworkers, Arizona State University, for the EM modeling algorithms and CAD tools described herein; and particularly W.L. Walters of Mayo for proofreading and helpful suggestions; and to E. Doherty, D. Jensen, L. Mason, and S. Richardson for preparation of text and figures. Special thanks to General Electric Corporate Research and Development Center, Honeywell, Texas Instruments, nCHIP, Georgia Tech, Hughes Aircraft, and 3M Corporation for support in the collaborative projects that resulted in the test coupons and active MCMs described in this chapter.

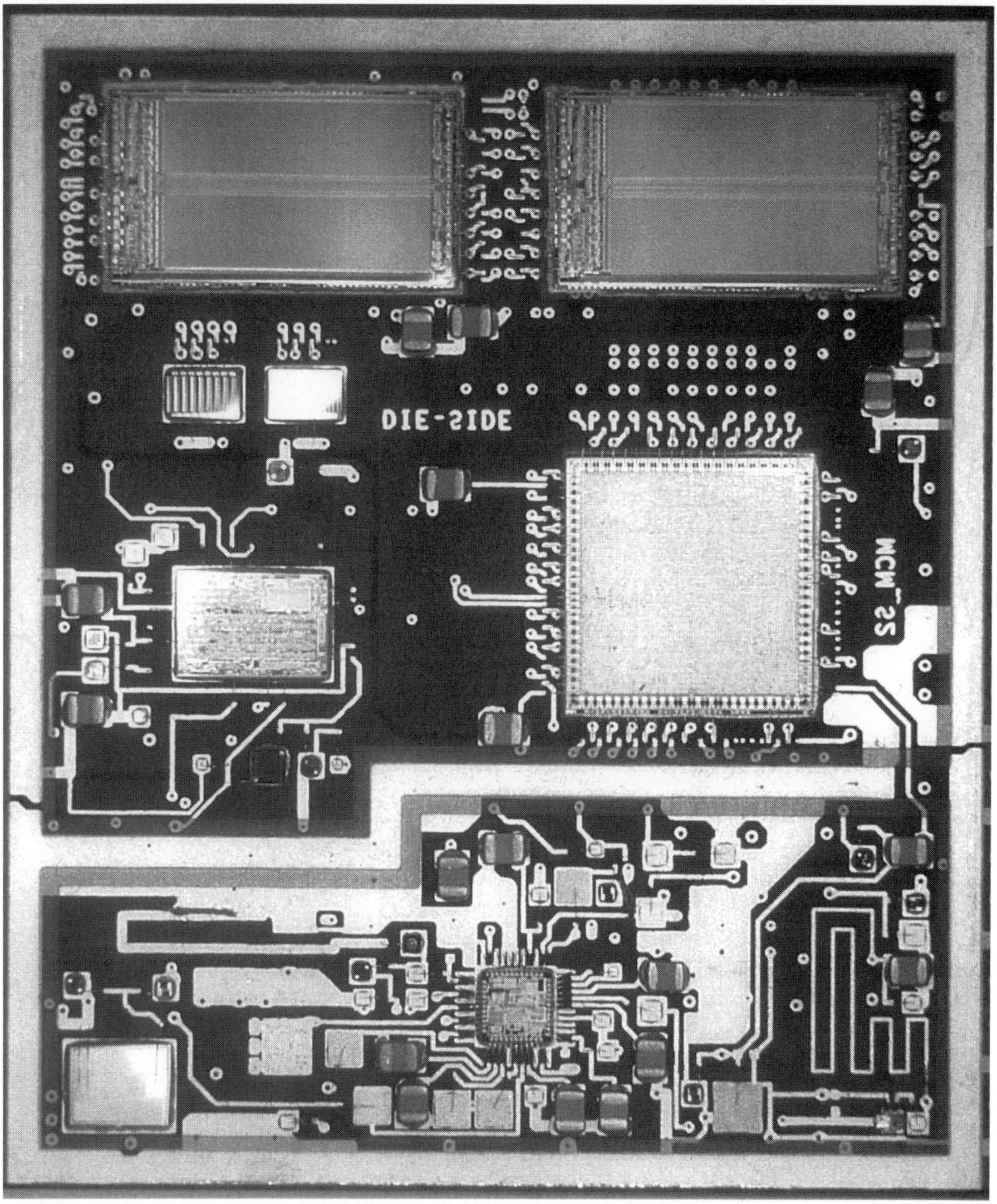

FIGURE 12.49 Photo of the "digital" side of a double-sided "mixed-mode" analog/digital MCM-L containing a microminiaturized global positioning system six-channel receiver. MCM dimensions are 1.37 × 1.43 inches. (MCM-L substrate fabricated by W.L. Gore & Associates.)

REFERENCES

1. Blood, W., *Motorola MECL System Design Handbook,* 4/e, HB205 Rev. 1, Motorola Semiconductor Products, Inc. 1988, Chapters 3, 4, and 7.
2. Gilbert, B.K., L.M. Krueger, and R.D. Beistad, Design of Prototype Digital Processors Employing Subnanosecond Emitter Coupled Logic and Rapid Fabrication Techniques, *IEEE Trans. on Components, Hybrids, and Manufacturing Technology,* CHMT-3(1): 125–134 (March) 1980.
3. Gilbert, B.K., Packaging and Interconnection of GaAs Digital Integrated Circuits. Chapter 8 (pp 289–331) in *GaAs Microelectronics,* Einspruch, N.G. and Wisseman, W.R., Eds., Vol. 11 in VLSI Electronics Microstructure Science series. Academic Press, Orlando, Florida, 1985 ISBN 0-12-234111-2.
4. Gilbert, B.K. and G.W. Pan, The Application of Gallium Arsenide Integrated Circuit Technology to the Design and Fabrication of Future Generation Digital Signal Processors: Promises and Problems. *Proc. IEEE* 76(7): 816–834 (July) 1988.
5. Hirsch, T., R. Miracky, and S. Sommerfeldt, Laser Customization of a Multichip Module Substrate," *IEEE Trans. on VLSI Systems* 1: 387–389 (September) 1993.
6. Muller, H., T. Rudwick, S. Sommerfeldt, and R. Miracky, Very Fast Prototyping of Multi-Chip Module Substrates," *Proc. ISHM/IEPS Int. Conf. on Multichip Modules,* Denver, April 1993, pp. 489–494. ISBN 0-930815-37-8.
7. Gilbert, B.K. and G-W. Pan, Packaging of GaAs Signal Processors on Multichip Modules. *IEEE Trans. on Components, Hybrids, and Manufacturing Technology* 15(1)15–28 (February) 1992.
8. Zabinski, P.J., D.J. Schwab, W.I. McNeff, and B.K. Gilbert, Performance Characteristics of Several Multichip Module Technologies for Application to High Clock Rate Digital Systems, *Proc. Government Microcircuit Applications Conf. (GOMAC),* November 1991, pp. 247–249.
9. Gilbert, B.K., W.I. McNeff, P.J. Zabinski, W.P. Kornrumpf, S.F. Tead, R.L. Maki, K.E. Carlson and J.A. Gerber, Development of Deposited Multichip Modules with Unique Features for Application in GaAs Signal Processors Operating Above 1 GHz Clock Rates. *Int. J. of Microcircuits and Electr. Packaging* 15(3): 144–159 (September) 1992.
10. Haller, T.R., B.S. Whitmore, P.J. Zabinski, and B.K. Gilbert, High Frequency Performance of GE High Density Interconnect Modules. *IEEE Trans. on Components, Hybrids, and Manufacturing Technology* 16(1): 21–27 (February) 1993
11. Gdula, M., W.P. Kornrumpf, and B.K. Gilbert, An Overlay Interconnect Technology for 1 GHz and Above MCMs. *Proc. 1992 IEEE Multi-Chip Module Conf. MCMC-92,* Santa Cruz, CA, March 18–20, 1992 pp 171–174. LC 91-77882, IEEE Catalog Number 92CH3124-5, ISBN 0-8186-2725-5.
12. Burgess, J.F., and C.A. Neugebauer, RWOH Integral Package for MCM Using the GE HDI Process with a Metal Barrier, *Proc. 1993 Electr. Components and Technology Conf.,* June 1–5, 1993, pp. 948–950. IEEE CH 33571; ISBN 0-7803-1432-8.
13. *Fairchild ECL Data Book,* Fairchild Camera and Instrument Corporation, 1977 (now National Semiconductor), Chapters 4 and 5.

13a. Haydl, W, Experimentally Observed Frequency Variation of the attenuation of Millimeter-Wave Coplanar Transmission lines with Thin Metallization, *IEEE Microwave and Guided Wave Letters,* Vol. 2, No. 8, (August) 1992.

13b. Lei, G. T., R. W. Techentin, and B. K. Gilbert, Joint Time and Frequency Domain Model of Power/Ground Plane Noise, *Proc. 29th Int. Symp. on Microelectronics (ISHM 1996),* Minneapolis, MN, October 8–10, 1996.

14. Randall, B., D. Schwab, B. Donham, D. Tuckerman, D. Benson, R. Hsu, N. Brathwaite, S. Enquist, K. Carr, S. Wendel, K. Seawright, P. Goodwin, and B. Gilbert, GaAs Digital MCM in $Si/Cu/Si0_2$ Technology for Clock Rates Up to 1 GHz, *Int. Electr. Packaging Society Conf.,* San Diego, CA, Sept. 12–15, 1993, pp. 615–627, IEPS Publishers, Wheaton, IL. ISBN 1-880433-15-X.
15. L.L. Hench and J.K. West, *Principles of Electronic Ceramics,* John Wiley & Sons, New York, 1990, pp. 237–280 (Chapter 6).
16. Carroll, K.R., J.M. Pond, D.B. Chrisey, J.S. Horowitz, R.E. Leuchtner, and K.S. Grabowski, Microwave Measurements of the Dielectric Constant of $SrBaTiO_3$, *Applied Physics Letters* 62(15): 1845–1847 (12 April) 1993.
17. Dey, S.K. and R. Zuleeg, Integrated Sol-Gel PZT Thin Films on Pt, Si, and GaAs for Non-Volatile Memory Applications, *Ferroelectrics* 108: 37–46, 1990.

17a. Hall, Stephen H. Hall, Wayne L. Walters, Larry F. Mattson, and Barry K. Gilbert, Multi-GHz VCSEL Electrical Packaging Analysis, *Proc. 46th Electr. Components & Technology Conf. (ECTC),* May 28–31, 1996, Orlando, Florida, pp. 605–611 (IEEE Catalog No. 96CH35931).

17b. Hall, Stephen H., Wayne L. Walters, Larry F. Mattson, and Barry K. Gilbert, VCSEL Electrical Packaging Analysis and Design Guidelines for Multi-GHz Applications. Submitted to *IEEE Trans. on Components, Packaging, and Manufacturing Technology –Part B,* June 1996.

18. Koopman, N.G., T.C. Reiley, P.A. Totta, Chip-To-Package Interconnections. Chapter 6 in *Microelectronics Packaging Handbook,* R.R. Tummala and E.J. Rymaszewski, Editors, Van Nostrand Reinhold, New York, 1989, pp. 366–391. ISBN 0-442-20578-3.

19. Lei, G-T, R. W. Techentin, W. L. Walters, B. K. Gilbert, Thermal Analysis of Multichip Modules Intended for Use with High Power, High Clock Rate Gallium Arsenide Digital Integrated Circuits, *Proc. 1992 ANSYS Conf. and Exhib.,* June 8–12, 1992, the Pittsburgh Hilton and Towers, Vol 2, 7.53–7.64.

20. Lei, G-T, R.W. Techentin, W.L. Walters, and B.K. Gilbert, Determination of the Thermal Characteristics of Complex MCM Systems Using the Finite Element Method and the Partial Element Equivalent Circuit Method. Submitted to *IEEE Trans. on Components, Hybrids, and Manufacturing Technology.*

20a. Logan, L. and D. Tuckerman, Statistical variation of TaN resistors in a chips first inorganic MCM-D process, after laser trimming. Personal communication, July 17, 1996.

21. Gilbert, B.K. and L.G. Salmon, Interconnection Design and Process Characteristics for Digital Multichip Modules Operating at High System Clock Rates. Chapter 5 in *Thin Film Multichip Modules,* G. Messner, I. Turlik, J. Balde, and P. Garrou, Eds., pp. 165–218, 1992. Int. Society for Hybrid Microelectronics Publishers, ISBN 0-930815-33-5.

21a. Wang, Gaofeng, Guangwen Pan, and Barry K. Gilbert, A Hybrid Wavelet Expansion and Boundary Element Analysis for Multiconductor Transmission Lines in Multilayered Dielectric Media, *IEEE Trans. on Microwave Theory and Techniques* 43(3): 664–675 (March) 1995 (ISSN 0018-9480).

21b. Lei, Guang-Tsai, Guang-Wen (George) Pan, and Barry K. Gilbert, Examination, Clarification, and Simplification of Modal Decoupling Method for Multiconductor Transmission Lines, *IEEE Trans. on Microwave Theory and Techniques: Part I* of two parts 43(9): 2090–2100 (September) 1995 (ISSN 0018-9480).

21c. Lei, Guang-Tsai, Robert W. Techentin, Paul R. Hayes, Daniel J. Schwab, and Barry K. Gilbert, Wave Model Solution to the Ground/Power Plane Noise Problem, *IEEE Trans. on Instrumentation and Measurement* 44(2): 300–303 (April) 1995 (ISSN 0018-9456).

21d. Pan, Guang Wen, Mikhail Toupikov, and Barry K. Gilbert, A Combined Finite Difference and Analytic Expression Approach to Crossover Capacitance in a Multilayer Dielectric Environment, *IEEE Trans. on Components, Packaging, and Manufacturing Technology Part B,* August 1996.

21e. Lei, G. T., R. W. Techentin, and B. K. Gilbert, Electrical Behavior of Multichip Modules in the Presence of Power and Ground Plane Noise, *Proc. 1995 Int. Conf. on Multichip Modules* (April) 1995 (ISBN 0-930815-42-4).

22. Gilbert, B. K. and G-W. Pan, The Application of Gallium Arsenide Integrated Circuit Technology to the Design and Fabrication of Future Generation Digital Signal Processors: Promises and Problems, *Proc. IEEE* 76(7): 816–834 (July) 1988.

23. Pan, G-W, K. S. Olson and B. K. Gilbert, Improved Algorithmic Methods for the Prediction of Wavefront Propagation Behavior in Multiconductor Transmission Lines for High Frequency Digital Signal Processors, *IEEE Trans. on Computer-Aided Design of Integrated Circuits and Systems* 8(6)608–621 (June) 1989.

24. Pan, G-W., G. Wang and B.K. Gilbert, Edge Effect Enforced Boundary Element Analysis of Multilayered Transmission Lines, *IEEE Trans. on Circuits and Systems, Part I: Fundamental Theory and Applications* 39(11): 955–963 (November) 1992.

25. Ramo, S., J.R. Whinnery, and T. Van Duzer, *Fields and Waves in Communications Electronics,* 2/e. John Wiley and Sons, New York, 1984. pp. 70–71, 124–125, 147–153, 178–183, 535–538.

26. L.N. Dworsky, *Modern Transmission Line Theory and Applications.* John Wiley and Sons, New York, 1979, Chapter 7.

27. Techentin, R.W., G. T. Lei, A. J. Staniszewski, G. W. Pan, D. J. Schwab, and B. K. Gilbert, Analysis of Heretofore Unsolved Electromagnetic Simulation Problems in High Performance Multichip Modules and Printed Circuit Boards: Simulation and Measurement of Power Plane Noise and Internal Inductance, *IEEE Trans. on Instrumentation and Measurement,* Fall 1994.

28. Pan, Guang-Wen, Jeffrey A. Prentice, Sharon K. Zahn, Andrew J. Staniszewski, Wayne L. Walters and Barry K. Gilbert, The Simulation of High Speed, High Density Digital Interconnects in Single Chip Pack-

ages and Multichip Modules, *IEEE Trans. on Components, Hybrids, and Manufacturing Technology* 15(4): 465–477 (August) 1992.

29. Pan, G-W, K.S. Olson and B. K. Gilbert, Corrections to "Improved Algorithmic Methods for the Prediction of Wavefront Propagation Behavior in Multiconductor Transmission Lines for High Frequency Digital Signal Processors," *IEEE Trans. on Computer-Aided Design of Integrated Circuits and Systems* 8: (11)1231 (December) 1989.
30. Gilbert, B.K. and W.L. Walters, Design Guidelines for Digital Multichip Modules Operating at High System Clock Rates, *Int. J. of Microcircuits and Electr. Packaging* 15(4): 171–182 (Fourth Quarter) 1992.
31. Zabinski, P.J., B.K. Gilbert, P.J. Zucarelli, D.V. Weninger, and T.W. Keller, Mixed-Signal MCM-L Design Using a Global Positioning System (GPS) Receiver as an Example, *Proc. Third Int. Conf. on Multichip Modules,* Denver, CO, April 13–15, 1994.

APPENDIX

A BRIEF DISCUSSION OF S-PARAMETERS

Figure 12.17 employs the concept of S-parameters, widely employed by the analog microwave design community but largely unfamiliar to the digital design community. Although a full treatment of S-parameters can be found in Refs. 25–26, we introduce the concept by stating that, for a simple two-port network (e.g., a structure as simple as a stripline or microstrip interconnect, with two ends or "ports"), S-parameters quantify the frequency-dependent behavior of the interconnect. For example, $S_{21}(f)$ and $S_{12}(f)$ (referred to as the insertion loss or transmission loss) quantify the amount of energy injected into either end of an interconnect which emerges from the opposite end, over the entire range of frequencies measured. An ideal transmission line would exhibit no losses at any frequency; that is, all energy injected from either end of the line would emerge unattenuated at the other end, over the entire measured frequency range. In that case the traces in the upper two panels of Fig. 12.17 would lie along the 0 dB line.

Alternatively, $S_{11}(f)$ and $S_{22}(f)$ (referred to as the return loss or reflection loss) quantify the amount of energy injected into either end of the transmission line that is actually reflected back at the injecting source itself. In an ideal transmission line, this energy should be extremely low and, in fact, the lower panels of Fig. 12.17 show curves that are below –20 dB; the closer the impedance of the line is to 50 Ω, the lower will be the average values of return loss and the flatter the curves. The "humps" on the return loss curves typically occur when the impedance of the line under test differs from 50 Ω. Reflections arise at the junction of the transmission line under test and at the 50-Ω test fixture, with standing waves appearing at the measurement head whose amplitudes wax and wane as a function of frequency, magnitude of the impedance mismatch, and line length.

A final comment will be presented regarding the relationship between the various S-parameters. The sum of the transmitted energy, the reflected energy, energy dissipated in resistive or dielectric losses, and any energy actually radiated from the interconnect (if any) should equal the total amount of energy injected into the line. Note that these individual energy components are *not* directly proportional to the voltages measured in the network. For a complete explanation of S-parameters, the reader is referred to Refs. 25–26.

CHAPTER 13

THERMAL MANAGEMENT

Dr. Iwona Turlik
Motorola
Robert F. Darveaux
Amcor Corp.

13.1 INTRODUCTION

In the race for superiority of the next-generation systems, engineers are packing dramatically more functions into integrated circuit chips, but at the same time they are increasing chip size nominally. This trend has generally been accompanied by increased power dissipation at the chip level.[3] These complex chips are packed very close to each other on an MCM. The net effect of these developments has been rapid increase in the module level heat flux.[1] Therefore, thermal performance has become an important design consideration that must be accurately predicted during the early design phase to ensure that the operating and reliability constraints will be met in the final product. Furthermore, heat removal becomes a critical factor in assuring high system reliability.[14] Few chip-type MCMs have greatly overcome many design and yield barriers and are usually assembled in QFP- or BGA-type packages. This type of MCM will use thermal enhancements appropriate for the package type selected. For MCMs that have more chips assembled, a more sophisticated thermal solution must be applied.

Basically, an MCM must be designed so that the semiconductor devices are maintained at or below their maximum rated temperature. Increased temperatures cause a variety of deleterious effects. For example, the reverse saturation current of a p-n junction increases with increasing temperatures,[7] reducing the electrical isolation provided by reverse-biased junctions in an IC. In a MOSFET, increased temperature leads to reduced switching speed.[8] At the system level, for example, electromigration, corrosion, and interfacial diffusion mechanisms increase with increasing temperature, leading to premature failures in metallization and bonded interfaces. The mean time to failure is decreased by a factor of two with approximately a 10° C increase in operating temperature.[9] Therefore, for quality performance and reliability, it is essential to properly select package design, process, materials, and cooling techniques to meet MCMs high power dissipation requirements.

13.2 THERMAL PERFORMANCE FIGURES OF MERIT

To compare various cooling technologies, it is necessary to establish a figure of merit (FOM) to assess the thermal performance of each design.[5] This is a difficult task because of great diversity in the designs. The most commonly used FOM is the junction-to-ambient thermal resistance, θ_{ja}:

$$\theta_{ja} = \frac{T_j - T_a}{P_c} \tag{13.1}$$

where T_j = device junction temperature
T_a = ambient temperature
P_c = chip power dissipation

In reality, the actual junction temperature is rarely measured in a thermal experiment. Instead, either the chip surface temperature is measured using an IR camera, or heater chips with resistors and temperature sensing diodes are utilized to measure θ_{ja}.

In most package designs, it is convenient to subdivide θ_{ja} into external and internal components. The internal resistance is composed of all the conduction and spreading resistances between the chips and the heat sink. The external resistance is due to spreading in the heat sink and convection off its surfaces to a moving fluid, such as air or water. The internal resistance is generally calculated in terms of the chip power dissipation.

$$\theta_{in} = \frac{T_j - T_{hs}}{T_c} \tag{13.2}$$

where T_{hs} is the average heat sink temperature. For MCMs, the external resistance is generally calculated in terms of the total module power dissipation.

$$\theta_{cx} = \frac{T_{hs} - T_a}{P_{mod}} \tag{13.3}$$

where P_{mod} is the total module power. This convention for calculating external resistance accounts for spreading effects due to unequal power dissipation by chips across a module.

The problem with using junction-to-ambient thermal resistance as an MCM FOM is that it neglects all geometric effects due to chip size, module area, and module volume. For example, the heat flux on the surface of a 2.5×2.5 mm^2 chip is 16× greater than that of a 10×10 mm^2 chip dissipating the same amount of power. Therefore, a more appropriate FOM would be to normalize θ_{ja} by the chip area.

$$\theta_{jaca} = \frac{T_{hs} - T_a}{P_{mod}}(A_c) \tag{13.4}$$

where θ_{jaca} = junction-to-ambient thermal resistance normalized by the chip area
A_c = chip area

Thus, packages with different size chips can be compared by using θ_{jaca}.

From a similar perspective, attaching a large heat spreader to each chip will reduce θ_{ja}, but chip packaging density is also sacrificed, so interconnection length between chips must increase. By using heat spreaders, improved thermal performance is achieved at the expense of increased size and signal delay. It would seem that a better FOM would be the maximum allowed power dissipation per unit area or per unit volume in the module. These FOMs would properly normalize the thermal performance to the packaging efficiency of the design. However, using maximum power dissipation as an

FOM can be misleading, because not all manufacturers have a consistent chip-to-ambient temperature rise. For some packages, the maximum power dissipation might be determined by structural reliability considerations whereas, in other packages, the maximum power might be determined by device temperature considerations. This type of detailed information is rarely reported, so the true maximum power dissipation for a given design is difficult to assess.

Bar-Cohen[2] suggests an alternative FOM by normalizing θ_{ja} to the area packaging density of chips on the module.

$$\theta_{jaapd} = \left(\frac{T_i - T_a}{P_c}\right)\left(\frac{V_{mod}}{N}\right) \tag{13.5}$$

where θ_{jaapd} = junction-to-ambient thermal resistance normalized by area packaging density
A_{mod} = projected module area
N = number of chips

This FOM properly accounts for geometric effects of heat spreading in the package, while it is independent of any specification on maximum junction temperature. In this sense, θ_{jaapd} is the best FOM to compare package structures from a thermal design perspective.

By extending the packaging density concept to three dimensions, a thermal resistance normalized by the volumetric packaging efficiency can be defined.

$$\theta_{javpd} = \left(\frac{T_i - T_a}{P_c}\right)\left(\frac{V_{mod}}{N}\right) \tag{13.6}$$

where θ_{javpd} = junction-to-ambient thermal resistance normalized by volumetric packaging density
V_{mod} = module volume

θ_{javpd} is probably the best FOM for a three-dimensional interconnection scheme, where modules are designed to be stacked. On the other hand, θ_{jaapd} is the best FOM for the planar interconnection schemes more commonly in use today.

13.3 MCM COOLING DESIGNS

A great number of MCM designs are currently being manufactured or are under development in the electronics industry. In this section, several MCMs will be described in detail, and the critical heat transfer design features will be discussed. This section is by no means an exhaustive list of technologies, but it does give a broad cross section of the various cooling strategies being employed.

13.3.1 Air-Cooled Modules

ALCOA Thermal Module. The MCM contains five IC chips on alumina substrate. Both cavity-up and cavity-down configuration of the MCM were employed (Figs. 13.1 and 13.2). Silver-filled epoxy is used as die attach material. The material for heat sink attach used is thermal grease. The materials used for the seal ring in the cavity-up is copper tungsten alloy. For the cavity-down design, Kovar™ is used for seal ring. Seal lid material is molybdenum and Kovar for cavity-up and cavity-down, respectively. The pins are made up of Kovar. Extruded aluminum heat sink is used to dissipate power in both MCMs.

For this application, cavity-down configuration is a better design from thermal management point of view, due to the shorter thermal path. For a typical workstation environment of 300 FPM, the maximum junction temperature difference between the two cases is approximately 10° C. The use of

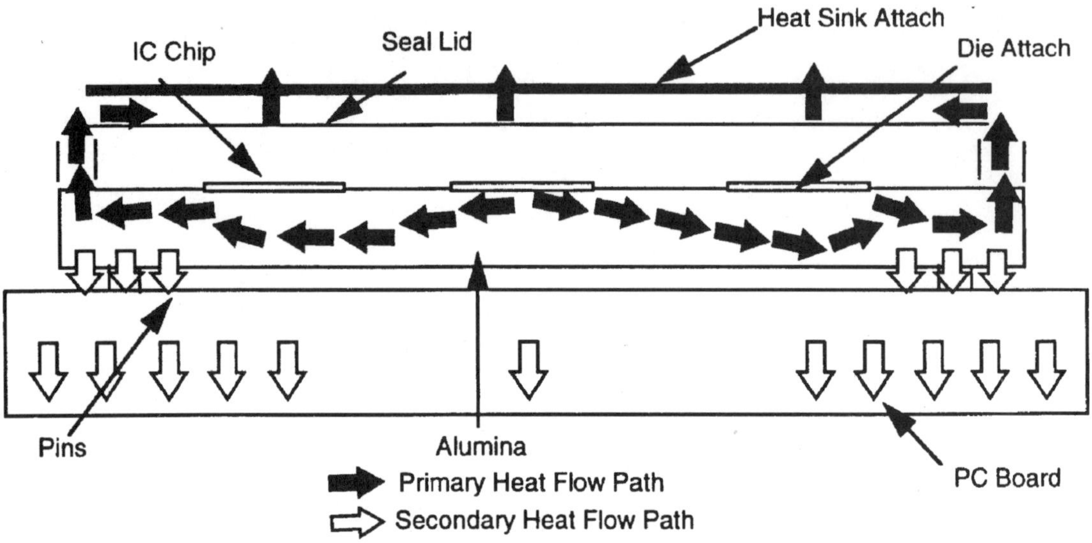

FIGURE 13.1 The primary and secondary heat flow paths for cavity-up MCM.[31]

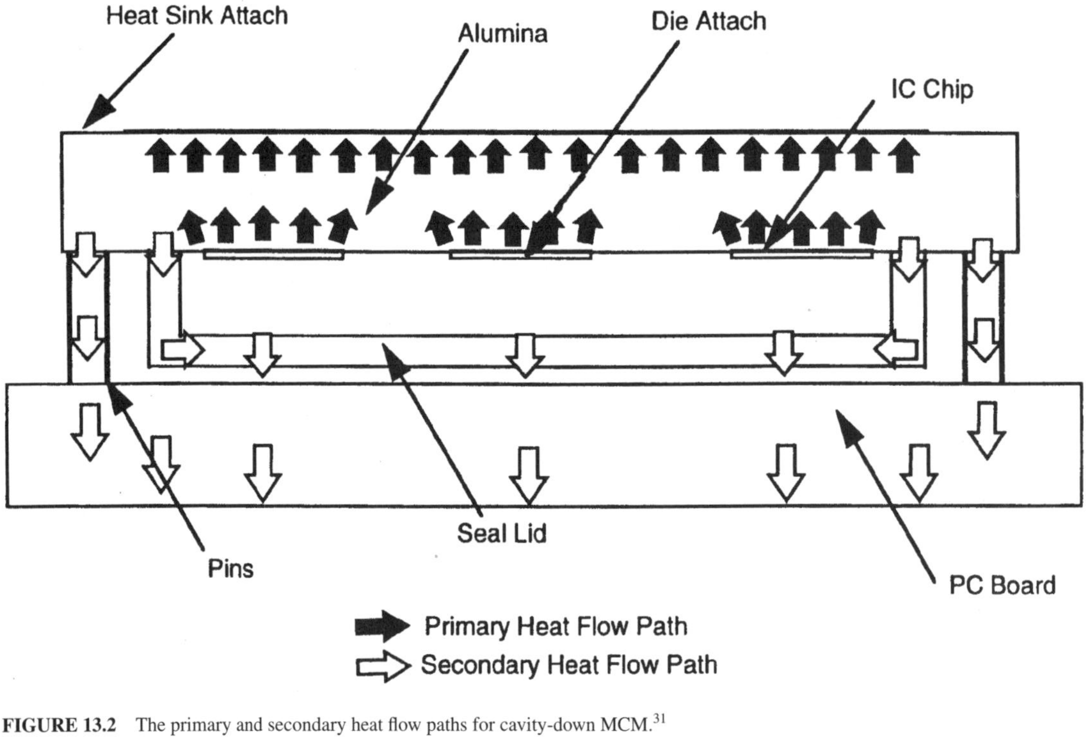

FIGURE 13.2 The primary and secondary heat flow paths for cavity-down MCM.[31]

molybdenum or Cu/W seal ring and seal lid is very efficient in reduction of the thermal resistance for cavity-up configuration. Another important result of this study was that the substrate material must have a thermal conductivity above 15 W/M°C, and that the materials with higher thermal conductivity over alumina do not much improve the junction temperature for the present design. Increased substrate thickness of cofired ceramic substrate allows a designer to use a larger number of individual layers for increased routing. According to Sullhan et al.,[25] the increase in thickness of 0.2032 mm in alumina ceramic adds only 0.5° C in maximum junction temperature for cavity-down design (Fig. 13.3) and decreases the maximum junction temperature by 10° C for cavity-up design. This is due to the reduced lateral thermal resistance to the flow of heat from chips to the sealing ring versus increased thermal resistance for cavity-down for thicker substrate.

AT&T Advanced VLSI Packaging (AVP). The first implementation of the AVP technology is shown in Fig. 13.4.[10] The 160 I/O pin grid array package consists of the WE 32100 central processing unit (CPU), memory management unit (MMU), and math accelerator unit (MAU). The chip I/Os range from 81 to 120 I/O. The silicon substrate measures 1.3 × 3.0 cm. The chips are spaced 500 microns apart. The substrate with the three chips flip chip mounted is assembled to the metal heat sink (aluminum) using a thick layer of compliant adhesive (silicone gel material). There are 130 vias for the CPU, 80 vias for the MAU, and 120 thermal vias for MMU. The heat sink structure is assembled to the PWB package.[10] In this case, the heat is dissipated from the chip active site to a heat sink through silicon substrate. A cross-sectional view of the heat transfer model configuration is shown in Fig. 13.5. The predicted and measured thermal resistances for this package are compared in Fig. 13.6. Good agreement between these two groups of results confirm the accuracy of the numerical analysis.

This study simulated steady-state internal thermal resistances affected by thermal vias, solder bump heights, high power I/O drivers, and chip sizes. The temperature distribution along a line passing through chip center is shown in Fig. 13.7.

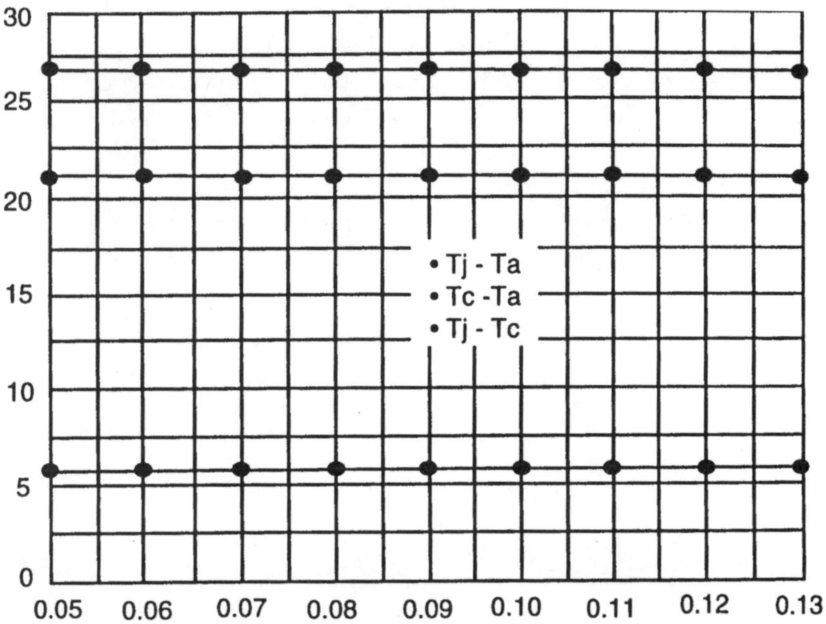

FIGURE 13.3 Substrate thickness vs. temperature (cavity-down).[31]

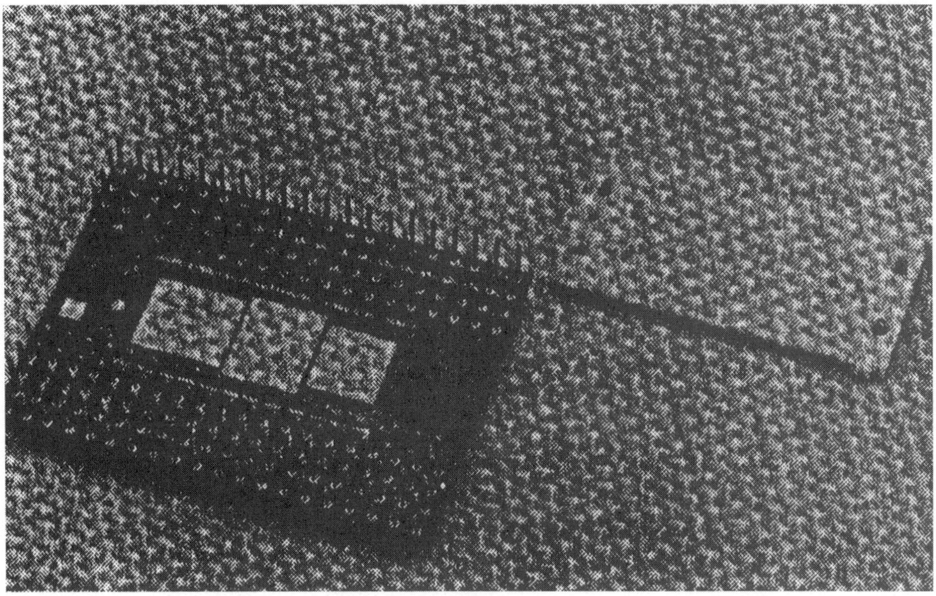

FIGURE 13.4 160 I/O pin grid array package for three-chip assembly.

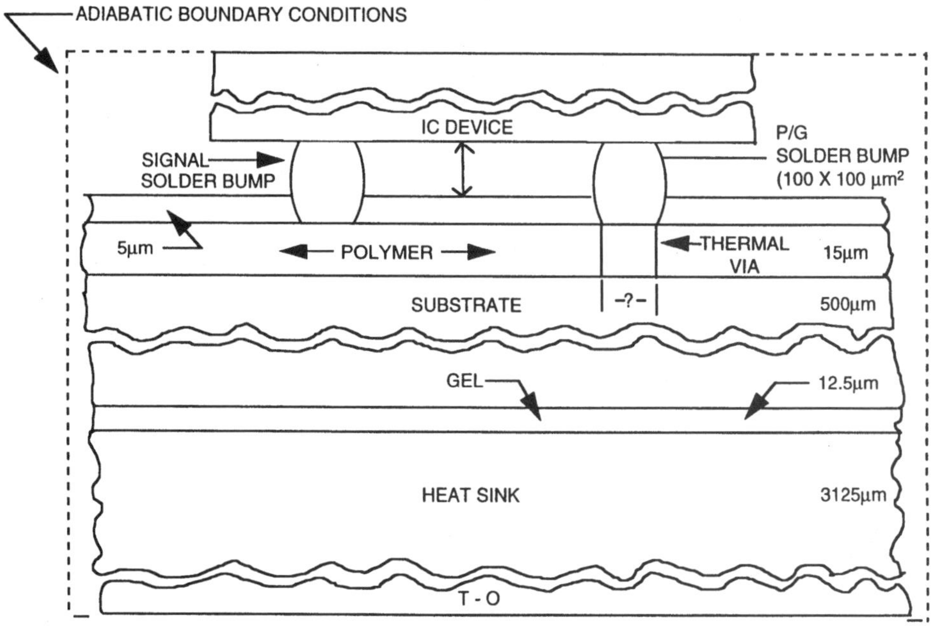

FIGURE 13.5 Cross-sectional view of the heat transfer for model configuration.

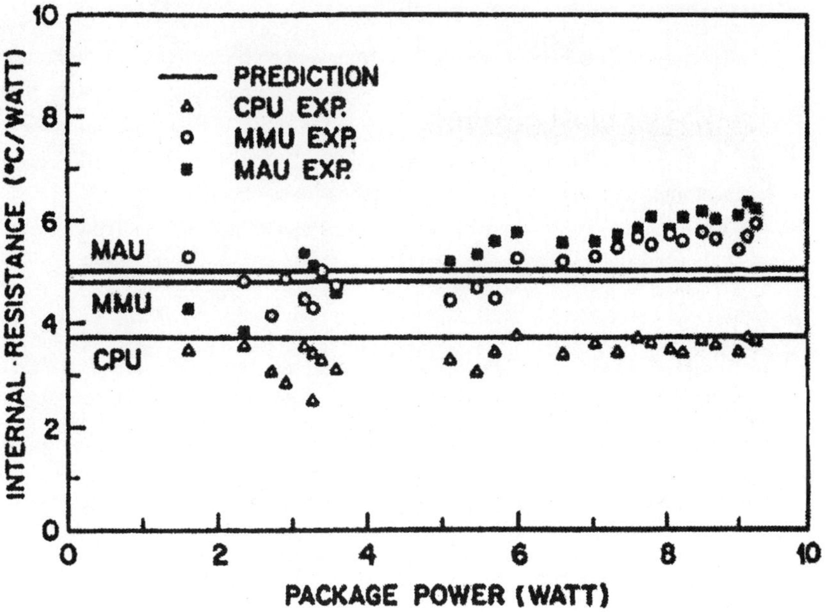

FIGURE 13.6 Measured and predicted thermal resistance for experimental package.

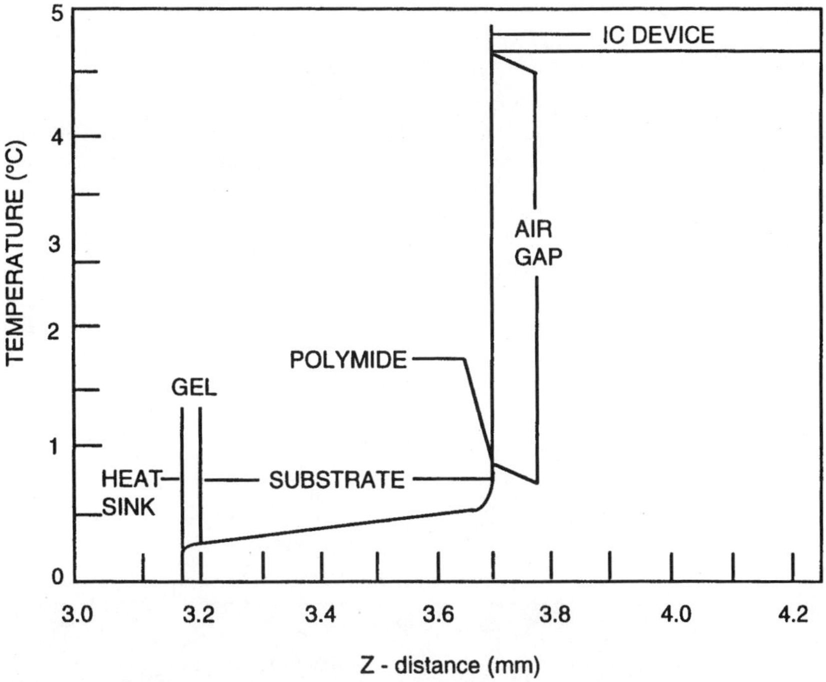

FIGURE 13.7 Temperature distribution along a line passing through chip center.

Several conclusions can be drawn from this study:[26]

1. Thermal performance of this MCM is improved by providing thermal vias to all solder bumps.
2. The number and size of thermal vias affect the internal thermal resistance (Fig. 13.8).
3. The distribution of thermal vias has negligible effect on heat distribution for this design.
4. The high-power I/O drivers cause nonuniform chip temperature distribution but have no significant impact on the maximum chip temperature, as shown in Fig. 13.9.
5. Solder bumps having thermal vias become the dominant dissipation elements when chip size is very small; e.g., 0.25 cm.[26]

IBM RISC Processor. The IBM thermal test module is shown in Fig. 13.10. The 59 × 59 mm module contains a RISC processor consisting of nine CMOS VLSI chips. The chip joining method chosen for this module was IBM's Controlled Collapsed Chip Connection (C4),[17,22] a type of flip chip bonding. The silicon substrate is glued to a silicon interposer using die-attach epoxies (to stiffen the substrate during thermal cycling), and its chip side is protected by a metal cap adhered on peripheral seal band. The whole assembly is bolted to an aluminum base with a layer of ZnO grease and a guard ring placed between interposer and the base. The module is connected to a printed wiring board through flexes on its four sides. A heat sink is clipped on the base after the module is mounted on the board. The thermal resistance from the base to the air inlet temperature and the pressure drop are 1.0 C/W and 0.4 Pa, respectively, at 1 m/s air flow and with a channel cross-sectional area of 32 ×

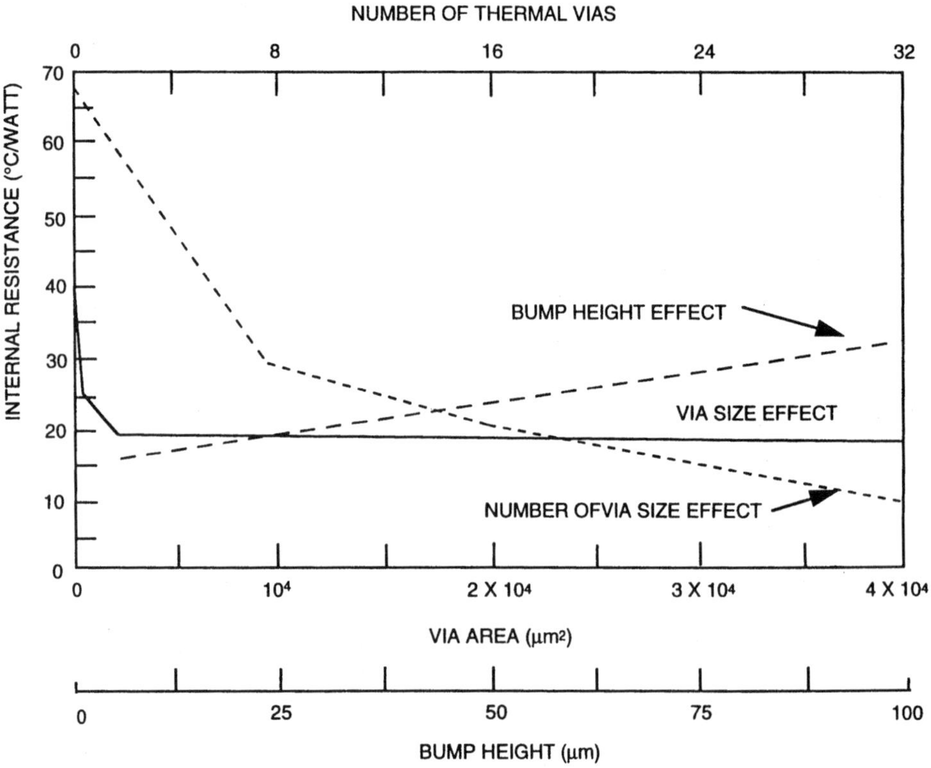

FIGURE 13.8 Effects of thermal via area, solder-bump height, and number of thermal vias (0.25 × 0.25 cm chip).

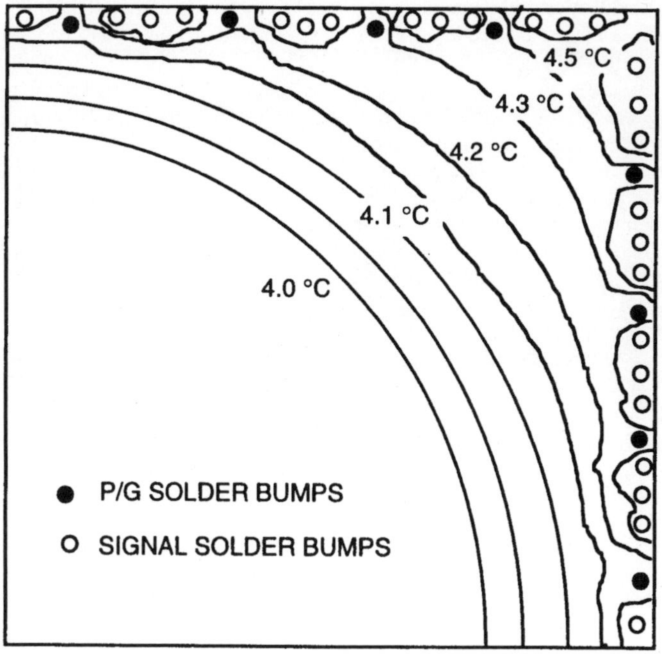

FIGURE 13.9 Effect of high-power I/O drivers (1 × 1 cm chip with 32 thermal vias and 25 mm bump height).

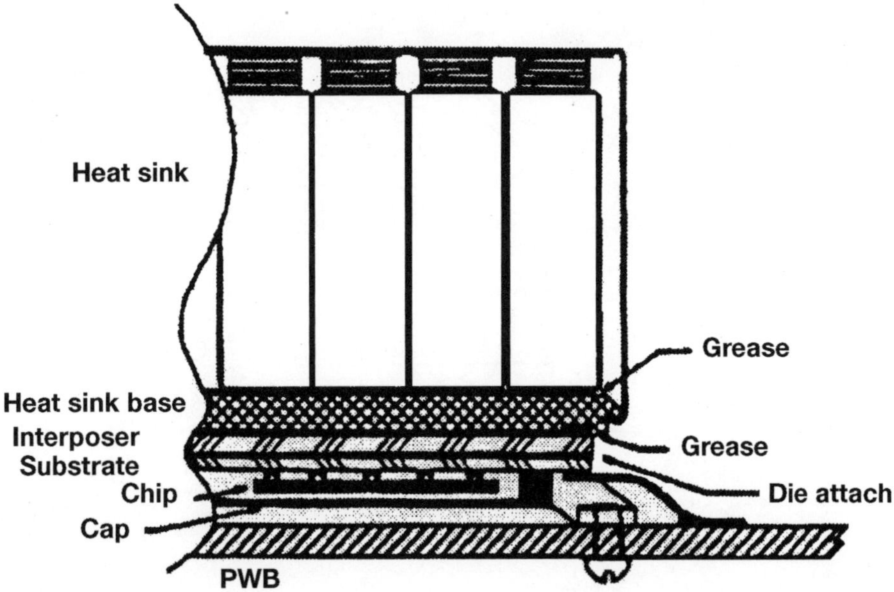

FIGURE 13.10 IBM thermal test module.

123 mm. The module can dissipate about 43 W at a chip temperature rise of 60° C when a heat sink with fin height of 25 mm is used at 1 m/s air flow. The module power will be reduced to 29 W if the fin height is 15 mm.[16,18]

IBM 4381 Processor. The IBM 4381 processor module shown in Fig. 13.11 is approximately 64 × 64 × 40 mm in size and is designed to accommodate 36 TTL chips with 704 circuits each.[27–29] The chips are 4.6 × 4.6 mm with 120 flip chip solder connections, so the maximum number of chip I/Os on the module is 4320. The substrate material is aluminum oxide with up to 32 layers of cofired molybdenum interconnections. The module is connected to a printed circuit board through 882 brazed pins. Twenty-two modules are accommodated on a PC board. A ceramic cap is soldered to the substrate by a 1.5 mm wide seal band around the perimeter. Each chip contacts the cap through a 0.1–0.35 mm thick layer of thermal paste. A unitary hollow pin aluminum heat sink is soldered to the ceramic cap. The serpentine design of the heat sink is such that most of the expansion mismatch between the aluminum and ceramic is absorbed by the flexible fin arrangement. The inlet plenum provides a parallel flow of air jets onto each module of a circuit board, so all modules are fed by fresh ambient air.

The thermal design goals were to dissipate up to 3.8 W/chip and 90 W/module. Approximately 45 percent of the generated heat is conducted off the back of the chips through the thermal paste, cap, and heat sink.[27] Another 30 percent is conducted down through the substrate and back up through the solder seal to the cap and heat sink. The final 25 percent of the heat is conducted through the substrate and I/O pins to the PC board. The chip-to-air thermal resistance was determined to be 17° C/W at an air flow rate of 9.4 L/s (20 cfm) per module at 16 mm of water head. Approximately 9° C/W resistance exists between the chip and the cap, and 8° C/W between the cap and the air.

Critical features for this design are dimensional control of the cap boss heights to determine the thermal paste thickness and reliable solder joints at the cap/substrate and cap/heat sink interfaces. Due to the relatively low thermal conductivity of the paste, its thickness must be kept to the minimum allowed by manufacturing tolerances. To minimize dimensional variation, and hence paste thickness, ultrasonic machining of the cap is employed.

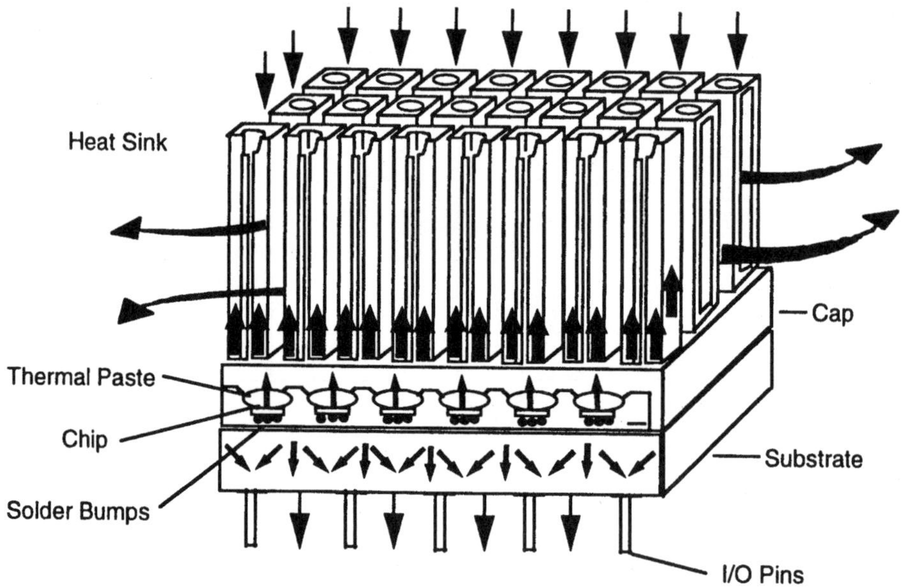

FIGURE 13.11 IBM 4381 processor module.[27]

Solder is used in the package to bond large structural components together. Reliability is an important consideration, because most of the heat flow is through these large area solder joints. In the IBM 4381 design, solder joint strains during power cycling are reduced by matching the expansion coefficients of the substrate and cap, and by using a compliant heat sink.[29] Nevertheless, some expansion mismatch exists in the structure due to temperature gradients. Apparently, the solder strains due to these gradients are low enough that the thermal resistance of the joints does not increase significantly during extended power cycling.

Mitsubishi High-Thermal Conduction Module. The Mitsubishi MCM shown in Fig. 13.12 houses nine 3,000-gate ECL chips.[30,31] The module size is 66 × 66 × 22 mm, and the chips are 8 mm on an edge. Each chip is solder-bumped to a cofired ceramic substrate by 223 connections, making a total of 2,007 chip I/Os on the module. The substrate has 624 brazed I/O pins. A ceramic cap is soldered to the substrate at its perimeter, and 13 × 13 × 0.25 mm copper heat spreaders are attached to each die using indium solder. The gas gap between the heat spreaders and the cap after the indium solder has cooled is nominally 30 μm. The gap provides mechanical compliance in the structure to minimize the reliability risk on the flip chip joint due to thermal and mechanical loading. The soldering operations are carried out under a hydrogen atmosphere, which results in a hydrogen filled module for enhanced heat transfer across the 30 μm gaps between the spreaders and the cap. Finally, a finned aluminum heat sink is attached to the ceramic cap by a soft, thermally conductive epoxy.

The maximum dissipation for the nine ECL chips is 4 W each, so the total module power can reach 36 W. For a prototype module with no heat removed through the I/O pins, the maximum chip-to-ambient thermal resistance was 7.7° C/W. The thermal resistances for the edge and corner chips were 5 to 10 percent less than for the center chip. Extensive finite difference analyses determine the amount of heat flow through various thermal paths in the module.[31] For the case where the package is soldered to a printed circuit board, 76.1 percent of the heat is conducted down through the substrate and back up through the cap and heat sink, and 10.4 percent is conducted through the substrate and off the backside of the PC board. Bar-Cohen[2] estimates that over half of the internal resistance is due to the chip/spreader, spreader/cap, and cap/heat sink interfaces.

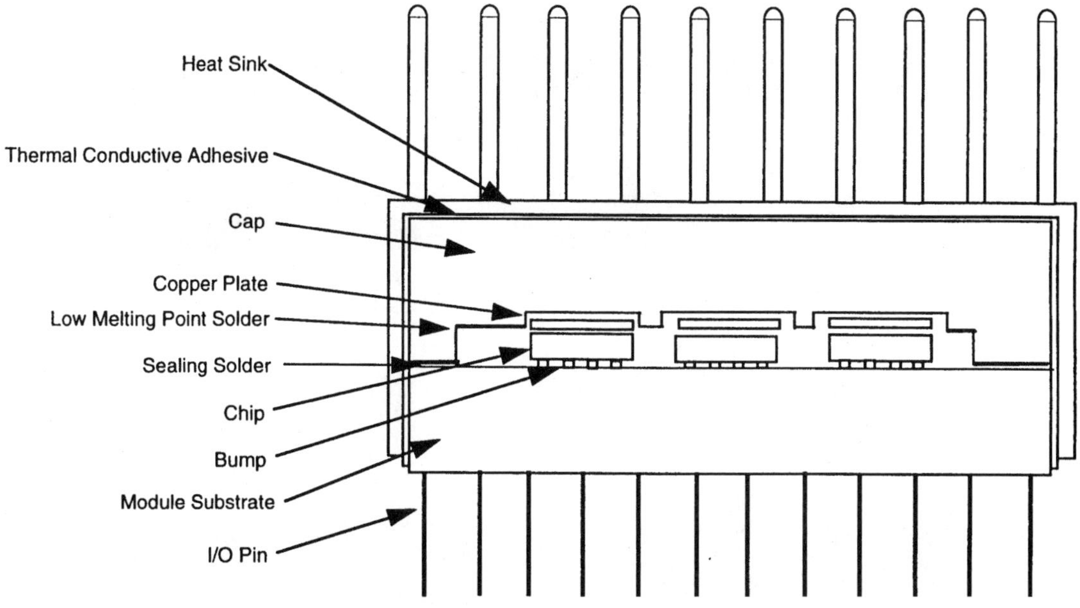

FIGURE 13.12 Mitsubishi high thermal conduction module.[31]

Hitachi Silicon Carbide RAM Module. The Hitachi SiC module shown in Fig. 13.13 is 27.4 × 27.4 × 16 mm in size, and it contains six 1 kb ECL RAM chips.[19] The chips are 1.9 × 4.0 mm in size with 77 flip chip connections. Twenty-five of the connections serve an electrical function, and the other 52 are only for enhanced heat transfer. The chips are attached to a silicon substrate, which is bonded to a silicon carbide base with gold/tin eutectic. Silicone gel is used to encapsulate for humidity and alpha particle protection. The module has 108 gull-wing leads for surface mounting to a PC board. Both the finned aluminum heat sink and the lid are attached by an adhesive silicone rubber.

The power dissipation for the six ECL RAM chips is 1 W each. For a 3 m/s air velocity, the measured chip-to-ambient thermal resistance was 5.5° C/W at the module level. This corresponds to 33° C/W for a single chip, which is about 5 percent less than the calculated value.[19] According to Hitachi's thermal analysis, approximately 25 percent of the thermal resistance is due to spreading in the chips and conduction through the solder bumps. Another 4 percent is due to conduction through the silicon substrate, gold/tin eutectic, SiC base plate, and silicone rubber adhesive. The final 71 percent of the thermal resistance is due to conduction in the heat sink and convection to air. It would appear that the optimization of the heat sink design would result in the greatest improvement in thermal performance. Bar-Cohen[2,40] estimates that the chip-to-air thermal resistance could be lowered to 27° C/W if the air velocity were increased from 3 m/s to a more typical value of 6 m/s.

DEC VAX 9000 Multichip Unit. The multichip unit of the DEC VAX 9000 computer is shown in Fig. 13.14.[24,32–34] The package is designed to accommodate a mix of 9 to 72 ECL chips. The MCA III gate array chips have 10k gates and 360 I/Os, and they measure 9.78 mm on a side. The self-timed random access memory (STRAM) chips are 2.54 × 2.54 mm, have 48 I/Os, and come in 1k × 4-bit and 4k × 4-bit densities. The clock distribution chip is 6.86 × 6.86 mm in size with 232 I/Os. The

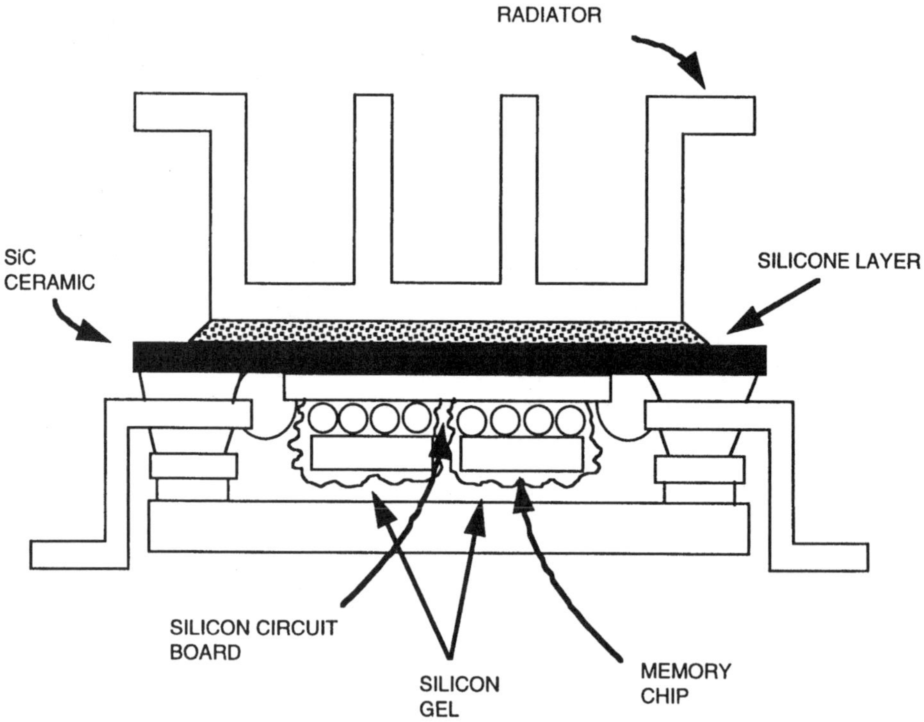

FIGURE 13.13 Mitsubishi high thermal conduction module.[31]

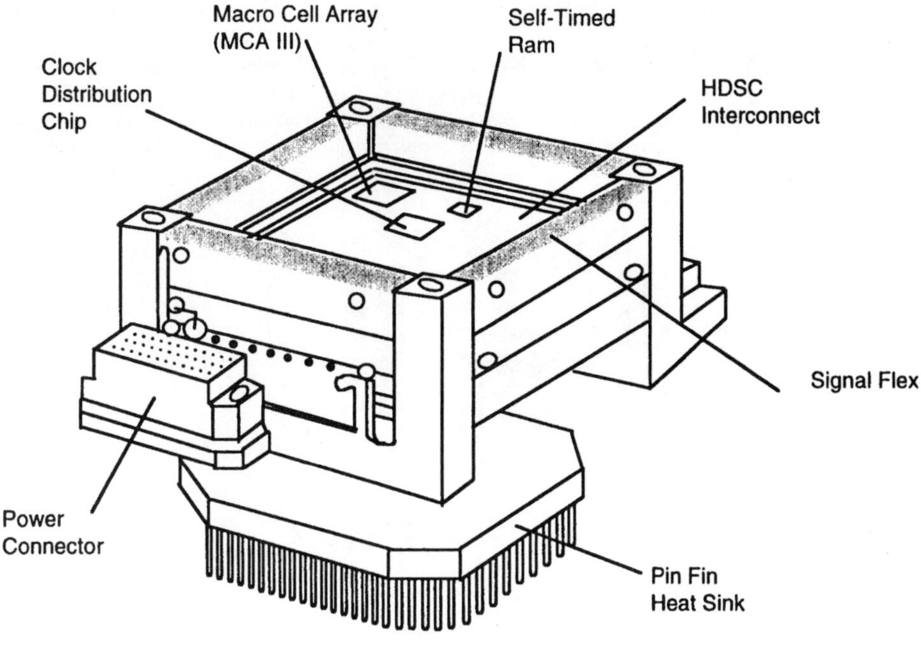

FIGURE 13.14 DEC VAX 9000 multichip unit.[34]

substrate is called a *high-density signal carrier* (HDSC) and consists of nine copper interconnection layers in a polyimide dielectric. Signal and power interconnection structures are made as two separate cores then bonded together. Lasers are used to cut holes in the HDSC at the die sites, and the structure is laminated to a chromium-copper base plate. A diamond-filled epoxy is dispensed at each chip site for die attachment to the base plate. Electrical connections are made by gang-TAB solder bonding to pads on the HDSC. Finally, the aluminum pin-fin heat sink is bolted to the chromium-copper base plate with nine screws. The heat sink is 102 × 104 mm in size with over 600 circular pins. The heat sink/base plate interface is dry (no thermal compound) and the surface specifications are a 0.025 mm total flatness over the entire mounting surface and a 0.2–0.4 μm RMS surface finish. The multichip unit is connected to a printed circuit board by flexible signal connectors on each of the four sides. The signal flex is gang bonded to the HDSC at one end and makes surface contact to the board at the other end. There are more than 800 signal I/Os per module.

The multichip unit is designed to accommodate a maximum of 300 W power dissipation while maintaining circuit temperatures below 85° C. The maximum heat flux on the chips is about 31 W/cm^2. For an air flow rate of 14.6 L/s (31 cfm), the maximum measured chip-to-ambient thermal resistance was 0.17° C/W on a module basis.[24] About 70 percent of this resistance is due to conduction through the heat sink/pins and convection off the surfaces. The tests indicate that the total thermal resistance can be decreased by about 10 percent if the flow rate was increased to 22.6 L/s (48 cfm).

The thermal resistance on a paper chip basis depends on the mix of chips on a module. For a module with one clock chip and eight gate array chips, the chip-to-ambient thermal resistance is about 1.5° C/W. For a module with one clock chip and 71 small STRAM chips, the thermal resistance would be 12° C/W. For a more typical mix of 1 clock chip, 40 STRAMs, and three gate array chips, the thermal resistance is 7.3° C/W on a per-chip basis. Clearly, the chip-to-ambient thermal resistance depends on the chip size and chip packaging density. To account for these effects, the thermal resistance normalized by the area packaging density (Θ_{jaapd}) can be used. The normalized value is 27° C/W (cm^2/chip) for all three of the module configurations discussed above.

NTT MCM. The NTT module shown in Fig. 13.15 is designed to accommodate 36 3 × 3 mm chips.[41] The module size is 70 × 70 × 22 mm with 288 brazed I/O pins around the perimeter. The substrate is cofired aluminum oxide containing power/ground and signal layers. Additional copper signal lines are built up on the substrate in a polyimide dielectric. Copper thermal vias are provided at each chip site for enhanced heat dissipation. The chips are attached using soft solder then wire bonded to pads on the substrate. Finally, the cap and cooling fins are attached.

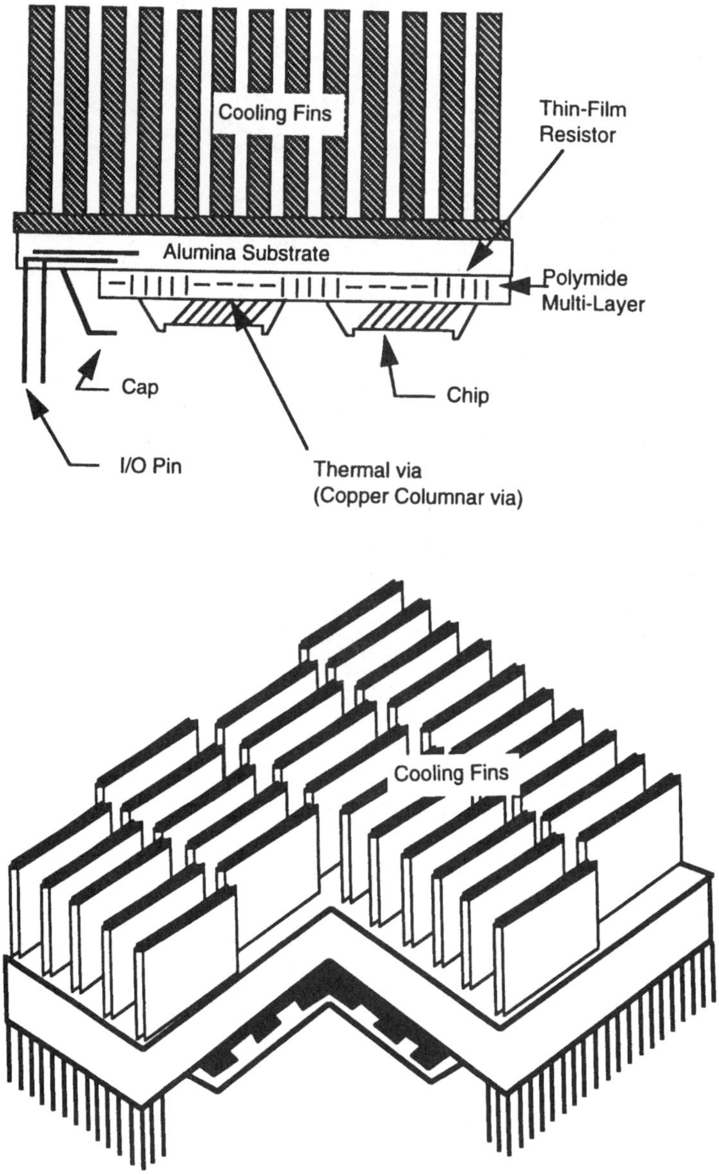

FIGURE 13.15 NTT MCM.[35]

The module contains 16 chips in the central area dissipating 1 W each and 20 chips around the periphery dissipating 0.5 W each, for a total module dissipation of 26 W. At an air velocity of 6 m/s, the measured chip-to-ambient thermal resistance was 28° C/W, which is about 10 percent lower than the calculated value.[35]

The incorporation of thermal vias under each chip reduced the overall thermal resistance by a factor of two compared to a module without thermal vias. Approximately 5 percent of the total thermal resistance was due to the copper/polyimide multilayer for the thermal via module compared to 90 percent for a module without thermal vias. The developers used a series of thermal and electrical simulations to determine the optimum number of thermal vias at each chip site. Increasing the number of vias obviously improves thermal performance, but it also subtracts from the available space for routing electrical interconnections, so an optimum number had to be determined.

MCC Microlaminar Module. The developmental MCC heat sink design shown in Fig. 13.16 was tested using a power transistor in a TO-127 plastic package which has a metal hat sinking surface.[23,36] Each heat sink is made of 250 μm thick copper fins attached to a 10 × 12 mm copper base. The spacing between fin times is 500 μm, and there are 26 total fins per heat sink. To simulate an MCM, 25 heat sinks were attached to a plenum and tube axial fan as indicated in Fig. 13.17. The test

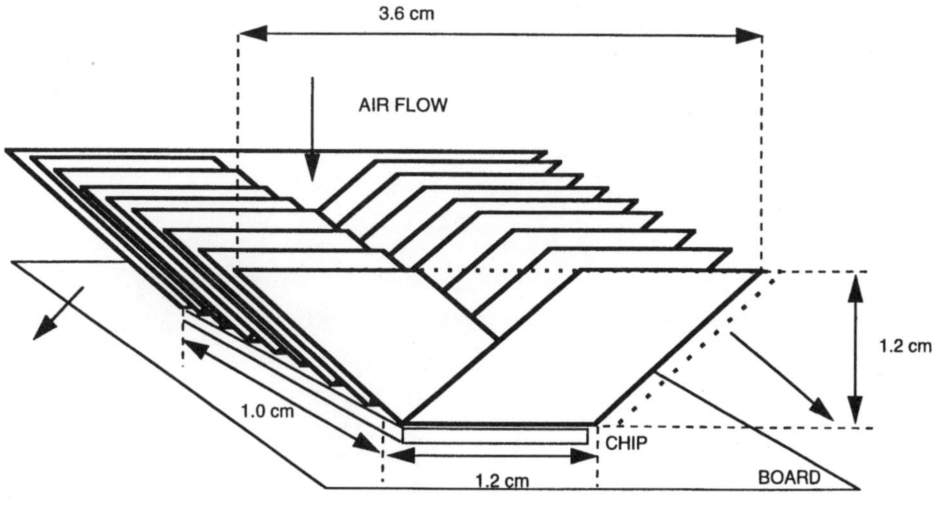

FIGURE 13.16 MCC microlaminar heat sink.[36]

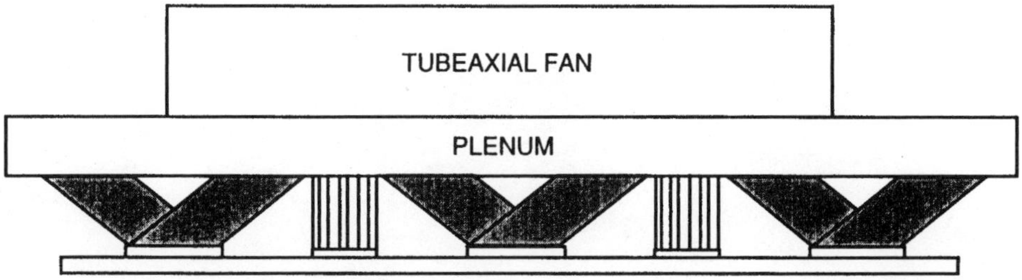

FIGURE 13.17 MCC microlaminar module.[36]

"module" was 150 × 150 × 26 mm in size, and the 25 "chips" were 10 × 10 mm copper plates soldered to power transistors.

A thermocouple was used to monitor the temperature of each heat sink base. At a total air flow rate of 23.6 L/s (50 cfm) and a pressure head of 5 mm of water, the maximum base-to-ambient thermal resistance was 2.2° C/W. In a real package, there would be an additional thermal resistance through the chip and die attachment material, so the chip-to-ambient thermal resistance is estimated to be 2.4° C/W. Hilbert, et al[36] showed that chip power densities of 10 to 20 W/cm^2 could be cooled by a novel heat exchanger design with volumes of less than 1 cm or 5 cm. As demonstrated in Ref. 36, 25 heat exchangers, closely coupled to a tubeaxial fan, are capable of cooling a multichip module dissipating over 500 W, similar to the volume and power dissipation of the early IBM TCM designs employing an MCM, using external cooling sources.[21] While demonstrating the efficacy of this heat exchanger design, the closely coupled geometry does not lend itself to application in conventional card cage mount systems. Moreover, structural support and contact of a single heat exchanger to a single chip were not addressed.

13.3.2 Liquid-Cooled Modules

IBM Thermal Conduction Modules. The IBM Thermal Conduction Module (TCM) was first used in the 3081 processor. Shown in Fig. 13.18, the TCM is 150 × 150 × 50 mm in size, and it houses 100 logic or 118 logic and array chips.[37–39] For the 3081 processor, the TTL chips were 4.57 × 4.57 mm in size. Each chip is attached to the substrate by an array of 120 C4 solder joints, so as many as 12,000 chip I/Os can be interconnected on the module. The 33-layer substrate is aluminum oxide/10 percent glass with molybdenum interconnections and 1,800 I/O pins brazed to the back surface.

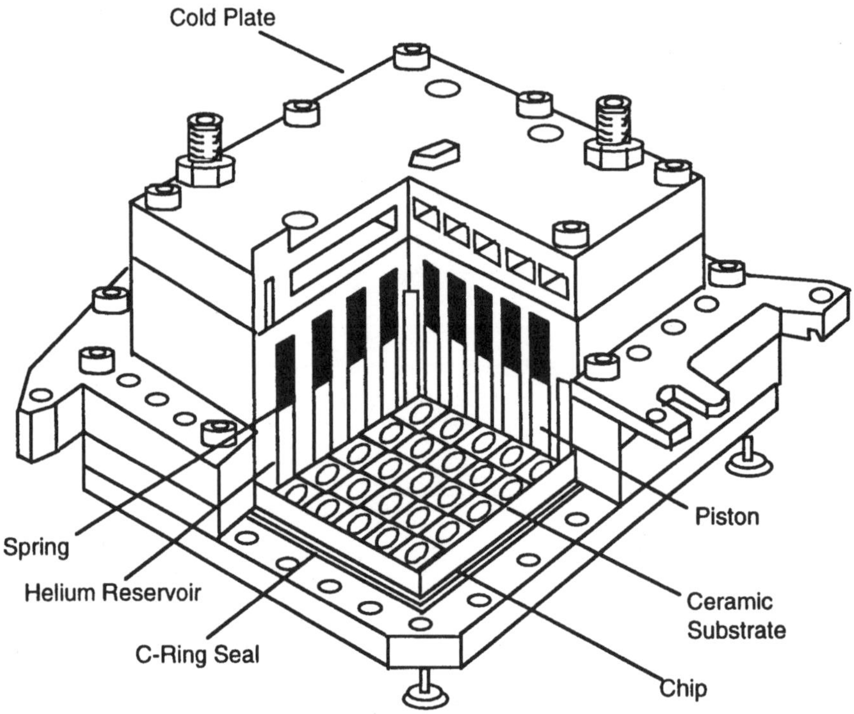

FIGURE 13.18 IBM thermal conduction module.[37]

Spring-loaded aluminum pistons contact the back side of each chip. The pistons provide mechanical compliance to accommodate chip height variation and thermal expansion mismatch between the heat sink and substrate. The heat is transferred from the hat assembly through a polycarbonate interposer to the water-cooled beryllium-copper cold plate. The interposer is used to provide a more uniform temperature distribution across the module. The entire module is hermetically sealed by means of a compressed C-ring, then it is back filled with helium to reduce the thermal resistance across the chip/piston and piston/hat gaps. Helium has a thermal conductivity approximately six times that of air. For the 3090 processor, 4.85 × 4.85 mm ECL chips are used, and design improvements were made for the water distribution system, pistons, and hat.[2]

The design goals for the 3081 processor were to accommodate up to 4 W/chip and 300 W/module with a junction temperature between 40 and 85° C. The measured internal thermal resistance from a device to the top-of-the-hat was 9.1° C/W for the 3081 technology.[38] This resistance value includes an assumed 0.75° C/W spreading resistance from the junction to the chip and a 0.70° C/W end-of-life degradation factor. The external thermal resistance between the hat and inlet water is 2.5° C/W on a per-chip basis, so the total thermal resistance is 11.6° C/W. It was found that 6 percent of the heat was dissipated through the PC board, so the junction-to-ambient thermal resistance is actually 10.9° C/W. For the 3090 technology, the peak power dissipation was increased to 7 W/chip and 500 W/module. The junction-to-ambient thermal resistance was estimated to be 8.7° C/W, which reflects design improvements to the TCM as well as slightly larger and more widely spaced chips.

Honeywell Silent Liquid Integral Cooler (SLIC). The Honeywell Bull SLIC micropackage (Fig. 13.19) was designed for the DPS-88 computer and typically supports 60 CML chips.[41] The 240 I/O multilayer ceramic substrate is 80 × 80 mm in size and housed in an aluminum frame. The total package dimensions are approximately 100 × 100 × 50 mm. The water-cooled cold plate has a flexible copper diaphragm that conforms to the back surface of the substrate. A 60 W micropackage con-

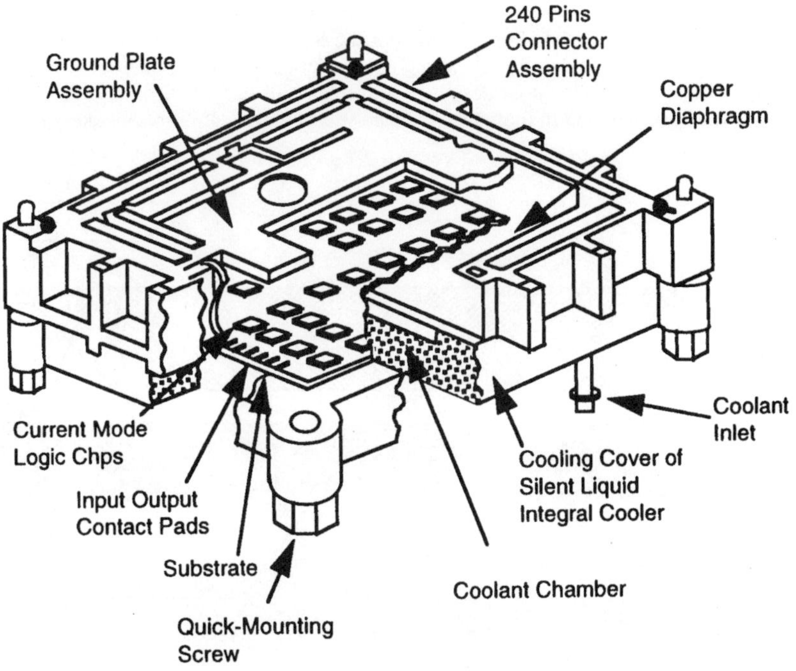

FIGURE 13.19 Honeywell silent liquid integral cooler (SLIC).

tains clusters of 6 to 7 chips dissipating 1.7 W each. There is a 20° C temperature differential between the high powered chips and the substrate back-surface, and a 10° C temperature differential between the substrate and the water. Based on the maximum chip power, the internal thermal resistance is 11.8° C/W. However, the external thermal resistance is generally based on the average power of 1 W/chip, the external resistance is 10° C/W. Therefore, the total chip-to-water thermal resistance is 21.8° C/W for the SLCI micropackage.

NEC SX Liquid Cooling Module. The liquid cooling module developed by NEC for its SX-1 and SX-2 supercomputers is shown in Fig. 13.20*a*–*c*.[41,42] The module is 125 × 125 × 60 mm in size, and it houses 36 flip-TAB carriers (FTCs). An FTC could contain one 1k gate CML logic chip or four 1 kb CML RAM chips. The logic chips are approximately 8 × 8 mm in size, typically dissipate 5.7 W of power, and have 1,761 I/Os. The RAM chips typically dissipate 1.35 W and have 52 I/Os. In either case, the chips are flip-TAB bonded to a multilayer ceramic substrate, then die attached to a copper tungsten cap. The logic FTCs are 12 × 12 mm in size, and the RAM FTCs are 14 × 14 mm. Thirty-six FTCs are solder-bumped to a 100 × 100 mm substrate. Each FTC has 169 solder connections. The base substrate is aluminum oxide with tungsten power and ground planes. Thin film signal layers are built on top of the substrate in a polyimide dielectric. The substrate has 2,177 I/O pins brazed to its back surface for attachment to a printed circuit board. Twelve modules are mounted on a board.

Each FTC is contacted by an adjustable stud mounted in a heat transfer block. To account for the FTC height variation across the module, each stud is adjusted individually to ensure a fine gap between stud and FTC. Then, the block is removed, thermal paste is applied to the studs, and the module is reassembled. Finally, a water cooled cold plate is attached to the heat transfer block. The maximum power dissipation per module is 250 W, and the reported chip-to-water thermal resistance is 5° C/W.[41]

The liquid cooling module technology was scaled up for the NEC SX-3/SX-X supercomputers.[42] The SX-3 substrate is 225 × 225 mm in size and it has 11,540 I/O pins. The substrate accommodates 100 FTCs, each with 604 solder connections. The logic chips have 20k gates, 485 I/Os, and dissipate 33 W of power. The memory chips have 40 kb capacity plus 7k gates, 487 I/Os, and dissipate 38 W. The maximum module power is 4 kW. For the SX-3/SX-X design, NEC reports a junction-to-water thermal resistance of 0.6° C/W,[42] which is an order of magnitude improvement over the SX-2 design. No details were given about design improvements in the SX-3, so it is difficult to determine how such a large reduction in thermal resistance was achieved. It is speculated that much higher thermal

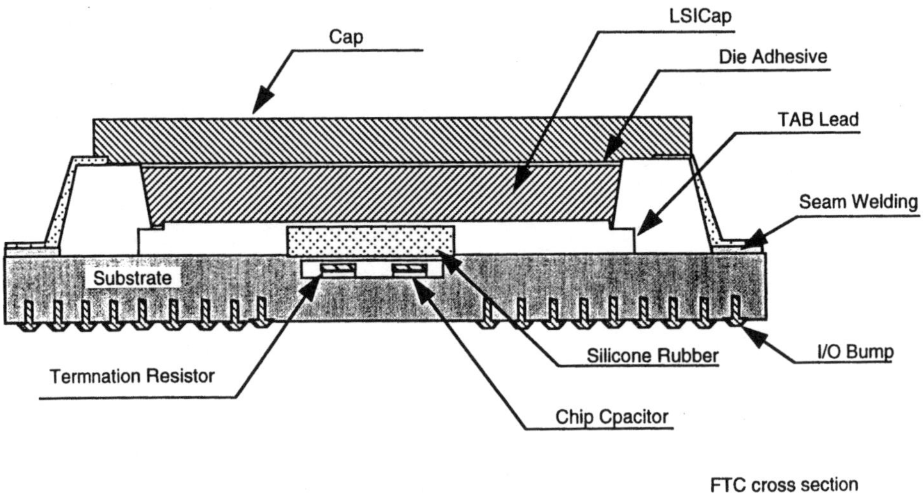

FIGURE 13.20a NEC flip-TAB carrier (FTC).[42]

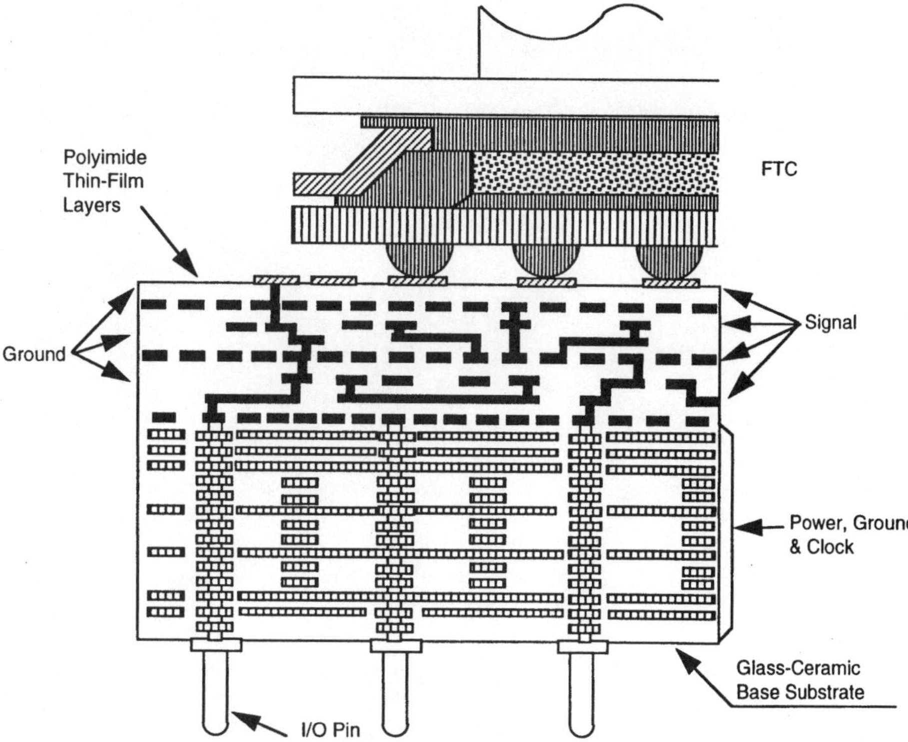

FIGURE 13.20b NEC multilayer substrate (MLS).[42]

conductivity materials were used for die attachment and FGC/stud contact to reduce internal thermal resistance, and improved water distribution and channel design were used to reduce external thermal resistance.

NTT Liquid Cooled Substrate Module. The developmental NTT liquid-cooled substrate module shown in Fig. 13.21 offers a unique approach in that the substrate also functions as a water cooled heat sink.[12,43] The module is approximately 105 × 85 × 3 mm in size, and it supports 25 8 × 8 mm chips. The prototype module had 900 I/O pins braced to the bottom of the substrate, but an alternative design would have I/O pads on both top and bottom surfaces to facilitate vertical stacking of modules in a single cabinet. The six-layer aluminum oxide substrate has 29 800× 400 μ^2 water channels formed by punching holes in the green sheet prior to firing. The chips are epoxy die-attached and wire-bonded to the substrate.

At 380 W module power dissipation and water flow rate of 0.8 L/min, the measured chip-to-inlet thermal resistance was 3.3° C/W. The developers estimate that thermal resistance can be reduced 33 percent by using Pb-Sn solder die attachment and 10 × 10 mm chips. On the other hand, increasing the substrate thickness from 1.2 to 6 mm to accommodate 35 conductor layers would increase the thermal resistance by 20 percent. Aluminum nitride is becoming more available as a substrate material. If the NTT module were fabricated from AlN, a 35-layer substrate would have about the same thermal performance as a six-layer Al_2O_3 substrate due to a 5× greater thermal conductivity of AlN over Al_2O_3.

MCNC Thermal Module. Several MCM technologies are being developed at MCNC Center for Microelectronics. One prototype design of the MCNC thermal module is shown in Fig. 13.22.[15,44–49] The 102 × 122 × 13 mm package contains 25 10 × 10 mm chips. The chips are reflow bonded to thin

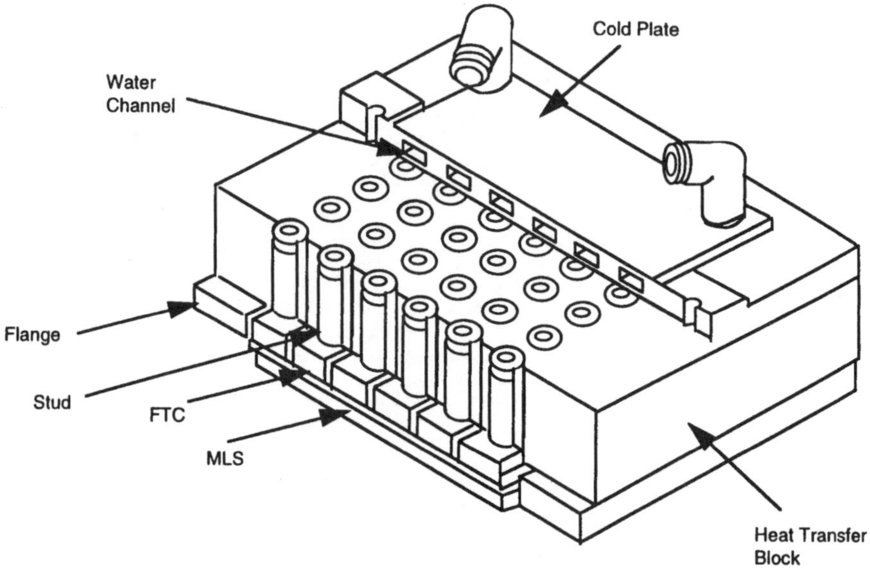

FIGURE 13.20c NEC multilayer substrate (MLS).[42]

film interconnections on the substrate with 95Pb/5Sn solder. Each 100k-gate chip has up to 1,356 connections, but only about 600 are needed for an electrical function with CMOS technology. The remaining solder bumps provide mechanical reinforcement and enhanced heat transfer between chip and substrate. If bipolar technology were employed, all solder bump connections would probably be needed for an electrical function. The thin film interconnection structure has up to seven layers of copper conductors in a polyimide dielectric. Electroless nickel vias are used to connect layers. The aluminum nitride substrate would have either molybdenum or tungsten power and ground planes. With current technology, up to 2200 I/O pins can be brazed to the substrate back surface. However, up to 10,000 second-level connections might be needed for a bipolar logic module, so alternative technologies are currently under development. The water-cooled silicon carbide heat sink is designed to accommodate a total flow rate of 500 cc/s (7.9 gpm). A layer of indium solder is used to attach each chip to the heat sink.

A copper prototype module was built to verify the heat sink design, and a silicon wafer with nine heater chips was bonded to the heat sink using gallium. At a power dissipation of 42 W/chip, the measured chip-to-water thermal resistance was 0.4° C/W,[44] which was about 15 percent lower than the calculated value. Extensive thermal stress modeling has been performed to analyze the potential reliability hazards for the module design shown in Fig. 13.22.[45–48] The simulations indicate that thermal stresses on the solder bumps during power cycling can be reduced by decreasing the substrate and heat sink thickness, and by clamping the substrate and heat sink together at the perimeter. Mechanical testing on indium die bond joints and 95Pb/5Sn solder bump arrays indicates that a fully populated array of solder bumps deforms at higher loads than an indium die bond.[48] This suggests that any heat sink/substrate expansion mismatch would be more readily absorbed by the indium, resulting in greater reliability for the Pb-Sn electrical connections. Prototype modules need to be tested to verify the simulation results and to expose other potential hazards such as thermomigration[50,51] and corrosion[52] in the indium die bonds, and stress-induced failure in the thin film interconnection structure.

Fujitsu Direct Immersion Cooling Module. The developmental Fujitsu immersion cooling design is shown in Fig. 13.23.[53] The chips are cooled by a boiling mixture of fluorocarbon fluids (3M Co.

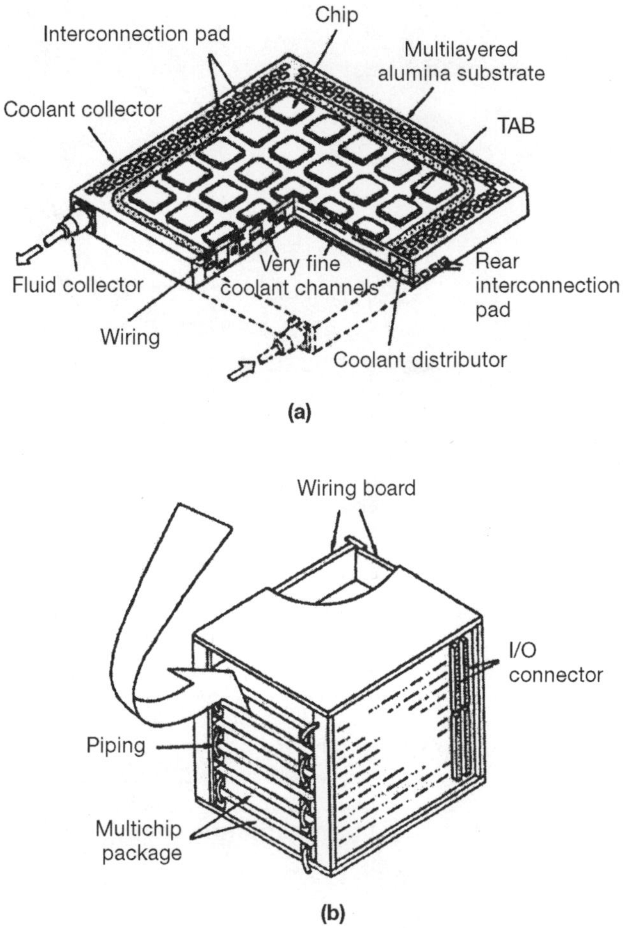

FIGURE 13.21 NTT liquid-cooled substrate module.[43]

FX-3300, boiling point 102° C, and FX-3250, boiling point 56° C). Two substrates with flip chip devices face each other and are separated by bubble guides and porous nickel bubble traps. In the prototype module, each glass-ceramic substrate supports 49 10× 10 mm chips. The total module size is 180 × 140 × 35 mm with the spacing between substrates being 20 mm. One water-cooled heat exchanger tube is set in the vapor layer to condense the vapor. One water-cooled heat exchanger tube is set in the vapor layer to condense the vapor, and another tube is submerged in the fluorocarbon fluid. The water flow rate was 500 cc/min in the vapor condensation tube and 1000 cc/min in the submerged tube.

At 1000 W power dissipation per module, the chip-to-water thermal resistance was 5.4° C/W. The thermal resistance was more than 3× higher for a module without bubble traps and guides. Also, the space between facing chips was important. Below a 1 mm space, the thermal resistance began to increase significantly due to interactions between bubbles leaving adjacent surfaces.

One problem with immersion cooling is temperature overshoot where the chip temperature rises beyond the point that nucleate boiling starts. The developers found that a 20:80 ratio of 3M Co. FX-3300:FX-3250 was able to minimize temperature overshoot while having little effect on cooling

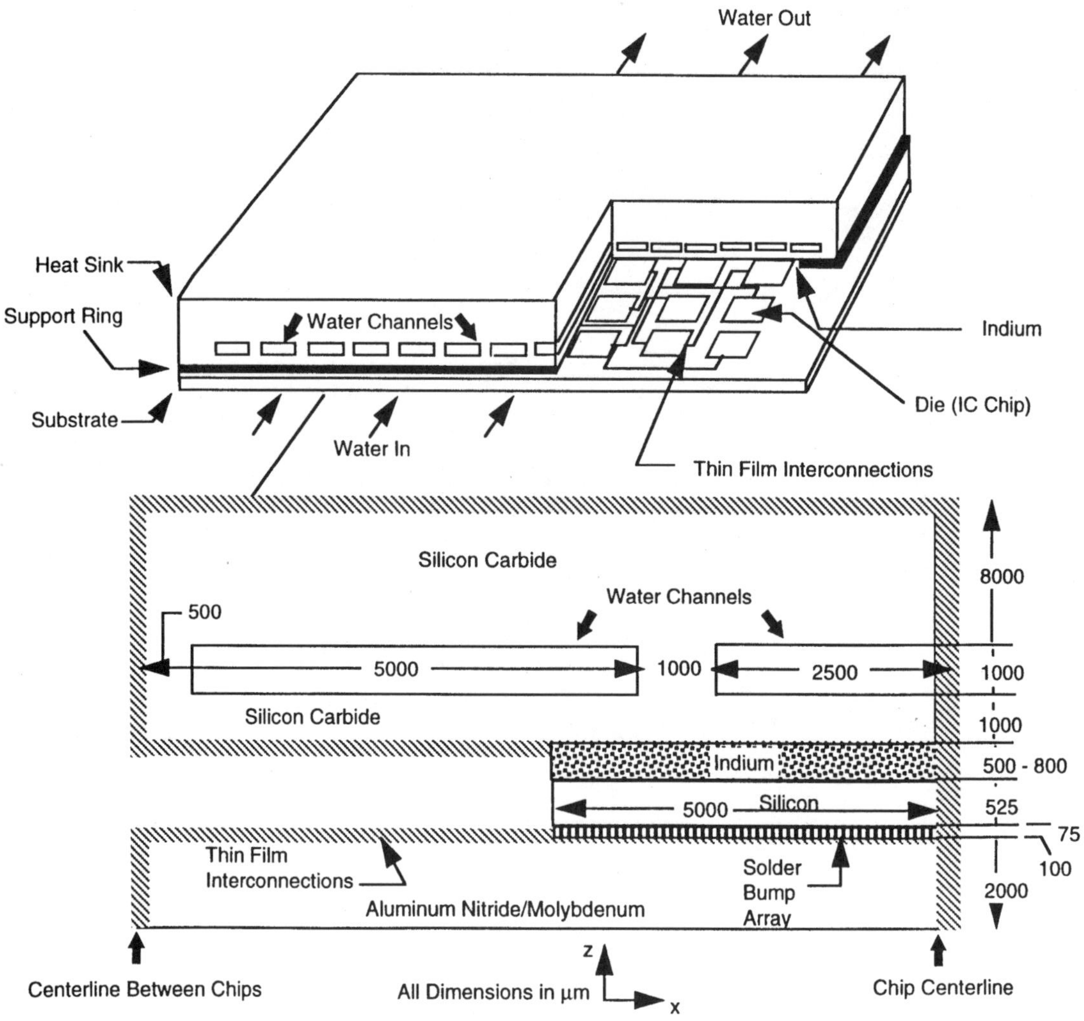

FIGURE 13.22 MCNC multichip package design.[46]

capability. Another potential concern for the design is reliability of the flip chip solder joints. Chu, et al.[37] reported that the IBM liquid encapsulation module had failures from contaminants being dissolved in the coolant and redepositing on the interconnection pads, resulting in corrosion-enhanced fatigue. Such a mechanism might be prevented by using a protective sealant on the solder joints.

DEC Integral Heat Pipe Module. A prototype MCM with an integral heat pipe has been developed by Fox et al. at Digital Equipment Corp.[54] The basic heat pipe structure is shown in Fig. 13.24. A porous saturated wick maintains the working fluid in contact with the back side of flip chip devices. Heat input results in evaporation of the fluid causing latent heat to be absorbed. As the vapor condenses on the cooler cap assembly, the latent heat is transferred. The condensed working fluid then returns by capillary action to the chip surface. The integral heat pipe design is a mechanically compliant, thermally effective technique for transferring heat off the back side of flip chip devices.

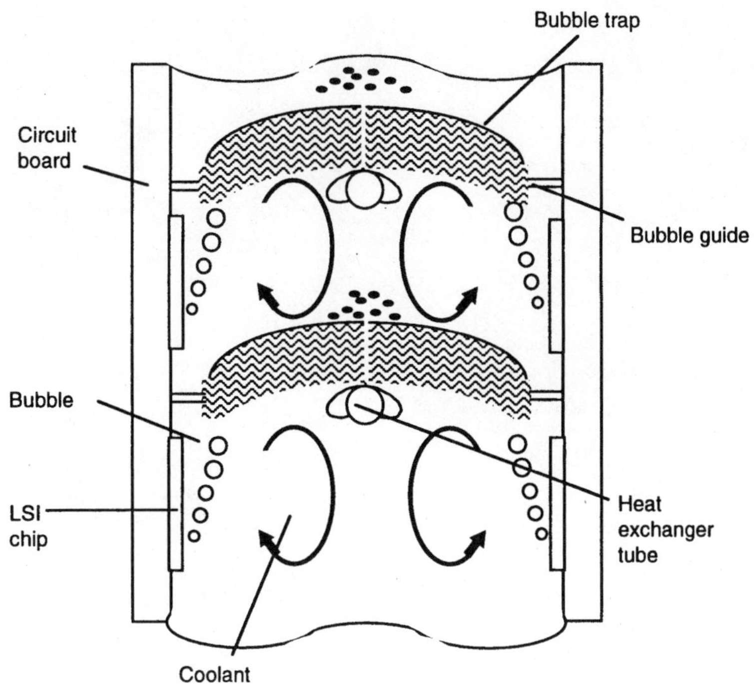

FIGURE 13.23 Fujitsu direct immersion cooling module.[53]

A 16-chip module approximately 76 × 76 × 38 mm in size was fabricated to test the heat pipe performance. Heater chips measuring 6.35 × 6.35 mm were bumped to a multilayer ceramic substrate. Nine sites had 256 I/Os, and seven sites had 400 I/O pins. A schematic of the multichip test vehicle is shown in Fig. 13.25. The condenser assembly was fabricated of brass, and a thermocouple was used to monitor the back surface temperature. The water flow rate was about 8 L/min.

At a power dissipation of 11 W/chip, the measured internal thermal resistance was 2.8° C/W with a 52 percent fill of pentane. Assuming a water channel convective heat transfer coefficient of 1 W/cm^{2}°C over a 36-cm^{2} area, the external thermal resistance is about 0.4° C/W on a per-chip basis. Therefore, the total chip-to-water thermal resistance is about 3.2° C/W for this design. Operating at 10° tilt of the package resulted in a 20–50 percent increase. The authors point out that accelerated life testing methodology for heat pipes is not well established, so reliability is difficult to assess over a short-term period. Contamination effects and interaction of the working fluid with the other package materials need to be investigated.

Fujitsu Computer Packaging Technologies Conformable Cold Plate. A cold plate typically consists of two relatively thick metal flat plates with coolant passages in between. To be able to remove heat efficiently, the cold plate has to be in a good thermal contact with the heat sources. In this design, the side of the cold plate which comes in contact with the chips is made of conformable thin metal sheet or a diaphragm.[20] The basic cold plate structure is shown in Fig. 13.26. The conformable cold plate contained a water channel that was 76 mm wide, 80 mm long, and about 2 mm deep. The pressure in the cold plate was below 3.4k PA (5 psi) which produced an average velocity range of 0.5 to 2 m/s. Spring-loaded plungers were located on the bottom side of the test section to improve the contact between the chip and stainless steel diaphragm. To further improve contact between the chips and a diaphragm, a thermal grease was used. Six Cu blocks simulated power dissipating chips

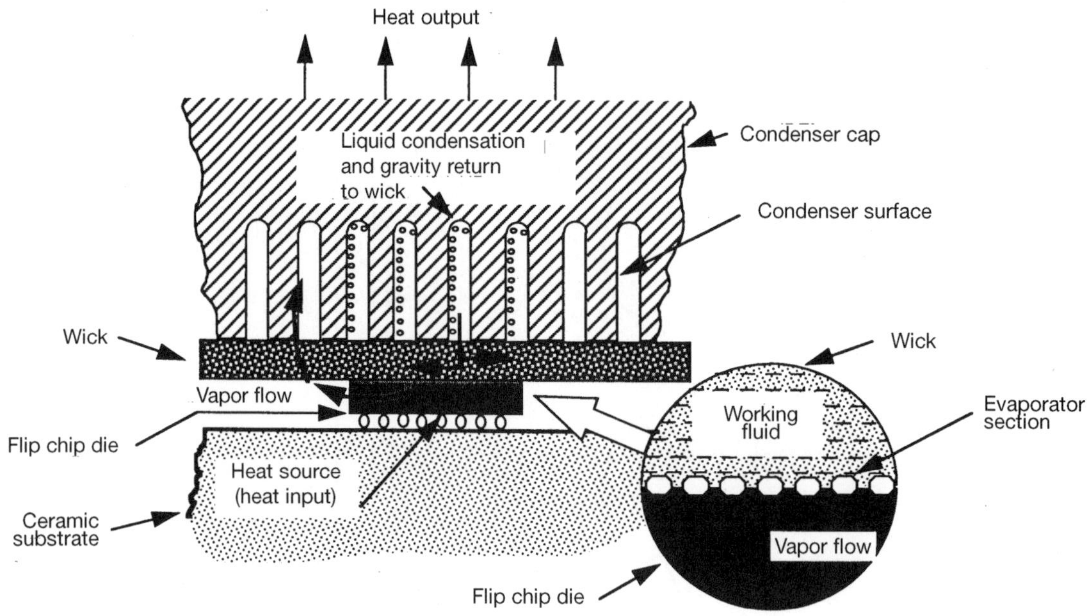

FIGURE 13.24 DEC integral heat pipe.[54]

in an MCM (the heater block, Fig. 13.27). The Cu chips were 15.5 × 15.5 × 16.3 mm in size. They were attached to a Lexan plate with a spacing of 4 mm. Thick film resistor heaters were soldered to the other side of Cu blocks and could be dissipated up to 40 W. External thermal resistance is in a range of 0.5 to 0.61° C/W for chips powered at 20–28.8 W with thermal grease and plungers used.

The simulation indicates that 80 to 100 W can be dissipated from the top side of the chips at water velocities of less than 2 m/s. (It was assumed that maximum chip temperature was 85° C, and the water inlet was 25° C.) The equivalent chip power dissipation for other liquid-cooled commercial systems are 27 W for IBM[37,4] and 30 W for Fujitsu.[53] Although this concept was not demonstrated on an actual MCM, it did show impressive results for liquid-cooled cold plates.

Fujitsu FACOM M-780 and VP2000. The Fujitsu FACOM M-780 also uses single-chip packages mounted on a PC board.[55,56] A total of 336 packages are surface mounted to both sides of a 540 × 488 mm board, with a maximum chip power dissipation of 9.5 W and a total board dissipation of 3 kW. A water-cooled cold plate with individual bellows contacting each package provides a chip-to-water thermal resistance of 2.4° C/W.

The Fujitsu VP 2000 computer[57] has 144 single-chip packages mounted on a 61-layer glass-ceramic substrate which is 245 × 245 mm in size. The total module size is approximately 325 × 325 × 70 mm. The logic chips are 13.5 × 13.5 mm with 462 TAB connections mounted in a pin grid array package which is 17 × 17 mm.

The water impingement cooling scheme is shown schematically in Fig. 13.28. The nominal chip power dissipation is 30 W, and the maximum power dissipation per substrate is 4,600 W. The average measured thermal resistance is 0.56° C/W with a maximum of 0.65° C/W.

CRAY-2. The CRAY-2 supercomputer uses arrays of single-chip packages on PC boards.[56,58] Eight PC boards are grouped to form a module, and 14 vertical columns with 24 modules each make up the computer. The entire system is immersed in 3M FC-77 Fluorinert fluid, which is pumped through the chassis at a rate of 265 L/min (70 gal/min). The total power dissipation of the system is 194 kW, which far exceeds that of any other supercomputer design.

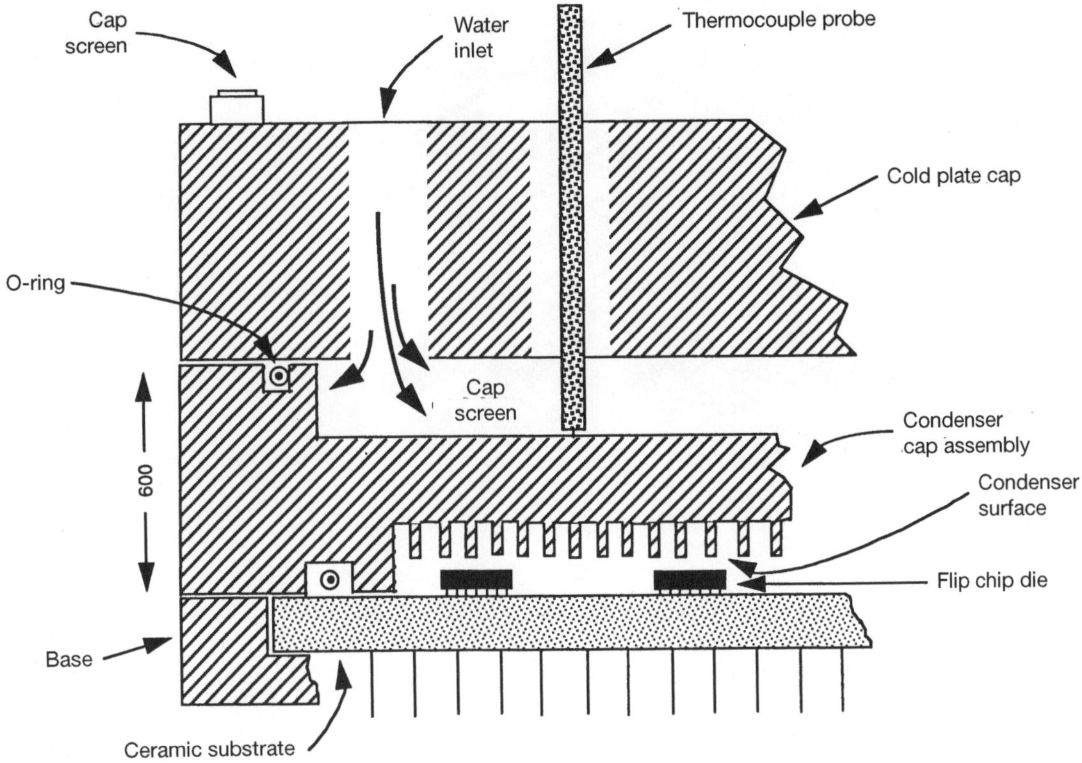

FIGURE 13.25 DEC integral heat pipe module.[54]

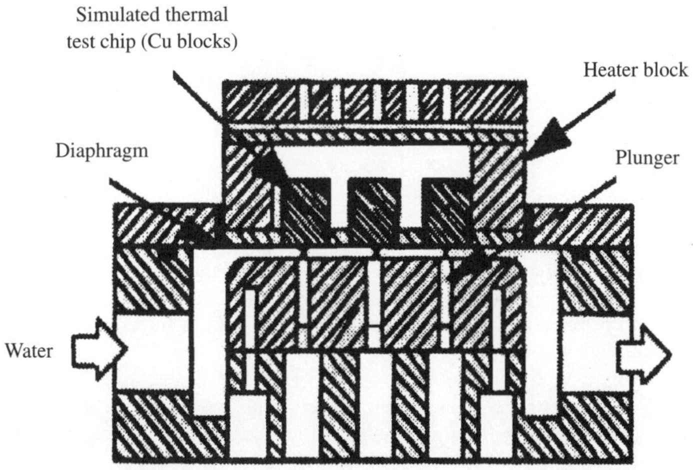

FIGURE 13.26 Conformable cold plate test section.

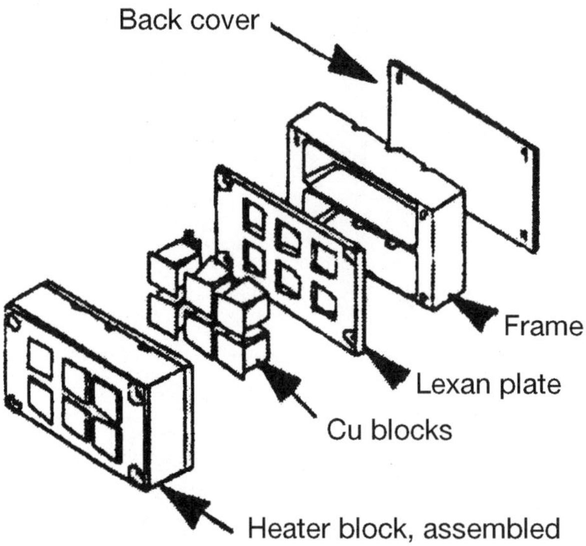

FIGURE 13.27 Heater block for simulating an MCM.

13.4 TECHNOLOGY COMPARISON

In comparing the MCM technologies described in this chapter, it must be pointed out that some designs are still in the developmental stages and are not being manufactured and shipped on a large scale. Clearly, more weight must be given to packages that have undergone full reliability qualification and are being used in the field. Nevertheless, the developmental modules have been included in the discussion to give a broad perspective of the various technologies being pursued at present.

A compilation of physical characteristics of the various MCMs is given in Table 13.1. It is seen that there is great diversity in the module size, chip size, and number of chips. Most of the modules in Table 13.2 are geared toward high-end computer. The number of chips ranges from 6 to more than 100, and the chip size ranges from 6.5 to 183 mm. The Hitachi SiC RAM module is the smallest package discussed in this chapter, but there are many other small-scale modules currently being produced. The NEC SX-3/SX-X and the Fujitsu VP2000 are by far the largest-scale modules built to date.

It is seen that the liquid-cooled modules are generally larger and have more chips and more I/Os than the air-cooled modules, although the DEC VAX 9000 and the MCC prototype module are comparable in size to the liquid-cooled packages. The design trends over time have been toward larger module size, larger chip size, and greater number of I/Os for the high-end systems. These trends are clearly pointed out by the evolution of the NEC SX-2 into the SX-3, which resulted in a 3× increase in the number of chips, a 4× increase in module size, an 8× increase in chip I/Os, and a 5× increase in module I/Os. In general, the scaling trends hold true for all of the newer technologies compared to the older technologies.

A comparison of MCM thermal performance characteristics is given in Table 13.2. The junction-to-ambient thermal resistance, shown in the first column, is seen to vary dramatically from module to module. This large variation is mainly due to differences in chip size; i.e., the packages with smaller chips show a larger θ_{ja}. Thermal resistance normalized by the chip size is shown in column 2. This figure of merit properly accounts for the hat flux on the chips, and it reveals a much smaller variation in performance among the technologies. With respect to θ_{jacs}, the DEC VAX 9000 multichip unit has

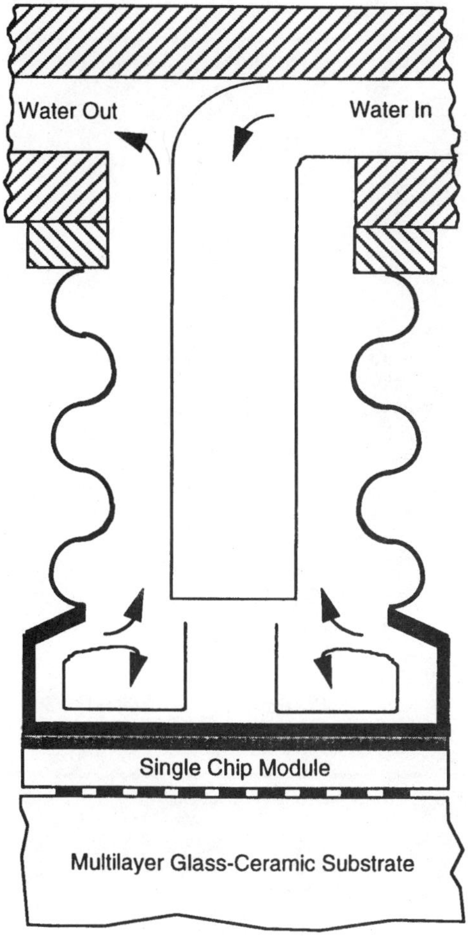

FIGURE 13.28 Fujitsu VP 2000 water impingement cooling structure.

the best performance of the air-cooled technologies, probably due to direct attachment of the chips to a large CrCu heat spreader. For the liquid-cooled technologies, the developmental MCNC thermal module shows the lowest value of θ_{jacs} due to direct solder attachment of the chips to a microchannel-type heat sink.

Probably the most relevant figure of merit for a MCM is thermal resistance normalized by the area packaging density, θ_{jaapd}. This FOM properly accounts for spreading effects and gives the best measure of the thermal packaging efficiency of a design. Two values of θ_{jaapd} are given in column 3 of Table 13.2. First, θ_{jaapd} is calculated on the basis of substrate area and, second, θ_{jaapd} is calculated on the basis of total module area, which usually includes a flange around the package perimeter. The value of θ_{jaapd} based on substrate area is most important from a chip-to-chip interconnection standpoint, and θ_{jaapd} based on module area is most important from a module-to-module and total system size standpoint.

For air-cooled technologies, it is seen that the IBM 4381, DEC VAX 9000, and the developmental MCC module have the best performance based on module area. Note that each of these designs uses

Table 13.1 MCM Physical Characteristics

Technology	Number of chips	Chip size, l × w (mm)	Module size, l × w × h (mm)	Number of chip I/Os	Number of module I/Os
			Air-cooled		
AT&T AVP	3	–10 × 10	48 × 30 × 17	330	160
Alcoa	5	11.5 × 10.7		—	—
		8.6 × 6.9	53 × 56 × 17		
DEC VAX 9000	9–72	2.54 × 2.54	127 × 127 × 50		
		6.86 × 6.86			
		9.78 × 9.78			
IBM 4381	36	4.57 × 4.57	64 × 64 × 40	4,320	882
Mitsubishi HTCM	9	8.0 × 8.0	66 × 66 × 22	2,007	624
HItachi SiC Ram	6	1.9 × 4.0	27 × 27 × 16	462	108
NTT	36	3.0 × 3.0	70 × 70 × 22	—	288
MCC	25	10 × 10	150 × 150 × 26	—	—
			Liquid-cooled		
IBM 3090	100–118	4.85 × 4.85	150 × 150 × 50	12,000	1,800
Honeywell SLIC	up to 60	—	100 × 100 × 50	—	240
NEC SX-3/SX-S	100*	—	240 × 250 × 70	48,500	11,540
SX-2	36	8.0 × 8.0	540 × 488 × 100	6,336	2,177
Fujitsu VP2000	144	13.5 × 13.5	325 × 325 × 70	66,528	8,640
FACOMM-780	336	—	540 × 488 × 100	—	—
NTT LCS	25	8.0 × 8.0	85 × 105 × 3	—	900
MCNC	25	10 × 10	102 × 122 × 13	15,000–31,000	2,000–10,000
Fujitsu DICM	49	10 × 10	180 × 140 × 35	—	—
DEC heat pipe	16	6.4 × 6.4	76 × 76 × 38	5,104	400

* Represents number of flip chip TAB carriers that contain one to four chips each.

impinging air normal to the plane of the package, but each module has a unique heat sink design. The IBM module has a serpentine-like unitary heat sink, the DEC module has a pin-fin design, and the MCC module has individual trapezoidal-finned heat sinks attached to each chip. These air impingement design shave about a 2× better thermal performance than the modules with air flow parallel to the plane of the package.

Based on substrate area, θ_{jaapd} is fairly consistent, with the Hitachi and NTT air-cooled modules having a slight advantage. Both of these modules have a substrate that is much smaller than the heat sink. This indicates that heat spreading measures are still effective at the module level. However, as module size increases, heat spreading to the package perimeter will become less effective.

For the liquid-cooled technologies, the Fujitsu VP 2000, NEC SX-3, and the developmental MCNC module have the best thermal performance with respect to θ_{jaapd}. The NEC liquid-cooled module had a 6× improvement in thermal performance between the SX-2 and SX-3 models, yet no changes in the design were reported. It is speculated that much higher thermal conductivity materials were used for die attachment and FTC/stud contact to reduce internal thermal resistance, and that improved water distribution and channel design were used to reduce external thermal resistance. The Fujitsu VP2000 had a similar improvement in θ_{jaapd} over the previous generation FACOM M-780.

The final FOM tabulated in Table 13.2 is the thermal resistance normalized by the volumetric packaging density, θ_{javpd}. This FOM is probably most appropriate for modules that were meant to be stacked vertically, such as the NTT liquid-cooled substrate design. It is seen that the air-cooled technologies have a fairly consistent value of θ_{javpd}, except for the DEC VAX 9000, which has a larger value due to its greater thickness. For the liquid-cooled designs, the developmental MCNC and NTT modules have the lowest value for θ_{javpd}.

Table 13.2 MCM Thermal Performance Characteristics

Technology	Thermal resistance θ_{ja} (°C/W)	Thermal resistance normalized to chip size θ_{jacs} (°C/W/cm^2)	Thermal resistance normalized to area packing density θ_{jaapd} (°C/W)(cm^2/chip) $A_{substrate}/A_{module}$	Thermal resistance normalized to volumetric packaging density (°C/W)(cm^3/chip)
		Air-cooled		
AT&T AVP	15	15	16/72	122
Alcoa	2.8	2.3	14/17	28
DEC VAX 9000*	7.3	0.98	18/27	134
IBM 4381	17	3.6	17/19	77
Mitsubishi HTCM	7.7	4.9	18/37	82
HItachi SiC Ram	33	2.5	12/40	64
NTT	28	2.5	14/38	84
MCC	2.4	2.4	20/22	56
		Liquid-cooled		
IBM 3090†	8.7	2.0	7.0/20	98
Honeywell SLIC	22		23/37	183
NEC SX-3/SX-S‡	0.6		3.0/3.8	23
SX-2¶	5.0	3.2	14/22	130
Fujitsu VP2000	0.65	1.2	2.7/4.8	33
FACOMM-780	2.4		19/23	
NTT LCS	3.3	2.1	5.6/12	3.5
MCNC	0.42	0.42	1.4/2.1	2.7
Fujitsu DICM	5.4	5.4	11/28	49
DEC heat pipe	3.2	1.3	7.4/12	44

* Packaged with 1 clock chip, 3 gate array chips, and 40 STRAM chips.
† Packaged with 100 logic chips.
‡ Packaged with 100 logic chips.
¶ Packaged with 36 logic chips.

13.5 FUTURE CHALLENGES

The primary goal of a cooling system is to enhance the performance and reliability of a module. As to performance, device characteristics are temperature dependent, so it is desirable to have all chips on a module operating at nearly the same temperature. With respect to reliability, many degradation processes such as electromigration, corrosion, and solder fatigue are thermally activated. Hence, the maximum operation temperature should be limited to reduce degradation rates. These thermal control goals must be met within the constraints imposed by manufacturing and assembly tolerances. Manufacturing constraints become more severe as the module size, number of chips, and number of I/Os increase. Hence, the technological challenge is more difficult for high-end systems. The primary challenge for low-end systems is reducing cost.

Significant gains in performance and reliability can be obtained through materials development and characterization. Typical desired properties for heat sink and substrate materials are high thermal conductivity, CTE matched to the chips, high fracture toughness, corrosion resistance, and relatively low cost. Substrate materials should also have a low ε_r for power/ground planes. Die attachment materials should also have a low ε_r for signal layers and a somewhat higher ε_r for power/ground planes. Die attachment materials should ideally be thermally conductive, compliant, fatigue resistant, low in process temperature, and reworkable. They might need to be electrically insulating or conducting, depending on the device technology. Obviously, the specific design will determine which mix of material characteristics is most important for a given application. In all applications, however, accurate material property characterization is needed, especially for modeling and reliability prediction.

Another aspect of thermal performance that needs to be characterized more fully is the machine duty cycle under use conditions. For example, to predict solder joint failure rates, accurate knowledge of chip and module power fluctuations is needed. It would be useful to design features that record upper and lower peak machine temperatures over time. For very high reliability applications, an active thermal control system might be used to provide power dissipation during low-utilization periods to maintain a uniform temperature across a module.

One major challenge for MCM design is the assembly of large structural components. Mechanical compliance is needed to accommodate dimensional tolerances of the components during assembly and to absorb thermal expansion mismatch during use. For TAB and wire bond technologies, there is mechanical compliance in the electrical connections to each chip, so less constraint is imposed on the thermal connection. For modules in which the chips are backside die attached to a substrate and wire bonded out, issues such as camber in the substrate and thickness variation in the chips are of less concern. The issue of thermal expansion mismatch between dissimilar materials is still a concern, however.

With flip chip technology, there is little compliance in the electrical connections to each chip, so compliance must be incorporated into the thermal connection. Mechanical compliance is important during both assembly and use. Designing a thermally conductive, mechanically compliant interface to cool the back side of flip chip devices is a major challenge in MCMs. Several different design approaches are listed in Table 13.3. Both compliant mechanical designs and materials are employed for back-side cooling. The MCNC module also depends on having CTE matched structural materials and a fully populated array of flip chip joints for mechanical reinforcement. The reinforcement of flip chip connections could be further enhanced by using an underfill epoxy, but the epoxy would have to be applied after all rework operations had been completed.

Another possibility for thermally conductive, mechanically compliant attachment is to use a liquid metal such as gallium or one of its alloys. A prototype module using gallium to attach a 100 mm silicon wafer to a copper heat sink has been tested at MCNC.[44] The main challenges to using a liquid metal for thermal contact are (1) developing methods of application, (2) containment of the material during extended use, and (3) prevention of liquid metal embrittlement. Gallium is known to embrittle solder alloys at the grain boundaries.[59]

Table 13.3 Back-Side Cooling of Flip Chip and Flip-TAB Devices in MCMs

Technology	Method to accommodate chip height variation during assembly	Method to absorb heat sink/substrate thermal expansion mismatch during power cycling
IBM 4381	Thermal paste	Thermal paste
Mitsubishi HTCM	Solder reflow attachment of heat spreaders	Gas gap between heat spreaders and cap
IBM 3090 TCM	Spring-loaded pistons	Spring-loaded pistons
Fujitsu VP2000	Bellows	Bellows
NEC SX series LCM	Adjustable metal studs	Thermal compound
MCNC	Solder reflow attachments of heat sink	Low yield stress die attachment solder (indium)
Fujitsu DICM	Fluorocarbon fluid	Fluorocarbon fluid
DEC heat pipe	Compliant wick	Compliant wick

Perhaps the ultimate cooling technology demonstrated to date has been microchannel water cooling.[6,11,13,60] Typical dimensions of the channels are 50–300 μm. One approach is to fabricate channels into the chip itself. Another approach is to fabricate a microchannel heat sink and attach it to the chip using solder or a microcapillary structure. For single-chip heat sinks, measured thermal resistance is an order of magnitude lower than any of the modules listed in Table 13.2. The remaining challenge is to incorporate such a technology into a large-scale MCMs under realistic manufacturing conditions. If individual heat sinks are employed for each chip, a compliant, low head-loss water distribution system must be developed. No such module has been reported to date, but the potential performance is very attractive.

REFERENCES

1. R.F. Chu, R.E. Simons, "Review of Thermal Design for Multichip Modules," *Proc. NEPCON West,* Anaheim, 1633 (1990).
2. A. Bar-Cohen, "Thermal Management of Air- and Liquid-Cooled Multi-chip Modules," *IEEE Trans. Compon. Hybrids Manuf. Technol.,* CHMT-10(2), 159 (1987).
3. S. Oktay, R. Habberman, A. Bar-Cohen, "High Heat from a Small Package," *Mechanical Engineering,* 36 (March 1986).
4. R.R. Tummala, E.J. Rymaszewski, Eds., *Microelectronics Packaging Handbook*, Van Nostrand Reinhold, New York, pp. 21–24 (1989).
5. R.W. Keyes, "A Figure of Merit for IC Packaging," *IEEE J. Solid-State Circuits,* SC-14, 265 (April 1978).
6. R.W. Pease, J.R. Brown, "High Performance Packaging for VLSI Systems," *SRC,* June 27, 1987.
7. M. Mahalingam, "Thermal Management in Semiconductor Device Packaging," *Proc. IEEE,* 73(9), September 1985.
8. W.N. Carr, J. Mize, *MOS/LSI Design and Application*, McGraw-Hill, New York, Chapter 3, pp. 85–89 (1972).
9. D.S. Peck, "The Analysis of Data from Accelerated Tests," *Proc. 9th Reliab. Phys. Conf.,* 69 (1971).
10. C.J. Bartlett, J.M. Segelken, N.A. Teneketges, "Multichip Packaging Design for FLSI-Based Systems, *Multichip Modules,* IEEE Selected Reprint Series, p. 116–129, (1991).
11. D.B. Tuckerman, R.F. Pease, "High Performance Heat Sinking for VLSI," *IEEE Electron Device Lett.* EDL-2(5), 126 (May 1981).
12. T. Kishimoto, T. Ohsaki, *Proc. 36th EIA/IEEE Electron. Compon. Conf.,* 596 (1986).
13. D.B. Tuckerman, *Heat-Transfer Microstructure for Integrated Circuits*, Ph.D. Thesis, Dept. of Electrical Engineering, Stanford Univ., February 1984.
14. R.C. Chou, "Heat Transfer in Electronic Equipment," *Proc. 8th Int. Heat Transfer Conf.,* 293 (1986).
15. L.T. Hwang, I. Turlik, A. Reisman, "A Thermal Module Design for Advanced Packaging," *J. Electron. Mater.,* 16(5), (1987).
16. M.F. Bregman, A. Kimura, T. Matsui, H. Nishida, K. Nishiyama, H. Ohkuma, A. Tanaka, C. Kovac, and D. McQueeney, "A Thin Film Multichip Module for Workstation Applications," *Proc. IEEE 42nd ECTC,* San Diego, CA, May 18–20, 1992, p. 968.
17. P.A. Totta, "Flip-Chip Solder Terminals," *21st EIA/IEEE Electron. Compon. Conf.,* 275 (May 1971).
18. L.S. Mik, "Thermal Management of Silicon-Based Multichip Modules," *Proc. Tenth IEEE Semi-Therm,* p. 59–63, (1994).
19. K. Okutani, K. Otasuka, K. Sahara, K. Satoh, Packaging Design of SiC Ceramic Multichip RAM Module, *Proc., IEPS Int. Electron. Packag. Conf.,* 299 (1984).
20. V. Mansingh, D. Horine, "Multichip Module Cooling with Conformable Cold Plates," EEP-Vol. 10-2, *Advances in Electronic Packaging*, p. 943–948, ASME 1995.
21. A.J. Blodgett, Jr., "Microelectronic Packaging," *Sci. Am.* 249, 86 (July 1983).
22. S. Oktay, Parametric Study of Temperature Profiles in Chips Joined by Controlled Collapse Technique, *IBM J. Res. Dev.,* 13, 272 (May 1969).

23. R. Nelson, T. Dolbear, K. Chen, and C. Chao, "Application of Air Cooled Microchannel Heat Exchangers to Card Cage Systems," MCC Technical Report, HVE-235-92.
24. J.M. McPhee, T.S. O'Toole, M. Yedvabny, "Cooling the Vax 9000," *Proc., ELECTRO/90 Electron. Conf.* (1990).
25. R. Sullhan, M. Fredholm, T. Monaghan, A. Agrawal, B. Kozarek, *Thermal Modeling and Analysis of Pin Grid Arrays and Multichip Modules,* Alcoa Electronic Packaging, Inc., (1991).
26. Y.C. Lee, H. T. Ghaffari, J.M. Segelken, Internal Thermal Resistance of a Multichip Packaging Design for VLSI Systems, *Proc. EIA/IEEE 38th Electron. Comp. Technol. Conf.,* Houston, (1989).
27. R.G. Biskeborn, J.L. Horvath, E.B. Hultmark, "Integral Cap Heat Sink Assembly for the IBM 4381 Processor," *Proc. IEPS Int. Electron. Packag. Conf.,* 468, (1984).
28. G.G. Werbizky, F.W. Haining, "Circuit Packing for Large Scale Integration," *EIA/IEEE Electron. Compon. Conf.,* 187 (1985).
29. N. Timko, Jr., "Blower Performance Enhancements in the 4381 Computer," *IEPS Int. Electron. Packag. Conf.,* 475 (1984).
30. M. Kohara, S. Nakao, K. Tsutsumi, H. Shibata, and H. Nakata, "High Thermal Conduction Module," *EIA/IEEE Electron. Compon. Conf.,* 180 (1985).
31. M. Kohara, S. Nakao, K. Tsutsumi, H. Shibata, and H. Nakaga, "High Thermal Conduction Package Technology for Flip Chip Devices," *IEEE Trans. Compon. Hybrids Manuf. Technol.,* CHMT-6(3), 267 (September 1983).
32. J.M. McPhee, Digital Equipment Corporation, Marlborough, MA, private communication, (December 1990).
33. J. Poi, S. Heng, R. Charlantini, and P. Gildea, "Cooling Components Used in VAX 9000 Family of Computers," *IEPS Int. Electron. Packag. Conf.,* 587 (1990).
34. E. Canty, "Understanding Digital's Multi-Chip Packaging Technology," Digital Equipment Corp., Marlborough, MA (1990).
35. S. Sasaki, T. Kon, and T. Ohsaki, "A New Multi-Chip Packaging Technology," Digital Equipment Corp., Marlborough, MA (1990).
36. C. Hilbert, S. Sommerfeldt, O. Gupta, D. J. Herrell, "A Low Pressure Air Cooling System for High Performance Modules," *Proc. NEPCON West, Anaheim,* 1667 (1990).
37. R.C. Chu, U. P. Hwang, and R.E. Simons, "Conduction Cooling for an LSI Package: A One-Dimensional Approach," *IBM J. Res. Dev.,* 26(1), 30 (January 1982).
38. S. Oktay, H.C. Kammerer, "A Conduction-Cooled Module for High-Performance LSI Device," *IBM J. Res. Dev.,* 26(1), 55 (January 1982).
39. A.J. Blodgett, D.R. Barbour, "Thermal Conduction Module: A High-Performance Multilayer Ceramic Package," *IBM J. Res. Dev.,* 26(1), 30 (January 1982).
40. A. Bar-Cohen, "An Addendum and Correction to Thermal Management of Air and Liquid-Cooled Multichip Modules," *IEEE Trans. Compon. Hybrids Manuf. Technol.,* CHMT-11(3), 333 (September 1988).
41. T. Watari, H. Murano, "Packaging Technology for the NEC SX Supercomputer," *EIA/IEEE Electron. Compon. Conf.,* 192 (1985).
42. D. Akihiro, W. Toshihiko, H. Hideki, "Packaging Technology for the NEC SX-3/SX-X Supercomputer," *EIA/IEEE Electron. Compon. Technol. Conf.,* 525 (1990).
43. T. Kishimoto, T. Ohsaki, "VLSI Packaging Technique Using Liquid-Cooled Channels," *EIA/IEEE Electron. Compon. Conf.,* 595 (1986).
44. D. Nayak, L. T. Hwang, I. Turlik, and A. Reisman, "A High-Performance Thermal Module for Computer Packaging," *J. Electron. Mater.,* 16(5), 357.
45. I. Turlik, A. Reisman, R. Darveaux, L. T. Hwang, "Multichip Packaging for Supercomputers," *Proc. NEPCON East,* Boston (1989).
46. R. Darveaux, I. Turlik, L.T. Hwang, and A. Reisman, "Thermal Analysis of a Multichip Package Design," *J. Electron. Mater.,* 18(2), 267 (1989).
47. R. Darveaux, I. Turlik, L.T. Hwang, and A. Reisman, "Thermal Stress Analysis of a Multichip Package Design," *IEEE Trans. Compon. Hybrids Manuf. Technol., CHMT-12*(4), 663 (December 1989).
48. R. Darveaux, "Mechanical Evaluation of Indium for Die Attachment in a Multichip Package," Ph.D. Thesis, North Carolina State University (1990).

49. R. Darveaux, E. Yung, I. Turlik, K.L. Murty, "Mechanical Characteristics of In and Pb5Sn Solders in a Thin Film Multichip Package," *Proc. MRS Meet.*, (Fall 1990).
50. B.N. Agarwala, J.S. Jaspal, "Mechanism of Microvoid Formation in Pb-In Solder Interconnections," AIME, (February 26, 1984).
51. W. Roush, J. Jaspel, "Thermomigration in Lead-Indium Solder," *EIA,IEEE Electron. Compon. Conf.*, 342.
52. R. Kossowsky, R.C. Person, L.C. Christovich, "Corrosion of In-Based Solders," *Proc. 16th Reliability Physics Symp.*, 200 (1978).
53. K. Yokuchi, N. Kamehara, K. Niwa, "Immersion Cooling for High Density Packaging," *EIA/IEEE Electron. Compon. Conf.*, 545 (1987).
54. L.R. Fox, G. Kromann, R.J. Hannemann, "Multichip Package Cooling with Integral Heat Pipe," *Proc. NEPCON West,* Anaheim, 1650 (1990).
55. H. Yamamoto, Y. Udagawa, M. Suzuki, "Cooling for FACOM M-780 Large Scale Computer," *Proc. Int. Symp. Cool. Technol. Electron. Equip.*, 110 (1987).
56. R.C. Chu, R.E. Simons, "Review of Thermal Design for Multi-Chip Modules," *Proc. NEPCON West,* Anaheim, 1663 (1990).
57. A. Kaneko, K. Seyama, M. Suzuki, "LSI Packaging and Cooling Technologies for FUJITSU VP2000 Series," 41(1), 12 (1990).
58. R.D. Danielson, N. Krajewski, J. Brost, "Cooling a Superfast Computer," *Electron. Packag. Prod.* 44 (July 1986).
59. J.K. Tien, B.C. Hendrix, P.O. Bretz, and A.I. Attarwala, "Creep-Fatigue Interactions in Solder," *EIA/IEEE Electron. Compon. Conf.*, 259 (1989).
60. A. Paal, R. F.W. Pease, "Extending Microcapillary Attachment to Rough Surfaces," Semiconductor Research Corp. Report C87268, (1987).

CHAPTER 14

KNOWN-GOOD DIE

Russell J. Wagner
Collins Avionics and Communications Division,
Rockwell International Corporation

John K. Hagge
Collins Commercial Avionics Division,
Rockwell International

14.1 INTRODUCTION

This chapter examines the need for *known-good die* (KGD) driven by the yield and reliability considerations of multichip modules. Various concepts and definitions of KGD are examined, and strategies for achieving high levels of "known-goodness" are discussed. Current methods used in industry are summarized and compared. Various initiatives working on solutions are examined, and likely future improvements are considered.

Competitive manufacturing technologies depend on six key elements:

- A supply of good materials entering the manufacturing process
- A supply of good parts entering the assembly process
- Well controlled processes that have little variation
- Robustly designed products that meet all customer expectations
- A product pricing structure that the customer finds affordable
- Short manufacturing cycles that meet both customer needs and market opportunities

These six elements combine to embody the familiar salesman's slogan of *good, fast, cheap.*

The requirement for having good parts entering the assembly process would be staunchly defended by most manufacturing engineers and is particularly relevant in the case of electronic assemblies using *bare* or *unpackaged* integrated circuits (ICs).

When ICs are handled in an unpackaged format, they are much more difficult to thoroughly test than after they have been sealed into single-chip packages. This is because of the difficulty of making electrical test contact; maintaining test access over temperature extremes; and problems in removing dissipated heat, providing adequate power and ground connections, and dealing with signal degradation from probe losses or contact resistance.

Consequently, most unpackaged ICs have historically received only limited probe testing, which typically resulted in probabilities of being good of 0.50 to 0.99, as shown in Table 14.1.[1,2] This results in several orders of magnitude higher fallout rate than for packaged ICs.[3] As seen in Table 14.1, different types of ICs tend to have higher KGD levels. Memory ICs in their mature stages can exceed 0.99 levels with simple DC probing. On the other hand, some categories typically have very low levels. CMOS ICs are reported[4] to have ten times higher early fallout rates than bipolar ICs.

Table 14.1 Cited estimates of the probability of being known-good for bare die obtained after DC wafer probe tests.

14.1.1 Concept of KGD

The term *known-good* was coined[11] to describe the needed probability of success that bare die must have to achieve reasonable assembly yields. The *known-goodness* needed for bare die can be quantified with simple probability equations as described in the next section. Several definitions of KGD have been formulated. The definition of ideal KGD[12] would state that it is known with high (such as 0.999) confidence levels that the ICs are free of any as-received defects and will remain defect-free through assembly, burn-in, environmental stress screening, and fielded infant mortality periods. This definition would result in fielded MCMs with high yield, a desired result for both the MCM fabricator and customer. A successful program to provide ICs meeting this definition to an MCM assembly process would require that at least 99.9 percent (or 99.99 in some applications) of all ICs would be fully functional at the following stages:

1. After MCM assembly processes (which follow IC vendors recommended handling procedures)
2. Over the design temperature range and operating speeds
3. After module burn-in or environmental stress screening (ESS)
4. After being fielded for a time approaching at least the minimum guaranteed field life (having been packaged in accordance with the IC vendor's recommendations)

Table 14.1 Cited Estimates of the Probability of Being Known-Good for Bare Die Obtained after DC Wafer Probe Tests

Description	Memory	Processor	ASIC	General	Reference
Early production	0.60	0.55	0.50	—	3
Early production	0.75	0.75	—	—	5
Second-year production	0.85	0.65	0.60	—	3
Mid-life production	0.90	0.90	—	—	5
Mature product	0.99	0.90	0.85	—	3
Mature product	0.93	0.95	0.90	—	5
Cited memory yields	0.978	—	—	—	6
Cited chip yields	0.95	0.90	0.85	—	7
Cited chip yields	—	—	—	0.80–0.95	8
Product start-up	0.99	—	0.90	—	9
Product in fifth year	0.99	—	0.98	—	9
State of the art	—	—	—	0.96	10

This ideal definition may be difficult to achieve. The latter two requirements are frequently ignored by declaring burn-in and infant mortality failures as a normal part of the manufacturing process. Nevertheless, the ideal definition is a useful goal, and six-sigma or even zero-defect levels are appropriate goals that will force the correct technology choices to be made. It should also be emphasized that known-good supplies of ICs can be achieved by two entirely different approaches: process controls during manufacture and screening tests after manufacture.

Other definitions of KGD are more pragmatic:

> A KGD is a bare die which has successfully passed the same electrical and reliability tests as its packaged counterpart.[13]

> KGD is a collective term that connotes bare die availability…with the same quality and reliability currently available in the equivalent single-chip packaged products.[14]

Or even more simply,

> A KGD is one that works in the system.[15]

These simple definitions are more useful for arranging purchase of KGD. However, they lack the definition of a metric to establish how good the die is. This uncertainty has led to the terms *possibly good die,*[16] *wishful thinking good die,*[17] *known-good die plus,*[18] *perfect die,*[19] *known-good/stays-good die,*[20] *good-enough die,*[21] and *almost-good die,*[22] which indicate the varied thinking on bare die reliability and yield that results without the guidance of metrics.

14.1.2 Impact of KGD on Yield

The yield which can be expected for an assembly of multiple parts which may have some small defect rate can be calculated from simple probability equations as follows:

$$Y_m = 100(P_c)^N \tag{14.1}$$

where Y_m = probable percent yield for assembled modules
N = number of ICs in the module
P_c = probability than an IC is good

This equation is plotted in Fig. 14.1, where it is evident that MCMs with more than a few ICs require ICs that are known-good to at least 0.999 probability levels to achieve high module yields.

When different chip types are used within a module, Eq. (14.1) can be modified for the number of chips ($A, B, C, \ldots$) of each type, and the probability for that type being good ($P_a, P_b, P_c, \ldots$) is as follows:

$$Y_m = 100(P_a^A \times P_b^B \times \ldots P_z^Z) \tag{14.2}$$

Low yields on module assemblies obviously have a large effect on module costs. Table 14.2 shows various comparisons cited for improvements in yield and cost due to institution of a KGD program.

14.2 BACKGROUND

Selected types of unpackaged ICs traditionally have been available to the hybrid microcircuit industry. However, there were legitimate concerns for low volumes, special processing, and warranty issues (Sec. 14.2.2). Today, with a growing market for MCM, COB, and PCMCIA applications, most IC suppliers offer some form of bare die either to the open market, to third party die processors, or to designated assembly houses. Industry initiatives have lead to the availability of a wide variety of methods (Secs. 14.3.4 and 14.3.5) for ensuring bare die quality and reliability. This section describes fundamental bare die quality and reliability issues.

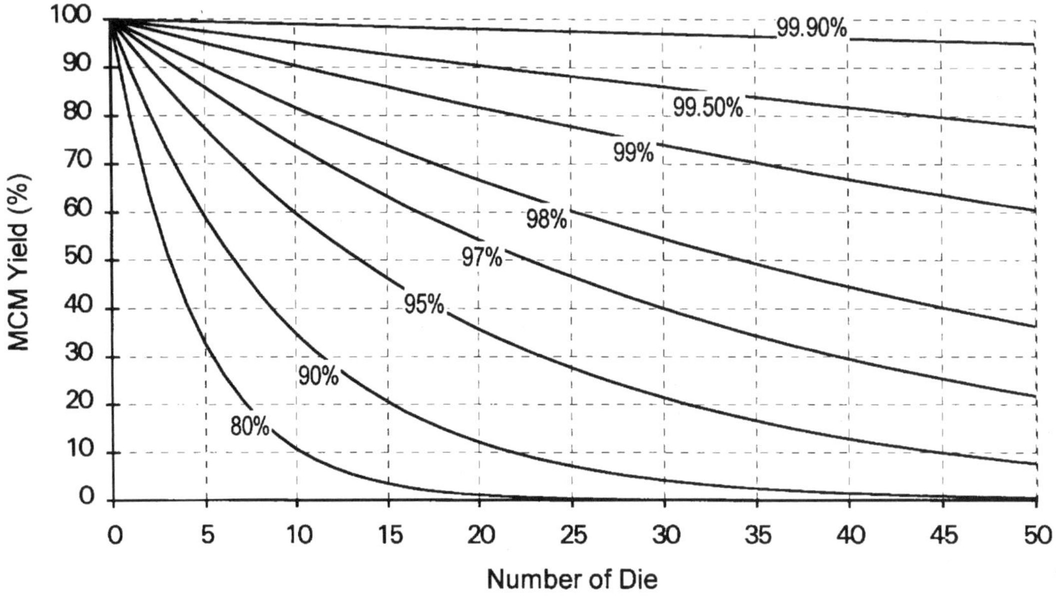

FIGURE 14.1 Module yield vs. die quality.

Table 14.2 Example Improvements with KGD

MCM Description	Without KGD: First-time yield (%)	Without KGD: Module cost	With KGD: First-time yield (%)	With KGD: Module cost ($)	Ref.
4-chip MCM-L	58	$158.00	100	92.00	8
5-DRAM MCM-L	77	$67.53	100	52.00	23
48-chip MCM-D	5	18.8×	94	1×	24
20-chip MCM	40	2.4×	95	1×	25
13-chip MCM	65	$4,980.00	100	3065.00	26

14.2.1 Traditional Integrated Circuit Manufacturing Flow

The semiconductor industry has routinely performed burn-in and full electrical testing of ICs *after* packaging them into single-chip packages as shown in Fig. 14.2. Existing industry test and burn-in infrastructure has supported the various standard package styles: dual in-line packages (DIPs), plastic leaded chip carriers (PLCCs), quad flat packs (QFPs), pin grid arrays (PGAs), ball grid arrays (BGAs), and so on. This extensive testing capability for packaged ICs has matured to the point at which the probability that the ICs are known-good can range to 0.99999 or even higher. Unpackaged ICs historically have not had the same 0.9999+ probability of being good, because they typically did not receive the benefit of full functional testing over the rated operating speed or temperature range. This meant that customers buying die were often buying a partially processed part.

When ICs are in an unpackaged format, they are much more difficult to thoroughly test than after they have been sealed into single-chip packages. Some testing occurs at wafer probe, but those tests have typically checked for only a limited set of parameters and have not included burn-in. This is because of the difficulty in maintaining test access over temperature extremes and problems related to removing dissipated heat, providing adequate power and ground connections, and signal degradation from probe losses or contact resistance.

When comparing the quality of unpackaged versus packaged chips, the cumulative effect of all the value-added tests determines the final product quality and reliability. Any incomplete or skipped tests may adversely effect the final results. Currently, most packaged integrated circuits pass through the following major manufacturing test steps.[8]

Wafer Probe Test. Once the semiconductor fabrication process is complete, the wafer is subjected to electrical tests to identify the potentially good bare dice on the wafer. A wafer probe card with

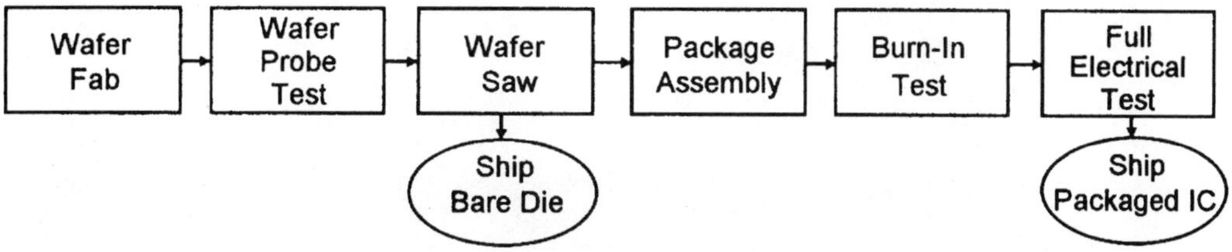

FIGURE 14.2 Traditional single-chip package manufacturing flow.

needle-like probes makes electrical contact to the microscopic bond pads of a single device (Fig. 14.3). Each die is electrically tested using a step-and-repeat process. Electrical tests are first performed on special process control structures or coupons strategically located on the wafer. The test coupons are used as a measure of quality for the overall fabrication process. They consist of unique process control features to test for mechanical, material, and electrical characteristics. Next, each individual die is tested at a minimum for basic functionality. These electrical tests typically consist of DC parametric test methods and may also include some amount of AC parametric and functional testing (Sec. 14.2.5). Each faulty device is marked with an ink dot. The wafer is then sawn into individual chips (diced), and the ink marked rejects are analyzed and discarded. The quality of bare die

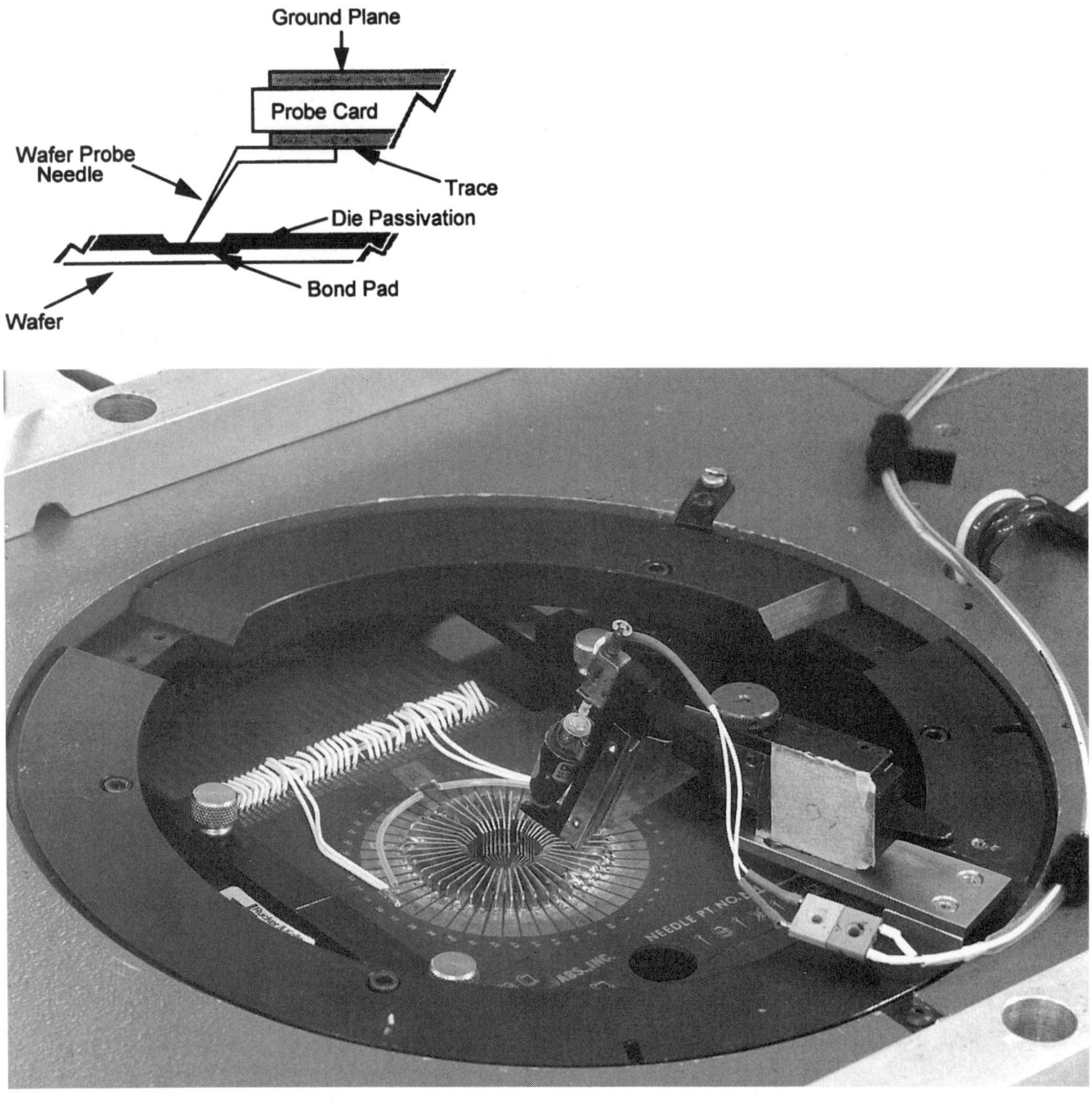

FIGURE 14.3 Typical wafer-level electrical testing with a probe card: (top) side view and (bottom) photo.

shipped at this stage of IC manufacturing will vary considerably, depending on the inherent maturity of the manufacturing process and the extent of wafer level electrical testing.

Burn-In Test. After the bare die have been assembled into single-chip packages, the parts are ready for burn-in and final electrical test. The packaged parts are plugged into sockets on special burn-in boards (BIBs) and transferred to precision controlled oven systems. The parts are then subjected to predefined profiles of elevated temperature and electrical bias. Burn-in is intended to accelerate semiconductor aging to expose infant mortality failures caused by process contamination or variability (Sec. 14.2.4). The need for integrated circuit burn-in (along with the specific burn-in test conditions) varies with the device technology (e.g., CMOS, bipolar), process maturity (e.g., 1.0 μm, 0.5 μm, 0.35 μm), by wafer lot, and by the intended products targeted reliability level. Mature devices intended for use in commercial or industrial may not require burn-in, but aggressive chip designs, high-end microprocessors and memories, or military and high-reliability products will typically require extensive burn-in.

Full Electrical Test. The final electrical test checks for full electrical functionality of the parts and weeds out chips damaged during package assembly or burn-in. These comprehensive electrical tests consists of DC parametric, AC parametric, and functional tests as discussed in Sec. 14.2.5. Many packaged parts are routinely tested and sorted into different categories based on speed, temperature, and performance attributes. Newer device types may undergo full electrical testing both before and after burn-in. High-end devices often require a multimillion dollar investment in automatic test equipment capable of high-speed testing with a comprehensive set of test vectors and algorithms. The voluminous test data collected is statistically analyzed and used to improve the chip fabrication, assembly, and test processes. After passing final electrical tests, known-good packaged parts are shipped fully warranted to meet the data sheet specifications.

14.2.2 Bare Die Supplier Concerns

Shipping bare die when burn-in and full electrical tests have not been run (see Fig. 14.2), creates justifiable warranty concerns for IC vendors. As a result, semiconductor manufacturers and third-party die processors utilize a variety of methods for assuring bare die quality and reliability. These methods include numerous temporary die carrier techniques as discussed in Sec. 14.3.4. Figure 14.4 shows the additional steps required for these KGD processing techniques.

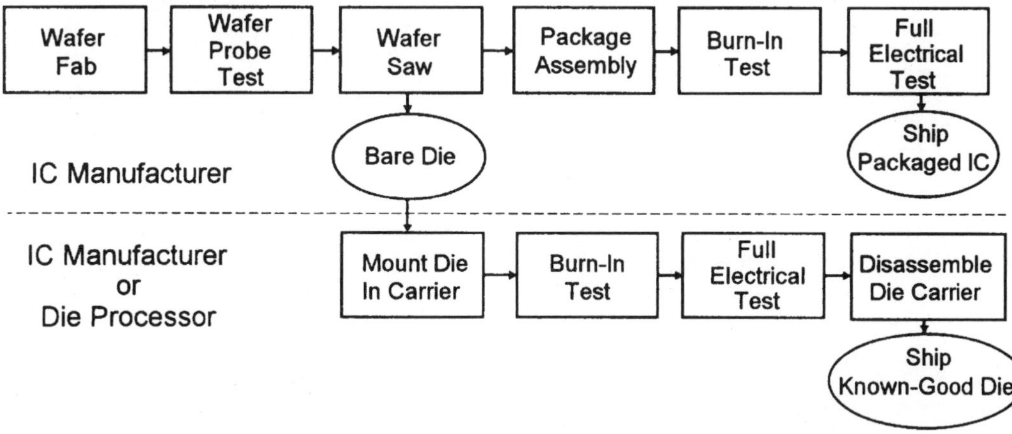

FIGURE 14.4 Modified manufacturing flow for KGD processing.

The sale of bare die products has historically accounted for only a very small fraction of the total worldwide integrated circuit sales. It was difficult to justify the cost of the additional processing steps required for KGD, and equally difficult to justify disruption of the established process flows.

There are also a wide variety of issues that are unique to supplying bare die products, including providing design data unique to bare die (die dimensions, die electrical parameters, die material properties, and so on), die change notification (for inevitable die shrinks), die packing and shipping, and die storage requirements. These issues are not trivial and have been addressed in detail by industry KGD specifications and guidelines as discussed in Secs. 14.3.2 and 14.3.3. The extent to which IC manufacturers or die processors warrant KGD products is highly dependent on the die users' capability and adherence to strict quality controls. KGD users take on a significant responsibility for proper application, handling, assembly, and long-term environmental protection of bare die products. Cooperation between suppliers and users is essential. Today, with growing demand for bare die products numerous KGD initiatives are underway to mature KGD technology.

14.2.3 Defect Activation

The effectiveness of sorting good chips from bad chips depends on having both (a) an appropriate applied environmental prestress sufficient to activate and reveal defective ICs, and (b) an adequate fault coverage electrical test to reveal operational defects. The prestress and pretest exposures must not significantly degrade the life of the remaining good ICs. The pretest can be applied either during or after the prestress exposures, and the prestress can be done with devices in a electrically biased or unbiased condition.

14.2.4 Reliability Prestress

IC failures generally follow a bimodal distribution in time commonly referred to as a bathtub curve as shown in Fig. 14.5. The life cycle of an IC is divided into three periods: infant mortality, operating lifetime, and wear-out. A small percentage of parts typically contain latent manufacturing defects due to contamination or process variability. Many of these devices fail in their early operating life—the infant mortality period. Once these infant mortality defects are exposed and removed, the remaining devices will exhibit a relatively stable rate of failure over the normal operating lifetime. The failure rate of the devices will eventually increase rapidly during the wear-out phase.

Integrated circuit reliability is typically expressed in failures in time (FITs) per billion device hours. With some mature products, defect rates during the normal operating life may be in the range of 10 FITs. For an aberrant (worst-case) lot of mature products, the rate might be 30 FITs or more. By comparison, a wafer lot of less mature ICs may be an order of magnitude higher in failure rate, as shown in the Table 14.3.[27]

Table 14.3 Effect of Process Maturity on Device Failure Rates, 1 MB DRAM

		FITs		
Process maturity	Wafer lot	Infant mortality (t_0)	Normal operating life (t_1)	Burn-in fallout (%)
Immature	Worst-case	100,000	300	>5
	Typical	1,000	100	≈5
Mature	Worst-case	1,000	30	>2
	Typical	100	10	<0.5

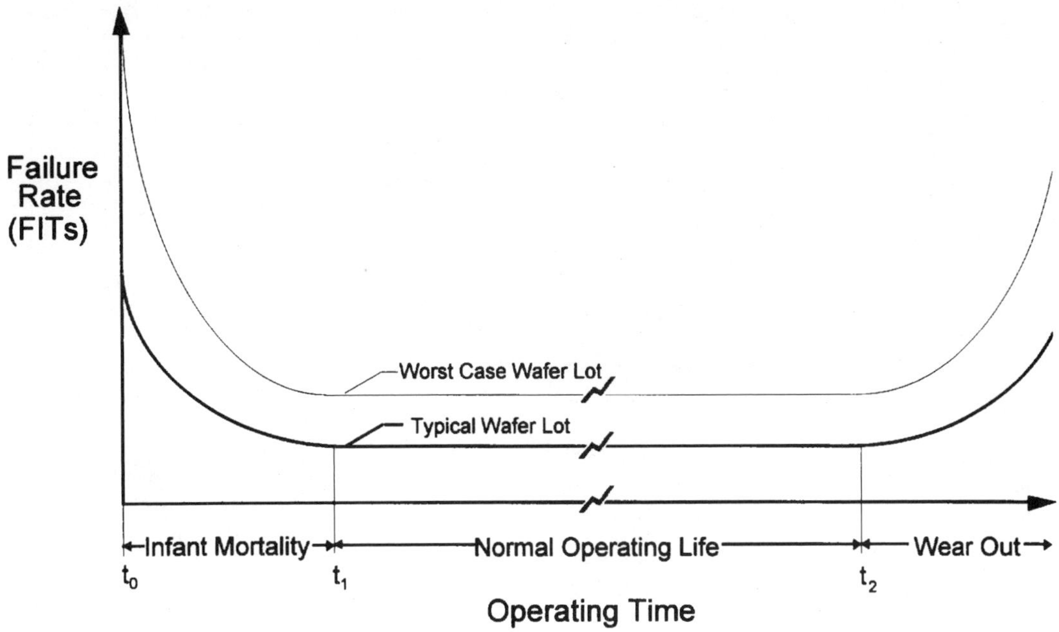

FIGURE 14.5 IC reliability bathtub curve.

Accelerated Prestress. To ensure IC reliability levels, semiconductor manufacturers use design for reliability methods,[13,28] fabrication process reliability monitors,[29] and accelerated prestress methods. The infant mortality period may exhibit failure rates in excess of 1,000 FITs and extend to thousands of device operating hours. This provides considerable incentive to accelerate the exposure of infant mortality failure mechanisms before shipment. Reliability calculations are used to predict the equivalent operating hours for every hour of accelerated prestress. First, the infant mortality period ($t_0 - t_1$) is established, based on curve fitting of available empirical data typically based on the Weibull distribution. Next, the acceleration factors and prestress conditions are developed using temperature, voltage,[30] current density, humidity, and so forth. The Arrhenius equation historically has been used as an empirical method to model temperature acceleration factors used in reliability calculations.

$$A_T = \exp^{-\left[\frac{E_a}{k_T}\left(\frac{1}{T_1} - \frac{1}{T_2}\right)\right]} \tag{14.3}$$

where A_T = acceleration factor
T_1 = baseline temperature (°K)
E_a = activation energy (eV)
T_2 = burn-in temperature (°K)
k_T = Boltzmann's constant, 8.61423E – 5eV/°K

Infant Mortality. Failure mechanisms associated with infant mortality failures include contamination, mask defects, and oxide film defects, as shown in Table 14.4.[13,30–33] These mechanisms tend to have lower activation energies (for temperature) and may be accelerated by elevated temperature (burn-in) and an appropriate electrical bias.

Table 14.4 Comparison of Early-Life and End-of-Life Reliability Failure Mechanisms for Bare Die

Fabrication and handling failure mechanisms	Activation energy (E_a)	Acceleration factors	Reliability prestress test methods
		Infant mortality ($t_0 - t_1$) failures	
Contamination/particulate	0. eV	Temp., voltage, humidity	Thermal shock (TS) Temp. cycle (TC) Static burn-in (SBI) Dynamic burn-in (DBI) Test during burn-in (TDBI)
Lithographs/masking errors	0.4 eV	Temperature, voltage	
Oxide failures	0.3–0.4 eV	Temp., voltage, elec. field	
Metallization	0.4 eV	Temp., current density	
Cracking, scratches	0.4–0.7 eV	Temp., voltage	
		End-of-life, wear-out (t_2) failures	
Electromigration	0.55–1.2 eV	Temp., current density	WLR, SWEAT
Ionic contamination	1.0–1.4 eV	Voltage, temp., elect. field	HTRB, SBI
Hot carrier injection	1.3 eV	Elec. field, temp.	Low-temp. burn-in, DBI
Oxide wear-out (TDDB)	0.3–0.6 eV	Elec. field, temp.	High voltage, temp.

Burn-In Issues. Burn-in is intended to accelerate semiconductor aging to expose defects contributing to infant mortality. The need for IC burn-in (along with the specific burn-in test conditions and profile) varies with product technology, process maturity, wafer lot, or the targeted FIT rate. Mature commercial or industrial devices may not require burn-in.[34] IC manufacturers achieving target FIT rates without burn-in may justifiably argue that burn-in handling and processing is a yield detractor. Nevertheless, burn-in is often a necessity for high-reliability products, devices with aggressive submicron process geometry, and leading-edge microprocessors and memories.

Burn-In Methods. Large sample sizes are generally required to provide enough data to characterize infant mortality because of the small percentage of devices failing. Static burn-in applies power to the device and keeps input and output pins at static or steady-state conditions under elevated temperature for a specified number of hours. Static burn-in can accelerate oxide defects, contamination and, silicon defects, but it does a poor job of accelerating metallization defects in CMOS. Bipolar processes can be screened with static burn-in, while CMOS processes are best suited to dynamic burn-in techniques. Dynamic burn-in applies test patterns to the device input signal pins to toggle the internal logic during temperature prestress. CMOS, by design, maintains its transistors in a quiescent condition unless actively switching. Thus, static burn-in will activate only half the device transistors on power up. The higher toggle coverage in dynamic burn-in significantly reduces burn-in escapes. For example, an order of magnitude reduction in FITs for ASIC devices is possible by improving the toggle rate from 50 percent to just over 90 percent.[35] Test during burn-in (TDBI) techniques provide monitored stimulus and response test patterns to the device during burn-in. This comprehensive method allows an IC manufacturer to tailor burn-in prestress to achieve reliability testing tailored to wafer lot conditions.[29]

Wear-Out Defects. A small sampling of parts is typically life tested to gain assurance that the majority of parts will avoid early wear-out related problems. Long test times are generally required to provide sufficient wear-out defect data. Failure mechanisms associated with the wear-out period include metal electromigration, hot carrier effects, ionic contamination, or gate oxide breakdown, as shown in Table 14.4. A number of these wear-out failure mechanisms can effectively be screened at the wafer level. Wafer-level reliability (WLR) techniques use discrete test structures and parametric monitors specifically designed to accelerate known failure mechanisms. One representative structure called *standard wafer level electromigration acceleration test* (SWEAT)[36,37] allows rapid in-process wafer fabrication monitoring for electromigration testing, based on the cross-sectional area of the metal trace being stressed, the applied current density, and self-heating temperature effects.

Next Generation Reliability Prestress. As technological barriers are neared in semiconductor fabrication, processes become more unstable, and reliability issues become more complex. Analysis of failure mechanisms must direct optimization of prestress conditions and provide feedback for continuous process improvement. Next-generation reliability prestress techniques may include concepts from innovative work such as I_{DDQ} testing (discussed in Sec. 14.2.5), *below-a-minute burn-in* (BAMBI),[24] physics of failure analysis, built-in self-test (BIST), and wafer-level burn-in (Sec. 14.5.1). For example, CMOS devices that were screened with I_{DDQ} test prior to burn-in showed a dramatic 50 percent reduction in burn-in fallout.[38]

14.2.5 Electrical Pretest

Integrated circuit quality is often expressed in *defects per million devices* (DPM). With some immature products, defects may be in the range of 3000 DPM. To ensure high quality levels, effective and comprehensive means are required for evaluating the overall IC manufacturing process. Wafer-level and final electrical tests play a major role in this evaluation process.

Electrical Test Methods. Basic types of IC electrical test methods include DC parametric, AC parametric, and functional tests, as shown in Table 14.5. DC tests are the most fundamental and easiest to perform at both the wafer level and final test. DC parametric testing provides a quick screen of gross device failures. AC parametric testing and functional testing become more challenging at the wafer level for higher-frequency measurements and under depressed temperature conditions.

DC Parametric. DC (direct current) parametric tests verify the non-time-varying analog voltage and current values at the device electrical connections (bond pads or package pins). These tests are performed at various conditions of electrical loading, temperature, and power supply voltage levels to verify that the device specifications can be met over the full operating range. Tests for shorted or open connections are the most fundamental and are performed first. Shorts or open circuit failures can quickly point out wafer probe misalignment, equipment setup errors, or catastrophic device failure. This test essentially uses the ability to forward bias and measure the p-n junctions of the input protection diodes and output drive transistors before the device to powered up. After the power supply voltage is applied to the device, tests are performed to verify switching levels, the device operating current, and leakage current values.

AC Parametric. AC (time-varying) parametric tests measure device signal timing relationships and values. These tests are performed at various combinations of supply voltage and temperature to give a full range of minimum, typical, and maximum timing values. Typically three combinations will give this full range: minimum temperature/maximum V_{DD}, typical temperature/typical V_{DD}, and maximum temperature/minimum V_{DD}. Propagation delay tests measure the time interval from a specified applied input signal to corresponding output change. Setup and hold tests are performed by varying the signal edge placement. Setup tests verify a specified input signal is valid at a certain number of time units before asserting a second signal. Hold tests verify that a specified signal output remains valid until after a stated input signal is asserted. Other specified timing relationships verified by AC testing may include signal edge placement, pulse width, and clock frequency. AC and DC parametric testing is greatly simplified for ICs designed with boundary-scan (see Chap. 16 and Ref. 3) because each I/O pin can be independently controlled by its boundary scan cell.

Table 14.5 Electrical Test Methods for Digital ICs

Type	Parameter	Description
		DC parametric
Shorts and opens	—	Application of constant current and measures forward voltage drop of diode p-n junctions on I/O pads/pins
Input-level tests	V_{IL}, V_{IH}	Switching voltage threshold levels for inputs
Output-level tests	V_{OL}, V_{OH}, I_{OL}, I_{OH}	Output voltage levels Output source current and sink current capacity
Static supply current	I_{DD} or I_{CC}	Steady-state device current, I_{DD} (CMOS), I_{CC} (bipolar)
Leakage	I_{IN}, I_{OZ}	Current flow through internal conduction paths
		AC parametric
Propagation delay	t_{pd}	Time interval from applied stimulus to output response
Setup and hold time	t_{su}, t_h	Time interval a given signal is valid before (setup) or after (hold) assertion of second signal
Clock frequency	f_{freq}	Range of input signal duty cycle and period
Signal timing	t_{we}, t_{oez}	Signal edge placement of critical timing relationships for specified input or output signals
Pulse width	t_{pw}	Period of specified signal assertion/deassertion
		Functional
Static	Verifies device structure; no signal timing attributes	
Dynamic	Verifies device logic or behavior with signal timing attributes	
At-speed	Verifies device behavior at full specified operational timing	
Built-in self-test	Verifies device logic or behavior through on-chip test circuitry through autonomous at-speed test	
		I_{DDQ}
External	Precision measurement of CMOS I_{DD} after settled state external stimulus	
Internal	Precision measurement of CMOS I_{DD} using built-in current sensor (BICS)	

Functional Testing. Functional tests verify that the device performs its intended function. These tests can be applied in a static, dynamic, or at-speed mode or autonomously by self-testing devices. Static functional tests check the circuit logic structure at lower than rated operating speed. This is also referred to as *settled-state testing,* where test patterns are applied, and the output response is allowed to settle before a measurement is made. Dynamic functional testing is concerned not only that the circuit structure is correct but also measures signal timing. Dynamic testing is not performed at-speed for a variety of reasons that often include tester limitations or test economics. At-speed testing verifies the circuit under test at the full rated device signal timing. All of these tests may be performed over the specified device operating temperature and supply voltage range. The trend of increased device size, complexity, and operating frequency accentuates both the need for full device operational verification and the difficulty in performing these tests. This requires careful test design to ensure that generated heat, high-frequency signal integrity, and full temperature range of operation are adequately accommodated. Industry data indicates that the functional test fault coverage of greater than 99 percent is necessary for producing high quality ICs.[3,39,40] Built-in self-test (BIST) and boundary scan access (see Chap. 16 and Ref. 3) eliminate the loading effect of external test equipment, allowing at-speed testing. Since ICs with BIST are self-testing, the problem of procuring or generating test vectors is eliminated.

I_{DDQ}. I_{DDQ} testing is used to augment traditional functional testing for CMOS devices, particularly ASICs. I_{DDQ} testing is performed by measuring the V_{DD} power supply current in the quiescent logic condition after the inputs have changed state and prior to the next input change. CMOS circuits draw negligible leakage current when not switching (steady state). However, fabrication defects (particularly gate oxide shorts) cause CMOS devices to draw a measurable quiescent current flow. Traditional voltage-based (logic level) functional test vectors must propagate faults to an observable output. I_{DDQ} testing requires considerably fewer patterns, since faults need only be activated by toggling the internal nodes. A comparison of devices tested using traditional functional test (52 to 82 percent fault coverage) showed a dramatic improvement in fault coverage (96 + to 99percent) using the same patterns for I_{DDQ} testing.[41] Industry studies have shown that device defects were reduced up to 11 percent by augmenting functional tests with I_{DDQ} testing.[41] RAM products benefit least from I_{DDQ} testing, since long-established RAM test algorithms have been optimized to identify leakage current problems. Custom products such as ASICs benefit most from I_{DDQ}, since product volumes and economics may limit the practical extent of test coverage or process optimization. I_{DDQ} can also provide a dramatic reduction in burn-in fallout (in excess of 50 percent).[38]

Test Yield Issues. *Yield* is the ratio of apparent fault-free units to total units coming out of a process. Without the benefit of complete pre-test and burn-in, a defective bare die may go undetected until the entire module is assembled and tested. At best, significant rework costs are incurred or, at worst, the entire module must be scrapped. Worse yet, the defective die may escape outgoing detection, and a product declared good will fail to meet the customers expectations. Figure 14.1 showed the effect of die yield on projected MCM yield using the familiar expression: $Y_{mcm} = 100(P_c)^N$. This formula assumes that the MCM test is able to screen out 100 percent of all the failures. Unfortunately, no test is good enough to detect every possible fault.

Defect Level. Overall quality is reflected by the defect level, i.e., the fraction of units that pass testing but may actually be faulty. Given the wide variation possible in wafer probe yield, quality is often highly dependent on the screening effect of upstream electrical testing. The ability of an electrical test method to detect faults is a measure of the test effectiveness or fault coverage. Defect level is a function of yield and fault coverage (see Fig. 14.6) that can be expressed with the following formula:[42]

$$DL = 1 - Y_{mcm}^{1-FC} \times 100\ \% \tag{14.4}$$

where DL = defect level, or the fraction of MCMs shipped that pass final MCM tests but may actually be faulty. Y_{mcm} = MCM yield, or the ratio of MCMs that pass MCM testing to total MCMs tested, or the projected yield based on chip yield and chip count (from Fig. 14.1). FC = fault coverage, or the ability of a test or inspection method to identify faults. Fault coverage is expressed as the ratio of faults detected to total faults possible, which ranges from zero (no test) to one (detects all faults).

The high fault coverage, low defect area of Fig. 14.6 is shown as an expanded logarithmic plot in Fig. 14.7. The defect level is expressed in defects per million (DPM) for various levels of die yield.

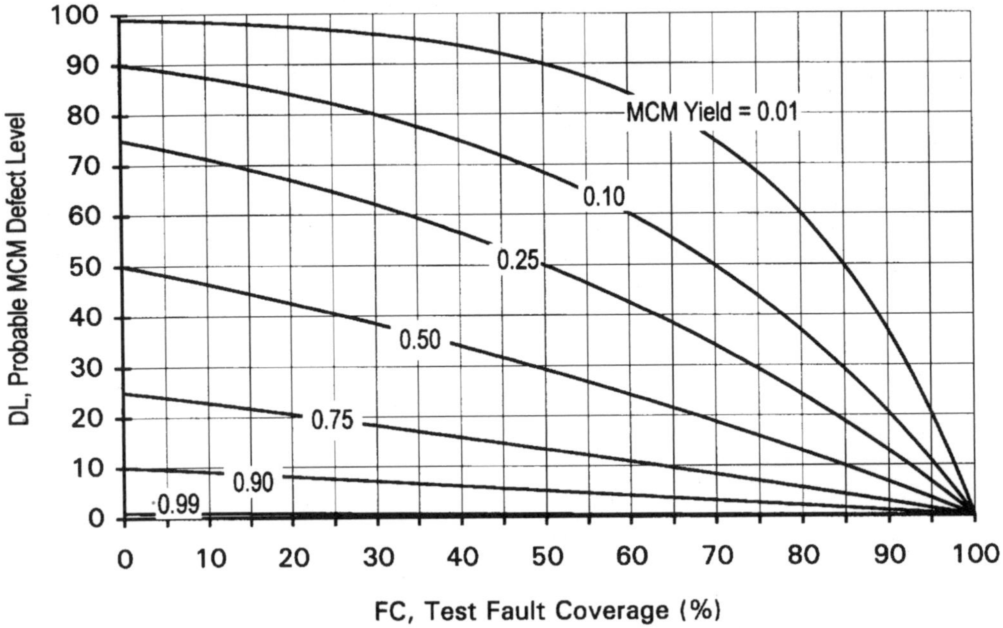

FIGURE 14.6 Defect level probability, a function of test yield and test effectiveness (fault coverage).

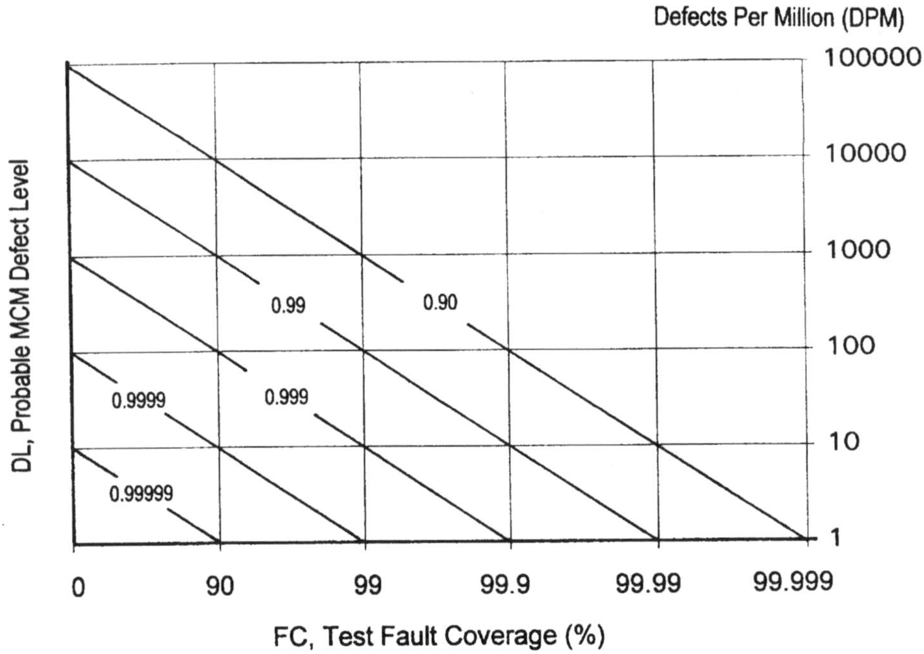

FIGURE 14.7 Defect level probability at high fault coverage.

For example, to ship better than 100 DPM MCM quality with a die yield of only 50 percent requires MCM Test fault coverage of nearly 99.99 percent. If the die yield is improved to 99 percent, identical DPM results are achieved with MCM test fault coverage of 99 percent. The significance of this relationship can be further illustrated by the following example of a six-chip MCM (see Fig. 14.8):

Case 1. For a six-chip MCM with chip yields of 60 percent (typical of some DC probed die), the MCM yield based solely on chip yield is projected as $Y_{mcm} = 0.60^{(6)} = 0.0466$, or approximately 5

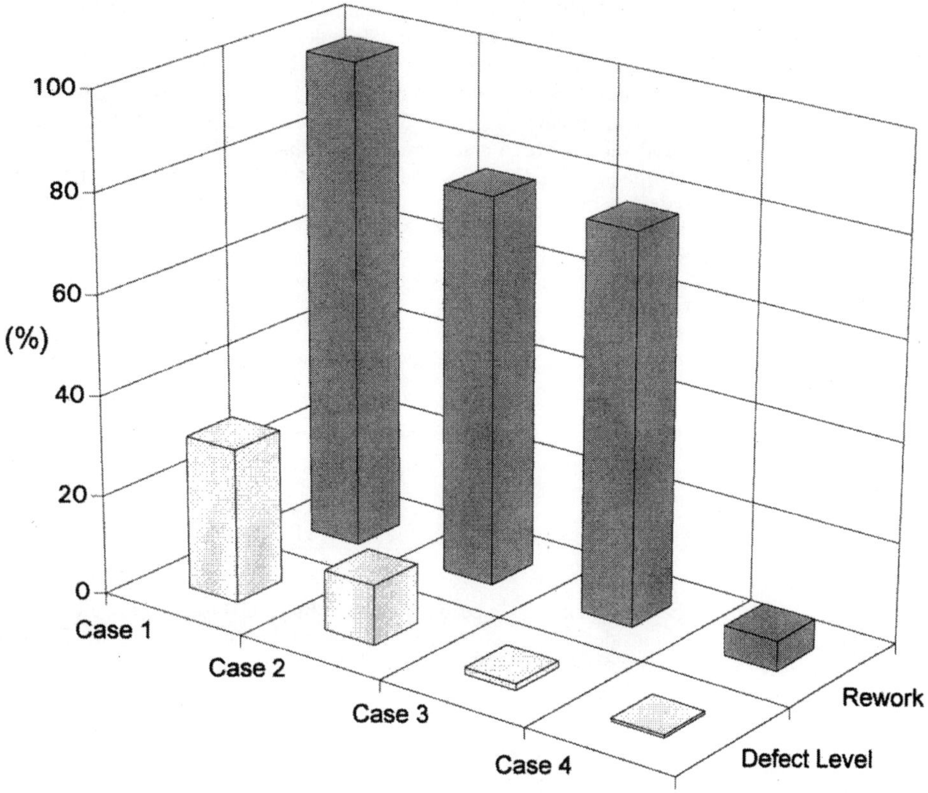

	Die Yield (%)	MCM Test FC (%)
Case 1	60%	90%
Case 2	80%	90%
Case 3	80%	99%
Case 4	99%	99%

FC = Fault Coverage

FIGURE 14.8 Low defect/rework level MCM manufacture requires very high quality in ICs and MCM testing.

percent. The magnitude of potential rework is equal to $(1 - Y_{mcm})$. Thus, about 95 percent of all MCMs going into MCM test may contain one or more defects. After MCM test with a relatively comprehensive 90 percent fault coverage, the defect level is estimated as $DL = 1 - 0.0466^{(1-0.90)} \times 100$ or approximately 26 percent. Thus, only 74 percent of assembled MCM that passed MCM testing are truly fault free.

Case 2. The MCM test fault coverage is maintained at 90 percent, while the chip yield is improved to 80 percent (typical of some hot chuck tested die). Y_{mcm} improves to a level of about 26 percent, and the defect level drops to about 12 percent. However, the level of rework is still nearly 74 percent.

Case 3. The die yield remains at 80 percent, and the MCM test fault coverage is improved to 99 percent fault coverage. Since the die yield did not change, the level of rework remains unchanged at nearly 74 percent. The defect level drops to about 1 percent (0.01329). This 13,299 defects per million (DPM) level may still be unacceptably high for manufacturers of high-reliability products.

Case 4. If, in addition to improving the MCM test effectiveness to 99 percent fault coverage, the die yield available for MCM assembly is also improved to 99 percent (typical of some hot chuck tested and burned-in die), the MCM yield improves to about 94 percent, and the defect level drops to about 600 defects per million (DPM). The level of potential rework drops dramatically to about 6 percent.

With further die yield improvements to 99.9 percent (typical of some die fully tested and burned-in to full data sheet specifications as KGD), the defect level would approach 60 defects per million (DPM). The die yield of 99.9 percent is not just 0.9 percent better than the 99 percent in Case 4; it is *ten times better*, since the defect level is 60 DPM compared to 600 DPM. To achieve six-sigma levels of 3.4 DPM will require further improvements in chip and MCM fabrication and testing methods. From Fig. 14.8, the following conclusions can be drawn:

- High MCM test coverage alone cannot adequately compensate for poor die yields.
- Improving die yields through KGD technology will significantly reduce both MCM rework and the defect level of MCMs shipped.
- It requires both high-quality KGD and highly effective MCM test to reduce defects to acceptable levels.

It should be noted that, for the purpose of presenting KGD issues, only the chip yield and number of chips were considered when projecting MCM yield. In actual practice, MCM yield is also dependent on other factors such as the yield of the MCM substrate and interconnect technique (wire bond, solder bump, and so on), and assembly process workmanship where:

$$Y_{mcm} = Y_{\text{chip}} \times Y_{\text{substrate}} \times Y_{\text{interconnect}} \times Y_{\text{workmanship}} \cdots \times Y_{xyz} \tag{14.5}$$

MCM packaging technology will be used only where it is cost and benefit competitive with the alternative packaging technology choices. Since it must compete against fine-line circuit board and chip-on-board technologies, assembly yields must approach the high yields obtainable in those technologies. The old approach of "assemble everything and then repair what doesn't work" simply will not be cost competitive in many sophisticated applications.

14.3 RECENT PROGRESS ON KGD

High-density multichip modules first emerged in the 1960s in pioneering work on silicon substrates at Philco-Ford.[43] Further development work was carried out in the 1970s on thick film ceramic constructions for mainframe computer applications. In the 1980s, many aerospace companies were developing multilayer thin film constructions of MCMs. By the late 1980s, it was evident that a number of "infrastructure issues" were barriers to exploitation of the potential benefits of MCM technology. In 1989, the U.S. Air Force Wright Laboratories began an initiative called the Electronic

Packaging and Interconnect Initiative (EPII) which brought industry experts together to chart a road map of technology developments needed. A core of eight aerospace companies (Hughes, Rockwell, IBM, Honeywell, Alliant Technology, Martin Marietta, GE, and TI) formed a subgroup called the Electronic Packaging and Interconnect Cooperative (EPIC) to provide continuing guidance to EPII. Following completion of the Air Force reports, this same group decided to continue as a voluntary advisory group known as the Packaging/Interconnect Technical Advisory Group (PI-TAG). In mid-1990, this group brought forth recommendations to the tri-services and ARPA that the highest-priority challenge for success of MCM technology was bringing known-good die to MCM assembly processes. This led to a series of workshops on KGD championed by the MCC consortium in Austin, Texas. Papers on known-good die began appearing at technical conferences.[44–46] Some semiconductor companies began delivering KGD to selected customers under strategic alliance agreements, and some die processors began offering KGD screening services for bare die.[47] Additional work soon followed supported by SEMATECH, SRC, SIA, and ARPA, as indicated in Table 14.6. Introduction of a variety of KGD hardware products soon followed, as shown in Table 14.7. Obviously, these lists do not cover all aspects of KGD developments. Since the relationship between KGD and yield is a generic issue in all MCM assembly technologies, the development of successful KGD approaches clearly had many roots from various companies' internal development programs.[48–56]

Table 14.6 The Genesis of KGD Technologies

Year	Milestone Completed
1989	EPII compiles list of MCM infrastructure issues
1990	EPIC/PI-TAG identifies KGD as highest priority issue
1991	MCC hosts first industry KGD workshop
1991	Die processor offers KGD screening service (chip supply)
1992	SEMATECH commissions MCC-led "KGD Task Force"
1992	KGD Task Force submits KGD Guidelines to JEDEC
1993	More processors offer KGD screening (ELMO, ETEC)
1993	JEDEC KGD guidelines published
1993	Micron advertises first IC supplier KGD availability
1993	ARPA funds various KGD programs
1993	MCC receives ARPA/AF contract, "Phase I, Infrastructure for KGD"
1993	MCC Phase I program develops "KGD Technology Assessment Guidelines"
1993	More KGD availability announcements (Motorola, National Semi., IDT, et al.)
1993	Die Information Exchange (DIE) electronic data format established
1994	More KGD availability announcements (Intel, Samsung, TI, et al.)
1994	MCC Phase I Report Released, Assessment of over 30 KGD methods
1994	MCC evaluates variety of KGD methods, Phase II contract
1994	Raytheon "Known Good Devices Phase I Report" Released
1995	Die Information Exchange (DIE) balloted by EIA
1995	MCC Wafer Level Burn-In (WLBI) Project
1995	ESPIRIT IV commission's GOOD-DIE Project for KGD database development
1995	EIAJ sponsors study group on KGD
1996	MCC Phase II Report released, assessment of KGD Methods

Table 14.7 Examples of Industry Availability of KGD Hardware

Year	Milestone Completed
1991	Various flexible membrane probe techniques applied for KGD (Packard-Hughes)
1993	ELM introduces silicon membrane probes for KGD
1993	Tessera offers TCC™ die grid array packaging solution for KGD
1993	IBM offers to sell KGD sockets, R3™ process, Harmonicair™ for KGD
1993	AEHR introduces "DiePak™ KGD sockets
1993	Fresh/Quest introduces "KGD probe card"
1994	TI/MMS introduce "Diemate™" KGD sockets
1994	3M/nCHIP/Micron/CYBEX introduce "KGD System" of carriers, sockets, tape and reel

By 1996, considerable work had been accomplished on identifying KGD issues and doing comparative studies of various KGD techniques,[13, 57–60] The variety of approaches still in contention is a clear indication of as yet immature status. However, both availability and quality of KGD are growing strongly.

The next several subsections discuss various methods available or being developed to improve known-good die probability levels.

14.3.1 Industry and Government KGD Initiatives, Evolution, and Accomplishments

Industry activities in KGD development cover a wide range of purposes, as shown in Table 14.8. Some companies have formed alliances to develop KGD products such as test sockets or assembly services. EIA is sponsoring an MCM division and KGD standards development. SEMATECH has sponsored the writing of documents for KGD guidelines and industry workshops, along with MCC. SIA has included KGD goals in their semiconductor industry road map efforts as shown in Table 14.9. SEMI has a group dedicated to development of equipment to support KGD technologies. The largest group has been the MCC-led KGD consortium, which has evaluated the cost and effectiveness of various KGD technologies. More recently, KGD initiatives have grown to include SEMATECH (USA), EIAJ (Japan), and ESPIRIT (European) sponsored KGD activities.

Several government agencies are active in the promotion of KGD technologies and infrastructure development. Support from the DOD agencies is largely aimed at a combination of achieving higher-reliability and lower-cost electronics, a strong benefit of MCM technologies that can be enabled by KGD availability. DARPA programs reinforce these objectives and have an additional agenda for rebuilding domestic electronics economies by infusion of competitive technologies and infrastructure development. A description of some of these programs and their principal participants is given in Table 14.10a.

14.3.2 Standards for Unpackaged Chips

Suppliers of bare die products have justifiable concerns about nonstandard process flows, increased risk, and warranty concerns, as previously discussed in Sec. 14.2.2. Users of bare die products have equally valid concerns about quality and reliability assurances, affordability, and the availability of die data. Designers of electronic assemblies now using standard surface mount technologies take for granted that full device documentation is readily available from component vendors for the various

Table 14.8 Some Industry Initiatives on KGD

Sponsor/lead	Program Description	Participants	Reference
3M	Bare die burn-in sockets	nChip, Micron, Cybex	61
Aehr	Bare die test sockets	Nitto Denko	62
SRC	Array probe card	Stanford, IBM	63
EIA	Multichip module division	various	—
EIA	Die information exchange	various	—
SEMATECH	KGD guidelines	various	—
Intel	Technical alliance program	Amkor/Anam, SCI Systems, Solectron, Valtronic, SMOS, Jabil Circuit, MMS, and others	64
MCC	KGD consortium	various	65
SIA	KGD road map	various	—
SEMI	KGD support equipment	various	—
Ericsson	Users group KGD concerns	various	—
ESPIRIT	GOOD-DIE database	IMEC, Lucas, CODUS, Philips, Rood, Elteck, GEC, Matra, Siemens, Mietec, TEMIC	—
EIAJ	KGD study group	NEC and others	—

Table 14.9 SIA Road Map Goals for KGD[66,67]

	1992	1995	1998	2001	2004	2007	2010
KGD probability level	0.90	0.99	0.999	0.9999	0.99999	0.999999	–
Fallout rate DPM	100k	10k	1k	100	10	1	–
Percentage of IC covered with BIST/DFT	–	25	40	50	70	90	90
Percentage of ICs with IEEE 1149.1 boundary scan	–	40	50	64	76	90	90

packaging types used (e.g., LCC, SOIC, PLCC, BGA). MCM designers have not enjoyed the same luxury. The MCM designer was often exposed to the following die-related problems:

- Die size changeability
- Limited, incomplete, or incorrect mechanical data for bare chips
- Limited or insufficient power dissipation information on bare die
- Limited or insufficient data on input/output buffer characteristics
- A bare die possibly having more electrical connections than a packaged IC

Table 14.10a Some KGD Initiatives at Government Agencies

Agency	Program description	Participants	Ref.
ARPA	Bare die testing	Univ. S. Florida	68
Sandia	Bare die testing	USC	69
ARPA	KGD processing	ETEC, Raytheon	70
Army	MCM/KGD survey	Ft. Monmouth	71
NSF	Wafer burn-in	Auburn Univ.	72
USAF/WL	Waver-level burn-in	nChip	73
ARPA	Flip chip burn-in and test	MCNC	74
USAF/Rome Labs	KGD for MCMs	RL/ERD	75
Rome Labs	KGD program	Raytheon, Chip Supply	13
ARPA	Enhanced DFT and FIST for KGD speed grading and dynamic fault identification	Rutgers, UT/Austin	76
ARPA	Wafer-level at-speed test	Hughes, Cascade	76
ARPA	Bare die burn-in for KGD	Aehr, MMS, MCC, TI, Motorola	76
ARPA	Wafer-level burn-in for KGD	Aehr, MCC	76
ARPA	Rapid burn-in of KGD	Insyte, HP	76
ARPA	Embedded test resources for KGD	Tektronix	76
MICOM	Low-cost bare die testing	Space Computer	—

Several industry, consortia, and government initiatives have addressed the issues of the supply, procurement, and development of KGD technologies, resulting in the specifications outlined below.

- KGD, "Standard for Known Good Die"[77] (now JEDEC Standard 49)
- FC-KGD, "Standard for Flip Chip Known Good Die"[78]
- TAB-KGD, "Standard for TAB Known Good Die"[79]
- TAG-KGD, "Technology Assessment Guidelines for Known Good Die"[80]
- DIE, "Die Information Exchange" Format[81]
- J-STD-012, "Implementation of Flip-Chip and Chip Scale Technology"[89]

The Microelectronics and Computer Technology Center (MCC) and SEMATECH jointly developed the first four specifications to address KGD procurement and KGD technology development issues. These specifications are intended to facilitate the procurement and supply of high reliability semiconductors in die form. The KGD described are intended to be *equivalent to* or *better than* their corresponding packaged parts. The major provisions of the first three standards address the following items:

3.2	General data	3.7	Change notification
3.3	Design data	3.8	Packing and shipping
3.4	Electrical test data	3.9	Storage
3.5	Quality assurance	4.0	Additional information
3.6	Reliability		

The TAG-KGD, "Technology Assessment Guidelines for Known Good Die," or TAG, provides guidance for organizations developing or evaluating KGD technologies. The TAG contains rules, recommendations, and evaluation methods, including the example test and burn-in attributes listed in Table 14.10b. The TAG-KGD was used for the test and evaluation of KGD carriers under the MCC/ARPA KGD Phase II contract.

The DIE Format was developed by the Logic Modeling Group of Synopsis Inc. with wide industry support and ARPA ASEM funding. The DIE format defines an EDA compatible data format for the exchange of KGD information as described in Sec. 14.3.3.

J-STD-012, "Implementation of Flip-Chip and Chip Scale Technology," was developed with wide industry support through IPC, EIA, JEDEC, SEMATECH, and MCNC. This document contains a wealth of implementation information, lessons learned, and suggested standardization for both chip scale packaging and flip-chip technologies.

14.3.3 Die Information Exchange Format

The Die Information Exchange (DIE) Format[82] was developed to convey the physical, functional, and administrative data of an unpackaged die in a computer sensible and human readable format. It provides the die characteristics needed for place and route, thermal analysis, electrical signal analysis, power distribution design, physical bonding, functional behavior, test, and timing analysis. Existing industry formats such as IBIS[83] or BSDL[84] are referenced to avoid duplicate or conflicting standards. The DIE information is intended to be directly processed by EDA tools to minimize entry errors and product cycle time. The format defines three levels of compliance, as show in Table 14.11.

Table 14.10b TAG KGD Test and Burn-In Criteria for KGD Technologies (≤100 I/O)[80]

Test attributes	Test rules	Test recommendations	Burn-in rules	Burn-in recommendations
Carrier operating temp.	0 to 125° C	–55 to + 150° C	die = 175° C amb. 150° C	die = 175° C amb. = 150° C
Time @ max. temp.	—	—	12 hr	168 hr
Contact resistance	<0.25 Ω	<0.25 Ω	<0.50 Ω	<0.25 Ω
Min. clock freq.	100 MHz	2 GHz	20 MHz	50 MHz
Array/peripheral pads	Peripheral	Both	Peripheral	Both
Pad metallizations	Al, Au, solder	Al, Au, solder	Al, Au, solder	Al, Au, solder
Pad pitch	≥200 μm	≥150 μm	≥150 μm	≥150 μm
Passivation opening	≥100 μm	≥75 μm	≥100 μm	≥75 μm

Table 14.11 Die Information Exchange Format Levels of Data Compliance

Level 0	Basic die information; the minimum set of die information (required by die user)
Level 1	Less common information; derived from or enhancements to level 0 data, e.g., tolerancing of level 0 data (desired by MCM designer)
Level 2	Highly proprietary, very unique, or MCM foundry-specific information, e.g., wafer lot processing data or lot-specific quality data (may require a special request or agreement)

Within each level of compliance, there are keywords called *settings* that identify and contain the detailed die information. The more than 86 settings listed by compliance level in Tables 14.12 through 14.14 reveal the level of detailed data necessary for robust design of MCMs and other bare-die products. The reader is referred to the DIE format standard document for a full explanation of the settings.[81]

DIE Structure. The DIE Format structure (Fig. 14.9) is composed of blocks, sections, settings (as listed in Tables 14.12–14.14), and setting values. Blocks represent the minimum unit of information that can be exchanged. Blocks are made up of sections that contain a major collection of related data. Common sections such as pad geometry or electrical definitions can be defined once and reused elsewhere. An optional external model section allows the inclusion of information defined by an existing standard such as an IBIS[83] or VHDL[85] file. Sections contain settings that are identified by the keyword they begin with, such as die_name, die_thickness, or die_BSDL. The settings are composed of the setting values or attributes that contain the bulk of the die specific data.

Future Standards Work. The specifications and guidelines listed identify the key issues and information required for the development, procurement, and use of KGD. This standardization effort has led to the development of die-level data sheets, die product applications support, and strategic alliances. The early MCC and SEMATECH KGD standards are continuing to mature and gain industry support in the from of JEDEC-49 and IEC-1360 standards. KGD database development through the

Table 14.12 DIE Format Level 0 Settings

block_level	die_pads	die_technology
block_notes	die_power_max	die_type
block_DIE_format_version	die_notes	die_thickness
block_disclaimer	die_packaged_part_name	pad_digital_circuit
block_version	die_section_version	pad_digital_circuit
block_source	die_power_nom	pad_digital_pull_down
die_junction_temperature	die_size	pad_digital_pull_up
die_lead_frame	die_solder_bump	pad_digital_threshold
die_mask_version	die_source	pad_geom_shape
die_manufacturer	die_substrate_connection	pad_geom_name
die_name	die_substrate_material	pad_supply_current_max pad_supply_voltage

Table 14.13 DIE Format Level 1 Settings

die_backside_finish	die_pads_tolerance	die_wafer_size
die_bonding_diagram	die_pads_VHDL_map	die_wafer_step_and_repeat
die_BSDL	die_passivation_material	model_name
die_conditions_storage	die_size_tolerance	model_type
die_fiducials	die_specific_heat_capacity	pad_digital_IBIS_model
die_IBIS	die_thermal_conductivity	pad_geom_metal_extent
die_packaged_part_bond	die_thickness_tolerance	pad_geom_passivation_extent
die_pad_pitch	die_VHDL	pad_geom_tolerance
die_pads_supply_grouping	die_wafer_scribe_line	

Table 14.14 DIE Format Level 2 Settings

die_bonding_sequence	die_manufacturer_cage	die_passivation_extent_size
die_conditions_bonding	die_measure	die_power_capacitance
die_conditions_process	die_military_spec	die_power_cell
die_conditions_sealing	die_packaged_part_attach	die_quiescent_current
die_conditions_special	die_pad_dielectric	die_saw_step_error
die_description	die_pad_metal	die_saw_width
die_flatness	die_pads_jumpers	die_separation
die_load_factor	die_pads_noncontact_area	pad_geom_bond_sites
die_lot		

ESPIRIT GOOD-DIE project will provide further access to needed KGD product data. The primary focus of the initial version of the DIE Format was on unpackaged digital integrated circuits. Early industry support for DIE turned to apathy as well conceived KGD product data sheets seemed to satisfy immediate needs. Eventually, a DIE like format is intended to become a *Device Information Exchange Format* to address all forms of unpackaged devices including passives, discrete semiconductors, and ICs. As the KGD market grows, and emerging information model standards such as EIA548 Electronic Data Interchange Format (EDIF) [87], the CFI CIR Electronic Data Book, and the ISO10303 Standard for the Exchange of Product Data (STEP) mature, the need for a DIE-like data format for the exchange of KGD will become a necessity.

14.3.4 Known-Good Die Technologies

Basically, there are three paths for ensuring that ICs entering an MCM assembly process are known-good to specific probability levels:

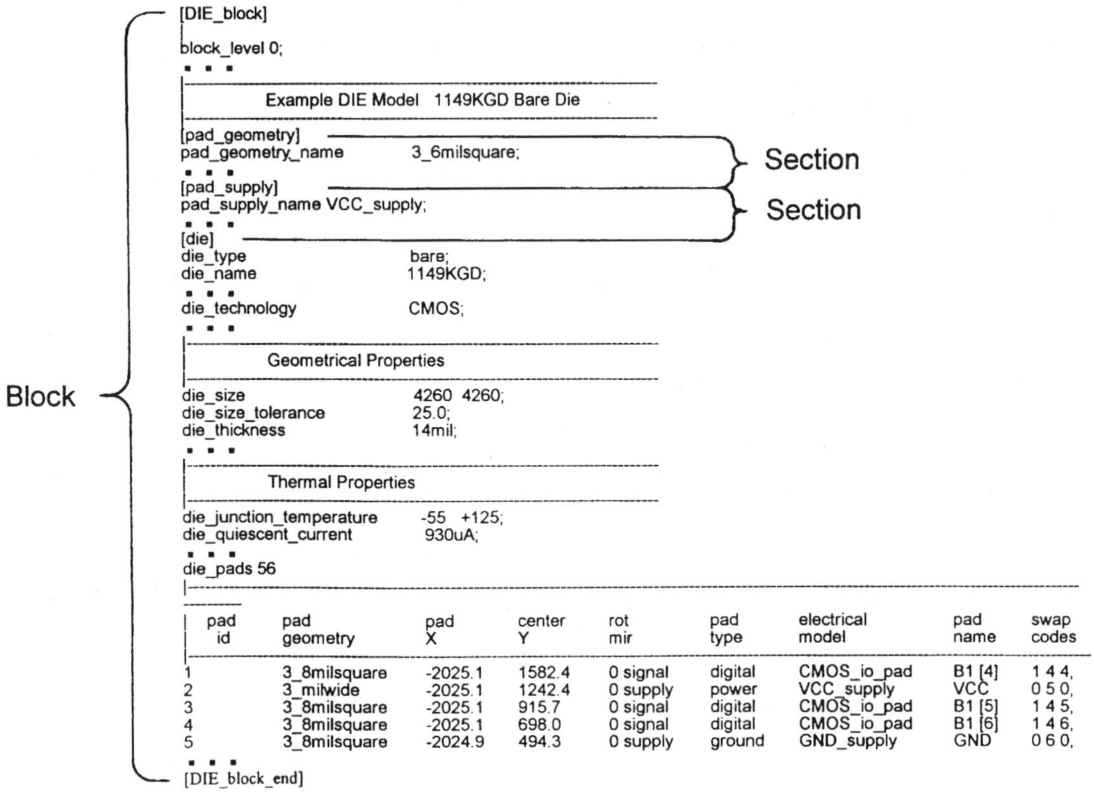

FIGURE 14.9 The basic DIE structure.

1. *Purchasing die from a semiconductor manufacturer who will guarantee a specific probability level.* This involves new efforts by semiconductor manufacturers in processing, configuration, and testing to raise statistical probability levels for unpackaged ICs.
2. *Purchasing die from a die supplier who will guarantee a specific probability level.* This involves new efforts by independent die suppliers to purchase unpackaged ICs and screen them to higher statistical probability levels.
3. *Bringing die testing in-house.* This involves new efforts by MCM fabricators to purchase unpackaged ICs and pretest them in-house prior to assembly into MCMs.

Initial efforts in all three approaches were hindered by IC vendor concerns about losing the "value added" they currently receive in single-chip packaged IC sales. However, as it has become evident to both vendors and customers that known-good unpackaged ICs are worth a price similar to packaged ICs, unpackaged ICs have become much more widely available.

Industry consensus is that the ideal long-term solution for ICs in high volume is the first approach, with the semiconductor manufacturer maintaining control of both the technology and the value added associated with ensuring known-good-die in some type of format (possibly bare die, TAB, or near-die-size "packages"). However, realistic industry consensus also acknowledges that, for leading-edge electronic products in low-volume or start-up phases, the first approach is not economically attractive for either the semiconductor manufacturer of the MCM producer, and the latter two approaches will probably be chosen until product delivery volumes justify the first approach.

General KGD Categories. Within the three paths identified above (semiconductor, third party, in-house), there are many different methods in use to raise the statistical probability level of ICs entering multichip assembly processes. These fall into four general categories:

1. *Process methods*—High yield through six-sigma processing, wafer-level screening, or improvements from special processing. This option is available only to semiconductor manufacturers and those who control custom IC designs.
2. *Design methods*—Inclusion in the IC design of boundary scan capability, built-in self-test features, sacrificial extra pads on the IC devoted to pretesting, and so on, followed by sorting chips by pretesting. Again, this approach is limited to those who control the IC design.
3. *Statistical sampling*—Sorting chip lots by packaging a statistically sufficient number of chips into single-chip packages and testing them sufficiently to certify the whole lot. This method requires the consumption of hundreds of test samples to ensure 0.99 KGD levels, and thousands of samples to ensure 0.999 KGD levels. Unfortunately, test lots of a few dozen samples are more typically used which results in KGD probability levels for 0.90 to 0.95.[1] Even when this approach is combined with aggressive yield improvement programs, KGD levels do not rise much above 0.95.[88]
4. *Testing every chip*—Sorting chips by pretesting all of them by wafer probing, IC socket, TAB, microchip carrier, or other methods.

Testing Every Chip. In excess of 35 different approaches to assure KGD quality and reliability have been developed. Wafer-level approaches are the long-term, potentially low-cost focus of industry initiatives, whereas near-term solutions offer single-chip test and burn-in approaches. Singulated KGD approaches fall into two general categories:

1. *Pressure contacts*—Miniaturized probes, die sockets, or temporary carriers that employ a variety of contactor interfaces including metal alloy bumps, membrane contact, conductive films or conductive adhesives.
2. *Metallurgical connections*—Solder, thermocompression, thermosonic, electrochemical, or welded low-impedance connections in a wide variety of forms including flip chip bumped die, repatterned die, TAB, bonded chip carriers, or chip-size packages.

A system that establishes electrical connection to a bare die for pre-test and burn-in will need to meet the following requirements:

- Provides stable connection throughout the full temperature range of testing
- Establishes low resistance and low impedance at each connection
- Able to connect to all die I/O
- Introduces no damage to the die that interferes with later assembly or field life
- Provides adequate current capacity for power and ground connections
- Introduces no significant performance penalties (e.g., crosstalk, reflections, attenuation)

In addition, the connection methods need to be compatible with the variety of different metallizations that may be present on the I/O pads of different chips to be tested. Aluminum is the most widely used termination, but gold and solder are other possibilities.

Pressure Contacts. Pressure contacts must provide low-resistance connections to the IC while avoiding damage to the die bond pads that would interfere with later assembly. Other issues include high-frequency response, fragility, probe compliance, and alignment. Since most die bond sites are depressed in wells below the passivation layer, alignment and planarity of the miniaturized contacting surfaces over temperature provides a significant challenge. Most ICs are designed with aluminum (Al) bonding pads that readily oxidize, forming a high-resistance aluminum oxide layer approximately 50–80 Å thick. Pressure contacts must break this native oxide layer through a wiping, scrubbing, or piercing action. Plasma or ultraviolet etching of bond pad sites is often a prerequisite for using nonpenetrating contact alternatives. Metal-to-metal connections may exhibit wiping action due to plastic deformation under high compressive forces, a concern for thin (<0.001 inch) or brittle (GaAs) die. Penetration of the pressure contacts must allow two or more touchdowns without unacceptable bond pad damage.[80]

Reference data for the wide variety of pressure contact approaches is presented in two comprehensive tables. Table 14.15 lists the temporary die carrier pressure contact techniques that have been a KGD thrust of many North American companies. Table 14.16 contains several unique pressure contact probe card technologies for KGD. Next-generation whole wafer-level pressure contact techniques are described in Sec. 14.5.1.

A sampling of bare die test sockets is shown in Figs. 14.10 through 14.15. The Die Mate™ sockets shown in Fig. 14.10 (and later in Fig. 14.17) were developed by Texas Instruments and Micro Module Systems and are used extensively by Intel in their KGD SmartDie™ product line. Contact with the IC pads and fanout to the socket I/O is accomplished with a thin film membrane substrate having compliant gold contact points to align with the IC bond pads.

Shown in Fig. 14.11 is a burn-in socket for bare die from TCL International (Japan) which is distributed by MarkeTech. It utilizes a field of probe points to contact the IC which are fabricated from 60 micron diameter wires with formed ends. Figure 14.12 shows the Fresh Quest Corporation (now CerProbe) bare die carrier for test and burn-in. Fanout and die contact are accomplished with a thin film substrate with photolithographic pattern plated probe (P4) contact points.

Another approach is shown in Fig. 14.13 from Aehr Test Systems, which utilizes a high-density polyimide thin film substrate. The interconnecting substrate accomplishes fanout from die contacting bumps on one side to the socket I/O on the other side. Another example of mechanical pressure contact sockets for bare die is shown in Fig. 14.14. This approach from IBM is designed for solder-bumped flip chip ICs. A ceramic PGA carrier plugs into a test board and has a field of die contacts coated with sharp palladium dendrites, which are reported to pierce through solder oxides. Further information on KGD for flip chip configurations is contained in a recent proposed standard developed by an industry working group.[89]

An interesting variation of the above techniques is shown in Fig. 14.15. In this example, a ceramic substrate accomplishes the fanout from the die bond pattern to the standard IC package. Connection between the die and the substrate is accomplished through compliant conductive bumps. A final pressure contact socket test and burn-in system is shown in Fig. 14.16, using a compliant polyimide-based membrane substrate that provides precision coplanarity.

Issues with Pressure Contacts. Extensive experience with connectors (which also depend on pressure contacts) in testing, burn-in, and field environments has established three basic requirements for success of pressure contacts:

1. Mating contacts must be noncorroding materials.
2. The process of making contact must provide a "wipe" action (or other effective motion) to move surface oxides or films out of the way.
3. A high Hertzian* contact stress must be applied and maintained throughout testing and temperature extremes.

Meeting these requirements is very difficult in the case of pressure contacts to a bare die. Both aluminum and solder terminations on an IC chip tend to oxidize forming an oxide layer that is poorly conducting or nonconducting. Most schemes for pressure contact testing of KGD do not provide the "wipe" lateral motion that so effectively moves surface contaminants out of the connection path in time proven connector and needle probe technologies. Some current approaches[120] attempt to substitute piercing contacts (dendrites or plated diamond grit) or burnishing (indenting) contacts for the

*Hertzian stress is the local stress experienced when a curved surface is deformed by pressure against another surface.

Table 14.15 Pressure Contact KGD Methods—Temporary Carriers

Trade Name	Description and references	Company
Particle Interconnect™	Metal-coated, irregular shaped diamond particles, piercing contact.[57, 90–92]	ACSIST Associates
DiePak™	Reusable family of hinged plastic carriers using Nitto Denko ASMAT thin film z-axis interconnect for nonscrubbing burnishing contact. Mechanical alignment of die in carrier with precision sawn die. Hinged lid applies controlled force and provides air tight seal. This solution also includes burn-in boards and equipment.[10, 57, 93–97]	Aehr Test
Microbeam™	Scrubbing controlled impedance contacts are individually cantilevered. Spring clip and pressure plate places die directly in sockets without the use of carrier.[13,57]	California Contacts
TDC	Temporary die carrier, temporary reusable ceramic carrier, self aligning scrub contact. Die is self-aligning.[13,57]	Consultar
QC²™	Quest chip carrier. Carrier with oxide penetrating probe set, carrier uses thin film membrane and P4 Photolithographic Pattern Plated Probe contactor, scrubbing contact with quick mechanical alignment for precision sawn die.[13,57,59,99]	Fresh Quest
TCA™	Dendrite temporary chip attach. Ceramic carrier with palladium dendritic structure for mating to C4 bumped die. A clamp hold die in place and serves as heat sink. Die is removed and bumps reflowed. Palladium dendritic structures provide piercing contact. Simple wash cleans dendrites between uses.[13,57,100,101,102]	IBM
TCL	TCL bare chip test and burn-in socket. Fine probe wires are embedded on the surface of a glass reinforced polycarbonate. Die is placed face down in carrier socket, hinged lid is snapped shut. Mechanical alignment is assured with precision sawn die. A similar TCL Bare Die Carrier uses spring loaded hold down instead of hinged lid to mate die to wire contacts.	MarkeTech International Tokyo Cathode Laboratory Co.
—	Coined Au bump (stud bump) formed on die pad. Die is mounted on carrier with conductive AgPd alloy powder distributed in epoxy resin on tip of bumps. Allows easy chip removal.	Matsushita
—	MCC tape-based KGD carrier. TAB lead frame with plated metallurgy at inner lead bond. Contact is temporary scrubbing contact. Die sits on pedestal and pads contact inner lead bond (ILB), allows compliance z-axis.[13,57,103]	MCC Harris
KGD*plus*™	Reusable, universal die carriers that emulate packaged device. Oxide penetrating probe with minimal bond pad damage. Non scrubbing Al bond pad contact, silicon insert fab by nCHIP.[13,57,104]	Micron 3M nChip
Gold Dot™	Membrane probe set, capable of variety of shapes, non scrubbing elastomer backed contact minimizes pad damage. Uses spring clamp and supporting plate and heat sink.[13,57]	Packard Hughes
—	Wire-based probing, compressed wires make scrubbing contact plastic base cavity to hold die on ceramic wiring plate for alignment. Designed to accommodate non planar surfaces.[13,57]	Plastronics
BTQI™	Burn-in test quality and inspection carrier. Thick film ceramic substrate temporary reusable package self aligning oxide penetrating scrub contact. ceramic substrate, thick film leads, gold bumps as contacts.[13,57]	Qualhi
DieMate™	Reusable carrier uses compliant bumped membrane, proprietary oxide piercing contact. Lid serves as heat sink. Thin film interconnect alignment fence for easy quick mechanical alignment on precision sawn die. Contacts hold die in place during align an clamp. Thermal management integral to design.[57, 105–114]	TI/MMS
T-L/P™	Temporary Lead Pack, modified ceramic package, thin film and proprietary non-scrubbing fine point oxide piercing contact mechanism. Minimizes retooling by using existing conventional packages. Compliance built in and provides air tight seal with assembly. Die position maintained during assembly disassembly by integral clamp.[13,15]	Tribotech
—	Carrier and socket using thin film contacts, non-scrubbing oxide piercing bumps, thin film interconnect, spring locating and alignment.[13,57,115]	Yamaichi

Table 14.16 Pressure Contact KDG Methods—Probe Cards

Trade Name	Description and references	Company
—	Four-segment probe card using individual sprint cantilever controlled impedance contacts. Extension of Microbeam interconnect concept.[57]	California Contacts
QPC™	Quest probe card features P4™ (photolithographic patterned plated probe) controlled impedance scrubbing contact. Provides 5 GHz bandwidth/500 MHz data rate.[99]	Fresh Quest
UMP	Universal membrane probe, TAB tape mounted with array pattern of nickel microbumps. Array patterns of 150 μm or 200 μm bump pitch will accommodate 80 percent of device pad patterns. Feature low-cost disposable TAB materials.[116,117]	HP Japan
Cobra™ Probe	Probe card for C4 bumped wafers. Individual probe beams deflect (buckle) providing scrubbing contact. Flat 5 mil dia. probe tip deflects 7 mils at 25 g.[118]	IBM
Gold Dot™	Membrane probe, extension of Gold Dot™ interconnect. Features 80 ps device edge speed, 350k contact life, 0.004 inch pitch.[119]	Packard Hughes

wipe action. Both techniques can be foiled by contaminant layers that deform but remain unbroken on the IC pads. There has been some good progress on these probe types, but variable contact resistance remains one of the trouble spots for pressure contacts.[15]

Typical contact resistance values for various approaches are shown in Table 14.17. The industry Technical Advisory Group (TAG) goal for contact resistance was established at 250 mΩ maximum[15] although a segment of industry believes that this value should be reduced to 50 mΩ maximum.

Caution should be used in comparing various reports of contact resistance, since some papers include wiring resistance in the reported number.[95,117] An excellent analysis of one KGD socket approach identifies the various elements of resistance introduced by the test socket system, as shown in Table 14.18. An exploded view of the test socket is shown in Fig. 14.17.

The challenges for pressure contacts include not only developing a low-impedance initial connection but maintaining connection integrity over testing at temperature extremes without either (a) relaxation of contact normal force or (b) losing alignment of probe-to-pad connections. Thermal expansion changes of different materials and stress relaxation of polymers, springs, and solders present significant challenges to pressure contact approaches.

Electrical Breakdown of Contact Resistance. An interesting study of high contact count membrane probes[63,128] has shown that there is a viable alternative to mechanical scrubbing for breaking down the interfacial oxide always present on aluminum bond pads. IC photolithography techniques were used to fabricate arrays of tiny tungsten probe tips that could be brought in contact with each IC pad. By sending a 50 mA pulse in one probe point and out a second probe point on the same IC pad, the initial contact resistance could be reduced by one to two orders of magnitude.

Metallurgical Connections. Metallurgical connections typically will have lower resistance and better stability than pressure contacts. Three types are used in KGD approaches. "Welded" joints are formed when the two connecting elements are joined through interdiffusion of atoms between the two (thermocompression, ultrasonic, and thermosonic wire bonding and TAB bonding). Soldered joints are formed when a low melting temperature metal is used to wet and bond the two connecting elements together (R3 and BATS). Plated/deposited joints are formed when atoms are deposited onto the IC pads to form the connection traces.[53] A summary of typical resistance values is shown in Table 14.19. Metallurgical connections will typically remain stable throughout KGD testing environments, provided they are designed to maintain low stress conditions.

Metallurgical connections used in creating KGD can be either permanent or temporary. Permanent connections (such as a TAB leadframe) remain with the die through KGD screening and into final assembly. Examples of KGD permanent metallurgical connection techniques are given in Table 14.20. Examples of two of the metallurgical connection techniques are shown in Figs. 14.18 and

FIGURE 14.10 DieMate™ test sockets for bare die. Courtesy of Texas Instruments and Micro Module Systems.

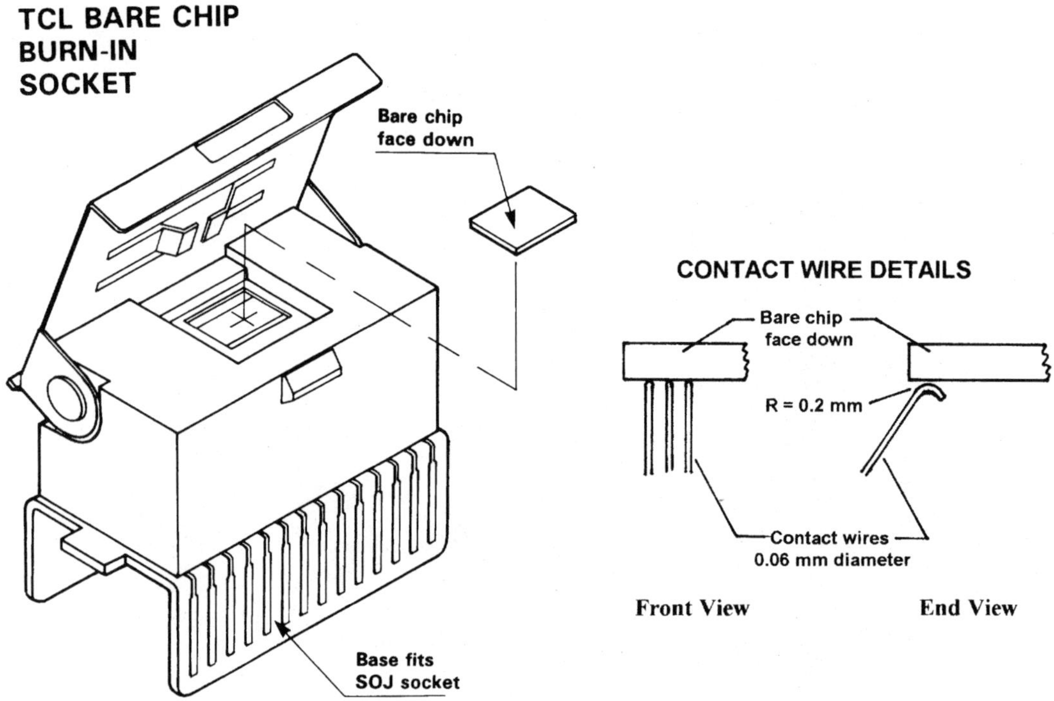

FIGURE 14.11 Burn-in socket for bare die. Courtesy of MarkeTech International.

14.19. Figure 14.18 shows an IC bonded into a TAB leadframe along with a typical test socket. Figure 14.19 shows a technique developed by ELMO to enlarge the bond pads on a die to permit wire bonding into a conventional package for test and burn-in. Afterward, the wire-bonds can be broken and the die removed from the package for assembly into the MCM.

Temporary metallurgical connections (sometimes called "soft" connections) are attached for the purpose of KGD screening but removed after testing to restore the original bare die configuration going into final MCM assembly. Examples of these KGD temporary connections are shown in Table 14.21 and in Figs. 14.20 through 14.23. Several of these methods utilize connections intentionally made to provide very low-strength bonds that can be removed after burn-in and test (e.g., SofTAB™ and SOFTOOL™). Such an approach is shown in Fig. 14.20 from a Sandia test program.[6] A similar approach for flip-chip ICs is shown in Fig. 14.21 from IBM. The device is soldered to undersize pads on a test substrate for burn-in and test. Afterward, the chip can be removed with low force by breaking the tiny area connections. Figure 14.22 shows another approach for flip-chip ICs. In this case, the IC is mounted to a special substrate with a sacrificial metal layer on the pads. After burn-in and test, special processing can cause the solder bumps to dejoin from the substrate, leaving a pristine solder-bumped die ready for assembly in the MCM. Figure 14.23 shows the temporary bond approach using TAB carriers.

Chip-Size Packages. An alternative packaging approach that avoids most of the bare die issues is to package the bare ICs in "near die size" or chip-size packages. A variety of approaches have emerged that combine the advantages of both single-chip package testability and bare die-size reduction. The approaches range from simply mounting a die on a very slightly larger chip carrier (the "miniature chip carrier") to coating the chip with a hermetic coating (the "sealed chip") to encapsulating the die in a very slightly larger package (the MiniBGA, MicroBGA, MicroSMT, and SLICC

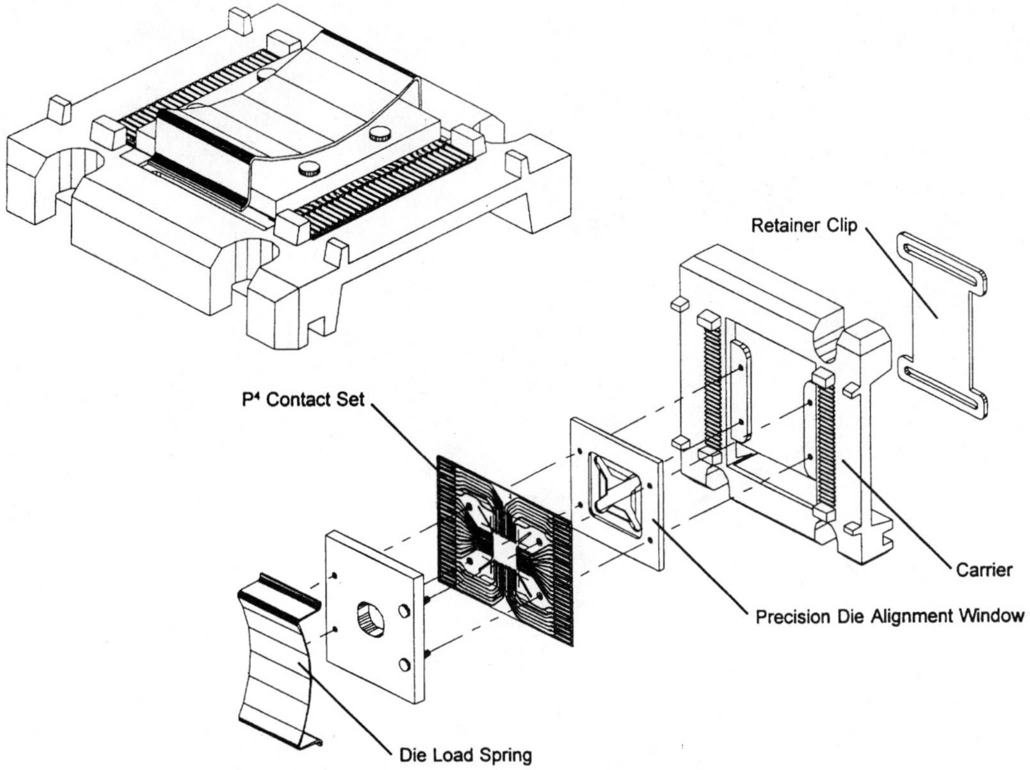

FIGURE 14.12 Bare die carrier for test and burn-in. Courtesy of Fresh Quest Corporation.

packages). Table 14.22 summarizes these approaches, which are also called *minimal packages, micro packages, chip-scale packages,* or *chip-size packages.* Figure 14.24 shows the Tessera chip-size package. Figure 14.25 shows a miniature device carrier being tested with a probe card. Table 14.23 shows the relative size comparison of some chip size approaches. The ideal chip size package will have the following attributes:

- Allows assurance that the IC is known-good
- Provides good mechanical protection prior to assembly into MCMs
- Has a footprint size equal to or approach the die foot-print size (no more than 1.2 times larger than die area)
- Has a height equal to or approaching the die height (no more than 2.5 times the die height)
- Has a lead pitch equal to or approaching the die pad pitch (no more than 1.2 times larger than die pad pitch)
- Introduces no significant performance penalties (electrical, thermal, mechanical, or environmental)
- Allows easy replacement of ICs or miniature chip carriers
- Allows options for use in either hermetic or nonhermetic MCM packages
- Meets desired low cost goals in both nonrecurring and recurring costs
- Provides desired producibility and reliability

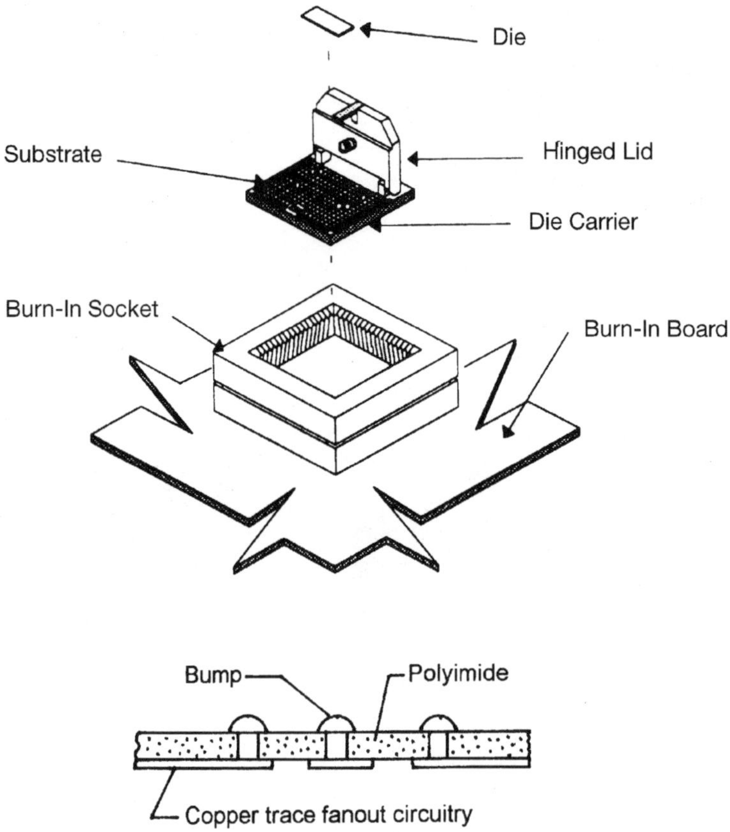

FIGURE 14.13 Burn-in and test socket/carrier for bare die. Courtesy of Aehr Test Systems.

Further information is available in a recent proposed standard for chip scale packaging developed by an industry working group.[89]

14.3.5 Levels of Test and Reliability

A number of the KGD approaches discussed in the preceding section have been successfully implemented as part of KGD product offerings by major semiconductor suppliers and die processors. Additional KGD product announcements are imminent, based on the development and evaluation results of government and industry sponsored programs. These product announcements foster the rapidly growing availability of bare die products.

To address the wide variety of customer needs, suppliers now offer bare die in an equally wide array of quality and reliability levels. The levels or degrees of die processing range from one to five different levels among the semiconductor manufacturers and die processors currently offering bare die products. The degree of processing ranges from none (relying on statistical data only) to full electrical test at the rated operating speed and temperature range with burn-in. Many of these die processing options exploit recent advances in wafer probe and die level KGD technology for at-speed and extended temperature testing.

Because the KGD market is still maturing, buyers and sellers of die must pay particular attention to understanding the expected and deliverable levels of die quality and reliability. For example, a DC

(a)

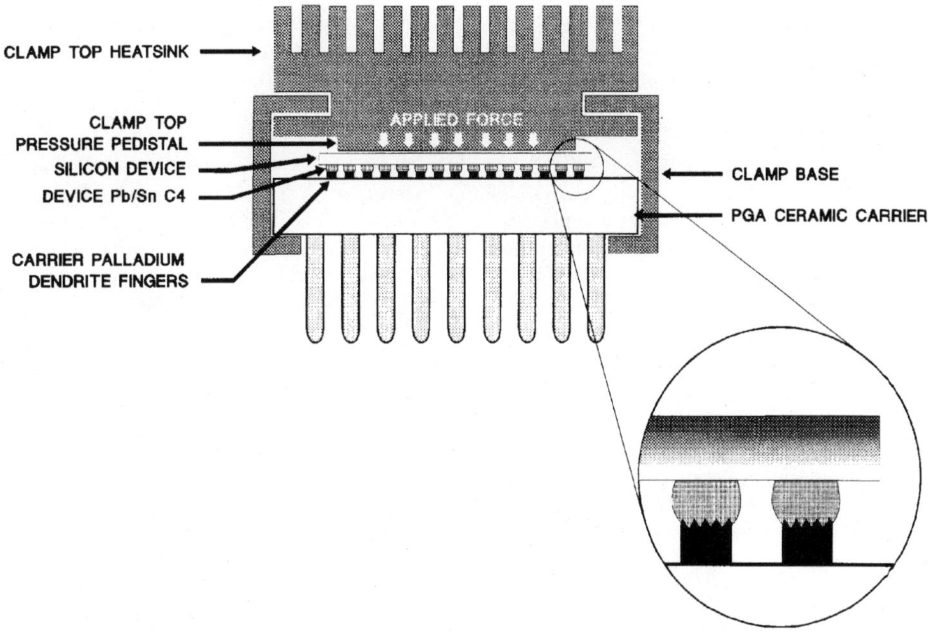

(b)

FIGURE 14.14 Bare die burn-in and test socket for flip chip ICs: (a) opened socket and (b) pressure contact detail. Courtesy of IBM.

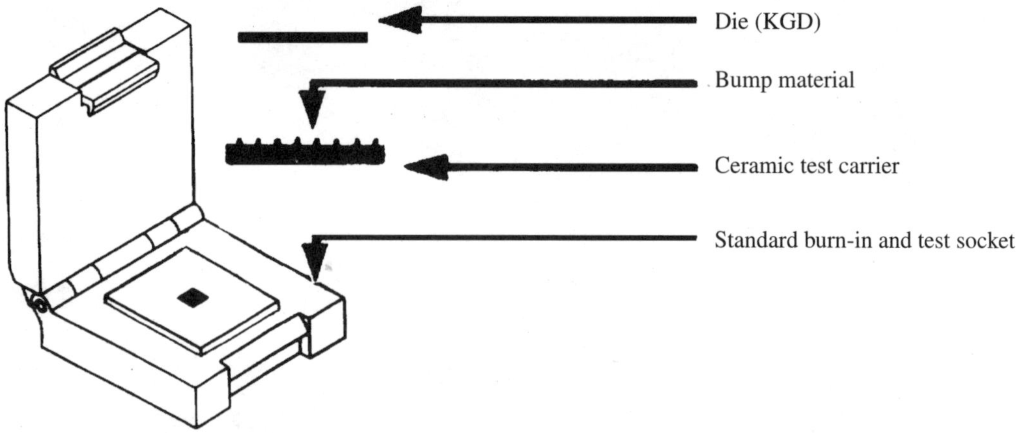

FIGURE 14.15 Bare die test and burn-in using conductive polymer bumps. Courtesy of EPI.

wafer probed die will result in die confidence levels as low as 0.50 for early production devices, while lot acceptance testing or statistical sampling will result in confidence levels of 0.90 to 0.95 at best for mature die. To achieve confidence levels in excess of 0.99 often requires full test and burn-in of every chip. Sections 14.4.2 and Table 14.1 provide additional metrics on these levels of testing. Table 14.24 below lists various degrees of testing currently employed by providers of bare die products.

KGD Product Offerings. Companies supplying some form of KGD include both semiconductor manufacturers, third-party die processors, and test houses. A representative sampling of KGD suppliers is summarized below.[13,157,158] This list captures the latest program information available at the time of publication; it is not intended to be all inclusive, since offerings will likely change with the rapid growth and maturation of KGD technologies.

Intel Corporation. Intel's SmartDie program offers one level of die processing—KGD—at the same cost, quality, and reliability as packaged devices. At the wafer level, these KGD are subjected to full electrical testing (DC, AC, functional). Smart Wafer Sort is used to ensure die operation over 0–80° C (junction). Intel utilizes the TI/MMS DieMate and other KGD technologies to supply fully tested and burned-in KGD. Quality (DPM) and reliability (FIT) rates are consistent with packaged product. The Technical Alliance program provides qualified assembly partners. Products include: flash memory, microprocessors, and microcontrollers.

National Semiconductor. National offers four levels of die screening through the Die Upscale program.

Option 1: DC wafer probed at 25° C and 100 percent visually inspected

Option 2: Option 1 plus sample lot testing for DC, functional, and burn-in

Option 3: DC/AC wafer probed at temperature (25 and 125° C) plus sample lot testing for DC, functional, and burn-in; four choices of increasing severity for sample lot testing

Option 4: Full 100 percent die-level testing at temperature and burn-in using KGD carrier technology

A fifth option for wafer level test and burn-in is proposed for future products. Die products requiring extended temperature operation are supported. Options include lots sizes ranging from 50-piece, 100-piece, and full-wafer minimums. Available products include logic, interface, SCAN, SRAM, and analog products.

Micron Semiconductor. Micron's DiePlus program offers four levels of bare die testing.

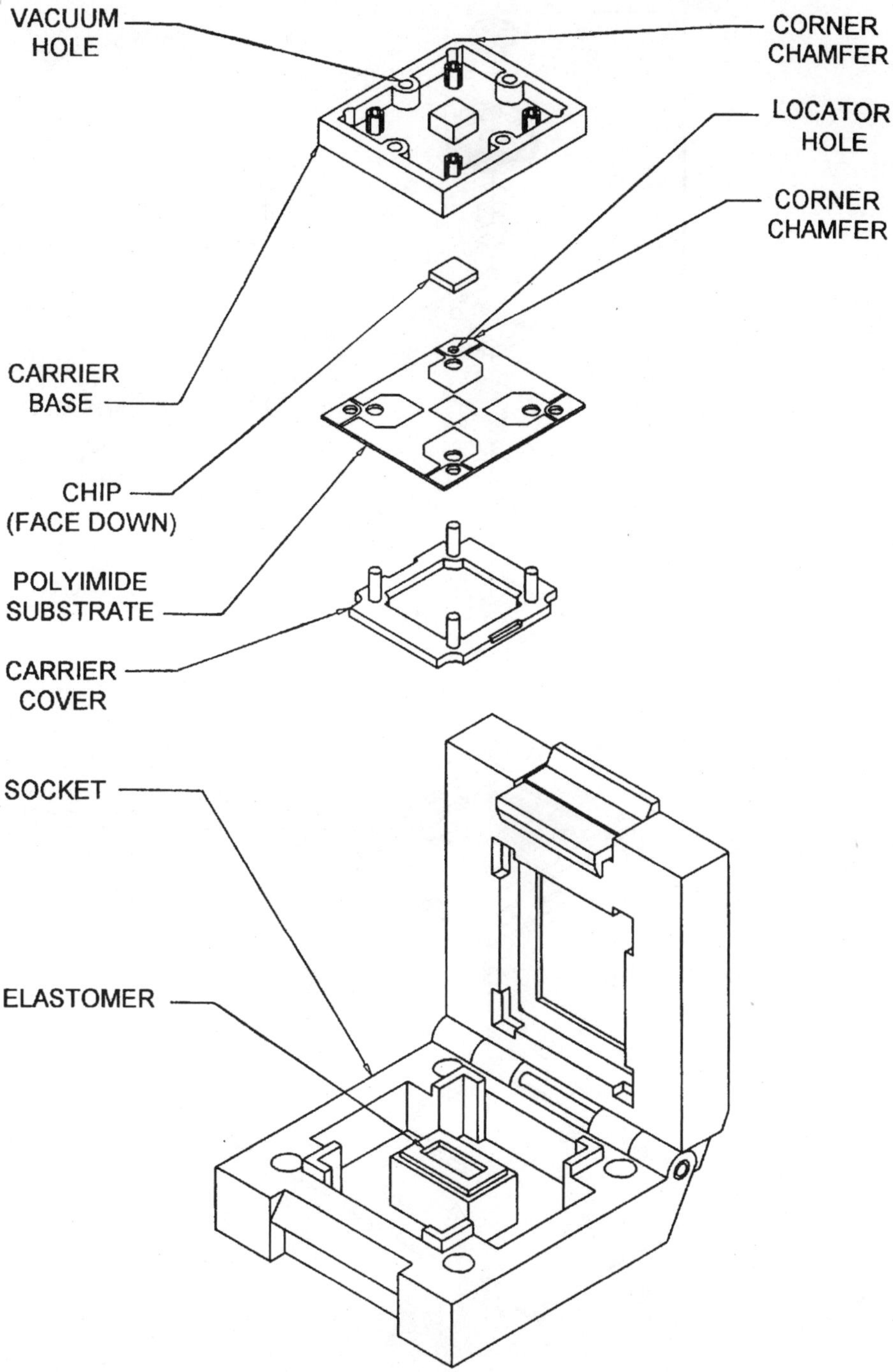

FIGURE 14.16 Yamaichi KGD test and burn-in socket and carrier. Courtesy of Yamaichi.

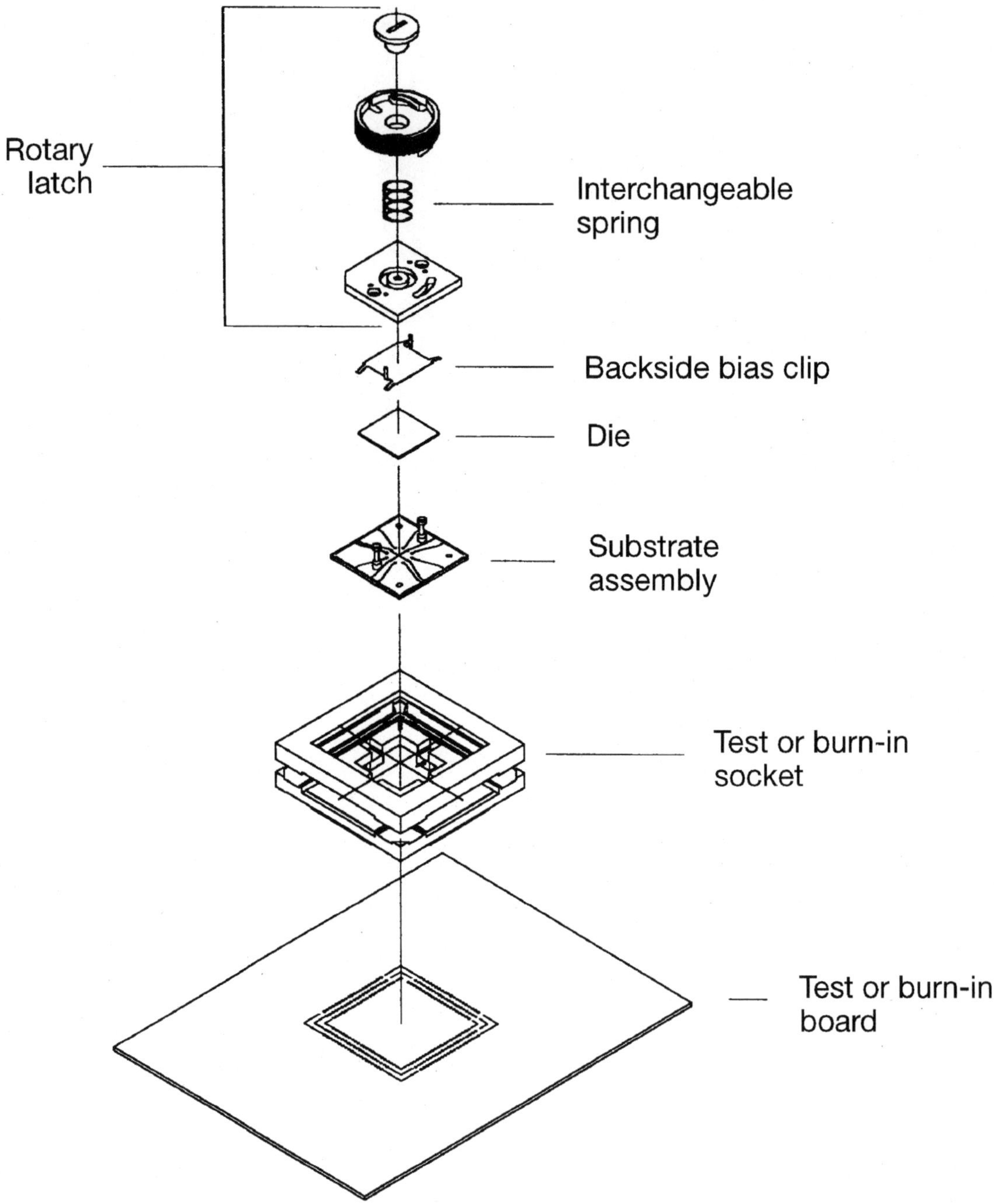

FIGURE 14.17 Exploded view of DieMate™ bare die socket. Courtesy of Texas Instruments.

Table 14.17 Values Reported for KGD Contact Resistance

			Contact resistance to IC (mΩ)		
Method	Organization	Force/contact (g)	Initial	After use	Reference
Flip chip probe	MCC	25	40	—	121
Membrane probe	Hughes	—	<250	—	122
Membrane probe	TI/Micro Module Sys.	—	30–80	40–275	123
Dendrite flip chip probe	IBM	10	9	—	124
Dendrite flip chip probe	IBM	100	2	—	124
TAB-lead probe	Harris/MCC	20	<50	—	103
ASMAT™ Z-conducting sheet, Au–Au	Nitto Denko	—	<20	—	125
ASMAT™ Z-conducting sheet, Au–Al	Nitto Denko	—	<100	—	125
DieMate™ socket	TI/Micro Module Sys.	7	<50	<100	111
DiePAK™ carrier socket	Aehr	—	80–190	—	95
Plated probe	Fresh Quest	—	140–850	140–850	126
Conductive adhesive	ETEC	—	<50	<50	70
Silicon membrane probe	ELM Technology	—	<100	<100	127
Carrier socket	TI	—	7	40–80	114

Table 14.18 Elements Contributing Resistance in a Particular KGD Socket System[114]

Element of Resistance	mΩ
Au-Al contact resistance, die to substrate	42–77
Fanout/routing traces in substrate	3,800–7,770
Au-Au contact resistance, substrate to socket	8–29
Fanout/routing conductors in socket	44–67
Solder joint, socket to PWB	14–29
Fanout/routing in PWB	21–28

Table 14.19 Summary of Connection Resistances with Metallurgical KGD Approaches

		Typical connection resistance			
Connection type	Base metal resistivity (μΩ-cm)	Joint (mΩ)	Lead length (mΩ/mil)	Total (mΩ)	Quoted range[13] (mΩ)
Gold wire bond	2.2	<3	0.5–2.0	<50	30–100
Aluminum wire bond	2.7	<3	0.6–2.4	<70	30–100
Copper TAB leads	1.7	<1	<0.1	<10	13–100
Flip chip solder ball	11–22	<3	none	<3	2–6
HDI overlay	1.7	<1	<1	<1	<1

Option C1: Standard Probe—Wafer probe tests include DC parametric and functional tests with hot chuck and guard banding to ensure commercial temperature operation.

Option C2: Speed Probe—Option C1 plus AC parametric and speed grades die at temperature. Unlike C1 which has no warranty, C2 guarantees access times.

Option C3: Probe*plus*™—Option C2 plus die are dynamically burned-in using SofTOOL carriers. Die is warranted for speed, and infant mortalities eliminated.

Option C7: KGD*plus*™—Die quality equals or exceeds that of packaged parts. Uses KGD*plus* carrier produced by 3M with the nChip silicon insert. Full electrical test and dynamic burn-in and fully warranted to data sheet specs. Products include SRAMs and DRAMs.

Texas Instruments. The Texas Instruments FlexSiDie program offers three levels of die processing that support both commercial and military options. KGD technologies include DieMate and a form of removable TAB.

KTD: Known-Tested Die, standard wafer sort (multiprobe). This can include standard die sort and accelerated voltage screen, lot acceptance testing, or hot chuck wafer probe testing at speed.

C-KGD. Commercial temperature processing using Die Mate or removable TAB.

M-KGD. Military temperature processing using Die Mate or removable TAB.

Products include: DSP, logic, SCOPE, DRAM.

IBM. One level of KGD processing is offered using the R3 process. IBM has shipped more than 1 million KGD since 1990. R3 allows processing of fully tested and burned-in flip chip C4 devices. Array sizes up to 625 I/O have been tooled. C4 KGD production will migrate to the dendrite process. Product: SRAM.

Integrated Device Technology, Inc. IDT die processing flows are the same as the delivered packed products for commercial, industrial and military products. IDT's temporary die mounting technique[13] is cost effective for small to moderate volume orders. Carrier methods are considered for larger volume orders. All die receive 100 percent visual inspection based on MIL-STD-883, Method 2010 Condition B, and are shipped in Gel-Pak containers within hermetically sealed ESD bags. KGD products include RISC processors, SRAMs, dual-port RAMs, and FIFOs.

Chip Supply Inc. Four major levels of die processing are available from Chip Supply.

Commercial die—as probed by the manufacturer, with no quality or reliability assurances

(a)

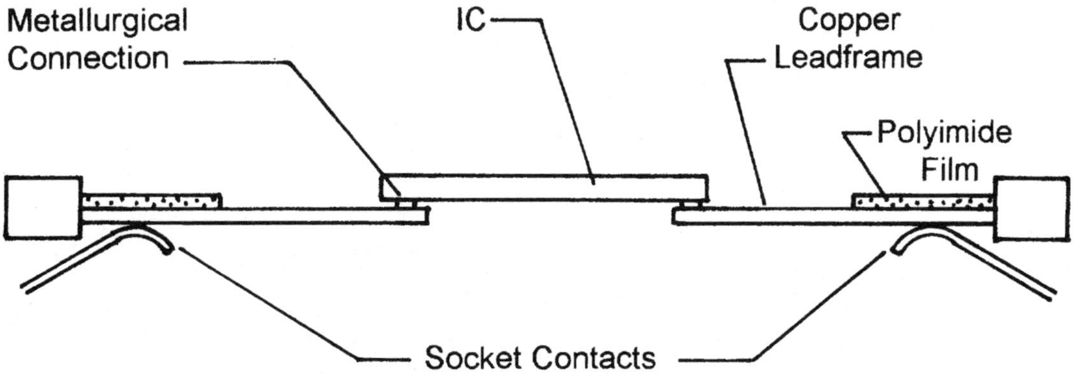

(b)

FIGURE 14.18 IC in TAB lead frame beside test socket: (a) TAB and test socket and (b) detail of TAB connections.

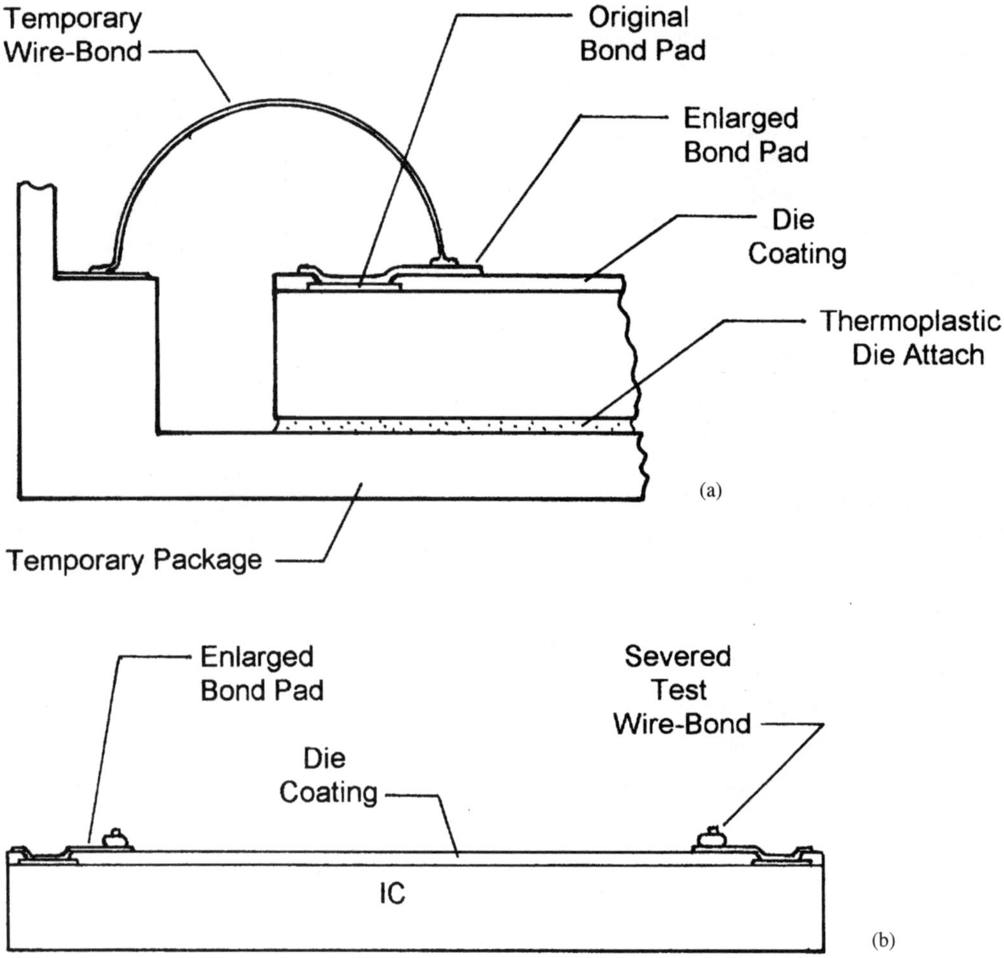

FIGURE 14.19 ELMO approach with enlarged wire bond pads; (a) die installed in temporary package and (b) KGD ready for assembly in MCM (after Ref. 129).

Probe and visual—commercial die probed at temperature for access time or speed and 100 percent visually inspected; increases the quality level of the die but has no effect on reliability

Lot acceptance testing—a small number of die from a wafer lot assembled into ceramic packages and screened to agreed upon conditions

Known good die—die 100 percent tested and burned-in using SofTAB, TAB, SofTOOL, chip on substrate (COS), or other KGD techniques. KGD supplied with certificate of compliance to predetermined test and burn-in levels and conditions.

Agreements with over 20 semiconductor manufacturers to purchase wafers for reprocessing and distribution in die form, including AMD, IDT, Micron, Motorola, Samsung, Xilinx, TI, and National.

Elmo Semiconductor. Elmo offers five levels of die screening, including level 5, which utilizes the Enlarged Bond Pad KGD technique.

Table 14.20 Metallurgical Contact KGD Methods

Trade name	Description and References	Company
Enlarged bond pad	Overcoat applied to wafer, contact windows opened over existing bond pads, enlarged bond pads are patterned and metallized over window. Die temporarily mounted into a TO package and wire bonded to enlarged bond pads. After test and burn-in, wirebonds are peeled off or clipped from bond pads. Allows multiple bonding.[13,57,129]	Elmo
HDI™	High Density Interconnect. An adaptation of the existing HDI MCM process. Die is overcoated with polymer, vias laser drilled down to bond pads. Vias are filled, temporary bond pads formed on overcoat, die attached to standard package and wire bonded. After burn-in and test, WBs are removed and bond pads are etched away.[13]	GE
TRB™	Testable Ribbon Bond. Alternative to TAB, die is conventionally ribbon bonded to disposable carrier with custom chip alignment tool. Die and inner lead bond stay intact, and the outer lead bond (OLB) is cut by laser. The ribbon leads are bonded with a themosonic wedge bonder face up or flip chip style.[13,57,130,131]	Hughes
IDT KGD	Au bumped die are mounted in a temporary package and Au ball bonded. After test and burn-in, wirebonds are severed just above gold ball bond and shipped with Au bump/ball on each pad.[13]	IDT
TAB	Tape Automated Bonding. A mature KGD technology suitable for highly automated production. Bumped die pads are gang bonded to photoetched metallized pads on tape. Tape is typically polyimide Kapton with sprocket holes similar to photographic film in 35 or 70 mm slide or reel formats.	TI, Harris
FAW	Frame Automated Wirebond. Wirebonded to temporary carrier, precision cut of wirebond to excise. Die with short wirebonds can be outer lead bonded (OLB) to substrate.[13,56,57]	Northern Telecom
—	Uses conventional Au ball bonding on a printed circuit board designed to test several chips simultaneously. Die are held mechanically by elastic spring mounts. After burn-in and test, a diamond blade excises gold wire bonds. The resulting flat gold bond pads are suitable for wirebonding.[138–140]	Samsung
SofTAB™	Uses standard TAB leadframe and gold bumped die. Inner lead bond is lightly attached. Modified pull test/shear removes the inner lead bond excising bumped die from TAB frame.[13,15,134,135]	Chip Supply
R3™	Reduced Radius Removal. Die is flip chip soldered to reduced radius pads on temporary reusable ceramic carrier. Carrier mounts in standard PGA socket. After burn-in and test, the die are removed from carrier by twist off shearing process leaving solder bump intact, ready for reflow.[4,13,54,100,102]	IBM
BATS™	Burn-in and Test Substrate. Die bumped with electroplated solder deposition process. Die attached to reusable ceramic carrier on solder pads with sacrificial metal. High joint strength during burn-in and test. After burn-in and test sacrificial metal is absorbed into solder bump with proprietary metallurgy process. Low dejoin force allows easy die removal.[13,60,74,136]	MCNC
SOFTOOL™	Temporary die attach and lightly wire bonded in a standard package. Bond wires are peeled/pulled from bond pads.[13,15,27,137]	Micron
—	Temporary die attach and light wire-bonding in temporary packaging for burn-in and test.[6]	Sandia et al.

Table 14.21 Metallurgical Contact KGD Methods—Soft Connection

Trade name	Description and references	Company
SofTAB™	Uses standard TAB leadframe and gold bumped die. Inner lead bond is lightly attached. Modified pull test/shear removes the inner lead bond excising bumped die from TAB frame.[13,15,134,135]	Chip Supply
R3™	Reduced Radius Removal. Die is flip chip soldered to reduced radius pads on temporary reusable ceramic carrier. Carrier mounts in standard PGA socket. After burn-in and test, the die are removed from carrier by twist off shearing process leaving solder bump intact, ready for reflow.[4,13,54,100,102]	IBM
BATS™	Burn-in and Test Substrate. Die bumped with electroplated solder deposition process. Die attached to reusable ceramic carrier on solder pads with sacrificial metal. High joint strength during burn-in and test. After burn-in and test sacrificial metal is absorbed into solder bump with proprietary metallurgy process. Low dejoin force allows easy die removal.[13,60,74,136]	MCNC
SOFTOOL™	Temporary die attach and lightly wire bonded in a standard package. Bond wires are peeled/pulled from bond pads.[13,15,27,137]	Micron
—	Temporary die attach and light wire-bonding in temporary packaging for burn-in and test.[6]	Sandia et al.

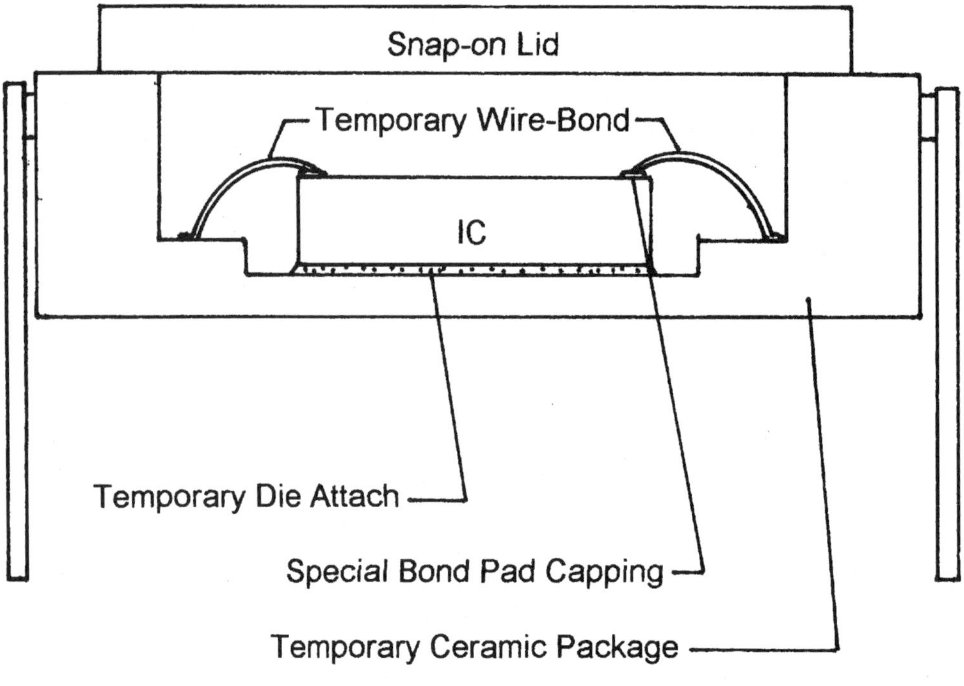

FIGURE 14.20 Sandia temporary wire bond connection for KGD (after Ref. 6).

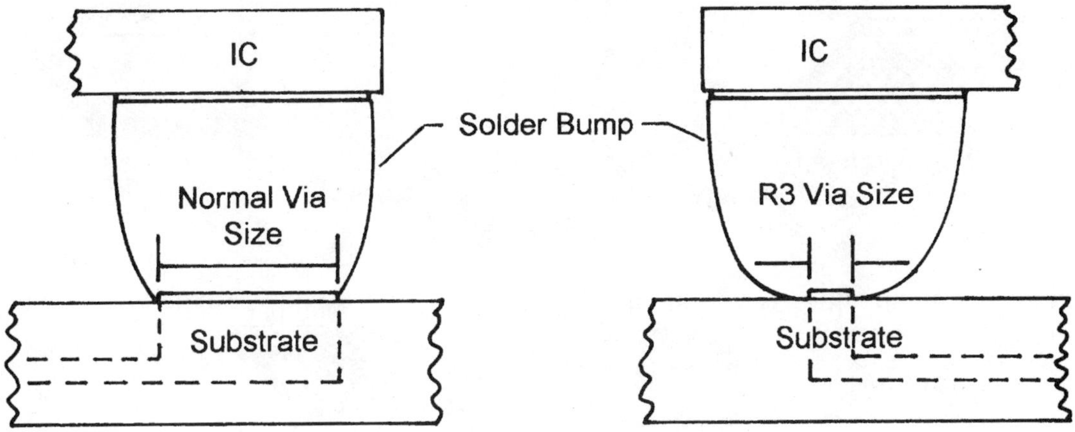

FIGURE 14.21 IBM R3 (reduced radius removal) technique for flip chip KGD: (a) normal MCM connection and (b) R3 connection for burn-in/test.

Option 1: Wafer probe, visual inspection, packaged sample, bondability and electrical tests

Option 2: Option 1 plus sample lot hot chuck tested

Option 3: Option 2 plus thermal cycling, pre/post burn-in electrical tests and burn-in

Option 4: Option 3 plus SPC probe, SEM analysis, and life test

Option 5: Option 3 plus Enlarged Bond Pad KGD wafer processing for full electrical test and burn-in

Major lines supported include: AMD,CDI, IR, NEC, Philips, Harris, BCK, Motorola, and TI.

Minco Technology Labs. Minco offers several solutions for die processing beyond the standard DC wafer probe test. The lowest cost solution involves 100 percent die probe (DC, Functional, and limited AC) at elevated temperature. The next level of processing options involve the use of temporary die carriers to allow full 100 percent test and burn-in. Several carrier solutions are available including DieMate. Minco's third solution includes supplying National Semiconductors' standard KGD products. Minco offers products from over fifteen semiconductor manufacturers including: AMD, Elantec, Harris, IR, LTC, Micro Semi, Motorola, TI, and National.

ETEC. Electronic Test Engineering Company (ETEC) is supplier of military and commercial component test services. ETEC provides a wide variety of test and burn-in options. KGD techniques include the ELMO enlarged bond pad and other temporary interconnect methods. Device testing includes SRAM, DRAM, ASICs, programmable logic, linear.

EPI Technologies Inc. EPI provides component testing services for both commercial and military products. EPI utilizes the EPIK™ KGD technique, which employs a temporary contact method using conductive polymer bumps on a ceramic carrier mounted in standard test and burn-in sockets. Products include SRAM, DRAM, microcontrollers, EEPROM, and logic.

Other companies including AT&T, Harris, HP, Motorola (five levels), Samsung, and Tektronix have developed die processing and KGD techniques for their products. Die users must chose the level of testing needed to deliver the best price and performance for a specific application.

14.4 REMAINING ISSUES

It is evident from the preceding discussion that use of unpackaged ICs with known-good probabilities much below 0.999 will lead to very low first-pass module yield, massive rework required to

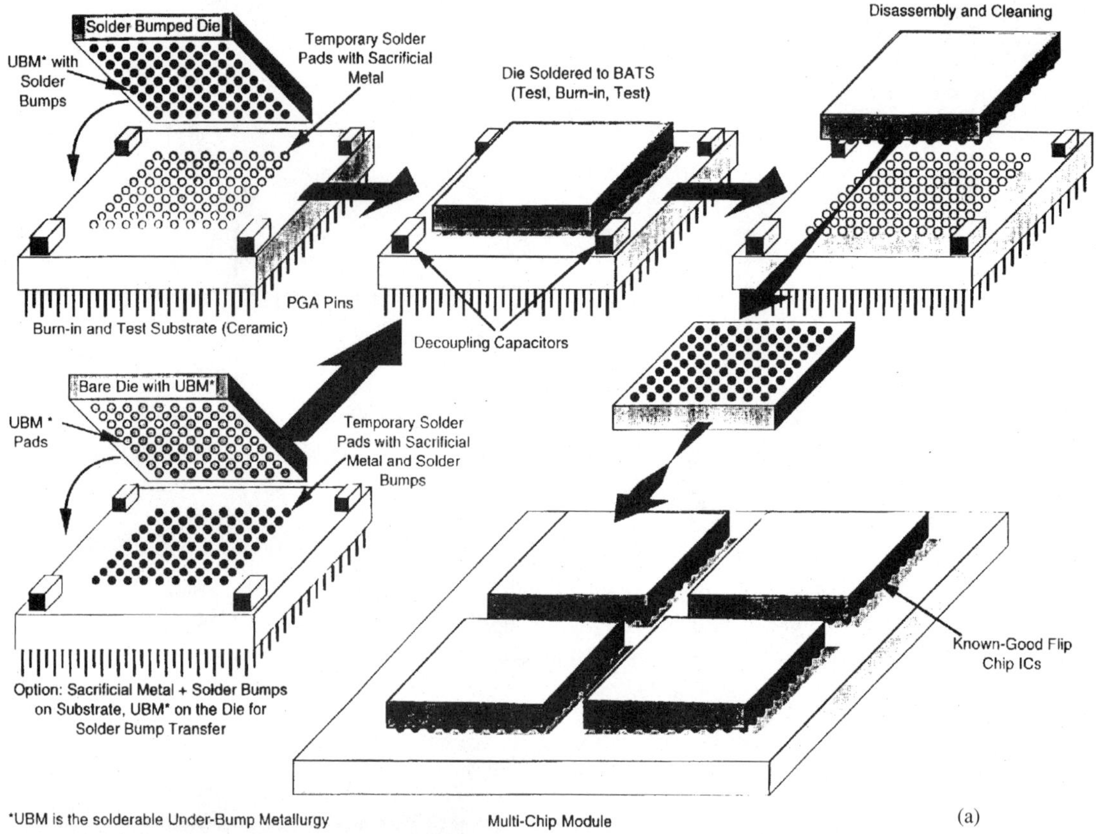

(a)

(b)

FIGURE 14.22 MCNC burn-in and test substrate (BATS) technique for flip chip KGD: (a) BATS process and (b) SEM photo of electroplating based bumps. Courtesy of MCNC.

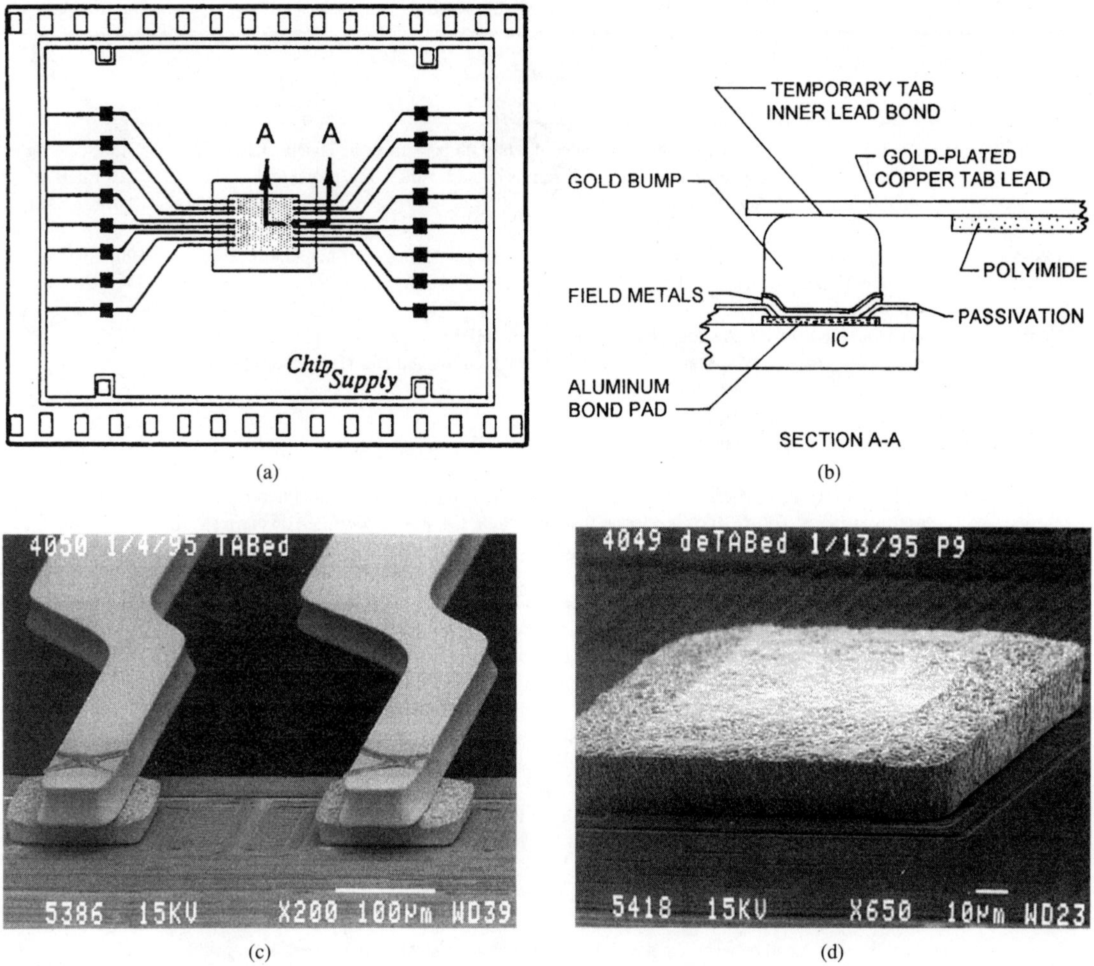

FIGURE 14.23 Temporary TAB approach: (a) TAB carrier, (b) section A-A enlarged, (c) SEM image of SofTAB™ leaded pads, and (d) SEM photo of electroplating-based bumps.

repair MCMs having faulty ICs, and very expensive resulting MCMs. Since MCM packaging technology will be used only if it is cost-effective (unless it provides critical performance advantages) compared with competing packaging technologies (hybrids, chip-on-board, circuit boards), it is essential that methods be developed for unpackaged ICs to more effectively screen out defective parts. The problem of ensuring that ICs entering MCM assembly are known-good is presently being addressed by task forces from SEMATECH, SRC, MCC, semiconductor manufacturers, MCM fabricators, and various government agencies. Key issues being addressed include:

- Establishing a broad coalition of IC manufacturers to develop approaches for improving availability of a wide spectrum of IC parts types of unpackaged ICs in standard formats suitable for multichip packaging
- Developing methods to quantify the known-good probabilities (0.9, 0.99, 0.999, …) of those ICs

Table 14.22 Metallurgical Contact KGD Methods—Chip-Size Package

Trade name	Description and references	Company
MSMT™	MicroSMT. Wafer post processing leads fabricated directly on pads then die encapsulated, back side of die is used for new contact points, and front is coated with epoxy, protective cap, and optional heat sink. Mates with Johnstech socket.[13,141–146]	Chip Scale Inc. (formerly Micro SMT)
COS™	Chip On Substrate. Die mounted on thick film ceramic substrate. After burn-in and test, trimmed to 0.030 milli inch greater than die.[13]	Chip Supply
TZOP™	Thin Zero Outline Package. An adaptation of the existing HDI process. Die is placed face down onto adhesive coated prefabricated flex, encapsulated, then imaged and vias etched. Thin film metalization provides surface for bump formation.	GE
CSP™	Chip Scale Package. A flip chip style carrier.[147]	Mitsubishi
SLICC™	Slightly Larger than IC Carrier. Combines flip chip and BGA technologies. Testable organic substrate carrier is 0.06–0.10 inches larger than die width. C4 die is underfilled on carrier with 0.020 inch diameter solder balls.[148]	Motorola
MDC, MCC	Miniature Device Carrier, Miniature Chip Carrier. A wide variety of approaches have been in use for over 20 years. The die is mounted to a patterned carrier slightly bigger than the chip, which allows fan out of the interconnect and ease of handling for test and burn-in. Interconnect schemes comprise a full spectrum of wire bond, TAB, and flip chip variations.[3,12]	Rockwell et al.
mBGA™	Mini BGA. Two polyimide and conductor layers are used to repattern peripheral pads to array pads of silver or gold. Carrier provides a common array pattern footprint to all die types. Z-axis film or particle interconnect membrane is used for die pad to carrier contact. The array pattern uses simple mechanical alignment with out the need for precision sawn die.[13,15,57,132,133]	Sandia National Labs
uBGA™	Micro BGA (Ball Grid Array), previously TCC (Tessera Compliant Chip). Thin film flex circuit reroutes peripheral bond pads to array pads of Ni/Au relaxed pitch. Compliance provided by elastomer sandwiched between die and flex. Mates to a PGA style carrier, or array socket.[13,15,16,149–153]	Tessera

- Establishing the technologies needed to result in a cost-effective price structure for known-good unpackaged ICs that is fair to both supplier and user

14.4.1 KGD Availability

Availability of Known-good bare die for MCM assembly has risen steadily from near zero in 1990 to the point where hundreds of millions of MCMs are annually being delivered having utilized some level of KGD processing. A dozen major semiconductor manufacturers and die processing companies have announced some level of KGD offerings. The more aggressive have provided their KGD data in Die Information Exchange format and have instituted some level of screening to ensure truly KGD parts (not just bare die that were DC wafer probed). Several dozen KGD approaches are being used or are in funded development. As the technologies mature and cost differentials become apparent, the number of surviving approaches will probably be reduced to less than half a dozen.

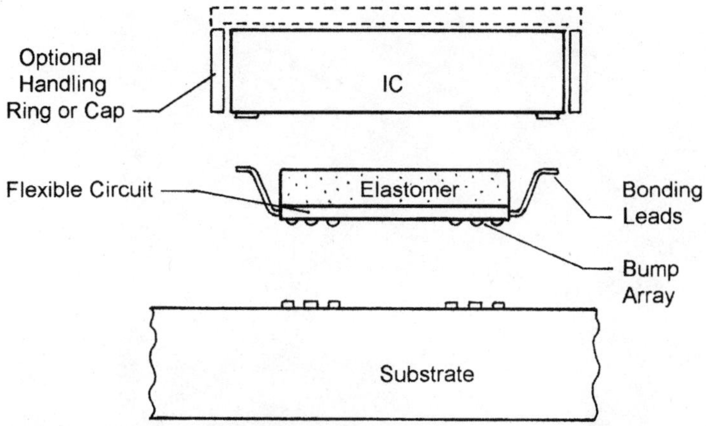

(a)

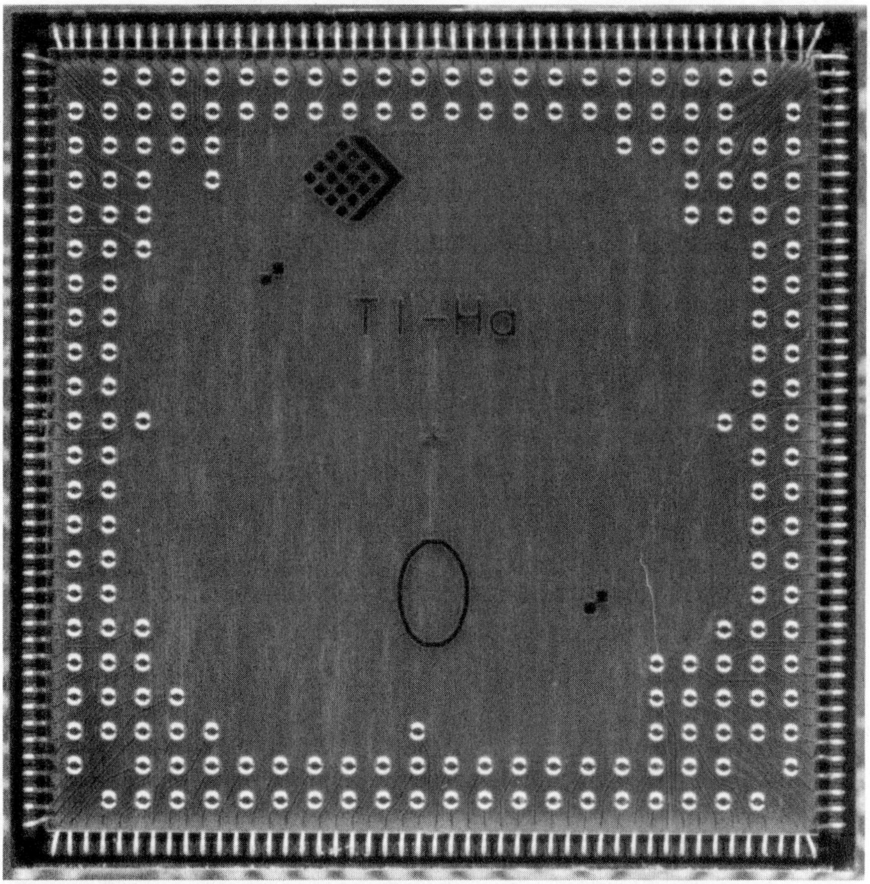

(b)

FIGURE 14.24 Tessera chip-size packaging approaches: (a) flex circuit providing fan-in from IC peripheral bond pads to bump grid array, (b) BGA flex adapter attached to IC perimeter pads, (c) chip and flex in Micro BGA format, and (d) Micro BGA attached to MCM substrate. Courtesy of Tessera, Inc. *(continues)*

(c)

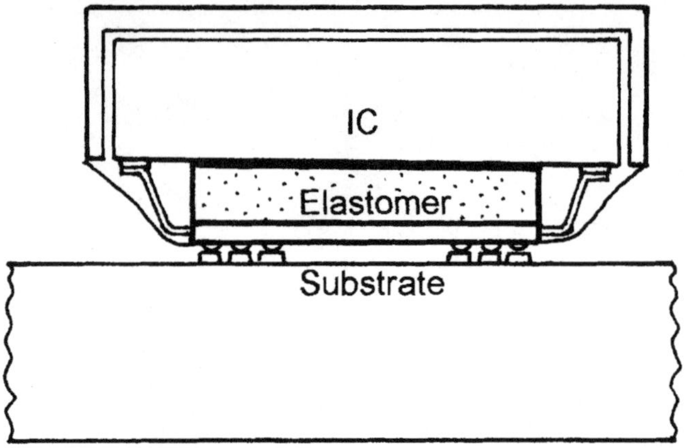

(d)

FIGURE 7.24 *(continued)*

14.4.2 Quantification and Measurement of KGD Level

The problem of defining and measuring the level of "known-goodness" of an IC chip was discussed earlier. From one point of view, the definition for a metric is deceptively simple: it is the probability that a die from a particular lot of purchased ICs will continue to perform properly through assembly, test, and fielded life. Unfortunately, industry efforts so far have not been able to reach agreement on a test method to measure this metric. Consequently, the excellent work done on establishing KGD guidelines is still missing a critical element.

As the often quoted statement from Lord Kelvin observes, one cannot fully understand and utilize what one is unable to measure.

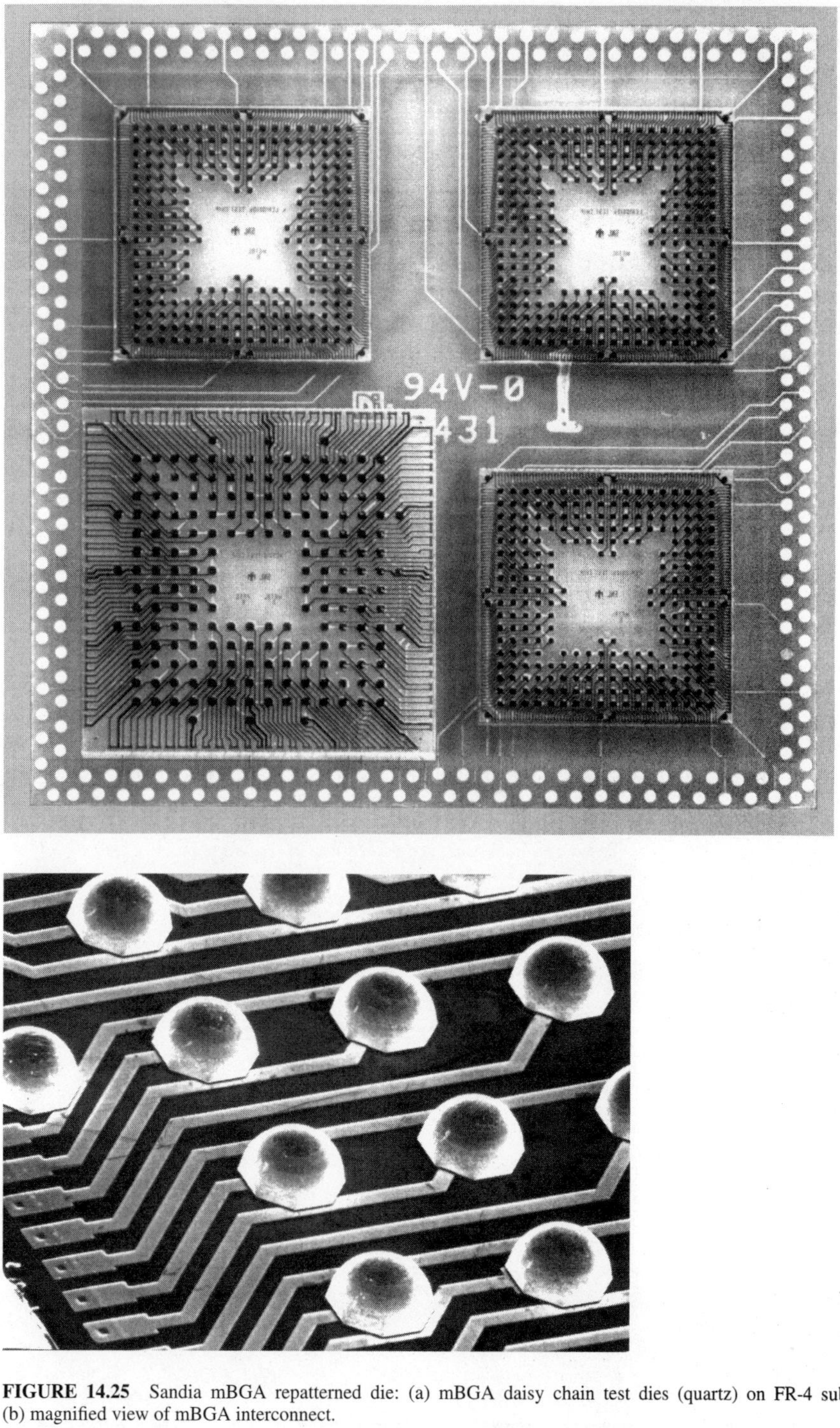

FIGURE 14.25 Sandia mBGA repatterned die: (a) mBGA daisy chain test dies (quartz) on FR-4 substrate and (b) magnified view of mBGA interconnect.

Table 14.23 Chip-Size Packages

Approach	Company	Area ratio*	Height ratio*	Reference
Miniature chip carrier	Rockwell et al.	2.0	2.1	12
Micro SMT	Micro SMT, Inc.	1.2	1.2	144
Micro BGA	Tessera, Inc.	1.2	2.0	151
Mini BGA	Sandia	1.0	1.2	154
SLICC	Motorola	1.8	2.0	148
Chip-scale package	Mitsubishi	1.2	1.6	147
Sealed chip	Dow Corning	1.0	1.0	155,156

* Compared to bare die

Table 14.24 Levels of Testing for Processing Bare Die

(a) Major Levels of Testing for Bare Die

Level	Method	Description	Est. probability limit	Comparative level of goodness
1	Historical data	Probability based on historical data from packaged part testing.	0.9	1×
1.5	Statistical sampling	Lot acceptance testing (LAT) or statistical sampling. Probability based on actual packaged part testing of a small population of die from the wafer lot.	0.95	1.5×
2	Wafer probe	Probability based on actual testing of each die at the wafer level.	0.99	10×
3	Individual die test	Probability based on test and burn-in of each die after wafer saw.	0.999	100×

(b) Sublevel Test Methods for Bare Die

A wide variety of possible combinations of temperature sort, speed sort, DC, AC, and functional test methods are used within each of the major levels of bare die testing shown in (a).

Method	Temperature sort			Speed sort		
	Ambient	Hot chuck	Cold (guard banded)	Static	Dynamic	At-speed
DC parametric	✔	✔	✔	✔		
AC parametric	✔	✔	✔	✔	✔	✔
Functional	✔	✔	✔	✔	✔	✔

The confusion caused by this missing metric is typified in Table 14.25, which shows several categories of KGD for one memory die product, along with informal estimates of the KGD level. As long as no metric is provided on KGD level by the IC vendor, an early customer of this particular bare IC will not be aware that:

- A probed lot of ICs will have 400 failures per 1,000 parts
- Screening Method B will have 190 failures per 1,000 parts
- Screening Method C will have 20 failures per 1,000 parts
- Screening Method D will have 1 failure per 1,000 parts

The difficulty in establishing a standard metric is related to the small defect rate involved. A KGD level of 0.999 suggests that 999 out of 1,000 chips will perform as expected. To verify this, clearly, a manufacturer will need to have tested at least 1,000 chips. To verify KGD levels to five nines, at least 100,000 representative chips will need to have been tested. Since the very large sample sizes required are impractical, it will be difficult to establish a standard metric in these early years of KGD technology. However, as bare die usage grows in volume, it will be possible for IC manufacturers to establish data bases based on feedback from fielded products, much the same way as has been done for packaged ICs. Such data bases for packaged ICs are published periodically, showing the defects per million (DPM) fallout rates and failures in time (FITs) rates being achieved by the ICs. Eventually, similar DPM and FIT data for bare die can provide the needed metrics for KGD level of goodness.

It is proposed that industry adopt a simple category nomenclature for describing the "known goodness" level in relation to the number of nines in the probability factor as shown in Table 14.26.

14.4.3 Cost, Cost, Cost

Interest in multichip module packaging (outside of a few computer companies) started in the early 1980s and grew rapidly as exciting demonstrations were made of impressive size reduction, performance improvement, and reliability robustness. Eventually, the emphasis gravitated to cost as companies struggled to make profitable businesses with this exciting new technology. As an industry expert[159] observed, the fascination with multichip modules has switched over to "multi-cheap" modules. Since one of the prime drivers on MCM cost is yield, KGD technology was recognized as an essential element in cost-effective MCMs. But KGD costs themselves become part of the issue as described in the following pages.

The cost that KGD will eventually have can be considered in three simple scenarios:

Scenario 1 considers reasonably complex ICs (such as microprocessors), which today have three roughly equal cost components: the chip, the package, and test.[160] Although a few people still argue that a bare die should cost less than a packaged one,[160] most packaging people recognize that (a) an IC manufacturer has no incentive to interrupt his normal product line with a *less* profit-

Table 14.25 Estimates of Differing KGD Levels for the Same Memory Die, Depending on Screening and Maturity[15]

Screening method	Die introduction	Mid-life	Fully mature
A. DC probed	0.60	0.88	0.95
B. Hot chuck, at speed	0.81	0.95	0.99
C. Hot chuck, at speed probe plus burn-in	0.98	0.99	0.995
D. Burn-in plus full data sheet testing	0.999	0.999	0.999

Table 14.26 Proposed Category Nomenclatures for "Number of Nines Known-Goodness Level"

Category	Probability*
1	0.9
1.5	0.95
2	0.99
2.5	0.995
3	0.999
3.5	0.9995
4	0.9999
5	0.99999
6	0.999999

* Probability that the bare die will be at least as good as the equivalent packaged part in DPM defect level and FIT reliability rate

able product line, and (b) at present, it is more costly to test and burn-in bare die than packaged die. In the limit case where bare die could be tested at cost parity with packaged ICs, only one of the three cost components (the package) could be eliminated. For this scenario, the lowest that KGD could ever cost would be 67 percent of the single-chip packaged IC cost.

Scenario 2 considers the case where packaging cost is considerably less than one-third of the total single-chip package cost. Low-cost plastic package parts might be examples of this case where packaging might be only 10 to 20 percent of the total cost. Obviously, there is even less potential for low-cost KGD in this case, since the cost of the package is even smaller than in Scenario 1.

Scenario 3 considers a single-chip packaged IC where the package cost dominates all other costs. ICs packaged in ceramic pin grid arrays with high pin count might be examples of this case, where package cost might be over 80 percent of the total cost. In this scenario, a good KGD solution might result in a bare die cost that is substantially less than the packaged IC.

In all three scenarios above it is evident that achieving low cost KGD depends upon developing screening methods for bare die that are at least as low cost as for single-chip packages.

Cost of KGD. The cost of obtaining known-good ICs varies widely at the present time. One major semiconductor company[26,64,73] has established a KGD product line with the same performance, reliability, and price as its equivalent single-chip packaged parts. Other companies offer their IC products[25,162–164] in several grade levels of increasing KGD "goodness," with corresponding increasing cost levels. Some long-term projections[18] show KGD eventually costing 80 to 90 percent of the price of the equivalent single chip packaged part. However, as those involved in module development know, KGD today typically cost from 1.2 to 10 times the equivalent single chip package price.[135] Figure 14.26 shows comparisons of multichip module costs with and without KGD. It is evident that the cost premium for obtaining KGD is probably not justified in many MCMs that have

- very low total cost

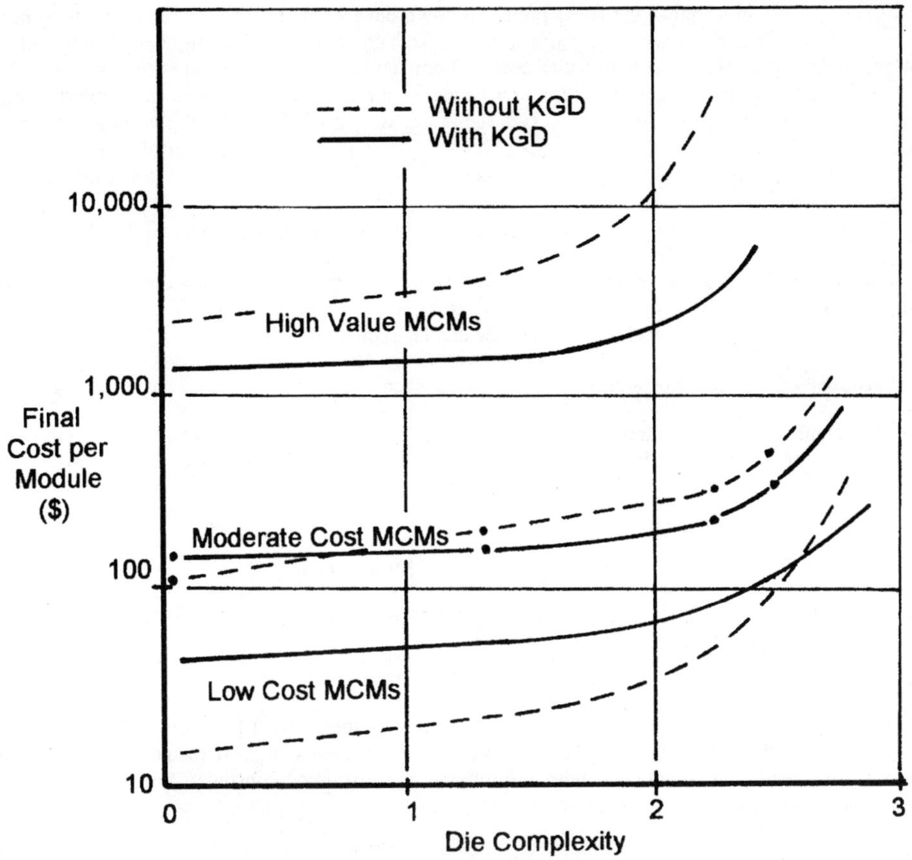

FIGURE 14.26 Estimates of the final cost of modules, including replacement of defective ICs (after Ref. 169).

- only a few ICs
- simple, low-complexity/cost ICs

On the other hand, it is also evident the KGD premiums are justified in

- High-cost MCMs
- MCMs with many ICs
- MCM using complex, expensive ICs

Several efforts to develop cost modeling and trade-off tools[21,165–167] for KGD are in progress. Maturation of these tools will help sort out which applications will benefit from KGD availability. From the studies so far, it is apparent that KGD can already be cost-effective in several situations.[166]

- When replacing ceramic single chip packages
- For complex, high-I/O ICs

This is really an acknowledgment that it is easier to justify a $20 premium for KGD on a $100 part than it is on a $10 part. For the lower-cost part, the KGD premium must be brought way down.

Two large drivers in bringing the KGD premium down have been shown[166] to be (1) the total non-recurring cost of the fixtures/carriers/burn-in boards/sockets divided by the number of uses and (2) the recurring costs associated with KGD testing. Items consumed in the testing must be very inexpensive for KGD to ever be cost competitive with low-cost plastic packaged devices. Similarly, items reused in KGD testing need to have a very large reuse life. At the wafer level, two large drivers in bringing the KGD premium down have been shown[168] to be (1) the total number of chips per wafer (small chip size or large wafer size is better) and (2) the time spent in burn-in and test (shortest times minimize amortization costs).

A summary of typical cost premiums needed today to obtain KGD is shown in Table 14.27.

Economics of KGD. As in every good business strategy, a decision to invest time and money on procuring known good die should be made only if it results in a positive return on investment. The theoretical cost of building an electronic module can be expressed as follows:

$$\text{Module build cost} = \text{parts cost} + \text{assembly cost} + \text{overhead cost} + \text{amortized development and tooling cost}$$

The actual cost of the module increases as yield losses occur, whether the losses are due to occasional defective components, handling damage, or occasionally inadequate assembly processes. The actual module cost is given by:

$$\text{Actual module cost} = \frac{\text{Module build cost}}{\text{Module yield}}$$

Therefore, the decision whether to use known good die is guided by the following design rule:

Design Rule 1. Reductions in actual module costs (due to KGD yield improvements) must be greater than the increased costs paid to get known good die.

The trade-off analysis of costs should consider three elements: cost premiums paid for known-good die, reduction in actual module cost from yield improvement, and (usually forgotten) the reduced overhead costs associated with minimizing the resources consumed with producing the unwanted scrap or devoted to rework and repair.

Table 14.27 Cost Factor to Obtain KGD, Compared to the Same Die in a Single-Chip Package

IC type	KGD level	Comment	Premium for KGD	Reference
Memory	0.999	Softool KGDPlus™ (low vol.)	1.5×	18
Memory	0.999	KGDPlus™ (high vol.)	0.85 to 1.5×	18
Various	0.93–0.97	—	1.0 to 1.5×	144
Various	0.999	—	8.0 to 12.0×	144
Memory	—	SmartDie™	1.0×	161
Microprocessor	—	SmartDie™	1.0×	161
Various	—	SofTAB KGD	1.2 to 10×	135
Various	0.0000	—	2.0 to 6.0×	144

Several points on module yield need to be considered. As previously discussed, the predicted module yield (Y_{mcm}) is given by the product of three probabilities:

- The probability that the IC chip set will be good (P_i)
- The probability that the other components will be good (P_c)
- The probability that the module assembly can be accomplished without any defects (P_{ma})

Thus, the module yield in percent is given by:

$$Y_{mcm} = P_i \times P_c \times P_{ma} \times 100 \tag{14.6}$$

In low to moderate volume purchases, customers are unlikely to pay much attention to price differentials of less than one percent, so a yield goal of 99 percent may be sufficient. On the other hand, if the volume of business is large enough that one percent extra cost over the year is a large dollar number, then yield improvement efforts may be warranted.

It is important to direct yield improvement efforts to areas where they can do the most good. For example, it has been shown that improving the KGD level in a module having two ASICs and 40 memory chips [21] is a far better investment if the memory KGD level is improved by one point from 0.98 to 0.99 than if the ASIC KGD level is improved by 10 points from 0.85 to 0.95. The first small action on all 40 memory chips raises the module yield by twice as much as the second larger action on only two chips. Obviously, all yield improvements are helpful, but the payback for investment in known good die will be well justified in some areas and not justified in others. It is helpful to examine the relative cost driver effect of each IC which can be calculated from:

$$\text{Cost driver effect of defective IC type } n = \frac{\text{module build cost}}{(P_n)^N - \text{module build cost}}$$

where N = quantity of component n used
P_n = KGD level for component n

This factor calculates the relative module dollar cost contributed by defective ICs. The factor should be calculated for each IC type and quantity and then compared to the other cost driver effects to determine the areas of highest potential payback. Once the identification of highest payback has been done, three approaches are available for trying to increase the KGD level of the ICs arriving for module assembly:

1. Switch the IC type to a more mature technology (It may be possible to purchase 0.5 micron technology devices at a much higher KGD level than 0.35 micron devices.[21])
2. Switch to a better IC vendor (some vendors have achieved much lower DPM defect processes than others and inherently have higher KGD levels).
3. Have the ICs screened to remove defective devices.

The Decision to Repair Defective Modules. The decision to set up repair procedures and facilities for modules containing defective ICs is a difficult issue. It is always preferable to "do it right the first time" and "do it right every time." The cost of tolerating anything other than "zero-defects" approaches is enormous in the "hidden factory" of repair operators, repair equipment, repair procedures, spare parts inventories, requalification, and so on. Arguments continue[170–172] about whether a business should use KGD approaches or simply pay the cost of replacement of defective ICs after assembly. Traditional hybrid microcircuit businesses have succeeded without KGD availability, but they have the opportunity for large savings from reduced rework as KGD devices become more economical and readily available.[17] The traditional approach has been, in effect, to achieve known-good die *within* the module by using the test and repair line to screen out defective ICs. As the cost of

KGD decreases, repair will become less attractive. Once again, the decision will be a business decision based on a simple design rule:

Design Rule 2. Repair of defective modules is appropriate only if the cost to replace defective ICs within a module is less than the cost to replace the entire module by building a new one. Full accounting should be given to all six elements of the cost to repair a module:

1. The cost of the replacement parts
2. The cost of the repair labor required
3. The overhead costs associated with parts and labor
4. The cost of qualifying and documenting that the repaired module will meet customer requirements
5. An appropriate share of the costs associated with having resources, equipment, and facilities devoted to repair
6. An appropriate share of the cost of lost productivity and capacity from the defective modules having been originally processed through the assembly areas

Clearly, it makes sense to do a \$10 repair action to save a \$500 module, and it does not make sense to do a \$50 repair and documentation action to save a \$20 module. One excellent set of recommendations for cost-effective MCMs[172] is to combine intelligent selection of low-risk die with careful partitioning and to package high risk die in forms that are pretestable and easily replaceable, as follows:

- Specify memory die that will have early maturity (and consequently high yield).
- Design custom and semicustom ICs on mature, high-yield fabrication processes.
- Use standard, high-volume microprocessor, DSP, and special-function ICs.
- Avoid low-maturity ICs and low-yield IC fabrication processes.
- Partition modules to contain high-yield ICs. Place low-yield ICs outside the module or keep them in single-chip packages.
- Use near die-size packages that can be pretested and burned-in for high-risk, low-maturity ICs.

One major producer of multichip module products has estimated that use of KGD strategies to eliminate the cost of rework saves from 25 percent to 75 percent of the net MCM costs.[173]

14.5 FUTURE IMPROVEMENTS

A KGD vendor has considerable flexibility in implementing a test and burn-in strategy. Some vendors will spend more time testing at probe and very little at final test. Others use the opposite strategy. Product failures at final test are painful for both the KGD supplier or MCM manufacturer. Industry consensus for a long-term, low-cost solution to KGD points to wafer-level test solutions. Nevertheless, near-term improvements in die-level KGD solutions, in terms of availability, quality and cost, will foster explosive growth in MCM, PCMCIA, COB, and other bare-chip products.

14.5.1 Wafer-Level KGD Assurance Technologies

A number of IC manufacturers have recognized the potential benefit of KGD, regardless of the final packaging destination. Wafer-level KGD solutions in particular are perceived as desirable for volume manufacturing. Significant cost advantages due to savings in set-up, handling, process control, and elimination of the cost penalties in packaging bad die are among the potential benefits.

Wafer-level KGD approaches can be performed using conventional and emerging probe card technologies (Sec. 14.3.4) for single or multiple die touchdowns. Probe cards are used for electrical testing at speed, through temperature, and for reliability testing. They can also be used in conjunction with short-duration prestress conditions of temperature or voltage that attempt to accelerate and expose reliability failure mechanisms.

Wafer-Level Reliability (WLR) Tests. The term *wafer-level reliability* (WLR) is associated with reliability tests and test structures for activation of wear-out defects and is the subject of intensive research. Reliability test structures have been designed and defined by current and upcoming JEDEC standards for each of the three main reliability concerns associated with wear-out defects: oxide breakdown,[174] hot carrier degradation,[175] and electromigration.[37] Future work will focus on in-line real-time process monitors and self-stressing test structures that provide autonomous stress stimulus and response monitoring and built-in reliability (BIR).

Wafer Level Burn-In (WLBI). While wafer level probe cards are capable of high-performance testing of a single die at a time on a wafer, whole-wafer technologies are being developed and evaluated for wafer-level burn-in tests. The issues of thermal management, signal and power distribution, mechanical alignment through temperature and over time, routability, and reliable contact are extremely challenging. Whole-wafer probing technology requirements can include:[168]

- Contact life >2000 burn-in cycles
- Power supply distribution in excess of 1000 W
- 150° C continuous operation, or 200° C for highly accelerated burn-in
- Contact targets ≤ 75 μm on ≤ 100 μm pitch (extendible to 50 μm pitch)
- Cost competitive in the current production burn-in process for packaged parts
- In excess of 15,000 die pad contacts across a 200 μm wafer, extendible to (300 μm wafer)

WLBI techniques encompass both on wafer interconnect redistribution techniques and off wafer approaches. On wafer techniques require additives or subtractive post processing steps (scribe line power grids, fusing) which may not be cost effective as a stand alone WLBI technique. Off wafer techniques can be inorganic (i.e. silicon) or organic membranes. The silicon approaches offer an exact thermal coefficient of expansion (TCE) match to the wafer under test (WUT) while organic membranes provide the advantage of compliance. Regular structures such as memories and chips incorporating built-in-self-test (BIST) features can minimize the required die pad contacts. Table 14.28 list some of the companies who have developed technology for whole wafer level burn-in and test.

Whole Wafer Testing. One WLBI approach is the silicon burn-in substrate (for convenience abbreviated SiBIS) illustrated in Figs. 14.27*a* and 14.27*b*. The SIBIS is fabricated using nChip's SiO_2 based multilayer thin film MCM-D technology, with solder bumps laid in a mirror image to the WUT (wafer under test) bond pads. The solder bumps deform under compressive force to form a gas tight contact of <100 mΩ. To aid bump uniformity, integral resistive heaters are being developed under the bond pads to allow reflow. The whole wafer test and burn-in assembly is housed in a steel vessel pressurized with air, nitrogen, or helium. An integral vacuum port maintains SiBIS to WUT contact. The high heat fluxes of the WUT require active cooling rather than a burn-in oven to maintain proper junction temperatures.

An alternate approach to wafer level test and burn-in developed by Elm Technology uses a full wafer flexible silicon membrane probe card (FW-SMPC) with metal oxide piercing contacts (Fig. 14.28*a* and *b*). The silicon membrane can be considered a very thin IC. This patented technology provides a single crystal layer of silicon and numerous layers of dielectric and metallization interconnect (Fig. 14.28*c*) forming a flexible silicon membrane typically less than 25 m thick. The silicon membrane is held to the wafer under test (WUT) under vacuum. The contacts allow at-speed functional test during burn-in.

Table 14.28 Wafer-Level KGD Testing, Pressure Contacts—Probe Cards

Trade name	Description and references	Company
SMPC™	Micromachines silicon membrane with iridium microfabricated silicon/dielectric membrane probe tips.[15,168]	Elm Tech. Corp.
—	Extension of die-level membrane probe system using Gold Dot™ bump technology.[57]	Packard-Hughes
—	An extension of die-level carrier polyimide membrane probe.[57,168]	MMS
SiBIS™	Silicon Burn-In Substrate. Silicon substrate multilevel interconnect probe card is mirror image of WUT pad patterns. Solder bumps are used to mate to the WUT under compression forming gas-tight seal. Integral decoupling caps and isolation resistors are built in.[57,73,168]	nChip
—	Needle probes contact multiple die, short-duration RAM test.[14,15,176]	Toshiba

A number of semiconductor manufacturers, MCM suppliers, and equipment vendors are cooperating in the evaluation of a wide variety of wafer-level technologies. Initiatives such as the ARPA-sponsored Wafer Level Test/Burn-In Project will provide future direction and advances in this revolutionary and potentially low-cost KGD technology. Wafer level burn-in will accelerate semiconductor process reliability learning through rapid feedback of failure data. The ultimate goal of this reliability learning is the fabrication of circuits that do not need burn-in.

14.5.2 Singulated Bare Die Testing

KGD methods for individual diced ICs will probably continue along the present four paths: sockets for bare ICs, temporary chip carriers, permanent chip carriers, and "near die-size" packages. Each method can be expected to provide cost reductions as usage grows. Bare die sockets will provide price reductions from current levels (several hundred dollars per socket) to a few dimes per I/O. The cost per use for temporary chip carriers and the cost of permanent chip carriers will both need to be reduced to dimes per die for wide acceptance. Near die-size packages will compete well when they reach cost parity with conventional single-chip packages.

Major improvements in temperature range capability and at-speed testing capability can be expected as thermal expansion, contact resistance, stress relaxation, and impedance control issues are resolved with new materials and new design configurations.

Standard Die Footprints. The standardization of any physical dimensions of semiconductor chips may appear to be counterproductive to the chip suppliers' desire shrink the size of a chip as it matures. However, with the growing interest in flip-chip attachment, industry must consider a bumped die as a *package* as opposed to a *bare die*. The recent development of a joint industry standard, "J-STD-012, Implementation of Flip Chip and Chip Scale Technology,'[89] proposes that suppliers observe standards for grid pitch based on a 0.5 mm (or 0.05 mm) grid. This allows for natural grid multiples that facilitate implementation of a common grid system. Most flip chip and CSP packages are currently designed for an I/O interconnect pitch between 0.5 and 1.0 mm, with 0.5 mm predominating. Future miniature array designs using 0.25 mm pitch are expected. Standard die footprints offer the opportunity for dramatic reductions in the cost and turnaround time of test hardware,[117] module design, and module assembly.

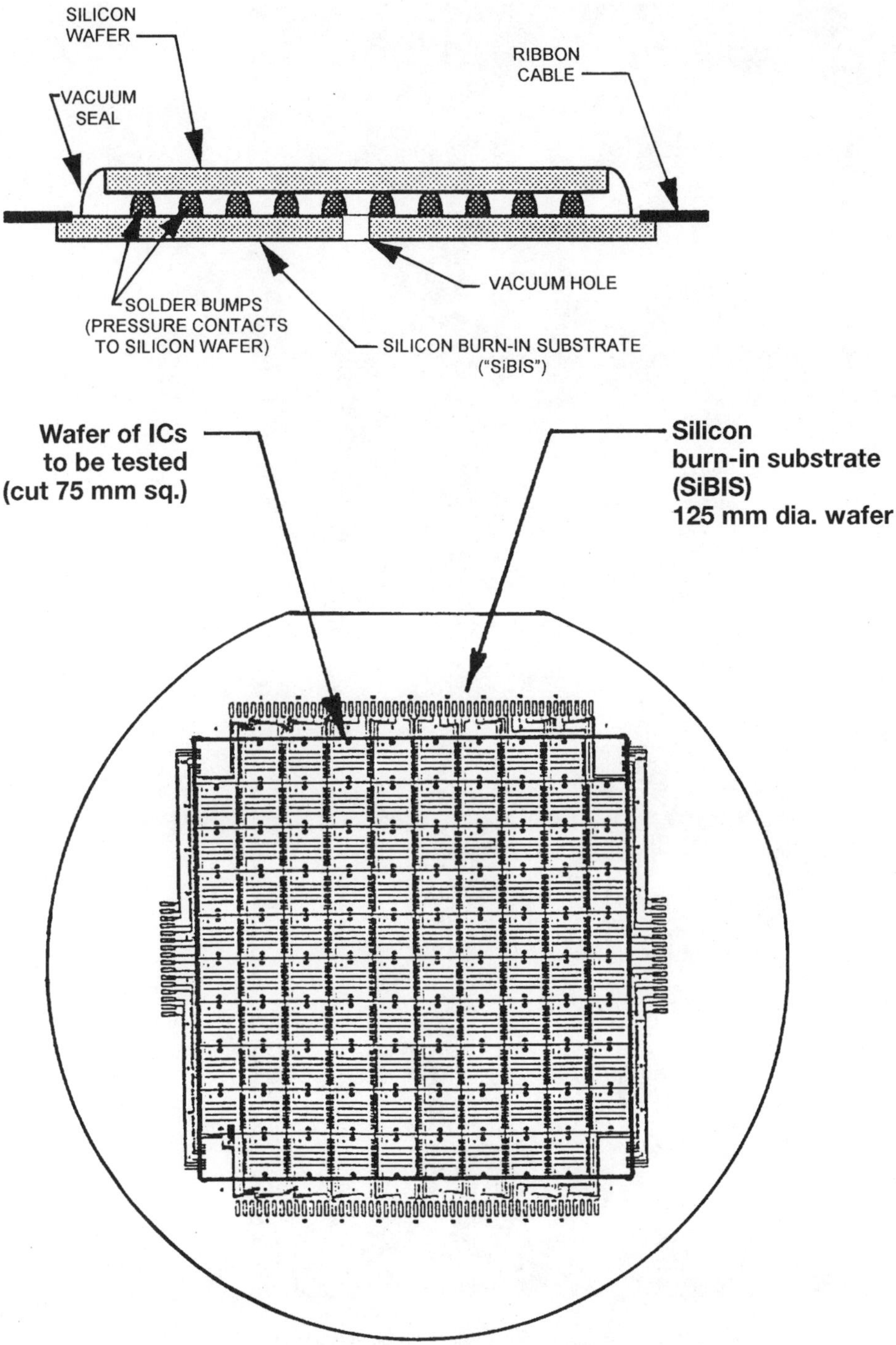

FIGURE 14.27 Silicon burn-in substrate: (a) SiBIS WLBI concept and (b) silicon membrane probe point, side views.

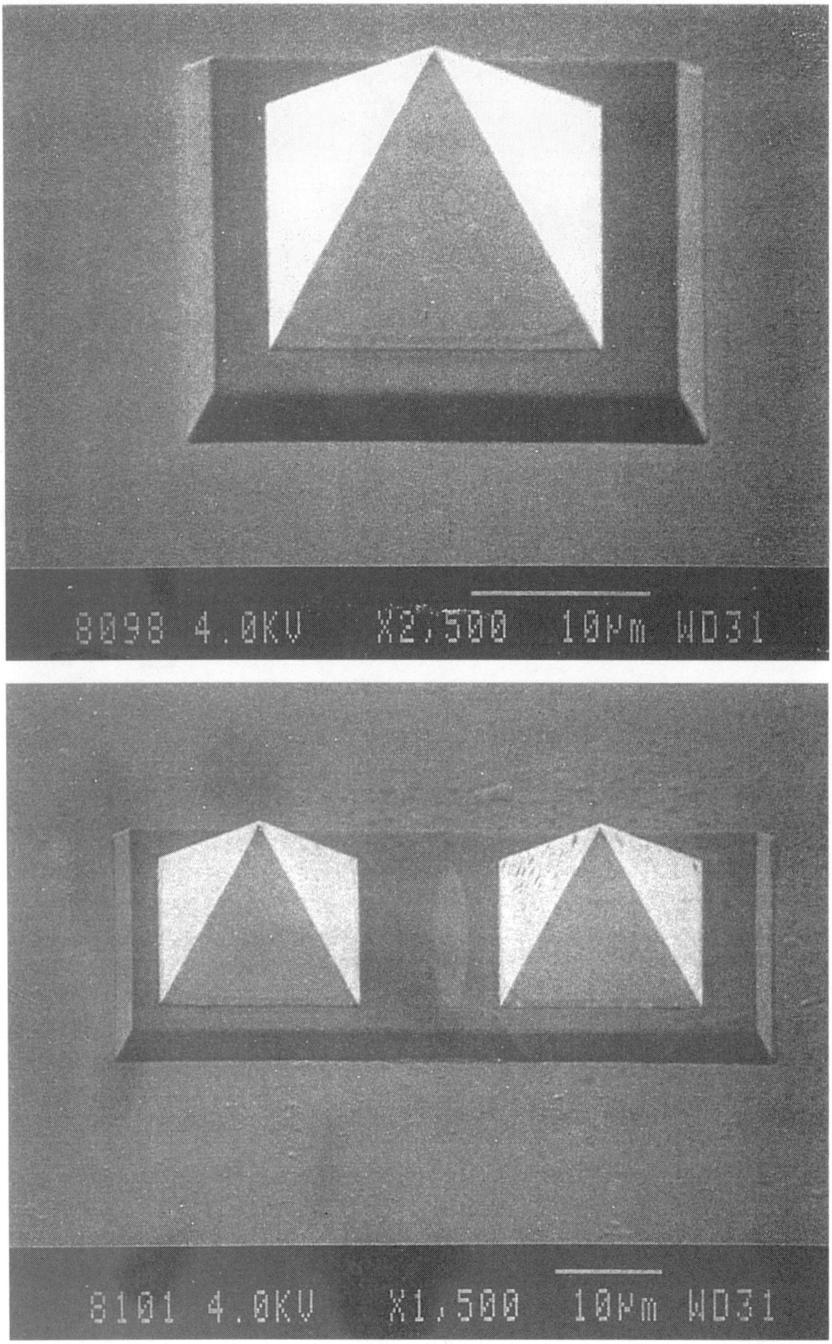

(a)

(b)

FIGURE 14.28 Micromachined wafer-level burn-in technique: (a) single and (b) double iridium pyramid probe point for silicon membrane probe card, and (c) silicon membrane probe point, side view. Courtesy of Elm Technology Corp. *(continues)*

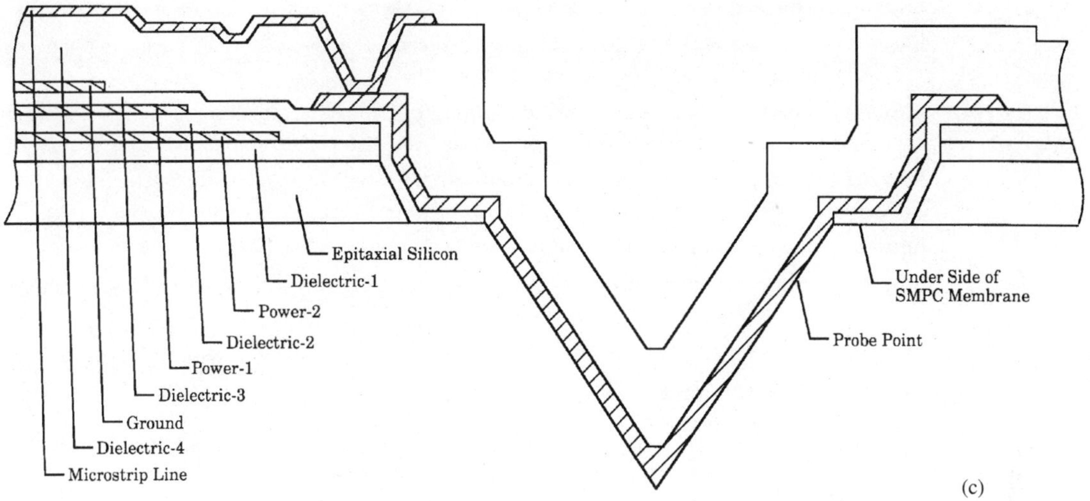

FIGURE 7.28 *(continued)*

14.5.3 Information Technology: Surfing the Internet for KGD

Today, with almost immediate access to computers worldwide, electronic information can be exchanged at the click of a button. As KGD technology matures, the Internet will provide access to a vast repository of KGD information. Several key World Wide Web sites containing KGD-related data are listed below. The reader should be alert to new or updated KGD web sites. A wide variety of Internet search tools are available to make this task relatively easy. However, when using these Internet search tools, the reader is advised to use the keyword "KGD" rather than "die" or "bare die" for the best (and the least shocking) results.

KGD and Related Web Sites

http://www.mcc.com/projects/kgd—Microelectronics Computer Technology Corp., KGD technologies and supplier data, wafer-level burn-in (WLBI) project briefs.

http://www.imec.be/kgd—ESPIRIT IV sponsored GOOD-DIE project information, KGD database.

http://www.gooddie.com—ESPIRIT IV sponsored GOOD-DIE project, KGD database.

http://www.isi.edu/midas/midas.html—USC Information Sciences Institute, MCM Interconnect Designer's Access Service, DARPA sponsored MCM brokerage, MCM, bare die, wafers, and KGD information for MCM designers.

http://www.isi.edu/fast/—DARPA-sponsored parts brokerage, die procurement.

http://www.mcnc.org—Microelectronics Center of North Carolina, wafer bumping services.

http://www.intel.com/design/smartdie—SmartDie information.

http://www.chipsupply.com—Chip Supply Inc. (USA), KGD supplier information.

http://www.eltek.com—Eltek Semiconductors (Europe), KGD supplier information.

http://www.yeu.com/tb140.htm—Yamaichi KGD test and burn-in sockets.

http://vhdl.org/pub/die—Die Information Exchange (DIE) Industry Group, specification updates, parser, DIE models.

http://vhdl.org/pub/ibis—I/O Buffer Information Specification, updates, guides.

http://vhdl.vhdl.org/vi/waves/—IEEE Waveform and Vector Exchange, specification updates.

http://esto.sysplan.com/ESTO/EL-Packaging/Projects—DARPA electronic packaging and interconnect KGD and related program briefs.

http://molothus.sysplan.com/ESTO/EI-Packaging/Presentation/Flipchip—DARPA electronic packaging and interconnect flip chip assembly and KGD and related program briefs.

http://erd.rl.af.mil/ER-News/Reliability.html—Rome Laboratory Reliability Sciences Newsletter, Overview of Wafer Level Reliability (WLR) development programs.

http://kegroup.com—ELMO KGD supplies information.

http://www.computers.org/tab/tttc/—Updates on IEEE 1149.X test busses, list of 1149.1 compatible components, I_{DDQ}, MCM test/KGD, wafer-level test.

14.5.4 KGD from PPM IC Processing

A variety of methods are being developed by various semiconductor manufacturers to increase known-good IC probability levels through yield improvements at the wafer level. These methods not only benefit MCM technology, they also reduce overall costs for single-chip packages. Wafer yield improvement eliminates wasted investment of labor, materials, and processing in faulty product. It also frees valuable assets and resources to increase plant capacity instead of consuming those resources in processing scrap. Successful efforts to improve IC manufacturing yields through process improvements also drive improved known-good IC probability levels. Hybrid microcircuit industry experience has shown that suppliers who have successfully implemented six-sigma or zero-defect yield improvement programs on their wafer fabrication lines subsequently deliver unpackaged ICs which consistently have higher survival rates through module assembly and burn-in steps.[1] Methods that incorporate improvements could include mask changes that increase device tolerance to processing variations, process changes that reduce defect levels, and design changes that increase device tolerance of defects. Programs that implement such improvements can raise wafer probe yields on new products from below 20 percent to above 50 percent, and on mature products from below 85 percent to above 95 percent. Such changes produce dramatic improvement in KGD levels of bare die without any additional KGD screening. Most of this chapter is devoted to discussions of screening approaches to sort good KGD from faulty ICs. This also reflects where the majority of industry activity on KGD is being directed. Hopefully, future efforts will be increased in wafer processing techniques to both "build it right the first time, and to build it right every time."

14.5.5 Full KGD Availability

After all this discussion, there probably remain two dominant questions in the mind of the reader.

1. Are all the bare ICs I need for my MCM, COB, or direct chip attach product available as KGD today?
2. Can I afford the KGD cost premium?

Answers, of course, will vary for different product applications, but the answer on availability is probably yes, and becoming more definitely yes with each year of KGD progress. IC manufacturers are rapidly increasing their KGD offerings, and what they do not offer is likely to be available from third-party die processors. The answer to the second question on cost is more complex. If the MCM or bare die application is a high-volume item, and only a modest KGD level is required, the KGD premium is probably easily justified. If the product application is a very high-value item, and rework is difficult, again, the KGD premium is probably justified. However, for other situations, successful businesses will need to work aggressively to bring the KGD cost premium down to affordable levels.

14.6 CONCLUSIONS

Other chapters in this book provide a fuller view of the exciting benefits and innovative opportunities possible with various MCM technologies. Miniaturization and performance demonstrations have been impressive. In select applications, MCMs have reached production rates of many millions per year. However, wider exploitation of MCM benefits will require continuing attention to cost reduction and yield improvement. Infrastructure development is proceeding but not yet mature. Ranking high on the list of remaining issues is the availability, "known-goodness," and affordability of all the bare die needed for a particular MCM design. Considerable progress has been made on these issues in the last few years, but much further development is needed to make MCM packaging technology cost competitive with other available packaging approaches. Many industry, consortia, and government initiatives are underway to address these infrastructure development issues, including the known-good die issue. Readers are encouraged to work with their suppliers, customers, and industry initiatives involved in MCM technology to assist in these infrastructure development issues. Significant progress is expected on these issues in the next few years.

ACKNOWLEDGMENTS

The authors would like to offer acknowledgments to those who have contributed to this chapter in several different ways, both indirectly and directly. First, there are several key individuals whose insight, persistence, and enthusiasm was fundamental to getting KGD technologies started and moving in productive directions: Stan Wagner, of USAF Wright Laboratory, who very early recognized the need for KGD and sponsored the seminal workshops that got things going; Don Orr, of MCC, who led the early industry efforts and organized much of the early cooperative multi company initiatives; and Nick Naclerio, of ARPA, who provided badly needed support of technology developments and infrastructure development[177] that have enabled significant KGD breakthroughs.

Second, there are industry colleagues who have worked very hard at enabling KGD availability to the great benefit of the electronics industry: Larry Gilg, of MCC; Jim Rates, of Chip Supply; Iwona Turlik, of MCNC and Motorola; Jose Avalos, of Intel; Gary Hill, of IBM; Randy Roebuck, of TI; Glenn Rinne, of MCNC; So-Soun Su, of IBM; Lt. Col. Mark Louis, of SEMATECH and USAF; Lt. Col. Ken Soda, of SEMATECH and USAF; Ray Fillion, of General Electric; Matt Salatino, of Harris; Barbara Vasquez, of Motorola; Kevin Wilson, of Insyte; Marshall Andrews, of MCC and ITRI; Monabu Bonkohara, of NEC; Mike Roughton, of Lucas Electronics; and the SEMATECH Low Cost KGD Project team.

Third, we express our sincere thanks to the following companies and organizations who provided materials that helped in writing this chapter: Acsist Associates, Aehr Test, California Contacts, Chip Supply, Dow Corning, ELM Technology, Elmo Semiconductor, ETEC, Fresh Quest, Harris Semiconductor, HP, Hughes, IBM, IDT, Intel, MCC, MCNC, Micro Module Systems, Micron Semiconductor, Micro SMT, Minco, National Semiconductor, nChip, Nitto Denko, Packard-Hughes Interconnect, Raytheon, Samsung, Temptronic, Tessera, Texas Instruments, 3M, Tribotech, and Gel-Pak.

REFERENCES

1. Hirschberg, A., "Multichip Module Die Issues," Personal Communication from Teledyne Microelectronics, Nov. 1991.
2. Condor, J., et al., "Dual TMS 320C30 Multichip Module, A Silicon-on-Silicon 3-D Memory Technology Prototype," *Proc. GOMAC,* Nov. 1991, pp. 243–246.
3. Hagge, J. K., and Wagner, R. J., "High Yield Assembly of Multichip Modules Through Known-Good IC's and Effective Test Strategies," *Proc. IEEE,* vol. 80, no. 12, December 1992, pp. 1965–1994.

4. Radke, C. E., et al., "Known Good Die and Its Evolution: Bipolar and CMOS," *Int. J. Microcircuits and Electronic Packaging,* vol. 16, no. 4, Fourth Quarter, 1993, pp. 302–312.
5. *Integrated Circuit Engineering,* "Mid-Term 1994 Status and Forecast of the IC Industry," 1994
6. Chu, D., et al., "Screening ICs on the Bare Chip Level: Temporary Packaging," *IEEE Trans. CHMT,* vol. 16, no. 4, June 1993, pp. 392–395.
7. Scannell, R. K., and Hagge, J. K., "Development of a Multichip Module DSP," *Computer,* April 1993, pp. 13–21.
8. Begay, M. J., "Bare-Die Test Strategies for the MCM Market," *Solid State Technology,* June 1994, pp. 65–75.
9. Gwinn, O., et al., "Unisyst MCM Experience and the Value of Known Good Die," *IEPS/ISHM Multichip Module Conference,* Denver, CO, April 1994, pp. 236–241.
10. Prokopchak, L., and Wrenn, J. M., "Development of a Burn-In and Test Process for Producing Known-Good-Die," *Proc. IEPS/ISHM Multichip Module Conf.,* Denver, CO, April 1994, pp. 23–28.
11. Hagge, J. K., "State-of-the-Art Multichip Modules for Avionics," *Proc. SPIE Int Conf on Advances in Interconnection and Packaging,* vol. 1390, November 1990, pp. 175–194.
12. Wagner, R. J., and Hagge, J. K., "Known-Good ICs for MCM Assembly," Chapter 13.2, *Thin Film Multichip Modules*, Editors G. Messner, et al., ISHM, Reston, VA, 1992, pp. 501–544.
13. Raytheon, "Known Good Devices Program," Interim Report, Contract F30602-93-C-0079, November 23, 1994, pp. 1–490.
14. Vasquez, B., and Lindsey, S., "The Promise of Known Good Die Technologies," *IEPS/ISHM Multichip Module Conference,* Denver, April 1994, pp. 1–6.
15. Gilg, L., "Known Good Die Tutorial," *IEEE Multichip Module Conference,* Santa Cruz, CA, January 1995.
16. Matthew, L., "Die Grid Array Package Provides KGD Solution," *Electronic Packaging & Production,* June 1994, p. 40.
17. Hagge, J. K., "The Road Towards Known Good Die for Hybrid Microcircuits/Multichip Modules," ISHM/IPC HMRC Meeting, Dana Point, CA, December 1992.
18. Johnson, J., "Known-Good Die Plus," ISHM/IPC HMRC Meeting, Dana Point, CA, December 1992.
19. Powell, T. and Gage, D., "Military Multichip Modules Need 'Perfect' Die," *Electronic Products,* October 1992, pp. 41–42, 51.
20. Chandra, G., "SPEC Coatings for Bare Die Protection," ISHM/IPC HMRC Meeting, Dana Point, CA, December 1992.
21. Bauer, C. E., et al., "Good Enough Die? A Cost Model for KGD Test Decisions," *Proc IMC,* Omiya, Japan, April 1994, pp. 266–269.
22. Wagner, R. J., "Known Good Die," Sematech KGD Workshop, Austin, TX, 1993.
23. Willet, M., "Multichip Carrier Modules," *Advanced Packaging,* March/April 1994, pp. 12–14.
24. Wilson, K. T., and Kleiner, M. H., "Testing Strategies for KGD," *Proc. NEPCON,* March 1994, pp. 2292–2301.
25. National Semiconductor, "KGD Technology Plan Presentation," Cedar Rapids, IA, July 1994.
26. Lineback, J. R., "Chip Makers Try to Add Quality While Removing the Package," *EBB,* November 1994, pp. 45–47.
27. Corbet, T., "A Process Qualification Plan for KGD," *(ICEMM) Proc.,* Denver, 1993, pp. 166–171.
28. Lavi, E., "VLSI Design for Burn-In," *Proc. 1991 Int. Symp. on Testing and Failure Analysis (ISTFA),* November 1991.
29. Micron Technology Inc., *Quality/Reliability Handbook,* 1991.
30. Signetics Company, *Signetics Reliability Handbook,* 1992.
31. Klinger, D., et al., *AT&T Reliability Manual*, Van Nostrand Reinhold, 1990, pp. 37–75.
32. Turner, T., and Donaher, H., "Reliability Problems A Brief History," *TEST, European Journal,* Sept. 1994, pp. S9–S11.
33. Pecht, M., *Handbook of Electronic Packaging Design*, Chapter 11: Design for Reliability, Marcel Dekker Inc, 1991.
34. Donlin, N. E., et al., "Is Burn-In Burned Out?," *IEEE International Test Conf. Proc.,* 1991, pp. 1113–1116.

35. Friedrich, L. W., "UNISYS System Burn-In Requirements and Burn-In Implementation," *Proc. NEPCON* 1993, pp. 745–757.
36. "Evaluation of Electromigration Using the SWEAT Procedure," Hewlett Packard application note 4156-7, 1994.
37. JEDEC, "A Guide for the Design of Straight Line SWEAT Test Structures," ballot JC-14.2-95-47, 1995.
38. McEuen, S. D., "Reliability Benefits of I_{DDQ}, I_{DDQ}, "*Testing of VLSI Circuits*, Kluwer Academic Publishers, 1993, pp. 43–48.
39. McCluskey, E., and Buelow, F., "IC Quality and Test Transparency," *IEEE Trans. Industrial Electronics,* May 1989, pp. 197–202.
40. Butler, K. M., and Mercer, R. M., *Assessing Fault Model and Test Quality*, Klewer Academic Publishers, 1993.
41. Kulkarni, V., "Increasing Fault Coverage with I_{DDQ} Testing," *Electronic Design,* August 8 1994, pp. 87–93.
42. Williams, T., Brown, N., "Defect Level as a Function of Fault Coverage," *IEEE Trans. Computers,* vol. C-30, no. 12, December 1981, pp. 987–988.
43. Kraynak, P., and Hill, Y. M., "Wafer-Chip Assembly for Large Scale Integration," *IEEE Trans. Electron Devices,* vol. 15, September 1968, pp. 660–663.
44. Hagge, J. K., and Wagner, R. J., "Assuring Known-Good ICs for MCM Assembly," *Proc. ISHM Advanced Technology Workshop,* Ogunquit, ME, June 1991.
45. Wagner, R. J., and Hagge, J. K., "Improving MCM Assembly Yields through Approaches for Known-Good ICs," *Proc. IEPS,* September 1991, pp. 882–897.
46. Hagge, J. K., "Progress on MCM Infrastructure Issues," *Proc. IEEE Symp. on High Density Integration in Communication and Computer Systems,* Waltham, MA, October 1991.
47. Smitherman, C. D., and Rates, J., "Methods for Processing Known Good Die," *Proc. IEPS/ISHM Multichip Module Conference,* Denver, CO, April 1992, pp. 436–444.
48. Bindra, A., "Prober Test Chips at Mil Temperatures," *Electronic Engineering Times,* p. 11, Feb. 23, 1987.
49. Holderfield, D. C., "Electrical/Thermal Prescreening of Semi-Conductor Chips," *Hybrid Circuit Technology,* pp. 37–40, June 1989.
50. Whitworth, W. C., and Rima, P., "A Complex TAB for Space Hybrids," *Proc. ISHM,* Baltimore, MD, Oct. 1989, pp. 614–619.
51. Lindberg, F. A., "Wirebonded Printed Circuit Assembly," *Printed Circuit Assembly,* pp. 10–13, Feb. 1989.
52. Scott, R. B., "Keys to the Coming TAB Technology Wave: Testability and Reliability," *Surface Mount Technology,* June 1990, pp. 15–17.
53. Fillion, R. A., et al., "Bare Chip Test Techniques for Multichip Modules," *Proc. EIA Electronic Components Technology Conference,* May 1990, pp. 554–558.
54. Chu, J. L., et al., "A 128kb CMOS Static Random-Access Memory," *IBM J. Res. Develop.,* vol. 35, no. 3, pp. 321–329, May 1991.
55. Maliniak, D., "Bare Chip Attachment Method Spreads I/O, Permits Pretesting," *Electronic Design,* pp. 32–33, May 23, 1991.
56. Walker, J. C., "Frame Automated Wirebonding: A TAB-Equivalent for Multichip Module Assembly," *Proc. IEPS,* 1992, pp. 389–398.
57. Andrews, M., *et al.,*"Final Report: Consortia for Known-Good-Die, Phase I," Wright Lab Technical Report WL-TR-94-5008, Air Force Material Command, WPAFB, Feb. 1994.
58. Williams, T. A., "Securing Known Good Die," *Advanced Packaging,* Fall 1992, pp. 50–59.
59. Falconer, E., and Lippold, M., "A Survey of Techniques for Producing Known Good Die," *Proc. 31st Annual ISHM Nordic Conference,* Gothenburg, Sweden, September 1993.
60. Rinne, G. A., et al., "Known Good Die Solutions for Flip Chips," *Proc. NEPCON,* Anaheim, CA, February 1994, pp. 2282–2290.
61. Anon., "3M Joins Bare Die Consortium," *EE Times,* September 19, 1994.
62. Anon., "Aehr Test, Nitto Denko Develop KGD Solution," *Elec. Pkg. & Prod.,* September 1993, p. 11.
63. Beiley, M., et al., "Array Probe Card," *Proc. IEEE Multichip Module Conference,* Santa Cruz, CA, March 1992, pp. 28–31.
64. Detar, J., "Intel Launches SmartDie Program," *Electronic News,* May 30, 1994, p. 56.

65. Gilg, L., Known Good Die (KGD) Industry Meeting, MCC, Austin TX, October 12, 1993.
66. Semiconductor Industry Association, "Semiconductor Technology Workshop, Working Group Reports," Irving, TX, November 1992.
67. Semiconductor Industry Association, "The National Technology Roadmap for Semiconductors," 1994.
68. Keezer, D. C., "Bare Die Testing and MCM Probing Techniques," *Proc. IEEE Multichip Module Conf.,* Santa Cruz, CA, March 1992, pp. 20–23.
69. Parker, R. H., "Bare Die Test," *Proc. IEEE Multichip Module Conf.,* Santa Cruz, CA, March 1992, pp. 24–27.
70. Delivorias, P., et al., "A Novel Technique for Known Good Die Processing Eliminates Failure Mechanisms Caused by Traditional KGD Processing," *Int. Reliability Physics Symp.,* Las Vegas, April 1995.
71. Erickson, J. M., and Hakim, E. B., "Survey Results on Multichip Module Test Issues," Fort Monmouth, June 1994.
72. Singh, A. D., "On Wafer Burn-In Strategies for MCM Die," *Proc. IEPS/ISHM Multichip Module Conf.,* Denver, CO, April 1994, pp. 255–260.
73. Tuckerman, D. B., et al., "A Cost-Effective Wafer-Level Burn-In Technology," *Proc. IEPS/ISHM Multichip Module Conf.,* Denver, CO, April 1994, pp. 34–40.
74. Adema, G. M., et al., "Flip Chip Technology: A Method for Providing Known Good Die with High Density Interconnections," *Proc. IEPS/ISHM Multichip Module Conf.,* Denver, CO, April 1994, pp.41–49.
75. Flynn, C. J., et al., "Rome Laboratory Initiatives for Assuring Known Good Die for MCM Applications," *Proc. ICMCM,* 1992, pp. 112–117.
76. ARPA, *Proc. Electronic Packaging and Interconnect Conf.,* Arlington, VA, February 27–28, 1995.
77. EIA/ANSI, "Standard Known-Good Die (KGD)," Proposed EIA/ANSI Standard, May 18, 1993.
78. Standard For Flip-Chip Known-Good Die (KGD), Working Document, July 22, 1993, Version 1.0.
79. Standard For TAB Known-Good Die (KGD), Working Document, April, 1993.
80. MCC/SEMATECH, Technology Assessment Guideline for Methods, Materials and Equipment Necessary to Prepare and Ship Semiconductor Devices as Fully Warranted Bare Die Product, July 29, 1993, Rev. 2.9.
81. EIA/ANSI, "Die Information Exchange (DIE) Format," Proposed EIA/ANSI Standard, 1995.
82. Harr, R. E., "Easing the Availability of Bare Die: Die Information Exchange (DIE) Format," ISHM, 1994.
83. IBIS Open Forum, I/O Buffer Information Specification, Intel Corporation, Hillsboro, OR., 1994.
84. IEEE, Boundary Scan Description Language, Supplement (B) to Standard Test Access Port and Boundary Scan Architecture: IEEE Std. 1149.1-b, Institute of Electronic and Electrical Engineers, New York, NY, 1993.
85. IEEE, VHSIC Hardware Description Language: IEEE Std. 1076, Institute of Electronic and Electrical Engineers, New York, NY, 1987.
86. IEEE, Waveform and Vector Exchange Specification: IEEE Std. 1029.1, Institute of Electronic and Electrical Engineers, New York, NY, 1991.
87. EDIF Steering Committee, Electronic Design Interchange Format: ANSI/EIA-548-1993, Electronic Industries Association, Washington, DC.
88. Simon, A. J., and Bailey, G. L., "Bare Chip Semiconductor Procurement System," *Proc. ISHM,* 1986, pp. 536–540.
89. Surface Mount Council, "Joint Industry Standard: Implementation of Flip Chip and Chip Scale Technology," Proposal, J-STD-012, JEDEC/EIA/IPC, January 1996.
90. ACSIST Associates, "Socketing Chip-On-Board to a Multichip Module Using Particle Interconnect," *Int. Conf. and Exhib. on Multichip Modules (ICEMM) Proc.,* Denver, 1993, pp. 335–340.
91. Reynolds, C., "Particle Interconnect to Probe Unpackaged Die," MCC KGD Technology Supplier Forum, Austin, May 13, 1993.
92. DiFrancesco, L., and Reynolds, C., "Socketing Chip-on-Board to a Multichip Module Using Particle Interconnect," *ICEMM Proc.,* Denver, April 1994, pp. 335–340.
93. Prokopchak, L., "Evaluating A Solution to Produce Known-Good Die (KGD)," *Proc. Semicon West,* San Francisco, July, 1995.
94. Burke, P., "Developing a Reusable Known Good Die Carrier," *ICEMM Proc.,* Denver, Apr. 1995.
95. Prokopchak, L., "Development of a Solution for Achieving Known-Good Die," *Proc. Int. Test Conf.,* Washington, D.C., September 1994.

96. Mochizuki, A., et al., "A New Ultra-High Density Bonding Technology," *ISHM Conf. Proc.,* 1992, pp. 342–347.

97. "Evaluation of New Micro-Connection System Using Microbumps," Yashuhiko Yamamoto, Masakazu Sugimoto, Kiyoshi Miyake, *ISHMConf. Proc.,* Nov. 16,1993.

98. Chrusciel, R. W., et al., "A New Technique for Known Good Die Processing Using a Novel Flip Chip Interconnect Technology," Submitted to International Test Conference, Washington, D. C., October 1995.

99. "Quest Chip Carrier & KGD Probe Card," MCC KGD Technology Supplier Forum, Austin, May 13, 1993.

100. Hill, G. R., and Ingraham, A. P., "Mechanical Interconnect System for Solder Bump Dice," *Proc. of 1st Int. Conf. on Flip Chip Technology,* San Jose, Feb. 1994.

101. Hill, G. R., and Ingraham, A. P., "Temporary Die Attach for Known-Good-Die," *Proc. ARPA Electronic Packaging and Interconnect (EPI) Principal Investigators Meeting,* Marina Del Ray, Feb, 1994.

102. Hill, G., and Bright, E., "Production of Known Good Die at IBM," *Proc NEPCON,* Anaheim, March 1994, pp. 780–786.

103. Salatino, M., et al., "Implementing a TAB-Based Known Good Die Method," *Proc. IEPS/ISHM Multichip Module Conf.,* Denver, CO, April 1994, pp. 16–21.

104. Cloud, G., et al., "Equipment, Processes, and Methods for High Volume KGD Production," *Proc. Semicon West,* Austin, 1994.

105. Roebuck, R., et al., "Known-Good-Die: A Practical Solution," *ICEMM Proc.,* Denver, Apr. 1993, pp. 177–182.

106. Roebuck, R., et al., "KGD Product: Integration into Semiconductor Manufacturing," *Proc. Semicon West,* San Francisco, June 1993.

107. Martin, S., et al., "A Practical Approach to Procuring Known-Good-Die," *ICEMM Proc.,* Denver, Apr. 1993, pp. 139–150.

108. Tech, H. B., and Rizzo, S., "Known-Good-Die Interconnect System—Design Verification Testing," *Proc. of ISHM,* Dallas, Nov. 1993, pp. 307–312.

109. Green, H., et al., "Testing Known-Good Die: Practical Solutions at Last!" *Semiconductor International,* September, 1994, pp. 48–52.

110. Agahdel, F., et al., "Known-Good-Die Carrier Reliability Evaluation," *Proc. IEEE Int. Integrated Reliability Workshop,* Fall, 1993.

111. Gucciardi, T., and Green, H. "Bare Die Testing," *Advanced Packaging,* March/April 1994, pp. 26–28.

112. "Temporary Packaging Yields Reasonably Priced KGD," *Electronic Design News,* Aug. 1994, p. 13.

113. Folaron, R., et al., "Process Integrated Equipment for Known Good Die," *ISHM/IEPS Proc.,* Denver, Apr. 1995.

114. Roebuck, R., et al., "Performance Testing of a Non-Destructive Burn-In Interconnect System for Known Good Die," *SEMI Assembly/Test,* Singapore, May 1995.

115. Fujishiro, I., "A Technology for Known-Good-Die Test and Burn-In, *Proc. IEPS/ISHM Conf. on Multichip Modules,* Denver, CO, April 1995.

116. Kondoh, Y. and Ueno, T., "Universal Membrane Probe for Known-Good-Die," *ICEMM Proc.,* Denver, April 1994, pp. 248–254.

117. Ueno, T., and Kondoh, Y., "Universal Membrane Probe for MCMs," *Proc. Int. Test Conf.,* Washington D.C., September 1994.

118. Motorola, "C4 Product Design Manual Volume I: Chip and Wafer Design," Issue A, Motorola, 1994.

119. Steele, M., "Shaped Contact Expands Horizon for Flexible Circuits," *Interconnection Technology,* October 1993, pp.26–29.

120. Lindsey, S., et al., "Comparison of Two Temporary Carriers for Test and of Memory-Die," *ICEMM Proc.,* Denver, Apr. 1994, pp. 29–33.

121. Nolan, R., "Bare Die Test and Burn-In," MCC Industry Meeting, July 1992.

122. Crumly, W., "Membrane Probe Technology," KGD Technology Supplier Forum, Austin, May 1993.

123. Teoh, H. B., and Rizzo, S. P., "Known Good Die Interconnect System, Design Verification Testing," *Proc. ISHM,* Dallas, November 1993, pp. 307–312.

124. IBM Brochure, "KGD, Temporary Die Attachment for Test and Burn-In," IBM, Endicott, NY, December 1993.

125. Miyake, K. "Connectivity Analysis of New Known Good Die Connection System Using Microbumps," *Proc. IMC,* Omiya, Japan, April 1994, pp. 156–161.

126. Corwith, A. E., "Probing at the Die Level," *Advanced Packaging,* Jan/Feb 1995, pp. 26–28.

127. Leedy, G., "KGD and At-Speed Functional Test Technology for Singulated IC," Elm Technology Corporation, private communication, February 1995.

128. Beiley, M. A., and Wong, S. S., "Micro-Machined Array Probe Card," Stanford University, SRC Pub. C93190, May 1993.

129. Duncan, L., "Known Good Die," MCC KGD Technology Supplier Forum, Austin, TX, May 13, 1993.

130. Congleton, H., and Root, R. E., "TRB Assembly—A Soft-Tooled Approach to Device Pre-Test and Interconnection," *ICEMM Proc.,* Denver, Apr. 1993, pp. 183–189.

131. Root, R., and Congleton, H., "TRB, A soft Tooled Alternative to TAB for Device Pretest," *Proc. Application Solutions for Known-Good-Die Symp.,* SEMATECH/MCC, Dec, 1992.

132. Treece, R. K., "mBGA Technology Overview," *Proc. ELECTRECON 1994,* May 1994, Indianapolis, IN.

133. Treece, K., "A Strategy for Obtaining KGD for MCMs," Applications Solutions for Known-Good Die, SEMATECH/MCC, Austin, Dec. 1992.

134. Smitherman, C. D., and Rates, J., "Methods for Processing Known-Good-Die," *Proc. International Symp. on Microcircuits,* Nov.,1993, pp.318–323.

135. Rates, J., "KGD, A Die Processors Point of View," KGD Suppliers Conf., Napa, CA, September 1994.

136. Rinne, G. A., "Burn-In and Test Substrate for Flip Chip ICs," MCNC Technical Report TR95-01, 1995, pp. 28.

137. Cloud, G., et al., "Known-Good Die: A Key to Cost Effective MCMs," *Electronic Packaging and Production,* September 1992.

138. Kim, I. U., et al., "A New Approach to Produce Cost-Effective Known-Good-Die," *ICEMM Proc.,* Denver, Apr. 1994, pp. 7–15.

139. Kim, I. U., et al., "A New Approach to Produce Cost-Effective Known Good Die," *Proc. IEPS/ISHM Multichip Module Conference,* Denver, CO, April 1994, pp. 7–15.

140. Kim, I. U., et al., "A New Method to Produce Cost-Effective Known-Good Die," VLSI Packaging Workshop, Kyoto, Japan, November 1994.

141. Marcoux, P., and Richmond, D., "Miniature SMD Packaging Simplified," *Elect. Packaging & Prod.,* January 1993, pp. 74–75.

142. Marcoux, P., and Richmond, D., "The Micro SMT Package for Surface Mount and Multichip Module Applications," *Proc. IEPS,* San Diego, Sept., 1993, pp. 457–470.

143. Marcoux, P., "Overview of Integrated Circuit Interconnect Options for Multichip Modules," *Proc. NEPCON,* Anaheim, March 1994, pp. 2275–2278.

144. Marcoux, P. P., "A Minimal Packaging Solution for Known Good Die and Direct Chip Attach," *Proc. Surface Mount Int.,* August 1994, pp. 19–26.

145. Young, J. L., "Chip Scale Packaging Provides Known-Good Die," *Proc. NEPCON,* Anaheim, CA, February 1995, pp. 52–59.

146. Richmond, D., "Micro SMT Discrete Device Technical White Paper," MCC KGD Technology Supplier Forum, Austin, May 13, 1993.

147. Costelow, T., "Mitsubishi Scales Down IC Package," *EE Times,* August 22, 1994, pp. 1, 82.

148. Banerji, K., "Development of the *S*lightly *L*arger than *IC C*arrier (SLICC)," *Proc. NEPCON,* Anaheim, March 1994, pp. 1249–1256.

149. Mathew, L. C., and DiStefano, T. H., "Future Directions in TAB: the TCC/MCM Interconnect," *Proc. Int. Flip Chip/TAB/Adv. Pkg. Symp.,* San Jose, Feb,1994.

150. Loo, M. C., "Surface Mount Studies of Fine Pitch Micro Ball Grid," *IEPS Proc.,* Sept. 1994.

151. Distefano, T. H., et al., "MicroBGA for High Performance Applications," *Proc. Surface Mount Int.,* August 1994, pp. 212–215.

152. Wakabayashi, S., and Distefano, T. H., "The (BGA as a Chip Size Package," *Proc. SEMI,* Japan, August 1995.

153. Faraji, B., et al., "Reliability of (BGA Packaging Technology, Intl Flip Chip, Ball Grid Array, TAB and *Advanced Packaging Symp. Proc.,* Feb. 1995.

154. Chanchani, R., et al., "A New Mini Ball Grid Array (mBGA) Multichip Module Technology," *Proc. ISHM,* Boston, MA, November 1994, pp. 196–201.

155. Camilletti, R. C., "Robust Protection of Bare Die," *Proc. ISHM Advanced Technology Workshop on Flat Panel Display Packaging,* Islamorada, FL, October 1994.

156. Camilletti, R. C., and Loboda, M. J., "Thin-Film Packaging for MCM Applications," *Proc. IEPS/ISHM Conf. on Multichip Modules,* Denver, CO, April 1995.

157. Blaisdell, B., et al., "KGD: Meeting The Many Needs Of The Market," *Proc. NEPCON West,* Anaheim, CA, February 27, 1995.

158. Gilg, L., *MCC/ARPA KGD Suppliers Conf. Proc.,* Napa, CA, September 1994.

159. Vardaman, J., "Multichip Packages," *Advanced Packaging,* January/February 1995, pp. 36–37.

160. Bartelink, D. J., "A Unified Approach to Chip, Test, and Assembly Technologies for MCMs," *Proc. IEEE Multichip Module Conf.,* Santa Cruz, CA, January 1995, pp. 221–228.

161. Tuck, J., "A New Debate: Packaged vs. Unpackaged," *Circuits Assembly,* August 1994, pp. 22–25.

162. National Semiconductor, "Known Good Die: Innovative Solutions," NSC Brochure JSM2.5M73, 1993.

163. Wood, A. G., "KGD Definitions Influence MCM Applications," *Proc. Surface Mount Int.,* August 1994, pp. 9–13.

164. Costlow, J., "Known-Good Dice Offer New Packaging Options," EE Times, August 22, 1994, pp. 4, 20.

165. Dislis, C., and Jalowiecki, I. P., "Economics Modeling for the Determination of Optimal Known Good Die Strategies," *Proc. IEEE Multi-chip Module Conference,* Santa Cruz, CA, January 1995, pp. 8–13.

166. Murphy, C. F., "Known Good Die Selection Tradeoffs: A Cost Model," *Proc. IEPS/ISHM Multichip Module Conference,* Denver, CO, April 1994, pp. 261–265.

167. Sandborn, P. A., and Moreno, H., *Conceptual Design of Multichip Modules and Systems*, Kluwer Academic, Boston, 1994, pp. 152–168.

168. Gralian, D., "Next Generation Burn-In Development," *IEEE Trans CHMT,* vol. 17, no. 2, May 1994, pp. 190–196.

169. MCC KGD Industry Meeting, Austin, TX, March 9, 1994.

170. Sanow, L., "Yield Considerations: Challenging the Necessity of Known Good Die," *Advanced Packaging,* March/April 1994, pp. 40–45.

171. Bechtel, J. S., "Testing MCMs: Solutions in Progress," *Electronic Packaging and Production,* November 1993, pp. 56V–56W.

172. Bauer, C. E., "Partitioning and Silicon Selection Strategies for Cost Effective MCM Designs," *Proc. Surface Mount Int.,* August 1994, pp. 14–20.

173. Strassberg, D., "Temporary Packaging Yields Reasonably Priced KGD," *EDN,* August 4, 1994, p. 13.

174. JEDEC, "Procedure for Wafer-Level Testing of Thin Dielectrics," JESD35, July, 1992.

175. JEDEC, "Hot Carrier Stress Specification," Proposed, September 7, 1993.

176. Furuyama, et al., "Wafer Burn-In (WBI) Technology for RAMs," *IEDM,* 1993.

177. O'Brien, K., and Pratser, S., "Establishing an MCM Infrastructure," *Advanced Packaging,* May/June 1994, pp. 34–37.

CHAPTER 15

MCM TEST AND DESIGN FOR TEST

Prof. Victor Nelson
Auburn University

15.1 MCM TEST ISSUES

Testing is a critical part of the manufacturing process for an integrated circuit (IC) device, multichip module (MCM), printed circuit board (PCB), or system. Each device, module, or system that leaves the factory must have an acceptably high probability of being fault free. Unfortunately, designers are

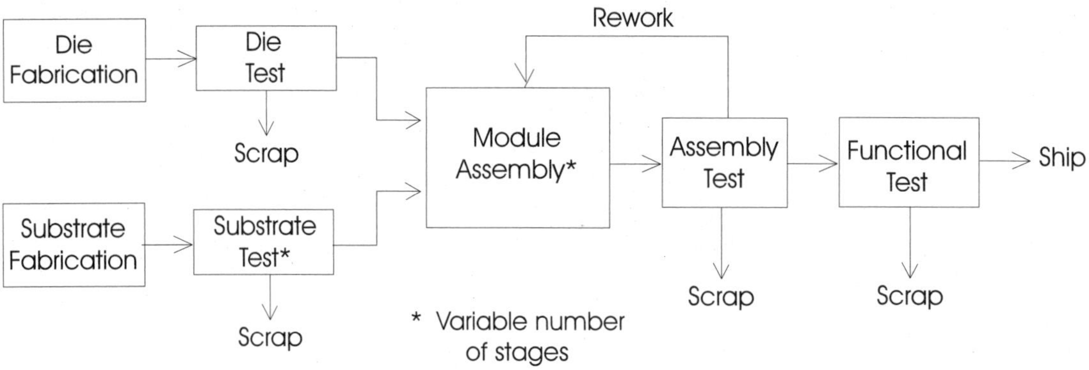

FIGURE 15.1 Testing in MCM manufacturing.

often overwhelmed by demands for increased functionality, smaller packages, higher speeds, and lower power consumption. As a consequence, testing is often made a secondary part of the process of designing and manufacturing MCMs.

Multichip modules present new and unique problems to test engineers. The features that make MCM technology attractive represent the biggest challenges to testing. With many high-speed, high pin-count devices and a high interconnect density on a single substrate, adapting traditional PCB test procedures and equipment for use with MCM technology is difficult. In addition, the complexity of an MCM is greater than that of a single integrated circuit, making established IC test methods and equipment difficult to adapt to MCMs.

The time and effort needed to develop and apply test procedures adds significantly to the overall cost of an electronic component over its lifetime and can lead to delivery delays that can be damaging in today's market. Consequently, both the process of developing a fault testing strategy and the actual application of tests to each circuit must be done as efficiently as possible while providing a high probability of detecting any circuit that is faulty.

As illustrated in Fig. 15.1, there are three basic processes in the creation of an MCM: die fabrication, substrate fabrication, and module assembly. Two levels of testing are performed: process-specific and comprehensive. A process-specific test detects specific classes of faults that can be introduced during that process. Test procedures for die, substrates, and MCM assemblies are radically different. The final comprehensive test step verifies the overall functionality and performance of the completed MCM, and detects any faults that escaped detection in the individual processes.

Development of a cost-effective MCM test strategy requires an understanding of potential faults in die, substrates, and MCM assemblies, as well as cost issues and effectiveness of test methods. This chapter examines these issues as they relate to MCM substrate and assembly testing. Testing of die used in MCMs is discussed in Chapter 14.

15.2 FAULTS AND DEFECTS IN MCMs

A *fault* can be informally defined as any condition that causes a device to function improperly. Digital logic networks may contain faults caused by broken or shorted interconnections, bad logic elements, improper power supply voltage, poor noise immunity, and others. Faults are usually categorized by their duration, extent, and behavior.[1]

The *duration* of a fault can be permanent, transient, or intermittent. A fault condition that does not change with time is referred to as a *solid*, *hard*, or *permanent* fault. A fault that appears and then disappears is referred to as a *transient* or *soft* fault. A fault that recurs periodically is called an *intermit-*

tent fault. Production testing typically focuses on permanent faults, with transient and intermittent faults more difficult to detect and more likely to occur during normal operation.

The *extent* of a fault indicates the amount of affected circuitry. A *local* fault affects a single device or component, while a *global* fault affects multiple circuits or components. Most test procedures are designed to detect isolated local faults, although most will detect global faults as well. The *behavior* of a fault describes its effect on logic signal levels or other electrical parameters in a circuit.

In this section, we examine physical defects and faults in die, substrates, and MCM assemblies, and then examine how the behavior of these faults is modeled for purposes of test development.

15.2.1 Physical Faults/Defects

Physical defects are primarily responsible for permanent and intermittent faults in electronic devices and assemblies, and eventually manifest themselves as incorrect circuit behavior. Most transient faults are induced by electromagnetic interference, environmental effects, and other outside agents, and are not related to defects. Defects in die, substrates, and MCM assemblies are typically considered separately for test generation, and are due to poor design, material variations, process variations, and poor workmanship.

Die Faults. Faults in integrated circuit chips have been well documented.[1–5] Most die faults are effectively open or short circuits that affect voltage levels. In digital circuits, the majority of opens and shorts result in input or output voltages effectively frozen (stuck) at values above the logic 1 threshold or below the logic 0 threshold.

Some die faults result in improper logic circuit input threshold voltages (V_{IH}, V_{IL}), incorrect output drive levels, abnormal power supply currents (static I_{CC} or dynamic I_{DDQ} in CMOS circuits), and excessive leakage currents at the outputs of tristate drivers. Other faults affect dynamic parameters such as propagation delays, flip-flop setup and hold times, and clock periods and widths. Die testing is not considered in this chapter; the reader is referred to Chapter 14 of this book for further information on die faults and known-good die (KGD) test programs.

MCM Substrate Faults. MCM substrates are susceptible to many different types of defects during fabrication.[6–8] Many of these are gross defects that can be identified by visual inspection, whereas others are more subtle and require complex electrical testing.

The most common concerns in substrates are open and short circuit conditions. A *short circuit* is the presence of a low-resistance path between conducting traces that should be electrically isolated. Shorts are the result of extra material deposited on a surface, traces formed too close together, and other mechanisms. Short circuits occur not only within substrate layers but also between traces on different substrate layers.

An *open circuit* is the presence of an abnormally high resistance in a conducting path, usually as a result of a broken or incomplete conducting circuit trace. Opens and high-resistance areas also occur in via holes, resulting in poor or missing connections between conducting traces in different layers.

Latent opens are imperfections in circuit traces, such as partial cracks or notches (*mouse bites*), indentations (*dishdowns*), reduced-width "necks," and other deformities.[9–10] Similar defects are also possible in vias. Latent opens are critical in that they may easily turn into open circuits during normal operation as traces are stressed. A region of a circuit trace affected by a latent open defect contains less conducting material and, therefore, carries higher current densities and has a higher resistance than the neighboring sections of the trace. Consequently, stress due to ohmic heating is more significant in the region of the defect, increasing the likelihood of stressing the trace to the point of mechanical failure. Latent opens are also stressed during soldering or wire bonding as die are attached

Latent faults are difficult to detect with simple electrical tests, because they do not interrupt conducting paths. Some of these defects can be identified through visual inspection, but others, such as cracks in vias, dishdowns (indentations in a trace), and other holes and nonuniformities on the surface of the substrate, in vias, or between layers, require special electrical tests.

Other substrate defects do not directly affect electrical performance but may affect the MCM in other ways during the assembly process. Examples include excessive warpage of the substrate and misalignment of via holes, pads, and traces.

MCM Assembly Structural Faults. The MCM assembly process consists of placing and attaching die and making electrical connections between die and substrate. Depending on the technology of the available die, electrical connections are made by wire bonding between pads on the ICs and the substrate, wire bonding TAB leads onto substrate pads, or soldering flip chip pads to substrate pads.

The most common structural defects in MCM assemblies are those related to electrical connections between die and substrate.[8] Typical defects include weak, misaligned, or missing wire bonds, and missing or cold solder joints between flip chip and substrate pads, resulting from contamination or corrosion of the wires or pads. Additional defects include bridging faults, which are shorts between conducting circuit traces caused by spots of conducting material or bent or misaligned leads.

In some cases, errors are made in the selection and placement of parts onto a substrate, including incorrect parts, wrong part values or speeds, and incorrect orientation of a part on the substrate. In addition, previously tested die and substrate traces and pads may be damaged during placement and attachment of die to a substrate.

15.2.2 Fault Models

The design of tests for specific faults in a die, substrate, or MCM assembly requires the use of fault models. Aside from optical inspection, testing does not detect defects directly but, rather, behavioral symptoms resulting from physical defects. For electrical testing to work, a fault must cause a circuit to behave differently from the fault-free circuit.

A *fault model* accurately represents the behavior of a physical fault or defect.[1–5,11] The type of model used depends on the type of testing to be performed and whether the circuit is digital, analog, or mixed. Most faults in digital circuits affect the logic functions realized by the circuits, while other faults affect various DC and AC circuit parameters.

Logical Fault Models. *Logical faults* alter the logic function realized by a digital circuit, causing it to function as an entirely different logic device. Most tests are designed to detect solid, logical faults. Other potential faults in digital circuits are not as simple to model and detect. Examples include *bridging faults*, which are shorts between wires that allow signals on one wire to affect signal values on another. Bridging faults also alter the logic function realized by the circuit, but in a manner different from stuck-at faults.

The most commonly used model for representing logical faults in digital systems is the *stuck-at* fault model. In this model, a fault in a circuit is represented as a wire whose voltage level is stuck-at logic 0 (s-a-0) or stuck-at logic 1 (s-a-1). The stuck-at model is justified by its simplicity and by its accurate representation of a large class of faults that occur in practical circuits. Open or shorted components, such as transistors and diodes, can be modeled as causing stuck-at faults in a circuit, as can broken wires and wires shorted to ground or to high voltage.

A faulty circuit realizes some logic function different from that realized by the fault-free circuit. For example, the circuit of Fig. 15.2 realizes the logic function:

$$f(A, B, C) = AB + C \tag{15.1}$$

A s-a-0 fault on logic line D, denoted by $D/0$, alters the logic function realized by the circuit to:

$$f(A, B, C) = C \tag{15.2}$$

A test for the fault $D/0$ consists of applying a test vector to the inputs and comparing the circuit output to that of the fault-free case. For example, the test vector $(A,B,C) = (1,1,0)$ applied to the fault-free circuit results in output $E{=}1$ but gives $E{=}0$ when fault $D/0$ is present.

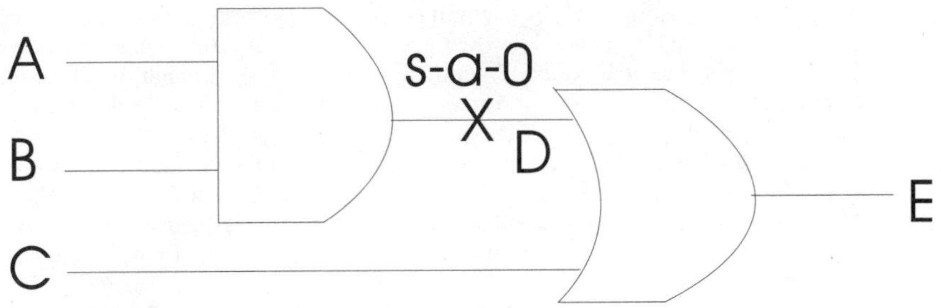

FIGURE 15.2 Stuck-at fault, *D* s-a-0.

When more than one wire is stuck, a *multiple fault* exists. Circuits with *r* wires have 2*r* possible single faults based on the stuck-at model. When multiple faults are considered, $3^r - 1$ faults can be enumerated. The latter number includes single faults as a special case of multiple faults. Many automatic test generation programs and fault simulators consider only single stuck-at faults, because they are much more likely to occur than multiple faults. In addition, it has been shown that tests for single stuck-at faults detect many multiple faults as well.[3–5]

DC and AC Parametric Faults. Many faults are *non-logical* in that they do not alter the logic function realized by a circuit. Instead, they affect various circuit parameters. *DC parametric faults* affect static currents and voltages, such as power supply current, I_{CC}, output voltage levels (V_{OH}, V_{OL}), logic thresholds (V_{IH}, V_{IL}), output loading, and leakage currents. CMOS circuits are also characterized by a quiescent power supply current, I_{DDQ}, that is unique during steady-state conditions since CMOS logic circuits contain no conducting paths between power and ground except during switching. Shorts are indicated by excessive I_{DDQ} levels that can often be detected by on-chip sensors to detect these faults inexpensively. Wafer probe of ICs and other preliminary electrical tests often check I_{CC} and other DC parameters so that faulty components can be discarded prior to performing more expensive tests.

AC parametric faults affect such timing parameters as propagation delays, signal rise and fall times, clock periods and pulse widths. Signal propagation delays often result in hazard conditions or violation of flip-flop setup and hold times or pulse width constraints, resulting in incorrect behavior. In addition, crosstalk, signal reflections, and other electromagnetic effects can be significant in high-speed MCMs. AC parametric faults are difficult to model and test. Most such faults can be detected only by functional testing at full operational speeds, with fault effects eventually causing errors that can be detected as incorrect logic values.

15.3 TEST METRICS AND ECONOMICS

The ideal goal of testing is to identify every faulty component and either discard or repair it before shipping it to a customer. However, in a production environment, cost per part and delivery delays must be weighed against the product quality and yield effect benefits of testing. The effectiveness of a test strategy must be weighed against the cost of developing and applying tests to allow the best strategy to be selected within cost constraints that will result in desired yields or defect levels. For this purpose, several parameters are used in determining the cost effectiveness of a test strategy.

15.3.1 Fault Coverage

The primary measure of the effectiveness of a test is *fault coverage*, or simply *coverage*. Given a universe of faults that could potentially be present in a component, the fault coverage of a test is defined as the percentage of those faults that will be detected by the test. The ideal goal of a testing strategy is 100 percent fault coverage. However, such perfection is rarely achievable at a reasonable cost or within a reasonable time period. Some faults are extremely difficult to detect, requiring costly and time-consuming test procedures. Therefore, test efforts are typically designed to cover only those faults most likely to occur.

The cost of designing and performing a test for a "difficult" fault may dictate that it be ignored during testing, or that the component be redesigned make that fault less difficult to detect. Many manufacturers are willing to accept somewhat less than 100 percent fault coverage to keep test costs and testing time within budgeted limits. Consequently, some faulty parts may be identified as good by a test. Such parts are referred to as *test escapes*.

The probability that a faulty part will be detected during testing is a function of the likelihood of occurrence of that fault and the fault coverage of the test, and it can be expressed as

$$\Pr(Faulty\ Part\ Detected) = \Pr\left(\frac{Test\ Fails}{Part\ Faulty}\right) \times \Pr(Part\ Faulty) \qquad (15.3)$$

The conditional probability Pr(*Test Fails/Part Faulty*) represents the fault coverage of the test. The probability that a defective part will escape the testing process undetected is

$$\Pr(Faulty\ Part\ Escapes) = \Pr\left(\frac{Test\ Passes}{Part\ Faulty}\right) \times \Pr(Part\ Faulty) \qquad (15.4)$$

where Pr(*Test Passes/Part Faulty*) = 1 – Pr(*Test Fails/Part Faulty*) and is the conditional test escape probability.

In general, the majority of stuck-at faults in a digital system can be detected with a relatively few test vectors. In some cases, acceptable coverage can even be attained with a small number of randomly generated test patterns, as illustrated in Refs. 3 and 5. Therefore, good coverage can be attained at a relatively low cost. However, some faults are more difficult to detect, requiring more expensive, specially designed tests. The result is diminishing increases in fault coverage with each additional test vector as shown in Fig. 15.3, with significant increases in test generation costs.

The expected coverage of a set of test vectors, or *test set*, should be evaluated early in the design process. Many CAD systems include tools that estimate or "grade" fault coverage by deterministic or statistical fault simulation. These will be discussed later under the topic of test generation methods.

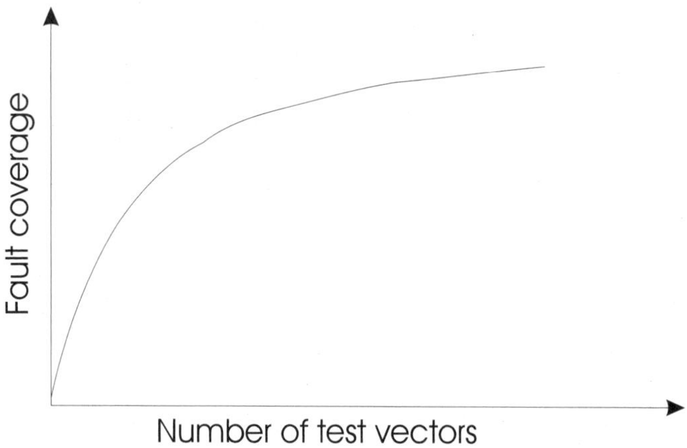

FIGURE 15.3 Test coverage vs. number of random test vectors.

15.3.2 Yield

The *yield* of a process is the number of parts accepted as good after completing the process, divided by the number of parts that entered the process.[12] Low yields result in a large amount of scrap and/or costly diagnosis and repair, which adds to the cost per part and increases the likelihood of test escapes. MCM yields are affected by faults in integrated circuit (IC) die, substrate traces, wire bonds, solder joints, and so on. Each component and each assembly step has some probability of introducing new faults into a module. A fault in any one of these usually results in the entire MCM being faulty.

To estimate the yield of an MCM manufacturing process, we begin with the ICs. For an MCM containing N ICs, the overall *chip yield*, Y_C, is the probability that all N ICs are good, which is given by

$$Y_C = \prod_{i=1}^{N} P_{C_i} \times 100\% \tag{15.5}$$

where P_{C_i} = the probability that IC_i is good

Die reliabilities are functions of IC yields, the testing performed by the die manufacturer, and any prescreening done by the MCM manufacturer. If all chips have approximately identical reliabilities equal to P_C, then the chip yield can be approximated as

$$Y_C = P_C^N \times 100\% \tag{15.6}$$

This is plotted in Fig. 15.4, showing that chip yields have a significant impact on MCM yield. Therefore, known-good die (KGD) programs, discussed in Chapter 14, are critical to achieving acceptable MCM yields.

Faults in substrates and assembly steps (wire bonds, et al.) likewise affect overall MCM yields. Assuming a substrate yield of Y_S and a yield Y_I for each interconnection between die and substrate (wire bonds, solder joints, and so on), the overall yield of an MCM can be written as

$$Y_{MCM} = Y_C \times Y_S \times Y_I^M \times Y_A \times 100\% \tag{15.7}$$

where M = the number of interconnections
Y_A = overall workmanship and other aspects of the assembly

High MCM yields require careful screening of incoming ICs to maximize P_C, and thorough substrate testing to maximize Y_S. In addition, considerable cost savings can be achieved by minimizing the probability of using a defective IC or substrate when assembling an MCM, thereby preventing good parts from being "wasted" in a bad assembly. Finally, structural testing of MCM assemblies is important to maximize interconnect yield Y_I by allowing rework or repair of interconnect faults.

15.3.3 Defect Level and Test Leakage

The ultimate goal of testing is to ensure high quality levels in outgoing products. The quality of a product can be expressed as the probability that it is fault free. Since most tests do not achieve 100 percent fault coverage, some test escapes can be expected. The *defect level*, *DL*, of a product is the fraction of test escapes in the outgoing parts:

$$DL = \frac{\text{Faulty parts passing test}}{\text{Total parts passing test}} \tag{15.8}$$

DL is usually expressed as a percentage or in parts per million.

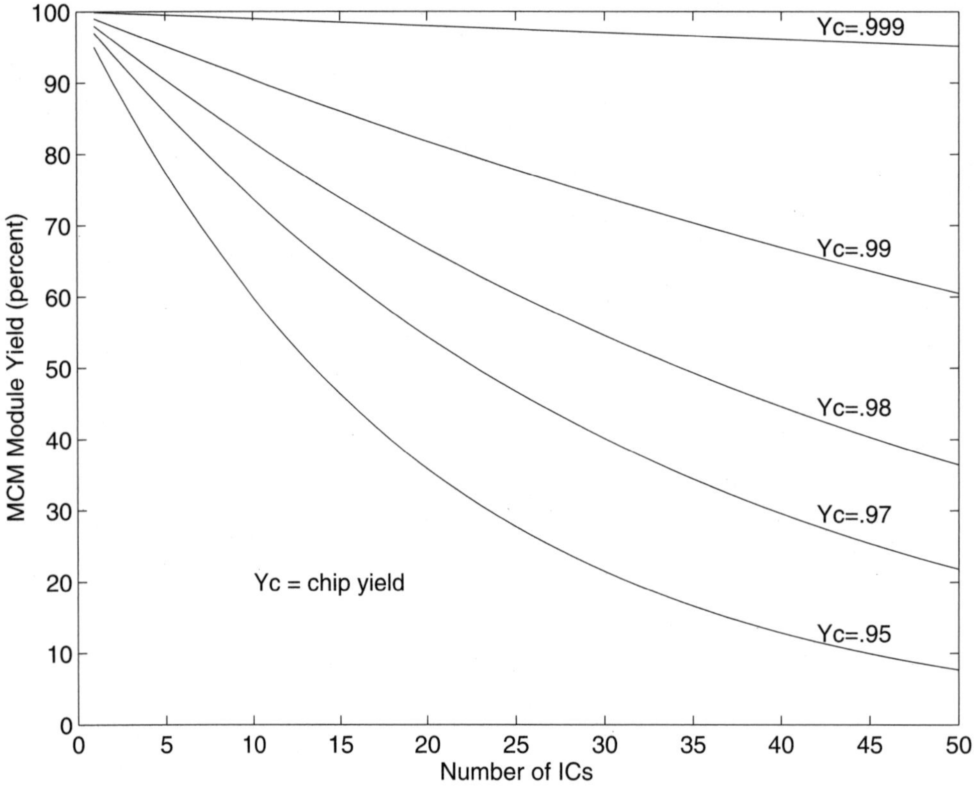

FIGURE 15.4 MCM yield vs. chip count.

Estimates of defect levels are used in developing test strategies to ensure that testing will result in acceptable levels of product quality. Estimates of MCM defect levels can be derived from predicted MCM yields and test procedure fault coverage values. A commonly used expression to estimate expected defect levels in MCMs is that proposed by Williams and Brown:[13]

$$DL = [1 - Y_{MCM}^{(1-FC)}] \times 100\% \tag{15.9}$$

where Y_{MCM} is the estimated yield of the MCM, and FC is the overall fault coverage of the test process. This expression is plotted in Fig. 15.5 to illustrate the effect of fault coverage on MCM yield for several values of Y_{MCM}. More detailed models that estimate the effects of fault coverage and other factors on defect levels are given in Ref. 14.

Reductions in defect levels require improving MCM yields and/or increasing fault coverage. Both approaches have significant costs associated with them and therefore represent trade-offs that must be weighed when deciding what defect levels will be acceptable.

15.3.4 Test-Related Cost Issues

The two primary factors considered in the development of an MCM test strategy are product quality and cost. While testing is critical for ensuring acceptable defect levels, the process of designing a test

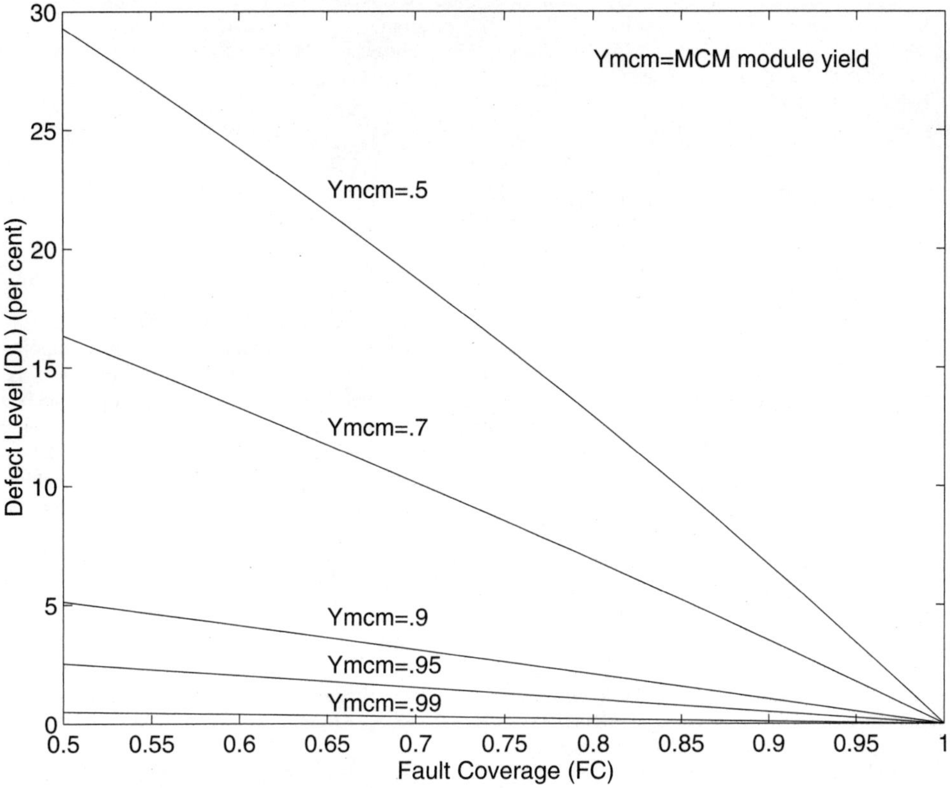

FIGURE 15.5 MCM defect level vs. fault coverage.

strategy and applying test procedures during production affects the overall cost of each MCM in several ways. Each manufacturer must decide what cost is acceptable to achieve desired defect levels in a given product.

Cost elements associated with testing can be classified as recurring or nonrecurring. Nonrecurring costs are those associated with product development, such as:

- Test design costs (design modeling, test pattern generation, fault simulation)
- Cost of designing testability features
- Test equipment and device fixturing costs
- Special prototype testing to identify potential problems

Recurring costs are associated with each manufactured part and include:

- Cost of applying test procedures
- Cost savings due to repairing versus scrapping parts
- MCM area overhead of testability features (circuitry, pins, and so on)

Testing may also affect product cost indirectly by affecting sales of an MCM. Examples are time-to-market delays and testability features that affect the performance of an MCM.

Several computer programs are available or under development to perform economic analysis of MCM test strategies. Examples include the *HiTEA* program developed by MCC and Savantage,[15] the *MCM Test Advisor Program (MTAP)* at Auburn University,[16] the *MCM Manufacturing Test Simulator (MTSIM)* at Colorado State University and Hewlett Packard,[7] and the Siemens *EVEREST* test strategy planner.[18–20]

15.4 TEST DESIGN AND TEST VECTOR GENERATION

Test design begins with developing a global test strategy and generating test vectors or other elements to implement that strategy. A global test strategy determines the stages of processing at which testing will be performed and the type of testing to be performed at each stage.[12,21–28]

It is generally acknowledged that the cost of detecting a fault in an electronic component increases by a factor of ten at each level of system assembly. For example, if the cost of detecting a fault in an integrated circuit chip is \$C, then the cost of detecting that fault after the chip has been mounted onto a board is \$10C, the cost to detect it after the board has been installed in a system is \$100C, and the cost to detect the fault after the system has been deployed in the field is \$1000C.

Consequently, it is advisable to expend considerable test energy in the early steps of assembly to minimize more expensive testing at later steps. In addition, testing in early stages of production allows faults to be detected where they can be repaired or discarded before adding value to them in the form of further processing or expensive die. This results in improved MCM assembly yields and lower cost per part.

To be cost effective, testing must be considered from the beginning of the MCM design process and throughout the design cycle. While this may increase the cost of developing a design, attempts to retrofit testability into a completed design or to generate test patterns for a design in which testability was not considered often results in significantly higher cost increases and/or acceptance of higher defect levels. Global test strategy decisions must be based on target defect levels, estimated die, substrate, and assembly yields, and available test facilities.

Once a global test strategy has been designed, generation of test programs, test vectors, and so on can proceed. A significant portion of the design cycle involves design verification. It is during this time that test vectors are best designed and verified. Many CAD systems include automatic test pattern generation (ATPG) tools that automatically produce test vectors for digital logic circuits.[29,30] Fault simulation tools provide indications of fault coverage. These will be discussed later in this chapter.

Test Application Costs. The cost of performing a test includes part handling to move the part to and from test stations and application of the test procedure. Optical inspection costs are related to the number of picture cells processed in each image and the number of features that must be extracted and identified. Electrical testing entails the application of test vectors and the capture and processing of circuit outputs. In-system testing involves exercising components within host boards or systems for some period of time and verifying correct operation. Costs associated with electrical and in-system testing are typically functions of the number of test vectors applied and the total testing time. Included in these costs are labor and the operation, maintenance, and depreciation of test equipment.

Test generation methods strive to derive minimum tests, i.e., tests that can achieve desired fault coverages with a minimum number of test vectors and test processing time while achieving desired defect levels.

Design for Test Costs. Design for testability (DFT) and built-in self-test (BIST) features represent both cost additions and savings. Testability features normally incur some area overhead; additional chip area, components, test pins, and/or connectors must be used to implement these features. In addition to the direct cost of the added components, increased circuit complexity may result in lower MCM yields. Also, DFT and BIST features may result in performance penalties due to greater propagation delays through test-related circuits.

Despite these drawbacks, the primary benefits of DFT and BIST are considerable reductions in test design efforts and even more significant reductions in test application times. These cost savings often offset the negative cost factors cited above. At the same time, DFT and BIST features typically result in significant increases in achievable fault coverage than can be obtained otherwise, thereby lowering defect levels.

Analyzing the cost benefits of DFT and BIST requires determining which are necessary for MCM testability and which components in the MCM will most benefit from these features. DFT and BIST features will be discussed later in this chapter.

Analyzing the cost benefits of DFT and BIST requires determining which are necessary for MCM testability and which components in the MCM will most benefit from these features. DFT and BIST features will be discussed later in this chapter. 11 1 1 Test Design and Test Vector Generation

After a global test strategy has been laid out, design of individual test procedures can begin. For digital circuits and modules, test design involves generating test patterns to stimulate circuit inputs and verifying that the set of test patterns will result in the desired fault coverage.

15.4.1 Test Vectors and Automatic Test Pattern Generation

A *test vector* is a combination of input values that will produce different outputs from a fault-free and faulty versions of a circuit. It is assumed that faults alter the logic function of the circuit. A *test set* is a collection of test vectors that will detect all possible faults in a circuit. Test generation is the process of designing a minimum test set for a circuit. Test generation methods differ according to whether the circuit is purely combinational or is sequential, containing memory elements.

Combinational Logic Circuit Tests. In a combinational logic circuit, the output at any time is a function of its inputs at that time. Testing consists of applying a test vector, waiting for propagation delays to elapse, and then comparing the output to the correct value. One approach to test set design is to exhaustively apply all possible input combinations. If all outputs are correct, then the circuit realizes the correct logic function and can be considered fault free. However, a combinational logic circuit with n inputs requires 2^n vectors for an exhaustive test. For large circuits, exhaustive testing is not practical, given the time it would take to apply all vectors and evaluate the outputs.

Test generation for combinational logic circuits involves identifying all test vectors for individual stuck-at faults in a circuit and then extracting from those a minimum set of vectors that will detect all faults. Fortunately, a given test vector may detect more that one fault in a circuit. For example, Table 15.1 lists test vectors that detect all ten of the possible stuck-at faults for the circuit of Fig. 15.6. From this list, we can see that a minimum test set of {*110, 010, 100, 001*} will detect all ten faults.

Note that faults *A/0*, *B/0*, and *D/0* have identical test sets and are therefore *equivalent*. In addition, tests for faults *A/1* and *B/1* also detect *D/1* and therefore *dominate* fault *D/1*. Likewise, *A/1* dominates *C/1* and *E/1*, and *C/0* dominates *E/0*. Complete test sets for a circuit can be derived while omitting dominated faults and all but one of each set of equivalent faults. Thus, a complete test of the circuit in Fig. 15.6 can be derived by considering only faults *A/0*, *C/0*, *A/1*, and *B/1*.

Table 15.1 Fault Table for Figure 15.6

Fault	Tests (A,B,C)	Fault	Tests (A,B,C)
A/0	110	*A/1*	010
B/0	110	*B/1*	100
C/0	0x1	*C/1*	000, 010, 100
D/0	110	*D/1*	000, 010, 100
E/0	xx1, 1x1, 11x	*E/1*	000, 010, 100

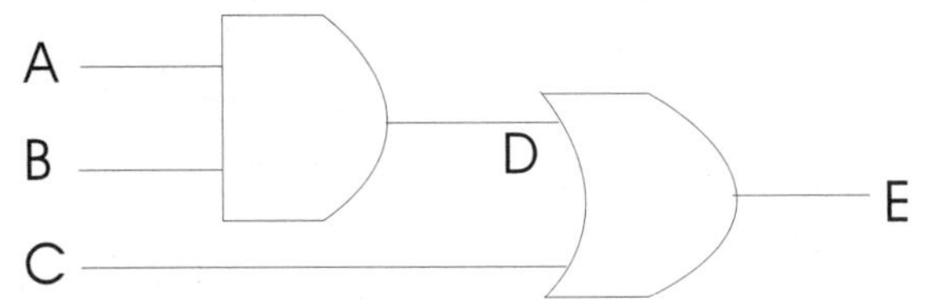

FIGURE 15.6 Circuit with ten possible stuck-at faults.

Many procedures have been developed for deriving test vectors for digital circuits.[3,4,30,31] Each test vector sensitizes a path from some fault to an observable output. For example, to test for the fault $D/0$ in Fig. 15.6, a logic 1 must be applied to net D to create an error, which requires that $A = 1$ and $B = 1$. In addition, input C must be set to 0 to allow the error to propagate through the OR gate to output E. Many CAD systems include automatic test pattern generation (ATPG) tools,[30] such as Mentor Graphics' *FlexTest*,[29] that process netlist descriptions of logic circuits, determine all possible stuck-at faults, and then generate minimum test sets for those faults.

Sequential Logic Circuits. A sequential circuit can be modeled as in Fig. 15.7, with n primary (external) inputs $\mathbf{x} = x_1 \ldots x_n$, m primary (external) outputs $\mathbf{z} = z_1 \ldots z_m$, and 2^r internal states, represented by r state variables $\mathbf{y} = y_1 \ldots y_r$.[31] The circuit contains r memory elements to store the internal state variables, and blocks of combinational logic to implement an output function $\mathbf{z} = \mathbf{f}_{\mathbf{out}}(\mathbf{x},\mathbf{y})$ and a next-state function $\mathbf{Y} = \mathbf{f}_{\mathbf{state}}(\mathbf{x},\mathbf{y})$. The output and next-state functions are derived from a state table that defines the behavior of the circuit.[31]

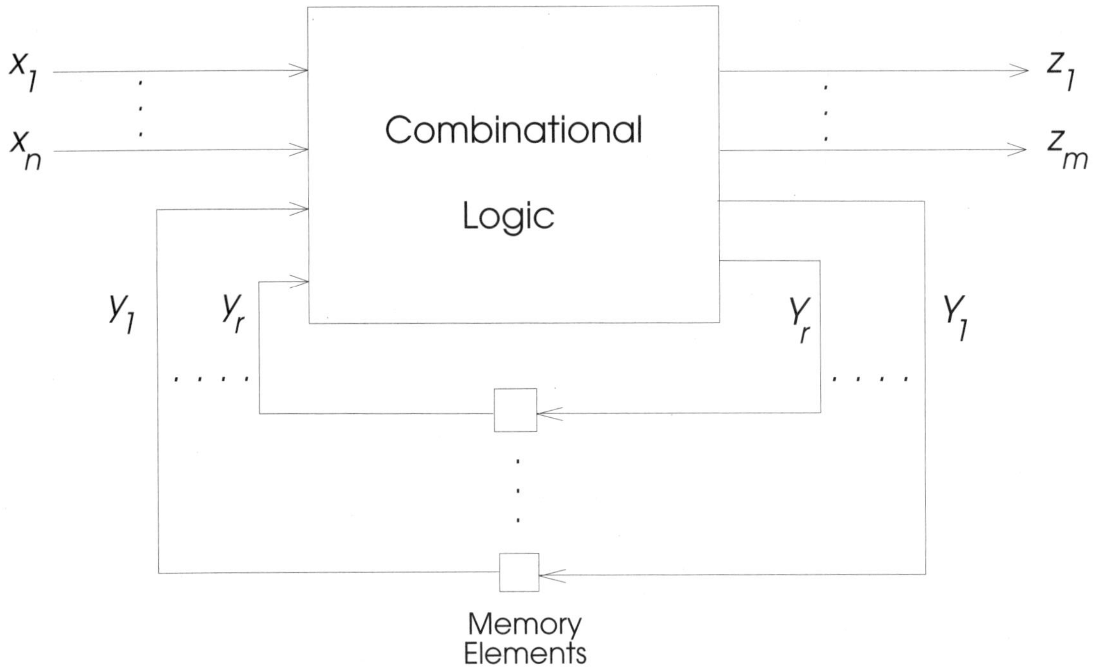

FIGURE 15.7 Generic sequential circuit model.

A complete test of a sequential logic circuit can be performed in two ways. The traditional method is to verify each state transition and output defined in the state table by the following procedure.

1. Place the circuit into a designated state.
2. Apply an input vector.
3. Verify correct primary outputs.
4. Apply a clock pulse to force a state transition.
5. Verify that a correct state was reached.
6. Repeat steps 1–5 for all possible combinations of states and inputs defined in the state table.

An alternative test method is to separately verify that the memory elements and combinational logic are fault free. Combinational logic test generation methods are used to derive test vectors for the combinational logic blocks, and these are applied via the primary inputs and state variables, with the outputs and next-state variables verified for correctness. This method allows fewer test vectors to be used than in full state table verification. Several methods for generating test vectors for sequential circuits are described in Refs. 3, 4, 30, and 31.

An exhaustive test of a sequential circuit requires a minimum of 2^{n+r} test vectors, representing all combinations of 2^n primary input vectors applied to each of the 2^r states. However, testing is hampered by the fact that only primary inputs and outputs are accessible to the tester. Forcing the circuit into a desired state may take many test vectors, and observing a next-state value may take many more test vectors, since state values may only be inferred from the outputs. Consequently, testing may require significantly more than 2^{n+r} test vectors. For complex sequential circuits, design for testability methods can be used to provide direct access to the state variables. These will be described later in this chapter.

15.4.2 Design Modeling and Simulation

Figure 15.8 illustrates the design cycle for an electronic component or module.[31] An integral part of this cycle is design verification and test vector generation. With the high cost of manufacturing MCM prototypes, there must be some reasonable assurance that the finished MCM will perform its intended function. This can be done by constructing a hardware prototype or by simulating the MCM assembly. Hardware prototypes of MCMs are usually wire-wrapped or printed circuit boards containing packaged ICs. However, it is hard to verify the complete functionality of a design from a prototype. Plus, signal delays and other characteristics of prototype boards may differ considerably from those of the MCM.

Design verification is usually done by modeling and simulation, with test generation closely tied to the simulation process. An MCM and each of its ICs are represented by software models. Simulation then involves applying input stimuli to the model inputs and examining the model outputs for correct behavior. As circuit details are added to the models, timing and other parameters can also be studied and verified.

Depending on the information desired from simulation, a model can range from an abstract model of the behavior of the IC, to a full gate- or switch-level model of the circuit. Behavioral models are most often written in a special hardware modeling language such as VHDL (VHSIC Hardware Description Language) and Verilog. VHDL was initially developed by the Department of Defense for the VHSIC (Very High Speed Integrated Circuit) Program and has since been adopted as IEEE Standard 1076.[32] Verilog was developed by Cadence and has recently been adopted as IEEE Standard 1364.[33] Both allow a digital circuit model to be represented at different levels of abstraction, ranging from a purely behavioral model with no implied circuit structure (which is useful when developing the basic functions of a circuit) to a full gate-level model of the actual circuit structure.

In some cases, hardware models of selected ICs are used for simulation. These are boards containing actual copies of components with interfaces to the simulation system that make them appear

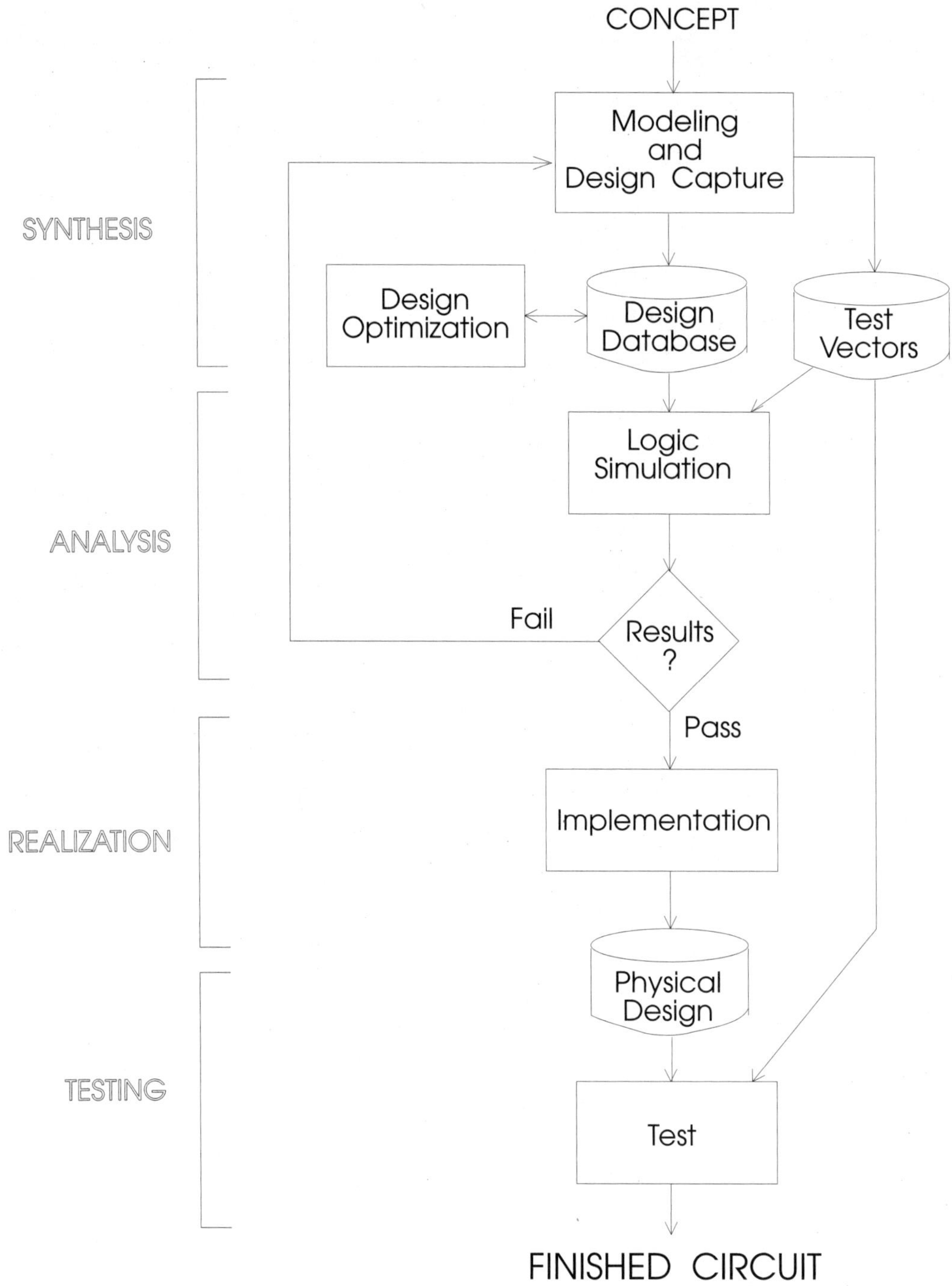

FIGURE 15.8 Design cycle for an electronic component.

as VHDL or Verilog models. Hardware models are faster, utilize very little computer memory, and are more accurate than software models. However, hardware modelers present a significant investment, and each component requires a separate model. In addition, minor differences may exist between a hardware model and the actual component, such as start-up conditions and signal drivers and receivers.[34]

15.4.3 Fault Simulation

In addition to verifying the functionality and parameters of a design, simulation is used to verify test vectors. Given a set of test vectors, generated manually or by an ATPG program, the coverage of the *test set* can be estimated or "graded" by a deterministic or a statistical fault simulator.[5] A deterministic fault simulator, such as Mentor Graphics' *QuickFault*,[35] automatically inserts each potential fault into the circuit and applies the vectors of the test set, determining the circuit outputs for each test vector and for each faulty circuit. The circuit outputs are compared to correct outputs derived by simulating the fault-free circuit. If they differ, the fault is considered detected by the test set. After repeating this for each vector in the test set and for each potential fault in the circuit, those faults that remain untested are noted. The percentage of undetected faults indicates the coverage of that test set. The list of undetected faults can be used to derive additional tests as needed to improve the coverage.

A statistical fault simulator, such as Mentor Graphics *QuickGrade II*,[35] estimates fault coverage by applying the vectors of the test set to the fault-free circuit and estimating the likelihood of detection of each fault, based on statistics related to the number of times each signal line is toggled. While not as accurate as deterministic fault simulation, a reasonably good estimate of fault coverage can be obtained in considerably less time, since simulations are not repeated for faulty circuits.

15.5 SUBSTRATE TESTING AND INSPECTION

Test strategies for MCM assemblies typically assume that the substrate and the dice are fault-free prior to attaching the dice to the substrate. The availability of known-good die (KGD) and known-good substrates (KGS) depends on the test strategies used by the die and substrate manufacturers. Known-good die issues are discussed in Chapter 14. In this section, we examine test and inspect strategies for bare substrates, i.e., for substrates before die and other components are mounted and electrically connected to them.

It is critical that substrates be identified as good prior to adding further value to them in the form of processing time and expensive components. In addition, substrate faults are more expensive to detect and diagnose after additional components are attached. For multilayer substrates, it is often possible to detect faults on one layer of a substrate before adding additional layers. Early identification of defects in a substrate may in some cases allow a substrate to be salvaged by some form of repair or rework process and can prevent good layers from being wasted by combining them with defective ones.

Substrates are susceptible to such faults as broken or incomplete traces, corresponding to open circuit conditions, and short circuits between traces. In addition, latent open defects may result in open circuits after the part has been stressed during operation. Other faults include design rule violations, such as lines too narrow or too close together, or unplanned islands of conducting material between circuit traces.

Current substrate testing methods include electrical checks for shorts, opens, latent opens, and affected performance parameters, and visual inspection, either manual or automated.[6,7,35–39]

15.5.1 Electrical Test Methods

Resistance Testing. The most straightforward way to check for continuity within a conducting circuit trace, or for isolation of two circuit traces, is to measure the resistance between test points. Stationary or moving probes make contact with the test points on the substrate, a voltage is applied between them, and the resistance between them is measured and compared to threshold values. Abnormal resistance measurements indicate irregularities between the test points.

Continuity testing identifies open circuits between two points on a circuit trace, which can be modeled by a high resistance, R_{open}, usually greater than about 10 to 500 Ω. If probes are positioned at opposite ends of a circuit trace and detect a resistance $R < R_{open}$, then it can be assumed that there is no open circuit in the trace between these points, i.e., there is a conducting path between them. $R > R_{open}$ indicates the absence of a conducting path between the test points due to an open circuit.

To perform a complete open circuit test, one resistance measurement must be made between each pair of end points on each trace on the substrate. For a substrate containing N traces with n end points per trace, the required number of open circuit tests, T_O, is:

$$T_O = N \times \sum_{k=1}^{n} (k-1) = N \times \frac{n(n-1)}{2} \tag{15.10}$$

If each trace has only two end points ($n = 2$), then the above reduces to $T_O = N$. However, $n > 2$ for many traces, since they may connect multiple IC pins or be routed through multiple substrate layers.

A short circuit is a low-resistance path between two points that were designed to be electrically isolated. A short circuit can be detected by probing test points on two circuit traces and measuring the resistance R between them. $R > R_{short}$ indicates that the traces are electrically isolated, while $R < R_{short}$ indicates the presence of a conducting path between the points, and thus a short circuit. Typical threshold values for R_{short} are in the range of 1 to 100 MΩ, while applied voltages are usually higher than those used in continuity testing; 100 to 250 V is common for isolation testing.

The detection of shorts requires probing and measuring the resistance between each pair of traces on the substrate. A substrate with N electrically isolated circuit traces would thus require T_S measurements, where

$$T_S = \sum_{k=1}^{N} (k-1) = \frac{N(N-1)}{2} \tag{15.11}$$

T_S usually can be minimized by identifying and testing only those pairs of traces that are in physical proximity, making them the most likely to be connected by short circuits.

Capacitance Testing. Resistance measurements provide direct indications of open and short conditions. However, resistance testing requires a significant number of measurements and, in the case of moving probes, many time-consuming mechanical movements. Time considerations in a production environment may preclude the use of exhaustive resistance testing on substrates with many nets.

Capacitance measurements provide indirect indications of opens and shorts and can be performed with far fewer probes than resistance testing. Consider the circuit traces in Fig. 15.9. Trace capacitance is determined primarily by its surface area. As illustrated in Fig. 15.9*a*, the capacitance between the end points A and B of a trace can be modeled as evenly distributed parallel capacitors along the line, with the total capacitance given by

$$C_{AB} = C_1 + C_2 + \ldots + C_N \tag{15.12}$$

The measured capacitance at any point on a trace should be the same as at any other point. If an open circuit exists at some point X on the trace as shown in Fig. 15.9*b*, then the capacitances between points A and X and between X and B are given by

$$C_{AX} = C_1 + C_2 + \ldots + C_k$$

$$C_{XB} = C_{k+1} + C_{k+2} + \ldots + C_N \tag{15.13}$$

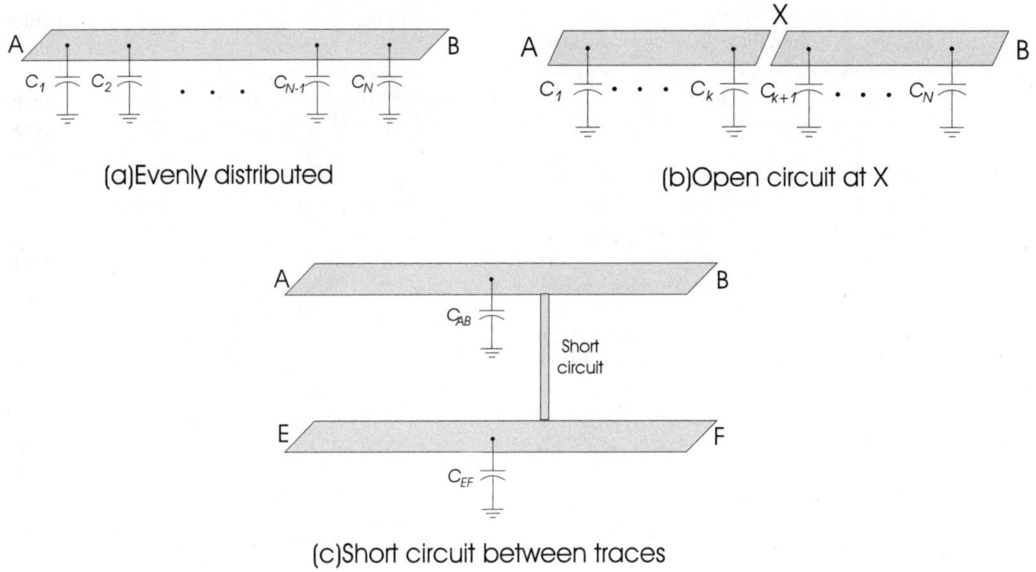

FIGURE 15.9 Distributed capacitance along circuit traces.

The open circuit can be detected by probing a point on each segment of the trace and noting that the capacitance values are not equal, with $C_{AX} < C_{AB}$ and $C_{XB} < C_{AB}$.

Likewise, if there is a short circuit between two traces *AB* and *EF*, as illustrated in Fig. 15.9*c*, the traces are electrically connected, and the capacitance at any point on either of these traces is the sum of the capacitances along both:

$$C = C_{AB} + C_{EF} \tag{15.14}$$

Capacitance testing entails measuring the capacitance between each test point and an internal or external conductive plane on the substrate or a metal chuck holding the part. The presence of opens and shorts is inferred from the measurements by comparing measured values to expected capacitance values derived from a model of the substrate layout or measured on one or more known-good substrates.

Note that both opens and shorts can be detected at the same time. Therefore, the total number of required probes is simply equal to the number of test points, resulting in a considerably shorter test time than for resistance testing.

Rise time testing produces indirect measurements of capacitance in some substrate testers.[40,41] A voltage pulse is applied to a test pad on a circuit trace and the signal rise time is measured. The time constant of the signal change is proportional to the capacitance of the signal line. Open circuits are detected by probing each end point on a trace and verifying that the rise times are identical and approximately equal to the expected value. A measured rise time at a node that is smaller than those measured at other nodes on the same net indicates a lower trace area, and therefore a discontinuity between the probed net and the rest of the trace. Short circuits are detected in a similar manner; a rise time value much longer than expected is an indication that the total trace area is greater than it should be and therefore must be the result of a short circuit between the probed trace and another trace or piece of metal on the substrate.

A significant shortcoming of capacitance testing is that it does not directly detect conducting paths between test points. Changes in the capacitance of a circuit trace may simply be the result of minor variations in the dimensions of the trace that would not normally be considered significant. In addition, a defect such as an unplanned island of conducting material shorted to a trace may result in a minimal net change in capacitance, allowing the defect to go undetected. Some substrate testers use capacitance measurements to identify possible locations of faults and then follow up with resistance tests at those locations to verify the presence or absence of conducting paths.

Latent Open Testing. Latent open faults are extremely difficult to detect with electrical testing. Such latent opens as cracks, mouse bites, and reduced line-width necks are circuit trace deformities that do not interrupt the continuity of the conducting path.

Many latent open defects can be identified through visual means, such as automatic optical inspection (AOI). However, visual inspection is applicable only to exposed areas of the substrate and is generally limited to defects in the x and y dimensions on the substrate surface. Latent open defects in the z dimension, such as *dishdowns* (indentations in a trace), cannot be easily identified through visual inspection. Also difficult to detect are cracks in vias, holes and nonuniformities either on the surface of the substrate or in vias or between layers, and separations between internal traces and via holes. Thermal cycling and other stress tests can be performed to induce mechanical fatigue failures in latent open defects, i.e., to force open circuits to occur, but this is too time consuming to be practical in a production environment.

Researchers at IBM have reported on a method for detecting latent opens in printed circuit board and MCM traces by exploiting nonlinear conduction characteristics of conducting circuit traces,[9] which should normally be linear passive components. The *LATEST* (latent-open test for electrical circuit line reduction detection) method has been implemented in the BSL LinearProbe tester.[10]

It has been shown that a metal circuit trace has nonlinear characteristics caused by temperature-induced changes in the DC resistance of the trace. Temperature changes result from ohmic heating of the trace. A region of a circuit trace corresponding to a latent open defect contains less conducting material and, therefore, higher current densities and higher resistance than the neighboring sections of the trace. Consequently, the defect region will heat and cool more quickly than the rest of the trace and will therefore exhibit different nonlinearities.

If a conductor is driven with a biased sinusoidal current of the form

$$i_{drive} = I_0 + I_1 \cos^2 \pi f t , \qquad (15.15)$$

resistance changes in the conductor result in distortion of the corresponding voltage waveform. As a result, the corresponding voltage waveform will contain many harmonics of the fundamental frequency. A pure sinusoid contains only odd harmonics, whereas adding a bias to a sinusoid produces both even and odd harmonics. If the drive current is of the form

$$i_{drive} = I_0 + I_1 \cos^2 \pi f_1 t + I_2 \cos^2 \pi f_2 t , \qquad (15.16)$$

then the resulting voltage will also contain intermodulation frequencies $f_1 - f_2$ and $f_1 + f_2$. The ideal frequencies f_1 and f_2 increase as feature sizes are reduced, with 1 *kHz* being typical for larger printed circuit boards and 1 *MHz* for MCMs.

A region of a circuit trace corresponding to a latent open defect exhibits different thermal properties from those of the surrounding regions, resulting in different distortions of the amplitude and phase of the test voltage. In particular, the phase of a good conductor lags that of the defective one. This difference can be identified easily by a phase detector, thereby detecting the defect.

The test requires two probes placed at opposite ends of the circuit trace, one to supply a drive current and the other to sample the return voltage. A single-frequency, biased sinusoid is applied, and the return voltage is filtered to remove all but the second harmonic, which is then compared to that of the drive oscillator. For MCMs with their smaller features requiring much higher frequencies, the harmonics of these high frequencies are subject to considerable distortion within the oscillator that is difficult to distinguish from the distortions due to device defects. Therefore, the second drive current form (containing two sinusoids of different frequencies) is used, with the difference frequency $f_1 - f_2$

isolated and examined. This high-frequency method has resulted in the detection of defect with a resolution as small as 0.01 Ω.[9]

15.5.2 Mechanical Contact Probe Testers

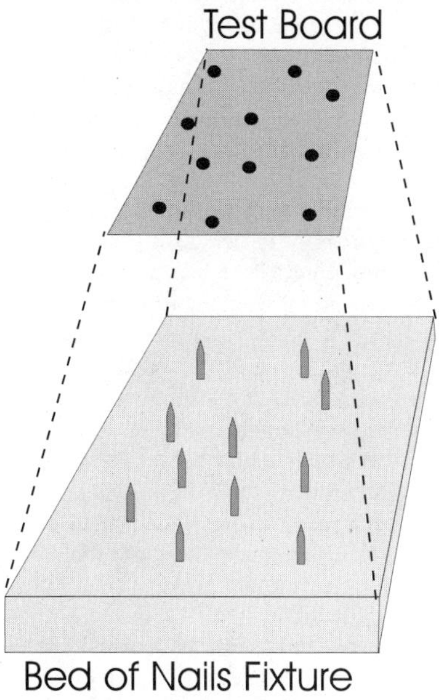

FIGURE 15.10 Bed of nails tester.

Bed of Nails Testers. A *bed of nails* is a test fixture in which vertical pins are arranged in a pattern corresponding to probe points on the substrate to be tested, as illustrated in Fig. 15.10. The substrate is placed on the bed of spring-loaded probe pins and held in place by vacuum or pneumatic pressure to ensure good electrical connection between the pins and the substrate. Test signals are then applied and measurements taken via the probe pins. No movement is needed during testing, allowing multiple tests to be performed concurrently involving any combination of the probes.

Bed of nails testers have been used for printed circuit board testing since the late 1970s[6,7,38,39] and are available in universal grid or fixed grid configurations. A *universal grid* tester has an array of thousands of spring-loaded pins arranged in a fixed grid pattern. Separate test pins are used to make contact with the test surface through two translator fixtures, as shown in Fig. 15.11. Pins are loaded through holes in the top plate that are drilled to match the pattern of test points on the substrate or board. These pins make contact with tester pins through holes in the base plate that conform to the tester pin grid. Each pin can be deflected slightly as needed. A typical universal grid tester configuration might contain 32,000 to 64,000 pins in a 16 × 20 inch grid and cost $250,000 to $750,000.

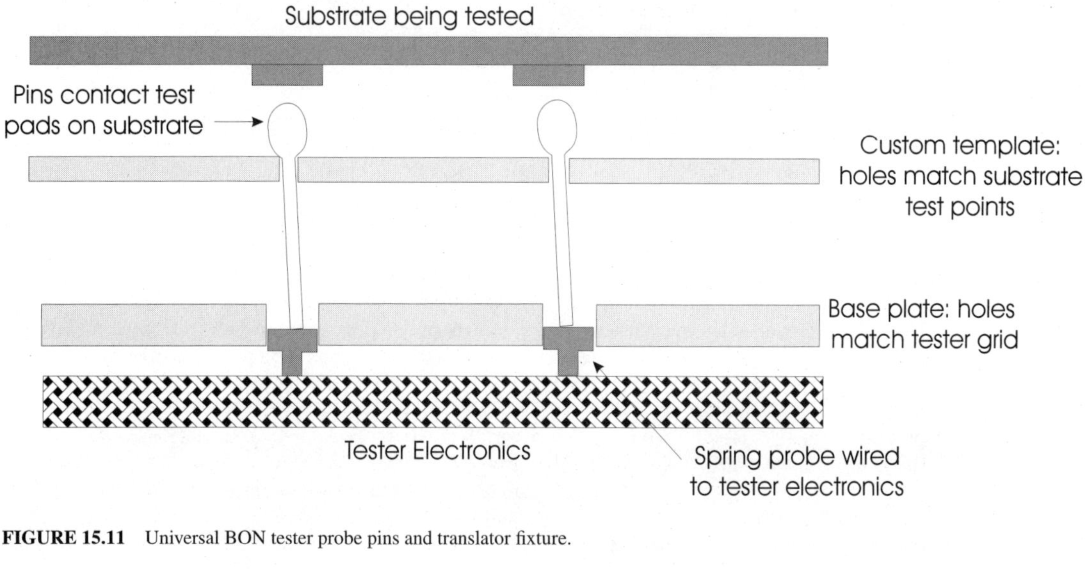

FIGURE 15.11 Universal BON tester probe pins and translator fixture.

A *dedicated* or *fixed-grid* tester contains a much smaller number of pins, typically 5,000 to 8,000, which are arranged on a special test fixture in the exact pattern of the test points on the surface to be tested. Each pin is a spring probe in which a probe shaft moves up and down in a barrel to allow it to adapt to the contour of the board or substrate. Because there are considerably fewer pins and associated pin electronics, dedicated testers are less expensive ($50,000 to $125,000) than universal grid testers. However, universal grid testers are more easily adapted to new boards, whereas dedicated testers require a custom test fixture to be developed for each new board, at a typical cost of $4,000 to $5,000. This is especially burdensome during substrate design, where engineering changes may require changes to the test fixture that result in delays and additional fixture costs.

Bed of nails testers are severely limited in achievable pin densities. The ability to develop a test grid with pitches comparable to test points on MCM substrates is very limited, both in terms of pitch [spacing between test points (15–20 mils)] and diameters of the test pins (12–16 mils). Typical universal grid testers might have 100 pins per square inch, with 100 mil spacing. Consequently, the use of bed of nails testers in MCM substrate testing is rapidly decreasing as feature sizes are reduced. Registration of substrate test point patterns to holes in drilled test fixture and pins is also difficult, especially in dedicated testers where pins are subject to a small amount of lateral movement.

Another significant concern about bed of nails testers is the mechanical force exerted by each probe on the substrate, which may be in the range of 50–100 g per contact and may damage delicate circuit traces. Research on probe design is striving to reduce these contact forces.

Variations of the traditional bed of nails testers are being developed to achieve higher pin densities. The most promising use membrane probes in which solder bumps are fabricated on a membrane in the pattern of the substrate test points. The membrane can then be placed on the substrate, conforming to the surface of the substrate with the solder bumps making contact with the test points. Electrical signals are then applied through the solder bumps in the same manner as in a bed of nails tester to perform resistance and/or capacitance measurements.

Flying Probe Testers. Because bed of nails pin-to-pin spacing is unable to keep pace with shrinking line widths and spacing on MCM substrates, and because of the high cost of bed of nails testers and test fixtures, the use of flying probe technology in MCM substrate testing is increasing.[6,7,38–42] A moving or "flying" probe is a wire that is mechanically positioned on a test point on the surface of a substrate. Two probes are typically used to perform resistance, capacitance, and other tests on the surface of the substrate, with some flying probe testers providing two pairs of probes to test both sides of a substrate or board without reloading the part.

Figure 15.12 is a photograph of the BSL LinearProbe tester,[41] which uses two flying probes for each surface of the substrate. Figure 15.13 is a photograph of the cantilever beam probes on the BSL LinearProbe, which are similar to the probes used in wafer testing. The probes are positioned under computer control by high-precision linear motors in the x,y dimension and then lowered in the z dimension by a stepper motor until contact is made with the pad. An optical system is often used to locate a reference mark on the substrate for the x,y coordinate system.

Solid contact with the pad must be made to perform accurate electrical measurements, and yet minimal contact pressure is desirable to minimize damage to the pads. A sensor in each probe detects and reports when contact has been made, and the pressure of the probe on the pad is programmable so that the pressure can be optimized for each application. The use of sensors also allows the probe to easily compensate for variations in planarity across the surface of the board.

The tester control computer processes an ASCII netlist description of the substrate, or some other layout information from the CAD database, to extract a set of x,y coordinates for the test pads. The control computer then determines the order in which the pads should be probed to allow testing to be performed with the least amount of probe movement. Resolution is typically much higher than in bed of nails testers. For example, the resolution of the probes in the BSL LinearProbe is 5 μm, with a probe-to-probe center of 75 μm.[41] The Integri-Test Micro-Prober 90 is capable of a resolution of 0.25 mil over substrate sizes that can range from 10×10 up to 24×24.[43]

The most straightforward method of testing for an open-circuited net is to probe two end points on the net and verify continuity between them by a resistance measurement. Alternatively, capacitance or rise time measurements can be made with one of the probes fixed at a reference point and

FIGURE 15.12 BSL LinearProbe flying probe tester.[41]

FIGURE 15.13 BSL LinearProbe cantilever beam probes.[41]

the other moved from pad to pad to make measurements. The BSL LinearProbe performs capacitance tests by applying short pulses with a probe and then measuring the rise time on the trace while the second probe is left stationary on an external ground plane or other reference point within the substrate.[41] In this manner, *N* probe movements are sufficient to detect opens on an *N*-net substrate. Capacitance testing also provides some indications of short circuits. However, these are usually verified by resistance measurements. If only resistance testing is used, each pair of nets must be probed to detect shorts; an N-net substrate would thus require a total of $N(N-1)/2$ probe movements and measurements.

Because probe positions are programmable, flying probe testers are more versatile than bed of nails testers in which test pins are configured into a permanent configuration. Flying probe technology also allows higher resolutions than bed of nails fixtures. However, mechanical probe movement is needed during testing, slowing down the speed at which multiple tests can be performed and limiting total throughput. Typically, 100–300 tests per minute can be performed by flying probes. This a significant limitation during high-volume production testing of complex MCMs.

Another concern with flying probe testers is that a test point might be probed multiple times during testing, resulting in wear and tear on test pads and circuit traces, whereas in bed of nails testing the pads are contacted only once.

15.5.3 Noncontact Methods

Given the expense and limited resolutions of bed of nails testers and the limited speeds of flying probes, alternative test methods are being investigated for MCM substrate applications. In addition, a mechanical probe can easily scratch or otherwise damage fragile pads and traces on an MCM substrate. Some flying probe testers allow contact pressure to be programmed to acceptably low levels. However, flying probes limit testing speed, because probes must be mechanically moved between test points.

Contactless test methods address these limitations. Because there is no mechanical contact with the test surface, the potential for damage to fragile substrates is virtually eliminated. In addition, computer-controlled movement of a test beam of charge or light provides the flexibility of moving probe testers and can be performed at much higher speeds to allow test speeds to approach those of bed of nails testers.

Electron Beam Testers. Electron microscope technology has been used for many years to perform nondestructive, noninvasive tests of VLSI circuits.[44] A scanning electron microscope scans the surface of a powered IC with a focused electron beam, making voltage measurements at selected test points. These measurements are organized into a profile of voltage values across the surface that is compared to that of a known good IC. Faults are indicated by locations in the profile at which voltages differ between the test and known-good ICs.

Electron beam IC testers have been designed to probe test points on the surface of relatively small IC chips whose features may be a fraction of a micron. Voltage-contrast electronic beam (VCEB) technology has recently been adapted to testing for opens and shorts on larger surface area fine-pitch bare circuit boards and MCM substrates.[38,44–46]

In MCM substrate testing, one electron beam is used to charge test pads on individual circuit traces to desired test voltages, and a second beam is used to read the charge on that and other circuit traces. The presence or absence of charge at a test point indicates a short or open circuit condition, respectively, between the initially charged pad and the test point.

A basic MCM substrate test setup is illustrated in Fig. 15.14*a*. Using scanning electron microscope technology, a thermionic electron gun generates a beam of electrons that are accelerated toward the test surface as illustrated in Fig. 15.14*b*. A blanking system switches the current density in the beam between values needed for charging a net and reading the charge on a net. The beam is then focused and deflected toward a test point on the substrate in a manner similar to that used to illuminate a picture cell on the face of a CRT. A computer controls the deflection, based on test point locations extracted from a CAD layout database, in much the same way as test points are located in flying probe testers. Test programs easily adapt to design changes without expensive hardware modifications to the test system.

A secondary electron energy analyzer completes the test setup, measuring voltages at probed test points. Voltage measurements in IC and bare circuit board or MCM substrate testing are made in a similar manner. A low-energy electron beam, called the *primary beam*, is focused on a test point on

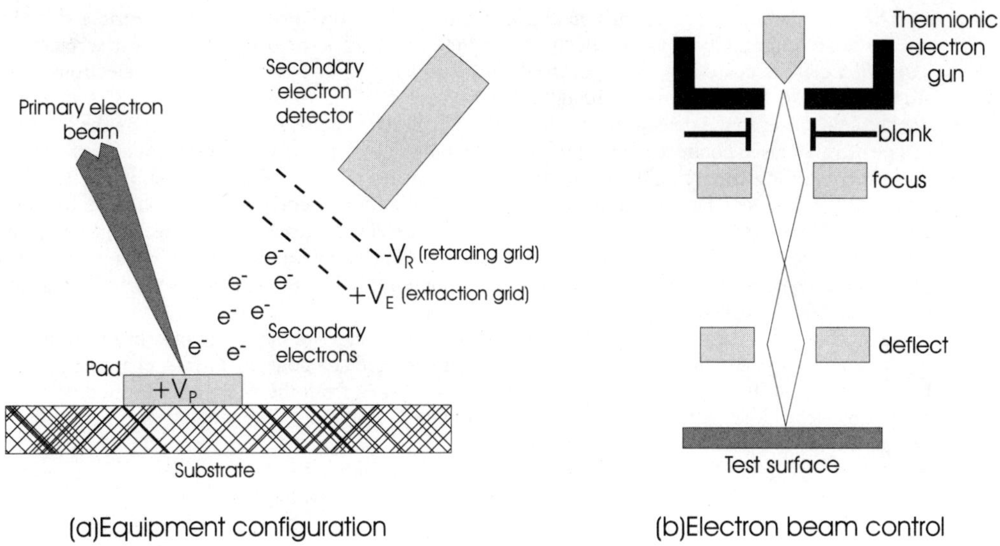

FIGURE 15.14 Voltage contrast electron beam test setup.

the surface that is charged to some voltage potential, V_P, as illustrated in Fig. 15.14*a*. Where the beam impinges on the test point, low-energy secondary electrons are released from the surface. The energy distribution function of the released secondary electrons is a function of the voltage potential of the material at the test point. When a voltage is applied to a test point, the secondary electrons gain or lose a proportional amount of energy as illustrated in Fig. 14.15.[44] This figure shows a 5 eV negative shift in the energy distribution function as V_P changes from 0 V to +5 V. This voltage contrast is the parameter measured by the test system.

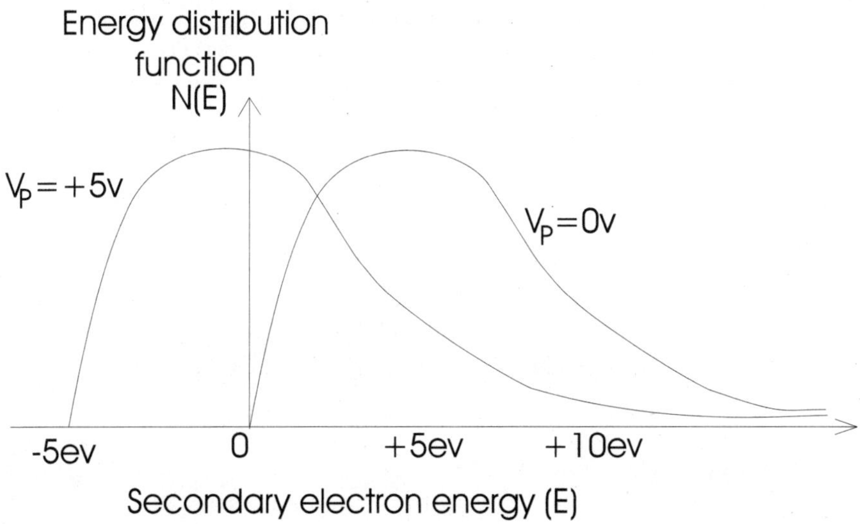

FIGURE 15.15 Secondary electron energy distribution function.[44]

The secondary electron energy analyzer, positioned nearby, produces two electric fields that affect the secondary electrons, as illustrated in Fig. 15.14*a*. The field nearest the test surface (the *extraction grid*) is positive with respect to the test surface, i.e., $V_E > V_P$, accelerating electrons away from the test surface and toward the detector. The second field (the *retarding grid*) is a retarding energy field closer to the detector and is biased at a negative voltage. If the test point on the substrate is a positively biased conductor, the low-energy secondary electrons have insufficient energy to overcome the retarding energy field and are turned back toward the surface. In contrast, if the test point potential is 0 V, secondary electrons are released with sufficient energy to overcome the retarding energy field and impinge on the detector. The number of electrons reaching the detector is therefore a function of the electron energy distribution which, in turn, is determined by the voltage at the test point. Since substrate testing is typically concerned only with the presence or absence of charge at a test point, a binary decision can be made by the detector.

In VLSI testing, the IC is powered, producing known voltages at the test points. In substrate testing, nets are charged by a high-energy electron beam directed at a node on a net. A charging electron beam contains sufficient energy that more primary electrons from the beam are captured than secondary electrons lost, resulting in the retention of charge in the net. The amount of time needed to charge the net is a function of the capacitance of the net and the charging current in the beam. For example, approximately 15 ms is required to charge a net with a 20 pF capacitance to 30 V, using a charging current of 4×10^8 amps.[46] After charging a test point, voltage measurements are made at each of the other test points associated with that net to check for open circuits. Charge will normally remain on the net long enough to allow a number of measurements to be made. Charge should have traveled through the conducting trace to each of these points. The absence of charge at a test point indicates an open circuit between that point and the pad that was charged.

To test for shorts, one test point is probed on each of the other nets on the substrate, which should be electrically isolated from the net containing the charged test point. Therefore, the charge at any point on these other nets should be negligible. The presence of charge at one of these test points indicates the presence of a conducting path, i.e., a short circuit, between the net containing the charged test point and the net containing the probed point. While the ideal case is an open circuit between two unconnected nets, in reality there will be some large resistance between them. Consequently, the disjoint net will still charge, although at a slower rate than a directly connected net.

Referring to the MCM substrate illustrated in Fig. 15.16, a typical MCM substrate test process would proceed as follows:

1. Select test point (1) and charge it to voltage V_P.
2. Probe nodes (2) and (3) to verify the presence of a voltage approximately equal to V_P along the net. The voltage at (2) will be significantly less than V_P, indicating the absence of charge at that node, and thus an open circuit between nodes and (2).

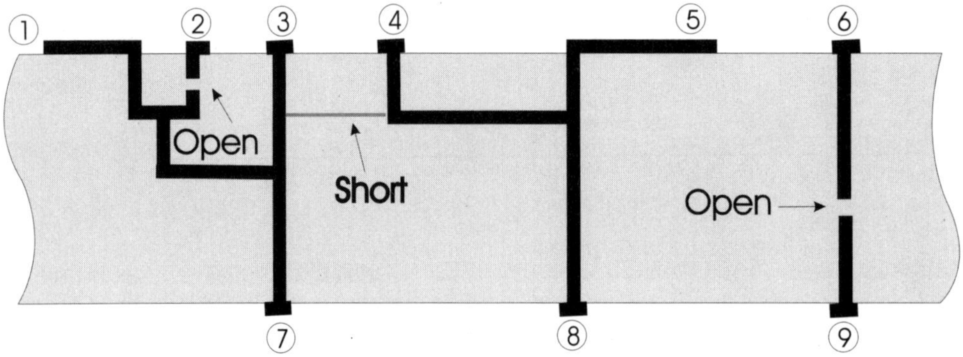

FIGURE 15.16 Electron beam probe points on a substrate.

3. Probe nodes (4) and (6) to verify the absence of charge on their respective nets. The presence of charge at node (4) indicates the presence of a conducting path, i.e., a short circuit, between the nets containing nodes (1) and (4).
4. Flood the substrate surface with electrons to return all traces to initial uncharged states.
5. Repeat steps 1–4 for each additional trace on the substrate.

After all nets on the substrate surface have been tested for opens and shorts, connections to the bottom of the substrate can be tested for open circuits by flooding the bottom of the substrate with electrons and verifying the presence of charge at connected test points on the top surface. For the substrate in Fig. 15.16, probing nodes (3) and (4) will verify continuity to points (7) and (8), respectively, but there will be no charge at node (6), indicating an open circuit between nodes (6) and (9).

Note that for a single charging operation, both open and short tests can be performed by the measuring beam. For a charged net with n test nodes, $n - 1$ probes of that net are needed to detect opens. Given N total nets on the substrate, one probe of each of the other $N - 1$ nets can detect shorts. The total test time is therefore

$$T = \sum_{i=1}^{N} \left\{ T_{charge} + T_{probe} \times \left[(i-1) + \sum_{k-1}^{n_k} (k-1) \right] \right\} \tag{2.17}$$

where T_{charge} = charging time
n_k = number of test nodes on net k
N = total number of isolated nets on the substrate

An electron beam can be focused on a smaller (submicron) area than can be accessed by a mechanical probe and is therefore an attractive technology for MCM testing applications. While sharing the flexibility of flying-probe testers in that test points can be derived automatically from the layout netlist, electron beams are significantly faster than flying probes and avoid mechanical contact with the substrate, resulting in less physical damage to fragile substrates during testing. Likewise, electron beam testing is noninvasive, i.e., the probe does not load down or otherwise affect the electrical characteristics of the material being probed. Electron beams also work well with surfaces that are not perfectly planar, adjusting to slight variations in height of the test surface. Finally, electronic beam testers can be used in vacuum of near-vacuum conditions that help reduce particle contamination of a substrate as compared to mechanical probe testers. This can be especially critical with small feature sizes.

Mechanical testers have high overhead associated with probing a large number of test points. Bed of nails testers are complex to set up, and flying probes are relatively slow in moving from point to point. While charge time can be long in electron beam testing, many voltage measurements can be made by probing beams after a single charging operation. The time needed to deflect a probing beam and make a measurement is considerably smaller than the charging time, making the overall set of measurements relatively fast.

Electron beam testers are not without drawbacks. They are still very expensive. In addition, it is more difficult to design a deflection system that can scan the large area of a circuit board or paneled MCM substrate than to scan a single IC. Placement of the detector is also difficult for large substrates and boards. However, test equipment manufacturers such as MCC, in collaboration with Hughes Research Lab[45] and Siemens,[46] are working on these problems and have begun to produce affordable equipment with acceptable performance for MCM substrates.

15.5.4 Automatic Optical Inspection

Visual inspection is used in one form or another in place of, or in addition to, electrical testing in the manufacture of printed circuit boards and MCMs. In particular, visual inspection is used to locate defects on bare boards and MCM substrates, defective wire bonds and solder joints, and other electronics manufacturing anomalies.

With rapidly shrinking feature sizes and increasing number and density of features on a single board, many printed circuit board manufacturers have taken advantage of automatic optical inspection (AOI) technology to identify many manufacturing defects. The same technology is being applied to MCM substrates, although the greatly reduced feature sizes and high interconnect densities of MCM substrates have presented significant challenges to the capabilities of AOI systems.[36,38,47,48] Fortunately, the availability of affordable high-performance computers and the development of efficient pattern recognition algorithms has begun to make AOI practical in MCM manufacturing environments. In addition, continuing improvements in video camera technology and image processing algorithms have made it possible to accurately capture images with very high resolution.

A significant advantage of AOI is that it can be applied during normal processing to individual substrate layers where electrical testing would not be practical or possible. In the early stages of processing, substrate layers are especially susceptible to damage from mechanical probes. In some cases, such as in MCM-C, circuit traces are poor conductors at these stages, making electrical tests difficult. AOI performed at early stages allows defects to be detected while repair is still possible, and thus reduces scrap. AOI also pinpoints the locations of defects and thus reduces repair time, whereas electrical testing can only identify that defects such as opens and shorts are present. Further inspection would be required to pinpoint their locations before repair can take place.

Finally, early detection of defects and in particular recurring defects can identify problems in the manufacturing process before there has been significant waste. In addition, substrate layers in MCM-C and MCM-L technologies are typically processed in parallel. This allows inspections to also take place in parallel. In contrast, MCM-D substrates must be processed and inspected serially. Inspecting individual substrate layers allows detection and elimination of individual defective substrate layers. This prevents defective layers from being combined with good layers later in the process and thus prevents good layers from being wasted, improving substrate yield.

A typical AOI test system hardware/software configuration is illustrated in Fig. 15.17. The test substrate is positioned, and then it surface is illuminated by a special high-intensity light source. A camera with high-quality lenses is focused on all or a portion of the substrate. Light from the image is sampled by optical sensors and transformed by analog to digital conversion into a matrix of grayscale data samples and stored in the memory of a computer. Each data sample is referred to as a picture cell or *pixel*.

The captured image is processed by the host computer in several steps. First, image processing algorithms are used to clean up and enhance the image, and then the pixels are converted to binary, black-and-white format. One level represents conducting material and the other represents spaces between conductors. Preprocessing of the image is critical because of the poor contrast between conductor and substrate materials in some MCM technologies. Such optical properties as color, transparency, and reflectivity must be handled via intelligent signal processing. MCM-D substrates present even greater challenges, because layers are processed and inspected serially. Some layers are transparent, allowing lower layers to be seen through upper ones, making it difficult to distinguish features on the top layer from those on lower layers.

Because of the large number of pixels in a single image, most AOI systems compress image data into a set of features rather than checking the image pixel by pixel.[48] *Feature extraction* is the process of locating and defining patterns of pixels in the array corresponding to circuit traces and other significant features. The result of this operation is that the image is represented by a feature set that is significantly smaller than the pixel array and therefore can be processed much more quickly during the pattern recognition phase.

Following feature extraction, the set of circuit traces is examined by pattern recognition algorithms to locate defects. Defects are identified in two ways: design rule checking and comparison to a reference image. Design rules for a particular substrate technology dictate minimum line widths, spacing between conducting lines, via hole diameters, annular ring widths, and so on. Design rules are specified for a given technology independently of the actual design and can thus be processed in the same manner for all substrates fabricated with that technology.

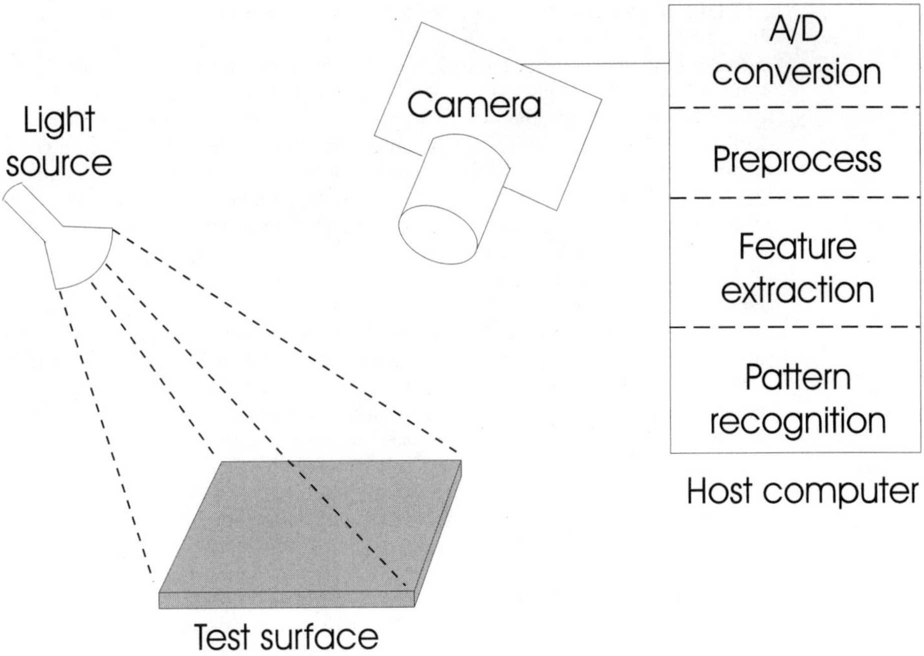

FIGURE 15.17 Automatic optical inspection system.

The detection of other defects requires comparison of the features of the acquired image to those of a *reference substrate* that is stored in the database of the host computer. A reference image is generated either by capturing and processing an image of a known-perfect substrate or by extracting features from the substrate layout data in the CAD database. Mismatches between the acquired and reference images are identified as faults. The result of the pattern matching operation are typically listed on an operator console or printed in a report, identifying the exact location and type of each defect. This information can then be passed along to inspection and repair/rework stations, allowing each defect to be quickly located and repaired. Typical defects detected by AOI software include:

- Opens
- Minimum line width violation
- Shorts
- Minimum spacing violation
- Cracks
- Minimum annular ring violation
- Mouse bites
- Protrusions
- Pin holes
- Islands (spurious material)
- Missing features
- Added features
- Misregistration
- Clogged via holes

AOI is limited by the resolution of the imaging system to defects whose sizes are greater than the pixel size. For example, the Orbot Models PT700 and PT716 systems have variable resolution between 0.25 and 1 mil, which implies that the smallest pixel size is 0.25 mil and the largest is 1 mil.[48] Current MCM projects are working with 4 mil and smaller lines and spaces. Therefore, AOI system resolutions must continue to decrease.

Pixel sizes are also limited by the ability to evenly illuminate the entire surface of the substrate and the ability of the optics in the imaging system to focus the image, which in turn is a function of the distance between the camera and the substrate (depth of field). AOI is also dependent on the ability to distinguish conducting traces from the substrate surface. The contrast between traces and surfaces differs between MCM technologies.

The size of the defect that can be identified is limited by the resolution of the video system, as measured by the pixel size. The selection of a pixel size is based on physical characteristics of the imaging system and consideration of speed versus precision trade-offs. As the pixel size is reduced, measurements are more precise, and smaller features can be recognized, including smaller defects.

However, as pixel size is reduced, more pixels must be stored and processed for a given image, increasing the time required to process each substrate. In general, the processing time is inversely proportional to the square of the pixel size. As an example, if the pixel size is reduced by a factor of 4 from 1 to 0.25 mil, the number of pixels increases by a factor of 16. Orbot has reported inspection times of 1 minute for a 9 in^2 board to 3.2 minutes for a 25 in^2 board with a pixel size of 1 µm.[48] These inspection times are reduced by a factor of approximately 10 when the pixel size is reduced to 2.5 µm.

Being a noncontact inspection method, AOI is a nondestructive testing method, i.e., there is no damage to the substrate. Since there is no mechanical movement, AOI is relatively fast compared to flying probes and electron-beam methods. AOI is computationally intensive; therefore, it is limited by the speed of the host computer and the efficiency of the pattern recognition algorithms executed. Typical are the Orbot Model PT700, which processes 5 ft^2 per minute, and Model PT716, which processes 16 ft^2 per minute.

A considerable advantage of AOI is that it can detect many defects that are extremely difficult to detect via electrical testing. Most latent opens in circuit traces, including mouse bites, cracks, and pinholes, cannot be detected by simple resistance and capacitance measurements, since a good conducting path exists throughout the length of the circuit trace. More complex tests are needed to detect these defects. However, AOI methods detect these faults routinely.

It is also impossible to detect islands of conducting material between circuit traces using electrical tests. Electrical tests are based on probing the substrate at predefined test points that are normally located on circuit traces. Detection of islands of material between traces would require probing the substrate at many points between traces, which is impractical because it would greatly lengthen test times.

There is no fixturing cost in AOI. Design rules are specified independent of the layout, including minimum line widths and separations, minimum annular ring widths, and so forth. The reference image may be captured from a known-perfect substrate or derived directly from the layout in the CAD database.

AOI has several shortcomings that are not characteristic of electrical testing. The primary limitation is that inspection is limited to visible areas on the surface of the substrate. Inner layers cannot be inspected, nor can the walls of via holes. Multilayer substrates must be tested one layer at a time, with electrical testing used once the layers have been combined to check the finished substrate assembly.

15.6 MCM ASSEMBLY TESTING

The MCM assembly process comprises placing die on a substrate, making electrical connections between pads on the die and the substrate, and performing various finishing operations. The assembled MCM must be tested for structural integrity and for correct functionality. Structural faults are those specifically related to the assembly process, with structural tests performed prior to encapsulating the MCM so that rework or repair can be performed. Testing is completed by verifying the overall functionality of the MCM, allowing die, substrate, or interconnection faults that escape earlier tests to be detected. Full functional testing is more complex and costly than structural testing, and it requires more work in test design, including generation and verification of test vectors.

MCM test strategies are often dictated by available test equipment, testability design features incorporated into the MCM, and potential for rework or repair.[21–23,25–28,34,49] The electronics industry has traditionally developed test procedures and equipment at three levels: single-chip packaged integrated circuit (IC), printed circuit board (PCB), and system. PCB and IC test equipment and methods are mature and relatively cost effective. ICs are tested by applying test vectors to their external input pins and observing their external outputs. Test points internal to an IC are generally not accessible. PCBs contain many single- and multichip packages and discrete components. PCBs are tested by probing test points on the board and pins in external connectors to provide stimulus or observe responses. If test equipment does not allow surface probing, then access to PCBs is restricted to external connectors, thereby limiting access to test points.

System testing involves installing MCMs into a host system, simulating normal operating environments, and exercising the system as a whole. Some manufacturers consider MCM testing too difficult or too costly and bypass testing of individual MCMs, relying on system testing to check out these modules within the context of their host circuit board or system.[50] For example, Intel Corporation tested an experimental Pentium processor MCM by plugging it into its host mother board and verifying that it would run Microsoft Windows and Flight Simulator programs.[51]

MCMs and other new electronic packaging technologies have challenged the test industry to develop new test methods and equipment for finished assemblies. Because of the number of components, MCMs are as complex as PCBs, suggesting the use of PCB test methods. However, reduced dimensions and new die attach methods have made internal test points virtually inaccessible, restricting test access to external I/O pins, suggesting the use of IC test methods.[22,23]

15.6.1 MCM Interconnect Testing

The primary objective of MCM structural integrity testing is to verify the presence or absence of defects introduced during die attach and interconnection processes by verifying electrical connections between ICs and substrates (wire bonds, flip chip solder joints, and so on). Where rework or repair is possible, testing is extended to diagnose individual faults. Interconnection problems include faulty wire bond or solder joint, shorts, damaged or open landing pads, and others. Additional assembly faults include wrong location or orientation of a part and damage to die or substrate during handling. Testing for structural faults can often be done without exercising the functionality of the MCM.

From the viewpoint of a tester, there are three types of assembly interconnections to be verified, as illustrated in Fig. 15.18.

1. Connections from IC outputs to external MCM pins
2. Connections from external MCM pins to IC input
3. Connections internal to the MCM between IC inputs and outputs

Internal connections between IC pins are generally not connected to external MCM pins. External MCM pins are directly controllable and/or observable by the tester. Internal interconnections, however, present a significant challenge since they can only be controlled and observed indirectly via the inputs and outputs of the ICs on the MCM.

Consider the two ICs wire-bonded to net N_1 in Fig. 15.19. Circuits on IC_1 drive net N_1 and circuits on IC_2 receive the signal. To test wire bonds W_1 and W_2 requires checking for electrical continuity between the driver in IC_1 and the receiver in IC_2 by verifying that IC_1 can properly drive N_1 and that IC_2 can receive correct values from N_1.

If N_1 is also connected to an external MCM I/O pin, then its logic state can be controlled and/or observed directly by the tester. However, to verify wire bond W_1, IC_1 must be made to drive N_1 to a

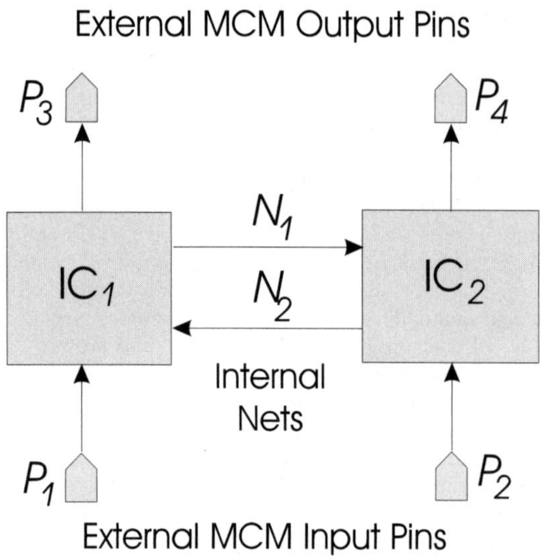

FIGURE 15.18 MCM assembly electrical connections.

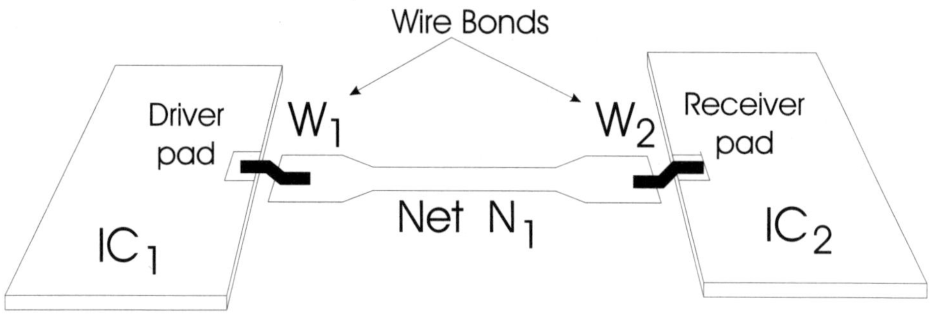

FIGURE 15.19 Testing wire bond connections.

known logic state that can be observed by probing N_1 with a test pin or propagating the value through IC_2. To verify wire bond W_2, N_1 must be forced to a known logic state, by IC_1 or a test pin, and propagated through IC_2 to some observable output.

If it is not necessary to determine which wire bond is faulty, W_1 and W_2 can be tested simultaneously by forcing IC_1 to drive N_1 to a known logic state and propagating the value of N_1 through IC_2 to an observable output. This requires IC_1 and IC_2 to be controlled concurrently, which may be difficult within a complex MCM.

Later in this chapter, we will examine design for testability methods that improve controllability and observability of nets within an MCM.

15.6.2 In-Circuit Test Approaches

Printed circuit board test methods and equipment have matured over the years.[6,7,49] PCB testing can be divided into surface probing methods, edge connector methods, and in-system tests, illustrated in Fig. 15.20. Surface probing methods utilize bed of nails or flying probe test equipment to gain access

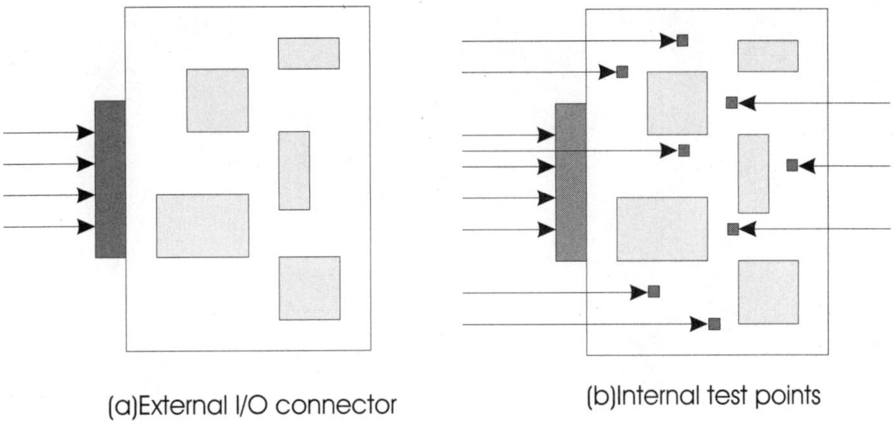

FIGURE 15.20 Access to PCB test points.

to internal nodes of the board to perform in-circuit testing. With proper design, in-circuit testing may also be performed from the external I/O pins of the MCM. Alternatively, external I/O pins may be used to perform limited functional testing of the MCM. The full functionality of an MCM is typically verified via in-system testing, in which all of the MCM's functions are exercised within the context of the MCM's host system.

An *in-circuit test* exercises a component within its host MCM circuit in isolation from the other components on the MCM. Since each IC is tested in isolation from the others, test vectors supplied by the IC manufacturer are often used to exercise and validate the operation of the IC, in addition to ensuring that all IC pins are properly connected to their respective pads on the substrate.

Connections between IC pins and substrate pads are tested by verifying that each signal trace can be driven by the IC outputs connected to that trace, and that each IC input connected to the trace can properly receive a signal from the trace. An in-circuit test of a component within an MCM comprises four operations.

1. Electrically isolate the component from other components in the MCM.
2. Stimulate the inputs of the component with test vectors.
3. Capture the outputs of the component.
4. Compare the captured outputs to expected values.

A component is isolated by forcing its inputs to be controlled directly by the tester. The ideal situation for in-circuit test is for all ICs to drive nets through tristate buffers that can be disabled by the tester. The alternative is to overdrive (backdrive) nets, overriding the effects of the on-chip drivers. Referring to Fig. 15.21, when Q_1 is conducting it acts like a low resistance. A sufficiently high current forced onto its output via a test pin can raise voltage on the net above the logic 1 threshold. When Q_1 is off, the test pin must pull the line to ground to create a logic 0 state. PCB testers designed for in-circuit tests typically have the capability to overdrive signal lines. However, most ASIC testers do not, precluding the use of in-circuit testing if tristate drivers are not used on IC outputs.

Figure 15.22 illustrates a partial configuration for in-circuit testing of a digital IC. To test module *B*, test vectors must be applied to inputs I_1, I_2, and I_3. I_3 is driven from the external connector, but test vectors must be applied to I_1 and I_2 via probes that make contact with test points on those nets, with the outputs of modules *A* and *C* disabled. Outputs O_1, O_2, and O_3 of module *B* are captured by the tester. O_3 is accessed at the external connector, while O_1 and O_2 are sampled via probes that make contact with these nets. When finished, the test pin drivers are disabled and the outputs of modules*A* and *B* reenabled to resume normal operation.

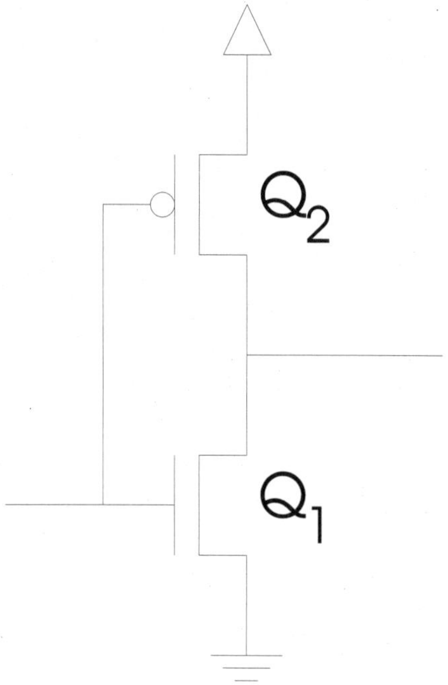

FIGURE 15.21 IC output driver.

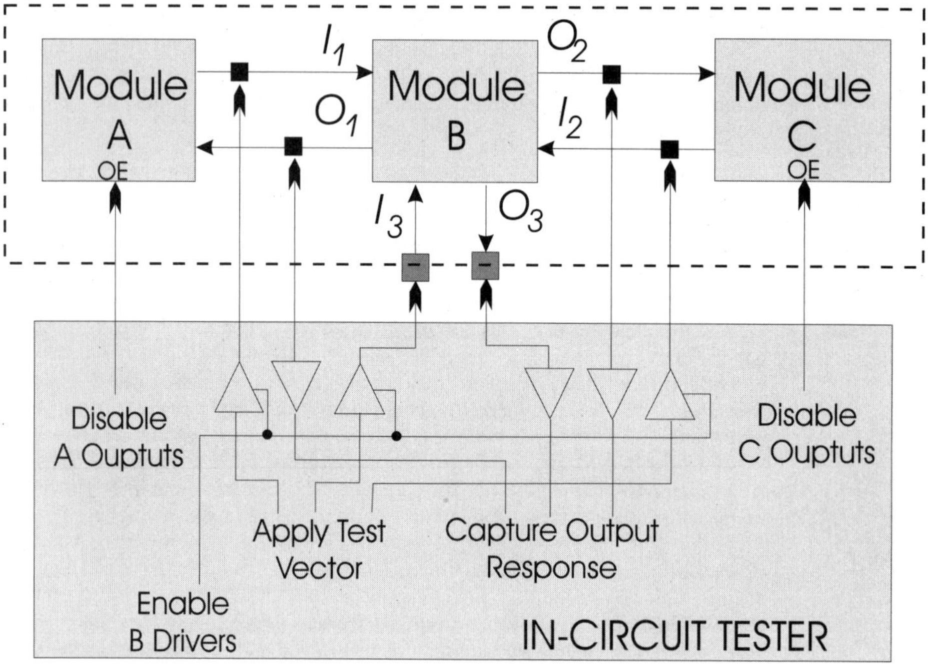

FIGURE 15.22 In-circuit test configuration.

The most difficult part of in-circuit testing for MCMs is gaining access to individual component inputs and outputs. This has been done routinely in PCBs with bed of nails or flying probe testers as described earlier in the substrate testing section. Test pins probe special test pads or circuit traces connected to IC pins without making contact with the pin itself, since pressure exerted by the probe could cause a faulty wire bond or solder joint to temporarily make connection with the pad and thus causing the test to pass. Reduced feature sizes and new die attachment methods have made surface probing a more significant problem in MCMs. This is especially true of flip chip technologies in which IC pads are soldered directly to substrate pads underneath the die.

An alternative to direct probing of the MCM surface is to incorporate extra pins to provide access to internal nodes. This is illustrated in Fig. 15.23, which shows the four internal nets I_1, I_2, O_1, and O_2, as well as the two output enable pins for IC_1 and IC_2 brought out to MCM pins, in addition to normal I/O signals I_3 and O_3. In this arrangement, the outputs of all ICs except the one under test must use tristate drivers that can be disabled by the tester so that it can drive the inputs to the isolated IC and capture its outputs.

Access to internal test points can also be increased by various design for testability (DFT) methods. In particular, boundary scan makes all IC pins controllable and observable from special external test pins. Boundary scan and other DFT methods will be discussed later in this chapter

In-circuit testing of analog components requires different approaches. The reader is referred to Ref. 49 for more information.

15.6.3 ASIC Approaches to MCM Assembly Testing

Because of high interconnect densities, PCB-oriented test equipment is often unable to probe interior test points within an MCM assembly. In such cases, all test vectors must be applied and all outputs observed via the I/O pins of the module, treating the MCM as a complex ASIC. Many manufacturers have adapted existing ASIC test equipment, used for single-chip packaged ICs, to MCM testing.

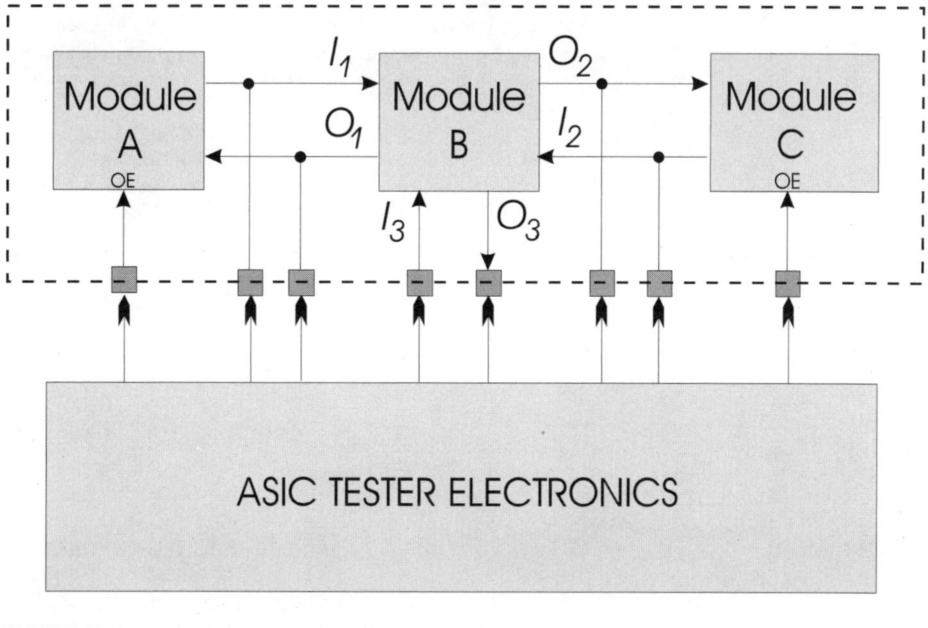

FIGURE 15.23 In-circuit test with an ASIC tester.

ASIC test equipment can be classified as comparison or stored pattern testers.[52] Both apply test vectors from a pattern memory to the device under test (DUT) and compare the DUT output responses to those of a correct device. A *comparison tester*, shown in Fig. 15.24*a*, applies test vectors concurrently to the DUT and a known-good device (KGD), and compares their responses. A *stored pattern* tester, shown in Fig. 15.24*b* applies test patterns to the DUT and compares its responses to expected patterns retrieved from a memory. The expected responses are usually generated via logic simulation, although in some cases expected patterns are obtained by exercising a known-good device and sampling its outputs.

ASIC tests are organized into test cycles. At the beginning of a typical test cycle, values are applied to logic (non-clock) inputs. After allowing for propagation delays and setup times, a clock is applied to trigger a state change. Outputs are sampled near the end of the cycle after allowing sufficient time for state changes to propagate to output pins. Many ASIC testers allow considerable flexibility in when inputs and clocks can be applied and when output sampling can take place within a test cycle.

Output responses are usually captured and stored in a memory for off-line error analysis. Most testers display input and output vectors and highlight differences between expected and actual outputs to assist test engineers in diagnosing the tested device. During production testing, it is often the case that only a "pass/fail" indication is needed. In such cases, a simple failure indicator is set whenever any output differs from its expected value.

One of the biggest drawbacks to using ASIC test approaches is obtaining test vectors and expected responses.[25,26,34] The most common method for generating test vectors and responses for an ASIC is through logic simulation of a model of the ASIC. Simulation results can usually be fed to the tester directly or through a translation program.[34] However, simulation models are often not available for all of the ICs used in an MCM. In these cases, tests must be designed by hand.

Limited Functional Testing for Structural Faults. If in-circuit testing is not practical, then functional testing can often be used to verify interconnections on an MCM without exercising its full functionality.[22,23,34] In limited functional testing, the application circuits on the MCM are exercised just enough to toggle all of their I/O pins and make the results observable to the tester, allowing interconnection faults to be detected.

Limited functional testing has been reported for a number of microprocessor- and microcontroller-based MCMs. Some of these are microprocessors with associated cache memory and bus control ICs,[53–57] and others are essentially single-board computers, comprising a microcontroller, programmable I/O circuits, and memory.[34]

The block diagram of Fig. 15.25 is that of an automotive engine control MCM, containing a microcontroller, programmable ASIC, and a flash memory.[34] An ASIC tester was used on this MCM with access limited to the external MCM I/O pins. Apart from power and ground, external MCM pins were connected to I/O pads on the microcontroller (μC) and the ASIC. Internal signal lines such as

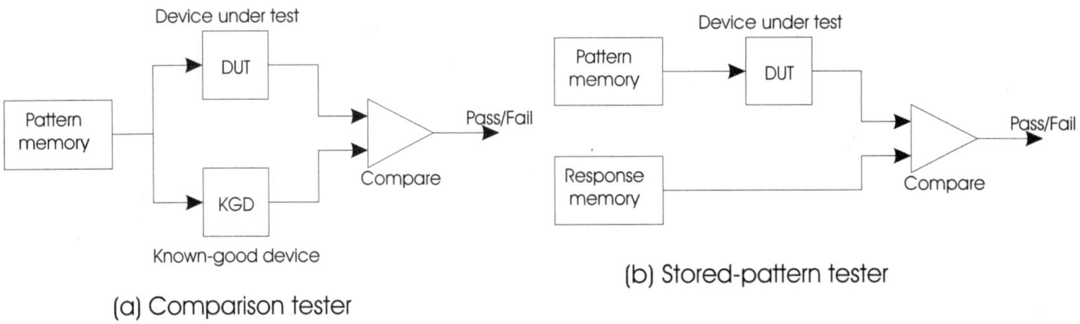

FIGURE 15.24 ASIC test equipment.

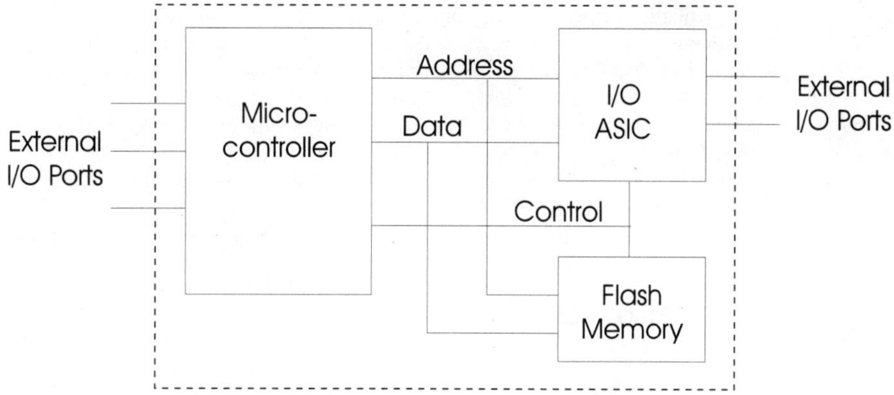

FIGURE 15.25 Microcontroller-based MCM.[34]

address and data buses and control signals used for data transfers between the μC and the ASIC and flash memory were not directly accessible by the tester. In addition, none of the components contained boundary scan or other testability features, nor could ICs be isolated via tristate drivers to perform in-circuit testing. Therefore, limited functional testing was the only available option.

In most microcontroller- or microprocessor-based MCMs, the on-board processor is able to control and observe most of the signal lines within the MCM, either directly or indirectly. Provisions are usually made in these MCMs to allow a short program to be downloaded from the tester into the processor's memory that can be executed by the processor to exercise itself and the other ICs on the MCM.[34] A typical test program would begin by exercising the processor itself and then the other components on the MCM.

For the MCM in Fig. 15.25, the test program performed five categories of tests to detect signal line stuck-at faults.[34]

1. The μC toggles its own outputs that drive external pins.
2. The μC samples its own inputs that are driven by external pins and writes the sampled values to one of its previously tested observable output ports.
3. The μC writes commands and data to the ASIC to toggle ASIC outputs that drive external pins.
4. The μC writes commands to the ASIC to sample ASIC inputs driven by external pins, reads the sample values from the ASIC, and writes them to a previously tested observable output port.
5. The μC writes commands and data to the flash memory, reads values back from the flash memory, and then writes them to a previously tested observable output port.

The test procedure comprised two stages: initializing and downloading the test program to the microcontroller, and execution of the test program by the microcontroller. By forcing one of its input pins to an initial state, this microcontroller can be made to bypass its application program and jump to a special boot loader program in ROM that reads test program code bytes via its serial communication interface, stores them in on-chip RAM, and transfers control to that program.

All multipurpose I/O ports on the microcontroller and ASIC were configured as general-purpose parallel outputs or inputs, and complementary values were written to them to toggle the corresponding MCM output pins. The address bus, data bus, and bus control lines interconnecting the microcontroller and the ASIC were tested indirectly during the process of reading and writing to the registers and I/O ports of the ASIC. Interrupt request lines from the ASIC to the μC and a few other special-function inputs/outputs require special sequences of instructions or tester actions, but detectable events associated with most of these signal lines can be created to complete the limited functional test.

Nonvolatile memory devices present unique testing problems in MCMs. The initial state of a flash memory must be assumed random, unless it can be initialized off-line before the MCM is tested. Flash memory erasure procedures require several milliseconds to complete, which often exceeds the number of available test clock cycles on high-speed testers. A simple test comprises sending commands to the memory device to read its built-in manufacturer and device identification codes, exercising many, but not all, of the bus interface between the μC and the memory.

The problems encountered in limited functional testing highlight the necessity and benefits of design for testability features, such as boundary scan, that would simplify interconnection testing for memory and other complex ICs without requiring complex and time-consuming test sequences.

15.7 DESIGN FOR TESTABILITY

Testability is an attribute of a circuit that determines the ability to create comprehensive, cost-effective tests for that circuit.[58] Testability measures the degree of difficulty associated with detecting the presence of a fault in a circuit, expressed in terms of controllability and observability of the signal lines to be tested. *Controllability* is a measure of the ease with which a signal line can be set to a desired logic state from the primary inputs of the circuit, and *observability* is a measure of the ease with which the logic state of a signal line can be propagated to an observable primary output.[3,5,59]

The high number and density of signal lines and the inability to directly access test points within a complex MCM greatly limit test effectiveness unless expensive, time-consuming procedures are used. Test costs and fault coverage can only be improved through the use of design for testability (DFT) methods applied to both ICs and MCM assembly designs. DFT improves signal line controllability and observability to increase fault coverage and reduces the cost of generating and applying tests. Unfortunately, DFT also increases MCM design time and cost and may degrade performance, discouraging designers from considering DFT from the beginning of a project. However, the negative effects of DFT are often outweighed by their benefits in test speed and fault coverage.[60,61]

DFT methods can be classified as ad hoc and structured. Ad hoc methods include the use of extra control and observation points brought out to external pins, partitioning a design to improve the testability of individual parts, the creation of test buses, isolating clocks from logic, and various built-in test methods. Structured methods allow systematic or automated methods to be used for test generation application, including boundary scan at the module level and internal scan design within ICs.

This section will examine ad hoc methods, boundary scan, and internal scan design methods. The following section will discuss special built-in self-test (BIST) features that incorporate test vector generation and test result analysis circuitry within ICs and modules.

15.7.1 Ad-Hoc Methods

There are a number of relatively simple methods for increasing testability of an MCM, with varying cost/performance trade-offs. A few of these methods require special provisions to be made in the ICs used in the MCM, while others can be implemented with standard off-the-shelf ICs.

If in-circuit test is to be used, the simplest way to improve controllability and observability of signal lines within an MCM is to provide the tester direct access to them. Where surface probing is used, this means adding special test pads on the surface with pad spacing consistent with test probes. Testing is further simplified if all test points are accessible on the same side of the substrate. Access is also improved by routing internal nets to external pins that can be accessed by the tester. Both options require sacrifices in MCM surface area and pin count that may not be acceptable.

In addition to making IC pins more accessible, control of IC pins can be simplified by including tristate drivers on all IC outputs that can be disabled during testing to allow the tester to control those lines.

A number of other design techniques are important to testability. These include isolation of clocks from other logic signals and minimizing the use of asynchronous logic to ensure well-behaved state transitions, and providing easy initialization of state elements.

Testability can also be improved by partitioning testable areas of the MCM into smaller modules. An exhaustive test of a combinational logic circuit with 32 inputs requires 2^{32} (about 4 billion) test vectors. If the circuit can be partitioned into two 16-input modules, then testing would require only 2^{16} (65,536) test vectors, assuming the blocks are tested concurrently.

15.7.2 MCM-Level Boundary Scan

The biggest obstacles to testing wire bonds and solder joints in dense circuit boards and MCM assemblies are associated with forcing individual ICs to drive nets to specified logic states and forcing other ICs to read these logic states in a way that can be verified. Simply increasing the number of test points or test pins does not solve these problems. At least a portion of the functionality of each IC must be exercised during testing, which is difficult when the MCM or board contains many complex ICs.

To address these problems, a number of manufacturers formed the Joint Test Advisory Group (JTAG), and in 1988 developed the JTAG Testability Bus Specification, which in 1990 was adopted as IEEE Standard 1149.1.[62–67] While developed initially for printed circuit boards, the Testability Bus is also being implemented in MCM designs.[12,68–71] Among the goals of the 1149.1 Testability Bus are

1. A standard interface between ATE systems and the devices on a PCB or MCM, allowing test data to be transferred to the devices and diagnostic information to be received from them.
2. A method for testing the interconnections between the ICs on a PCB or MCM without exercising the logic in the ICs. Tests are based only on the number of ICs and the types of pins on each IC.
3. A method for testing and locating individual faulty ICs.

These goals are accomplished by incorporating a programmable test controller within each IC and special logic at each I/O pin. The basic architecture of an IC containing a Testability Bus interface is shown in Fig. 15.26. Two special features are included on each module a *boundary scan register* (BSR) and a *test access port* (TAP).

The BSR is a chain of special register cells, positioned one per I/O pin around the boundary of the chip. Each cell contains a flip-flop placed between a circuit input/output line and the corresponding external pin as illustrated in Fig. 15.27. During normal operation, BSR cells are transparent to the chip, passing on-chip logic inputs and outputs to and from the I/O pins. During test operations, the BSR cells can drive IC inputs or outputs, or they can be configured into a serial shift register, called a *scan chain*, to transfer test data from the tester into the I/O cells via the dedicated TDI (*test data in*) pin, and from the I/O cells back to the tester via the TDO (*test data out*) pin.

The BSRs of the individual components on an MCM or circuit board are typically configured into a single scan chain by connecting the TDO pin of one device to the TDI pin of the next, as illustrated in Fig. 15.28. One TDI pin and one TDO pin are connected to the ATE system at the external connector. Through these pins and the scan chain, the ATE system can supply test instructions and data to all chips on the MCM or circuit board and receive test results and other diagnostic data from them.

Instructions from the ATE system are received and executed in each IC by the TAP controller and associated logic, illustrated in Fig. 15.29. The TAP controller has two input signals supplied by ATE system: TMS and TCK. TMS carries command bits broadcast from the ATE system to the ICs, synchronized to a clock signal on TCK. These commands cause each TAP to route incoming bits from the TDI pins to either an *instruction register* (IR) or one of several *data registers* (DRs). The latter include the boundary scan register, a bypass register, and optional user-defined registers. The *bypass register* is a one-bit data register that routes TDI directly to TDO within the IC, bypassing the on-chip BSR and thus temporarily eliminating that IC from the overall board or MCM shift register. The bypass registers allow an ATE system to isolate selected ICs during testing, bypassing the others.

Figure 13.30 presents the state diagram of the TAP controller, which shows how bit sequences on TMS control the shifting of bits into the IR and DR. Note that separate state sequences control the IR and DR. When loading the DR, the current instruction in the IR is decoded to select the BSR, the bypass register, or one of the user-defined registers to receive values from TDI and to drive TDO.

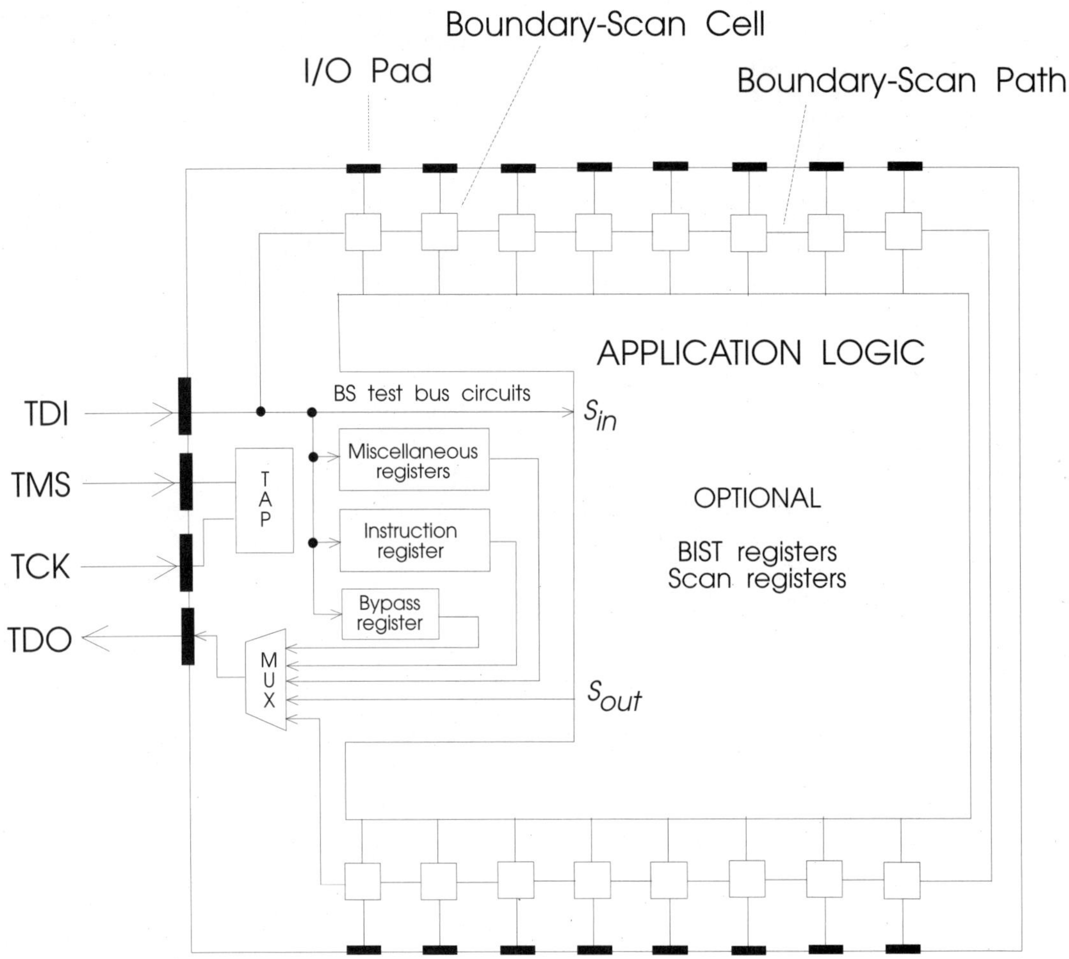

FIGURE 15.26 Circuit for use with 1149.1 testability bus.

Referring to Figs. 15.29 and 15.30, a typical test operation is performed as follows:

1. A bit sequence is broadcast on TMS to force all ICs to load their IRs with instruction bits from their TDI inputs.
2. A second bit sequence is broadcast on TMS to force the ICs to route a test data sequence from their TDI inputs to their BSRs.
3. Each TAP configures its IC according to the received test instruction.
4. The test instruction is executed, with results captured in the BSR.
5. Test results are shifted from the BSR out to the ATE system via the TDO outputs.

Two general types of test operations can be performed: IC-to-IC and IC-to-MCM pin interconnection tests, and tests of on-chip circuits internal to the ICs. Test vectors are shifted serially from the ATE system into the BSR, and test results are captured and shifted serially out of the BSR to the ATE system. The TAP controller is required to implement at least three test instructions: BYPASS,

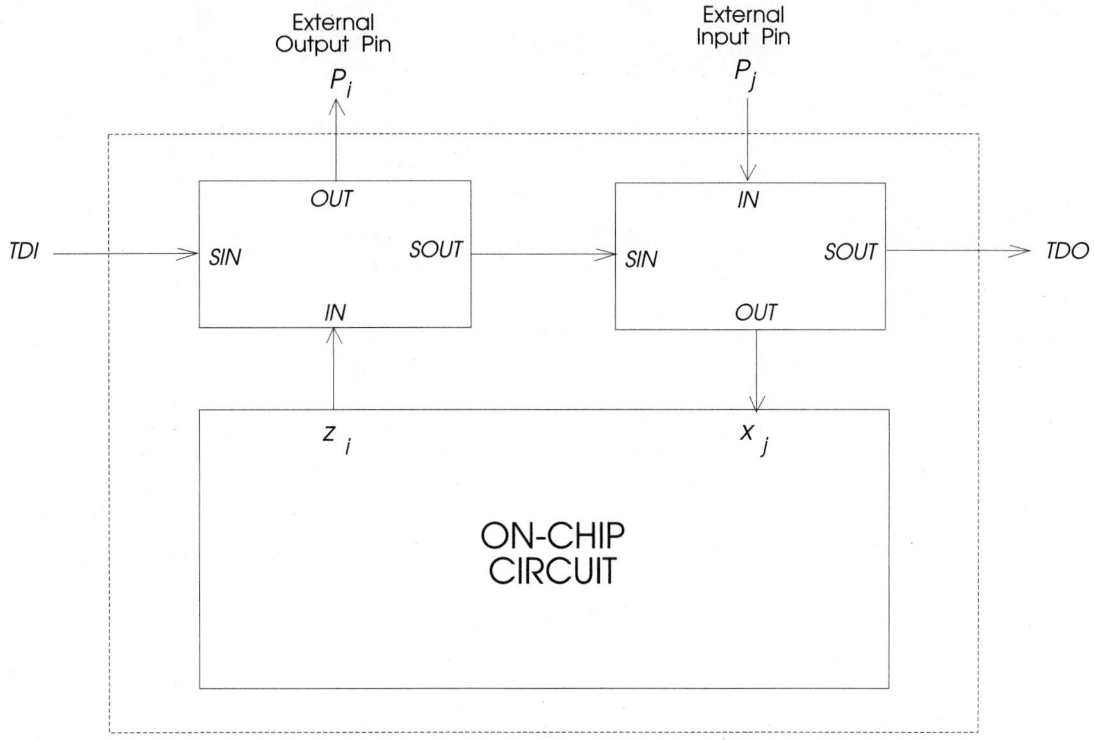

FIGURE 15.27 Boundary scan cells between circuit I/O lines and external pins.

EXTEST, and SAMPLE; it may also include a number of optional instructions such as INTEST and RUNBIST. These instructions are defined as follows:

BYPASS. Bypass the BSR by routing TDI directly to TDO as shown in Fig. 15.31*a*. This is the default power-on state of the instruction register.

EXTEST. Drive data from BSR cells onto IC output pins, and capture values in BSR cells from IC input pins as shown in Fig. 15.31*b*. This implements interconnection tests by broadcasting test vectors from IC outputs to other IC inputs and external MCM pins, where they are captured in the BSR and shifted out for verification.

SAMPLE. Capture in the BSR all IC inputs destined for on-chip logic inputs and all on-chip logic outputs destined for IC output pins as shown in Fig. 15.31*c*. This takes a snapshot of the current state of all I/O pins on the IC.

INTEST. Apply test vectors from the BSR to on-chip logic inputs and capture on-chip logic outputs in the BSR as shown in Fig. 15.31*d*. This performs a test of the internal logic of the IC without using its external pins.

RUNBIST. Initiate operation of on-chip BIST circuits. Initial data needed for BIST operations is shifted into special user-defined test registers, and test results are captured in special output registers that can be shifted out to the ATE system.[72]

EXTEST, SAMPLE, and INTEST. These all utilize the BSR. The logic diagram of a boundary scan cell is shown in Fig. 15.32. An IC input pin drives cell input IN, with cell output OUT driving the corresponding on-chip circuit input. An IC output pin is driven by cell output OUT, with input IN of that cell driven by the corresponding on-chip circuit output. To form the boundary scan chain, SOUT

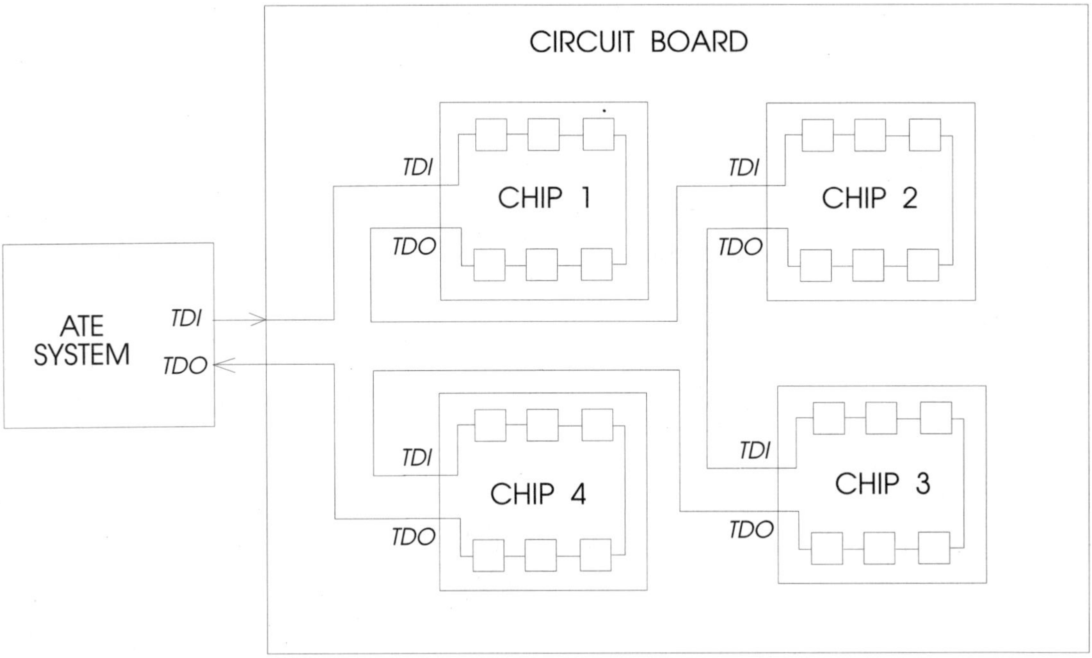

FIGURE 15.28 Circuit board configured as a single scan register.

of one cell is connected to SIN of the next, with TDI driving SIN of the first cell and SOUT of the last cell driving TDO.

The operation of each BSR cell is dictated by the cell control signals shown in Fig. 15.32. During normal circuit operation, *Mode-Ctrl* = 0, connecting IN directly to OUT, bypassing the registers within the BSR cell. During scan operations, *ShiftDR* = 1, causing SIN to be shifted to SOUT through the data register. For the EXTEST and INTEST instructions, *Mode-Ctrl* = 1, sending data from the shadow register to OUT. For EXTEST, SAMPLE, and INTEST, *ShiftDR = 0*, capturing the state of IN in the data register. *ClockDR* and *UpdateDR* are activated by the TAP controller to load the data and shadow registers, respectively.

The use of the EXTEST instruction for interconnect testing is illustrated by Fig. 15.33. In scan mode, test vectors are shifted into the output cells of chips 1 and 2. When EXTEST is executed, these values are moved to the corresponding shadow registers from which they drive internal net NET_1 and the external MCM output pad. Data applied to the external MCM input pad and the logic state of NET_1 are then captured in the data registers of the input cells in chips 1 and 2, respectively. Finally, scan mode is selected to shift the captured data out to the ATE system for analysis.

Boundary scan is not yet supported in many of the ICs used in MCMs, and therefore partial scan designs are often used, with the ICs supporting scan design organized into a scan chain.[12,28,65] Figure 15.34 illustrates how non-scan ICs fit into the test picture. The best option for non-scan ICs is the configuration of Fig. 15.34*a*, which allows in-circuit testing to be performed on the non-scan IC. The inputs of the non-scan IC are supplied with test vectors from the external MCM inputs and scan IC boundary cells, and its outputs drive external MCM pins or are captured in scan IC boundary cells. The configuration of Fig. 15.34*b* is less desirable since some of the IC inputs and outputs can be neither controlled nor observed directly from scan cells or external inputs.

Test development for boundary scan designs is supported by an increasing number of CAD tools. A special *Boundary Scan Description Language (BSDL)*, a subset of VHDL, has been developed that describes the testability features of components, MCMs and boards that comply with IEEE/ANSI

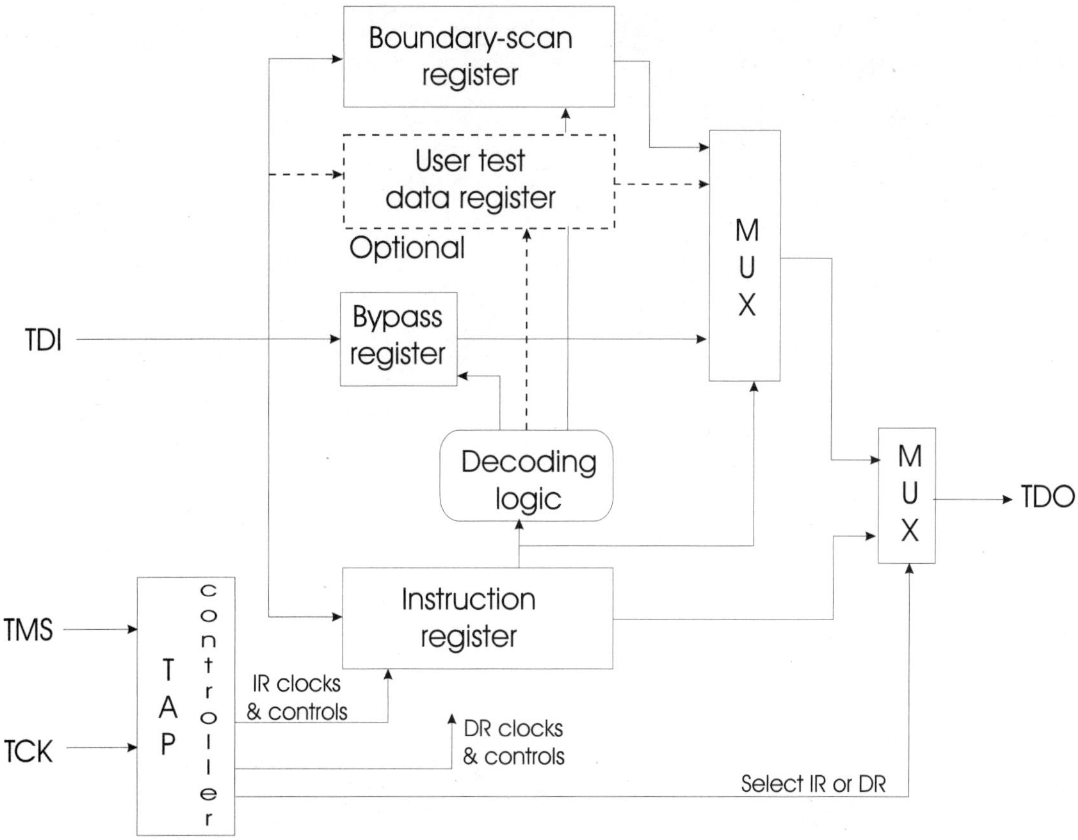

FIGURE 15.29 On-chip test bus circuitry.

Standard 1149.1. BSDL describes IC boundary scan and instruction registers, TAP inputs and outputs, and TAP instructions implemented in the controller. From a BSDL description, CAD tools can automatically generate test programs for MCMs and PCBs.

In addition to automatic test generation, synthesis tools, such as Mentor Graphics' *BSDAdvisor*, are available that automatically insert boundary scan logic into an IC design, represented by a BSDL description. This allows boundary scan to be added to a design with minimal design effort.

With the success of boundary scan in board and MCM testing, a number of related testability standards are also under development by IEEE/ANSI. These include:

1149.2 Extended Digital Serial Access Port. This is similar to 1149.1 but with no TAP controller. Extra control pins are added to each IC to select and control testing.

1149.4 Testability Bus: Real-Time Analog Interface. This is a strict superset of 1149.1 that supports testing of analog ICs and other and discrete components.

1149.5 Module Test and Maintenance Bus. This is a bus protocol focusing on high-level systems testability, partitioning systems into subsystems, modules, and boards.

15.7.3 Internal Scan Design

The logic ICs on an MCM can be considered as a collection of sequential circuits. Testing a sequential circuit within an IC on an MCM is significantly more difficult than testing MCM structural faults

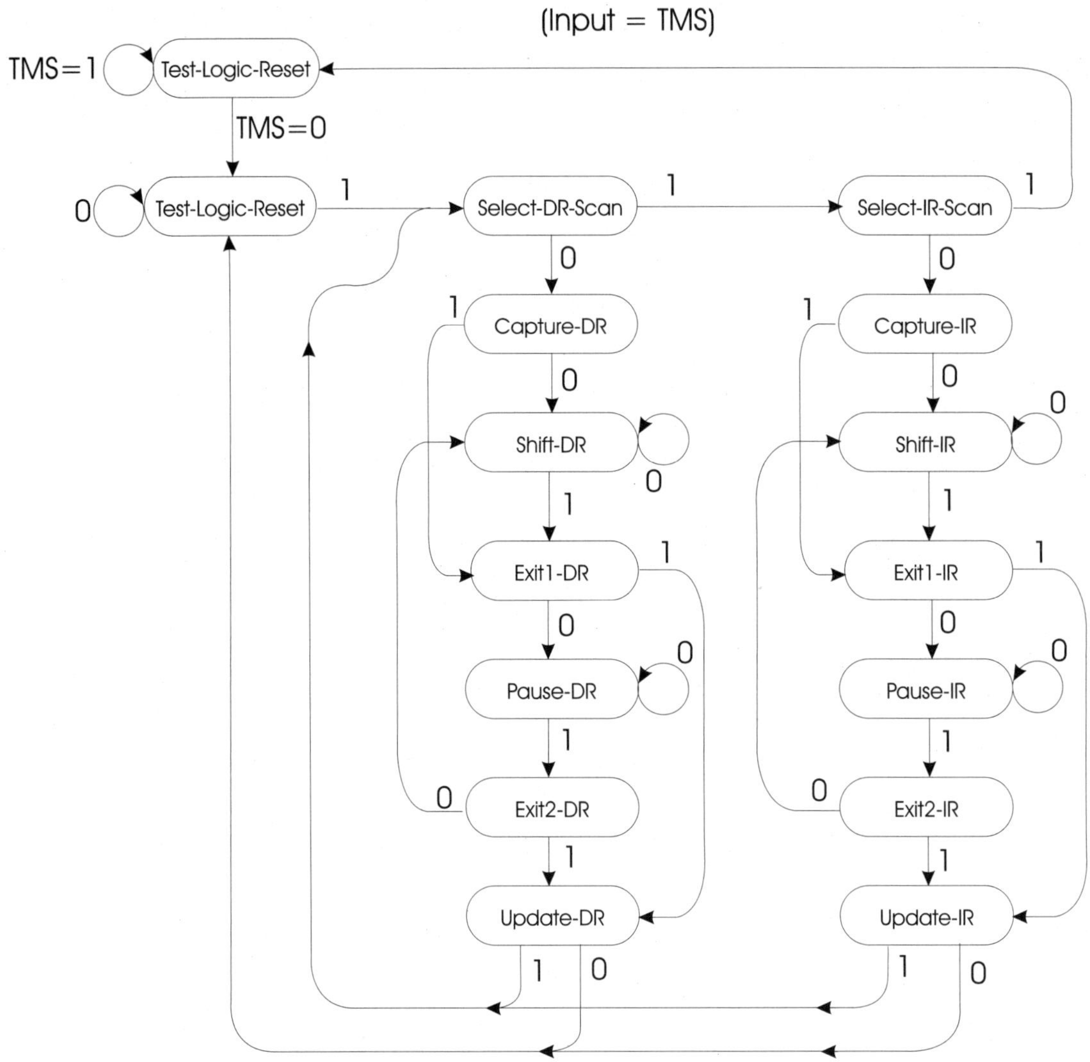

FIGURE 15.30 State diagram of the TAP controller.

because of the inability to directly control or observe the internal states of the circuit. Test vectors may be applied only to primary inputs of the IC, and only primary outputs may be observed. Internal *scan design* addresses this problem by providing a mechanism to directly set the internal state of the circuit to a desired value and to directly observe the internal state.[3–5,31,75–77] This is accomplished by isolating the memory elements during test operations and providing special inputs and outputs to control and observe their values as illustrated in Fig. 15.35.[31]

Scan design methods alter the design of the memory elements to operate in two modes: normal and scan. As shown in Fig. 15.35, a special external control signal, *Select*, is added to the circuit to select either the scan or normal operating mode of operation for the flip-flops. In scan mode, the memory elements are configured into a single shift register, called a *scan chain*. The input of the first flip-flop in the scan chain is driven by a special test input, $Scan_{in}$, and the output of the last flip-flop

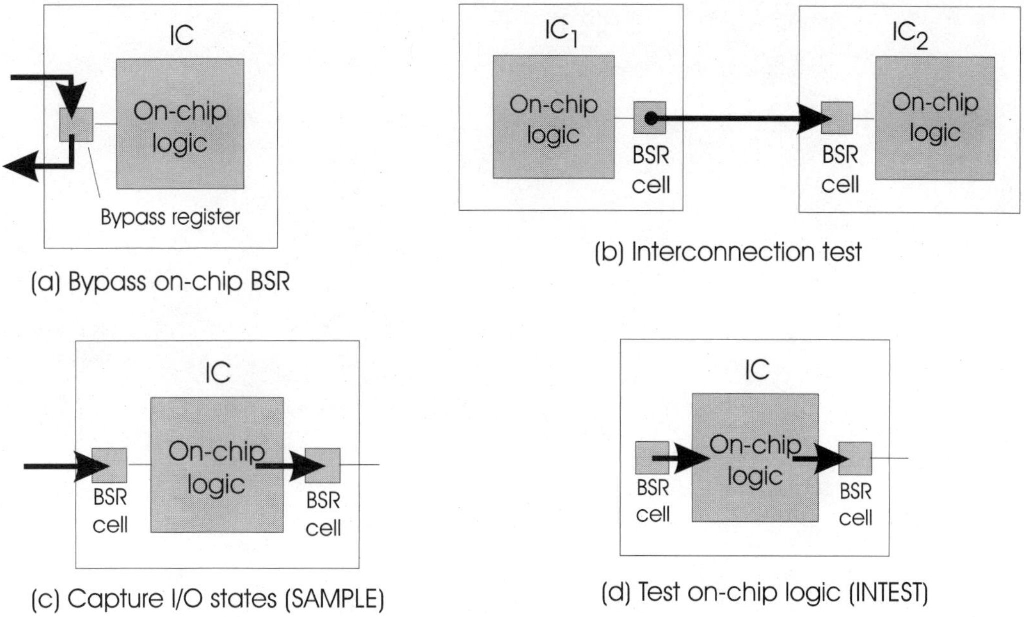

FIGURE 15.31 Configurations for boundary scan test instructions.

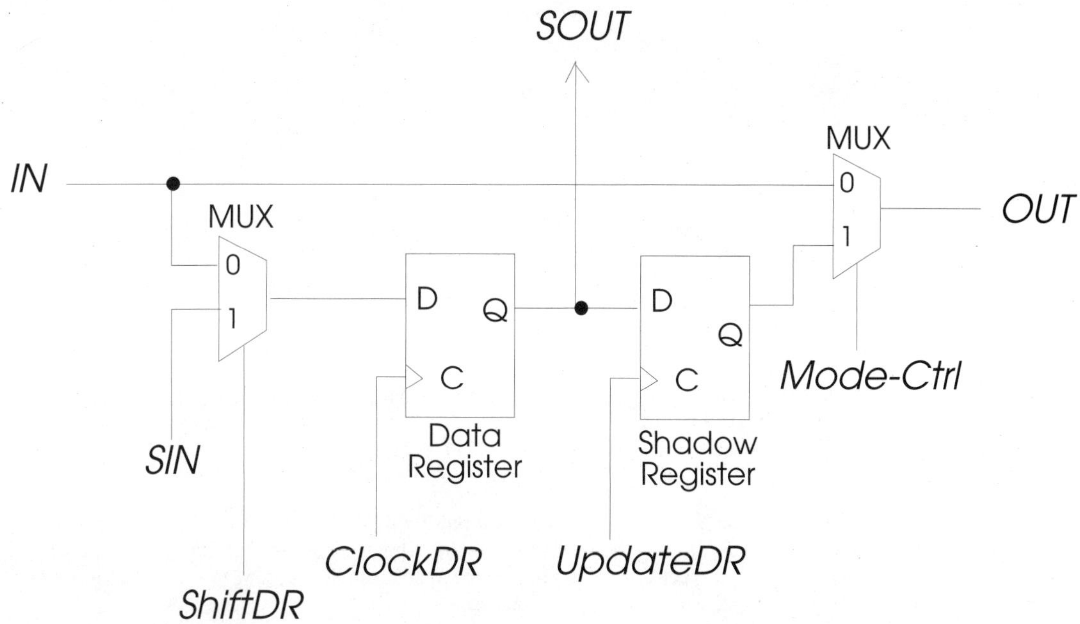

FIGURE 15.32 Boundary scan cell.

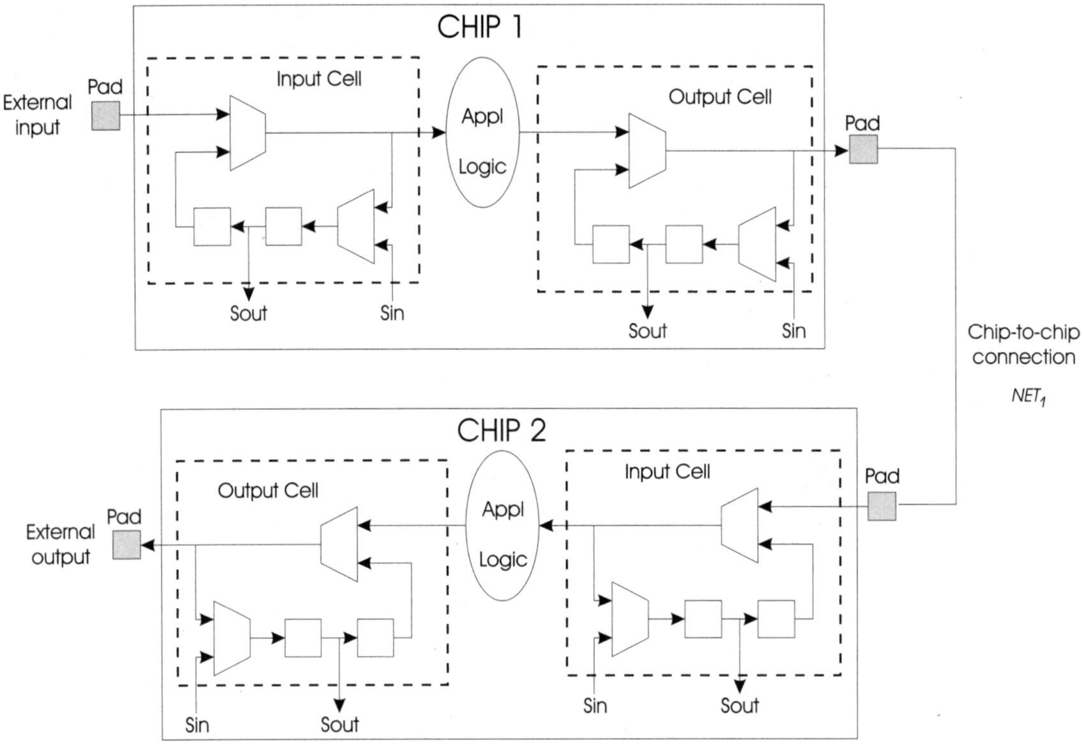

FIGURE 15.33 IC-to-IC and IC-to-MCM pin interconnect test configuration.

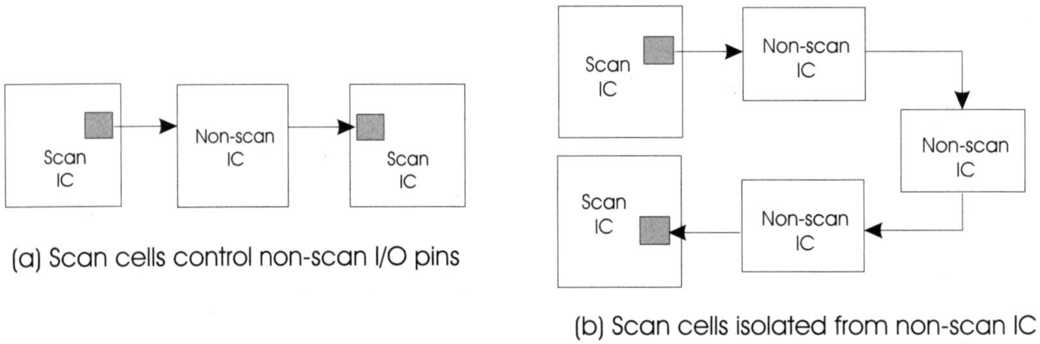

FIGURE 15.34 Partial boundary scan configurations.

in the scan chain is connected to a special test output, $Scan_{out}$. While configured in scan mode, a pattern may be shifted serially into the memory elements while shifting out the current pattern stored in them. This allows a test vector to be applied to the state variables $y_1 \ldots y_r$ in r clock periods, while making the previous state variables $Y_1 \ldots Y_r$ accessible to the tester.

A number of methods have been used to implement scan designs. One is to add a multiplexer to each flip-flop excitation input, as illustrated in Fig. 13.36, with the multiplexer controlled by the

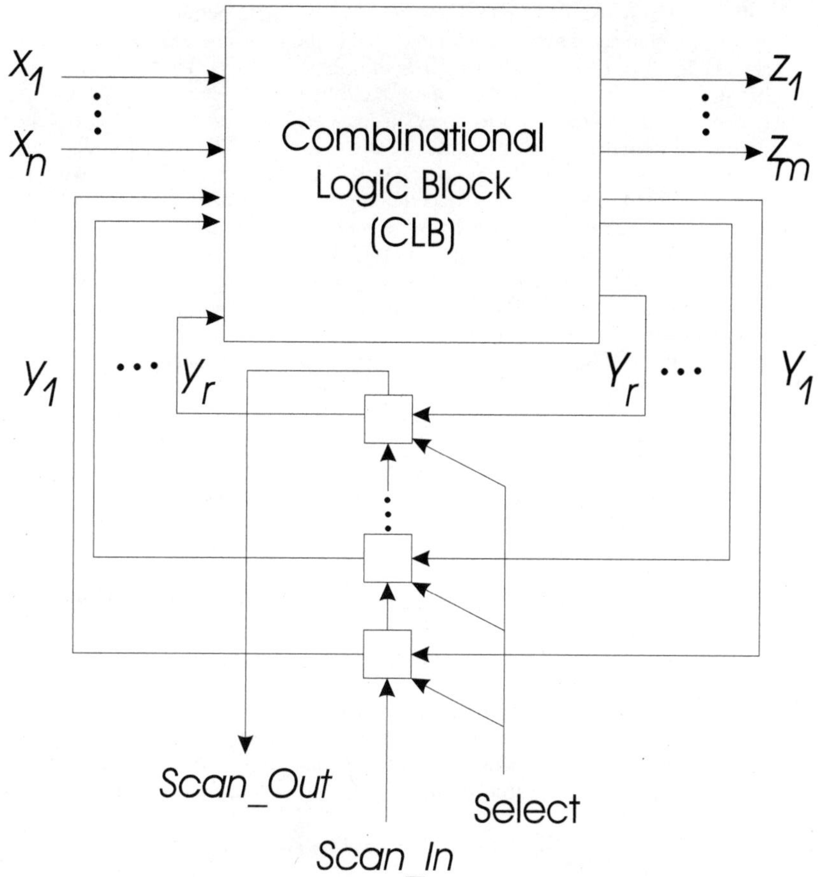

FIGURE 15.35 Incorporating scan design into a sequential circuit.

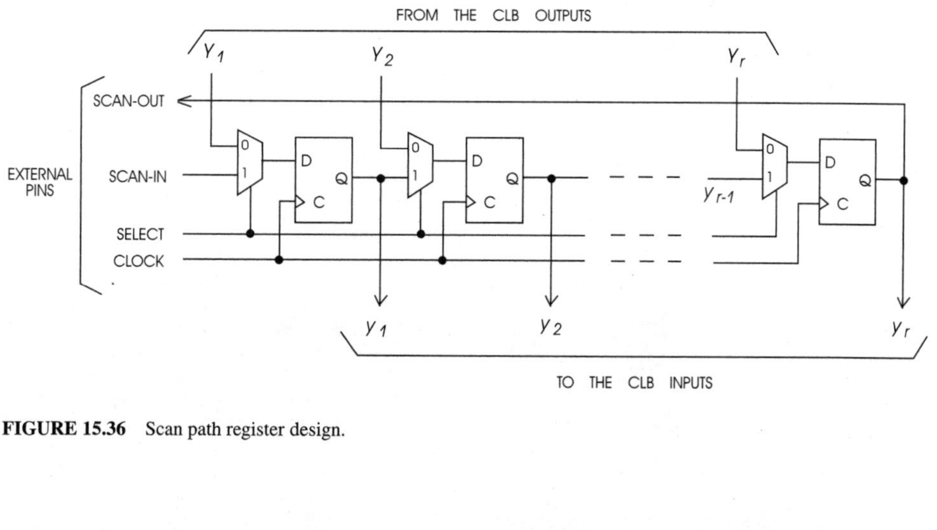

FIGURE 15.36 Scan path register design.

Select pin. In *Normal Mode* (*Select* = *0*), the next-state outputs of the combinational logic block, $Y_1 \ldots Y_r$, are connected to the excitation inputs of the flip-flops and the circuit operates according to its state table. In *Test Mode* (*Select* = *1*), the flip-flops are connected into a serial shift register with each flip-flop output connected to the input of the following flip-flop. The input to flip-flop y_1 is connected to $Scan_{in}$, and the output of flip-flop y_r is connected to $Scan_{out}$. In this mode, the flip-flop inputs are decoupled from the combinational logic, allowing them to be driven serially from $Scan_{in}$.

Level sensitive scan design (LSSD) has been used extensively within IBM and in other designs.[53,76,77] In LSSD, each scan cell comprises a pair of latches, *L1* and *L2*, as illustrated in Fig. 15.37. During normal operation, *L1* is the storage cell and is clocked by clock *C*. During scan operations, *L1* and *L2* form a master-slave flip-flip that is incorporated into the scan chain. All master and slave latches in the scan chain are controlled by separate test clocks, *A* and *B*, respectively, to ensure hazard-free operation. LSSD also specifies a number of design rules to prevent hazards and other timing problems in sequential circuits.[76]

Referring to Fig. 15.35, the basic test process for a circuit built with scan design principles involves the following steps:

1. In *Scan Mode,* shift alternating 1s and 0s through the scan chain to verify correct operation of the memory elements.
2. Serially shift a test vector into $y_1 \ldots y_r$ while concurrently applying a test vector to the primary inputs $x_1 \ldots x_n$.
3. Verify that the primary outputs $z_1 \ldots z_m$ are correct.
4. Change to *Normal Mode* and clock the circuit to force a state transition, i.e., to load the memory elements with the next-state values $Y_1 \ldots Y_r$.
5. Change back to *Scan Mode* and shift out the state values $Y_1 \ldots Y_r$ to verify their correctness.

Steps 2–5 of the above process are repeated for each combinational logic test vector, although step 5 of one test can be combined with step 2 of the next, i.e., the next test vector can be loaded while previous next-state values are shifted out via the scan chain. In general, for a sequential circuit with *n* inputs and *r* state variables, the number of clock cycles needed to verify a state table is:

$$N_{clks} = 2^{n+r} \times (r+1) + r \tag{15.18}$$

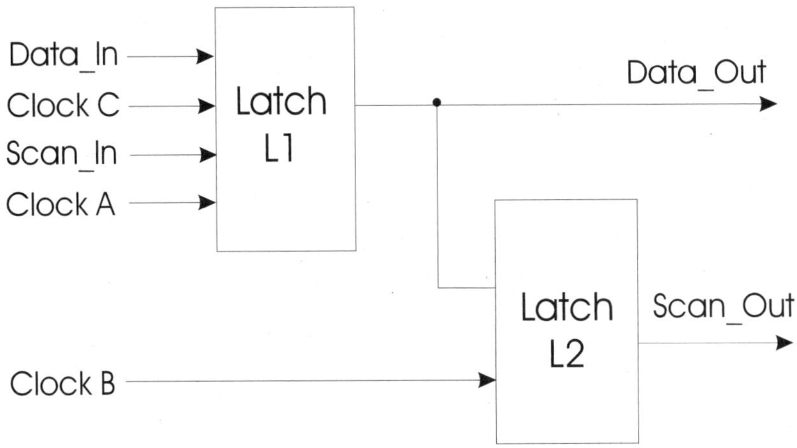

FIGURE 15.37 Level sensitive scan design (LSSD) scan cell.

where $(r + 1)$ cycles are needed to shift in each state vector and clock the circuit for each of the 2^{n+r} state/input combinations, and r additional clock cycles are needed to shift out the final result.

Since the combinational logic is isolated from the flip-flops during testing, combinational methods can be used to derive a minimal test set for the combinational logic block that will, in general, require far fewer than (2^{n+r}) test vectors. Consequently, test time can be greatly reduced as compared to simply verifying the state table of the circuit.

15.7.4 Built-In Self-Test

Even with the use of scan design and other DFT features, testing ICs within an MCM is expensive, because many test vectors must be applied by the automatic test equipment (ATE) to the circuit under test (CUT), with each response captured and verified. In addition, it is often difficult to get test vectors and responses to and from individual ICs within a complex MCM. A typical ATE test setup is shown in Fig. 15.38*a*.

Built-in self-test (BIST) improves testability by embedding test pattern generation and/or response capture and evaluation circuitry within the circuit under test, as depicted in Fig. 15.38*b*.[31] Test evaluation is done by signature analysis,[3–5,31,78] in which test response values are combined into a single value called a *signature*, which can be compared to the signature of a known good circuit to detect faults. By building BIST circuits into the ICs of an MCM, testing can be performed autonomously, and at full operational speed, with minimal guidance by the tester. The result is increased testing speed and lower test cost. At the same time, fault coverage is increased by placing BIST circuits where they have maximum controllability and observability of signals within the ICs.

Figure 15.39 illustrates a standard sequential circuit model modified to include BIST circuit elements. *Autonomous linear feedback shift registers* (ALFSRs) generate pseudo-random test patterns and apply them to the combinational logic block and the scan path input. The states of the primary

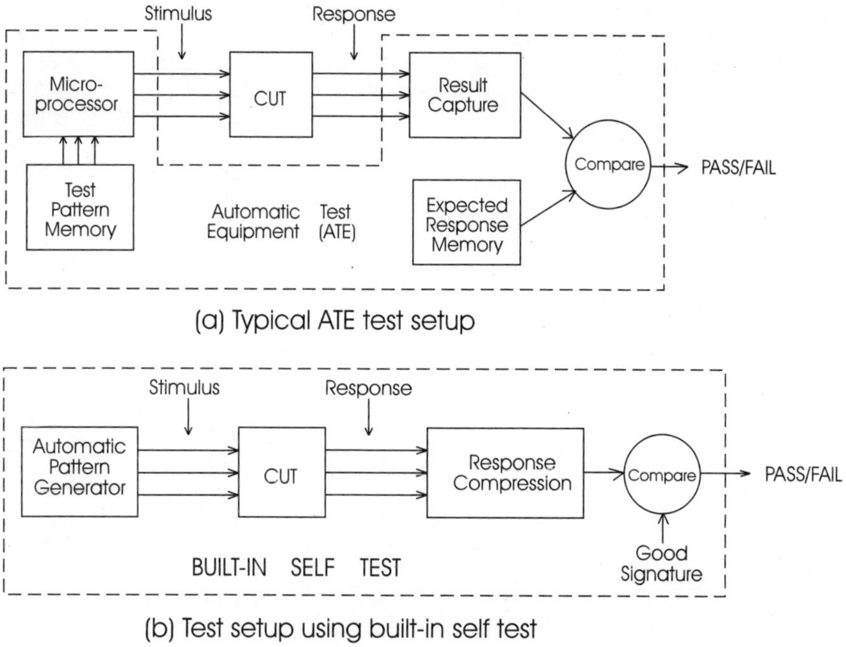

FIGURE 15.38 Test setups with and without built-in self-test (BIST).

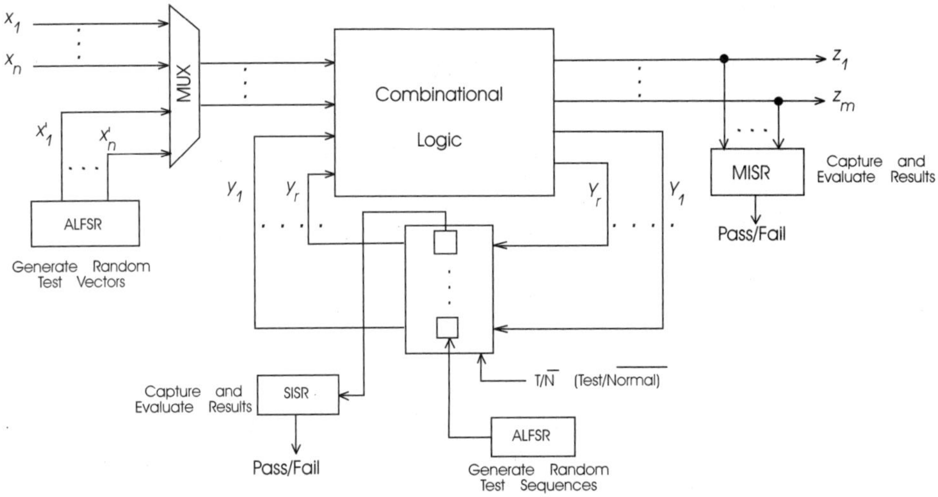

FIGURE 15.39 Sequential logic circuit with built-in self-test (BIST).[31]

outputs are captured and *compressed by a multiple-input signature register* (MISR), while serial data from the scan path is captured and compressed by a *serial-input shift register* (SISR). This section will describe the design and operation of these BIST elements.

Pseudo-random Test Vector Generation. BIST circuits used to generate pseudo-random test vectors and compute signatures are typically based on the use of a *linear feedback shift register (LFSR),* which is a series configuration of D flip-flops and EXCLUSIVE-OR (XOR) gates whose design and operation are based on principles of polynomial arithmetic in cyclic coding theory.

For generating test vectors, an *autonomous LFSR (ALFSR)* is used, which is an LFSR with no external inputs. An n-stage ALFSR produces a periodic pseudo-random sequence of n-bit binary numbers according to a special *generating function* realized via feedback lines and XOR gates within the ALFSR. If the sequence contains all $2^n - 1$ non-zero values, resulting in a period of $2^n - 1$, the generating function is called a *primitive polynomial.*

Figure 15.40 shows two general ALFSR circuit structures. In structure 1, the feedback path realizes the logic equation

$$a_n = a_0 c_0 \oplus a_1 c_1 \oplus \ldots \otimes a_{n-1} c_{n-1} \tag{15.19}$$

The feedback coefficients $c_0 \ldots c_{n-1}$ are selected to create the desired primitive polynomial, with $c_i = 1$ if flip-flop output a_i is fed back to the shift register input via an exclusive-OR gate, and $c_i = 0$ if a_i is omitted from the feedback circuit.

A four-bit ALFSR circuit is illustrated in Fig. 15.41. At output a_0, the value a_1 will be present after one clock period, a_2 after two clock periods, and so on. This sequence of outputs is represented in polynomial form as:

$$f(x) = a_4 x^4 + a_3 x^3 + a_2 x^2 + a_1 x + a_0 \tag{15.20}$$

where x^k represents a time delay of k clock periods. In the feedback path, $a_4 = a_1\ r\ a_0$, and therefore $a_4 r a_1 r a_0 = 0$. This corresponds to the primitive polynomial

$$+p(x) = x^4 + x^1 + x^0 = x^4 + x + 1 \tag{15.21}$$

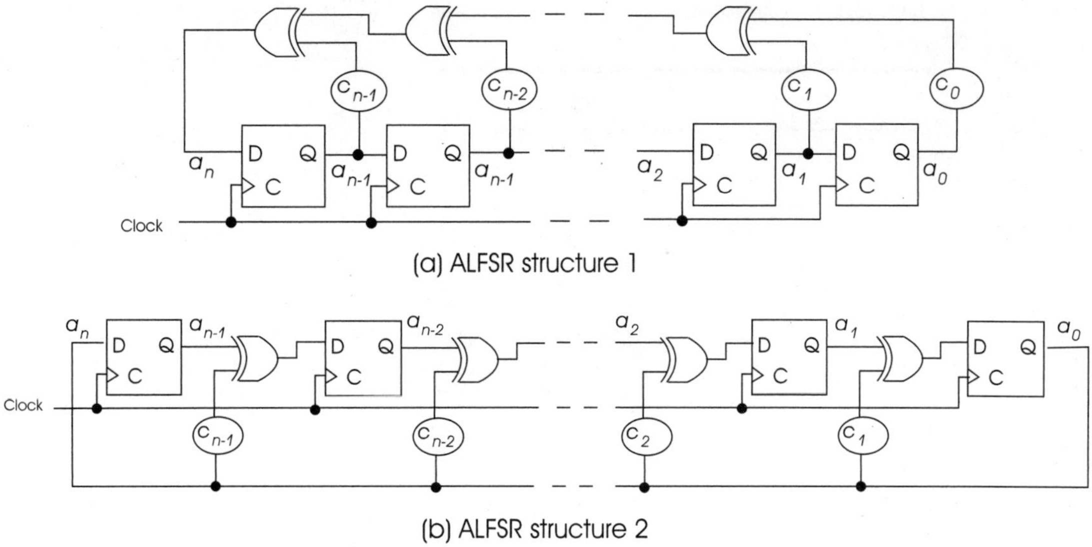

FIGURE 15.40 General ALFSR circuit structures.

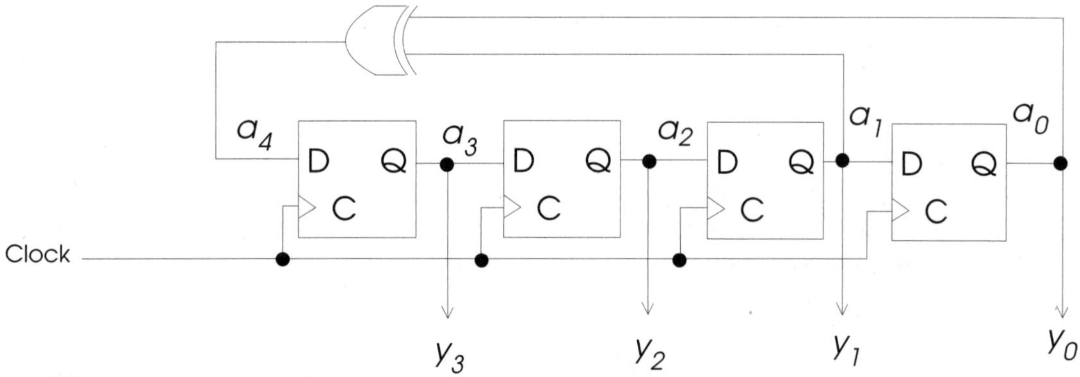

FIGURE 15.41 ALFSR structure 1 with $p(x) = x^4 + x + 1$.

If the register is initialized to a "seed" state of 1000, the pseudo-random sequence given in Table 15.2 results. Primitive polynomials for sequences of length $2^n - 1$ are published in a number of sources.[4]

ALFSR structure 2 of Fig. 15.40*b* is typically used where there are two or more feedback taps. As with structure 1, the c_i values indicate the presence or absence of feedback taps, with an XOR gate placed at the input to each flip-flop to which there is a feedback connection. As a result, at most one XOR gate delay occurs between clock cycles, whereas in structure 1 all of the XOR gates are in the feedback path, resulting in $(k - 1)$ gate delays for k feedback lines. Structure 1 can be built from a standard shift register with XOR gates added externally, while structure 2 requires XOR gates between the flip-flops.

Signature Analysis. The capture and analysis of circuit responses to each individual test vector of a test sequence is often impractical, especially if test circuits are built into the circuit module. In such

Table 15.2 Pseudo-random Sequence for an ALFSR with $p(x) = x^4 + x + 1$

Clock	$a_3a_2a_1a_0$
1	1000
2	0100
3	0010
4	1001
5	1100
6	0110
7	1011
8	0101
9	1010
10	1101
11	1110
12	1111
13	0111
14	0011
15	0001
16	1000

cases, response data for an entire sequence of tests is usually compressed into a single data value called a *signature*.[3–5,31,78] The signature of a fault-free circuit is unique. If the signature obtained for a given test is incorrect, then the circuit is known to be faulty. If the signature is correct, the circuit is assumed to be fault-free.

A number of data compression methods are relatively simple to implement. Examples include counting the number of 1s in a sequence of output values, which detects all odd numbers of errors and some even numbers of errors; counting the number of 0-to-1 and 1-to-0 transitions on the output lines; and computing parity over a sequence of outputs, which detects all single errors and all odd numbers of errors. These methods are easily implemented, although a number of output error patterns may go undetected due to aliasing, which will be defined below.

Higher fault coverage is achieved through the use of error-detection coding methods implemented with an LFSR.[3–5,31,75] Let us assume that a test response sequence on an output line z can be represented by a polynomial $z(x)$:

$$z(x) = z_n x^n + z_{n-1} x^{n-1} + \ldots + z_1 x^1 + z_0 \tag{15.22}$$

where x^k represents a time delay of k clock cycles, and z_k represents the data value at clock cycle k, as produced by the kth test vector. Error detection is performed in cyclic coding theory by dividing the data sequence polynomial $z(x)$ by the generator polynomial of the LFSR, $p(x)$, producing a quotient $q(x)$ and remainder $r(x)$. This can be expressed as:

$$z(x) = q(x)p(x) + r(x) \tag{15.23}$$

Polynomial division is performed serially by the LFSR as the response data sequence arrives. The $(n - k)$ bits of the quotient are shifted out of the LFSR, leaving the k-bit remainder $r(x)$ in the LFSR after the last step, which is the signature of the circuit.

If a fault results in one or more output errors, then the output sequence $z^*(x)$ differs from $z(x)$ of the fault-free circuit, and the polynomial division performed on $z^*(x)$ by the LFSR produces:

$$z^*(x) = q^*(x)p(x) + r^*(x) \tag{15.24}$$

Remainder $r^*(x)$ is the signature of the faulty circuit. The circuit is assumed to be fault-free if $r^*(x) = r(x)$.

Since information is lost by discarding the quotient, it is possible for certain faults to go undetected, resulting in less than 100 percent fault coverage. There are 2^{n-k} different $(n - k)$-bit quotient values out of 2^n possible response sequences that have the same k-bit remainder $r(x)$. Only one of these quotients corresponds to the correct sequence resulting from fault-free circuit operation; $2^k - 1$ other sequences are produced by faulty circuits, having the same remainder/signature as the fault-free case. This situation is referred to as *aliasing*, with the signature referred to as an *alias* when equal to that of the good circuit. The probability that an fault is masked is the probability that $r^*(x) = r(x)$ for an incorrect sequence $z^*(x)$, which is given by:

$$P_M = \frac{2^{n-k} - 1}{2^n - 1} \tag{15.25}$$

For large values of n, this reduces to:

$$P_M \approx \frac{2^{n-k}}{2^n} = 2^{-k} \tag{15.26}$$

Therefore, for long response sequences, the probability of missing an error is reduced as the number of stages, k, in the LFSR increases.

Figure 15.42 shows a LFSR configuration with a single input, called a *serial input signature register* (SISR). The generating function of a SISR is realized in the same manner as described above for ALFSR elements. For the SISR in the figure, because the feedback taps are at stages 1 and 0, the generating function is

$$p(x) = x^4 + x + 1 \tag{15.27}$$

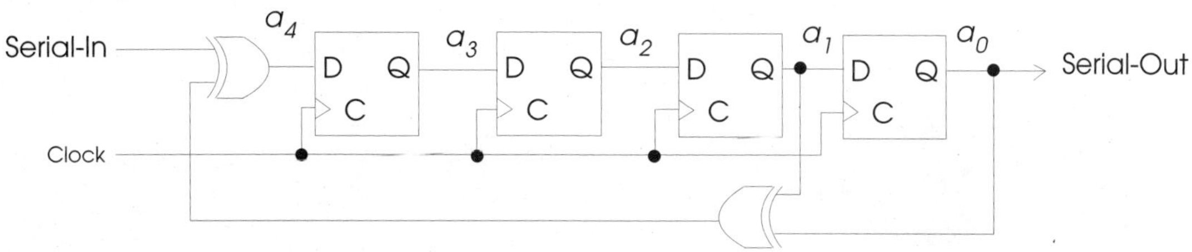

FIGURE 15.42 SISR logic diagram.

The operation of the SISR is demonstrated in Table 15.3*a* for the uncorrupted input sequence $z(x)$ = 010001101110. Table 15.3*b* lists the response obtained when an erroneous input sequence occurs; in this case, one bit has been altered in the input sequence. Note that the error response, $r^*(x)$, differs from the expected response, $r(x)$.

For a logic circuit with *m* parallel outputs, *m* SISR elements can be used to compute individual signatures for each output. A more efficient method is to use a *multiple-input signature register* (MISR), which effectively computes the signatures of parallel input sequences concurrently within a single LFSR circuit. A four-bit MISR is shown in Fig. 15.43, which is based on the same generating function as the SISR of Fig. 15.42. The number of parallel inputs to the MISR, *m*, may be less than the number of MISR stages, *k*.

In general, an SISR or MISR of length *k $ 2* is capable of detecting all single-bit errors. If the LFSR generator polynomial $p(x)$ has an even number of terms (which occurs when $p(x)$ is divisible by $x + 1$), then all odd numbers of errors can be detected. In addition, any *burst error,* i.e., any group of consecutive errors, of length up to *k*, can be detected. Therefore, the fault coverage is a function of the number of stages *k* in the LFSR, the number of parallel inputs *m*, and the length of the inputs

Table 15.3 SISR Operation

(a) Response to correct sequence		(b) Response to erroneous sequence	
Serial in $z(x)$	LFSR contents $a_3a_2a_1a_0$	Serial in $z(x)$	LFSR contents $a_3a_2a_1a_0$
0	0000	0	0000
1	0000	1	0000
0	1000	0	1000
0	0100	0	0100
0	0010	1*	0010
1	1001	1	0001
1	0100	1	0000
0	1010	0	1000
1	1101	1	0100
1	0110	1	1010
1	0011	1	0101
0	1001	0	0010
r(x) =	1100	r*(x) =	1001

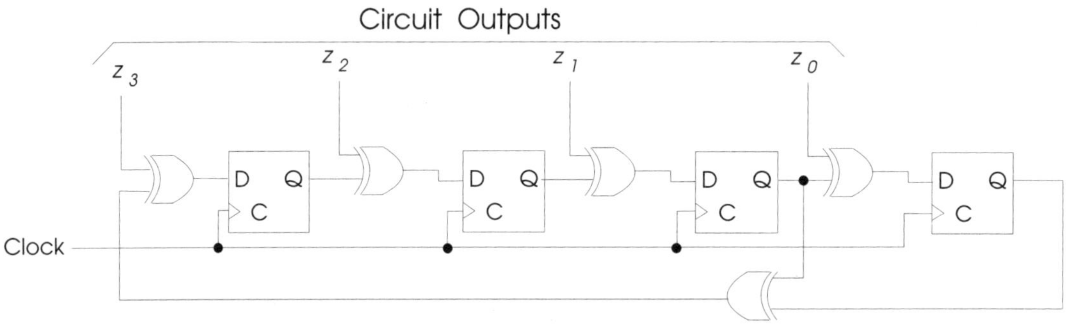

FIGURE 15.43 Multiple-input signature register (MISR).

sequence n, as well as the LFSR generating function $p(x)$, and the initial state of the LFSR. More detailed descriptions of signature analysis fault coverage properties are presented in Refs. 4, 31, 74, and 78.

Built-In Logic Block Observer (BILBO). The addition of LFSR elements to a circuit to perform pattern generation and signature analysis adds to the complexity and expense of a circuit module. One approach that can minimize the total number of flip-flops in the circuit is to combine the pattern generation and signature analysis functions with the normal state flip-flops of the circuit. The resulting register structure is referred to as a *built-in logic block observer (BILBO)*.[79]

A four-bit BILBO register is shown in Fig. 15.44. The BILBO has two control lines, B_1 and B_2, which select the mode of operation as summarized in Table 15.4. Normal mode is selected by $B_1B_2 = 00$, causing the flip-flops to operate as a parallel-load register whose inputs, $I_1 - I_4$, come from CLB outputs $Y_1 \ldots Y_4$, while outputs $F_1 - F_4$ represent the current state $y_1 \ldots y_4$. $B_1B_2 = 01$ sets $D_1 = SCAN\text{-}IN$ and the BILBO is configured into a serial shift register to be used as a scan path. $B_1B_2 = 11$ configures the BILBO as an ALFSR for test pattern generation, with D_1 connected to the feedback signal. When $B_1B_2 = 10$, the flip-flop inputs become $D_i = I_i r F_{i-1}$, and thus the LFSR functions as an MISR to perform signature analysis.

When using BILBOs, the circuit is typically partitioned into modules, as shown in Fig. 15.45. For testing CLB #1, BILBO #1 is operated as an ALFSR to apply test vectors while BILBO #2 is configured as an MISR to perform signature analysis. The two BILBOs reverse roles to test CLB #2. After testing is complete, the BILBOs are returned to parallel-load mode to allow them to operate as normal-state variable flip-flops.

Table 15.4 BILBO Operating Modes

B_1B_2	Input MUX	D_i	Function	Test function
00	SCAN-IN	I_i	parallel-load register	normal (non-test) mode
01	SCAN-IN	F_{i-1}	linear shift register	scan path mode
10	Feedback	$I_i r F_{i-1}$	MISR	signature analysis
11	Feedback	F_{i-1}	ALFSR	pattern generation

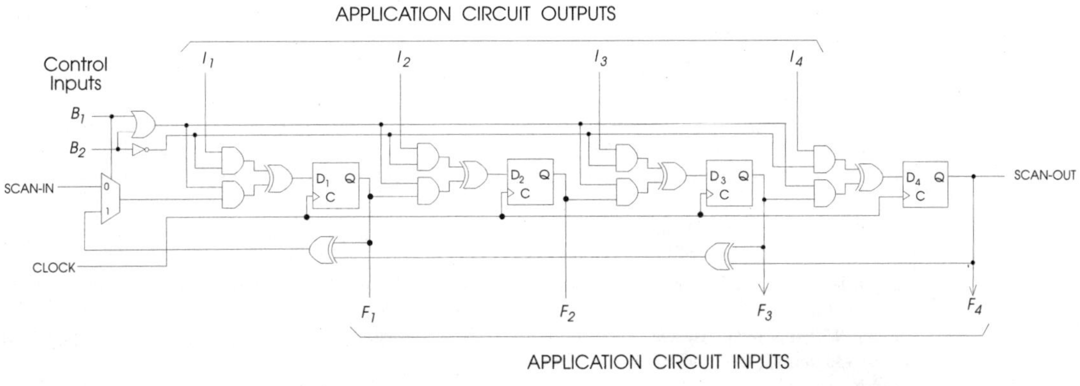

FIGURE 15.44 Built-in logic block observer (BILBO).

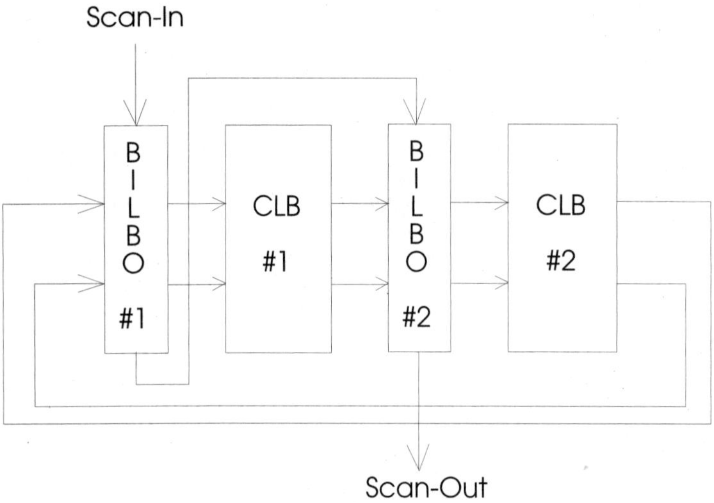

FIGURE 15.45 Built-in logic block observer (BILBO)

REFERENCES

1. V. P. Nelson, "Fault-Tolerant Computing: Fundamental Concepts," *IEEE Computer,* vol. 23, no. 7, July 1990, pp. 19–25.
2. J. P. Hayes, "Fault Modeling," *IEEE Design & Test of Computers,* April 1985, pp. 88–95.
3. A. Miczo, *Digital Logic Testing and Simulation,* Harper & Row, Publishers, Inc., 1986. ISBN 0-06-044444-4.
4. M. Abramovici, M. A. Breuer, and A. D. Friedman, *Digital Systems Testing and Testable Design,* Computer Science Press, New York, NY, 1990.
5. M. A. Breuer and A. D. Friedman, *Diagnosis & Reliable Design of Digital Systems,* Computer Science Press, Inc., Potomac, MD, 1976.
6. G. Hroundas, "Bare Board Testing," Chapter 17, *Printed Circuits Handbook,* 3rd Ed., C. F. Coombs, Jr., Ed. in chief, McGraw-Hill, Inc., New York, NY, 1988, pp. 17.1–17.26.
7. J. Grasso, "Electrical Testing of Bare Fine Pitch Printed Circuit Boards," Chapter 20, *Handbook of Fine Pitch Surface Mount Technology,* J. H. Lau, Ed., Van Nostrand Reinhold, New York, NY, 1994, pp. 567–597.
8. S. Trigwell, "Failure Mechanisms of Wire and Die Bonding," *Solid State Technology,* May 1993, pp. 45–49.
9. A. Halperin, T. DiStefano, and C. Shinwu, "Latent Open Testing of Electronic Packaging," *Proc. 1994 IEEE Multichip Module Conf.* (MCMC-94), March 15–17, 1994, Santa Cruz, CA., pp. 83–88.
10. "Latent Open Detection With Flying Probes," Bath Scientific Ltd., UK, BSL Transconductance Option Brochure.
11. W. Maly, "Tutorial: Realistic Fault Modeling for VLSI Testing," *Proc. 24th Design Automation Conf.,* June 1987, pp. 173–180.
12. J. Hagge and R. Wagner, "High-Yield Assembly of Multichip Modules Through Known-Good IC's and Effective Test Strategies" *Proceedings of the IEEE,* vol. 80, no. 12, December 1992, 1965–1994.
13. T. Williams and N. Brown, "Defect level as a function of fault coverage," *IEEE Trans. on Computers,* vol. C-30, no. 12, Dec. 1981, pp. 987–988.
14. P. Maxwell and R. Aitken, "Test Sets and Reject Rates: All Fault Coverages Are Not Created Equal," *IEEE Design & Test of Computers,* March 1993, pp. L42–51.

15. M. Abadir, et. al., "Analyzing Multichip Module Testing Strategies," *IEEE Design & Test of Computers,* vol. 11, no. 1, Spring 1994, pp. 40–52.
16. V. P. Nelson and A. D. Singh, "MTAP: A MCM Test Advisor Program," *Proc. ISHM Advanced Technology Workshop on MCM Test,* Napa Valley, CA. September 10–13, 1995.
17. M. Tegethoff and T. Chen, "Manufacturing Test Simulator: A Concurrent Engineering Tool for Boards and MCMs," *Proc. Int. Test Conf.,* 1994, pp. 903–910.
18. I. Dear, C. Dislis, A. Ambler, and J. Dick, "Economic Effects on Design and Test," *IEEE Design & Test of Computers,* December 1991, pp. 64–77.
19. C. Dislis, J. Dick, I. Dear, I. Azu, and A. Ambler, "Economic Modeling for the Determination of Test Strategies for Complex VLSI Boards," *Proc. IEEE Int. Test Conf.,* 1993, pp. 210–217.
20. C. Dislis, J. Dick, and A. Ambler, "Algorithms for Cost Optimized Test Strategy Selection," *Proc. IEEE Int. Test Conf.,* 1993, pp. 383–391.
21. J. Trent, "Test Philosophy for Multichip Modules," *Proc. Int. Conf. on Multichip Modules,* Denver, CO, April 1992, pp. 444–452.
22. A. Flint and W. Blood, "MCM Test Strategy: Board Test in an IC Environment," *Proc. Int. Conf. on Multichip Modules,* Denver, CO, April 1993, pp. 429–434.
23. A. Flint, "Testing Multichip Modules," *IEEE Spectrum,* March 1994, pp. 59–62.
24. E. Trischler and M. Johansson, "TEN: A Concurrent Test Engineering Environment," *IEEE Design & Test of Computers,* Fall 1994, pp. 6–16.
25. L. E. Roszel, "MCM Test Methodologies and Project Experiences," *Proc. Int. Conf. on Multichip Modules,* Denver, CO, April 1995, pp. 424–427.
26. L. E. Roszel, "MCM Foundry Test Methodology and Implementation," *Proc. IEEE Int. Test Conf.,* 1993, pp. 369–372.
27. N. Shen and J. Yang, "Reliability By Design for MCM Manufacturing: The Roadmap and an Example," *Proc. 1993 IEEE Multichip Module Conf.,* Santa Cruz, CA, March 1993, pp. 100–105.
28. M. Lubaszewski, M. Marzouki, and M. H. Touati, "A Pragmatic Test and Diagnosis Methodology for Partially Testable MCMs," *Proc. 1994 IEEE Multichip Module Conf.,* Santa Cruz, CA, March 15–17, 1994, pp. 108–113.
29. ASIC/IC Design For Test Process Guide, Mentor Graphics Corp., Wilsonville, OR, 1994.
30. "Automatic Test Generation for Digital Electronic Circuits," *AT&T Tech. Journal,* vol. 73, no. 2, April 1994, pp. 19–29.
31. V. P. Nelson, H. T. Nagle, B. D. Carroll, and J. D. Irwin, *Digital Logic Circuit Design and Analysis,* Prentice-Hall, Inc., Englewood Cliffs, NJ, 1995.
32. IEEE Standard 1076-1987, "IEEE Standard VHDL Language Reference," IEEE Standards Board, 345 East 47th Street, New York, NY 10017, March 1987.
33. M. D. Edwards, *Automatic Logic Synthesis Techniques for Digital Systems,* McGraw-Hill, New York, NY, 1992.
34. V. Nelson and A. Flint, "Test Development of a Microcontroller-Based MCM for Automotive Applications," *Proc. Int. Conf. on Multichip Modules,* Denver, CO, April 1995, pp 114–119.
35. *Fault Analysis User's Manual,* Mentor Graphics Corp., Wilsonville, OR, 1994.
36. H. Davidson, "Inspection and Testing," Chapter 8, *Multichip Module Design, Fabrication & Testing,* J. Licari, Ed., McGraw-Hill, Inc. 1995, pp. 325–351.
37. T. Russell and Y. Wen, "Electrical Testing of Multichip Modules," *Multichip Module Technologies & Alternatives,* Van Nostrand Reinhold, 1993, pp. 615–660.
38. R. Fillion, et. al., "MCM Testing," Chapter 13, *ISHM Handbook on Thin Film Multichip Modules,* pp. 487–592.
39. S. DeFoster, M. Mancini, and J. Zalesinski, "Automatic Testing of Metallized Ceramic-Polyimide (MCP) Substrates," pp. 548–553.
40. C. Shipley, "Flying Probes," *Advanced Packaging,* Fall 1992.
41. BSL LinearProbe: System Description, Bath Scientific Ltd., UK, 1995.
42. "A Multi-Probe Approach for MCM Substrate Testing," *IEEE Trans. on Computer-Aided Design of Integrated Circuits and Systems,* vol. 13, no. 1, Jan. 1994, pp. 110–121.
43. Micro-Prober 90 Technical Data Sheet, Integri-Test Corp., Commanck, NY 11725.

44. J. T. Thong, Ed., *Electron Beam Testing Technology,* Plenum Press, New York, NY, 1993.
45. O. C. Woodard, et. al., "Voltage contrast electron beam testing experiments on very large scale integrated circuit packaging substrates," *J. Vac. Sci. Technol. B,* vol. 6, no. 6, Nov/Dec 1988, pp. 1966–1970.
46. M. Brunner and R. Schmid, "Electron-beam MCM substrate tester," *Proc. 1993 IEEE Multichip Module Conf.* (MCMC-93), March 15–18, 1993, Santa Cruz, CA, pp. 62–68.
47. O. Hecht and G. Dishon, "Automatic Optical Inspection (AOI)," *Proc. IEEE,* pp. 659–661.
48. Orbot, Inc., Application Training Kit, V.07.88, 600 W. Cummings Park, Woburn MA, 01801.
49. F. J. Langley, "Testing In Assembly," Chapter 21, *Printed Circuits Handbook,* 3rd Ed., C. F. Coombs, Jr., Ed. in chief, McGraw-Hill, Inc., New York, NY, 1988, pp. 21.1–21.26.
50. C. A. Dean and Y. Zorian, "Do You Practice Safe Test? What We Found Out About Your Habits," *Proc. IEEE Int. Test Conf.,* 1994, pp. 887–892.
51. Discussion at Intel Exhibit, Int. Conf. on Multichip Modules, Denver, CO, April 1995.
52. R. J. Feugate, Jr. and S. M. McIntyre, *Introduction to VLSI Testing,* Prentice-Hall, Englewood Cliffs, NJ, 1988.
53. T. Bardsley, J. Drummond, and J. Eastman, "Integrating Wire Bond Die From Multiple Vendors on a C4-Based MCM," *Proc. Int. Conf. on Multichip Modules,* Denver, CO, April 1995, 286–291.
54. S. Berke, D. Golden, and E. Rogers, "Test Strategy for a Microprocessor Based Multichip Module," *Proc. Int. Conf. on Multichip Modules,* Denver, CO, April 1995, 419–423.
55. R. Wagner and J. Jorgenson, "Design-For-Test Techniques Utilized in an Avionics Computer MCM," *Proc. IEEE Int. Test Conf.,* 1993, pp. 373–382.
56. J. Jorgenson and R. Wagner, "Analyzing the Design-For-Test Techniqu*ç*s in a Multiple Substrate MCM," pp. 360–365.
57. F. Chong, et. al., "The Evolution of MCM Test from High Performance Bipolar Mainframe Multichip Modules to Low Cost Work Station Multichip Modules," *Proc. Int. Conf. on Multichip Modules,* Denver, CO, April 1993, pp. 404–410.
58. T. W. Williams and K. P. Parker, "Design for Testability–A Survey," *Proc. of the IEEE,* vol. 71, no. 1, pp. 98–112, Jan. 1983.
59. L. H. Goldstein, "Controllability/Observability Analysis of Digital Circuits," *IEEE Trans. Computers,* vol. CAS-26, no. 9, Sep. 1979, pp. 685–693.
60. J. Gabay, "Modern Design-for-Test Concerns: How Far Should I Go?," *ASIC & EDA,* October 1994, pp. 60–77.
61. B. Caldwell and T. Langford, "Is 1149.1 Boundary Scan Cost Effective: A Simple Case Study," *Proc. Int. Test Conf.,* 1992, pp. 106–109.
62. Test Access Port and Boundary-Scan Architecture, IEEE Standard 1149.1-1990, IEEE Standards Board, 345 East 47th St., New York, NY 10017, May 1990.
63. Parker, K.P., *The Boundary-Scan Handbook,* Kluwer Academic Pub., Norwell, Mass., 1992, ISBN 0-7923-9270-1.
64. M. Tegethoff and K. P. Parker, "IEEE Std. 1149.1: Where Are We? Where From Here?" *IEEE Design & Test of Computers,* Summer 1995, pp. 53–59.
65. R. Tulloss, "IEEE 1149 Standards–Changing Testing, Silicon to Systems," *Proc. IEEE Int. Test Conf.,* 1993, pp. 399–408.
66. C. W. Yau, et. al., "Boundary-Scan Testing for Electronic Subassemblies and Systems," *AT&T Technical Journal,* vol. 73, no. 2, March/April 1994, pp. 40–48.
67. M. Robinson, F. Mailhot, J. Konsevich, "Technology-Independent Boundary Scan Synthesis (Technology and Physical Issues)," *Proc. IEEE Int. Test Conf.,* 1993, pp. 157–166.
68. R. Bracken, B. Kraemer, L. Paradiso, and C. Jensen, "Multichip Modules Die and MCM Test Strategy: The Key to MCM Manufacturability," *Proc. Int. Conf. on Multichip Modules,* Denver, CO, April 1992, pp. 456–460.
69. K. Posse, "A Design-for-Testability Architecture for Multichip Modules," *Proc. IEEE Int. Test Conf.,* 1991, pp. 113–121.
70. S. Hilla, "Boundary Scan Testing for Multichip Modules," *Proc. Int. Test Conf.,* 1992, pp. 224–231.
71. R. Vutukuru, P. Subbarao, and R. Vemuri, "Boundary Scan Test Structures and Test Bench Compilation in a Multichip Module," *Proc. 1992 IEEE Multichip Module Conf.,* March 18–20, 1992, Santa Cruz, CA, pp. 44–47.

72. T. Langford, "Utilizing Boundary Scan to Implement BIST," *Proc. IEEE Int. Test Conf.,* 1993, pp. 167–173.

73. K. P. Parker and S. Oresjo, "A Language for Describing Boundary-Scan Devices," *Proc. IEEE Int. Test Conf.,* 1990, pp. 222–234.

74. Products for Testability, Product Brochure, Texas Instruments, Inc., Dallas, TX, 1991.

75. V. D. Agrawal, et. al., "Built-In Self-Test for Digital Integrated Circuits," *AT&T Tech. Journal,* vol. 73, no. 2, April 1994, pp. 30–39.

76. E. B. Eichelberger and T. W. Williams, "A Logic Design Structure for LSI Testability," *Proc. Design Automation Conf.,* June 1977, pp. 462–468.

77. T. Storey, "A Test Methodology for VLSI Chips on Silicon," *Proc. IEEE Int. Test Conf.,* 1993, pp. 359–368.

78. J. E. Smith, "Measures of Effectiveness of Fault Signature Analysis," *IEEE Transactions on Computers,* vol. C-29, no. 6, June 1980, pp. 510–514.

79. Koenemann, B., J. Mucha, and G. Zwiehoff, "Built-in logic block observation techniques," *Digest of Papers, 1979 Test Conf.,* IEEE Pub. 79CH1509-9C, pp. 37–41, Oct. 1979.

INDEX

U

V